WATER SUPPLY

Fifth Edition

WATER SUPPLY

Fifth Edition

Alan C. Twort
BSc, FICE, FCIWEM

Don D. Ratnayaka
BSc, DIC, MSc, FIChem E, FCIWEM

Malcolm J. Brandt
BSc, MICE, MCIWEM

BUTTERWORTH
HEINEMANN

IWA
Publishing

BINNIE BLACK & VEATCH

Butterworth-Heinemann
Linacre House, Jordan Hill, Oxford OX2 8DP
225 Wildwood Avenue, Woburn, MA 01801-2041
A division of Reed Educational and Professional Publishing Ltd

A member of the Reed Elsevier plc group

First published by Arnold 1963
Second edition published by Arnold 1974
Third edition published by Arnold 1985
Fourth edition published by Arnold 1994
Fifth edition published by Arnold 2000
Reprinted by Butterworth-Heinemann 2001

Co-published by IWA Publishing
Alliance House, 12 Caxton Street, London, SW1H 0QS
Tel: +44 020 7654 5500; e-mail: publications@iwap.co.uk; website: www.iwap.co.uk

Whilst the advice and information in this book are believed to be true and
accurate at the date of going to press, neither the authors nor the publisher
can accept any legal responsibility or liability for any
errors or omissions that may be made.

British Library Cataloguing in Publication Data
A catalogue record for this book is available from the British Library

Library of Congress Cataloguing in Publication Data
A catalogue record for this book is available from the Library of Congress

ISBN 0 340 72018 2

For information on all Butterworth-Heinemann publications
visit our website at www.bh.com

Produced and typeset by Gray Publishing, Tunbridge Wells, Kent
Printed and bound in Great Britain by The Bath Press, Bath

FOR EVERY TITLE THAT WE PUBLISH, BUTTERWORTH-HEINEMANN
WILL PAY FOR BTCV TO PLANT AND CARE FOR A TREE.

Contents

Preface

A number of key specialists in the water industry have contributed to the production of this fifth edition of *Water Supply* and as a result most chapters have been re-written, extended and updated.

The text gives up-to-date national and international standards for drinking water quality set by the UK and EC regulatory bodies plus the World Health Organization and the US Environmental Protection Agency. It describes the incidence and significance of the main chemical constituents found in raw waters, and the types of bacteria, viruses and protozoan organisms which present a hazard to human health. New matters of concern with respect to chemical and microbiological contaminants are listed. The design is given of treatment works and the equipment used for chemical coagulation, clarification methods including dissolved air flotation, rapid gravity filtration, slow sand filtration and membrane filtration. Sludge disposal methods are presented with design parameters and tables of relevant data.

The new 1999 UK regulations with respect to monitoring for cryptosporidium oocysts are given, and treatment practices recommended to reduce the risk of oocysts passing into supply are described.

Advanced and specialised treatments are described for iron, arsenic and manganese removal, plumbosolvency control, defluoridation and fluoridation, nitrate and ammonia removal. Taste and odour causes and removal; and reduction of volatile organic compounds and micropollutants by use of granular activated carbon (GAC) and advanced oxidation processes are dealt with. Further material is added on desalination by ion exchange, electrodialysis, reverse osmosis and thermal processes.

The new approach to the management of distribution systems is described. Zoning of supplies, telemetered monitoring of district flows, computer modelling of flows and water quality modelling, together with geographic information systems are increasingly being used to provide data on system performance, levels of service to consumers and the condition of assets for the development of asset management plans. Methods of rehabilitating pipelines are discussed. An extended chapter on pipes and pipelines gives additional information on the design of steel, polyethylene and PVC pipes according to numerous international and in-country standards.

Material on the yield of sources has been remodelled to emphasize the important role played by underground water supplies throughout the world. New approaches to definitions of 'yield' are discussed; and reference is directed to the new *Flood Estimation Handbook 1999* for estimating flood magnitudes.

Experience with privatization of the water industry in England and Wales is reported together with a consideration of the growth of private sector participation in public water supply overseas. Levels of staffing of waterworks undertakings in UK and other countries are given. New information is presented on the water demand experienced by USA water undertakings and the revised approach of the US Environmental Protection Agency to drinking water quality control.

The authors are grateful to the many contributors and reviewers who have aided the production of this fifth edition and to the firm of Binnie, Black & Veatch who have made this co-operative venture possible. The text also benefits from the contributions of F. M. Law, F. W. Crowley and Dr R. C. Hoather to previous editions. However we must make clear that responsibility for the statements and opinions expressed lies with ourselves.

Alan C. Twort
Don D. Ratnayaka
Malcolm J. Brandt

Contributing Authors, Reviewers and Advisors

Contributing authors from Binnie, Black & Veatch

Peter B. Clark
BA, MA, MSc, MICE

Technical Director (Hydraulics)

Tony N. Coe
MA, FIMechE, FCIWEM

Chief Mechanical Engineer (Pumping Plant)

Neville A. Cowton
BSc, MSc, MICE, MCIWEM

Technical Director (Service Reservoirs)

Ken J. Edworthy
BSc, FGS, MCIWEM

Consultant Hydrogeologist (Groundwater Supplies)

Mike J. Little
MA, MICE, MCIWEM

Technical Director (Pipelines)

David E. MacDonald
BSc, MSc, MCIWEM

Chief Hydrologist (Surface Supplies & Floods)

John C. Maunder
BSc, MIEE

Consultant, Control & Instrumentation
(Control & Automation Systems)

Alvin J. Smith
BSc, MIBiol, MIWEM

Chief Biologist (Water Biology & Storage)

Peter J. Speight
BSc, MIEE

Chief Electrical Engineer (Electrical Systems & Motors)

Contributing author from Drinking Water Inspectorate

Claire R. Jackson
MRSC, FCIWEM

Principal Inspector, Drinking Water Inspectorate; formerly
Senior Chemist, Binnie & Partners (Water Quality
Issues and Standards)

Technical reviewers and advisors
Binnie, Black & Veatch

John Ackers
BSc, MICE, MCIWEM

Roger Brown
BSc, MSc, DIC, FICE, FCIWEM

Ken Harper
BA, MICE, MCIWEM

Terry Heard
BSc, MRSC, MCIWEM

Peter Mason
BSc, MPhil, PhD, FICE

Chris Scott
BSc, MSc, MICE

Others

David Drury MIBiol, MCIWEM   Drinking Water Inspectorate
Bob Hulsey BS, MS   Black & Veatch
Owen Hydes OBE, BSc, MRSC   Drinking Water Inspectorate
Frank Law BSc, MICE, FCIWEM   Institute of Hydrology
Jon McLean BSc, MBA, MCIM   Hanovia Limited
Pierre Mouchet PM, Eng.Agr.& Forestry   Degremont
Trevor Peploe B.Tech, MRSC, MIWEM   Paterson Candy Limited
Mark Smith BSc, MRSC   Drinking Water Inspectorate
Neil Wade MA, FIMechE   Mott, Ewbank Preece

Abbreviations used in bibliographies

ASCE	American Society of Civil Engineers, Reston, VA, USA
ASME	American Society of Mechanical Engineers, New York
AWWA	American Waterworks Association, Denver, USA
BHRA	British Hydromechanics Group, Cranfield, UK
CIRIA	Construction Industry Research & Information Association, London
CIWEM	Chartered Institution of Water & Evironmental Management, London
DoE	Department of the Environment, London
FAO	Food and Agriculture Organization, Rome
HMSO	Her Majesty's Stationery Office, London
IAHS	International Association of Hydrological Sciences, Wallingford, UK
ICE	Institution of Civil Engineers, London
I Chem E	Institution of Chemical Engineers, Rugby, UK
IoH	Institute of Hydrology, Wallingford, UK
IWE/IWES	Institution of Water Engineers (& Scientists), London
IWSA	International Water Supply Association, London
NEWA	New England Waterworks Association, Massachusetts, USA
NERC	Natural Environmental Research Council, Swindon, UK
OFWAT	Office of Water Services, Birmingham, UK
SWTE	Society for Water Treatment and Examination, London
USGS	United States Geological Survey, Washington, USA
WHO	World Health Organization, Geneva
WMO	World Meteorological Oganization, Geneva
WRc	Water Research Centre, Medmenham, UK

1

Public water supply requirement and its measurement

1.1 Categories of consumption

It is useful to divide public water consumption into the following categories.

(1) Domestic
In-house use – for drinking, cooking, ablution, sanitation, house cleaning, laundry, patio and car washing.
Out-of-house use – for garden watering, lawn sprinkling and bathing pools.
Standpipe use – from standpipes and public fountains.

(2) Trade and industrial
Industrial – for factories, industries, power stations, docks, etc.
Commercial – for shops, offices, restaurants, hotels, railway stations, airports, small trades and workshops, etc.
Institutional – for hospitals, schools, universities, government offices, military establishments, etc.

(3) Agricultural
Agricultural use is for crops, livestock, horticulture, greenhouses, dairies, farmsteads.

(4) Public
Public use is for public parks, green areas, street watering, water mains and sewer flushing, fire-fighting.

(5) Losses
Distribution losses – leakage from mains and service pipes upstream of consumers' meters or property boundary; leaks from valves, hydrants and washouts, leakage and overflows from service reservoirs.
Consumer wastage – leakage and wastage on consumers' premises and from their supply pipes, misuse or unnecessary use of water by consumers.
Metering and other losses – source meter errors, supply meter errors, unauthorised or unrecorded consumption.

Many domestic supplies are not metered. In the UK about 14% of domestic supplies in England and Wales were metered in 1999, but none in Scotland and Northern Ireland. A DoE report[1] mentions that in Denmark only town houses are normally metered; in the Netherlands about 24% of houses are unmetered; and in Germany only 30–40% of households are individually metered, the rest being block-metered. In the USA, whilst metering of domestic supplies is widespread, it is not universal. New York City did not start a programme to meter all domestic supplies until 1987. This was in response to the Delaware River Basin Commission's requirement that the major cities it supplies in the states of Delaware, New Jersey, New York and Pennsylvania, should achieve metering of all services within a ten year period.[2] A survey by the Asian Development Bank (ADB) in 1996[3] showed that, of 27 Asian cities serving over 1 million people, only 15 were fully metered and six metered less than 7% of their connections (Calcutta 0%: Karachi 1%).

Trade and industrial supplies are usually metered because they are a major source of income to a water undertaking. In the UK many small shops and offices occupied only in the daytime used not to be metered, but now generally are, even though their consumption is small. Overseas standpipe supplies are not metered and are usually given free. The city of Bombay, for instance, supplies 400 Ml/day (megalitres per day) to some 6 million people in its slums.[4] In many countries large quantities of water are used for watering public parks and green areas and supplying government offices and military establishments, etc. They are often not metered nor paid for if the government (or state or city) supplies the water. The result is that reported water distribution losses depend on the accuracy with which the unmetered consumption is estimated.

1.2   Levels of total consumption

The usual measure of total consumption is the amount supplied per head of population; but in many cases the population served is not known accurately. In large cities there may be thousands of commuters coming in daily from outside; in holiday areas the population may double for part of the year. Other factors having a major influence on consumption figures are:

● whether the available supplies and pressure are sufficient to meet the demand, 24 hour or intermittent;
● the number of population using standpipes;
● the extent to which waterborne sanitation is available;
● the undertaking's efficiency in metering and billing, and in controlling leakage and wastage;
● how much of the supply goes to relatively few large industrial consumers;
● the climate.

Some countries, such as India, rarely have any cities with a 24-hour supply. The ADB survey of 1996 already referred to[3] showed that 40% of 50 Asian cities surveyed did not have a 24-hour supply, and about two-thirds had street standpipe supplies. Hence comparison of average total consumption between undertakings is not informative. High consumption can be caused by large industrial demand and low consumption by a shortage of supplies. However the general range of total supplies per capita is:

● from 600 to 800 lcd (litres per capita per day) in the big industrial cities of USA;
● from 300 to 550 lcd for many major cities and urban areas throughout the world;

● from 90 to 150 lcd in areas where supplies are short or there are many street standpipes, or many of the population have private wells.

In England and Wales the average total supply was 288 lcd (1998/99). In Scotland it was 460 lcd and in Northern Ireland 407 lcd in 1997/98, mainly because these areas have a high rainfall providing plentiful supplies of good quality water.

1.3   Consumption surveys

A consumption survey is necessary when losses appear to be large, or consumers in some areas cannot get an adequate supply, or metering and billing practices appear to be inefficient. This situation often occurs on many undertakings throughout the world where lack of money and technical resources has resulted in water supply systems where leakage and consumer wastage is high, and records of consumption are unreliable. In such a situation, it should be noted that:

> Total supply =Total legitimate potential demand
>
> > *plus* consumer wastage and distribution losses
> >
> > *minus* unsatisfied demand.

Hence the total supply can seem to be adequate when expressed as the water available per head of population, but this may conceal the fact that, due to excessive wastage and leakage, there is much unsatisfied demand because some consumers do not get the water they need. It is then necessary to conduct a consumption survey to find the state of the system. The steps involved are the following.

(1) Log the initial state of the undertaking before any remedial work is started, by marking on a map of the distribution system areas where water pressure is too low for consumers to get what they need (a) at peak demand times, and (b) during the whole of the daytime.

(2) Check the accuracy of source meters, e.g. by diverting the source output over a temporary measuring weir, or by measuring the input to a tank.

(3) Check the general validity of supply meter readings by, for instance, check-reading a number of meters over a period and comparing with the readings billed.

   Find the typical number of supply meters found stopped at any one time, and investigate what billings are made when meters are found stopped.

   Find the average age of meters and how frequently they are brought in for testing and repair.

   Test some meters of typical size and age for accuracy. Where more than 15% of meters are found stopped at any one time, or many meters are over 10 years old and not brought in for testing and repair, a substantial amount of under-recording must be suspected.

(4) Assess typical domestic consumption per capita by classifying dwellings into five or six classes, and test metering 30–35 dwellings typical of each class. (See Section 1.4 below.)

(5) Test meter a few typical standpipes and, by estimating the population reliant on each, estimate the typical standpipe consumption per capita.

(6) Examine the supply meters on all large trade and industrial supplies; check the accuracy of those found in poor condition.

List the largest potential trade consumers and check their billing records to see whether they seem reasonable having regard to the size of their supply pipe, their hours of take, and amount of water likely to be used for their production. It can be found that some major consumer is missed from the billings, or an establishment may have two supply feeds of which only one is metered.

(7) Meter, or by some other means estimate, the amount of water supplied unmetered to such as government or municipal offices, and also to public parks and gardens to get a measure of their probable consumption.

From the foregoing an estimate can be made of the probable total potential demand on the system in the following manner. On a map of the distribution supply districts, mark areas of the different classes of housing. Using appropriate population densities per hectare and measuring the areas of each class of housing within a district, estimate the total domestic demand per district by using the appropriate consumption per capita derived from Step 4. Add an allowance for unavoidable consumer wastage which will not have been registered by the test metering under Step 4.

In each district any standpipe consumption should be added, and the trade and other non-domestic demands apportioned according to the character of the district and the location of major trade consumers. An allowance for a reasonable degree of unavoidable distribution leakage should be added, usually expressed as a percentage addition to the total domestic and trade demand for each district. This gives the total average daily demand on the whole system, broken down into sufficiently small supply districts for the demand in each district to be distributed to 'nodal points' of the mains layout, i.e. key junctions of mains. These 'nodal demands' can then form the basis for an hydraulic analysis of flows in the distribution system, as described in Section 14.14. From this can be ascertained the adequacy of the distribution system to meet the demands.

1.4   Test metering in-house domestic consumption

To assess average domestic consumption per person when supplies are not metered or records of metered domestic consumption are not reliable, it is necessary to test meter a sample of properties. It is best to rely on the results of test metering 30–35 households from each of five or six classes of households. Larger samples are difficult to conduct accurately because of the need to ensure all meters work accurately, all properties are leakfree, and the difficulty of keeping check of the number of people in each household. The households test metered should be typical of their class; they should not be chosen at random because the sample size is too small for random selection and could result in a bias of the sample mean towards the higher or lower end of the range within the class. Only about five or six classes of household should be adopted because it is difficult to distinguish between a larger range of households with any certainty. The test period should be 2–4 weeks, avoiding holiday times and, if possible, extremes of weather. Meter readings and occupancy rates should be ascertained weekly. Theoretically a sample size of at least 30 households is required to provide a reasonable estimate of the mean consumption in a given class of households. In practice 35 properties per class will need to be test metered to get a minimum of 30 valid results because of mishaps – a stopped meter, a leak discovered, or occupants gone away, etc.

Unfortunately all such tests show a wide scatter of results, as illustrated in Fig. 1.1. The mean of a sample can therefore be substantially influenced by a few households where the

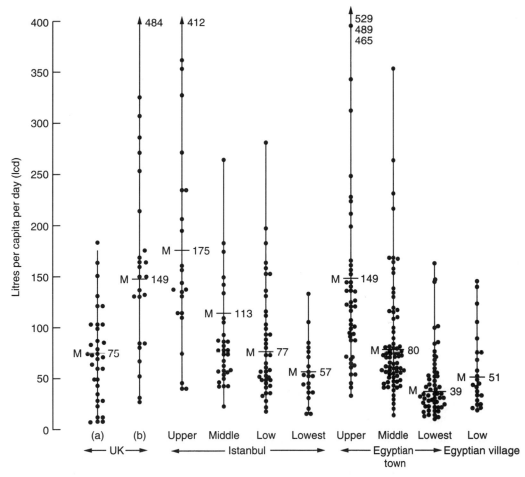

Fig. 1.1 Variation of domestic consumption per capita within a given class of dwelling. UK: (a) 1–2 occupancy and low number of water consuming appliances; (b) 2–4 occupancy in mainly detached houses with high number of appliances. Data from Russac D. A. V., Rushton, K. R. and Simpson, R. J. Insights into domestic demand from a metering trial. *JIWEM*, June 1991, pp. 342–351. Istanbul and Egyptian data from Binnie and Partners' Reports.

consumption seems extraordinarily high. Provided such high consumption is not due to meter reading error, these figures should not be excluded because domestic consumption is so highly variable. The problem is, however, that the sample size is too small to evaluate the incidence of such high consumers which may be, for example, one in 20, so that a sample size of 30 may contain no such high consumers, or one or two. However, if five classes of housing are adopted there will be at least five separate samples from which to judge, roughly, the frequency of such exceptional consumption.

The mean value of per capita consumption should be the total consumption in the 30 or so households tested in a given class, divided by the total occupancy during the test period, because the total domestic demand is estimated on the basis of the population in each class of housing.

An alternative to test metering individual properties is to meter the flow through a main supplying properties of the same class. But this method does not reveal high or low household consumption that may need investigation, and may also include some leakage from the supply main and service pipes downstream of the metering point. However it is a useful supplementary method of assessing mean consumption if reasonably leak-free conditions can be assured. Relying on analysis of existing billing records is inadvisable because the purpose of test metering is to check the validity of billings.

1.5   Confidence limits for sample means

When the mean $\bar{x}$ of a sample is obtained, the probability that the mean μ for the whole population lies within a range of values about $\bar{x}$ is given by the formula $\mu = \bar{x} \pm Z\,s/\sqrt{n}$, where:

- Z is the probability coefficient, value 1.96 for 95% probability, 2.33 for 98% probability, and 2.58 for 99% probability;
- s is the standard deviation of the sample values:

$$s = \sqrt{\frac{\sum(x - \bar{x})^2}{n - 1}}$$

- n is the number of values in the sample.

The formula is based on the assumption that the means of random samples from a population are normally distributed. The term $s/\sqrt{n}$ is called the 'standard error of the mean' or 'standard error' (SE) for short.

The formula is not as useful as might be expected because consumption per capita varies so greatly that the standard deviation, s, of samples is seldom below 30 lcd and often higher. This results in large samples being necessary to get a useful degree of accuracy in the mean. For example, with an s value of 30 lcd a sample size of 865 is needed for 95% probability the population mean lies within ±2 lcd of the sample mean. To test meter such a large sample is impracticable. If, alternatively, existing billing records are used for the analysis, the weakness of this procedure is that the mean will include metering and billing errors, consumer wastage and leakage, and it will be difficult to know accurately the population in residence during the billing period.

1.6   Components of domestic in-house consumption

Table 1.1 gives some analyses of in-house domestic consumption in litres per resident per day. Consumption is influenced by differences of climate, ablution habits, and the number and capacity of water fittings installed. The figures are not strictly comparable because they are obtained by different methods and are from samples inevitably limited in size. In the 'diary' method, residents book down daily the number of times they use each fitting for a period, and these are multiplied by the average consumption for each type of fitting used. In the 'data logging' method, pulsed output consumer meters record flows at very frequent time intervals and these are analysed by computer, the different uses being identified by differences of flow pattern. In a few instances, notably work undertaken by Anglia Water in UK, as many as 14 meters were installed in each household at every point of use.

Table 1.1 Some breakdowns of domestic in-house consumption (in lcd)

| | UK | | | NORWAY | MALDIVES | USA | | | | | | | | |
	Anglian Water Co. 1993 (a)	S. West Water Co. 1985 (b)	Scotland 1991 (c)	Inst. of Water Research 1983 (d)	Male Study 1983 (e)	Dept. Housing Estimate 1983 (f)	MWD of S. California average 1993 (g)	Phoenix, Arizona pre-1980 (h) / 1990 Code (i)		Denver, Colorado (j)	Boulder, Colorado (k)	San Diego, California (l) 1996–1998	Seattle, Washington (m)	Tampa, Florida (n)
Toilet use	48	33	43	30	28	82	113	83	24	80	75	60	65	63
Ablution														
– baths	19			40	nil	26		26	26	6	5	2	4	4
– showers	6	41	50		85	61	95	87	48	49	50	34	43	39
– washbasin	13													
Clothes washing	30	18	37	25	35	64	64	63	42	59	53	62	45	54
Kitchen use and cleaning	23	31	5	28	27	35	49	34	32	40	44	41	33	45
Dishwasher	2		12			9	15	9	5	5	5	3	4	2
Car washing and patio use	4	9	1	7	nil	excl.	excl.	excl.	excl.	excl.	excl.	excl.	excl.	excl.
TOTAL	145	132	148	130	175	277	336	302	177	239	232	202	194	207

Notes and sources.

(a) Data log of 100 properties 1992–93. *JIWEM*, Oct 1995, pp. 477–485

(b) Diary log of 863 properties. *JIWEM*, Dec 1988 pp. 626–631

(c) Wright P. Water Resources Management in Scotland, *JIWEM*, April 1995, pp. 153–163

(d) *Water Bulletin*, 18 March 1983, pp. 12–13

(e) Diary log of 63 properties with private wells, Binnie & Partners Survey.

(f) Estimate for non-conserving households with 19–26 l/flush toilets and 18–30 l/min showers. *JAWWA*, Mar 1987, pp. 52–58

(g) Metropolitan Water District of Southern California (MWD), Supplier to undertakings serving 14.9 million population. Figures are estimated average for multi-family residences and may include leakage, etc. *Report on Urban Use Characteristics*, MWD, April 1993 by courtesy T. A. Blair, Senior Resource Engineer, MWD.

(h), (i) Phoenix 1990 Code estimate is for households having 6 l/flush toilets 9 l/min showers, 139 l/load washing machines and 32 l/load dishwashers. Data by courtesy T. M. Babcock, Water Resource Specialist, Phoenix City Water Dept.

(j)–(n) Logs of 99–100 households for each city. Part of *Residential End Uses of Water Study 1996–98*, funded by AWWA Research Foundation and 12 municipalities. Data copyright Aquacraft Inc. and AWWA Research Foundation, by courtesy P. W. Mayer, Study project engineer, Aquacraft Inc. Water Engineering & Management, Boulder, Colorado.

The UK analyses shown in Table 1.1 are for relatively low domestic consumption, partly because of their relatively early date and partly because they relate to housing areas outside the most affluent parts of England. As Section 1.8 shows, average domestic consumption has now risen to 146 lcd, but in the more affluent parts of England it averages about 160 lcd.

In USA the higher domestic consumption is primarily due to the larger capacity toilet and shower fittings at present still in use. These are gradually to be replaced by lower consumption fittings under the US Energy Policy Act 1992 as set out in Table 1.2. It will take some years before sufficient numbers of the new 6-litre flush toilets are installed to have a significant effect on consumption; but some water authorities in USA are encouraging this by offering rebates on water charges to customers who change to 6-litre toilets. US toilets, however, are prone to leak because they use a horizontal flap or drop valve which is raised to start the flush, the valve re-seating when the toilet tank is nearly empty. New York reported that, of 80 000 6-litre flush toilets installed, 1.5% were found leaking due to faulty flap valves.[5] Consequently some US water suppliers allow 15 lcd for 'toilet leakage' in their estimates of consumption.

Air conditioner and humidifier usage is excluded from the US figures in Table 1.1. Evaporative or 'desert' coolers are used in some parts of western USA where the climate is hot and arid. Such coolers use a fan to draw air through a vertical porous pad of cellulose fibre, down which water is trickled. One type recirculates the surplus water and uses 12–15 l/h; another type bleeds off part of the surplus water to reduce deposits on the porous pad and uses up to 40 l/h. The consumption effect of such coolers depends on the percentage of dwellings equipped with them and the length of the hot season. Phoenix City in Arizona,[6] where the summer climate is exceptionally hot and dry (July average 40°C and 29% daytime humidity) estimates evaporative coolers add 86–98 lcd to annual average daily domestic consumption, based on 48% of Pheonix residences possessing them. The MWD of Southern California[7] estimates that only about 1% of the average residential water use is used for cooling, representing about 5 lcd; but this is an overall average for coastal, inland valley, and desert climate regions with the greater part of the population residing in the coastal region. In hot and wet climates, as in the tropics with a high humidity, evaporative coolers are not used. The electrical air conditioners used in such climates do not consume water.

1.7 Ex-house use for garden irrigation and bathing pools

Garden watering in UK can increase daily consumption by 30–50% during a prolonged dry period, but the total amount used in a year depends on whether a 'dry' or 'wet' summer is experienced. In the UK's changeable climate, dry summer periods are often of relatively short duration and the time-lag between the start of a dry period and the build-up of garden watering demand means that the peak of the latter is short lived so the amount used on garden watering in a year of 'average' weather, when expressed as an average annual daily amount, is not large enough to be separately quoted. In the north of UK prolonged dry periods are rare; but in the drier south-eastern part of England, garden watering has been estimated to account for nearly 5% of the total supply during recent years of prolonged low summer rainfall, such as in 1995–96.

In the USA the water used for irrigation of household lawns and gardens is very substantial – exceeding the in-house consumption. In the drier western states, summer

Table 1.2   Sizing of domestic water fittings

	USA Energy Policy Act 1992		UK Water Supply (Water Fittings) Regulations 1999
	Before	After	
	Note (a)		Note (b)
WC toilets			
– size of flush	13–26 litres	6 litres	7.5 litres until 1 Jan. 2001, 6 litres from 1 Jan. 2001 – but cisterns installed before 1 July 1999 can be replaced by same volume
– type of flush	flap-valve	flap-valve	siphonic until 1 Jan. 2001 from 1 Jan. 2001 flushing or pressure flushing cistern
Showerhead flow	11–30 l/min	9.5 l/min max. at 8 psi	as BS 6700: 1997, i.e. 12 l/min (13 mm supply)
Faucet, i.e. tap	10–26 l/min	9.5 l/min max. at 8 psi	as BS 6700: 1997, i.e. washbasin 6–9 l/min (13 mm) kitchen 6–12 l/min (13 mm) bath 12–16 l/min (19 mm)

Typical consumption of other water using appliances

	As Phoenix Water Undertaking USA		UK required for efficiency as Water Supply (Water Fittings) Regulations 1999
	Prior to new codes	Under new codes	
	Note (c)		Note (d)
Clothes washing machines	208 l/load	139 l/load	27 litres per kg of load (av. actual 56 (range 42–77) litres per 2.5 kg load)
Dishwashing machines	53 l/load	32 l/load	4.5 litres per place setting (av. actual 23 (range 18–30) litres per full load).

Notes.
(a) The US Energy Policy Act 1992 applied to the manufacture and installation of new fittings as from 1 Jan. 1994.
(b) The Regulations imply that a 7.5-litre siphonic flushing stern installed after 1 July 1999 could not be renewed by one of the same size after 31 Dec. 2000. The mode of flushing for 6-litre cisterns is not stated, thus flap-valve discharge is permitted and siphonic discharge is not precluded although it is unlikely to be effective with 6-litre cisterns unless new designs are possible.
(c) The code figure examples are those adopted by Phoenix Water Undertaking, 1997.
(d) Values are from *Which* publication reports Feb. 1998, p. 42 (clothes washers); Dec. 1996, p. 42 (dishwashers).

irrigation use may range from 300 lcd (e.g. Boulder, San Diego) to 600 lcd (Eugene) or even 800 lcd (Denver).[8] Similar high consumption for garden irrigation is experienced in south Australia, the summer use ranging from 150–300 lcd in metered supply areas to 400 lcd in unmetered areas.[9] The total amount used depends on the length of the dry season and the mix of housing in an area. The MWD of southern California[7] estimates average annual household irrigation use amounts to about 175 lcd for people living in single family dwellings and about 70 lcd for those living in multi-family dwellings, i.e. blocks of flats, etc. These MWD averages conceal a wide variation according to climatic location; residences in desert regions having nearly twice the average consumption of those in the coastal regions, with those sited in inland valley regions coming intermediately.

Water used for swimming pools depends on the incidence of such pools in an area. Coupled with car washing and miscellaneous outdoor uses, the MWD of Southern California estimates the consumption this represents is about 3% of the total average residential consumption, or 15–20 lcd.

1.8   General levels of in-house domestic consumption

In the UK in-house consumption is strongly related to the class of property served and, in many other countries this also applies, as shown in Figure 1.2. Excluding USA where large capacity water fittings are still in use, domestic in-house consumption for average middle class properties having a kitchen, a bathroom or washroom, and some form of waterborne sanitation, falls into a fairly narrow range of 120–155 lcd irrespective of climate or country. Other factors, such as occupancy and household income, influence consumption; but these are unstable factors impracticable to ascertain. The ACORN socio-economic, property type and location, etc. classification which is available in UK, although used by

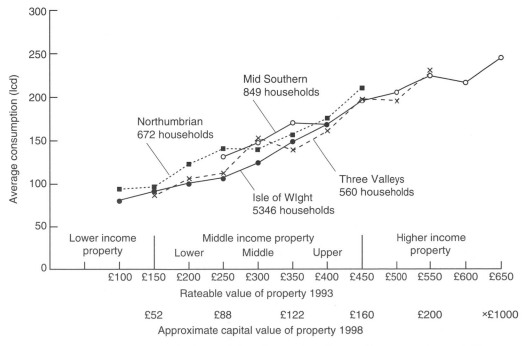

Fig. 1.2   Relationship between in-house domestic consumption and property class – UK water companies. Source: National Metering Trials 1989–1993, Final Report 1993.

some water companies, is not entirely satisfactory,[10] but has the advantage of being a 'ready-made' classification. The class of dwelling occupied, in terms of type (flat or house, etc.) and value (size, age, etc.), is the most practicable basis to use for estimating domestic consumption, since it permits visual identification of the predominant class in an area, and the average occupancy therein can be estimated with reasonable accuracy.

The influence of dwelling class can be seen in the figures of average domestic consumption in England and Wales given below.[11] The higher domestic consumption reported by the Water-only Companies reflects the fact that 87% of the population they serve resides in the more affluent southern parts of England where housing standards are generally higher than in the north.

1998–1999	10 Regional water and sewerage companies	16 Water-only companies*
Population in households	39.38 million	11.34 million
	Reported average domestic consumption excluding supply pipe leakage	
Un-metered households	145 lcd	159 lcd
Metered households	135 lcd	140 lcd
(Percent popn metered)	(11.8%)	(10.3%)
Weighted means	143 lcd	157 lcd
Overall mean	146 lcd	

*Figures exclude Cholderton Water Co.

The lower figures for metered consumption cannot be taken as due to metering because metering is optional for most householders, and householders choosing to have a metered supply will therefore tend to be those expecting to pay less for their supply because of their low consumption, such as single occupants or elderly retired people. This is reflected in the household occupancy figures which average 2.0–2.2 people/household for metered properties and 2.5 people/household for unmetered properties.

Occupancy affects consumption per capita in individual households. The consumption per capita in a one-person household tends to be 15–25% more than that in a three-person household; while occupancies higher than three per household show a decline in per capita consumption. But to estimate consumption for a whole area of a distribution system, it is the average occupancy which is important. In England and Wales, average occupancy according to census figures[12] showed a decline from 2.70 in 1981 to 2.47 in 1991 and is expected to continue to fall slowly. The 1991 figures varied from 2.10–2.20 in retirement areas, to figures of 2.55–2.67 in the denser populated urban areas. In the USA mean household occupancy reported by the US Census Bureau was 2.63 people in 1990 and this also is expected to decline slowly. Individual US water suppliers report average occupancies varying from 2.4 in multi-family complexes to 3.1 in single family residences; the usual range being 2.5–2.8 people per household. In many other countries, however, such as Egypt, India, etc. the average occupancy is five or six people per household.

Domestic consumption reported by various countries is not necessarily comparable mainly because there is no assurance that the figures quoted are produced on the same basis. In the USA the typical in-house consumption (excluding cooling) is 180–230 lcd, but this can be expected to reduce gradually with increased installation of low flush toilets and reduced rates of consumption by other fittings. Average domestic plus small trade consumption reported by European countries for 1995[13] centred about 150–160 lcd (range 113–260 lcd), Switzerland, Italy and Norway being the only ones above 200 lcd. The ADB survey of 1996[3] showed that, of 27 Asian cities serving over 1 million population, the middle third had 135–186 lcd domestic consumption: selected individual domestic consumption being – Hong Kong 112 lcd; Jakarta 135; Shanghai 143; Colombo 165; Singapore 183; Kuala Lumpur 200; Bangkok 265.

Table 1.3 gives reasonable water supply allowances to make for average in-house consumption according to the type of property served. The figures are not to be taken as exact or always applicable, but they represent the average supply that should be sufficient to meet usual in-house domestic requirements, given reasonably efficient control of consumer wastage and leakage, and reasonably efficient metering, billing, and income collection practices. The figures exclude water used for cooling, bathing pools and the irrigation of lawns and gardens.

1.9   Standpipe demand

Standpipe consumption is influenced by the distance over which consumers must fetch water, the usages permitted from the standpipe, the degree of control exercised over use at the standpipe, and the daily hours of supply. At a WHO conference[14] standpipe consumption of 20–60 lcd was reported from various parts of the world. The range of uses can be as follows:

(1) water taken away for drinking and cooking only;
(2) additionally, water taken away for household cleaning, clothes washing, etc.;
(3) additionally for bathing and laundering at the standpipe;
(4) plus for watering animals at or near the standpipe;

together with

(5) spillage, wastage and cleansing vessels at the standpipe.

Where the supply is for drinking and cooking only, the consumption in terms of water taken away may be only 7 lcd, but spillage and wastage at the tap will cause the minimum consumption to be 10 lcd. Take-away consumption for items (1) and (2) is about 15–20 lcd, but wastage and spillage will raise this to 25 lcd which is the minimum design figure that should normally be allowed. Where there is little control exercised over consumers' usage of water at a standpipe and, in any case where bathing and clothes washing takes place at or near the standpipe, then 45 lcd needs to be provided. If purpose-built bathing and laundering facilities are provided the consumption will rise to 65 lcd. In India 50 lcd is the usual standpipe design allowance; in Indonesia where the water is sold from standpipes, 15–20 lcd occurs. In rural Egypt where uncontrolled all-purpose usage tends to occur, including watering of animals, test figures of 46, 51 and 71 lcd were obtained.[15]

In low income communities in the lesser developed countries it is often the practice that one householder has an external 'yard tap' which he permits his neighbours to use, usually charging them for such use. These are, in effect, 'private standpipes' and, because carrying distances are short, the consumption can be about 90 lcd based on numbers reliant on the yard tap water.

A public standpipe should supply at a rate sufficient to fill consumers' receptacles in a reasonably short time, otherwise consumers may damage the standpipe in an attempt to

Table 1.3 In-house domestic water and standpipe demand – suggested design allowances

Class occupancy and type of property	UK and Europe (lcd)	Elsewhere in warm climates (lcd)
A1 Highest income groups: villas, large detached houses, large luxury flats	190	230–250
A2 Upper middle income groups: detached houses, large flats	165	200–230
B1 Average middle income groups: two- or three-bedroomed houses or flats, with one or two WCs, one kitchen, one bath and/or shower	155	180–200
B2 Lower middle income groups: generally small houses or flats with one WC, one kitchen, one bathroom	140	
– ditto overseas block metered		160+
– ditto overseas individually metered		130
C1 Low income groups: small cottages; flatlets; bedsits with kitchen and bathroom	90–110	
Overseas:		
C1 Tenement blocks, high density occupation with one shower, one Asian toilet, one or two taps		
– block metered or free		130+
– individually metered		90
C2 Lowest income groups: above poverty line: low grade tenement blocks with one- or two-roomed dwellings and high density occupation		
– with communal washrooms (unmetered)		110
– with one tap and one Asian toilet per household: block metered		90
C3 Lowest income groups: one tap dwellings with shared toilet or none; dwellings with intermittent supplies		50–55
D Standpipe supplies		
– in urban areas with no control		70+
– in rural areas under village control		45
– rural but with washing and laundering facilities at the standpipe		65
– minimum for drinking, cooking and ablution		25
– drinking and cooking only		8–10

Note. Figures exclude lawn and garden irrigation, bathing pool use and use of evaporative coolers; but include an allowance for unavoidable consumer wastage.

get a better flow. In Rangoon the water undertaking uses short standpipes, permitting a typical vessel to be stood on the ground below the tap, so water is not wasted. The water pressure is low but the tap is 19 mm (3/4 in) size so that it gives a good flow. The tap is press operated, the body being made of cast iron so that it does not present a temptation for theft. Brass taps often get stolen from standpipes. The hours of supply need to be adequate morning and evening, and proper drainage should be provided at the standpipe to take away spillage.

1.10 Industrial and commercial demand

Industrial demand for water can be divided into four categories.

(1) *Cooling water demand* – usually abstracted direct from rivers or estuaries and returned to the same with little loss. None is used from the public supply.
(2) *Major industrial demand* – factories using upwards of 1000 m^3/day for such industries as paper making, chemical manufacturing, production of iron and steel, oil refining, etc. Such large supplies are frequently obtained from private sources. A few public water undertakings supply 'non-potable' water for such purposes. Such a supply is not intended for drinking or culinary uses and may undergo no treatment by the water supplier, or it may receive precautionary disinfection in case it is inadvertently used by someone for drinking. Dependent upon the type of industry served, a user may apply extensive treatment to the supply because some industries require very low suspended or dissolved solids in the water. These non-potable supplies are always shown separately from the 'public' water supply statistics.
(3) *Large industrial demand* – factories using 100–500 m^3/day for food processing, vegetable washing, drinks bottling, chemical products, etc. These demands are often met from the public supply.
(4) *Medium to small industrial demand* – factories and all kinds of small manufacturers using less than 50 m^3/day, the great majority taking their water from the public supply.

Since all factories and manufacturers have staff on their premises, they will all take some water from the public supply for the use of such staff, irrespective of whether they have a private source of supply. Some more consistent figures of demand for various types of industry are given in Table 1.4. Often it will be found that about 90% of the total industrial demand in a large industrial area is accounted for by only 10–13% of the industrial consumers. Hence it is important to check the accuracy of metering these few large industrial consumers when conducting a consumption survey of the type outlined in Section 1.3 above, because an error in measuring their consumption can have a major effect on assessment of losses.

Figures for light industrial demand in industrial estates are given in Table 1.5. Only a few service industries on such estates, e.g. milk bottling, concrete block making, laundries, etc. use much process water. Many light industries, such as those involved in printing, timber products, garment making, etc. use water only for their staffs.

Commercial and institutional metered demand for shops, offices, schools, restaurants, hotels, hospitals, small workshops and similar activities common in urban areas, appears to run at an average of about 25 lcd over the whole population served in England. This includes the domestic use by people living in such premises or in attached living quarters. But in the USA it can be much higher: the Metropolitan Water District of Southern California estimates it is 125–133 lcd for their region, of which almost 30% is consumption for cooling systems and outdoor irrigation in the hot climate.[7] Typical allowances made for demand in certain types of commercial and institutional premises in UK are given in Table 1.6.

Table 1.4 Typical magnitudes of water demand by various manufacturing industries (note individual factory consumption varies widely and recycling of water can reduce consumption substantially)

Process or product	Data source	Consumption quoted
Automobiles	(a),(b)	12 m³/vehicle; 5 m³/car
Bakery	(b),(c)	2 m³/t of product; 1.4 m³/t
Brewery	(b),(d)	7 m³/m³; 7 m³/m³ +2 m³/m³ for malting
Canning		
– generally	(a),(e)	25 m³/t, 20 m³/t
– meat, vegetables	(f)	30–35 m³/t
– fish	(f)	60 m³/t
Confectionery	(c),(f)	12 m³/t; 35 m³/t
Chemicals; plastics	(a),(e)	9–23 m³/t; 30–80 m³/t
Concrete products	(c)	1 m³/t
Concrete blocks	(c)	1 m³ per 200 blocks
Dyeing	(b)	83 m³/t of fabric
Fish processing and packaging	(d)	8.5 l/kg
Food processing and packaging:		
– biscuits, pet foods, cereals, pasta	(f)	8–15 m³/t
– jams, chocolate, cheese, cane sugar	(f)	20 m³/t
– frozen vegetables, poultry	(f)	45–50 m³/t
Iron castings (small production)	(c)	0.4 m³/t used in sand moulds
Laundry	(b),(c)	20 m³/t of laundry; 26 m³/t
Leather production	(a)	70 m³/t
Meat production, slaughtering	(a)	5 m³/t livestock;
	(b),(f)	40 m³/t fresh meat; 16.5 m³/t
Milk bottling	(b),(d)	3 m³/m³; 2 m³/m³
Paper production (wood pulp)		
UK	(e)	135–150 m³/t; 90 m³/t av. UK
USA; South Africa	(g)	240 m³/t av.; 44 m³/t
fine quality paper	(g)	800 m³/t or more
Rubber, synthetic	(a)	12–13 m³/t
Soft drinks	(b),(d)	7 m³/m³; 3 m³/m³
Terrazo tiles	(c)	1 m³ per 10 to 20 m³ of tiles
Textile processing	(d)	300 m³/t av.

Information sources.
(a) Weeks C. R. and Mahon T. A. A Comparison of Urban Water Use in Australia and US. *JAWWA*, Apr. 1973, p. 232.
(b) Thackray J. E. and Archibald G. G., The Severn-Trent Studies of Industrial Use. *Proc. ICE*, August 1981.
(c) Binnie & Partners. Survey of Water Use in Riyadh Factories, 1977.
(d) Schutte C. F. and Pretorious W. A. Water Demand and Population Growth, *Water South Africa*, 1997, pp. 127–133.
(e) Rees J., *The Industrial Demand for Water in SE England*. Weidenfeld, London, 1969.
(f) Whitman W. E. and Holdsworth S. D. *Water Use in the Food Industry*. Leatherhead Food Research Association, 1975.
(g) Van der Leeden F. *Water Resources of the World*. Water Information Centre Inc., 1975.

Table 1.5 Light industrial estate water consumption

Usage	Consumption allowance
Basic factory requirement for cleaning and sanitation	0.05 m³/day per worker
Average consumption in light industrial estates with no large water-consuming factories	0.25–0.50 m³/day per worker
Average consumption in light industrial estates which include a proportion of factories engaged in food-processing ice-making and soft drinks manufacture	0.90–1.10 m³/day per worker
Typical factory consumption SE England:[a]	
– clothing and textiles	90% under 6 m³/day per factory
– leather, fur, furniture, timber products, printing, metalworking and precision engineering.	70–80% under 25 m³/day per factory
– plastics, rubber, chemical products, mechanical engineering, and non-metallic products	70–85% under 125 m³/day per factory

(a) Source: Rees J. *The Industrial Demand for Water in SE England*, Weidenfeld 1969.

Table 1.6 Allowances frequently used for water consumption in commercial and institutional establishments

Usage	Consumption allowance
For small trades, small lock-up shops and offices in urban areas	In the UK – 3–15 lcd. Elsewhere – up to 25 lcd (applied as a per capita allowance to the whole urban population)
Offices	65 l/day per employee (but actual consumption can be three to four times this if waste is not attended to).
Department stores	100–135 l/day per employee
Hospitals	350–500 l/day per bed
Hotels	350–400 l/day per bed; up to 750 l/day per bed luxury hotels in hot climates
Schools	25 l/day per pupil and staff for small schools; rising to 75 l/day per pupil and staff in large schools

Note. The figures for offices, department stores and schools apply to the days when those establishments are open.

1.11 Agricultural demand

Most water for crop irrigation, horticulture, greenhouses, etc. is taken direct from rivers or boreholes because it does not need to be treated. The principal use of the public supply is for the watering of animals via cattle troughs, for cleaning down premises, and for milk bottling. Table 1.7 gives estimates of such consumption.

Table 1.7   Agricultural water demand

Usage		
Dairy farming	(a)	av. 70 l/day per cow in milk at 22 l/day milk plus 20–50 l/day per cow for dairy cleaning
	(b)	4.5 l per litre of milk, but less depending on dry matter content of feed
	(c)	38–95 l/day per cow
Beef cattle	(a)	25–45 l/day per animal for drinking
	(b)	27–33 l/day per animal winter and 33–36 l/day summer
Beef – abattoir use	(c)	av. 1500 litre per animal
Sheep	(a)	3–5 l/day per animal for drinking
Pigs	(a)	2–9 l/day per animal for drinking
Sows in milk	(a)	18–23 l/day per animal
	(b)	23 l/day per animal
Poultry	(a)	20–30 l/day per 100 birds for egg production; 13 l/day per 100 birds for meat production
	(c)	20 l/day per 100 birds for meat production plus 17–20 l/bird for abattoir and packaging
Crop irrigation	(a)	dependent on crop and rainfall; rough rule used for SE England – 13 mm of water per week in growing season ($= 130$ m^3/ha per week)
Glass house production	(a)	20 m^3/day per hectare or more in growing season 12.4 m^3/day per hectare in winter rising to three times this or more in spring and summer

(a) Soffe R. J. ed. *The Agricultural Notebook*, 19th edn. Blackwell Science, Oxford, 1995, p. 532.
(b) Dalal-Clayton D. B. *Blacks Agricultural Dictionary*, 2nd edn. A. & C. Black, London, 1985.
(c) Schutte C.F. and Pretorious W. A. Water Demand and Population Growth. *Water South Africa*, April 1997, p. 127–133.

1.12   Public and miscellaneous use of water

Public use of water in overseas countries can be substantial, especially in hot dry climates where the water is used to maintain parks, green areas, ornamental ponds, fountains and gardens attached to public buildings. The quantities used can be prodigious and have to be assessed for each particular case in relation to the area to be watered and the water demand of the type of cover planted, e.g. grass, date palms, etc. Often the quantity of water used for such purposes is only limited by the available supply. In addition, in many overseas countries, supplies to government owned properties such as government offices, museums, universities, military establishments, are often considered as 'public usage' and are not metered and paid for.

In the UK supplies to public parks, government or local authority offices and similar would be metered and thus form part of the metered consumption. The only unmetered usages of water not paid for would be for firefighting, and for routine maintenance purposes such as for fire hydrant testing, sewer cleansing, and flushing dead ends of mains.

Minor unquantified usages will be for building supplies, usually via a temporary standpipe attached to a fire hydrant. The total of these is estimated by undertakings in England and Wales to be about 1.2% of the total input, equivalent to about 3 lcd on the total population.[11]

1.13   The accuracy of water meters

British Standard 5728: Part 1 specifies the accuracy required for four Classes of cold water meters used in UK. Table 1.8 summarises them. The smaller meters used for domestic supply are usually Class C or D and are 'semi-positive meters' in which an eccentrically pivoted plastic cylinder is caused to rotate by the through-flow of water (see Chapter 15). Larger meters, predominantly used on trade supplies, are usually Class B, and are 'inferential' meters of the rotating vane type. Household domestic supply meters on a 13 or 19 mm diameter service pipe would normally have a maximum capacity of 3 m^3/h because the maximum drawoff rate is unlikely to exceed 40 litres/min, i.e. 2.4 m^3/h (see Section 15.5). Figure 1.3 shows the BS accuracy limits for Class B, C and D meters of 3 m^3/h maximum capacity, and some test results on Class B and C meters obtained by WRc.[16] The latter show that, for meters of 3 m^3/h maximum rating, Class B meters are likely to

Table 1.8   Meter accuracy requirements to BS 5728

Accuracy	Percentage of *nominal flow rate* of a meter to which the accuracy quoted left applies		
	Class B	Class C	Class D
For meters less than 15 m^3/h nominal rating:			
for ±2% accuracy	8% and above*	1.5% and above*	1.15% & above*
for ±5% accuracy	2%–<8%	1.0%–<1.5%	0.75%– <1.15%
For meters of 15 m^3/h nominal rating and above:			
for ±2% accuracy	20% & above*	1.5% and above*	Not specified
for ±5% accuracy	3%–<20%	0.6%–<1.5%	

*'And above' means – to *maximum flow rate*. See Notes below.
Notes.
1. The *maximum flow rate* is defined in BS 5728 as – 'the highest flow rate at which the meter is required to operate in a satisfactory manner for a short period of time without deteriorating'.
2. The *nominal flow rate* is defined as half the maximum flow rate and also that at which 'a water meter is expected to operate in a satisfactory manner under normal conditions of use, i.e. under steady or intermittent flow conditions'.
3. Manufacturers' literature should be consulted for the nominal rating of meters which varies according to the type and size of meter. Some manufacturers quote a nominal rate which is more than half the maximum flow rate.
4. BS 5728 requires the 'starting flow' at which a Class D meter begins to register to be not more than half the minimum flow at which ±5% accuracy is obtained, but does not specify a starting flow for Classes B and C meters.
5. Manufacturers' literature should be consulted for actual accuracies, starting flows and loss of head through meters.

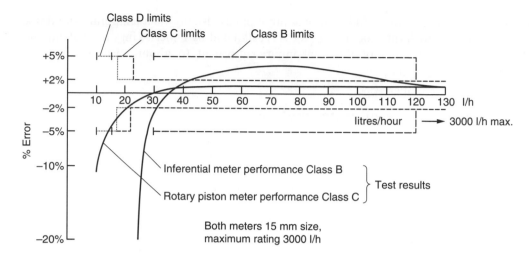

Fig. 1.3   Performance of two types of small water meters at low flows (as WRC Report TR 221 1984).

exceed 5% under-registration at flows below 30 l/h and Class C meters at flows below 15 l/h. The specified limits for similar sized Class D meters show they may exceed 5% error at flows below 11.25 l/h.

Consequently domestic meters tend to under-record consumption due to their inability to measure low flows accurately. A relatively fast dripping tap (4 drips/s) will waste 3–4 l/h, and the thinnest continuous stream about 6 l/h. Both these flows will tend to be under-measured, if recorded at all. The WRc found that low flows caused by near-closed float valves to WC cisterns and storage tanks were also seriously under-recorded, resulting in a mean under-registration of 2.5% for 'direct' supply systems and 6% for 'indirect' systems.[16] In a direct system all cold taps and WC cisterns are fed direct from the mains. In indirect systems only the cold water drinking taps in the kitchen and bathroom are fed direct from mains, the rest are fed from a float valved roof storage tank. The accuracy of all meters also deteriorates with age. Most tests show the great majority of domestic meters under-record; rarely does a meter over-record.

In the National Metering Trials in England 1989–1992 (see Section 1.19) it was found that, of 200 meters withdrawn annually for testing, approximately 20% had failed, most by under-recording; about one-sixth of them due to blockages from particles in the flow.[17] Estimates of the under-recording of supply meters reported by undertakings in England and Wales for 1998–99 are mostly 3–4%. The two water companies having the largest number of domestic meters installed, namely Anglian and Severn Trent, estimated their household meters under-registered by 2.9 and 4.1%, respectively, and their trade meters by 2.6 and 6.6%, respectively.[11] Generally 3% under-recording would be considered an average for semi-positive meters and 5% under-recording for inferential meters; but if meters are over 10 years old and have not been regularly removed for testing and refurbishment, substantially greater under-recording must be suspected.

Trade meters of the inferential (vane) type usually conform to Class B accuracy. One manufacturer claims an accuracy of 1% down to one percent of the maximum flow for meters on 80–150 mm diameter pipework. But such meters have to be carefully sited because an adjacent upstream bend or tee can seriously affect their accuracy. Because of this WRc thinks that their under-recording in practice is greater than that of domestic supply meters. The 'multi-jet' meter (dividing the flow into several streams) was developed to improve the accuracy of such meters.

Source meters and meters on large mains have hitherto been of the Venturi or Dall tube type so that the majority of existing meters are of this type. But in future increasing use is likely to be made of the electromagnetic meter which is simpler to install and more accurate. Venturi and Dall tubes can have an accuracy of ± 1.25%, but must be carefully sited to prevent turbulence caused by an upstream bend or valve, etc. affecting their performance. Even the downstream length of main must be free of obstruction for several pipe diameters (see Section 10.16). Electromagnetic meters have an accuracy of ± 1% but recent developments have enabled some manufacturers to quote ± 0.25% accuracy for their meters (see Section 15.15).

1.14 Consumer wastage

The term 'consumer wastage' is used to mean all leakage and waste of water on consumers' premises and from their supply pipes. Where supplies are plentiful or water is cheap, or waste prevention measures are slack, consumer wastage can be 50 lcd or more. Block metering also tends to result in high consumer wastage because, under this system where one meter measures the supply to a block of flats or households and the landlord pays the water charges (recovering the cost through the rents charged), individual householders do not pay for their own wastage. Table 1.3 shows that 25–40% extra domestic consumption needs to be allowed for block metered premises.

In England and Wales wastage on consumers' internal plumbing systems is not high because for many years the quality and design of water fittings and plumbing has been controlled by byelaws, exercised by every undertaking. These byelaws require, among other things, that all WC and storage cisterns are float valve controlled and the cistern overflow pipe must discharge outside the premises. Hence overflows can be easily noticed or heard by waste inspectors checking premises at night; and the nuisance created by the overspill at the premises may (sometimes) motivate the occupier to take remedial action. Some wastage caused by dripping taps is, however, unavoidable. Some undertakings in UK at one time re-washered taps free of charge as part of their waste reduction routines.

In the UK a major cause of consumer wastage is leakage from consumers' underground supply pipes because many are over 50 years old and made of galvanised iron. In the installation of 50 000 meters on household supplies in the Isle of Wight during the National Metering Trials, 1989–92, it was reported that 8000 service pipes were either repaired or replaced in part or in total, most defects being found on the customer's supply pipe downstream of the boundary stopcock. This represents 1 in 6 of such pipes being found faulty.[18] For 1998–99, undertakings in England and Wales reported that estimated leakage from consumers' unmetered underground supply pipes averaged about 46 l/day per property (range 20–61 l/day); but on properties where meters were installed at the boundary of consumers' premises, supply pipe leakage was reported as averaging 18 l/day per property (range 0–28 l/day).[11]

1.15   Minimum night flows as an indicator of leakage and wastage

The minimum night flow (MNF) to a section of the distribution system can act as an indicator of distribution leakage and consumer wastage in the section. There is, of course, some legitimate demand for water during the hours 01.00 to 04.00 when flow testing takes place and this has to be deducted from the MNF recorded. Table 1.9 shows the lowest night flows found to small residential areas which can be taken as comprising legitimate night domestic demand plus unavoidable consumer wastage from dripping taps. However, the larger the test area metered the more likely it will include night-time consumption for such as – hospitals, nursing homes, police and fire stations, railway stations, airports, clubs, etc. Measuring their consumption during the period of night testing, by reading their meters at the start and end of a night test, is usually not practicable. Instead their usual rates of night consumption have to be found before the test takes place and deducted from it.

Although MNF tests do not measure quantities of water lost, they are a good indicator of the state of a system, and Table 1.10 shows the usual interpretation of results obtained. An MNF test is, however, impracticable if the supply is intermittent or houses have large storages which fill at night. In the UK house-storages are relatively small and therefore are usually full before an MNF test takes place. Increased pressure at night can, however, result in some more water going into storage. Scott reported increased water pressure at night caused his storage tank to fill by a further 11 litres.[19]

Table 1.9   Minimum night flows found on groups of domestic premises

Reference	MNF (l/h per connection)	Test conditions, etc.
Gledhill	1.2	Supply mains tested before and after for leaks; test repeated three times
Reid	1.0	Tests on leak-free portions of distribution systems serving 750–1000 dwellings in Manchester
Cook	1.5–2.1	Deduced from night consumption tests on five systems supplying 200–2200 people after a moderate leak detection exercise
Shaw Cole	1.6	Quoted as usual USA experience
WRc	1.7	Average minimum night use per household (or 0.6 l/h × No. of people)
Edwards	2.2–2.5	Anglian Water Co. test on 100 metered properties 1992–1993

Sources.
Gledhill E. G. B. An Investigation of the Incidence of Underground Leakage. *JIWE*, 1957, p. 117
Reid J., *Proc Symposium on Waste Control*. IWES, 1974, p. 105.
Cook R. G. Discussion on Paper by Thackray *et al. Proc. ICE*, August 1978, Table 15.
Shaw Cole E. Water Losses and Leakage Control, *Proc. IWSA Congress*, 1978.
WRc, *Managing Leakage*; Report E, WRc, Water Services Assn & Water Cos Assn, 1994, pp. 35–36.
Edwards K. and Martin L., A Methodology for Surveying Domestic Water Consumption. *JCIWEM*, Oct 1995, 477–488.

Table 1.10 Figures for minimum night flow per connection

MNF per connection	Interpretation
5 l/h	About the lowest found in practice on parts of systems in good condition
7 l/h	A frequent 'target level' for distribution districts, indicating good control over leakage and wastage
9 l/h	Experienced on large systems where there is a fair amount of nocturnal demand and/or some distribution leakage and consumer wastage
11 l/h	Indicative of substantial night demand and/or considerable distribution leakage and/or consumer wastage

1.16 Distribution losses from 24-hour supply systems

Distribution losses comprise leaks from mains, joints, valves, hydrants and washouts, and leaks from service pipes upstream of consumers' meters or boundary stopcocks. These distribution losses cannot be measured directly but have to be estimated by deducting estimated consumers' consumption (including estimated average leakage from their supply pipes and plumbing systems) from the total input to a system. The ferrule connections of service pipes to mains are often a major cause of distribution leakage. Hence distribution losses are influenced both by the length of mains needed to serve consumers and the number of service pipe connections per kilometre. The following are three estimates of what has been found on UK distribution systems.

(1) Research by WRc in 1978 suggested distribution losses were in the range 100–200 l/h per km for 'newer' mains and 150–300 l/h per km for 'older' mains.[20] As an example Scott estimated leakage from 1780 km of mains in Bradford was 212 l/h per km.[19]

(2) Further research by WRc in 1994[21] collating data from numerous tests, suggested the following formula for estimating the aggregate of small leaks from distribution systems in England and Wales. The formula attempts to distinguish between mains losses and service pipe connection losses, as well as the influence of pressure.

For mains in 'average' condition:

$$\text{Leakage} = (40 \text{ l/h per km of main} + 3 \text{ l/h} \times \text{No. connections/km})$$
$$\times (P/50)^{1.3} \text{ approx.*}$$

where P is the average hourly pressure (m) over 24 h; *plus* or *minus* 50% for mains in 'poor' or 'good' condition, respectively.

(3) The 1997–98 reports from water undertakings in England and Wales[11] showed the following estimated distribution losses.

---

*The formula $(P/50)^{1.3}$ is used in place of WRc table of values and formula in Reports F and G[21] because it is simple and sufficiently accurate bearing in mind the difficulty of judging the condition of the main.

Company reports 1998–99	l/h per km	
	Average	(Range)
Regional water and sewerage companies (excluding Thames)	291	(156 – 421)
Thames Water Co.	773 (a)	
Water-only companies	269	(156 – 387)

(a) Value reported from Thames appears unreliable.

Some of the difference between the regional and water-only companies distribution losses may stem from differences of approach in estimating losses, or because the larger regional companies have a larger scale of problems to deal with. But physical factors may also contribute to the difference. The regional companies supply the largest urban areas in the country which tend to have older systems than the water only companies; some include coal-mining areas where ground settlement has disturbed mains, and several have to supply hilly areas requiring high distribution pressures.

Leakage from service reservoirs can be found by direct static testing; hence it should not comprise an ongoing loss. The WRc,[20] collecting information on experience with 123 service reservoirs, reported three per 100 had experienced gross leaks, five per 100 had leaked 2–4% their capacity a day, and 10 per 100 had leaked 0.5–1.5% capacity per day. Acceptance figures for a new concrete service reservoir would normally be a drop in water level of 1.5–2 mm in 24 hours. If the depth of water is 4 m (about the shallowest likely) and the reservoir holds one day's supply, this would represent leakage not exceeding 0.05% of the average supply per day, which is a negligible amount. Overflow discharge pipelines should be so designed that any overflow can be seen and therefore stopped.

Leaks on distribution systems break out continuously so the total leakage from a system for a period is the aggregate sum of each leak-rate multiplied by the time it runs before repair. Hence the frequency with which all parts of a system can be tested for leaks influences the level of leakage experienced. Obviously there is a practical limitation to that frequency, so some level of leakage is unavoidable. There is also a need to determine the economic level of resources which should be put into leak detection and repair. This is a complex problem which is discussed in Chapter 15.

1.17   Total unaccounted-for water, or total losses

Total losses for a period have to be estimated by measuring the total input to a system and deducting the amount supplied to consumers. Many different figures for losses are quoted by water undertakings because 'loss' figures are influenced by many factors such as – the age of mains, supply pressures, efficiency of leak and waste prevention measures, and how unmetered supplies are estimated. However from experience on projects to reduce losses on water systems in UK and overseas, it is possible to list levels of loss and the circumstances commonly found to give rise to them as shown in Table 1.11.

Expressing total losses as a percentage of the supply can result in false comparisons, because some undertakings may provide a large proportion of their supply through a few connections to industry, whilst others have no large industrial supplies. Percentage losses can also decline on account of a rise in consumption, and not because the losses have actually been reduced. For this reason some authorities (e.g. OFWAT) prefer to quote

Table 1.11 Typical figures of unaccounted for water

Percentage of total supply	Typical circumstances applying
6–9%	Small residential areas with no leakage and all supply meters in good condition
10–13%	Small systems with little leakage; residential parts of large systems with little leakage
16–17%	Usual lowest reported for whole cities, often immediately following some intensive leak eradication programme
20–22%	Achievable in large systems with reasonably efficient leakage and waste control methods
25%	The average level attained by large systems with mains and service pipes in moderate condition
26–35%	Systems with old mains or where ground conditions are poor; poorly metered systems; systems needing attention
35–55%	Systems with many old mains and service pipes in poor condition; systems with inefficient metering and lack of attention to leaks and consumer wastage

The percentages include both distribution leakage and leakage on consumers' supply pipes and plumbing systems.

average losses per km of mains, or per connection. But these measures do not provide a common basis for comparison because rural undertakings have long mains serving few connections per km, and urban undertakings have a high number of connections per km of main. Hence losses as a percentage of input are still widely quoted because they provide a rough common measure for comparison – provided bulk supplies given are excluded and the proportion of the supply taken by large industrial consumers does not vary greatly from one undertaking to another. The most reliable indicator of losses is the minimum night flow per connection for discrete areas of the distribution system as shown in Table 1.10.

The age of a distribution system is a major factor influencing losses. High losses quoted by several UK water undertakings are primarily due to the advanced age of many of their mains and service pipes. Thames Water reported the average age of its mains was 70 years, whereas many European cities report 40–45 years as the average age of their mains.[22] To renew all old mains and service pipes in UK would take many years because of the cost, and the need to undertake renewal of mains piecemeal to avoid unacceptable traffic disruption.

Different interpretations of the term 'unaccounted-for water' (UFW) or 'losses' can also give rise to confusion. Zurich water undertaking stated in 1995 that 'water losses currently amount to 6% of the production', but then added – 'in addition to a further 5% loss from unaccounted-for continuous drippings of household installations'.[22] Quite a few low loss figures have later been revealed as excluding consumer wastage. In the USA, as already mentioned, 15 lcd 'toilet wastage' is commonly included in domestic consumption instead of appearing under 'losses'. Maddaus reported that San Francisco and Los Angles, targeting high consumption properties, found an average of 115 lcd unnecessary consumption in them due to leaking fixtures.[23] World-wide figures for losses reported by Geering[24] ranged from 28–27% (Hong Kong, Taiwan), through 22–20% (Portugal, Lithuania, Malaysia, Sweden, Czech Republic) and 17–13% (Finland, New Zealand, Italy, Spain, France) down to 7% (Germany, Switzerland, Singapore). Later figures for 1996 reported by ADB for 50 Asian cities reported they had unaccounted-for water ranging from 6–63%.

The wide range of figures for UFW probably reflects the variety of methods used to estimate it as well as the range of actual losses themselves. Apart from metering errors there is always some unmeasured consumption that has to be estimated. High figures in excess of 45% UFW may be partly due to leakage from pipes, and partly due to lack of data about consumption. Low UFW figures of 10% or less are possible if a system has only recently been built or extensively renewed; but they can result from more liberal estimates of unmetered water supplied than other undertakings adopt, or inclusion of consumer wastage in consumption. The most plausible estimates of UFW come from water authorities where full-scale attempts have been made to discover and rectify leakages and losses from most of their distribution system.

On the last mentioned basis, a realistic target for unaccounted-for water in cities and large undertakings is 16–17% of the supply. This is usually achievable if adequate resources are applied to leakage and waste prevention, and an undertaking does not have many mains over half a century old. As an example of what can be achieved, the Delaware River Basin Commission in US (referred in Section 1.1 above) reported that major undertakings supplying 4630 Ml/day from the Delaware Basin had reduced unaccounted-for water from 20.9% in 1989 to 17.6% in 1993.[2] For a UK total supply of about 288 lcd, 16–17% losses would represent 48 lcd. Taking an average of 2.5 people per dwelling this would imply an average loss figure of 120 l/day per property, or 5.0 l/h per connection. This relates reasonably well to the minimum night flow likely to be experienced where there is good control over leakage and wastage, as shown in Table 1.10.

The water from leaks is not actually 'wasted'. Most of it percolates underground so the hydrologist often needs to take account of it when assessing groundwater flows.

1.18   Effect of price on water demand

When water is sold by meter the theoretical price-demand relationship applying is:

$$Q = kP^e$$

where Q is the demand at price P per unit of consumption, k is a constant for the particular units used and e is the coefficient which measures the 'elasticity' of the demand. Since price increases will tend to cause a reduction in the demand, Q is proportional to the inverse of P, hence e is negative. When $e = -1.0$, Q is proportional to $1/P$, i.e. changes of P cause almost proportional changes in Q. A low value of e indicates a high degree of inelasticity, e.g. at $e = -0.2$ a 29% price increase of P would only cause Q to decrease by 5%.

Measurements of e are difficult to make accurately because conditions before and after some price rise are often not the same because of weather or economic changes. Also the elasticity value must be influenced by the size of the price rise, a large one-off price rise having a greater influence than smaller rises annually. A time lag also occurs between a rise in price and any observable effect on demand. Consequently e values quoted show a

Table 1.12 Price elasticity of demand

e value range	Type of supply for *e* value quoted	No. of *e* values quoted
zero to −0.10	In-house demand; lower elasticity of municipal demands	6
−0.11 to −0.25	In-house demand; household annual demand; some municipal demand *e* values	11
−0,26 to −0.50	Household summer demand; some municipal demands; lower range industrial demand *e* values	12
−0.51 to −1.00	Usual upper range municipal demand *e* values; some industrial demands; some household summer demands (a)	19
−1.01 to −1.25	Highest *e* value of municipal demand; lower elasticity range for UK industrial demand	6
−1.26 to −1.58	Out of house summer demand; higher *e* values for UK industrial demand	3

Note.
The household values are mostly for USA and Perth in Australia; none are for UK. Municipal demand *e* values are for USA and Canada, the *e* values being usually higher in the western drier states.
Sources.
Hanke S. H. A Method for Integrating Engineering and Economic Planning, *JAWWA* Sept. 1978, p. 487.
Thackray J. E. and Archibald G. G. The Severn Trent Studies of Industrial Water Use. *Proc. ICE*, Aug. 1981, p. 403.
DoE, *Water Resources & Supply: Agenda for Action*, Stationery Office, Oct 1996, p. 65.
(a) University of Delaware, Center for Energy and Environmental Policy, *e* value reported 0.605 for 'summer price elasticity; rising block rates'.

wide variation for apparently similar situations. Table 1.12 shows the most common findings which come mostly from the USA and for periods of the 1960s and 1970s.

Despite the difficulty of getting an accurate measure of *e*, the elasticity of demand is important to an undertaking. Industrial and trade consumers and the higher income householders are often a major source of income to many water undertakings overseas. Hence the elasticity of their demand is an important factor, because a price rise may cause them to reduce their take, thereby not producing a proportionate increase of income.

1.19 The question of metering domestic supplies in UK

With rising demand for water, increasing difficulty in developing new supplies, and recent periods of low rainfall, the question has been raised whether universal metering of the 85% unmetered domestic supplies in England and Wales should be adopted. The questions centres mainly on cost, fairness to the consumer, and what reduction of consumption could be expected.

The National Metering Trials 1989–92[25] showed the cost of installing a meter was £165 inside a property and £205 outside at 1993 prices. Only about 30% of properties could be metered inside and 5% could not be provided with a meter at all due to plumbing difficulties or high cost. In many older housing areas up to 20 properties can be fed by one common supply pipe: one undertaking was said to have 650 000 properties fed through joint service pipes.[26] Such common supply pipes are often laid at the rear of terraced properties, even occasionally by a supply pipe laid through the roof space of a terrace of properties. Multi-occupancy buildings, such as blocks of flats, are mostly fed by one metered pipe and inserting meters on individual households would cause unacceptable disruption of kitchen fitments.

The savings in consumption achieved by metering are difficult to assess. The National Metering trials gave erratic results; the best information coming from the Isle of Wight where 50 000 meters installed resulted in 5–9% reduction.[27] Probably it is safe to assume that metering can achieve at least 6% reduction, equivalent to about 10 lcd on current domestic consumption in England and Wales of 150–160 lcd. In addition a reduction of similar magnitude in supply pipe leakage can eventually be expected where meters are installed outside. This is a useful benefit, but whether universal domestic metering is justified by the cost is a complex matter. About half of the water companies in England and Wales have installed domestic meters free when requested by the householder and all water companies will be required to conform to this practice under a new Water Industries Bill expected to be enacted by Parliament in 1999.[28] It is also usual practice to meter all newly built houses. The aim is, of course, to restrain rises in domestic consumption, especially in areas where additional sources of water are difficult or expensive to procure. Reducing demand for fresh water is, of course, an environmental benefit to all.

The Director General of Water Services has estimated the cost of metering will be 'no more than £30 a year ... split down as of two-thirds for water (£20) and one third for sewerage (£10)'.[28] The estimated cost, which is an average, covers meter provision, meter reading and extra billing costs and is presumably per household metered. This represents a substantial increase on current charges to consumers (see Section 2.19) and any offsetting saving of cost due to the reduced production of water achieved by metering is unlikely to be significant (less than £1 per annum), unless the reduction in consumption is sufficient to avoid the need to develop a new source of supply. However the Water Industries Bill 1999 will also extend indefinitely the right of companies to continue using rateable values as a basis for unmeasured charges for water and sewerage. Hence substantial extension of domestic metering is likely to occur only in those areas where it is economic to restrain rising demand for water.

The question of fairness of metering to householders also causes debate. Metering obviously causes the large family to pay more than the small family; but if a tariff is adopted which allows a large family a sufficient basic water allowance at low cost, then the single occupant can take several times his basic need at the same low cost. At present charging for unmetered supplies according to property value seems reasonably fair, since there is a strong relationship between property value and consumption per capita as Fig.1.2 shows and low income families generally reside in low value property. However, the metering of domestic supplies is so widely practised in many countries that it is difficult to maintain that domestic metering creates an unacceptable injustice.

1.20   Growth trends of consumption and forecasting future demand

The increase in public water supply consumption in England and Wales since 1970 is shown in Fig. 1.4. Over the period 1970–99 total per capita consumption rose from 277 lcd to a peak of 331 lcd in 1995–96 which has been reduced in the last three years to 288 lcd in 1998–99 due to increased measures for reducing leakage.[11]

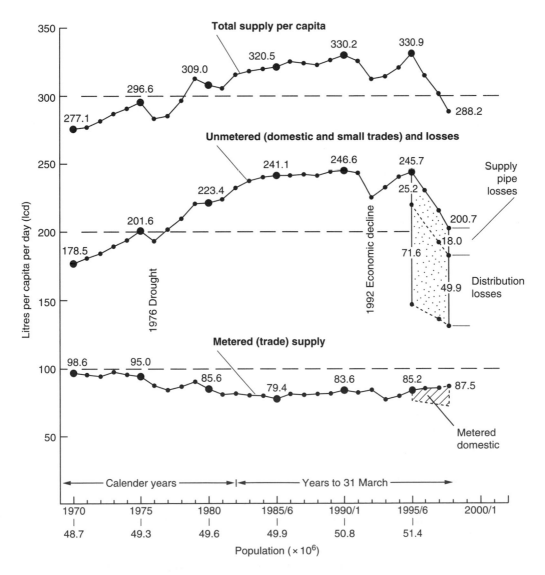

Fig. 1.4 Growth of consumption per capita, England and Wales.

	1995–96	Change lcd	1997/98
Domestic consumption:			
reduction in unmetered supplies	−12.9		
less increase in metered supplies	+6.0		−6.9
reduction in supply pipe leakage			−7.2
Trade and industrial consumption:			
reduction in unmetered supplies	−3.2		
reduction in metered supplies	−3.7		−6.9
Distribution leakage – reduction			−21.7
Net change			−42.7

The figures are the aggregate of 27 water company estimates whose individual estimates vary. The reduction is primarily due to increased work on leakage reduction and measures to restrain domestic consumption, but part may be due to variation of the summer climate which affects household and garden water use.

Long-term forecasting of water demand presents problems. Figure 1.4 shows how a period of sharp economic decline in 1992 reduced domestic consumption immediately and trade consumption the following year. Such incidents interrupt previous trends of increase which, in the case of domestic consumption in developed countries, tend to be asymptotic to some future maximum demand per capita. The ultimate maximum level of domestic consumption in the developed countries depends on the standards of housing and installation of water fittings householders can generally afford, coupled with the policy of the water supplier. Where additional supplies are difficult or expensive to procure, or where construction of any new water supply scheme would meet with strong environmental opposition, measures to restrain the rise of domestic demand may be adopted. In the UK metering of domestic supplies and for garden watering is being used for this purpose. In the USA, the previous high levels of domestic consumption are being reduced by the introduction of low consumption water fittings. Where, however, plentiful supplies exist, as in Scotland at present, there may be no need for restraint.

In under-developed countries future rises of consumption are often limited to the amount of water available, with the result there can be much unsatisfied demand. Estimating the long-term potential demand therefore involves a different approach. Population growth forecasts may be available, but they can be unreliable because of unexpected changes in birth rates. A United Nations manual[29] gives mathematical methods of forecasting population growth based on fertility and mortality rates, sex ratio and age distribution of population, etc.; but such data may not be available in a given case or necessarily relevant. A major proportion of the population growth in many cities in under-developed countries is often caused by migration from rural to urban areas. Hence estimating future water demand of such a city involves the following steps.

(1) Plotting the population trend for the past 10–20 years, and investigating the likely proportion due to immigration and that due to natural increase of the existing population.
(2) Dividing the supply area into different classes of housing and assessing in which classes of housing the main rises of population have occurred and are likely to occur in the future.

(3) Assessing typical rates of domestic consumption per capita in the different classes of housing.
(4) Seeking likely figures for future immigration and natural increase, and allocating these to appropriate classes of housing.
(5) Finding what provisions are being made, or are likely to be made, for new housing development.

On the foregoing basis, forecasts of future demand can be built up, but these may need to be adjusted to ensure they represents a realistic continuation of past consumption trends to date. Generally speaking, consumption forecasts for more than 10 years ahead tend to be unreliable.

Commercial and institutional rises of demand are often estimated as a per capital allowance on the population growth, because these are activities tending to relate to the size of the population served. Rises of manufacturing and industrial demand are, however, usually dominated by the needs of a relatively few major manufacturing concerns whose developments plans should be ascertained. The sum of their estimated individual future requirements needs to be written down to allow for the probability that not all such developments are likely to be achieved in the period of forecast.

1.21 Demand constraint methods

Much emphasis is being placed in the USA and in the drier parts of south-east England on the adoption of demand constraint measures. In England emphasis is being placed on extending the metering of domestic supplies and on increasing measures to reduce leakage and wastage. In the USA, as required by the Safe Drinking Water Act 1996, the USEPA published draft guidelines in 1998 to water suppliers for 'conservation planning', i.e. measures to induce economy in the use of water. The guidelines propose three sequential levels of approach. The first comprises universal metering, loss control (i.e. leakage and wastage reduction), and public education. The second and third levels include such measures as water audits, pressure management, re-use and recycling, and integrated resource management.

However, emphasis on demand constraint does not necessarily apply in all countries, where some undertakings may be reluctant to curb the demand from metered industrial consumers and from metered households occupied by higher income groups because the payments made by them comprise a major part of the undertaking's income needed to cross-fund supplies given free by standpipes or below cost to low income groups. The position varies according to the circumstances applying.

In many countries restricted hours of supply have to be adopted in order to prevent consumption and losses exceeding available supplies. Metering is adopted for the same reason but it must be reasonably efficient to be effective. Intermittent supplies bring many problems. Consumers store water when the supply is on, but throw away the unused balance when the supply next comes on, believing the new supply is 'fresher'. Consumers may leave taps open so as not to miss when the supply comes on again, hence storage vessels overspill. Intermittent supplies make leak detection and prevention of consumer wastage very difficult. In fact the hours of supply have to be reduced to at most 4 hours in the morning and 4 hours in the evening and frequently less, to gain control of consumption. To some extent intermittent supplies are self-defeating – more consumer wastage and more distribution leakage occurs because of the difficulty of maintaining the system in a good state. The situation is often exacerbated by loss of income due to difficulties with metering and income collection. Furthermore, if mains become emptied, contaminated groundwater may enter mains and endanger the health of consumers. Nevertheless many undertakings world-wide have to adopt intermittent supplies.

On 24 hour supplies temporary reduction of domestic consumption during a shortage of supplies is widely adopted by banning the use of water for washing vehicles, refilling bathing pools, and the use of hosepipes and sprinkler equipment for the watering of gardens. Good publicity in times of temporary shortage can effect a modest temporary reduction in demand, perhaps as much as 10 percent. Metering of 24 hour domestic supplies can curb excessive consumption, especially water used for lawns and gardens, provided a two-part tariff is adopted which imposes a financial penalty if consumption exceeds a reasonable amount. Promoting the use of low consumption dish and clothes washing machines, toilets and showers makes a valuable reduction. The maintenance of steady distribution pressures makes the operation of float-controlled valves and other fittings more reliable so that less malfunctioning and wastage of water occurs. Wessex Water found 5% and 10% consumption reductions in areas where static minimum pressures were maintained day and night by means of remote pressure sensors and control valves operated by telemetry under computer control.[30] Keeping pressures to the minimum necessary reduces water taken unnecessarily, and also achieves a substantial reduction of the amount of water lost from leaks. According to the WRc formula given in Section 1.16 a reduction of pressure from 50 to 30 m nearly halves the flow from leaks. Flow limiters have sometimes been used to curb domestic consumption but have not always been effective. If set too low, consumers leave taps open to fill containers which overspill, or they are tempted to by-pass the limiter in an effort to get a better supply.

To restrain commercial and institutional demand it is important to meter both large and small shops, offices and other businesses, because wastage in them from plumbing fittings is frequently high because no one is present who is responsible for paying the water charges and the premises are unoccupied outside working hours. Many cases have been reported of night and weekend flows to unoccupied premises being nearly as high as daytime flows when staff are present – especially in government offices in some countries. Manufacturers are also often unaware of the potential financial savings they can achieve by adopting water conservation measures, not only by reduction of their water purchase costs but also by reduction of their effluent discharge costs due to reduction of its volume and improvement of its quality. Yorkshire Water Company has demonstrated the considerable savings that can be achieved when manufacturers permit a water company to investigate their processes and advise how economies can be achieved.[31]

1.22   Maximum day's consumption

The maximum consumption for a day is usually expressed as a percentage of the average annual daily supply. Some figures are given in Table 1.13. In temperate climates a design figure of 140% is often adopted.

1.23   Maximum week's consumption

The maximum weekly demand (expressed as the average daily demand during the peak week) is usually only a few percent below the maximum day demand, but it is of

Table 1.13 Maximum day's consumption

Location, etc.	Ratio: maximum day to average annual daily consumption
UK	
Rural areas in which water is used for spray irrigating crops	140–150%
Seaside and holiday resorts	130–140%
Residential towns, rural areas	122–125%
Industrial towns	117–122%
Peak due to garden watering in prolonged hot dry weather	150–170%
World-wide	
USA typical peak domestic demands due to lawn sprinkling:	
– western state	215–340%
– eastern states	195–295%
As above – but excluding lawn sprinkling (i.e. in-house only):	
– western states	180–185%
– eastern states	130–140%
Cities with hot dry summers	135–145%
Cities in equable climates	125–135%
Cities with substantial industrial demand (e.g. Singapore 124%; Hong Kong 122%; Penang 116%; Damascus 115%; Toulon 115%; Marseilles 111%)	110–125%

Note. USA figures from Linaweaver F. P. *et al.*, *A Study of Residential Water Use*, Johns Hopkins University, Baltimore, MD, 1967.

importance because the daily overdraw above the average for seven consecutive days cannot usually be met from the amount of service reservoir storage provided. This means that the maximum output of the source works must be at least equal to the average daily demand for the 7 days of the peak week.

1.24 Maximum hourly rate of consumption

The maximum hourly demand depends upon the size of the population in the area served and the nature of the demand. For mainly domestic areas, excluding garden watering and sprinkler demand, the peak flow factors shown in Fig 1.5 are applicable in UK. The domestic peak period for residential areas is usually between 07.30 and 09.00 hours or earlier for commuting areas; and about noon for mixed residential and industrial areas. However garden watering demand can create an evening peaking factor of 3.0 or more. In the USA with its hot summers, the peak hourly factor for in-house demand was reported as about 3.0 in eastern states and between 4.0 and 5.0 for western states. Sprinkler demand can increase the factor to 6.0. All these factors apply to the average annual daily demand.

In practice the peak hourly demand does not affect the sizing of the smaller distribution mains which are usually at least 100 mm diameter in UK, and 150 mm diameter in USA to

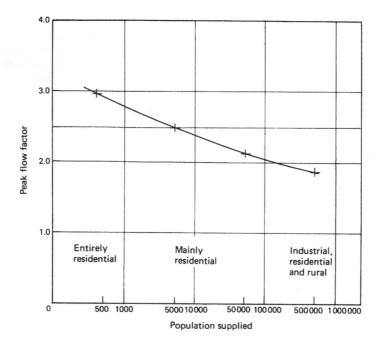

Fig. 1.5   Ratio of peak hourly flow to annual average flow in an undertaking (Adams, *JIWE*, 1955).

meet fire demand. The larger, principal feeder mains, and those supplying an outlying area may have to be sized according to the peak hourly demand. However, a diversity factor applies, i.e. the peak hourly flow factor reduces as the size of the population served increases.

References

1. Department of Environment. *Water Resources and Supply: Agenda for Action*, 1996, Annex C, pp. 62–63.
2. Featherstone J. Conservation in the Delaware River Basin, *JAWWA*, Jan. 1996, pp. 42–51.
3. McIntosh A. C. and Yniguez C. E. (eds). *Second Water Utilities Data Book: Asian and Pacific Region*. Asian Development Bank, Manila, Oct. 1997.
4. Unvala S. P. Bombay Water Supply. IWSA Congress 1995. *Water Supply,* **14**, Nos 3/4.
5. Vickers A. Implementing the US Energy Policy Act. *JAWWA*, Jan. 1996, pp. 18, 20, 112.
6. Babcock T. M. Water Resources Specialist, City of Phoenix Water Services Dept. Letter and data, 1997.
7. Metropolitan Water District of Southern California. *Report on Urban Use Characteristics*. April 1993, pp. 8, 10, 11.
8. AQUACRAFT Inc. and AWWA Research Foundation. *The Municipal End Uses of Water Study, 1996–99*. (Funded by AWWA and 12 participating municipalities.)

9. Weeks C. R. and Mahon T. A. A Comparison of Urban Water Use in Australia and US. *JAWWA*, April 1973, p. 232.
10. Russac D. V. A., Rushton K. R. and Simpson R. J. Insights in Domestic Demand From a Metering Trial. *JAWWA*, June 1991, pp. 342–351.
11. OFWAT. *1998–99 Report on Leakage and Water Efficiency.*
12. UK 1991 Census. *Key Statistics for Local Authorities,* Table 8. pp. 141–155.
13. Water Services Association. *Waterfacts '97,* p. 63.
14. World Health Organization. Public Standposts for Developing Countries, *Bulletin No. 11.* WHO International Reference Centre, The Hague, May 1978.
15. Binnie-Taylor. *Report on Egyptian Provincial Water Supplies,* Oct. 1979.
16. Welton R. J. and Goodwin S. J. *The Accuracy of Small Revenue Meters,* Report TR 221, WRc, 1984.
17. Hall M. Technological Developments in Metering, IWEM Symposium 'Paying for Water', Jan 1992. *JIWEM,* Aug. 1992, p. 517.
18. Smith A. The Reduction of Leakage Following Water Metering on the Isle of Wight, IWEM Symposium 'Paying for Water', Jan. 1992. *JIWEM,* Aug. 1992, p. 516.
19. Scott J. W. Private communication, Jan 1991.
20. WRc, *Waste Control and Leak Detection in Water Distribution Systems,* WRc regional meeting, Autumn 1978.
21. *Managing Leakage,* Report F p.7 and Report G p. 28. WRc, Water Services Assn and Water Cos. Assn, 1994.
22. Skard B. C. Methods of Diagnosis and Performance Indicators for Rehabilitation Policies. Proceedings IWSA Congress 1995. *Water Supply Review Jnl,* **14,** No. 3/4, 1996, pp. 351–355.
23. Maddaus W. O. The Effectiveness Of Residential Water Conservation Measures. *JAWWA,* March 1987, p. 52.
24. Geering F. and Lohner E. Design for Continuity and Reliability in Distribution Systems. *Water Supply Review Jnl,* **14,** No. 3/4, 1996, p. 78.
25. Water Services Assn. *Water Metering Trials: Final Report,* 1993, p. 4.
26. Roberts K. F. Requirements for Trial Programme. *Proc WRc Seminar on Water Metering,* April 1986.
27. Water Services Assn. *Water Metering Trials: Final Report,* 1993, p. 67.
28. OFWAT. *1998 Annual Report of the Director General of Water Services,* June 1999, Stationery Office, London.
29. United Nations, *Methods of Population Projections by Sex and Age.* 1956.
30. Howarth M. Minimum Need for Water when Available Resources Fail. *IWSA Rev Jnl,* **2,** No. 2, 1984, p. 191.
31. Edwards A. M. C. and Johnston N. Water and Wastewater Minimization: the Aire and Calder Project. *JCIWEM,* Aug 1996, pp. 227–234.

2

The organisation and financing of public water supplies

2.1 The control of public water supplies

Every government or state is forced to exercise some overall control over public water supply because piped water and waterborne sewerage are the two universally essential requirements for urbanised and city life. Four special features of water supply determine how much control the state must exercise over it:

- the water supplied must be free of contaminants injurious to health and be palatable;
- all citizens should have access to it;
- the service is nearly always a monopoly;
- the development of public water supplies entails the use of large amounts of capital.

The first two requirements reflect the basic public health nature of public water supply. The water has to be free of contaminants that could cause the spread of disease or injure health, it has to be palatable so that people use it, and it has to be available to all people in a community to prevent unsafe supplies being used which could bring disease into the community. In many countries the supply to the lowest income groups is therefore given free, usually by standpipe. To ensure a safe supply many countries adopt the water quality standards recommended by the World Health Organization (WHO); but limited financial resources may prevent individual undertakings achieving full compliance with every standard. In the richer countries extensive national water quality standards are set which are legally enforceable, and systems of inspection may be used to ensure they are achieved.

The monopoly characteristic of piped water supply forces most governments to control the charges made for it, so that all people can afford a supply. Where domestic supplies are metered or charged according to the value or size of a dwelling, a government can limit the charge made for a basic water allowance sufficient for all householders, sometimes below the cost of production. In the latter case, trade and large household consumers may have to pay more to make the undertaking financially self-sufficient, or the government itself may have to deficit fund the undertaking.

Water supply systems are 'capital intensive', i.e. the cost of developing a new supply is usually too large to be met from the income received from current consumers, so the money for it has to be borrowed. Reservoirs, dams, treatment works and bulk supply mains, have to be sized large enough to cater for future rising water demand. Hence it is reasonable to borrow money for such works, repaying it gradually from rising sales of water. Raising the necessary capital is one of the key problems facing governments. If a country is rich enough the government may lend money from its own funds. Alternatively a government can apply to an international lending agency such as the World Bank or Asian Development Bank for a loan, or take a loan from another country under 'tied aid'. Alternatively a water undertaking may be permitted to borrow money directly from the 'money market', i.e. from commercial sources of money; but the amount so borrowed may have to be controlled to ensure the total 'public sector borrowing' does not exceed the government's need to maintain its own financial stability. If, however, a government allows water to be supplied by commercial companies, the companies are like any other commercial concerns and can borrow from the money market without being constrained by the government's public sector borrowing limits.

Hence public water supply, however it may be provided, always comes under some form of government control. In most developed countries the government's controls are set out in Acts and Regulations. Additionally in countries belonging to the European Community, EC Directives have to be complied with. However, in many of the less developed countries the government's control may extend only to the publishing of water quality standards which should be achieved, and to controlling the expenditure of undertakings and their charges to consumers. In such circumstances water undertakings have to work within the expenditure limits set for them annually, and may only be able to meet the published targets for water quality and levels of service to the extent that the permitted expenditure allows.

2.2   The approach to privatisation of water in England and Wales

Prior to 1973 there were 198 separate water supply undertakings in England and Wales owned by 64 municipal authorities, 101 joint boards, and 33 statutory water companies.[1] The water companies were 'statutory' because each was authorised to be set up by a private Act of Parliament. By regulation, the maximum profits the statutory companies were allowed to make was limited to 10% on their share capital.

There were also 29 River Authorities which controlled abstractions of water from rivers and lakes by means of a licensing system. The River Authorities also licensed the discharge of effluents to rivers, controlled navigation on rivers, and promoted land drainage, flood protection, fishing, and sea defences. There were, however, some 1300 sewage works owned by municipalities and other smaller local government bodies which discharged their effluents to watercourses, many of these effluents being of poor quality through lack of expenditure on sewage treatment works.

With the growing demand for water it became evident there was a need to control all sewage and other wastewater discharges to rivers and streams and all abstractions of water, on a whole river-catchment basis. Consequently the 1973 Water Act transferred all the River Authorities in England and Wales, together with all the water undertakings (*except* the statutory water companies), together with all the sewerage functions of the municipal and other authorities, to 10 new regional *Water Authorities* covering England and Wales. Each new Water Authority covered one or more whole river basin catchments, so that it became possible to plan water abstractions, towns' sewage and industrial wastewater discharges to make the best use of available water resources and to preserve the hydrological environment for environmental and recreational purposes. This radical restructuring of water, sewerage and water controls by government was along the lines that the water industry had suggested was necessary.[2]

The 10 new Water Authorities faced considerable expenditure on improving sewage effluents and therefore needed to raise funds for this purpose, but by the early 1980s the government was also faced with inflation problems and was consequently reluctant to allow the new Water Authorities to increase their borrowings and charges to consumers to finance the necessary improvements. As a short-term measure the government modified the quality conditions for Water Authorities' sewage works effluent discharges and postponed provisions of the Control of Pollution Act 1974 which otherwise would have revealed increasing pollution of rivers.[3] Nevertheless capital expenditure by the Water Authorities nearly doubled in the period 1975–85. But an even larger financial worry for the UK government was that much more money was needed by the Water Authorities to meet new EC Directives stipulating the quality of bathing waters, conditions for disposal of dangerous wastes to the environment, and new quality standards for drinking water. Facing these financial problems it occurred to government ministers in 1985 that one way to ease the difficulty of increased government funding was to privatise the Water Authorities so they could tap the commercial money market to raise the additional capital required.

'Privatisation' of the Water Authorities meant changing them from publicly owned bodies to commercial companies owned by shareholders. Capital raising by these companies to improve water and sewerage standards would not then come under the government's restraints on public sector borrowing. Also the new companies would then be responsible for raising charges to consumers, relieving the government of direct responsibility for this unpopular duty. A 'one-off' advantage to government was that selling the water undertakings to shareholders, would bring in money to the Treasury. However the main arguments put forward by the government for privatisation were that it could not find the capital money needed to improve water and sewerage services, and also that the profit-motive of the new companies should tend to make them more efficient than public bodies. There was some criticism that the first argument was invalid, since both private and public funding made the same demand on the country's economic resources, and the Authorities could have been permitted to raise more foreign loans as they had been allowed to in the past.[4] The government also admitted that the publicly owned Water Authorities could not be criticised as inefficient.

Government ministers first considered transfering the Water Authorities with all their functions to new companies; but it was quickly pointed out that commercial companies could not be given powers to license abstractions and discharges, and to undertake pollution prevention, flood control, and coastal protection, etc. because these were public control measures. This and a number of other problems were resolved as follows.

- Only water and sewerage services would be transferred to the new companies.
- The balance of the work – planning of water resources development, licensing abstractions and discharges, pollution control, inland navigation, amenity protection, flood protection, land drainage and coastal defences, etc. – would be taken over by a new public body, the '*National Rivers Authority*' (NRA).
- The water and sewerage service companies would need to come under strict financial control to satisfy the public that they were giving proper service at a justifiable cost. This control would have to be exercised by a '*Director General of Water Services*' responsible to government.

- If the companies were to provide only water and sewerage at prices limited by the Director General, their potentiality for profit might be insufficient to attract many purchasers of their shares; there was no opportunity to expand water services because 99% of the population were already on a public water supply and 96% were connected to a sewer.[5] The issue of shares might therefore fail. The companies needed freedom to undertake other commercial activities which offered the prospect of better profits to shareholders. To unravel this predicament, water and sewerage work would have to be undertaken by '*Water Service Companies*' which would be subsidiary to and wholly owned by '*holding companies*' which would be *public limited companies (plcs)* free to undertake other commercial activities. It would therefore be the plc shares which would be put on the market, and only their subsidiary Water Service Companies would come under the Director General's control.
- To make shares in the plcs more attractive, much of the long-term debt still owed by the Water Authorities would be written off by government.
- The existing statutory Water Companies, supplying water only, would remain, but would be allowed to set up a plc holding company if they wished.

2.3   General provisions of privatisation

The consequent arrangements under the Water Act 1989 are shown in Figs 2.1 and 2.2. Later four Consolidation Acts were passed in substitution of the complex 1989 Act. These

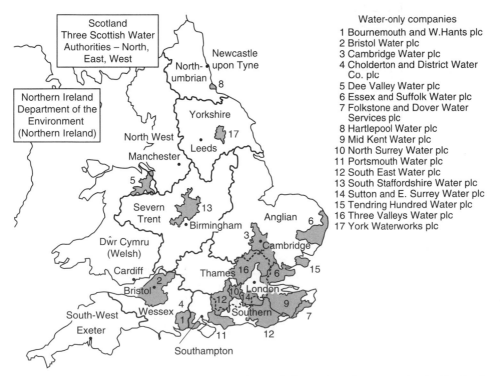

Fig. 2.1   The ten water and sewerage companies and 17 water-only companies in England and Wales in 1999.

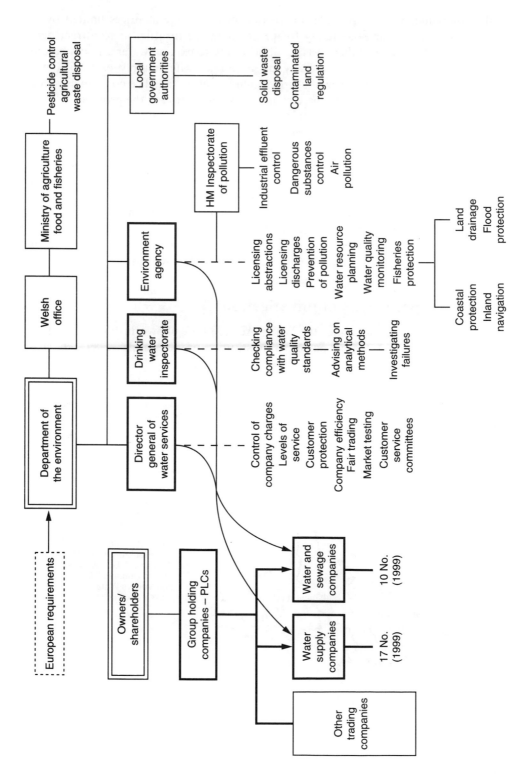

Fig. 2.2 Organizations involved in water supply, England and Wales 1999.

Consolidation Acts repeated all past legislation still in force so that previous Acts need no longer be consulted. The four Acts were as follows.

- The *Water Industry Act 1991* set up the new plcs and their subsidiary *Water Service Companies*, transferring to the latter the water and sewerage functions of the previous Water Authorities, and also set up the *Director General of Water Services* defining his powers to control the Water Service Companies.
- The *Water Resources Act 1991* set out the functions of the new National Rivers Authority (NRA).
- The *Statutory Water Companies Act 1991* applied specifically to the statutory (Water-only) Companies keeping them in being.
- The *Land Drainage Act 1991* transferred the functions of previous internal land drainage boards and the land drainage powers of local authorities to the NRA.

The 1989 Act resulted in three types of water service companies whose numbers in 1999 are as follows.

(1) Ten Water Service Companies providing sewerage services for the whole of England and Wales and water supplies in those areas not supplied by the Water-only Companies; each owned by a public limited company (plc).
(2) Eleven 'Water-only Service Companies' (former statutory Water Companies) providing water supplies in certain areas, who have chosen to set up plc holding company.
(3) Six 'Water-only Service Companies' (former statutory Water Companies) providing water supplies in certain areas, but who had not chosen (by 1999) to set up a plc holding company.

The powers of the Director General relate only to the Water and Sewerage Service Companies or Water-only Service Companies, whose work is said to be 'the core business' of the plcs where these have been set up. (Both types are referred to as the 'Water Service Companies'.) The main provisions of the Water Industry Act 1991 in respect of the Water Service Companies are as follows.

(1) The DoE Secretary of State can issue Water Quality Regulations which set the quality standards required for drinking water.
(2) A Drinking Water Inspectorate under the Secretary of State monitors and reports on the water quality achieved by the Water Service Companies.
(3) A plc cannot sell its 'core business' of water and sewerage, or water-only, nor alter its area of supply without the permission of the Secretary of State. The Director General can advise whether he thinks such permission should, or should not be granted; or if he thinks the core business should be taken over by others.
(4) The Director General monitors the quality of the services given to customers by the Water Service Companies and can direct their improvement where he thinks this necessary.
(5) The Director General controls the charges made to consumers by the Water Service Companies. These charges cannot be increased over a 5-year period by more than a percentage expressed as – 'the current rate of inflation $\pm$ K' known as 'the K-factor'. Where K is positive, this represents sufficient additional income, per 5-year period, which (in the Director General's estimation), is needed to meet additional approved expenditure to improve the core services. Where K is negative, this means the Director

General considers a company has already sufficient resources to meet its 5-year commitment without using the whole of the inflation allowance.

(6) The Director General has power to require the Water Services Companies to give him all information he requires to keep under review their performance in respect of their economy, efficiency of service to consumers, proper financing, adequacy of service and fair treatment of consumers.

2.4   Experience with water privatisation in England and Wales

Released from Treasury constraints on borrowing, the new companies were able to finance and therefore progress with a number of much-needed improvement works. The higher cost of market borrowing than use of government loans was not of any particular significance because the subsidies on interest rates on government loans through the Public Works Loan Board were in any case being gradually withdrawn.[6]

The plc holding companies had, however, some initial difficulties in pursuing other commercial ventures successfully.[7] They were in the strange position of being companies set up before they had decided what to do. Some bought, or set up, commercial ventures, but later sold them. Some made heavy losses on diversification into other lines of activity.[8] This could reflect the hybrid nature of privatisation; running the plc commercial enterprises requires business experience and entrepreneurship; whereas providing water and sewerage on a 24-hour basis is a primarily a technical service of a routine kind.

There were problems of 'cross-funding'. Many plcs transferred the water engineering design staffs, laboratory staffs and laboratories they took over into separate subsidiary companies. They used these companies, and other equipment supply companies they had purchased, to supply their Water Service Company with services or goods. The Director General decided that supply of such services or goods by plc-owned companies to the core business of water and sewerage must be market tested by competitive bidding to ensure a competitive market price was paid for them,[9] and a plc must not favour a bidder because he or she intended to sub-contract work to a plc-owned company. The Director General also found expenditure on new capital works was, in some cases, substantially less than a plc had forecasted and on which the Director had based his 'K values' controlling the Water Service Companies' charges.[10] Some of this shortfall was due to new works not proceeding as fast as had been forecasted, but it also raised a doubt as to whether the reduced expenditure was due to improved efficiency or to the plc's forecasted expenditure being unnecessarily high. Where large savings appeared to have been made on the estimates, the Director persuaded all but one of the plcs to give rebates to customers or to limit the increase of their charges to less than the K percentage he had previously authorised.

Privatisation also changed the public perception of water supply. Many consumers now tend to regard it as a profit-making commercial enterprise and consequently are critical of any restrictions on use (especially for garden watering) when this is forced on an undertaking by an exceptionally dry period. This has meant that water companies have had to work more closely to source-failure risks regarded as acceptable by the Director General (see Section 3.15), and have pursued a policy of increased metering of domestic supplies in some areas which will inevitably increase domestic charges for water (see Section 1.19).

Privatisation has worked successfully, but accompanying it there has been a need for strict regulation and external monitoring. This has not proved easy to apply because of the difficulties associated with establishing measures of efficiency, ensuring fair allocation of costs and checking that services are paid for at market-tested prices. Monitoring has become necessary of the terms under which plc subsidiary companies provide services to the core business, and to check how far a plc's core business is used to provide support to the other subsidiary companies it owns. The Director General has required that, where directors or managers of a Water Service Company also hold directorship of another plc subsidiary company, procedures must be followed which make clear such directors or managers can have no influence on transactions between the two. Where multi-utility companies have been formed, such as those providing both water and electricity, so that common services are possible, there will be more complicated problems in ensuring such transactions are fair to both water and electricity consumers.[11] The take-over of a number of the plcs by large foreign firms who own many other companies providing goods and services for the water industry will add to problems of monitoring. Inevitably privatisation has meant that more controls have had to be applied to public water services than were necessary when these services were publicly owned.

2.5   The Environment Agency

The Environment Agency was set up under the Environment Act 1995 in order to achieve a more co-ordinated approach to the use of fresh water and to protect the environment against various forms of pollution. It took over the work of the National Rivers Authority (NRA) which had been set up when privatisation took place (see Section 2.3). Consequently the former functions of the NRA now exercised by the Environment Agency in England and Wales include:

- the licensing of water abstractions and wastewater discharges;
- the development and management of water resources;
- monitoring and controlling pollution of water resources;
- protecting any surface or underground waters from which water undertakers are authorised to take water;
- promoting environmental and recreational benefits of water;
- flood protection, land drainage and coastal defences;
- fisheries protection and control of inland navigation.

The management of water resources is the most important of the Agency's duties in relation to water supply. There are many conflicting demands for fresh water. A difficulty inherited from earlier 1963 legislation when the River Authorities were set up, lies in the continued existence of 'licences of right' which entitled abstractors of 1963 to take as much as their plant could then abstract in 24 hours. This inhibited the Agency's ability to plan best use of water resources because it has to pay compensation if the issue of a new licence reduces the quantity available under an existing licence of right, even if the holder thereof uses much less than his entitlement or perhaps takes no water at all because he has ceased some business which used the water. However the Agency is endeavouring to overcome this problem by negotiating reduction of licensed amounts.

Although the Environment Authority issues abstraction licences, this has made no difference to the procedure which has to be followed when a Water Service Company requires a new source of supply. The company must, of course, take account of the Environment Agency's views as to whether or not it would be willing to issue a licence for the proposed new abstraction, but the company still has to apply to the DoE Secretary of State for a Ministerial Order sanctioning the construction of new sourceworks. A public inquiry has then to be held at which objectors to the proposed sourceworks can state their case. If the proposed Order is objected to in Parliament, it has to go to a Parliamentary Committee which hears the objections again and makes recommendations to Parliament which can decide whether or not to approve the Order. If objection to an Order in Parliament is thought to be likely, the company may decide to promote a Bill instead of an Order, because the Bill goes straight to a Parliamentary Committee for examination which saves a certain amount of time. In either case it takes at least a year to get through the procedure.

Another major role of the Environment Agency is that of monitoring the quality of surface waters, and taking remedial action against pollution. Under the Water Resources Act 1991 the DoE Secretary of State can set 'Water Quality Objectives' for surface waters taking into account the uses to which a surface water is put. He can issue Regulations for the control of dangerous substances, and for procedures to be adopted by people holding poisonous or polluting materials. By means of Orders he can designate 'water protection zones' for safeguarding underground supplies, restricting activities within them. The role of the Environment Agency is to see such objectives are met. In practice the Agency takes the lead on many matters, developing proposals to put to the DoE for approval.

2.6   Other pollution control measures

The Drinking Water Inspectorate (DWI) plays a key role in England and Wales in ensuring that water supplied by all the companies complies strictly and continuously with the many quality requirements described in Section 6. The role of the DWI is not only that of monitoring company performance but also in giving guidance on the best techniques available for ensuring the use of reliable quality monitoring procedures and for advising where new or improved techniques need to be adopted to deal with new or unresolved problems of pollution. However the DWI is primarily concerned with ensuring that unacceptable pollution does not pass into public water supplies; the control of sources of pollution is exercised by other authorities.

The principal legislation dealing with prevention of environmental pollution generally is:

● the *Control of Pollution Act 1974* (under which *inter alia*, HM Inspectorate of Pollution was first set up);
● the *Environmental Protection Act 1990* (which widened the scope of the 1974 Act);
● the *Environment Act 1995* (which set up the Environment Agency and further widened the scope of previous Acts).

Wastewater discharges, including both industrial and sewage effluents, are controlled by the Environment Agency which can refuse a licence for an effluent which does not meet its conditions. It can prosecute polluters and carry out works to remove or mitigate pollution, recovering the cost from those liable. It has to keep registers, open to inspection by the public, of discharge consents given, water quality objectives set, and sampling results.

'Point source' agricultural pollution can occur from slurries, silage effluents, yard washings and vegetable processing wastes. These all have high BODs (Biological Oxygen Demands), silage effluents exceptionally so, and the slurries and yard washings have a high ammonia content as well. Guidance to farmers, and grants, are available through the Ministry of Agriculture, Food and Fisheries (MAFF) which published a 'Code of Good Agricultural Practice for the Protection of Water' in 1991. The legal requirements on farmers are set out in the Control of Pollution (Silage, Slurry and Agricultural Fuel Oil) Regulations 1991.[12] Many treatment methods have been tried on farm wastes, but the simplest of them is to store slurries and sludge drainage for 9–12 months in open ponds, after which they can be sprinkled evenly over grassland in the right weather conditions, avoiding any direct runoff to a watercourse.[13]

'Diffuse pollution' of the environment occurs from the use of pesticides (including herbicides, etc.) and therefore control has to be applied to the usage. Not all such pollution comes from farming. Increasing detection of the presence of the herbicides atrazine and simazine in groundwater was attributed mainly to their use by road and rail authorities. Dealing with the many types of pesticides and herbicides used is a complex matter, as illustrated by the fact that MAFF lists about 400 approved compounds. Some compounds have been banned, including DDT, aldrin, dieldrin and chlordane. An added complication is that farmers rotate the use of different pesticides from year to year to avoid build-up of resistance; this increases the difficulty of monitoring for such contaminants.

The DoE Secretary of State has designated 'nitrate sensitive areas' within which, in co-operation with MAFF, farmers are assisted to adopt practices designed to limit, and possibly reduce, the amount of nitrates found in water. A Directive of the EC in December 1991[14] made a similar policy mandatory on member states, stipulating the need to identify 'vulnerable zones' where underground or surface waters used for drinking would contain more than 50 mg/litre nitrate if no protective action were taken. Among the practices required by the EC Directive are measures to prohibit the application of nitrogenous fertilisers during certain periods of the year, and to limit the application of nitrogenous fertilisers and manures.

As part of the Environment Agency, HM Inspectorate of Pollution deal more especially with controlling the discharge of industrial wastes (solid, liquid, and gaseous) including trade wastes discharged to public sewers. EC Directives[15] set limits for the discharge of certain dangerous substances to the aquatic environment. In response, the UK DoE produced an initial 'Red-List' of 23 substances ranking for priority control, to which further substances are being considered for later inclusion. Except for the metals mercury and cadmium, all the Red-List substances are organic compounds, mostly pesticides and herbicides, although some are particular wastes from industry. Two quality standards have to be applied to the discharge of such substances to surface waters:

- the application of an 'environmental quality standard' (EQS); or
- the application of 'best available technology not entailing excessive cost' (BATNEEC) – sometimes alternatively denoted as a 'uniform emission standard' (UES);
- whichever is the more stringent.

An EQS is set by estimating the effect the discharge of a substance has on the environment, so this involves research and monitoring of such discharges by the Environment Agency, followed by the setting of an appropriate EQS by the DoE. The BATNEEC standard or UES involves HM Inspectorate of Pollution approaching an industrialist producing or discharging a Red-List substance in order to assess what process should be applied to the substance to render it suitable for discharge and setting limits for the amount discharged. Obviously co-ordination between the two different procedures must take place to ensure the most stringent requirement is applied. HM Inspectorate also have to deal with radioactive wastes under the Radioactive Substances Act 1993.

Solid waste disposal sites are the responsibility of the County Council Waste Regulation Authorities.[16] Before licensing a site the Authority must obtain agreement of the Environment Agency and of any water undertaking likely to be affected. Compliance with EC Directive 80/68[17] ('Protection of groundwater against pollution caused by certain dangerous substances') is required, which prohibits certain substances from waste disposal sites entering groundwater, and limits the amount allowable for others. The substances prohibited are – cadmium, cyanides, mercury; mineral oils and hydrocarbons; organohalogen, organophosphorous and organotin compounds; and those possessing carcinogenic, mutagenic or teratogenic properties. Substances which must be limited (although no limit criteria are stated) include the inorganic substances normally limited in drinking water supplies, e.g. arsenic, chromium, etc. (see Table 6.1(B), Section 6.47) plus biocides, taste or odour producing substances, and toxic compounds of silica.

Contaminated land has to be identified and registered by the local authorities who must then inform the owner and the Environment Agency. The local authority can require the owner to carry out remediation of the land, or can undertake remediation itself, charging the cost to the owner. If, however, the Environment Agency decides the land is a 'special site', it takes over responsibility for enforcing or undertaking the necessary remediation measures. A 'special site' as defined by the DoE Secretary of State DoE is one which would or might cause serious harm, or serious pollution of controlled waters.

2.7   Organisation of public water supplies in Scotland and Northern Ireland

In Scotland there are now only three regional Water Authorities which provide water and sewerage services: North, East, and West Scotland Water Authorities. They were set up in 1996 under the Local Government (Scotland) Act 1994 which reorganised Scottish local authorities into single tier Unitary Authorities. Prior to this there had been 12 regional Water Authorities which had taken over some 200 water supply and 230 sewerage undertakings owned by local authorities or joint boards. The three new Water Authorities have the additional duty of promoting conservation and effective use of water resources, a duty previously held only by the Secretary of State for Scotland.[18] They also have powers to seek private finance for capital investment projects, in line with the UK government's Private Finance Initiative.

Under the Environment Act 1995, a Scottish Environment Protection Agency (SEPA) took over the work of the previous 10 River Purification Authorities (RPAs) which had been set up under the Rivers (Prevention of Pollution)(Scotland) Act 1951. These RPAs had been responsible for authorising effluent discharges, and for monitoring their quality and that of the receiving waters. With their early start the RPAs had been successful in maintaining the quality of Scottish rivers. However, they did not have general powers to license abstractions, mainly because of Scotland's liberal availability of water supplies in relation to demand. They could control irrigation abstractions for commercial, agricultural and horticultural use under the National Heritage (Scotland) Act 1991, and could apply to the Secretary of State for a control order permitting them to control abstractions from a specific length of river. But only one such order was ever issued because of the lengthy procedure involved. Although the Scottish EPA has the same limited powers for controlling abstractions that the RPAs had, it is expected to have the ability to extend control order systems to cover all abstractions above a certain amount.[19]

In Northern Ireland, water and sewerage services formerly run by the local authorities and joint boards, were taken over by the Northern Ireland Department of Development under the Water & Sewerage Services (NI) Order 1973. But in 1996 the work was transferred to the Northern Ireland Water Service, which acted as a separate government Agency under the Department of the Environment in Northern Ireland. Its headquarters in Belfast carries out overall control and planning, and so on, while four Divisions perform the day-to-day operational functions. The intention was that ultimately the four Divisions would be privatised, but the change has not yet come about. Control of pollution and effluent discharges comes under the Environment & Heritage Service of the DoE (NI), as set out in the Water (Northern Ireland) Act 1972. DoE consent has to be obtained for all effluent discharges to inland and coastal waters and for the discharge of material at sea.

2.8   Public water supplies in USA

Public water supplies in the USA are characterised by the many small supply systems that exist as shown in Table 2.1. The majority of small systems are privately owned; the larger systems are predominantly municipal and publicly owned. The physical isolation of many small communities in a large country is partly a cause of this fragmentation, but an additional factor according to Okun[20] is that developers have often preferred to site new residential communities outside the limits of major urban areas where land is cheaper and property taxes lower, and consequently have installed individual small water supply systems in preference to connecting the community to the nearest existing large system.

Table 2.1   Number of water utilities in USA in mid-1993[21]

USEPA designation	Population served	No. of water utilities	Total population served (million)
Very small	25–500	36 515 (62%)	5.569 (2%)
Small	501–3300	14 516 (25%)	20.053 (8%)
Medium	3301–10 000	4251 (7%)	24.729 (10%)
Large	10 001–100 000	3062 (5%)	85.035 (35%)
Very large	over 100 000	326 (1%)	109.797 (45%)
Total		58 670	245.183

2.9   Functions of the US Environmental Protection Agency

The US Environmental Protection Agency (USEPA) sets drinking water quality standards under the Safe Drinking Water Act (SDWA) 1996. It can issue Regulations stipulating:

- an MCLG (maximum contaminant level guide) which is not mandatory;
- an MCL (maximum contaminant level) which is mandatory;
- a 'Treatment Rule' which is mandatory and which sets out the type(s) of water treatment to be adopted where an MCL is not appropriate or sufficient, e.g. for protection against bacteriological contamination, *Cryptosporidium* and *Giardia* (see Sections 6.57), and for minimising the incidence of disinfection by-products.

An 'MCLG' is defined under the 1996 Act as 'a level at which no known or anticipated adverse effect on human health occurs and that allows for an adequate margin of safety'. An 'MCL' is defined as 'a level as close to the MCLG as feasible'. 'Feasible' is defined as 'practicable according to current treatment technology, provided no adverse effect is caused on other treatment processes used to meet other water quality standards'.[22] Under the previous Safe Drinking Water Acts of 1974 and 1986, the USEPA had issued Regulations covering some 85 organic or inorganic substances, and 16 others covering radionuclides and microbial levels. When issuing such standards the USEPA had not been required to assess the cost to water utilities of implementing a regulation nor to estimate the health benefits a regulation would achieve. This gave rise to widespread criticism that relative risks, costs, and benefits had not been taken into account when setting standards, thereby imposing some costs on public water suppliers which could not be justified in relation to the incidence of such contaminants and the numbers of individuals at risk.

Under the 1996 Act the USEPA is now required to support any proposed new Regulation by publishing a report on:

- the risk the contaminant presents to human health;
- the estimated occurrence of the contaminant in public supplies;
- the population groups and numbers of people estimated to be affected by the contaminant;
- the benefits of reduced risks to health the proposed Regulation should achieve;
- the estimated costs to water utilities of implementing the Regulation;
- estimated changes to costs and benefits of incremental changes to the MCL value proposed;
- the range of uncertainties applying to the above evaluations.

The USEPA has to publish and keep updated every 5 years a list of contaminants likely to require regulation, choosing at least five of them to Regulate every 5 years, giving priority to those posing the greatest health risk first. A period of 18–27 months for public comment and USEPA's consultation with certain authorities must be allowed before a proposed Regulation containing an MCL is promulgated, i.e. formally put into operation.

To follow these new procedures USEPA had first to collect country-wide data revealing the incidence of various types of contaminants in public water supplies and the effectiveness of current treatment processes in reducing such contaminants to required levels. All systems supplying 100 000 population or more were required to submit detailed reports on the results of 18–24 month comprehensive sampling programmes, with lesser sampling programmes being required by smaller systems serving upwards of 10 000 population. Also to help small water systems serving 10 000 population or less, the 1996 Act required USEPA to provide a list of treatment systems that would assist these small systems achieve the required quality standards. USEPA must also provide small systems with 'variances' which permit the adoption of treatment processes that will achieve nearest compliance with an MCL, taking into consideration a system's resources and the quality of its source water. Government funds made available to States are to be used to assist in training waterworks operators to standards set by the USEPA. The funds are also intended to help these small systems obtain technical assistance with water quality compliance problems and source protection.

The radical changes made by the 1996 SDWA should achieve a more realistic approach to establishing the setting of water quality standards, and make it more practicable for utilities to comply with them. But the performance of the many small undertakings may continue to present a weakness, unless they can be provided with sufficient day-to-day technical and laboratory assistance to ensure they achieve compliance with the many sophisticated drinking water standards in force. The evaluation procedures the USEPA must undertake to support any proposed new Regulation are complex and are likely to raise problems of interpretation and differences of approach concerning the methods to be adopted which may need clarification and agreement.

2.10 Private sector participation in water supply

Since the later 1980s there has been increasing interest in involving the private (i.e. commercial) sector in providing some or all of the activities required to provide a public water supply. While the public sector must always retain ultimate control because water supply is an essential service and also a monopoly, the advantages of private sector participation are:

● commercial funds for capital and improvement works are easier to obtain than government or state funds;
● specialised technical and managerial skills can be brought in to benefit a water undertaking;
● improved efficiencies of service can be obtained by setting them as contractual obligations on a private company.

Private companies have, of course, always been used by publicly owned water undertakings to provide such services as the design and construction of new works, repair of burst mains, analysis of water samples and so on. But a wider variety of contractual arrangements for private sector inputs have now been developed and used successfully. The two main classes of private sector input are 'build contracts' which cover the development of new infrastructure, and 'operational agreements' which aim to improve the efficiency of one or more water supply operations by putting them for a period under the direction of a private company. In general, the greater the responsibilities undertaken by a private company, the longer is the period of contract.

A type of *build contract* which has been developed for the design and construction of new works, such as new source works or treatment works, is the 'build, own, operate, and transfer' (BOOT) contract. A private company backed by a bank or some other financial institution, or perhaps using some bi-lateral loan provided by its government, finances and builds some new works, operates them for a term of years, and transfers them to the public water authority at the end of the contract. With private finance the duration of such contracts is often 20 years or more. The water authority repays the cost of the works over the contract period or as otherwise defined, and pays for the operation of the works on some time-related or works output basis. A 'build, operate, transfer' (BOT) contract is very similar, but is usually used to cover the case where the contractor does not provide funds to construct the works, being paid for the works as they are constructed, then operating them for a given period before transferring ownership to the water undertaking.

Operational agreements are now the main type of private sector participation methods adopted. They comprise a private company taking over or controlling for a period some or all of the operational activities of a water undertaking. Often they involve both the public and private sectors working closely together under a 'public–private partnership' arrangement, with risks allocated and shared in some agreed manner. The principal types of operational agreements which can be distinguished are as follows.

Contracting out. A private organisation takes responsibility for a specific package of work for a fixed duration, such as meter reading; or billing and revenue collection; or operation of a water treatment plant. The responsibility of the private company is limited to its contracted task.

Management Contract. A private company contracts to provide a service to improve certain water supply operations for a fixed fee, or a fee partly based on achievement in meeting some specified performance target, such as reduction of leakage, betterment of service levels to consumers, etc. The management contractor has to work closely with the water authority, advising and directing its staff in the use of new or improved techniques to gain the end-result desired. The contractor does not take on any significant risks or responsibility save that of ensuring the provision of the quality of managerial and technical staff that he or she has contracted to supply and they are effective in achieving the objective of the contract. The water authority retains responsibility for executing the operations necessary including their necessary funding. The duration of such contracts is normally only a few years and variety of different arrangements are possible.

Leasing (or affermage). Under leasing a private company takes over all, or some of a water authority's operations for a fee, or for a full or part share of the income collected from consumers. Specific targets to be achieved by the operation taken over will usually be set. Leasing does not necessarily involve the lessee, i.e. the private company, in providing large finance for the construction or improvement of capital works, because the government or state may find that it can get better capital financing terms from international or bilateral funding agencies, who may provide 'soft' loans and grants. Compared with a management contractor, however, the lessee is more directly responsible for management and technical aspects, needs a larger financial input to cover his or her day-to-day working capital, and bears more financial risk. On the other hand, if the lessee takes a share of the undertaking's income, there are greater opportunities for profit through adoption of improved efficiencies. Duration of leasing agreements is normally 10–15 years.

In many French cities, mainly for the operation of treatment works, leasing (termed 'Affermage') has been widely used for many years for fixed periods. One of the reasons for this approach is that, whereas in the whole of the UK there are only 31 suppliers of public water (27 in England; three in Scotland; one in Northern Ireland), France has over 3000 separate water undertakings. Likewise Germany has 1500, and Spain 8000.[23] Hence the use of large firms able to provide a wide range of specialised technical services involved in water treatment, is an advantage where many small undertakings exist with limited technical resources.

Concession. Under a concessionary agreement a private company takes over full responsibility for a whole system, including planning and funding all capital works needed for rehabilitation and expansion of the system, taking on most risks in return for receiving all, or part of the total income generated from water sales and other charges. This motivates the private company to use its expertise to develop the system in the most efficient manner in order to maximise its income. The duration of such agreements is normally 20–30 years.

All types of private sector input contracts need to be robust, and the longer term contracts need to provide for adjustment to meet future conditions that cannot be foreseen sufficiently accurately to be fair to the undertaking's consumers and the private company. For instance, the capital works programme may have to change to meet a different growth of water demand from that expected, and tariffs may need to be adjusted for inflation. Some of the most important matters which contracts have to cover are as follows.

- The source of funds and the terms of their repayment have to be defined.
- Detailed targets for achievement have to be set, such as standards of water quality, development of a new source for a given output, distribution performance, levels of service to consumers, and measures of efficiency for operations taken over.
- The contractor's reimbursement has to be defined. Under most contracts an independent regulator or regulatory body sets water tariffs to consumers, or tariffs can be fixed through a competitive bidding process; but if the contractor is to be reimbursed partly or wholly from tariff income there may need to be provisions assuring the contractor of some minimum rate of return on his or her financial outlay.
- The question of staffing has to be decided. The water authority's staff will expect their employment to continue under the private contractor, but the contractor may not find it possible to take on all staff because of the obligation to achieve greater efficiency. Negotiations have to take place to solve this problem.
- The allocation of risks is a key issue to be clarified. Too many risks placed on the contractor mean that the price must be increased substantially. Too few risks on the contractor can reduce the motivation to improve the services taken over.

There are some disbenefits to public sector participation. There may be a social cost if some water authority staff have their employment terminated. If the contract is undertaken by an international contractor, more offshore currency expenditure may fall on the government if the contractor requires a proportion of the charges to be paid in this way. The problem for poor countries is that of funding the contractor and any improvement works necessary, before the improved supplies or greater efficiency make it possible for the undertaking to gain an increased income through more sales of water or the raising of tariffs. Drawing up a contract is a complex matter on which it is essential to take experienced advice. It may take up to 2 years to produce a contract, evaluate bids received, and negotiate a contract, and bidding costs for contractors are high. As a result there may be a tendency for contractors to favour only the less onerous and less risky contracts which offer the best opportunity for a return on their expenditure. When the contract starts, the water authority will need to set up an efficient regulatory body, able to check all aspects of the contractor's work to the end of the contract and negotiate any financial adjustments necessary. In addition it may be advisable for government to establish a framework setting out procedures which must be followed when letting contracts of a major kind to the private sector, in order to ensure that all such transactions are properly carried out.

With the availability of public funds for improvement of water services inevitably limited by the many other demands made on the public sector, private sector participation with its access to commercial funds is a potential alternative world-wide. The details of how this is best achieved are evolving rapidly as experience is gained in many countries.

2.11   Organisation of a water undertaking

The typical organisation of a large public water supply undertaking is similar to that shown in Figure 2.3. Smaller undertakings may merge some departments; the very large ones may set up regional offices whose organisation replicates the head office organisation on a reduced scale. Sometimes in the large organisations there may be a separate 'Resource Planning' or 'New Works' department, and also a 'Records Department'. If the undertaking has a large pumping and mechanical plant there may be a separate 'Electrical and Mechanical Engineering Department'. A 'Training Department' is essential in the larger undertakings because it can make a major contribution to the efficiency of the undertaking.

From long experience there is a need to centre the day-to-day responsibility for maintaining the quality and sufficiency of the supply upon one person. This is usually the Chief Engineer, as advised by the Chief Scientist. The smaller undertakings often have a combined 'Engineer and Manager' in charge. The larger undertakings may have an 'Operations Manager' in charge of the Distribution and Supplies Departments, who reports to the Chief Engineer. In this key position, the Chief Engineer or Chief of

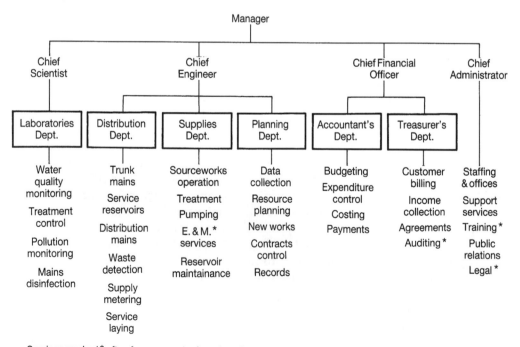

Services marked * often form separate departments.
Stores and stockyards will come under Distribution Dept. (for pipes etc.) and Supplies Dept. (for chemicals, spares, etc); replacements being ordered through the Accountant's Dept. and stock usage being checked by the Audit Section.

Fig. 2.3   A typical division of functions in a moderately sized water undertaking.

Operations has to decide what is to be done in an emergency, such as when a source fails; a major burst occurs; or there is evidence of contaminated supplies.

The large modernised water undertaking will normally have a central control room, manned 24 hours a day, into which out-station conditions are reported on instruments. The data flowing into this control station will show the outputs of sources, what plant they have running and what is available on standby, water levels in service reservoirs, records of flow in some key mains, and some key distribution pressures. Into this control room will also come warnings from the fire service of the whereabouts of fires, or of major road accidents that might involve discharge of chemicals that could endanger the quality of any source. Arrangements are usually also made for messages from the public to flow into this control room because the public are often the first to notice signs of a burst main. Rotas have to be established whereby key personnel, such as valve operators familiar with sections of the distribution system and mobile repair gangs, are on standby to carry out any night-time emergency work required.

A key part in maintaining a good service to consumers is played by the local engineer, technician, or 'inspector' in charge of a distribution area. This local manager is responsible for managing a limited area of the distribution system and becomes familiar both with its detailed layout and the consumers in the area. The local knowledge so obtained is invaluable in operating valves when burst mains or fires occur, in detecting signs of waste and in reporting defects in the distribution system. Local distribution managers are the water undertaking's employees who most frequently makes direct contact with consumers, hence they can monitor and report back any difficulties that consumers are experiencing.

2.12 Staffing levels

Many factors affect levels of staffing in a water undertaking. Those using surface supplies which require treatment will need more staff than those using underground supplies, since treatment works usually have to be staffed 'around the clock' and more staff will also be required for maintaining buildings, dams or intakes, and catchment grounds. Borehole supplies often require little treatment and can be operated by automatic or semi-automatic control. A multiplicity of supplies involves more staffing than when a single major source gives the principal supply.

Some selected figures of staff employed per 1000 connections are given in Table 2.2. In England it is the larger water companies or those supplying compact urban areas who can achieve 1.0–1.1 employees per 1000 water connections. The smaller English water companies or those who have to supply extensive rural areas utilise 1.4–1.5 employees per 1000 connections. Overseas water undertakings have much higher ratios of staff per 1000 connections in most cases because of the low cost of labour and consequent use of manual processes for many operations, and the need to use more staff on income collection from metered domestic consumers who are required to pay in cash monthly.

2.13 Charging for public water supplies

Water supplies often have to be given free or below cost to those on very low incomes. Most standpipe supplies in low income countries are given free. The consequent financial loss has to be made good either by the water supplier charging other consumers more, or by the state or government meeting the deficit. Even in the UK the extension of water

Table 2.2   Some selected figures of waterworks staffing

	Staff per 1000 connections	No. of connections $\times 10^3$	No. of staff	Notes
England:				
Water-only Companies				
1995/96 (av. 18 companies)	1.16	4731	5486	
1996/97 (av. 18 companies)	1.14	4695	5353	
1997/98 (av. 17 companies)	1.12	4791	5376	
Overseas:				
Selected undertakings 1995 or 1996				*Work put out to private sector*
Kuala Lumper, Malaysia	1.12	157		Prodn
Taipei, Taiwan	1.14	1289	1465	B & C + leak repairs
Johor Bahru, Malaysia	1.2	223		Prodn + leak repairs
Selangor State, Malaysia	1.4	932	1322	Prodn
Singapore	2.0	910	1865	B & C
Seoul, Korea	2.3	1873	4332	Meter reading
Chonburi, Thailand	2.6	47		Prodn
Hong Kong	2.8	2100	5830	
Johore State, Indonesia	2.9	534	1544	
Chiangmai, Thailand	2.9	35		
Penang State, Malaysia	4.0	263	1058	
Penang Island, Malaysia	4.4	133		
Bangkok, Thailand	4.6	1241	5736	Prodn
Thai Provincial W.A.	5.5	1195	6547	
Lahore, Pakistan	5.7	372	2106	
Jakarta, Indonesia	5.9	362	2131	B & C
Shanghai, China	6.0	1828	11 060	

Notes.
Prodn = operation of sourceworks. B & C = billing and income collection.
Data sources: England – 'Waterfacts' '96, '97 and '98.
Overseas – McIntosh C.A. and Yniguez C.E. (eds) *Second Water Utilities Data Book; Asian and Pacific Region*, Asian Development Bank, Manila.
(Other cities in India, Pakistan, Bangladesh, etc. frequently show 15–25 staff per 1000 connections.)

mains to rural areas had to be funded by government or local government through a series of Rural Water Supply Acts in the 1940–60s because the cost of the long water mains required for rural dwellers did not give an economic return for the water undertakings and was too expensive for rural dwellers to finance.

Where domestic supplies are unmetered, it is common practice, as in UK, to apply a charge related to the value of a householder's dwelling, on the basis that most low-income householders occupy low-value property. Where domestic supplies are metered, a basic quantity can be allowed for each household (per charge period) at a rate sufficiently low for the low-income householders to meet. Neither system is perfect since there are some low-income householders who occupy a large property because of a large family size and, with metering, the lowest rate of charge has to cover an amount sufficient for a large family, which gives the single occupant a liberal supply at the cheapest rate. However, any system for charging according to the number of occupants or household income is impracticable.

When fixed charges according to property value apply, it is relatively simple for the water undertaking to set the charges at a level which will provide a required income because the number of properties in each valuation range can be known. But if all supplies are metered, the setting of an appropriate tariff to produce a given income has to take the elasticity of the demand into account (see Section 1.19). This is particularly important because the margin of an undertaking's annual income over expenditure to pay loan charges and to provide a profit, may be heavily dependent on the amount of water sold at the higher rates of charge. A graph of the percentage of households taking less than a given amount needs to be produced, as shown in Figure 2.4. The area under this graph represents the quantity of water (per 100 households) in each charge band. Assuming the elasticity of demand for the lowest band is -0.3, and 31.5 m³/day is taken within that band by every 100 consumers when the price is £0.50/m³; then the amount taken when the

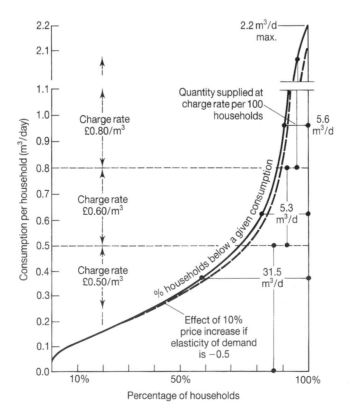

Fig. 2.4 Cumulative frequency graph of household consumptions.

price is raised by 10% to £0.55/m^3 is given by:

$$\text{Reduced volume taken} = 31.5 \times [(0.50)/(0.55)]^{0.3} = 30.6 \text{ m}^3$$

Assuming the other band prices are increased by 10% and the elasticities are −0.6 for the middle band and −0.9 for the upper band, it can be shown that raising the tariffs by 10% only increases the total income from every 100 domestic consumers by 5.3%. The income from metered supplies to trade and industrial consumers under an increased tariff has likewise to be computed taking an appropriate value for the elasticity of their demands.

2.14   Financing of capital works

A water undertaking usually borrows money to finance construction of major new capital works, because the cost of such works is too large to meet from current income and can reasonably be spread over future consumers who will benefit from the works. Loan repayments are typically 10–15 years for plant and machinery; 20–30 years for buildings; and 50–60 years for dams and land. Routine capital works expenditure on extension of mains can normally be met from current income. Three methods of loan repayment are possible – capital plus interest: annuity; and sinking fund. If a loan of £X is repayable in n years at r% interest rate (where r is expressed as a decimal) the annual amounts payable under the three different methods are as follows.

	Annual payment
Capital plus interest	$X/n + r$ interest on the loan balance outstanding at the start of each year
Annuity	$Xr(1 + r)^n/[(1 + r) - 1]^n$
Sinking fund	$Xr/[(1 + r) - 1]^n$ to sinking fund $+ Xr$ interest on loan.

The annuity repayment method is most usual. The capital plus interest method results in reducing yearly payments. The second and third are equivalent if the interest received on the sinking fund is the same as that charged on the loan.

Public water authorities may be able to borrow from government, from an international lending agency, or may be authorised to borrow from the money market. Private companies usually borrow from the money market. In theory they can also raise more money by offering further shares for sale on the market if their full share capital has not been issued; but this would rarely be adopted for new works. Of course if an undertaking has an excess of income over expenditure, part of the excess can be set aside to build up a fund to finance, or partly finance, new capital works. If a variety of capital projects have to be funded, a separate loan may not be raised for each. Instead short-term borrowing from current funds or from the bank may be adopted to fund the capital outlay, until the time is considered appropriate for raising a single long-term loan. Much depends on interest rates prevailing and whether these are expected to move up or down in the future. Clearly if interest rates are extremely high, there can be hopes they will later fall, so that postponing the raising of a large long-term loan could be advantageous, even if current short-term borrowing is costly. But things do not always work out as expected.

2.15 Depreciation and asset management planning

'Depreciation' is an accountancy term for the practice of writing down the initial cost of an asset annually. The amount an asset is depreciated is debited against income, so the equivalent amount of money (in cash or securities) is set aside and allocated to a depreciation fund. This fund is used later to meet the cost of renewing the asset when it becomes worn out. Depreciation has been defined as representing 'the consumption of assets by current users thereof'. The amount of the depreciation depends on accountancy practice. The depreciation must, in the first instance, be sufficient to write off the whole cost of the asset before it has to be replaced. But the replacement cost may be higher due to inflation, or because a better quality replacement is required. This has to be allowed for in the amount of the depreciation. Also tax considerations may influence the depreciation. In the UK a 25% initial depreciation of plant and machinery is allowed against tax, thereafter 10% of the reducing balance. An accountant may therefore wish to benefit from the large initial tax allowance.

An 'asset management plan' consists of assessing the condition of each asset, and estimating the length of its remaining life, how much it will cost to maintain and ultimately replace. The process has to be applied to all assets of a water undertaking – the distribution system, sourceworks, treatment works, pumps and machinery, etc. For a large undertaking this is a large task, and systems for simplifying it may need to be developed. Usually a long-term plan for asset renewal for 20 years or so ahead is developed, and a 5-year rolling programme of renewals is produced in more detail with estimates of the year-by-year expenditure involved. The longer term plan is necessary because worn-out assets are seldom replaced by identical ones. Old mains may be renewed in larger size or by a differently routed main in order to cater for increased future demands. Newer standards of water quality may make it necessary to replace old treatment plant by different or more extensive plant. Older equipment needs to be replaced by more efficient or more sophisticated apparatus. Thus asset management planning has to based on a technical plan for the future development of the undertaking, and estimated costs have to be updated at intervals.

Some assets, such as dams and large trunk mains, last so long, 60–100 years or more, that their loan cost has long since been written off. Hence no 'value' for them may appear in the books, so depreciation is not applicable. Valuation of such assets is only necessary in the rare and unlikely event of the undertaking being sold. On the other hand such assets present a liability because, in due course, a dam may need extensive repairs and water mains cannot last indefinitely. Hence, it is prudent to build up a 'contingency fund' to meet possible repair or renewal costs. This is particularly important in the case of water mains, in order to avoid getting into the position of finding it necessary to replace a large number of mains of a certain age simultaneously at some future. This could pose a large unexpected financial burden on an undertaking. Hence depreciation, contingency funds, and part of maintenance expenditure, are all financial measures with the same aim of making timely provision for the cost of preserving and renewing assets.

2.16 Cost comparison of proposed capital projects

Discounting to compare present values of projects. If proposed projects differ in phasing of capital expenditure and running costs, etc. they can be compared by 'discounting' their costs to obtain the 'total present value' cost of each project. This allows the timing of expenditure to be taken into account. The present value, P, of £X due to be paid in n years' time is taken as $P = £X/(1 + r)^n$, where r is the discount rate expressed as a decimal. (This is the inverse of compound interest calculation, i.e. P invested at r% compound interest accumulates to £X in n years' time.) To compare projects, the costs have to be for the same year-by-year water outputs for the same period – usually 20, 25 or 30 years. The estimated capital expenditure and works renewal costs, plus the running costs, for each year are estimated and discounted* to give the equivalent present value cost and the total is summed for the chosen period. The project having the lowest total present value is the cheapest at the chosen rate of discount. Usually projects are compared over a range of discount values; 8, 10 and 12% being commonly used. The higher the discount rate, the smaller are the 'present values' of future costs. Hence high discount rates tend to favour schemes which can be built in stages, or which have cheaper initial capital outlay despite having higher running costs (see Section 14.7).

A government or lending agency may stipulate the discount rate to be used when comparing proposed capital projects. This is a way of putting projects on a par basis with respect to their demands for capital; it prevents one scheme using a disproportionate amount of capital as compared with another. A discount rate fixed in this manner is said to represent – 'the social opportunity cost of capital' – being used to assess all capital demands made on a government or fund for social improvement projects, such as housing, roads, hospitals, etc. From time to time the UK government Treasury stipulates the rate of discount to be used when public sector schemes are put forward for funding.

Inflation of prices is not usually taken into account when discounting, on the assumption that if inflation occurs it affects all prices proportionately. History, however, shows otherwise. As standards of living have increased, labour costs have inflated more than material prices due to increased productivity of machines. Fuel oil prices doubled in 1973, then fell back, then rose again later. Hence foreseeable differences of inflation need to be allowed for – if foreseeing them with any reliability is possible.

Shadow pricing of costs is sometimes adopted when market prices do not represent true costs. Thus, if unemployment is high, the shadow price for labour is the wage paid (including oncosts) less the cost to society of that person when unemployed (e.g. the unemployment pay, etc.) Taxes are excluded from shadow prices because they represent only a transfer of money from one section of society to another. Where the unofficial exchange rate for offshore currency in the market is higher than the official exchange rate, the shadow price for purchase of offshore goods may need to be based on the unofficial rate because this represents the real cost of offshore purchases. Such shadow pricing has the benefit of showing in a more favourable light schemes which use more local labour or more inshore goods than another, which is an advantage to the country in which the project is to be built. However, carried to extremes, shadow pricing is complex, because each item price has to be broken down into its component parts, e.g. labour, fuel, plant,

---

*A useful additional formula for obtaining the present value P of a regular sum £x paid annually for n years at r% discount is

$$P = \frac{£x(1 + r)^n - 1}{r(1 + r)^n}$$

materials, further broken down into offshore and onshore elements, and taxes, and so on, before a shadow price for each item can be estimated. Consequently shadow pricing tends to be adopted only when a funding agency requires it, and the agency may stipulate that only a few major items, such as labour, should be shadow priced.

2.17 Other aspects of project assessment

The *internal rate of return (IRR)* is another way of comparing projects. In this case the 'value' of 'the outputs' has to be assessed. The IRR on the project is then defined as:

> the discount rate at which the value of the outputs from a project for a term of years equals the value of the costs of the project for the same term of years.

The international funding agencies and the World Bank often require a project to show not less than 10% IRR before they rank it for funding.

The weakness of the internal rate of return approach is the difficulty of setting 'a value' for the outputs of water when a public water supply serves so many purposes, e.g. health, comfort, recreation, trade, industry, food production, and so on. The tariff charges made to consumers cannot be used as a 'value' because they only reflect the money return required (after paying running expenses and any profit) which equals the interest rates charged on capital monies. Hence the 'value' put on the water output is bound to be somewhat arbitrary, such as, say, a cost per cubic metre for domestic supplies which would 'normally' be charged or which represents, say, 1% or maybe 2% of average family household income; and a cost for trade and industry supplies might be judged from the elasticity of their demand. The funding or loan agency can, of course, stipulate what 'value' is to be placed on the water output.

A private water company, it should be noted, does not need to 'value' its water output because it is not a funding agency. A funding agency or government which has a limited amount of funds for social benefit schemes, must endeavour to apply its funds to best effect, so it must assess and compare the benefits given by its funding of various schemes. A commercial company is only concerned with the cost of borrowing and the need to set consequent charges sufficient to give a satisfactory rate of return on its share capital. The Director General of Water Services in UK has stated that the cost of equity capital for the Water Service Companies in England and Wales 'might be 6–7% in real terms allowing for inflation, business taxes and cost of debt finance in the range 3–5% in real terms'.

Full *social cost-benefit analysis* (SCBA) of projects gives rise to many subtle (and at times confusing) differences of view. Merrett[24] considers that using a 'social time-rate of discount' (STRD) to evaluate cost and benefits is reasonable since it reflects human appreciation that a benefit now has more certainty than one in the future, but it must not be too high. He suggests 3% is appropriate because it does not devalue a future cost or benefit too rapidly, taking 23 years to reduce a value by half. The much higher discount rates of 8–12% used to express the 'opportunity cost of capital' he considers are only appropriate for comparing the *cost* of alternative schemes under 'social cost-effectiveness analysis' (SCEA) procedures. They are inappropriate for SCBA analyses because they 'ignore the huge importance of environmental and distributional effects in decision making'. In effect he distinguishes between assessing projects purely on financial cost, and the very different matter of assessing the benefits and disbenefits of a project (in terms of social improvement, health, enjoyment, etc. versus environmental losses, etc.) some which should not be devalued by discounting to negligible amounts in a short period of years because they are costs paid, or benefits enjoyed, in perpetuity.

To the extent that monetary values can be put on environmental gains and losses they can be included in a SCBA of a project. For instance, what people will pay for fishing rights, or for a visit to a nature reserve or open country, including their travel costs, can act as a crude monetary evaluation of an environmental benefit. This can put a useful comparative measure on at least a few of the many other considerations that have to be the subject of an Environmental Impact Assessment (EIA). An SBCA is an economic assessment taken as far as it is reasonably possible to attribute monetary values to outcomes; an EIA deals with other matters not expressible in economic terms. Both are only aids to decision making which may have to be taken on entirely different grounds, such as political expediency, public reaction, etc.

2.18   Short- and long-term marginal costing

When an undertaking has spare capacity at its source works, it can meet additional demand at little extra cost. Over 90% of a water undertaking's costs are fixed, being loan repayments for works and the operational costs involved in the day-to-day running of a system which supplies water to tens of thousands of premises. Less than 10% of the cost varies with the amount of water sent out by the source works. But if new major works have to be built, the cost of the supply rises sharply because the loan repayments on the new works have to be met even though the immediate rise in consumption is small. To avoid this sudden rise in charges, 'long-term marginal costing' can be used, under which account is taken of the cost of the next stage (or stages) of capital expenditure required. This means that, before the new source works are actually required, consumers pay more than current costs, so money is saved which can be set aside to partially or even wholly fund the cost of the new works. Price rises are thus smoothed out, and the early increase in charges may induce metered customers to curb their demand. This may postpone the date when the new works will be required.

How much current consumers should be asked to pay for works that benefit future consumers is a debatable point; but often current consumers themselves benefit from works still standing and productive but long since paid for by previous consumers. Arguments about the monetary value to be put on paid-off assets still in use are only relevant in the relatively rare case when a water undertaking is put up for sale. However, a much more important point is that an undertaking should not become so heavily debt burdened by long-period loans, that it finds itself in a financial difficulty if new, unexpected heavy financial commitments come upon it. This situation occurred in England, where many new schemes had been built in the 1960s and 1970s to meet a swiftly rising demand for water. As a consequence the water authorities were already burdened with many long-term loans when new demands came upon them in the 1980s to meet new standards of water quality as well as improved sewage effluents. Hence, when the Government decided to privatise the water authorities in the 1980s, it wrote off £5.2 billion of the long-standing debt of the water authorities[25] because it feared the size of the debts might not make the authorities sufficiently attractive financially to make their sale to the private sector a success.

2.19 Typical prices charged for water

Average household bills for water 1997/1998 were as follows.

	Unmetered median (range) per annum	Metered median (range) per annum
England and Wales		
Water service companies (10)	£103 (£94–138)	£94 (£85–111)
Water-only companies (17)	£114 (£72–168)	£89 (£65–116)
Scotland	£72 (£66–77)	not applicable

Tariffs for metered domestic supplies in England and Wales averaged £0.67/m^3 (range £0.43–1.00) plus a standing charge in most cases of £24 per annum.

References

1. Evans H. R. The Structure and Management of the British Water Industry 1945–91. *Yearbook 1993*, IWEM.
2. IWE. *Evidence to the Central Advisory Water Committee*. Feb. 1970.
3. Kinnersley D. *Troubled Water*. Hilary Shipman, London, 1988, pp. 122–123.
4. Hill S. W. *Taking Stock: Privatization of Water Services*. Private publication. June 1989, pp. 16–17.
5. Water Cos. Assn. *Waterfacts '92*, p. 52.
6. Henley D. *et al*. *Public Sector Accounting and Financial Control*, Chapman & Hall, 1993, p. 122.
7. Isack F. Money Down the Drain, *Construction News*, 31 Aug. 1995, pp. 16–17.
8. cf. *Construction News* 1 Dec. 1994, and *New Civil Engineer*, 1 Dec. 1994, p. 7, and 9 Feb. 1995, p. 6.
9. Director General of Water Services, *Report for the Period 1996/97*, p. 42.
10. Director General of Water Services, *Report for the Period 1996/97*, pp. 32–34.
11. Director General of Water Services, *Report for the Period 1996/97*, p. 43.
12. Statutory Instrument No. 324, 1991.
13. Barker P. Agricultural Pollution Control and Abatement in the Upper Thames Region. *JIWEM*, June 1991, pp. 318–325.
14. EC Directive 91/676.
15. EC Directive 76/464 and Directive 86/280, with additional substances listed in EC Directives 88/347 and 90/415. [UK Regulations in response were Statutory Instruments 2286 (1989); 337 (1992) and 2560 (1997).]
16. DoE regulations 1994, under Part II of the Environmental Protection Act 1990, in response to EC Directive 75/442 and in accordance with DoE Circular 11/94.
17. EC Directive 80/68.
18. Anderson P. Demands and Resources for Public Water Supplies in Scotland from 1991 to 2016. *JCIWEM*, June 1997, pp. 164–169.
19. Adeloye A. J. and Lowe J. M. Surface Water Abstraction Controls in Scotland. *JCIWEM*, April 1996, pp. 123–129.

20. Okun D. A. Addressing the Problems of Small Water Systems, *IWSA Conf Proceedings 1995*, Special Subject 9, pp. 439–442. Also MacDonald J. A. *et al.* Improving Service to Small Communities. *JAWWA*, Jan. 1997, pp. 58–64.
21. AWWA Small Systems Research Committee, Research Needs for Small Systems: A Survey. *JAWWA*, Jan. 1997, pp. 101–113.
22. Pontius F. W. Future Directions in Water Quality Regulations. *JAWWA*, March 1997, pp. 40–48.
23. Report on Proceedings of Conference on 'Supply and Demand: a Fragile Balance'. *JCIWEM*, Aug. 1996, p. 298.
24. Merrett D. *Introduction to the Economics of Water Resources*, University College London Press, London, 1997, p. 92.
25. Kinnersley D. *Coming Clean: The Politics of Water and the Environment*. Penguin Books, Harmondsworth, 1994, p. 78.

3
Hydrology and surface supplies

Part I Hydrological considerations

3.1 Introduction

When a new water supply is required to meet anticipated increased demand it is common practice to undertake an assessment of the water resources of the region, to consider all possible sources of supply and then to define an appropriate development plan. The range of development possibilities are as follows.

Surface water sources – (a) direct supply from an impounding reservoir or lake, supplemented if necessary by gravity feed from an adjacent catchment or pumped inflow from another source; (b) abstraction from a river or canal, supplemented if necessary by releases from a storage reservoir; (c) collection of rainfall runoff from the roofs of buildings or bare catchments and feeding to storage tanks.

Ground water sources – (a) springs, wells, and boreholes; (b) adits and collecting galleries driven underground; (c) extraction of ground water by riverside wells or sub-surface extraction wells sunk in the bed of a river course or 'wadi'; (d) using ground water extraction to supplement abstraction from rivers or reservoirs; (e) use of artificial recharge of aquifers.

Water reclamation schemes – (a) desalination of brackish water or seawater; (b) re-use of acceptable wastewater discharges by appropriate treatment; (c) demineralisation or other treatment of a minewater, including blending with a freshwater supply.

Other types of schemes – (a) integrated (conjunctive) use of surface water, ground water, or water reclamation schemes according to availability; (b) obtaining a bulk supply from some other water authority.

An assessment of the hydrology of a catchment is necessary to ensure that any proposed additional abstraction will not detract from the rights of other water users or cause unacceptable harm to the environment. For this purpose a 'water balance survey' of the catchment is often desirable. This quantifies for a given water year of average or minimum rainfall:

- rainfall on catchment *less* losses from evaporation, transpiration, and natural surface runoffs (i.e. after correcting for abstractions and discharges);
- resultant percolation to underground:
  plus inflow *less* outflow through underground water boundaries;

less abstractions from wells and boreholes;

$+/-$ change in soil moisture content, underground water storage and water in transition zone;

● balance remaining – water unaccounted for.

Theoretically if all inflows and outflows have been accounted for, a balance of zero will be obtained. But if the calculations have been conducted monthly the balance remaining will tend to be positive for wet periods and negative for dry periods because of time lag effects. However these should largely cancel out over a full year. The balance achieved is largely dependent upon adequate local hydrological data being available; if not, values have to be estimated from relationships derived from catchments elsewhere with similar characteristics. But once a reasonably satisfactory balance has been achieved there can be assurance that the hydrology of the catchment has been properly appraised.

3.2   Catchment areas

The accurate definition of the catchment boundary is usually a prerequisite for the assessment of a source. In theory the topographic catchment to a source is found by examining a contour map, but in practice difficulties occur. Many parts of the world are still unmapped at an adequate scale; even where maps exist, the contours on them may be largely extrapolated or out of date because of subsequent constructional work. Personal knowledge of a catchment is therefore important, and a visit to key sections of the topographic divide is desirable. When larger scale maps cannot be obtained, those at 500 000 scale by the USA Air Force in their 'World Tactical Pilotage' series (obtainable in UK from the Director of Military Survey, Ministry of Defence or major mapsellers) are recommended. Where possible an aerial survey followed by detailed photogrammetric contouring is the best policy. Associated ground control survey checks will be essential as well, but a rapid mapping programme measured in weeks rather than months can be completed for a modest cost. Direct examination of stereo pairs of air photographs may resolve some uncertainties. Large inaccessible basins can be examined from satellite photographs, libraries of which are now available internationally. However the best sources are those giving digital descriptions of terrain and rivers, now on CD-ROMs; the most detailed global coverage is emerging from ESRI, Redlands, California, the major American geographic information software company.

Computer programs exist which will define a catchment boundary from any gridded data set. That of the UK Institute of Hydrology operates on a 50×50 m digital terrain model derived from 1:50 000 scale contours,[1] and has been used throughout the United Kingdom to define the catchment boundaries to streamflow measurement stations. Such programs can generate catchment boundaries quickly and objectively, but may require manual adjustments in some cases, particularly in areas of low relief where the stream drainage network is poorly defined.

Once the catchment boundary has been located by defining the direction of downhill flow at right angles to the contour lines, one may come across water-courses marked on maps that appear to cross the catchment divide. If not the top pound of a canal, this watercourse may well be a contour leat. A leat is an open channel gravity-flow catchwater, often constructed to augment the flow into a reservoir or to bring water to a mill or mine. Leats may either collect all the drainage from land on their higher side, take only the flows of the major streams which are intercepted, or can be derelict and carry no flow in or out the catchment. A survey is required to assess the condition of the leat and its capacity relative to the local runoff as the leat may be sized to contain all but the largest floods or may be so small that it is overtopped many times a year.[2]

The catchment to a ground water source is not readily defined, even where contoured maps of the water table are available. In the majority of cases the water table contours represent a scaled-down version of the surface topographic contours, but cases can occur where ground water enters some distance outside the surface catchment limits. Natural water table levels tend to rise and fall together in any one region but, particularly where pumping abstraction is taking place, the ground water boundary divide may migrate seasonally. This illustrates the fact that there is a continual balancing effect proceeding in an aquifer between steady abstraction and intermittent replenishment. Satisfactory confirmation of such conditions may require as many as one observation well per km² once away from flat terrain, a density which is rarely practicable except under research funding.

3.3   Data collection

The collection, archiving and dissemination of hydrological data is expensive and is usually a government funded activity. As governments try to limit public spending, data acquisition budgets tend to be cut on the basis that the data does not produce immediate or obvious benefits. But it has been demonstrated that the availability of reliable hydrological data results in benefits an order of magnitude larger than the costs of collection.[3] The need for more data is particularly important in countries where changing climatic conditions are causing serious depletion of water supplies. In Nigeria, for example, it was reported in 1998 that failure of a number of large scale resource developments to achieve their forecasted output was partly due to the absence of accurate data on rainfall and river flows.[4] If large capital sums are not to be wasted, the basic hydrological data on which water schemes are designed must be adequate and reliable. But where good hydrological data is available and subjected to experienced analysis, very large savings can be made. On the Bhatsai water scheme for Bombay in the 1970s, a re-assessment of the hydrological data based on the Tansa river flows showed a river regulating scheme was possible, increasing the potential yield of the source from 1140 Ml/day for a direct supply scheme to 1800 Ml/day and saving a large proportion of the cost of a pipeline.[5]

Good hydrological data forms the fundamental basis on which water schemes need to be devised. The first requirement is to check the validity of all basic data by ascertaining how such data was produced, including the need to visit all flow measuring structures to check their condition and probable degree of accuracy. If relevant data does not exist, systems must be set up for acquiring it. Once data is obtained it must be carefully filed or computer archived, so that it is permanently available for recomputing potential yield when catchment or other conditions change.

To assess a potential surface water source the prime need is for long streamflow records which show the variations of flows with time, while for ground water sources, records of changing aquifer water levels are most important. Nowadays many catchments are already partially developed so that, in order to make full use of the available streamflow data, it is necessary to distinguish between measured and natural river flow. In order to naturalise a flow record (see Section 3.9), details are required of all abstractions, effluent returns and reservoir storage changes within the catchment. Similarly in order to set recorded aquifer levels in context, details should be kept of pumped output from wells together with pumping and rest water levels.

3.4   Streamflow measurement

Riverflow or streamflow records taken in the vicinity of an existing or proposed intake or dam site are an invaluable aid to the assessment of the potential yield of a source. In general the longer the period of records the more reliable any yield estimates based on them are likely to be, but even a very short record often provides a significant improvement over estimates derived from generalised regional relationships.

Flows are often obtained by measuring the 'stage level' of a river, i.e. the elevation at some location of the water surface above an arbitrary zero datum. The majority of streamflow records are obtained in this manner by converting continuous or regular stage measurements to discharge by means of a rating curve. The simplest way to measure stage level is by means of a permanent staff gauge set so that its zero is well below the lowest possible flow. Although such gauges are simple and inexpensive they must be read frequently when the water level is changing rapidly to define the shape of the streamflow hydrograph. It is preferable to construct a stilling well to house a computer-compatible solid state level recorder and a chart recorder with an unlimited natural level scale operating from a float. The solid state level recorder has marked advantages in terms of data processing but the chart record gives an important visual check of existing conditions and can provide a backup record should the digital data from the solid state recorder become corrupted.

A rating curve for a site can be obtained in a variety of ways of which the two most common are by means of velocity-area methods using a current meter,[6] or by means of weirs or flumes which are mostly permanent structures. They include:

● sharp edged plate weirs;[7]
● broad crested weirs;[8]
● triangular profile weirs;[9]
● critical depth flumes.[10]

Dilution gauging[11] and ultrasonic gauging[12] can also be used to help define the stage/ discharge relationship at sites where conditions are difficult for current meter measurements, and also to check an existing rating curve. The choice of gauging method depends on channel and streamflow characteristics, staff time availability and cost.

Current meter measurements are most often used when large flows have to be measured and the available fall is small. They are also often desirable for smaller rivers with sediment laden flows. Current meter gauging stations are relatively easy to set up in that they often require little modification of the existing channel. Since each potential gauging site is unique, each requires a careful pre-assessment of the width and depth of the channel, likely flood velocities and alternative ways of current meter measurement. Whenever possible current meter gauging stations should be located in straight uniform channel reaches with relatively smooth banks and a stable bed. For many rivers it is difficult to locate accessible sites with these characteristics and particular care must be taken to find a site that will give satisfactory results.

Current meter measurements can be made by wading with a current meter attached to a graduated wading rod to measure depth. Wading can usually be carried out safely when water depths are about a metre or less and the maximum stream velocity is no more than about 1 m/s. When wading is not practical, current meter observations may be carried out by lowering a current meter, ballasted with a weight, from the deck of a bridge. A crane may be necessary because heavy ballast may be needed if the current is strong. Measurements are best carried out from single span bridges as the turbulence caused by bridge piers can cause considerable errors. If no suitable bridge is available near the gauging site it may be necessary to construct a cableway from which to suspend a current meter and ballast weight, or to carry out measurements from a boat.

At current meter stations discharge measurements are usually made by subdividing the river cross-section into vertical sections. The mean velocity in each vertical section is measured and applied to its area. The river discharge is computed from the sum of the individual section values. The number and spacing of the verticals should be such that no section accounts for more than 10% of the river flow. The velocity measurement points are normally located by means of a tagged tape or wire stretched across the river or from graduations painted on the deck of a bridge. Detailed observations for many different types of river have shown that the mean velocity of a vertical can be closely approximated from the mean of two observations made at 0.2 and 0.8 of the depth. If the water depth in the vertical section is less than about 0.6 m, or if time is limited, then one observation at 0.6 of the depth will approximate to the average over the whole depth.

The Water Resources Division of the United States Geological Survey has produced some excellent publications on current meter gauging techniques[6] while standard practice in the United Kingdom is contained in BS 3680 of which parts 3A, 3C, 3G and 3Q are most relevant to the use of a current meter.

Despite the relatively low capital cost of velocity-area stations, the need to make sufficient current meter measurements over a wide range of river levels to establish the station rating curve, followed by repeat measurements each time a major flood is thought to have shifted the river bed profile, makes heavy demands on staff time. Also as a result of the difficulty of getting to the gauging site at the time of peak flood flows and of arranging for the necessary measurements by current meter, it may take a long time to obtain a complete rating curve for a river section. The alternative of a standard gauging structure, therefore, is often more attractive to the engineer, particularly so for catchments of less than 500 km^2.

Sharp edged plate weirs, of rectangular or vee-shape depending upon the sensitivity required, are generally only suitable for spring flows or for debris-free small streams. The need to keep the weir nappe aerated at all stages limits their use, but they are frequently used for low flow surveys as they can be rapidly placed in small channels.

Of all the weirs that designers have tried, perhaps the most successful has been that of Crump[9] (see Section 10.14). It has a simple and efficient flow characteristic $Q = 1.966\,H^{1.5}$ m^3/s, whilst operating up to the total head at which tailwater reaches 75% of that upstream (relative to crest height). Using the crest tapping designed by Crump it is possible to go further and attain reasonable results up to 90% submergence, but this versatility is marred by a tendency for the crest tappings to block in floods carrying sediment. A minimum head of 6 cm on the weir is necessary for accuracy, but with compounding of the weir crest this can be achieved. The flat-vee variant is another possibility.

The accuracy of streamflow data obtained from weirs can often be poorer than generally realised. Laboratory rating conditions rarely appear in real rivers where deterioration of the weir crest or siltation upstream from the weir resulting in non-standard approach conditions can cause errors in excess of 10%. Drowning out of the weir at high flows can also be a problem and may result in the gross over-estimation of flood discharges.

Critical depth flumes are appropriate on smaller catchments of, say, under 100 km^2 which have a wide flow variation and where sensitive results are required. Essentially a contraction of the channel forces flow to attain critical depth over a fixed section whatever the upstream head. A unique upstream head–discharge relationship is then created such that for a rectangular throat

$$Q = 1.705 C_V b h^{1.5} \text{ m}^3/\text{s}$$

(see Section 10.9), where b is the effective width (m), h is the effective head (m) and C_V is the approach velocity coefficient. Introduction of a shape factor is possible for trapezoidal or U-shaped throats.

Calibration of an existing sluice structure may be achieved using formulae derived from laboratory model tests, but these should be checked by current meter measurements wherever possible. Sometimes long, relatively homogeneous sluice keeper's records exist,[13] and the calibration of such sites can then produce records of flow of several decades in length for a modest cost.

Dilution gauging[11] is a flow measurement technique that is particularly well suited to small turbulent streams with rocky beds where the shallow depths and high velocities are unsuitable for accurate current meter gaugings. The approach can also be used to calibrate non-standard gauging structures. With dilution gauging the discharge is measured by adding a chemical solution of known concentration to the flow and measuring the dilution of the solution some distance downstream where the chemical is completely mixed with the stream flow. Sodium dichromate is the most commonly used chemical although dyes such as Rhodomine B have the advantage that they can be easily detected at very low concentrations. With the commonly used 'gulp injection' method a known volume of chemical is added to the stream flow as quickly as possible in a single 'gulp' and downstream samples are used to construct a graph of concentration against time. It follows that if a known volume of chemical V of concentration C_1 is added to a streamflow and the varying downstream concentration C_2 is measured regularly then

$$V C_1 = Q \int_{t_1}^{t_2} C_2 \, \mathrm{d}t$$

A graph of C_2 against time is drawn and the area below it between t_1 when the chemical just starts to be detected in the stream, to time t_2 when it ceases to be detectable is measured. This gives the integral on the right, hence Q can be found.

Ultrasonic gauging uses the transmission of sound pulses to measure the mean velocity at a prescribed depth across a river channel. Two sets of transmitters/receivers are usually located on either bank of a rectangular channel, offset at an angle of about 45° to the direction of flow. They send ultrasonic pulses through the water, mean water velocity at pulse level being a function of the difference in pulse travel times in upstream and downstream directions. In essence ultrasonic flow measurement is a velocity-area method requiring a survey of the channel cross-section at the gauging site. If used in conjunction with a water level recorder, ultrasonic measurements can provide a complete record of stream flows. However the method is most frequently used to check the stage discharge relationship of an existing station.

3.5   Rainfall measurement

Precipitation is measured with a rain gauge, the majority of which are little more than standard cylindrical vessels so designed that rainfall is stored within them and does not evaporate before it can be measured.[14] In an effort to ensure that consistent measurements of the precipitation reaching the ground are obtained, observers are recommended to use standard instruments which are set up in a uniform manner in representative locations. Many national meteorological institutions provide pamphlets designed to ensure good standard observation practice and the World Meteorological Organization plays an effective co-ordinating role.

The standard daily rain gauge in UK is the Meteorological Office Mark II instrument which consists of a 127 mm diameter copper cylinder with a chamfered rim made of brass. Precipitation which falls on the rain gauge orifice drains through a funnel into a removable container from which the rain may be poured into a graduated glass measuring cylinder. The Bradford rain gauge is a daily storage gauge which was first made for Bradford Waterworks who required a larger capacity instrument to deal with the higher rainfalls in the Pennines in the north of England. The greater depth of the container below the ground surface also reduces the risk of the collected rain freezing. Monthly storage gauges are designed to measure the rainfall in remoter areas and are invaluable on the higher parts of reservoir catchments. The Seathwaite gauge is a monthly storage gauge developed for use in the Lake District in north-west England.

Ideally rainfall should be measured at ground level but this gives rise to problems due to rain splashing into the gauge. The higher the rim is placed, the more some rain will be blown away from the gauge orifice and go unrecorded. All standard storage gauges in the United Kingdom are set into the ground with their rim level and 300 mm above the ground surface which should be covered by short grass or gravel to prevent any rainsplash. Many international gauges are set with their rim one metre high: these can be expected to read 3% lower than the standard British gauge.

In the United Kingdom daily storage gauges are inspected each day at 09.00 hours and any rainfall collected is attributed to the previous day's date. If the inner container of a rain gauge should overflow as the result of exceptional rainfall, or possibly because of irregular emptying, it is important that the surplus water held in the outer casing should also be recorded. Monthly storage gauges are usually inspected on the first day of each month to measure the previous month's rainfall total. Corrections may need to be made to the measurements taken at any gauges visited later than the standard time during spells of wet weather.

Specialised problems occur in snow prone areas. Small quantities of sleet or snow which fall into a rain gauge will usually melt to yield their water equivalent, but if the snow remains in the collecting funnel it must be melted to combine with any liquid in the gauge. If there is deep fresh snow lying on the ground at the time of measurement, possibly burying the gauge, a core of the snow should be taken on level ground and melted to find the equivalent rainfall. Countries that each year experience a snow cover throughout the winter months require regular snow course surveys[15] to monitor the amount of precipitation.

Continuously recording rain gauges are invaluable for flood studies. The original type gives a daily chart recording of the accumulated contents of a rain-filled container: this empties by a tilting siphon principle each time 5 mm has collected. As staff time costs rise for chart analysis this type falls out of favour. A more recent development is the tilting bucket gauge linked to a logger which will run for at least 1 month. Each time the bucket tilts to discharge 2 mm the event is recorded in a computer compatible form. A daily or monthly storage gauge is often installed on the same site as a recording gauge to serve as a check gauge and to avoid the possibility of loss of data due to instrument failure.

Particular care must be taken when siting a new rain gauge from which the resulting records are to be published. The gauge should be placed on level ground, ideally in a sheltered location with no ground falling away steeply on the windward side. Obstructions such as trees and buildings, which affect local wind flow, should be a distance away from the gauge of at least twice their height above it. In particularly exposed locations, such as moorlands, it used to be standard British practice to install a turf wall[15] around the gauge. With staff reduction to reduce costs, however, it has proved difficult to sustain the level of turf maintenance that is needed.

The 150 cm^2 'mouth' area of a standard rain gauge is an almost infinitesmally small part of the catchment area it will be taken to represent. Nevertheless by judiciously siting the gauges in typical catchment settings a relatively small number of gauges can provide a surprisingly accurate representation of catchment rainfall. The number of gauges required to give a reliable estimate of catchment rainfall increases where rainfall gradients are marked. A minimum density of one per 25 km^2 should be the target, bearing in mind that significant thunderstorm systems may be only about 20 km^2 in size. In hilly country, where orographic effects may lead to large and consistent rainfall variations in short distances, it can be necessary for the first few years to adopt the high densities suggested in Table 3.1. Thereafter high densities are only required where control accuracy necessitates it.

In large areas of the tropics there is great variation in rainfall from place to place on any one day, but only a relatively small variation in annual totals; in such areas the rain gauge densities of Table 3.1 will be excessive and it is better to concentrate on obtaining homogeneous records of long duration at a few reliable sites. Rainfall (and snowfall) totals are best computed monthly and summed each year. Weekly or 10 day values should be avoided because they neither correspond to the duration of significant hydrological events nor precisely fit an annual calendar.

Table 3.1   Rain gauges required in a hill area[16]

Catchment area (km^2)	Number of gauges
4	6
20	10
80	20
160	30

Measurement of catchment rainfall

There are several methods for computing catchment precipitation from rain gauge measurements ranging from simple numerical procedures, interpolation from isohyetal maps or Thiessen polygons, and from numerical interpolation procedures of which Kriging[17] and trend surface are most frequently used. The simplest objective method of calculating the average monthly or annual catchment rainfall is to sum the corresponding measurements at all gauges within or close to the catchment boundaries and to divide the total by the number of gauges. The arithmetic mean provides a reliable estimate, provided the whole catchment is of similar topography and the rain gauge stations are fairly evenly distributed. Also if accurate values of area rainfall are obtained first from a large number of rainfall stations within a catchment by one or other of the more time consuming methods described below, then it may be found that the mean of the corresponding measurements from a smaller number of stations may provide equally acceptable results. In the Thames Basin, for example, it was found that the annual catchment rainfall for the 9980 km^2 area could be derived by taking the arithmetic mean of 24 well-distributed representative gauges, to within $\pm 2\%$ of the catchment value computed by a more elaborate method using measurements from 225 stations.

The isohyetal method is generally considered to be the most accurate method of computing catchment rainfall. The method is, however, laborious and subjective, and dependent on the analyst having a good understanding of the rainfall of the region. The monthly or annual rainfall total recorded by each gauge within or close to the catchment boundaries is plotted on a contour base map. Isohyetal lines, i.e. lines joining points of equal rainfall, are then drawn on the map taking into account the likely effects of topography on the rainfall distribution. If any isohyetal shows an unusual feature, the gauge readings must be checked and the shape of the plotted isohyets critically reviewed. When the isohyets have been adjusted to give the most likely rainfall distribution, the total precipitation over the catchment for the period considered is obtained by planimetering the areas between isohyets. The mean catchment rainfall is calculated by summing the products of the areas between each pair of isohyets and the corresponding mean rainfall between them, and then dividing by the total catchment area.

When long-term average rainfall values are available for most stations in a region for a common period of 30 or more years, it is often useful to plot isopercental lines instead of isohyetal lines. Isopercental lines connect locations on a map which have the same percentage of the average annual rainfall or the average monthly rainfall – the choice being one of convenience. Once defined they are a reliable means of estimating missing individual gauge readings at sites where the average is known, and also of determining the area rainfall for a given event if the long-term average annual catchment rainfall is already known. If, for example, the rainfall total for each station for a given year is plotted as a percentage of the station's standard period long-term average annual rainfall, then isopercental lines can be drawn. The areas between pairs of isopercental lines can then be planimetered and applied to the mean percentage between isopercentals. The overall catchment percentage is obtained by dividing the sum of these products by the catchment area. This is applied to the long-term average annual rainfall to provide the average catchment rainfall for the given year. The advantages of the isopercental approach are that isopercental values for a particular period tend to be much less variable spatially than the actual rainfall totals and thus the plotting of isopercentile lines can be computerised more readily. The isopercental technique is particularly applicable in regions with many long

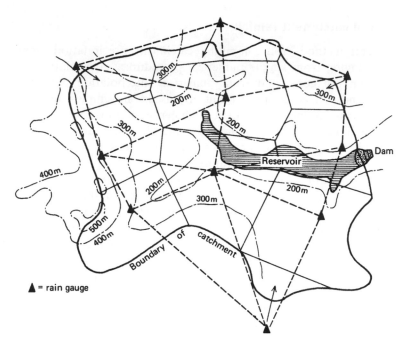

Fig. 3.1 Thiessen's method of estimating general rainfall over an area.

rainfall records where reliable long-term average annual catchment rainfall values are available. The technique is much less useful in regions with limited or fragmentary records.

The most popular method of weighting gauge readings objectively by area has been that of Thiessen. An area around each gauge is obtained by drawing a bisecting perpendicular to the lines joining gauges, as shown in Fig. 3.1. The portion of each polygon so formed lying within the catchment boundary is measured and the rainfall upon each is assumed to equal the gauge reading. The total precipitation is the weighted average of these values. One drawback is that, if the gauges are altered in number or location, major alterations to the polygonal pattern ensue. To maintain homogeneity it is better to estimate any missing individual gauge values. The gauges must also be reasonably evenly distributed if the results are to lie within a few per cent of the isohyetal method. The approach is not particularly good for mountainous areas because no account is taken of the effects of altitude on rainfall when deriving the Thiessen coefficients for individual polygons.

In mountainous areas where there may be few stations the main difficulty is to allow for the influence of topographic effects. One widely used approach for such areas is to develop a multivariate regression model using parameters such as elevation, orientation, exposure or distance from the sea, and then using a numerical interpolation procedure such as Kriging[17] to smooth out residual discrepancies from the regression correlation.

Progress has been made in rainfall estimation by both weather radar and satellite. The strength of both lies in the spatial view they afford with the former being particularly good for flood forecast purposes. Radar averaging is normally 2×2 km or 5×5 km over 5 or 15 minutes: the lowest altitude scan needs to be calibrated against concurrent autographic rain gauges. Satellite estimates are far more approximate, being related to cloud top temperature and only indirectly to actual rainfall amount. For large basins in the tropics it is now possible to obtain public domain estimates of $0.5 \times 0.5°$ grid satellite 'monthly rain estimates' from the Climate Analysis Center, Washington DC.

3.6 Evaporation and transpiration measurement

Evaporation is a key part of the hydrological cycle in that, on a global basis approximately 75% of total annual precipitation is returned to the atmosphere by the processes of evaporation and transpiration. Water evaporates to the air from any open water surface or film of water on soil, vegetation, or impervious surfaces such as roads and roofs. The rate of evaporation varies with the colour and reflective properties of the surface (the albedo), and with climatic factors of which the most important are solar radiation energy, windspeed, the relative humidity of the air and the temperature of the water and air.

Transpiration is the water used by plants to produce a mature crop. A small part of this water is retained in the plant tissue, but the bulk passes through the roots to the stem or trunk and is transpired into the atmosphere through the leaves. As it is almost impossible under field conditions to differentiate between evaporation and transpiration when the ground is covered with vegetation, the amounts of water used by both processes are usually combined and referred to as 'evapotranspiration'.

Evapotranspiration losses vary with the same meteorological factors as evaporation, but also depend upon the incidence of the precipitation, the characteristics and stage of development of the vegetation, and the properties of the soil.

It is difficult to measure evaporation with any certainty by means of an evaporation tank because of the edge effects associated with such an instrument. The standard approaches to the problem are either direct or indirect.[18] For direct measurement of evaporation the USA Class A pan is of galvanised iron or monel metal, 1.21 m in diameter and 255 mm deep. It is set on a standard wooden framework 100 mm above ground level, thus allowing air to circulate all round it. As a result measured evaporation is higher than that of a natural water surface and a reduction factor must be applied. This is generally taken to be 0.7, but it can vary between 0.35 in areas of low humidity, very strong wind and surrounded by bare soil, and 0.85 where high humidity and light winds prevail.[19] The British Symons sunken tank is 1.83 m square and 610 mm deep, with the rim 75 mm above ground level. It is more nearly a model of reservoir evaporation but suffers from inconsistent results if it is not in tight contact with the surrounding ground. The heat storage of a small tank is correspondingly small, whereas a large lake takes time to warm up or cool down. As a consequence tanks results do not quite match the evaporation of a nearby lake in regions with strong seasonal temperature variations. Peak lake evaporation rates in the Kempton Park experiment[20] occurred up to a month after peak tank measurements and this was explained by heat storage theory. Annual open water evaporation ranges from 700 mm in Northern Europe, through 1500 mm in much of the tropics, to more than 2500 mm in hot arid zones.

Unfortunately no standard percolation gauge to measure evapotranspiration by a given type of vegetation has yet been devised. Most are formed by a large diameter pipe sunk about one metre into the ground and carefully filled with the original soil layers. A surface cover of short grass is irrigated by a trickle hose if potential evapotranspiration is to be measured. Otherwise any water percolating through the soil layers during rainfall is drained from the bottom of the pipe to a nearby access manhole where it is measured.[21] Deduction of the measured percolate from rainfall measured by an adjacent rain gauge gives the evapotranspiration loss from a non-irrigated percolation gauge, but only between dates of equal soil moisture content in the gauge will this figure be meaningful.

A lysimeter measures water-balance elements to get a measure of evapotranspiration on a larger scale. The term normally covers a small catchment plot underlain by impermeable geology, with a single type of vegetation and from which all overland flow and shallow drainage can be measured. Rainfall is measured above the canopy of vegetation on the plot so that the total evaporative losses from the canopy and at ground level can be inferred. The technique is particularly useful for showing the relative water consumption characteristics of moorland, forest, arable crops, and pasture.

Penman's formula[22] for predicting water surface evaporation indirectly is recognised as the most accurate, being based on physical principles, but it involves use of data which may not always be available, e.g. on measurements of radiation (or sunshine duration), wind run at 2 m above ground, vapour pressure and air temperature, all of which should be taken at the same site. At altitudes above about 1000 m, McCulloch's fuller version of Penman's equation[23] should be used as it makes express allowance for the corresponding pressure drop as well as adjusting the radiation term for latitude.

Penman also showed how simple coefficients could be applied to his open water evaporation figures to obtain the evapotranspiration rate from a grassed surface. The use of the latter became standardised as being the evapotranspiration (designated ET_o) – 'from green grass surface cover 80–150 mm high, actively growing.' Crop coefficients k_c could then be applied to give the evapotranspiration of various types of crops at various stages of development and meteorological conditions, e.g. $ET_{(crop)} = k_c \times ET_o$.

Thornthwaite's formula for evapotranspiration from short vegetative cover, widely used in USA, is empirical and simpler than Penman's, being dependent on sunshine hours and mean monthly temperature. Thornthwaite's method has been widely used, but it is strictly valid only for climates similar to that of eastern USA where the method was developed. The method tends to give potential evaporation estimates higher than those produced by the Penman formula, particularly so during the summer months.

The most recent guide to assessing the irrigation requirements of growing crops is the FAO's Irrigation and Drainage Paper 46 – 'CROPWAT: A Computer Program for Irrigation Planning and Management' 1992. This computes evapotranspiration Et_o values according to the Penman–Monteith method, to which quoted crop coefficients, k_c as mentioned above, are then applied according to type of crop and its stage of development. An earlier FAO publication of 1976 – 'Crop Water Requirements: Irrigation and Drainage Paper 24' was based on the Blaney–Criddle formula which uses temperature and length of daylight hours to give ET_o values, which are then modified by values given for a range of humidities and wind speeds.

It has to be borne in mind that evapotranspiration formulae assume no shortage of water to meet crop growth and potential evaporation. Hence, in dry periods with insufficient precipitation, lesser evapotranspiration figures must apply. A good explanation of all the foregoing indirect methods of estimating evapotranspiration is given by Wilson.[24]

3.7   Soil moisture measurement

The study of soil moisture has always been of vital interest to agriculturalists, but more recently the potential impact of soil moisture content on runoff has been understood to a greater extent. The temporary storage of rainfall in the soil and aquifer layers of a catchment can be significant in the overall catchment water balance. Soil moisture variations occur predominantly in the first metre below the surface. From the driest (Wilting point) condition to the wettest drained state (Field capacity) may mean a rise from 3 to 10% water content in very sandy soil, or from 20 to 40% in a clay soil. Thus the maximum range of water storage in one metre of soil may be as much as 200 mm. Markedly higher values apply to peat. Additional water may be held under waterlogged conditions whenever the drainage rate is lower than the rainfall intensity.

A knowledge of the types and distributions of soils over the area of interest is an essential prerequisite for the selection of sampling points at which to measure soil moisture. For England and Wales this information can be obtained from maps published by the Soil Survey and Land Research Centre at Silsoe, while for Scotland soil maps can be obtained from the Macaulay Land Use Institute, Aberdeen. In developing countries it may be necessary to engage a soil scientists to carry out soil surveys in the catchments where detailed hydrological measurements are required in order to ensure that the sampling points cover the main soil types and vegetation zones.

A common method of measuring soil moisture is by means of a gravimetric determination whereby a soil sample of known volume removed from the ground with a soil auger, is first weighed, then dried in a special oven and finally reweighed. The method is accurate provided care is taken with the measurements and is often used to calibrate other techniques. However, the method is time consuming and requires laboratory facilities. It is also a destructive process and has obvious limitations where regular sampling is required. The neutron measurement probe was developed in the 1950s to provide direct field measurements, but because it used a radioactive source, it has been superseded by the soil capacitance probe[25] which poses no safety management problems.

3.8 Catchment losses

A significant proportion of rainfall is lost by immediate evaporation or by the later transpiration of growing vegetation. In some cases there will also be deep infiltration that eventually emerges in the sea without ever appearing in surface drainage channels. Catchment losses are best estimated from a water balance conducted over a number of years on the catchment concerned, or on one of similar rainfall, geology and land use.

A typical loss rate in England would be 450 mm per annum, with significantly higher figures above 500 mm per annum generally only occurring where afforestation predominates in a high rainfall area. Table 3.2 gives some idea of the variation in loss in different regions of the world. Although it might be thought that losses would be higher in years hotter than average, this is often more than offset by the concurrent dryness of the weather, which leads to a deficit of moisture in the soil which in turn leads to a limit on the transpiration of water by the growing vegetation. As soil dries out and approaches wilting point it has been shown that evapotranspiration rates can drop to only about one-tenth of those to be expected from weather data. Some plants are much more successful than others at their control of water use in drought conditions, for instance pine trees have a marked ability to conserve water in this way.

Losses do not always drop with cooler altitudes because advected wind energy together with higher radiation gains may intervene. Detailed measurements at experimental catchments by the Institute of Hydrology have shown that losses increase with the density

Table 3.2   Typical catchment losses in various parts of the world

Country	Location	Catchment cover	Annual rainfall (mm)	Annual loss (mm)	Marked seasonal variation
Nigeria	Ibadan	Rain forest	2500	2350	No
Malaysia	Johor basin	Forest/oil palm	2320	1240	No
Sri Lanka	Kirindi Oya	Mixed forest	1650	1230	Yes
Hong Kong	Islands	Grass	2100	1050	No
Zaire	Fimi	Rain forest	1700	1040	No
Thailand	Chaa Phraya	Forest/rice paddy	1130	1000	No
Japan	Ota	Conifer forest	1615	890	Yes
Australia	Perth	Mixed grass/forest	875	760	Yes
S. Africa	Transvaal	Mixed grass/forest	870	760	Yes
Kenya	Tana	Forest/savannah	1100	730	Yes
India	Bombay	Rain forest	2550	700	Yes
Zimbabwe	Low Velt	Mixed grass/forest	655	560	Yes
Lesotho	Maseru	Grassland	600	530	Yes
Holland	Castricum	Low vegetation	830	450	Yes
Britain	South England	Pasture/arable	600–900	450–530	Yes
	Midlands	Pasture/arable	650–850	440	
	Central Wales	Moorland/forest	1500–2300	480–530	
	Pennines	Moorland/forest	1150–1800	410–460	
	NE England	Moorland/pasture	700–1250	380	
	S. Scotland	Moorland/pasture	600–1800	360–410	
	N. Scotland	Moorland/pasture	1250–2500	330–380	
Algeria	Hamman Grouz	Scrub	420	400	Yes
Russia	Moscow	Agricultural	525–600	375	Yes
Iraq	Adheim basin	Scrub/grassland	420	350	Yes
Sth. Korea	Had basin	Forest	1180	320	Yes
Oman	Oman	Rock	160	130	Yes
Iran	Khatunabad	Bare ground	150–550	50–200	Yes

and height of natural vegetation growth and crops; this is particularly so with mature coniferous forests. However, predicting forest annual loss compared with that from grazed pasture is still not easy. It has been shown that the forest loss is due to intercepted raindrops being evaporated back into the atmosphere at rates up to five times normal transpiration values for short grass. This is because water laid out in thin films on vegetation can take up available heat in the atmosphere more readily; the sight of a forest steaming gently in a short spell of sunshine between showers is not uncommon. To quantify the extra loss to be expected requires[26] an idea of the depth of water the canopy of vegetation can hold during a shower and the frequency of showers. One point to note is that whilst rain is being evaporated off the outside of leaves, water will not be transpired through them. As a result perhaps only 90% of forest interception losses will be an addition to the catchment losses that would prevail anyway.

A review[27] of 94 catchment experiments in Africa, Asia, Australia and North America concluded that conifer forests could be expected to reduce catchment water yield, on average, by 40 mm per 10% forest cover. More recent studies[28] at Plynlimon in the UK found that the magnitude of the reduction in water yield was 29 mm per 10% forest cover which equates to a 15% reduction in the water yield for a completely forested upland catchment. It should be noted, however, that while heavy afforestation of catchments reduces their overall water yield, the results of the Plynlimon study and from a study[29] of catchment data from 40 European agencies found no statistically significant relationship between the proportion of forest cover and measures of low flow. These findings support the view that low flows in upland headwater catchments are primarily influenced by drainage from minor sources of groundwater which remain largely unaffected by forest interception losses.

3.9   Streamflow naturalisation

Nowadays the majority of catchments are already partly developed, so that it is rarely sufficient to evaluate a potential resource directly from the as-gauged stream flow records. Most surface water resource studies are preceded by some form of naturalisation to determine the sequence of streamflows which would have occurred with the river basin in its current state but with no abstractions or discharges of water. If there are no reservoirs within the catchment then the flow naturalisation involves the use of the basic equation given below over a daily or, more commonly, a monthly time step.

> Natural flow = gauged river flow
>
> *plus* sum of all upstream abstractions
>
> *minus* sum of all upstream discharges and return flows to river

The naturalisation of a stream flow record is critically dependent upon the availability of good quality data for both the stream flows and for all abstractions from, or return flows to, the river system upstream of the gauging station over the whole period for which naturalisation is required. The main types of abstraction or losses from a river system which may need to be considered are:

- public water supply abstractions
- irrigation abstractions;
- power station cooling water abstractions;
- industrial (non-cooling) water abstractions;

and the more common gains are from:

- sewage and industrial effluent returns;
- irrigation return flows.

For catchments which contain a reservoir, allowance also needs to be made for the storage and attenuation of flows, and for additional evaporation losses from the reservoir surface which can be a significant item in warm semi-arid environments. Although relatively simple in outline, the naturalisation of a flow record often proves to be a complex and time consuming process in practice, usually because of the limited availability of the necessary data. Where the required data, either for the abstractions and returns, or for changes in reservoir storage are not available, appropriate estimates must be made. For England and Wales, the Water Archive and Monitoring system, and the National Abstractions Licence Database, operated by the Environment Agency are important repositories of the types of data required for naturalisation. These data bases do not, however, solve all the problems of naturalisation.

There has been a considerable growth in spray irrigation usage in the United Kingdom in the last two decades in order to enhance crop yields and to meet the tight product controls imposed by supermarkets and other purchasers. Details of licensed annual volumes and maximum daily abstractions are normally readily available for all abstractions, but the actual pattern of usage is much more difficult to obtain, as data on actual abstractions is often limited to the annual values for the larger abstractions. As a result irrigation abstractions are often assumed to follow the same pattern each year although in some studies soil moisture and rainfall data have been used to refine estimates of actual usage.

Diversions of water into channels for subsurface irrigation are more difficult to evaluate because they are unlikely to be controlled; they may be made through a variety of diversion canals or by sluices ('slackers') from embanked main rivers. The water is used to ensure that local ditch levels are kept high enough to permit crop roots to draw water from the associated water table, but low enough to encourage good root growth. Obtaining an exact estimate of irrigation consumption of this type on a major basin while it is happening is rarely possible. Some indication of the quantities involved in English conditions are shown in Table 3.3. Larger use occurs where piped under-drains (i.e. land drains) exist as well.

The impact of groundwater abstractions on river flows is another problem area. The amount by which a particular groundwater abstraction reduces the river flows over a given length of river depends on the complex inter-relationships of a number of variables. Abstractions from 'bankside' wells where the river bed is in direct contact with the

Table 3.3 Subsurface irrigation quantities in England

	Average rainfall (mm)	Consumptive water use for subsurface seasonal irrigation in the growing season (Ml/day per km^2)						
		Mar	Apr	May	Jun	Jul	Aug	Sep
South levels (fens)								
Cambridgeshire (arable)	225	–	0.3	0.8	1.3	1.2	0.6	–
Somerset moors	290	–	–	0.3	0.2	0.4	0.4	0.1

Note. Subsurface irrigation normally occurs only in well-watered flatland areas used for arable crop farming.

groundwater table can be expected to cause a direct matching reduction in river flows, whereas the impact of abstraction from wells more remote from the river may be time-lagged, particularly at times of low flows when groundwater table levels may be below river bed levels. The results of regional groundwater models may help the engineer to make an appropriate judgement about how best to allow for the effect of groundwater abstractions in the naturalisation process.

River flows are likely to be directly affected wherever inland power stations[30] are built. A 2000 MW station on full load can lose about 65 Ml/day by cooling tower evaporation, but this reduces with lower load factors. The recent change to more efficient combined cycle gas turbine power stations is likely to lessen the overall impact on rivers as older power stations are closed down. It should also be recognised that there are some significant abstractions which do not require licences. British Waterways, for example, take water to replenish their canals for navigation. Frequently these flows are not measured. In urbanised areas there will be abstractions by infiltration to sewers laid below the water table level; and there may be additions of leakage from water distribution systems.

3.10 Long-term average catchment runoff

Wherever possible average catchment runoff should be calculated from a long streamflow record which, if subject to artificial influences, has been naturalised as described in the previous Section. For those sites where streamflow data are either short term or non-existent, long-term average runoff can be estimated:

(1) by correlating the brief records available for the study catchment with those of a long record station in a catchment with similar characteristics;
(2) by deducting loss estimates from catchment rainfall figures;
(3) by the use of a rainfall/runoff model.

Correlations between long and short record stations are best carried out using monthly data. The use of daily figures is more time-consuming and often produces a large scatter while annual values provide too few points. An initial mathematical 'best fit' relationship between the stations may be derived by computer but a graphical plot should always be produced to ensure that the computed relationship provides a good fit to the bulk of the data. A manual adjustment should be made if, for example, the mathematically computed relationship is unduly influenced by the values for a few high flow months, or if the relationship implies unreal intercept values.

An estimate for average runoff for a catchment for which no streamflow records exist is often obtained by deducting a value for average annual catchment losses due to evapotranspiration from the average annual catchment rainfall obtained from an annual isohyetal map of the region for a selected standard period, such as 1961–90. If possible, the value for average annual catchment losses should be based on the typical average annual loss value obtained from similar gauged catchments throughout the region. In the UK estimates of actual evapotranspiration can also be obtained from the MORECS estimates[31] produced by the Meteorological Office for defined 40 × 40 km grid squares, or from the estimates of catchment average annual potential evapotranspiration[32] and the adjustment factors listed in Table 3.4.

Table 3.4   Adjustment factor for estimating actual evaporation in the United Kingdom[33]

Standard average annual rainfall (mm)	500	600	700	800	900	1000	>1100
Adjustment factor		0.88	0.90	0.92	0.92	0.94	0.96	1.00

For most regions rainfall stations are more numerous and have longer records than streamflow measurement stations. Rainfall/runoff models are therefore commonly used to extend short-term stream flow records. Typically a chosen model is calibrated by adjusting the model parameters so as to produce the best possible match between predicted and measured flows. The calibrated model is them used to extend the short-term stream flow data to cover a particular standard period. One of the usual checks on the synthetic stream flow data generated by rainfall/runoff models is whether they accurately reproduce the long-term mean runoff estimated by other means. However for a catchment where only a short-term stream flow record is available and there are no representative long-term stream flow measurement stations in the region, an approximate estimate of the long-term catchment runoff can be obtained using a rainfall/runoff model such as HYSIM[34] or HYRROM[35] (see Section 3.16).

3.11   Minimum rainfalls

Experience of past recorded droughts and low rainfall is an important factor in assessing probable future conditions that may be encountered. In the variable climate of the UK the longest known spell without any recorded rain at all[36] was for 73 days from 4 March 1893 at Mile End in London. At the other extreme, in desert climates many years may have no rainfall at all. At Calama which is near the desert region of northern Chile, it is believed that virtually no rain fell for 400 years until a sudden storm fell in 1972.

The most notable droughts in England and Wales this century were the following.

1921	Annual rainfall lowest in over 100 years in SE England. Spring sources hard hit as the autumn rainfall was insufficient to prevent flow recession which began in a dry spring and continued until January 1922 in many parts.
1933–34	Two dry summers with a remarkably dry winter intervening occurred in Wales and mid-England.
1943–44	A similar pattern to 1933–34 in S England. Very low flows experienced in spring-fed rivers because preceding years were also dryish.
1949	Exceptionally low summer rainfall and high temperatures affected sources reliant on river flows or with little storage.
1959	Similar to 1949.
1975–76	Many low flow records broken because of low summer rainfall.
1988–92	A succession of dry winters taxed groundwater supplies in southern and eastern England. Runoff deficits in parts of east England were the largest for 150 years.
1995–96	Two dry summers with a very dry winter intervening experienced in the Pennines.

The point about these diverse low rainfall experiences was that they followed no predictable pattern. Hence in estimating the minimum yield a source will provide, it is necessary to bear in mind the types of drought which past experience shows are possible.

To estimate drought rainfall for catchment modelling either one can use a knowledge of recorded minimum rainfall for a region, expressed as a percentage of the average, or one can carry out a statistical analysis[37] of available rainfall measurements.

3.12 Minimum rates of runoff

In temperate climates with variable rainfall, when minimum runoffs are expressed as rates per unit catchment area, it can often be seen that geology and topography of the catchment are the major influences, except where human activity has interfered. Clearly the dry weather flow of many small catchments is zero; and bournes, which are streams flowing strongly when the water table is high, dry out gradually from their headwaters as the water table level falls away from the stream bed. But in temperate zones where rainfall occurs throughout the year, large rivers do not dry up.

Several studies have attempted to predict minimum flows of specified severity after regional analysis of flow records. Notable examples include those for Malaysia,[38] Europe,[39] New York State[40] and that produced in 1992 by the Institute of Hydrology (IoH) for the UK.[41] The last named provided formulae for estimating drought flows based catchment characteristics. A summary national equation for UK was also quoted which accounted for 62% of the variations encountered and is as follows.

$$\text{1 day mean flow exceeded 95\% of the time } (Q_{95(1)}) = 44B^{1.43} \, S^{0.033} \, A^{0.034}$$

where: $Q_{95(1)}$ is the 1 day flow, expressed as a percentage of the long-term average daily flow; B is the base flow index (BFI) (see Table 3.5); S is the Standard Average Annual Rainfall 1941–70 in mm; A is the catchment area in km^2.

The mean annual 7-day minimum flow, $MAM(7) = 6.40 \, Q_{95(1)}^{0.953} \, S^{-0.0342}$, where $MAM(7)$ is expressed as a percentage of the long-term average daily flow, S, is as defined above, and $Q_{95(1)}$ is derived as shown.

The BFI represents the proportion of river flow which it is estimated is derived from underground storage. A way of evaluating it was developed by the IoH which used a computer program to analyse a long sequence of daily flows by locating the minima of consecutive non-overlapping 5-day flow totals. The program then searched for the 'turning points' in this sequence of minima, connecting them together to form the estimated baseflow hydrograph (Fig. 3.2 shows a graphical representation of the procedure). The BFI is then the volume of flow below the baseflow hydrograph, divided by the total flow for the same period.

The BFI values for over 1300 gauged catchments in UK are given in the latest issue of the IoH/British Geological Survey publication *Hydrological Data UK: Hydrometric Register and Statistics*.[42] For any ungauged catchment an estimate of the BFI can be obtained either by interpolation between the values for upstream and downstream gauging stations quoted in the Hydrometric Register for the same river, or by transposing the value for a gauged catchment with a similar average annual rainfall, surface geology and soil type. The approach has been adopted by a number of countries for analysis of their river low flows, the advantage being that annual values of BFI tend to be more stable than other low flow variables. Details of the method are given in the IoH's Low Flows Study.[41] The range of BFI values typically applying is shown in Table 3.5.

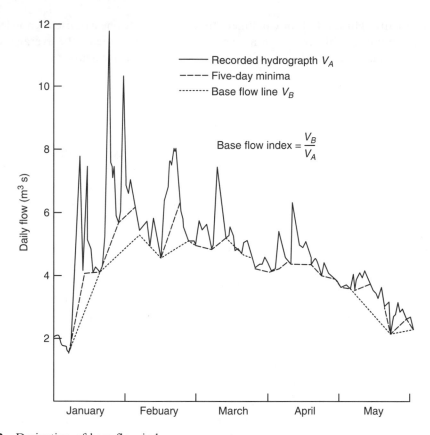

Fig. 3.2   Derivation of base flow index.

Table 3.5   Typical baseflow indices for various rock types

Dominant permeability characteristics	Dominant storage characteristics	Example of rock type	Typical BFI range
Fissured	High storage	Chalk	0.90–0.98
		Oolitic limestones	0.85–0.95
	Low storage	Carboniferous limestone	0.20–0.75
		Millstone grit	0.35–0.45
Intergranular	High storage	Permo-Triassic sandstone	0.70–0.80
	Low storage	Coal measures	0.40–0.55
		Hastings beds	0.35–0.50
Impermeable	Low storage at shallow depth	Lias	0.40–0.70
		Old Red sandstone	0.46–0.54
		Metamorphic-Igneous	0.30–0.50
	No storage	Oxford & London clay	0.14–0.45

3.13   Maximum rainfalls

Figure 3.3 is a plot of the maximum measured rainfalls recorded at individual points in the world. It must be stressed that the highest falls are precipitated only in the most unusual hill areas in certain climatic zones, once the duration exceeds one day. In addition the lines do not represent any continuous event, except possibly for a storm lasting for up to, say, four hours. The maximum figures for Britain are seen to be about 20–30% of world maxima, with the lowland easterly part of the country suffering less severely in long duration storms. The greatest is the Martinstown, Dorset, storm of 1955 in which 280 mm (11 in) of rain were officially noted in 18.5 hours, with an unofficial estimate at the heart of the storm claiming 350 mm (14 in).

Maximum rainfalls vary with the season of the year because thunderstorm intensities are associated with high sea and air temperatures, However high rainfalls over a day or so may occur at any time of year wherever the weather system can bring in moist air steadily, and conditions (often orographic) exist to cause precipitation.

It is quite possible for many years to elapse without any outstanding maximum rainfall event and then for several unusual maximums to be clustered close together, probably due to high concurrent sea temperatures and dominant weather system movement routes. Many countries have compilations of extreme meteorological events and these should not be overlooked. Also many national meteorological agencies make available long period rainfall measurements and duration-intensity-frequency estimates. These are often adequate for estimating storm magnitudes that can be expected more frequently than once every 50 years; but possible rarer storms need to be investigated by thorough regional

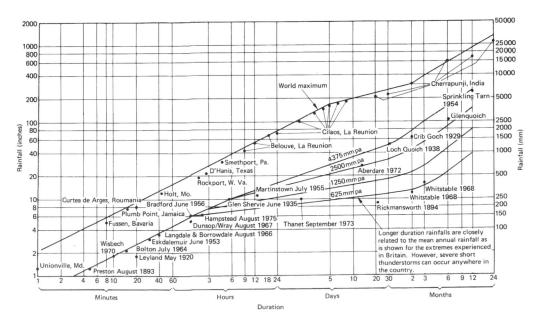

Fig. 3.3   Maximum recorded world and British rainfalls. World maximum data as given by Paulhus J. L. H. in Indian Ocean and Taiwan Rainfalls Set New Record. *Monthly Water Review*, **93**(5), 1965, pp. 331–335 – except for Cherranpunji reported by Dhar O. N. and Farooque, S. M. T. in A Study of Rainfalls Recorded at Cherranpunji *Int Assoc Hydrol Sci*, **XVIII**(4), 1973, pp. 441–450.

studies. In the UK the most up-to-date information is contained in Volume 2 of the *Flood Estimation Handbook*, Dec 1999[43] and shows how rainfall frequency calculations can be applied to UK catchments, leading to estimation of either the rainfall depth for a given return period and duration, or of the return period corresponding to a given depth and duration of rainfall. The parameters of the rainfall frequency model are provided digitally on CD-ROM(FEH) to a 1 km grid.

3.14   Maximum runoffs

Peak runoff rates are often difficult to estimate because of the damage occasioned by many major floods and the debris and sediment brought down by the floodwater. Figure 3.4 gives some recorded maximum runoff data experienced in UK. In lowland areas peak runoff rates are normally much lower than Fig. 3.4 suggests because of the temporary storage available in side channels and drains or in adjacent low lying land and the greater likelihood of drier soil conditions before a storm. For example in the fenland area of eastern England entire catchments may be drained quite adequately by pumping stations capable of pumping no more than 13 mm of runoff from their catchment per day. This is very much less than the maximum precipitation rates shown in Fig. 3.3 where rainfalls up to about 250 mm in 24 hours can occur in Eastern England, as shown by the plot for Thanet. These figures are not, however, large compared with experience elsewhere in the world where ten times higher rates can be experienced.

Four methods of estimating maximum runoffs are dealt with below, of which the fourth applies specifically to the UK.

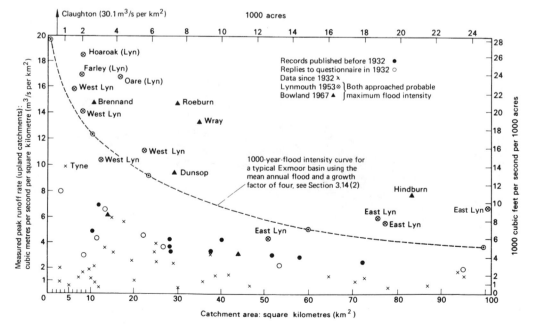

Fig. 3.4   Flood data in Britain in and since the ICE 1933 Report (*Proc ICE*, 1960).

(1) Probability analysis of existing flood records

This method plots recorded peak floods on probability paper, as described for the analysis of droughts in Sections 3.16 and 3.17 below, in order to estimate the probability of occurrence of a flood of given magnitude. The type of probability paper used is that which gives a best straight line plot for the recorded floods. Usually the peak flood for each year of record is plotted; but sometimes all peak flows above a given level are plotted, in which case care has to be taken to ensure that the events plotted are truly independent and that they are counted per 'water year' (i.e. summer plus winter) and not per calendar year.

The method is of limited value because a probability plot of past annual maximum floods is only useful for return periods up to about twice the length of the period of record. Thus 30 years record of annual maximum floods cannot be safely used to estimate the magnitude of a 100-year return period flood. In addition the accuracy of past flood records cannot be checked and catchment conditions may have altered since records were taken or may alter in the future.

(2) Use of regional flood probability curves

A different approach applicable world-wide is to use 'regional flood probability curves' which are published for many regions. These are curves as illustrated in Fig. 3.5. In this approach the mean annual flood for a given catchment is first obtained from the period of

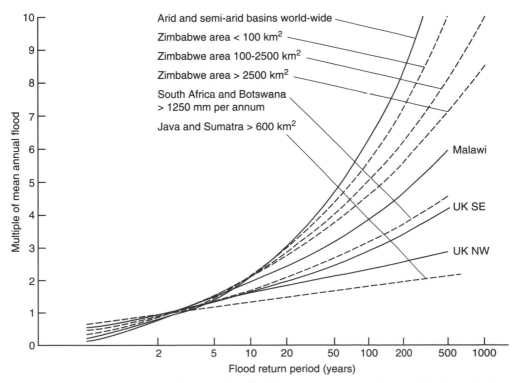

Fig. 3.5 Examples of regional dimensionless flood frequency curves. Source: UK, Law, F. Inst. Hydrology. Elsewhere, Meigh J. R., Farquharson F. A. K. and Sutcliffe J. V. *Hydrological Sciences Journal*, **42**(2), April 1997, pp. 225–244.

historic record and is termed 'the index flood'. A 'growth factor' (i.e. multiplier) taken from the appropriate regional flood probability curve is then applied to the index flood to give the probable magnitude of a flood for a given return period, T, longer than the period of record. The published regional flood probability curves are derived from analyses of flood magnitude frequencies for catchments in a region possessing similar characteristics. In the absence of an adequate historic record, the mean annual flood may be estimated from quoted formulae (relevant to a particular region) which are based on the catchment area and its average annual rainfall, together with other parameters as necessary to allow for other catchment characteristics, such as average slope, etc.

However, in the UK the more recent approach as set out in Volume 3 of the *Flood Estimation Handbook*,[43] is to take the *median* annual flood for a catchment as the 'index flood', rather than the mean. The *Flood Estimation Handbook* also introduces the concept of a 'pooling group' of catchments with similar characteristics to replace geographical regions. Full details of the method are given in that volume.

(3) Flood estimation derived from maximum precipitation

In this method the maximum precipitation that has occurred, or may occur on a catchment, is converted to consequent flood flow using either (a) a unit hydrograph derived from the catchment, or (b) a 'synthetic' hydrograph used to derive the probable maximum flood under the most unfavourable conditions. The method is of wide application because use of the unit hydrograph for the catchment permits assessment of flood flows likely from given rainfall storms; and use of a synthetic hydrograph permits an estimate to be made of the maximum flood possible under conceivable extremes of rainfall incidence and catchment conditions. Both approaches are similar, but as the second is principally used for designing impounding reservoir overflows, it is described in Chapter 5, Section 5.6.

(4) Flood estimation in UK

The principal guidance for estimating flood magnitudes in UK and Ireland is now Volumes 3 and 4 of the *Flood Estimation Handbook 1999*.[43] Alternative methods for flood frequency estimation for any catchment, gauged or ungauged, are presented. Volume 3 provides statistical procedures intended principally for use for return periods between 2 and 200 years. However it should be mentioned that this statistical procedure should not be used for assessing the required design capacity of impounding reservoir flood overflow works, except in the case where no loss of life can be foreseen as a result of a dam breach and very limited additional flood damage would be caused (see Section 5.6). Volume 4 uses a unit hydrograph rainfall-runoff method and is applicable to a wider range of return periods than the statistical approach given in Volume 3. It uses Volume 2 procedures for estimating depth and duration of rainfall for a given return period, or vice versa. This permits estimation of floods for return periods between 2 and 10 000 years and derivation of the probable maximum flood from estimates of the probable maximum precipitation. In addition up-dated summaries of flood peak data and flood event data are presented in Appendices to these volumes.

The procedures set out in Volumes 3 and 4 are relatively complex by reason of the number of influencing and limiting factors it is advised should be taken into account. Although the *Flood Estimation Handbook* provides much useful guidance on the choice and use of method, the application of the procedures to any particular case in UK or Ireland is best undertaken by an experienced hydrologist.

Part II Yield of surface sources

3.15 Introduction and definitions of yield

The term 'yield' does not have a precise meaning; its definition varies according to the context in which it is used. Definitions commonly used are:

(1) 'average yield' – the average abstraction taken from a source over a number of years;
(2) 'failure yield' – the steady supply that could be maintained for a given percentage of days in a year (as averaged over two decades or more);
(3) 'historic yield' – the steady supply that could just be maintained through a repetition of the worst drought on record;
(4) 'probability yield' – the steady supply that could just be maintained through a drought of specified severity and probability;
(5) 'operating yield' – the output that could be deployed with existing system constraints and a fixed set of operation rules.

Waterworks practice in many countries has been to adopt definition (4) when planning a new source, but to move to definition (5) for the final design of some new scheme and also when reassessing the yield of an existing source. Probability yields, (4), for major sources where failure would seriously disrupt industry or city life are generally set by the once-in-100 year drought (1% yield in annual risk terms). A whole supply system is typically checked out against the once-in-50 year (2%) drought, whilst minor sources may well be run at 3 or 4% yield if support measures are available. Although the old-fashioned terms of 'safe' or 'reliable' yield linger on (often equated with a standard 2% yield), they no longer serve a good purpose. All water authority managers have to be aware that almost every source has a quantifiable risk of partial failure.

Standards vary widely, but they generally rise with the economic development of a country. Developing countries may improve intermittent supplies until they are reliable through a 20- or 30-year drought, but the advent of industrialisation normally leads to the need for even greater reliability, e.g. Malaysia and Singapore have accepted the need to provide city water supplies that are adequate through a 50-year drought. For the majority of countries the choice of risk must depend upon:

- the social and economic consequences of supply failures of different magnitudes;
- the extent to which demand might be lowered (voluntarily or by legislation) during a drought crisis;
- the available reserves elsewhere in the undertaking or the possibility of temporarily reducing residual river flows;
- the uncertainty about population numbers served and unit consumption growth.

Sections 3.18–3.21 provide details of how the probability yields of surface sources may be assessed. It is often necessary to compute, not simply the yield of an individual source, but also the amount (usually larger) by which this source raises the yield of the water supply system of which it is part. In order to determine this increase in system yield the water engineer responsible for planning any new development must first be able to evaluate the yield of the existing system. Section 3.24 provides guidance on the conjunctive or integrated use of resources from a number of sources.

No source or group of sources should be considered as having a fixed yield. Yields may vary with time as the result of:

- changes in design standard or accepted methods of yield estimation;
- an improved understanding of the actual variations in runoff within the source catchment as the result of longer records;
- changes in catchment characteristics or level of development which may increase or reduce runoff during drought periods;
- a change in the amount and/or seasonal distribution of runoff within the catchment due to climate change.

In England and Wales the original approach to yield estimation was to adopt historic yields, followed in the late 1960s and 1970s by a widespread change towards probability yields based on the analysis of historic or synthetic design drought sequences. The severe droughts which occurred in parts of England and Wales during 1975/1976, 1984 and in 1995/1996 focused the attention of water authorities on operational requirements to a greater extent than previously; particularly so on the need to determine how much supply could be drawn from a source during a drought without running too great a risk of failure of supply. Although it has always been recognised by water authorities that a reservoir is not actually operated at a constant drawoff until it empties and that measures to reduce demand (e.g. the imposition of hosepipe bans, drought orders or other restrictions) are taken when the contents fall below some critical level, little was published on the subject.

Added impetus for the development of techniques to determine the operational yields of sources came with the privatisation of water utilities which has led to growth in the number of countries where a regulatory agency sets 'levels of service'. Since the privatisation of the water utilities of England and Wales in 1989 the Director General of Water (OFWAT) has required the water companies to report to him annually on the levels of service attained in respect of raw water availability ('DG1'), continuity of supply ('DG3') and water usage restrictions ('DG4'). Under DG1 the water companies must specify not only the yield of their resource systems but also the estimated frequency of restrictions imposed on consumers during droughts. The Director General considered a reasonable reference level for raw water availability would be:

(a) a hosepipe ban on average not more than once in 10 years;
(b) a need for a major public campaign requesting voluntary savings of water on average not more than once in 20 years;
(c) Drought Orders imposing restrictions on non-essential use not more frequently than one year in 50 on average;
(d) Drought Orders authorising standpipes or rota cuts not more frequently than one year in 100 on average.

Such levels of service are not statutory requirements, but are bound to exert an influence on risk policy. In some case monetary compensation of consumers may be stipulated if their supply is cut off for longer than a stipulated period for reasons within the supplier's ability to control. However, it is only in countries where living standards are high that stipulated levels of service have any meaning. In many countries lack of money for water development means that sources may have to be run to the limit of their current capacity, periods of shortage being faced when they come.

In 1995 the National Rivers Authority (NRA) published suggestions for a standardised method for assessing the yield of sources taking into account issues such as seasonal demand variations, levels of service and realistic drought management procedures. The Environment Agency, taking over the work of the NRA, then published in 1996 'Standards of best practice' for groundwater and surface water yield assessment 1996[44] which were adopted by the water companies in 1997. The methodology proposed for surface water yield assessment was to 'simulate the realistic operation of the water resources systems ... over as long a period as possible' on the basis of:

- historic data (generating flow records back to earlier critical periods if necessary);
- system capacity;
- control rules used to initiate demand constraints;
- the maximum licensed abstraction amount; and
- 'outage', i.e. temporary losses of output due to maintenance, system failure or water quality problems.

3.16   River intake yields

Usually abstraction from a river has to be agreed and authorised by some river controlling authority, or perhaps a department of state or regional government, or by an Irrigation Authority charged with looking after farmer's interests. In the UK the controlling authority is now the Environment Agency. Sometimes the authorised abstraction simply stipulates the maximum that can be taken in 24 hours. This kind of abstraction might be allowed if the proposed abstraction represents only a small proportion of the lowest flow in the river. More often the permitted abstraction is partly or wholly proportional to the river flow or to the amount left in the river, and may vary seasonally. Whatever conditions apply an analysis of the river flows is the first requirement.

Obtaining river flow data

By far the most reliable estimate of yield is obtained by analysis of a reliable long series of flow measurements at or near the abstraction point. If the period of record is short, it may be possible to extend it by correlation with a longer record of flows on a nearby catchment with similar characteristics of rainfall, topography and geology. Attempts to create a flow record from rainfall can raise difficulties, although HYSIM,[44] Sacramento,[45] Standford[46] and the IoH's HYRROM model[35] have improved the art. Assuming rainfall and potential evapotranspiration data exist, it may be possible to synthesise a flow record by use of a set of equations which forms a model of the catchment hydrology. This has to take account of how the catchment soils, and the aquifer below, take in and drain away water. Average surface flows have to be adjusted by changes in the soil moisture content stored within the range of the vegetation roots; and minimum surface flows will depend strongly upon the chosen recession constant which denotes aquifer drainage. The modelling becomes complex when the catchment is partly permeable and partly impermeable and, in that case, is best left for the hydrological specialist to undertake. Computer routines improve the speed of modelling, but optimising the constants necessarily involved has to be a matter of experienced judgement and compromise. The program MIMIC from the IoH can automate the optimisation for a rainfall-runoff model, but it is always necessary to test the validity of a model by comparison of the synthesised flows with measured flows over a period when daily rainfall is also recorded.

If, however, there is a reasonable period of recorded runoff concurrent with a long rainfall record, it can be possible, particularly in high rainfall areas, to correlate the two records directly, provided both wet and dry seasons are covered. This gives a longer runoff record.

Once the flow record has been obtained, a number of adjustments may be necessary to make the record represent natural flows (see Section 3.9). The adjustments to the gauged flows may include additions for irrigation consumption and for water turned into canal feeders or for other abstractions which take water out of the catchment. Deductions may be necessary for effluent discharges where these originate from water obtained from outside the catchment. The timing and growth of these changes will need to be ascertained so they can be applied correctly to the past records. Some abstractions, such as irrigation, will be seasonal.

Estimating minimum intake yield and its risk of failure

It is preferable to have a flow record for 20 years, taken at or near the point of abstraction. Weekly records of flow are best used because daily flows may be subject to erratic changes due to operation of river gates and diurnal changes in abstractions and discharges. Monthly runoffs may have to be used if weekly or daily records are not available.

To estimate the intake yield on the basis of risk of failure the following procedure is recommended.

(1) List the lowest flows in each year of record in order of magnitude, starting with the lowest.
(2) Compute the ranking of flows on a percentage basis using the formula

$$\% \text{ rank} = \frac{(m - 0.44)}{(n + 0.12)} \times 100\% \tag{3.1}$$

where m is the 'rank number' of the flow, e.g. the lowest flow is ranked 'No. 1', the next lowest 'No. 2', and so on; n is the total number ranked.

Thus, if there are 23 years of record, the percentage ranking of the lowest flow experienced will be $(1.0 - 0.44)/(23 + 0.12) = 2.42\%$

(3) Plot the magnitude of each flow against its percentage ranking on a type of probability paper (see below) which best gives straight-line interpretation, as exampled on Fig. 3.6.
(4) Draw a 'best fit' straight line through the plotted points from which can be read off the probabilities of failure of given flows.

Equation (3.1) is used for percentage ranking because the ranking needs to be unbiased in a statistical sense. If it were taken as $m/n \times 100\%$, the largest of the flows listed would have 100% probability, implying no flow could be larger – which is an unjustified assumption. The formula is that which is considered most satisfactory out of several which have been suggested. It allows for the probable error that a small sample may not be truly representative of the larger population from which it is drawn.

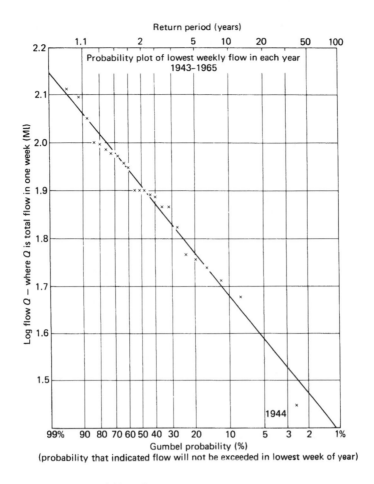

Fig. 3.6 Dry weather flow probability plot.

The straight line through the plots must avoid implying that some low flow event is more, or less frequent than appears reasonable. For example the plot labelled '1944' in Fig. 3.6 has a ranking probability of about 2.4% (1 in 40 year event) but with only 23 years of record available it is quite possible for it to be '1 in 60 year event', i.e. of 1.5% probability as inferred from the line which fits the other points well.

From Fig. 3.6 the probability of failure of different minimum flows can be read off. But because the points have been plotted for the estimated natural flow, adjustments are necessary to allow for deductions and additions to the flow caused by other abstractors and dischargers, both currently and as may be expected in the future.

3.17 Choosing appropriate probability paper

Probability paper is ruled according to the frequency distribution of events assumed by some probability function. The aim is to choose a probability paper on which the observed events, when plotted, form a straight line, or approximately so. This aids extrapolation of the line to estimate frequencies, and indicates how far the events plotted conform to the probability function on which the paper is based. There are various frequency distributions that can be assumed, but the commonest used for analysis of hydrological data are as follows.

(a) *Normal probability paper* which assumes normal frequency distribution, i.e. equal distribution of events about the mean.
(b) *Log-normal probability paper* which plots the log of the event magnitude against normal frequency distribution. It may give better straight line results than (a) for hydrological data, being useful for skewed distributions.
(c) *Gumbel (or log-log) probability paper* which has ordinates for magnitude of events spaced linearly, and abscissa values spaced proportional to $-\log_e [\log_e(1 - 1/T)]$, where T is the return period of the event. It is principally used to assess the frequency of extreme events (hence sometimes called 'extreme probability paper'), such as for annual maximum rainfalls or runoffs.

Hydrological data for short time-periods under 1 year is quite often skewed; for longer time periods it more nearly approaches normal distribution. But where average runoffs are very low, such as less than 250 mm per annum over the catchment, even long time-period droughts may be skewed such that the log-normal or log-Gumbel paper may be needed to obtain straight line plotting. Some guide rules suggested by Hardison[47] as to what type of paper to use, briefly put, are as follows.

(1) Use normal distribution if the coefficient of variation of flows is less than 0.25.
(2) Use log-normal distribution if the coefficient of skew is algebraically greater than 0.2.
(3) Try Gumbel distribution if neither of the above gives satisfactory straight lines.

The coefficient of variation

$$V = \frac{100 \times \text{standard deviation}}{\text{mean}}$$

The coefficient of skew

$$\gamma_1 = \frac{\mu_3 (\text{3rd moment})}{\text{cube of standard deviation}}$$

or, if the skew is not great

$$\gamma_1 = \frac{3(\text{mean} - \text{median})}{\text{standard deviation}}$$

(3rd moment $\mu_3 = \sum(x-\bar{x})^3/n$ and standard deviation $= [\sum(x-\bar{x})^2/n]^{1/2}$)

3.18   Yield of direct supply impounding reservoirs

Assessing reservoir storage volume

To assess the storage volume of an existing reservoir, the water surface area at different levels of drawdown is measured so that a curve of area against water level can be drawn. Reading areas for each unit of depth from the curve, these can be converted to volumes and added to give the total volume. Water areas can be measured from aerial or satellite photographs, or by instrument survey. Areas for at least four water levels are needed. Often the area increases rapidly for the upper 20–25% of depth, so two of the area measurements should cover this depth to assist in drawing the area/water level curve accurately, one being at top water level. To calculate the storage volume for a proposed reservoir, a contoured map can be used for measuring areas, but the contours need to be spot checked by instrument survey. Areas are usually taken to the centreline of a proposed dam, a deduction being made later for the volume occupied by the upstream shoulder of the dam – except in the case where fill for an earth dam is taken from a borrow pit within the proposed reservoir below the intended impounding level.

The volume/water level curve should be extended to cover surcharge volumes above the spillway crest because this volume will be required for flood routing calculations. Allowance must be made for unusable 'bottom water', normally taken as the soffit level of the lowest existing or proposed drawoff pipe. If silting of an existing reservoir is suspected, it may be necessary to take some soundings to find the silt level. This can be higher than the level of the scour pipe which may only draw the silt down locally.

Estimating historic yield

A simple graphical method as shown in Fig. 3.7 can be used to find the historic minimum yields for various storage volumes. The catchment runoff is plotted cumulatively year by year. Abstractions of a uniform quantity are then straight lines, the maximum vertical intercept between such a line and the cumulative runoff then shows the amount of storage need to support that uniform abstraction.

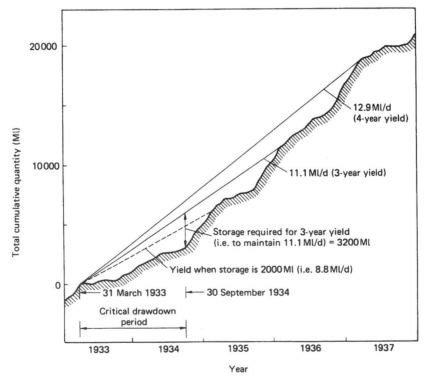

Fig. 3.7 Five-year mass flow diagram.

A more satisfactory and accurate numerical method of finding the historic yield with the aid of a computer is to use the formula:

$$\text{Yield} = \frac{\text{inflow over a period} + \text{storage}}{\text{length of critical period}} \tag{3.2}$$

Moving catchment runoff totals for periods of 2, 3, ... , n consecutive months are summed by the computer starting at months 1, 2, 3, ... , etc. to find the lowest total runoff for each period. Weekly or daily flows can be used instead.

The minimum runoff found for each period is plotted on a cumulative minimum runoff diagram as shown in Fig. 3.8. This, in effect, compresses onto one diagram the driest segments of the mass curve shown in Fig. 3.7. As a result adjacent points on Fig. 3.8 may be for entirely different low runoff periods, and the graph will normally have an irregular shape as the period lengthens to include a second dry season and then a third. By marking the storage on the negative ordinate and striking a tangent to the plotted curve, the minimum historic yield can be found from the slope of the line because it expresses the above eqn (3.2). An advantage of the method is that it is simple to find the minimum yield for other storage values if, say, enlargement of a reservoir is under consideration.

Check computations using eqn (3.2) above can be used for greater accuracy in computing the minimum yield read from the graph in Fig. 3.8. Alternatively the computer may be programmed to do this and print out the minimum yield for each period of consecutive months. But the graphical method is a visual aid to understanding the runoff

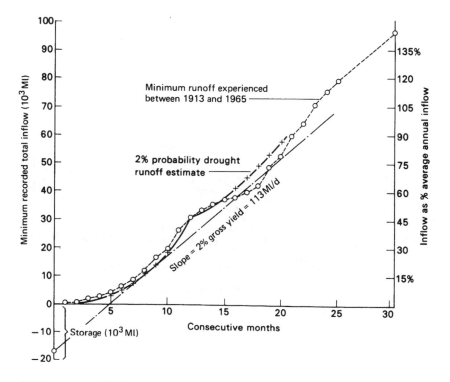

Fig. 3.8 Minimum runoff diagram.

pattern and can reveal if there is any unusual feature which may indicate an error in the computations or in data used.

Corrections for evaporation from the reservoir water surface may have to be made to the catchment runoffs. The deduction is minimal for UK impounding reservoirs but in hot climates it can be large and must be taken into account. Sometimes the evaporation is more than the catchment inflow, causing a 'negative inflow' to be registered.

Corrections for compensation water releases from the reservoir may be needed if the amount to be released is not a fixed amount daily (which can be deducted from the gross yield), but is required to vary seasonally, or if spates (i.e. increased releases for short periods) are required. These can be dealt with in the same manner as corrections for evaporation.

Estimating minimum yield and failure risk by probability analysis

Impounding reservoirs frequently have enough storage to even-out the catchment inflow for 2 or 3 years, and some for longer. To find the minimum steady yield it is therefore necessary to find the minimum inflow for periods varying from one dry season to those which include three dry seasons or more. However, the distribution of rainfall and runoff for short periods of under 12 months tends to be skewed, whereas that for longer periods more nearly approaches the normal distribution (see Section 3.17). It is therefore best to approach the problem of finding the minimum yield in two parts, so that appropriate probability paper can be used for each.

For periods of under 1 year containing one dry season, the runoff totals for periods of 1, 2, 3, ... , 11 consecutive months are obtained for each year of record, They are tabulated in order of magnitude for each period. An example for a 4-month period is given in Table 3.6. The tabulation can cease when the number of runoff totals obtained is nearly equal to the number of years of record i.e. totals are well above the mean runoff for the period. The probability, expressed as 'return period', is obtained by eqn (3.1) in Section 3.16 where n is the number of years of record.

The totals and their return periods are plotted on log-log paper as shown in Fig. 3.9 which is likely to give straight line results. (If not, some other type of probability paper must be tried.) The lines drawn through the plotted points for each period must be

Table 3.6 Example of ranking of drought flows for a 4-month period

Runoff 10^3 cu. m	Starting month	Return period years
28.40	June 1934	50
41.50	July 1959	25
50.50	June 1949	16.7
56.30	July 1947	12.5
60.90	June 1933	10.0
64.00	June 1921	8.3
66.90	June 1929	7.1
72.20	July 1955	6.25
72.80	April 1938	5.55

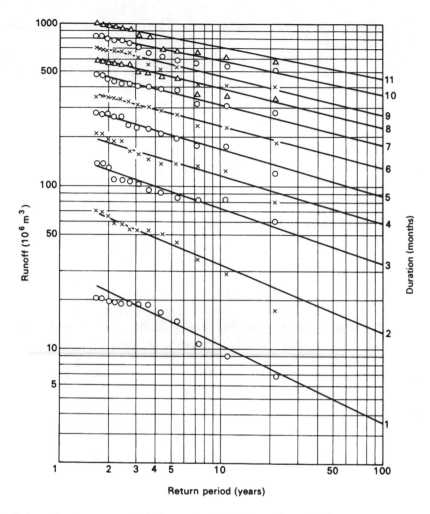

Fig. 3.9   Independent events probability analysis (1–11-month periods).

reasonably positioned; they should show steadily increasing increments of flow as the period increases to include the first wet season.

For longer time periods exceeding 1 year, the number of independent period events grows fewer, and the extent to which periods overlap increases. The periods will also contain different proportions of wet and dry season runoffs, according to their start date. This can result in plots that are confusing to interpret. It is preferable therefore to sum flows for these long periods beginning on a fixed date each year, which is the usual starting date for the dry season. If inspection shows this does not pick out the lowest flow, an alternative rank list should be formed with an adjusted start date. The ranking of the runoffs and calculation of their return period is carried out as described above for the short periods, again using eqn (3.1) in Section 3.16. The runoffs are plotted on normal probability as illustrated in Fig. 3.10 with straight lines drawn through the plots for each period.

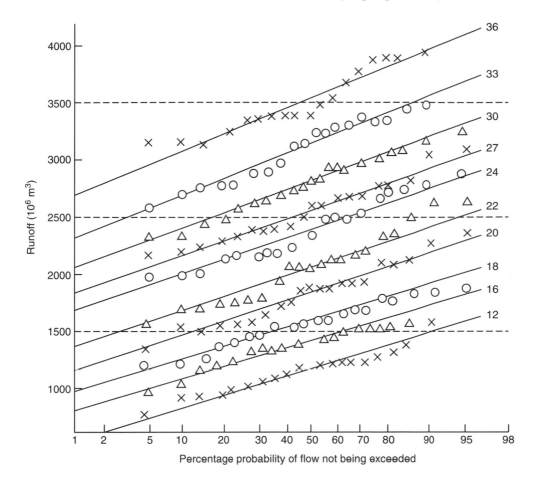

Fig. 3.10 Overlapping events probability analysis for 12–36-month periods.

From these diagrams a minimum runoff diagram as Fig. 3.8 can be produced for, say, a 2% risk of failure, as described above when estimating the minimum historic yield. From this diagram minimum yields for different volumes of storage can be estimated and checked by calculation.

A further development possible is to express the minimum yield as a proportion of the mean catchment flow over the period of record, and the reservoir storage as a proportion of the mean annual flow volume, as shown in Fig. 3.11. The slope of the curve on this type of diagram changes as the critical period for minimum yield changes from one dry season to two dry seasons, etc. However, if daily or weekly runoffs have been used instead of monthly runoffs the change of slope will be less perceptible.

3.19 Yield of a pumped storage reservoir

To estimate the yield of a pumped storage reservoir it is more accurate to work on the basis of daily flows. The use of mean monthly flows is inexact and necessitates reducing the computed yield by an arbitrary amount to allow for part of the flows in excess of the

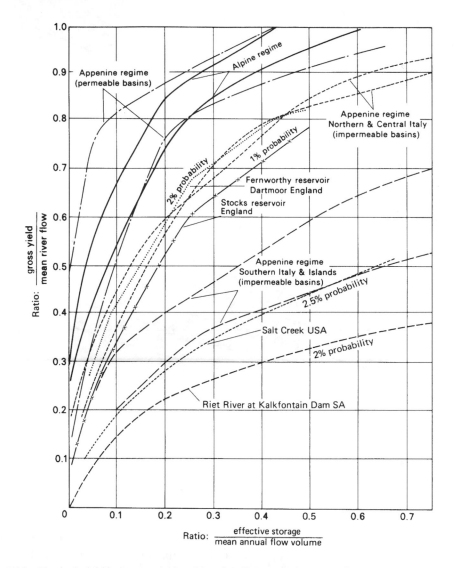

Fig. 3.11   Typical yield/storage relationships for direct supply reservoirs.

mean being uncollectable. If some daily flow records are available these may assist in estimating what allowance should be made; otherwise a frequent practice is to assume that only 90% of the potential abstraction is possible. The better alternative is to develop a series of daily flows which represent flows during a period of minimum runoff (see Section 3.20)

With daily flows available, the basic calculations are as shown in Table 3.7. However, the sub-calculation required to calculate potential abstraction will depend upon:

(a) the rule(s) laying down the abstraction conditions;
(b) the assumed maximum pumping capacity; and
(c) the time lag between a change of flow and the consequent change of pumping rate.

Table 3.7 Pumped storage reservoir calculation(Abstraction condition – two-thirds flow above 30 $\times$ 10³ m³/day)

Day	Flow to intake 1000 m³ per day	Quantity available at intake 1000 m³ per day	90% pumped to store 1000 m³ per day	Amount supplied ex store 1000 m³ per day	Reservoir contents	
					Net change of storage 1000 m³ per day	In store end of day 1000 m³ per day
15.8	42.0	8.0	7.2	5.5	+1.7	151.7
16.8	39.0	6.0	5.4	5.5	−0.1	151.6
17.8	38.0	5.0	4.5	5.6	−1.1	150.5

The last two factors need careful practical consideration before any computations take place. With appropriate control equipment and variable speed pumps, the abstraction may closely follow flow variations; but other arrangements are often adopted. A range of fixed speed pumps may be started and stopped automatically to give such as 0.5Q, 1.0Q, 1.5Q or 2Q outputs according to their combination. Also if manual operation of pump outputs is used, the time lag between change of flow and change of abstraction will depend on the manning pattern and whether variable or fixed speed pumps are used.

On a simple basis computations can assume an unlimited volume of storage reservoir is available so the maximum drawdowns occurring each year can be calculated and plotted on probability paper to find the drawdown probable once in, say, 50 years. But this does not solve the problem of what yield is available with a reservoir of a lesser capacity than the maximum drawdown. Calculations then have to proceed for a range of drawoffs to produce a yield-storage relationship with an associated risk in order to find the yield for a given storage for any degree of risk. This process is difficult to carry through successfully unless the flow record is long enough to make the probability plot one for interpretation only, because points from an unnatural distribution of this type to do not permit confident extrapolation.

In view of these difficulties it is more satisfactory to work directly on an estimate of the daily flows likely during, say, a drought of 2% risk, as shown in Section 3.20 below. It is then possible to compute the yield of 2% risk according to the size of storage reservoir adopted and factors (a), (b), (c) mentioned above. A further limiting factor to be taken into account is the need to ensure refilling of the reservoir after some maximum drawdown. The most secure provision is to ensure refilling of the reservoir in any single wet season following a dry season, but this is not always possible. Some pumped storage schemes accept that refilling will only occur once every few years. Such schemes have to be tested to make sure they can achieve an initial filling sufficient to meet the anticipated initial demand.

All such problems can be dealt with by applying appropriate computer calculations to the record of daily flows. Among the most useful result is that of finding the most economic size of pumps to install. Increasing pump capacity beyond a certain level may increase the yield by only a small amount when flows fall rapidly and critical drawdown periods are short. On the other hand the need to ensure refilling of the reservoir during a wet period may be an over-riding factor determining maximum pumping capacity.

3.20   Producing daily flows representative of a given drought

The use of daily flows facilitates computation of the minimum yield of a pumped storage scheme, or of a regulating reservoir as described in the next section. But daily flows are rarely available for a long enough period to include a 2% drought period. Statistical techniques exist for generating long sequences of daily flows but it has yet to be shown they can be made to produce storage requirements matching those required in practice. However to provide a tool for rapid and consistent testing of a variety of schemes, the following method for compiling a 2% design drought of daily flows from a record of monthly flows has proved satisfactory in giving results consistent with experience.

(1)   From a diagram such as Fig. 3.9 the 2% minimum runoffs for 1, 2, ... , *n* months are listed as percentages of the long average annual flow (AAF). The monthly incremental percentages are derived, and these are rearranged to represent a realistic calendar order, retaining the sequence of minimum runoffs for 1, 2, ... months, as shown below.

(A) Driest periods from 2% analysis			(B) Re-arranged monthly sequence	
Period	Runoff % AAF	Monthly increment	% AAF	Month
1 month	0.9	0.9	5.2	April (as 7th)
2 months	1.9	1.0	2.5	May (as 5th)
3 months	3.1	1.2	1.3	June (as 4th)
4 months	4.4	1.3	1.0	July (as 2nd)
5 months	6.9	2.5	0.9	August (as 1st)
6 months	11.5	4.6	1.2	September (as 3rd)
7 months	16.7	5.2	4.6	October (as 6th)
...			...	

(2)   The procedure is continued for 8, 9, 10, ... , *n* months using such as Fig. 3.10. However, when the longer periods of 12, 13, or so months include the wet seasons, it will not be found possible to achieve exactly equivalent totals for each of the driest 12, 13, 14, ... , *n* month periods; but this will not affect the drought yield in practice.

(3)   To convert the monthly flows to daily flows, some record of daily flows (however short) is needed. Then, to derive daily flows for, say, April in the table above, the record of daily flows is examined to find an April which has a total flow nearest to the total given for April in the table above, and its values are scaled down proportionately to give the required 5.2% AAF monthly total. Some minor adjustments may be necessary. (i) Daily flows from the end of 1 month to the beginning of the next may need adjustment to give continuity, retaining the correct totals for each month. (ii) In the absence of any really dry months on record, it may be best to apply a straight line recession of flows below the historic record to derive the required minimum flow for the month.

(4)   The daily flows so produced need checking to see they have a proper balance of spates and recessions. To do this the following procedure is used.

    (i)   From the historic record, daily runoff deficiencies below some arbitrary value (such as 40% of the average daily flow, ADF) are summed monthly and plotted against the volume discharge for the month – as Fig. 3.12.

    (ii)   A mean curve is plotted through the points.

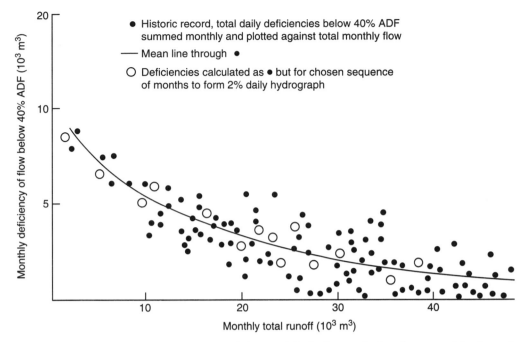

Fig. 3.12 Check of 2% daily hydrograph. If the 2% daily hydrograph plots lie reasonably about the mean line for the historic record plots, the sequence of months chosen for the 2% daily hydrograph is satisfactory.

(iii) From the 2% hydrograph of daily flows, daily deficiencies summed monthly are similarly calculated and plotted against volume for the month.

If the 2% hydrograph values under (iii) all fall below the line (ii), the 2% hydrograph has not chosen a severe enough sequence of months for the longer periods of 12, 13, ... , n months, and needs adjusting.

(5) To estimate the amount of adjustment necessary, seasonal deficiencies below the same 40% arbitrary flow for (say) 8, 20, 32, ... consecutive months (e.g. covering one dry season; two dry + one wet; and so on) are taken from the historic record and plotted on log-Normal probability paper as Fig. 3.13 and extrapolated to give the 2% deficiencies. The difference between the values for 2% probability gives the individual seasonal deficiencies, hence the months used for the 2% daily hydrograph can be altered to get agreement for each season.

The above procedure may seem complex but is not difficult in practice, the computer calculations involved being simple. The method has the advantage that it permits subsequent work to proceed more rapidly, and direct analysis for the 2% yield of any given size of reservoir storage is possible.

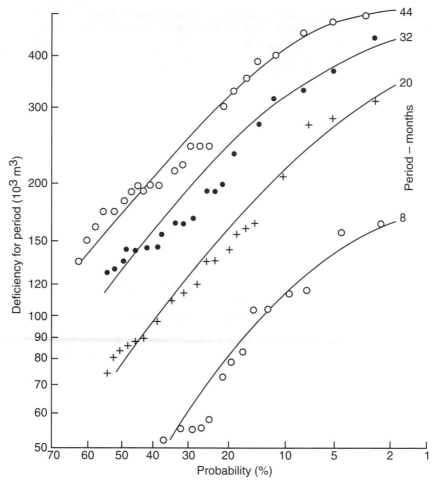

Fig. 3.13   Seasonal deficiencies plotted on log-Normal probability paper to check 2% daily design hydrograph.

3.21   Yield of regulating reservoirs

A regulating reservoir (see Fig. 3.14) impounds water from a catchment A, and releases water to support an abstraction at some location B downstream, when flows at B are not sufficient to meet the required abstraction. This means that the yield obtainable is greater than that provided by catchment A alone. Usually some compensation water has to be released from A to maintain a flow in the stream below it; and at B various abstraction conditions can apply. The latter fall into two main groups. One group requires maintenance of a given flow continuously below the intake B. The other requires abstraction at B to be fully supported by equivalent releases from A until the natural flow at B reaches a certain figure; thereafter, as flows at B continue to rise, releases from A are cut back until the natural flow at B is sufficient to support the whole abstraction. This procedure preserves the natural low flow regime of the river below B which may be desirable on environmental grounds.

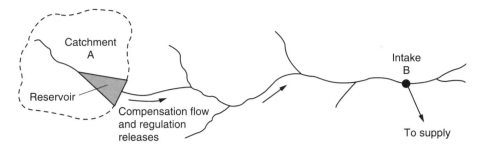

Fig. 3.14 Regulating reservoir.

As with pumped storage schemes it is best to calculate minimum yield by first producing daily flows for, say, a 2% drought period for the natural flows at the intake point B. The proportion of this flow diverted into the regulating reservoir A will have to be assessed. It will probably bear a varying relationship to the natural flow at B. The calculation can then proceed according to the rules laid down for compensation releases at A and the abstraction conditions at B. An example is given in Table 3.8.

Some points have to be borne in mind. Releases from the reservoir may need to include an allowance for evapotranspiration and other losses en route to the abstraction site (hence the 10% addition to Col. 4 in Col. 5 of Table 3.8). Also the time taken for released water to travel to the abstraction point must be taken into account. Dependent on the distance involved, the time lag may range from several hours to 1 or 2 days or more. This can mean that an increase of release in expectation that flow at the intake will decline may be wasted if rain should come and increase the flow. Hence a further allowance of 10% or 15% may have to be added to the releases to cover discrepancies between actual and

Table 3.8 Regulating reservoir calculations. Conditions: Abstraction required 28 Ml/day. Reservoir compensation release 2.0 Ml/day. Residual flow below intake 10 Ml/day

Day	(1) Natural flow at intake units 1000 m^3	(2) Reservoir inflow	(3) Net flow at intake	(4) Intake flow deficiency	(5) Reservoir release	(6) Reservoir change/day	cum.
1	45.0	3.3	41.7	Nil	2.0	+ 1.3	+ 1.3
2	41.0	3.0	38.0	Nil	2.0	+ 1.0	+ 2.3
3	38.0	2.8	35.2	0.8	2.9	− 0.1	+ 2.2
4	35.5	2.6	32.9	3.1	5.4	− 2.8	− 0.6
5	33.0	2.4	30.6	5.4	7.9	− 5.5	− 6.1
6	etc.						
	Col (3)	= Col (1) − Col (2)					
	Col (4)	= Abstraction + Residual flow − (3) − Compensation release					
	Col (5)	= Col (4) × 110% + Compensation release					
	Col (6)	= Col (2) − Col (5)					

theoretical release requirements. A dry weather recession curve for the natural flow at the intake, converted into a guiding rule, can be used to aid release decisions. Use of such a decision rule in UK has shown that actual releases tend to be up to 20% more than the theoretical requirement in wet years and about 3% more in dry years.

3.22 Catchwater yields

A catchwater is usually a channel which leads water from some remote catchment into an impounding reservoir. The catchment would otherwise not contribute any flow to the reservoir. The flow from the catchwater increases the yield of the reservoir, but the problem is to find out by how much, taking into account the fact that the catchwater is seldom big enough to take all of the higher runoffs from its catchment.

The maximum flow the catchwater will carry may need to be estimated by hydraulic analysis (according to its grade and size, etc.) or by installing a temporary gauge to measure it directly during some period of heavy rainfall. The catchment runoff will also need to be estimated on a time basis to produce a flow duration curve for, say, a 2% dry year, of the type shown in Fig. 3.15, from which can be deduced the amount collectable for a catchwater of a given size. Figure 3.16 shows a relationship between catchwater size and percentage of runoff collectable, derived from hydrographs obtained on small British catchments with average rainfalls of 1500 mm per annum or more. It was found the relationship checked equally well against data from a tropical catchment, such as one in Singapore.

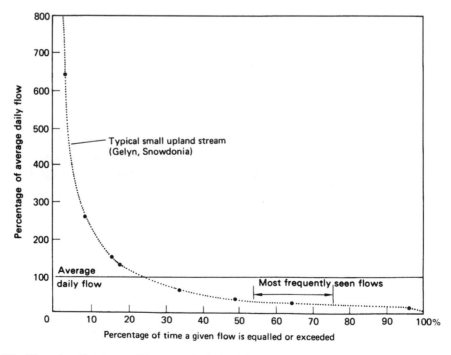

Fig. 3.15 Flow duration curve. The curve is derived by measuring the number of hours the stream flow exceeds given level of flow.

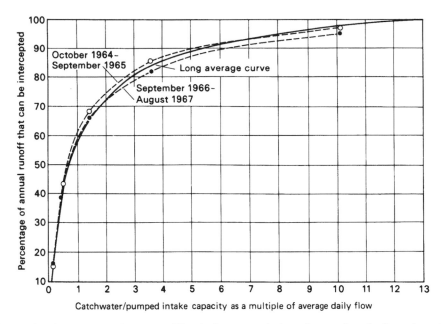

Fig. 3.16 Catchwater transfer curve. The design curve is based upon hourly flow duration data from four small mountain catchments in England and Wales.

Because many catchwaters are simply open unlined channels cut to a gentle gradient in a hillside, they often contribute no inflow during the dry season, and will not contribute flow to the reservoir until initial precipitation is sufficient to wet their bed and banks to saturation level. Hence their main contribution to the reservoir is during the wet season. Although Fig. 3.16 strictly applies to average annual flows, it can be used without too much error to assess the yield during a given season of the year, provided the appropriate average daily runoff for that season is used in place of the average annual flow.

The catchwater contributions to the reservoir are added to its direct catchment inflow, so that the minimum yields for various consecutive months can be plotted on a diagram such as Fig. 3.17. This shows the extra yield a catchwater provides for a given reservoir storage and the change of critical drawdown period.

In some instances large catchwaters picking up quite large streamflows may be built, in which case it may be necessary to allow for any minimum flow that must be left in streams. A simple structure on a stream may allow diversion into the catchwater only when the streamflow rises above a certain level. The calculations of yield must accordingly take this into account.

3.23 Compensation flows

Compensation water is the flow that must be discharged below an impounding reservoir to serve riparian owners and other abstractors downstream. Each country tends to have its own water law to preserve water rights and the setting of compensation water can involve much legal dispute. In Britain the compensation water to most impounding reservoirs is set by some Parliamentary Act. In the early part of the century compensation water was

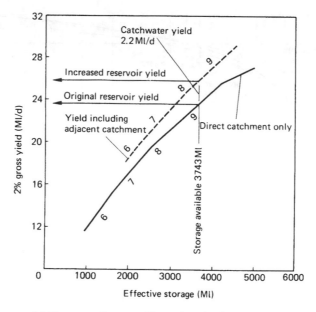

Fig. 3.17 Catchwater yield/storage diagram. Note that the figure by each straight line denotes the length of the critical period in months.

often set at one-third of the gross yield of the reservoir, but this proportion tended to reduce to one-quarter in later years.[48] Nowadays the compensation water is often required to be varied seasonally, and extra discharges as 'spates' may be stipulated at certain times of the year to meet fishing interests.

The discharge of a fixed amount every day has also been criticised on environmental grounds as being 'unnatural' and not conducive to the maintenance of fauna and flora which need periods of varying flow. Considerable progress has been made in quantifying the water requirements at different stages in the life cycle of fish, invertebrates and macrophytic vegetation.[49] American studies of physical habitats have been followed in France, Norway, Australia and UK as a means of defining environmentally acceptable flows. The software calculations with the PHABSIM program, available in the public domain, depend on field measurement of river velocity, depth, substrate and tree cover. They determine ecological preferences and hence seasonal variation of compensation water, but inevitably not all the requirements can be met if a reasonably economic yield is to result. Hence some compromise solution has to be found. Nevertheless the technique gives a far more satisfactory means of engineering water resource developments to achieve minimum environmental damage.

3.24 Conjunctive use and operation rules

When a water supply undertaking has several sources, conjunctive or integrated use of them may be a means of improving the total yield or of reducing costs, or both. Thus extra water from an underground source when the water table is high, or from a river in flood, may permit a cut-back in the supply from an impounding reservoir with a 2-year critical drawdown period, enabling it to store more water. Similarly it may be possible to keep storages with short critical periods in continuous full use to avoid overspills and so maximise their supply, at the same time reducing drawoff from a larger reservoir with a longer critical period, thereby gaining larger reserves to meet critical drawdown conditions. In a similar fashion it may be possible to reduce costs if the source producing the cheapest water can be over-run for part of the year, whilst dearer sources are cut back.[50,51]

However, there can be physical conditions which limit possibilities for conjunctive use, such as:

(a) isolation of sources and their supply areas;
(b) sources supply areas at different elevations;
(c) incompatibility of one source water with another;
(d) the need of certain manufacturers to use only one type of water.

It is not always possible to change frequently from one type of water to another, particularly if one is a 'hard' water from underground and the other a 'soft' river or impounding reservoir supply. Domestic consumers may complain that the taste and colour of the water has changed: tea will taste markedly different. Deposits in plumbing systems and water mains may be affected. Some industrial consumers, having installed water treatment plant to cope with one type of water, cannot change to a different supply without long forewarning and considerable expense.

To ascertain potentials for conjunctive use the whole system of sources needs drawing out in diagrammatic form, showing:

- source outputs (average day critical yield; maximum day plant output);
- impounding or pumped storage capacity; length of critical drawdown period;
- area served, line of trunk feeders, key service reservoirs fed;
- elevation of supply at sources; high ground areas in the supply area;
- any legal or other restrictions on source outputs.

It is helpful to allocate a different colour for each source and its associated works. The possibilities can then be examined for conjunctive use. Key factors will probably be the need for major interconnecting mains and extension of treatment works capacity. The cost of these must be roughly assessed to see whether they are likely to be worthwhile having regard to the possible gain in yield. Once a possible scheme for conjunctive use has been clarified, this can be tested by computer calculations on a month-by month basis to check the combined yield during a chosen critical dry period.

Operational rules can be developed to assist in judging when storage reservoirs can supply more than their minimum yield for a given risk. Monthly reservoir drawdowns over a long period of simulated inflows can be used to develop a control curve as shown in Fig. 3.18. This shows the minimum storage required at the beginning of every month to ensure maintenance of a given supply rate. To produce a control curve of this type involves calculating backwards in time, from an assumed zero storage at the end of each month of the year. By applying this process to droughts of every duration, it is possible to locate the maximum storage required at any time of the year to ensure the reservoir never quite empties at the assumed abstraction rate. Different abstraction rates will require different control curves. A 'family' of such curves is therefore produced, each for a different level of supply. Hence reference to the curves and a storage/water level chart, can show whether the water level in the reservoir permits an increased abstraction or not.

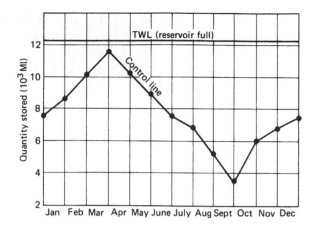

Fig. 3.18   Reservoir control curve. The control line represents the level to which the reservoir could have been drawn down during any month in the period 1910–64 (55 years) and still maintain a total outflow of 46 Ml/day to supply and 13 Ml/day to compensation. When the contents are above the control line the draw off may exceed 46 Ml/day to supply (up to the limit of the treatment works capacity). If the contents fall to or below the control line the draw off must be limited to 46 Ml/day to supply.

Reservoir control curves have to be based on the most severe droughts historically recorded, or on a 'design' drought of specified probability. Neither can forecast the magnitude of some drought. Hence a control curve tends to be of more practical use in permitting extra water to be supplied when storage is high, than when storage is low. If a reservoir is three parts empty with the dry season not yet ended, most engineers would attempt to restrain demand, in preference to relying solely on a 'control' curve which poses a significant risk that it might not apply to what the future may bring.

3.25   Use of computer models for hydrological computations

A wide variety of computer software programs are available for undertaking hydrological computations. The most common tasks they are used for comprise:

- statistical analyses of precipitation and flow data;
- computation of surface and/or underground flows through a catchment;
- hydraulic calculation of channel flows, flood analyses and reservoir routing;
- simulating water quality changes down channel flows allowing for time of travel.

Initial parameters entered into the computer form a 'model' of the system to be analysed. Hence fieldwork may be necessary to obtain the necessary data, and the model should be tested to see it replicates reasonably accurately conditions that have been experienced.

'Dynamic' programs trace the effect of changing inputs with time, such as rainfall, on catchment flows. Other models can be 'single event' or 'spatial', such as tracing the surface water profile down a river course under steady-state flow conditions. A computer program breaks down a problem into a lengthy series of relatively simple steps of calculation, pursued in logical order, using 'feedback' connections where the result of one step modifies an earlier calculation. The procedure is the same as that adopted for manual calculation and uses the same theoretical or empirical formulae. The exception is the assessment of groundwater flow, which is through a three-dimensional porous medium, for which the computer uses the finite element or finite difference mathematical approach (see Section 4.5) – a method that would be impracticable by hand.

Software packages vary according to the size of problem to be handled and the extent and complexity of the sub-calculations involved. For example, evapotranspiration values may be simple approximations, or be worked out using one or other of the formulae mentioned in Section 3.6. The more complex programs are suited for the use of specialists or researchers, and are often modified by the user to suit particular conditions applying. Others are 'easy-to-use' packages which do not require the user to have a knowledge of the computer operating system. They cater for the normal range of hydrologic inputs required, but may not be modifiable by the user to incorporate some unusual feature. Packages are often combined, so that a wider range of inter-related phenomena can be handled in one program, such as the combination of computed flows with consequent water quality changes. At a simpler level much can be achieved by computer spreadsheet water-balance models written for each new task.

A good guide to the types of programs available, with addresses for their procurement, appears in *Handbook of Hydrology*, published by McGraw-Hill, 1992. The *Hydrology Handbook* published by the American Soc. Civil Engrs, 345 East 47th Street, New York, 1996, also has information on models and their application. Programs are available from many sources – universities, scientific institutions, specialist government departments, consulting engineers and so on. Some programs are free, others need to be purchased. Four useful sources of hydrological programs are:

World Meteorological Organization, HOMS Office, Case Postale No. 2300, Geneva 2, Switzerland.

The Institute of Hydrology, UK, Wallingford, Oxfordshire OX10 8BB.

US Geological Survey, WRD, 415 National Centre, Reston, VA 22092.

US Army Corps of Engineers, Hydrologic Engineering Center, 609 Second Street, Davis, CA 95616–4587.

The catalogues of commercial firms supplying software, such as that of Scientific Software Group, P.O. Box 23041, Washington DC 20026–3041, are also useful since they give details of a range of programs they can supply from many organisations and establishments.

3.26 Rainwater collection systems

Rainwater tanks collecting runoff from roofs or impervious surfaces form a useful source of drinking water where daily rainfall is frequent, as in equatorial climates and in the monsoon periods of monsoon climates. The supply is particularly useful if local sources are polluted because only simple precautions are necessary to keep the rainwater free of pollution.

For individual house rainwater collecting tanks the principal constraint is usually the size of tank which can be afforded or which it is practicable to install. Tanks up to 600 or 800 litre can generally be made of one piece material and are transportable. Tanks of 1000 litre capacity are usually more economical if constructed *in situ*. 'Ferrocement' tanks made of layers of cement rendering applied to wire mesh are particularly economical if the requisite skill to construct them is available or can be trained. Otherwise simple reinforced concrete tanks can be constructed. Such tanks have the advantage they are repairable if they leak or can be rendered inside if poorly constructed. Tanks made of plastic plates bolted together tend to fracture under the repeated bending caused by changing water levels and the fracture is usually unrepairable. Steel tanks made of plates bolted together tend to rust at the joints and are difficult to repair.

Roof areas are not usually a principal constraint provided houses are one storied. There is usually at least 20 m^2 of roof area which can discharge to guttering along one side of the roof with, perhaps, a return along another side. Traditional roofing for low income communities is galvanised iron, but sometimes asbestos cement roofing will be found or the more modern types of roof will be tiled or asphalted. With house occupancies varying from six to 12 people, about $1\frac{1}{2}$–$2\frac{1}{2}$ m^2 of developable roof area per person is usually available.

Daily rainfall records need to be tabulated. The daily runoff to the collecting tank is usually calculated according to the roof area available per person. The runoff is taken as 95% (daily rainfall – 1 mm), the 1 mm deduction being for initial evaporative loss on the roof, and the 95% allowing for guttering overspill. Some field tests need to be undertaken to assess the size of guttering required, where it should be positioned relative to roof eaves, and what allowance should be made for overspill during intense rainfalls. Local practice and experience, where available, can act as a guide. Although UK rainfall conditions are not likely to apply overseas where roof rainfall collection systems are mainly used, some UK publications can be of assistance.[52,53] If large roofs, e.g. of commercial premises, etc. are used, the use of siphonic outlets to guttering may achieve economy in downpipe sizing. The siphons become primed by air-entrainment of initial flows, thus making the whole head between gutter outlet and the bottom of the downpipe available for the discharge, reducing the size of downpipe required.[54,55]

Calculations proceed by using trial abstractions over typical recorded dry periods, commencing when the collecting tank can be assumed full. It is convenient to work on a units of roof area and storage available per person, sizing this up later according to the average number of people per household for which the design should cater. Operating rules for householders should be simple, e.g. either a fixed amount 'X' per person per day, or 'X' reduced to '$\frac{1}{2}$X' when the tank is half empty. An appropriate minimum amount drawoff rate would be 2.5 litres per person per day, sufficient for drinking and cooking purposes. The theoretical tank size so calculated needs to have allowances added to it for 'bottom water' and 'top water'. The top water allowance – for example 6 in or 150 mm – means the householder does not have to restrict his or her take until the water level is that much below overspill level. The level can be marked clearly inside the tank. The advised abstraction 'X' has to be expressed in terms of commonly available vessels. A standby supply must also be available to afford householders a supply when longer dry periods occur than assumed for the calculation and also for rescuing householders who run out of water for any reason.

Although house tanks will not give a sustainable amount in dry seasons of the year, they will be well used in the wet season because they relieve the householder of having to carry water from a distance. The rainwater in the tanks does not need to be chlorinated, but tanks do need cleaning out annually. Mosquito breeding in tanks can prove a nuisance in some climates and specialist advice may need to be sought as to how to control it.

References

1. Morris D. G. and Heerdegen R. G. Automatically Derived Catchment Boundaries and Channel Networks and their Hydrological Applications. *Geomorphology*, **1**(2), 1988, pp. 131–141.
2. Mansell-Moullin M. The Application of Flow-Frequency Curves to the Design of Catchwaters. *JIWE*, **20**, 1966, pp. 409–424.
3. Simpson R. P. and Cordery I. A Review of Methodology for Estimating the Value of Streamflow Data. *Inst Engrs Aust Civil Eng Trans*, **24**, 1987, pp. 79–84.
4. Tor S. M. Strategies Towards Sustainable Development in Nigeria's Semi-Arid Regions. *JCIWEM*, June 1998, pp. 212–215.
5. Binnie & Partners, *Binnie News*, Nov. 1967, pp. 37–38.
6. United States Geological Survey. *Techniques of Water Resources Investigations of the USGS, Book 3*. Chapters A6–A8 on stream gauging procedure. US Government Printing Office, Washington, 1968.
7. BS 3680: Measurement of liquid flow in open channels. *Part 4A: Method using thin plate weirs*, BSI, 1981 as amended 1987.
8. Bos M. G. (ed.). Discharge Measurement Structures. ILRI Publication 20, 3rd edn. Reprinted 1990.
9. Herschy R. W., White W. R. and Whitehead E. *The Design of Crump Weirs. Technical Memo No. 8*, DoE Water Data Unit, 1977.
10. Ackers P., White W. R., Perkins J. and Harrison A. J. M. *Weirs and Flumes for Flow Measurement*. John Wiley, Chichester 1978.
11. United States Geological Survey, *Techniques of Water Resources Investigations of the USGS. Book 3*, Chapter A16, Measurement of discharge using tracers. US Government Printing Office, Washington, 1985.
12. BS 3680: Measurement of liquid flow in open channels. *Part 3E: Measurement of discharge by the ultrasonic (acoustic) method*. BSI, 1993.
13. Sargent R. J. and Ledger D. C. Derivation of a 130 Year Run-off Record from Sluice Records for the Loch Leven Catchment, South-East Scotland. *Proc ICE Water, Maritime and Energy*, **96**(2), 1992, pp. 71–80.
14. World Meteorological Organisation, *Guide to Meteorological Instruments and Observing Practices No. 8*, 1983 edn.
15. Hudleston F. A Summary of Seven Years Experiments With Rain Gauge Shields in Exposed Positions 1926–1932 at Hutton John, Penrith. *British Rainfall*, 1933, pp. 274–293.
16. Institution of Water Engineers, Report of Joint Committee to Consider Methods of Determining General Rainfall Over Any Area. *Trans IWE* **XLII**, 1937, pp. 231–299.
17. Creutin J. D. and Obled C. Objective Analysis and Mapping Techniques for Rainfall Fields: An Objective Comparison. *Water Resources Research*, **18**(2), 1982, pp. 413–431.
18. World Meteorological Organisation. Measurement and Estimation of Evaporation and Evapotranspiration. *Note No. 83*. WMO Geneva 1966.
19. Doorenbos J. and Pruitt W. O. Crop Water Requirements. *Irrigation and Drainage Paper 24*. FAO 1976.
20. Lapworth C. F. Evaporation From a Reservoir Near London. *JIWE*, **19**(2), 1965, pp. 163–181.

21. Rodda J., Downing R. A. and Law F. M. *Systematic Hydrology*. Newnes-Butterworth, Oxford, 1976, pp. 88–89.
22. Penman H. L. Vegetation and Hydrology. *Technical Communication N.53*. Commonwealth Bureau of Soils, Harpenden, 1963.
23. McCulloch J. S. G. Tables for the Rapid Computation of the Penman Estimate of Evaporation. *East African Agricultural and Forestry Journal*, **XXX,** No. 3 1965.
24. Wilson E. M. *Engineering Hydrology,* Macmillan, Basingstoke, 4th Edn. 1990.
25. Dean T. J., Bell J. P. and Baty A. J. B. Soil Moisture Measurement by an Improved Capacitance Technique. Part 1. Sensor Design and Performance. *J Hydrology*, **93,** 1987, pp. 67–78.
26. Calder I. R. *Evaporation in the Uplands*. John Wiley, 1992.
27. Bosch J. M. and Hewlett J. D. A Review of Catchment Experiments to Determine the Effects of Vegetation Changes on Water Yield and Evapotranspiration. *J Hydrology*, **55**.
28. Kirby C., Newsom M. D. and Gilman K. Plynlimon Research: The First Two Decades. *Report No. 109*, Institute of Hydrology, Wallingford, 1991.
29. Gustard A., Roald L. A., Demuth S., Lumadjeng H. S and Gross R. *Flow Regimes from Experimental and Network Data (FREND) I and II*. Institute of Hydrology, Wallingford.
30. Ege H. D. Management of Water Quality in Evaporation Systems and Residual Blowdown. *Water Management by the Electric Power Industry*. University of Texas, Austin, TX, 1975.
31. Thompson N., Barrie I. and Ayles M. *The Meteorological Office Rainfall and Evaporation Calculation System: MORECS*. The Meteorological Office, London, 1981.
32. Smith L. P. and Trafford B. D. Climate and Drainage. *MAFF Technical Bulletin 34*. HMSO, London, 1975
33. Gustard A., Bullock A. and Dixon J. M. Low flow estimation in the United Kingdom. *Report No. 108*, Institute of Hydrology, Wallingford, 1992.
34. Manley R.E. Simulation of flows in ungauged basins. *Hydrolog Sci Bull*, **3**, 1978, pp. 85–101.
35. Blackie J. R. and Ecles C. W. O. Lumped Catchment Models, in *Hydrological Forecasting* (eds Anderson M. G. and Burt T. P.), Chapter 11. John Wiley & Sons, Chichester, 1985.
36. Holford I. *The Guinness Book of Weather Facts and Feats*. Guinness Superlatives Ltd, London, 1977.
37. Tabony R.C. *The Variability of Long Duration Rainfall Over Great Britain,* Scientific Paper 37, Meteorological Office, London, 1977.
38. Enex/Drainage and Irrigation Dept. *Magnitude and Frequency of Low Flows in Peninsular Malaysia*. Hydrological Procedure No. 12, Malaysia, 1976.
39. Gustard A. and Gross R. Low flow regimes of Northern and Western Europe, *Proc FRIENDS in Hydrology Conf*. IAHS Publ, **187**, p. 205, 1989.
40. Darmer K. I. *A Proposed Streamflow Data Program for New York*. USGS Water Resources Division, Open File Reports, Albany, New York, 1970.
41. *Low Flow Studies Report*. Institute of Hydrology, Wallingford, 1992.
42. *Hydrometric Register and Statistics 1991–95*. IOH/British Geological Survey, Wallingford, 1998.

43. *Flood Estimation Handbook*, Vol. 1 – Overview; Vol. 2 – Rainfall Frequency Estimation; Vol. 3 – Statistical procedures for Flood Frequency Estimation; Vol. 4 – Restatement and Application of the FSR Rainfall-Runoff Method; Vol. 5 – Catchment Descriptions. Institute of Hydrology, Wallingford, Dec. 1999.

44. *Water Resources and Supply: Agenda for Action*, DoE 1996. Annexes D and E contributed by Environment Agency.

45. Codner G. P. and Ribeny F. M. J. The Application of the Sacramento Rainfall Model to a Large Arid Catchment in Western Australia, *Proc Hydrology Symposium* 1976, Institute of Engineers, Australia.

46. Fleming G. *Computer Simulation Techniques in Hydrology*, Elsevier, Oxford, 1975.

47. Hardison C. H. Storage to Augment Low Flows, Paper 8, *Proc Symposium on Reservoir Yield*, WRA, 1965.

48. Gustard A. *et al. A Study of Compensation Flows in UK*, Report 99. Institute of Hydrology, Wallingford, 1987.

49. Bullock A., Gustard A. and Grainger E.S. *Instream flow requirements of aquatic ecology in two British rivers*, Report 115, Institute of Hydrology, Wallingford, 1991.

50. Lambert O. A. *An Introduction to Operational Control Rules Using the Ten Component Method*, Occasional Paper No. 1, British Hydrological Society, 1990.

51. Parr N. M. *et al.* (eds). *Water Resources and Reservoir Engineering*, British Dam Society, Thomas Telford, London, 1992, pp. 11–40.

52. Code of Practice BS 6367: 1983 *Drainage of Roofs and Paved Areas.*

53. Sturgeon C. G. (ed.). *Plumbing Engineering Services Design Guide.* Inst. of Plumbing, 1983.

54. May R.W.P. Design of Gutters and Gutter Outlets. *Report IT 205*, Hydraulics Research Station, 1982

55. May R. W. P. The Design of Conventional and Siphonic Roof Drainage Systems. *JCIWEM*, Feb. 1997, pp. 56–60.

4

Groundwater supplies

4.1 Groundwater and aquifers

The most prolific sources of underground water are the sedimentary rocks, sandstones and limestones, the latter including the chalk. They have good water storage and transmissivity, cover large areas with extensive outcrops for receiving recharge by rainfall, and have considerable thickness but are accessible by boreholes and wells of no great depth.

The porosity of a rock is not an indicator of its ability to give a good water yield. Clays and silts have a porosity of 30% or more, but their low permeability due their fine grained nature makes them unable to yield much water. Solid chalk has a similar high porosity and low permeability, but is a prolific yielder of water because it has an extensive network of fissures and open bedding planes which store large quantities of water and readily release it to a pumped well or borehole. Limestone, on the other hand, is so free draining that, though it may yield large quantities of water in wet weather, the recharge rapidly drains away in dry weather when the yield is low.

In England the main aquifers used for public supply[1] comprise the following:

(1) The Chalk and to a lesser extent the Upper and Lower Greensands below it, both of the Cretaceous Period.
(2) The Bunter Sandstones and to a lesser extent the Keuper Sandstones of the Triassic-Permian Period.
(3) The Magnesian and Oolitic Limestones of the Jurassic Period, and to a lesser extent the Carboniferous Limestones and Millstone Grits of the Carboniferous Period.

The chalk and greensands, and the limestones of the Jurassic Period are widely spread over southern and eastern areas of England; the Triassic sandstones and Carboniferous limestones occur in the Midlands. Similar formations occur in northern France and across the lowland northern plains of Europe, where they give good yields.

On a world-wide scale, some of the very largest aquifers are listed in Table 4.1. The most extensively used formations for water supply are the following.

(4) The shallow alluvial strata, sands and gravels of Tertiary or Recent age, which are so widespread that, despite their varying yields, they form the principal source of supply in many parts of the world because of the large areas they cover and ease of access.
(5) The many areas of Mesozoic to Carboniferous age sedimentary rocks – chalk, limestones and sandstones as listed under (1)–(3) above which mostly give good yields, subject to adequate rainfall.
(6) The hard rock areas of Paleozoic to Cambrian or pre-Cambrian age including igneous and metamorphic rocks which are relatively poor yielders of water unless well

Table 4.1 Major aquifer systems world-wide

	Strata	Basin area (km^2)	Estimated reserves (MCM)	Estimated recharge (MCM/year)	Estimated use (MCM/year)	Source
Nubian Aquifer System Egypt, Libya, Chad, Sudan	Cambrian & Tertiary	2.0 m	150 m	small	460	(a), (b)
Great Artesian Basin Australia	Triassic-Cretaceous sandstone	1.7 m	20 m	1100	600 (1975)	(c), (d)
Hebei Plain China	Quaternary alluvium	0.13 m	−0.75 m	−35 000	10 000	(c)
Algeria, Tunisia, N. Sahara	L-U Cretaceous alluvium	0.95 m.	v. large	v. small	> recharge	(c), (e)
Libya	Cambrian-Cretaceous sandstones	1.8 m	24 m	small	> recharge	(f)
Ogallalah Aquifer USA	Jurassic sediments	0.075 m	0.35 m	60	3400	(g), (h)
Dakota Ss Aquifer USA	Upper Mesozoic sandstones	> 0.4 m	> 4.0 m	> 315	725	(h), (i)
Umm er Radhuma Aquifer Saudi Arabia	Tertiary-Palaeocene sediments	> 0.25 m	0.025 m	1048	n.a.	(j)

(a) Idris H. and Nour S. Present groundwater status in Egypt and the environmental impacts. *Env Geol Wat Sci*, **16**(3), 171–177, 1990.
(b) Lamoreaux P. E. *et al*. Groundwater development, Kharga Oasis Western Desert of Egypt: a long term environmental concern. *Env Geol Wat Sci*, **7**(3), 129–149, 1985.
(c) Margat J. and Saad K.F. Deep-lying aquifers; water mines under the desert. *Nature Res*, **20**(2), 7–13, 1984.
(d) Habermehl M. A. The Great Artesian Basin, Australia. *BRMJ Austr Geol Geophys*, **51**(9), 1980.
(e) De Marsily G. *et al*. Modelling of large multi-layered aquifer systems: theory & applications. *J Hydrol* **36,** 1–34, 1978.
(f) Mayne D. *The Libyan Pipeline Experience*, Brown & Root Ltd., Leatherhead, UK 1991.
(g) American Society of Civil Engineers *Groundwater Management*. 1972.
(h) Johnson R. H. Sources of water supply pumpage from regional aquifer systems. *Hydrogeol Jnl*, **5**(2), 54–63, 1997.
(i) Helgeson J. O. *et al*. Regional Study of the Dakota Aquifer, *Ground Water* **20**(4), 1982, pp. 410–414.
(j) Bakiewicz V. *et al*. Hydrogeology of the Umm er Rhaduma aquifer Saudi Arabia, with reference to fossil gradients. *Q J Geol*, **15**, 105–126, 1982.

fractured and fissured, but which must perforce be used because they occupy large areas of the continents where alternative supplies are scarce.

Over vast areas of Africa and India reliance has to be placed on the relatively small yields which can be drawn from the ancient hard rocks under (6) above. The Deccan Traps of

India,[2] for example, cover almost 0.5 million km^2 and are used by a large rural population. Across Africa almost one half of the continent is underlain by hard basement rocks, and although these provide relatively poor yields, they are the main source of underground water for the rural populations.[3,4] From such rock formations, supplies typically in the range 10–100 m^3/day per boring are sufficient for basic domestic use by small populations, but are normally inadequate for any large-scale agricultural or manufacturing development.

The advantages of groundwater are substantial. The wide area typically occupied by aquifers makes it possible to procure water close to where it is required. Many aquifers provide water that requires no treatment other than precautionary disinfection – though in many developing countries even disinfection is not adopted. The cost of a borehole is relatively modest if water is at no great depth; and the supply can be increased to meet additional demands by drilling additional boreholes so far as the local groundwater resources permit this.

But these advantages often give rise to consequent disadvantages through failure to realise the need to apply measures to conserve and protect underground supplies. For example, the large cities of Bangkok, Jakarta, Calcutta and Manila were initially able to gain supplies by drilling and pumping from boreholes close to, or even within their urban areas. But in each case the benefits have been largely reversed by urban coverage reducing the recharge of the aquifer. Subsidence can result from over-pumping. Intensive pumping of groundwater from the thick series of alluvial aquifers beneath Bangkok has resulted in their partial dewatering and consolidation, giving rise to ground subsidence at the surface.

There is often a failure to monitor a groundwater resource and the effects of overdraw remain unseen until the resource is seriously depleted. The European Environment Agency estimates that about 60% of European cities with more than 100 000 inhabitants are located in or near areas with groundwater over-exploitation, as shown by recent severe problems of supply in parts of Spain and Greece.[5] There is also commonly a failure to understand the importance of aquifer protection, which is particularly important in urban areas overlying an aquifer where poor sanitation methods or badly maintained sewerage systems result in many shallow aquifers becoming thoroughly polluted.

Sometimes the physical or chemical quality of a groundwater can act as a constraint on its use. Proximity to the sea poses a salinity hazard affecting boreholes in coastal areas and on oceanic islands. Development of freshwater resources then requires great care because, if seawater is once drawn in, it may prove difficult or impossible to reverse the process (see Section 4.15). In areas where water circulates to great depth, the groundwater which eventually emerges at the surface may be warm or even hot. Whilst air cooling methods can be applied, the warm or hot groundwater may have taken up an undesirably high concentration of minerals or gases which may be difficult to deal with.

Cessation of large scale abstractions of groundwater can bring rising groundwater levels which present many serious problems. In Paris, London, Birmingham and Liverpool, rising groundwater levels are occurring because of reduced industrial demand for water consequent upon the demise of the major water-consuming industries and their replacement by other types of manufacturing which have no large water demand. These rising groundwater levels threaten the flooding of tunnels and deep basements, can cause chemical attack on the structural foundations of buildings, and decrease the ability of drainage systems to dewater surface areas, Similar rising groundwater levels can occur where excessive irrigation is applied or where large scale river impounding schemes for hydro-power development or irrigation raise water levels upstream, causing soil salinisation and sometimes land-slope instability.

4.2   Yield uncertainties and types of abstraction works

The yield of a well or borehole is dependent on the following.

- The aquifer properties of the strata from which the water is drawn, and the thickness and extent of the aquifer and area of its outcrop.
- The extent to which the well or borehole intercepts a water storing and transmitting network of fissures, cracks and open bedding planes in the aquifer.
- The depth, diameter and construction details of a borehole or well, and the screen type or gravel packing used.
- The extent to which the abstraction affects other water users drawing water from the same aquifer or using surface water fed by spring discharges from the aquifer.

Uncertainty arises from the fact that any borehole intercepts only a small volume of the strata so that, in aquifers where the water is only able to move through open fissures or bedding planes, there is always a possibility that a hole intersects no fissures or planes large enough to give a good supply. Many cases can be quoted where a boring in one location gives a poor yield, and a second boring only a few feet away provides a yield four or five times as much. Clearly the more fissures or cracks that exist, the more is it likely that a reasonable yield will be obtained. Thus boreholes are favoured in the well-fissured layers of the Upper Chalk or in the looser Pebble Beds of the Bunter Sandstone. In loose gravels and sands there is less uncertainty as to the yield of a boring.

   Estimation of the probable yield of a proposed underground development is difficult. A knowledge of the hydrogeology in the area, and records of what other boreholes or wells in the same or similar formation have yielded, can be of help. But where little is known about the hydrogeology of a proposed borehole site, it is advisable to sink a 'pilot hole' on the site, usually about 150 mm diameter, to provide information from the samples withdrawn. If the results seem promising and it is decided to adopt the site, the pilot hole can be reamed out to form a borehole large enough to accommodate a pump. However it must be borne in mind that, if the small pilot bore appears to give a good yield for its size, this may not be a reliable indicator that a larger yield should be obtainable from a larger hole in the same or at a nearby location. A suggested method of approach for locating a borehole for optimum yield, using mapping data from lineations, bedrock geology, vegetation and drainage to produce probabilities of yield, and used to assist rural groundwater development in poor aquifers, has been described by Sander.[6]

Need for hydrogeological survey

Where an aquifer is already supporting abstractions and feeds rivers and streams flowing through environmentally sensitive and recreationally important areas, it can be essential to conduct a full hydrogeological investigation of an aquifer before a new abstraction is proposed. An environmental assessment of the possible effect of a proposed new abstraction for public supply is often a requirement of an abstraction licensing authority or government department. A hydrogeological investigation may also be needed to make sure possible sources of pollution on the catchment, such as leachates from solid waste tips or abandoned contaminated industrial sites, will not render water from a proposed abstraction unsuitable for public supply. The work involved in such an investigation may be extensive and is best carried out by an experienced hydrogeologist who will be familiar with the range of geophysical and other techniques available (see Section 4.7).

Types of abstraction works

The range of abstraction works which are used for various different ground conditions include the following.

(1) Boreholes can be shallow or deep, sunk by different methods, and have many different screening methods used to prevent the ingress of unwanted fine material and keep the bore stable, or to draw upon the best quality or quantity of water available underground.
(2) Large diameter wells, 15–25 m deep, can have a boring sunk from their base to considerable depth to intercept some main water-bearing strata, the well lining being sealed to prevent entry of surface waters.
(3) Although not built now, many older wells have an adit or adits driven from it (see Plate 1); an adit being a small (1 m wide by 2 m high) unlined tunnel driven at some level below ground surface where it is expected that more water bearing fissures or bedding planes will be intercepted. Some adits have been several kilometres long.
(4) A 'well field' can be adopted, i.e. the sinking of several moderately sized boreholes, spaced apart in some pattern, their yields being collected together. The system is used to develop a good yield from an area of an aquifer where a single well could not be expected to give a large enough yield.
(5) Collector wells and galleries of various design using porous or unjointed pipes sunk in river bed or in river bankside deposits of sand and gravels can be used for abstraction of shallow groundwater, or galleries can be driven into a hillside to tap the water table (see Section 4.16).

These methods of development are dealt with in more detail in the following Sections.

4.3   Potential yield of an aquifer

It is often necessary to quantify the limiting yield of an aquifer for water supply purposes. Formerly this used to be taken as equal to the long-term average recharge from rainfall percolation, provided the storage capacity of the aquifer is large enough to even-out year to year variations of the recharge. But further studies of climate variability have shown there can be long runs of years of below-average recharge so that aquifer yields rarely exceed 90% of the long average recharge. Additionally, however, an allowance may have to be made to prevent the reduction of aquifer-fed spring flows causing environmental damage. As rough guide, therefore, abstraction of about 75% of the average long-term recharge is a safer estimate of the likely maximum sustainable development.

To assess the possible yield of an aquifer in more detail, the groundwater catchment must first be defined from the contours of the water table. It is usually assumed that the water table reflects the ground surface to a reduced scale, but variations from this can occur in asymmetric scenery containing features such as escarpments, or where one valley has cut down deeper than its neighbours. It is important to check this, because the underground water table catchment 'divide' may not coincide with the topographical surface catchment divide above. It has to be assumed that underground flow is in the direction of the major slope of the water table. This may not always be true, for instance where there are karst limestone fissures; but it is best to follow the general rule in the absence of any evidence to the contrary.

Assessing the amount of recharge

Sometimes the catchment outcrop may be remote from the point at which the wells tap the strata concerned. The average recharge is estimated as average rainfall minus evapotranspiration over the outcrop area (see Sections 3.6–3.8). Sometimes clay covered areas of an aquifer discharge surface runoff to rivers which, in turn, leak a certain amount of recharge back to the groundwater through the river beds. This amount is difficult to estimate and can be near guesswork until there is a long experience of successful pumping to confirm some mathematical model of the aquifer hydrology. Fortunately, recharge from river bed leakage is usually only a significant amount in arid areas. In temperate wetter areas the hydraulic gradient between the water level in the river and the adjacent groundwater table level is usually too small to cause significant river bed leakage.

To assess aquifer recharge, percolation formulae[7] can be of use, but they need confirming before being used outside the area for which they were derived. Examples from semi-arid areas like Jordan or Western Australia are based on a simple percentage of long average rainfall, say 3 or 5%. Research by tracking the tritium contents of chalk-limestone pore water and fissure water has demonstrated that, whereas water passing through fissures may travel down to the water table at more than 0.3 m/day, the pore water recharge front may move down at only 1–2 m/year. Distinguishing between the volumes travelling by these alternative routes is fraught with difficulty because it depends on the size of the recharge event and the possibility that pore water will drain out to fissures during any major drawdown of the water table. A more helpful analytical approach lies in recession analysis of dry-weather flows from groundwater catchments, as the next paragraph describes.

River flow recession curves can be used to determine that part of river flow, termed 'base flow' which is fed from underground aquifer storage (see Section 3.12). In prolonged dry weather the natural flow of a river will comprise only aquifer drainage through springs. At any instant of time, spring flow, Q, is related to the volume of stored water, S, in the aquifer by the relationship:

$$Q = kS$$

If Q_o is the spring flow at time $t = 0$, and Q_t is the flow at later time, t,

$$Q_t = Q_o e^{-kt}$$

where k has the unit day^{-1} if t is in days. (Typical values for k lie between 0.01 per day in a good aquifer, to 0.10 per day in a relatively impermeable aquifer.)

Consequently if log Q_t is plotted against t (days) for a prolonged dry period, this should give a straight line of slope k, as shown in Fig. 4.1. If a single long period free of rain is not available for analysis, it is possible to link together shorter dry period flow recessions which, plotted as shown in Fig. 4.2, will be asymptotic to the natural recession curve. Section 3.12 shows how this recession curve can be used to estimate the base flow for a given period of record.

Base flow is percolation routed through storage. Hence by summing the baseflows over a period and adding a correction for the change of storage between the beginning and end

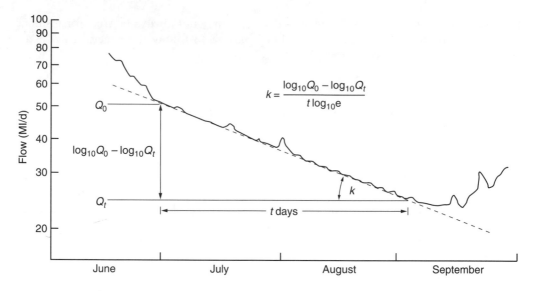

Fig. 4.1　Groundwater recession graph.

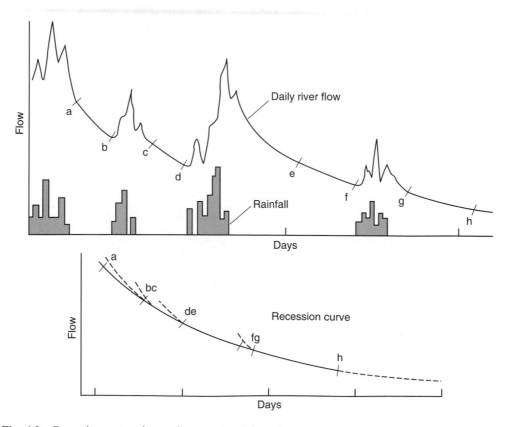

Fig. 4.2　Recession curve drawn from parts of daily flow record.

of the period the percolation for the period can be estimated. The correction for change of storage is obtained by taking the change of water table level between the beginning and end of the period and multiplying this by the aquifer storage coefficient (see Section 4.4). Ranked and plotted on probability paper (Fig. 4.3), these seasonal percolation values can be used to estimate recharge probabilities.

Percolation gauge or lysimeter drainage readings (see Section 3.6) can also be used, but it is difficult to be sure they represent actual average catchment conditions. Their rims prevent surface runoff and, being on level ground, local runoff does not occur so the measured percolate may be an optimistic estimate of the amount that reaches the water table under natural conditions. Practical difficulties occur when such gauges are kept in use for a long period. They can become moss covered, or the soil may shrink away from the edge of the container. Alternatively percolation can be estimated by a soil moisture storage balance method of the type demonstrated by Headworth.[8] Once the readily available moisture in the root zone is used up by evapotranspiration in a dry period, the subsequent build-up of soil moisture deficit must be made up before excess rainfall can percolate to the aquifer once more. As Fig. 4.4 shows, the average percolation produced by this method depends strongly upon the amount of moisture stored within the root zone. These storage or 'root constant' values are known for several crops, ranging from 25 mm for short-rooted grassland, to over 200 mm for woodland. Figures such as Fig. 4.4 require specific calculation for each climatic regime, preferably with daily rainfall data.

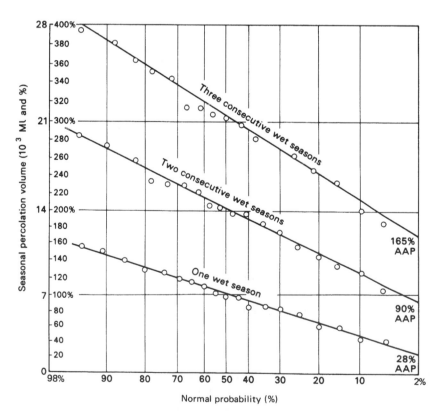

Fig. 4.3 Seasonal percolation probability plot.

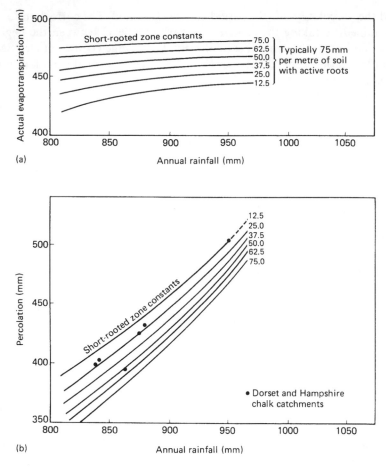

Fig. 4.4   Relationship between rainfall and (a) evapotranspiration and (b) percolation using various root constants (after Headworth[8]).

Yield constraints

Having found the potential yield of a groundwater catchment, it is possible to compare it with the total authorised (or licensed) existing groundwater abstractions. This gives the upper limit to what extra may be obtained, or, if too much is already being taken, the extent of groundwater mining that must be taking place. It should be noted, however, that mining can occur in one part of an aquifer whilst elsewhere springflows may indicate no mining is apparent. This indicates low transmissivity within the catchment or parts of it, suggesting a better siting of abstraction wells or a different approach to estimating the maximum yield.

  Where a group of wells exists within a catchment, the average maximum drawdown they can sustain without interfering with their output can be known. Using the average storage coefficient, S', for the aquifer supplying the wells, the amount of storage they can draw upon to reach this drawdown can be estimated. From this it is possible to calculate the minimum yield for various time periods for a given severity of drought, i.e.

> Abstraction possible = storage available for given pumping level;
>
> *plus* recharge during the given time period;
>
> *less* loss from springflows.

This calculation would be carried out for, say, a 2% drought for periods of 6–7 months (1st summer); 18 months (two summers + one winter); and so on. This can reveal what is the length of the dry period which causes the lowest yield, and whether pumps are sited low enough to average out percolation fluctuations.

Seawater intrusion into wells close to the sea may limit abstraction. The pumping has to be kept low enough to maintain a positive gradient of the water table to the mean sea level. However, since seawater is 1.025 times denser than freshwater, under equilibrium conditions the freshwater–saltwater interface is $40h$ below sea level, where h is the difference in height between the freshwater surface level and the sea level. Thus if the rest water level in a well is 0.15 m above mean sea level, the saline water interface will be about 6.0 m below, or 5.85 m below mean sea level. (In practice the interface will be a zone of transition from fresh to salt water rather than a strict boundary.) This means a pump suction can be sited slightly below mean sea level without necessarily drawing in seawater; but this is heavily dependent on the aquifer's local characteristics, because a stable interface may only be formed in certain types of ground formations. A more reliable policy is to use coastal wells conjunctively with inland wells. In the wet season the coastal wells are used when the hydraulic gradient of the water table towards the sea is steepest, reducing the risk of drawing in seawater. In the dry season the inland wells are used. By this method a higher proportion of the water that would otherwise flow to the sea is utilised than if only the coastal wells were used.

Small low-lying oceanic islands face special seawater intrusion problems described in Section 4.15.

4.4 Assessment of aquifer characteristics

The two principal characteristics of an aquifer are its horizontal transmissivity, T, which is the product of its permeability times its wetted depth; and the storage coefficient, S. Transmissivity, T, is the flow through unit width of the aquifer under unit hydraulic gradient. Its units are therefore m^3/m per day, often abbreviated to m^2/day.

The storage coefficient, S, is defined as the amount of water released from an aquifer when unit fall in the water table occurs. Where free water table conditions occur it is the volume of water released from a unit volume of an aquifer (expressed as a percentage of the latter) that will drain by gravity with a unit fall of the water table level. But when the aquifer is confined under pressure because of some impervious layer above, it is the percentage of unit aquifer volume that must be drained off to reduce the piezometric head by unit depth. The difference between these two meanings, although subtle, is vital. Whereas the former may be in the range 0.1–10.0%, the latter may be 1000 times smaller (demonstrating the incompressibility of water).

By considering a well as a mathematical 'sink' which creates a cone of depression in the water table, it has been shown by Theis[9] that drawdown in a homogeneous aquifer due to a constant discharge, Q, initiated at time $t = 0$ is

$$h_o - h = \frac{Q}{4\pi T}\left(-0.5772 - \log_e u + u - \frac{u^2}{2.2!} + \frac{u^3}{3.3!} - \frac{u^4}{4.4!} + \dots\right)$$

$$= \frac{Q}{4\pi T} W(u) \tag{1}$$

where $u = r^2 S/(4Tt)$, h_o is the initial level and h is the level after time t in a well distance r from the pumped well. Any consistent set of units can be used. For example, if Q is in m^3/day and T is in m^2/day, then h_o and h must be given in metres. S is a fraction. $W(u)$, the 'well function' of the Theis equation, can be obtained from tables.[10] Although both the Theis equation and the Jacob simplification of it (see below) are derived for a homogeneous aquifer they are found to work in well-fissured strata, such as Upper and Middle Chalk. The essential need is to be able to assume reasonably uniform horizontal flow on the spatial scale being considered, together with the absence of any impervious layer that interferes with drawdown.

In the Theis method of solution a type curve of $W(u)$ against u is overlaid on a plot of the pump test drawdowns versus values of log (r^2/t). Where a portion of the type curve matches the observed curve, coordinates of a point on this curve are recorded. With these matchpoint values the equations can be solved for S and T.

However Jacob's less exact method[11] is easier to apply and meets most situations that confront an engineer. He pointed out that if time t is large, as in most major pumping tests, then u is small (say less than 0.01), and so the series in the Theis equation can be shortened to:

$$h_o - h = \frac{Q}{4\pi T} - 0.5772 - \log_e(r^2 S/4Tt)$$

$$= \frac{Q}{4\pi T} 2.30\log_{10}(4Tt/r^2 S) - 0.5772 \qquad \text{Note (a)}$$

$$= \frac{2.30Q}{4\pi T} \log_{10}(2.25Tt/r^2 S) \qquad \text{Note (b)} \tag{2}$$

(a) $2.30 \log_{10} = \log_e$ and inverting the log changes the sign.
(b) anti-$\log_e(-0.5772) = 1/1.78$ and $4/1.78 = 2.25$

Plotting drawdown against time at an observation borehole within the cone of depression thus produces a straight line (Fig. 4.5).

If the drawdown for one logarithmic cycle of time is read off, the value of $\log_{10} (2.25 \, Tt/r^2 S) = 1.0$ in eqn (2).

Hence

$$h_o - h = 2.30Q/4\pi T \times 1.0 \text{ where } h_o - h \text{ is in metres and } Q \text{ is in } m^3/\text{day}.$$

i.e. $T = 2.30 \, Q/4\pi(h_o - h)$

Then reading off the time intercept, t_o days, for zero drawdown

$$S = 2.25Tt/r^2 \text{ where } T \text{ is in } m^2/\text{day and } r \text{ is in metres}.$$

Certain qualifications apply as follows.

(1) If the regional water table rises or falls as a whole during the test then the drawdown should be adjusted by the equivalent amount.

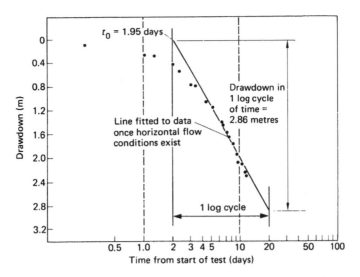

Fig. 4.5 Jacob's pump test analysis for the test pumping of a well at 4000 m^3/day. Data are plotted from an observation well 200 m from the test wall.

Hence $T = \dfrac{2.3 \times 4000}{4\pi \times 2.86} = 256$ m^3/day metre, and $S = \dfrac{2.25 \times 256 \times 1.95}{200^2} = 0.028$ or 2.8%

(2) Early time data should not be used because there will be substantial initial vertical flow as the storage is evacuated. Boulton suggests the necessary horizontal flow conditions will exist when $r > 0.2$ days and $t > 5dS/K_t$ days; where d is the wetted aquifer depth and K_t is the vertical hydraulic conductivity which can be taken as T/d if K_t is not otherwise known. (In horizontally layered strata K_t may be one-third or less of the horizontal conductivity.) Some reiteration with values of S and T is required to use this guide.

(3) Where the pumped well only partially penetrates the aquifer,[12] it will be necessary to adjust the drawdown values for the non-standard flow lines that then ensue unless the observation well is at least 1.5 times the aquifer depth away from the source. Rather than make complex adjustments it is preferable to use the Jacob rather than the Theis method on a long-term pumping test.

(4) Where drawdown is large compared with the aquifer thickness and the aquifer is unconfined, the measured drawdowns should be corrected[11] by subtracting from them

(drawdown)^2/2 × wetted aquifer depth

as shown in Table 4.2 which gives a layout for pumping test results.

Analyses for S and T can also be made during the recovery part of the test at an observation bore. If only levels in the pumped well are obtainable, the information that can be achieved is usually limited to an estimate of T on recovery.[13]

Once S and T are established it is possible to predict with the above equations what drawdown below current rest water level will result at different pumping rates, different times and other distances. Where more than one well can create a drawdown at a point of interest, the total effect can be calculated by the principle of superposition, i.e. the drawdowns due to individual well effects can simply be added. Analytical solutions[10,12] exist for many aquifer

Table 4.2   Layout of pumping test results

Average pumping rate (Q)_____

Results from observation well No. _____ Radial distance (r) _____

Aquifer thickness _____

Regional water table change during test, δh_o per day _____

(1)	(2)	(3)	(4)	(5)	(6)	(7)	(8)
Time from start of ppg t	r^2/t	Level in well h above datum h_o	Drawdown $S_1 = h_o - h$	Correction for region-change water table $\delta h_o \times t$	Corrected drawdown Cols $(4) \pm (5)$	Correction for large drawdown $\dfrac{(\text{Col. 6})^2}{2d}$	Final drawdown Cols $(6)-(7)$

conditions, including boundary effects from impermeable faults and recharge streams. Care is needed to adopt the solution appropriate to the lithology and recharge boundary.

It will be appreciated that the engineer has no control over the values of S and T found at a well site. The water drawn from the hole may come from local aquifer storage after a long residence time. However, resiting, deepening or duplicating a bore are all options that may be called upon once S and T are known.

4.5   Groundwater modelling

Early groundwater movement models used an electrical analogue of Darcy's law of groundwater flow which states:

Flow $Q = Tiw$

where T is transmissivity, and i is hydraulic gradient through an aquifer cross-section of width w. Transmissivity was defined in Section 4.4 as flow per m width, i.e. m^3/m per day under unit hydraulic gradient. With the introduction of computers, mathematical modelling of groundwater flows became possible. The most common models use the finite difference or finite element approach. Under the former, and assuming flow is near enough horizontal, a grid is superimposed on a plan of the aquifer to divide it into 'nodes'. Between nodes the flow is related to the hydraulic gradient and the transmissivity of the aquifer in directions 'x' and 'y' (the transmissivity sometimes being taken the same in either direction). The finite difference method adjusts the calculations so that 'boundary conditions' between each flow stream and its neighbours match.

Equations (with many terms) can therefore be set up connecting node-to-node flows and head changes, and to which overall limiting boundary conditions apply, such as the assumed (or known) upstream initial head applying, the lateral boundaries of the aquifer, and the downstream conditions applying, such as the outcrop of the aquifer. The equations can then be solved by the computer to match the boundary constraints by reiterative methods or matrix formulation, to a specified degree of accuracy, resulting in the computer providing water table contours and field flows. Three-dimensional flows and both transient and steady-state condition of flows can be dealt with, together with such matters as consequent aquifer-fed spring flows, the effect of pumped abstractions and so on.

There are many computer models available for simulating groundwater flows, and a resume of their characteristics and capacities is given by Maidment.[14] MODFLOW developed by US Geological Survey in 1988 is quoted as 'popular and versatile'. It can deal with two- and three-dimensional flows and incorporates numerous ancillary facilities. Another useful model is PLASM, originally written for the Illinois State Water Survey and described as 'recommended as a first code for inexperienced modellers because it is interactive and easiest to operate'. Use of a groundwater model requires a sound knowledge of hydrogeology and some practical experience of modelling techniques. Also in order to obtain the necessary data for the model, a hydrogeological investigation of the aquifer is essential (see Sections 4.2 and 4.7).

4.6   Test pumping of boreholes and wells

The aim of test pumping is not only to find how much water a well or borehole will presently yield, but also to find:

(a) the effect, if any, of the pumping on adjacent well levels, spring and surface flows;
(b) the sustainable amount that should be abstractable through possible dry periods of different severities;
(c) the drawdown/output relationship whilst pumping in order to decide the characteristics required for the permanent pumps installed;
(d) the quality of the water abstracted;
(e) data sufficient to derive estimates of the key characteristics of the aquifer penetrated, i.e. its transmissivity and storage coefficient.

An initial problem is to decide what size of pump to use for test pumping. If the pump's maximum output is much less than the well is capable of yielding, the test will not prove how much more water the well could give and what effect this would have on adjacent sources. On the other hand a test pump with an output larger than the well can yield, means a waste of money. A major cost of test pumping is the temporary discharge pipeline required. For a large output the discharge line may need to be laid a considerable distance to ensure there is no possibility of the discharged water returning to the aquifer and affecting the water levels in the test well and in other wells kept under observation.

The test pump size has to be estimated from experience in drilling the well and on what other holes in similar formations have given. Water level recovery rates after bailing out (see Section 4.9) and upon stopping drilling are important indicators of possible yield; the drill cores can indicate where well fissured formations have been encountered, and an experienced driller should be able to notice at what drilling level there have been signs of a good ingress of water to the hole. The most usual type of pump used for test pumping is the electrically driven submersible, its output being adjusted by valve throttling. In rare instances a suction pump (i.e. one above ground having its suction in the well) may be used if the water level in the well is very near ground level; and for small holes an airlift pump is possible and cheap (see Figure 4.6) although rarely used. The discharge must be accurately measured and continuously recorded. A venturi meter, orifice meter, or vane meter can be used for measuring discharge; or for best accuracy the discharge may be turned into a stilling tank equipped with a V-notch measuring weir outlet with a float recorder to log the water level over the notch. The advantage of a stilling tank and weir is that it provides a nearly constant outlet head for all outputs and is a positive method of measurement. If

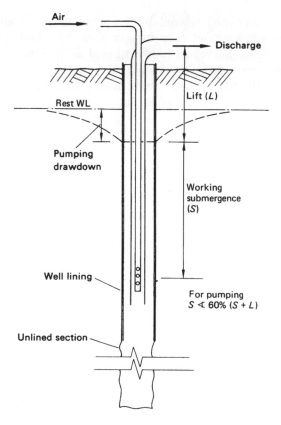

Fig. 4.6   Air lift pump.

meters in the discharge pipeline are used, the pump outlet head will vary with the flow making it more difficult to maintain a constant test pump output, and the meter accuracy must be checked before test pumping and at least once during the test.

Test pumping regimes

Every effort should be made to carry out the test pumping when groundwater levels are near their lowest seasonal decline towards the end of the dry season. Whilst dry weather during the testing is an advantage, it cannot be guaranteed in a variable climate, and some rainfall during the testing does not invalidate the observations taken, provided it is not extreme. Test pumping outside the dry season when groundwater levels are high should be avoided if at all possible, because it gives uncertain estimates of the yield during the dry season.

The test pumping is usually carried on for at least 3 weeks. Three stages of increasing output, such as 1/3rd, 2/3rds, and full output are tried first without stoppage, each for at least 24 hours. The pump is then stopped and the recovery rate of the well water level is carefully measured. When sufficient of the recovery curve against time has been obtained to show it's trend towards starting rest level, the pump should be restarted at maximum output and kept at that rate for 14 days, the recovery again being measured when pumping is stopped.

If the maximum pump output is greater than the well can yield, there will be a continuous dropping water level, and the output must be throttled back in an endeavour to get a steady drawdown. In other cases the output, after causing an initial drop of water level, may either show a continuous slow decline of water level as pumping continues, or a diminishing decline showing a trend towards some maximum drawdown for the given output. It is not always possible to reach stable conditions within a 14-day pumping test.

Problems can occur. If the pump fails in the middle of the test, the important thing then is to get the well water level recovery recorded. After re-starting the pump, the output should be adjusted to that previously. It is not easy to keep the pump output constant because it will reduce as drawdown increases, and fluctuations in the voltage of the electrical supply to the pump can cause quite marked changes of pump output. Adjustment of the pump output by valve operation has therefore to be done carefully, to avoid over-adjustment and re-correction which will cause fluctuating water level readings difficult to interpret. Measuring the water level accurately in a pumped borehole can present difficulties. Using an electrical contact device to detect the water level must be done manually. An air pressure measuring device sited below the water level is not very accurate. Using a submerged electronic pressure transducer can record levels, but it must be calibrated *in situ*.

The pumping water levels and pump outputs should be plotted as shown in Fig. 4.7 from which a 'type curve' of output against drawdown can be derived. It is also useful to replot the stepped drawdown results as a Bruin and Hudson curve[15] as shown in Fig. 4.7. This can reveal the proportion of the drawdown due to the characteristics of the aquifer,

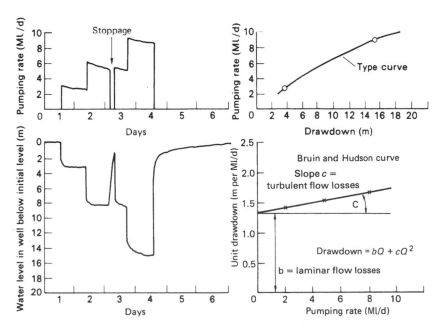

Fig. 4.7 Step-drawdown test results (Bruin J. and Hudson H. E. Selected Methods for Pumping Test Analysis. Report 25, Illinois State Water Supply, 1955).

and the proportion due to hydraulic characteristics of the well or borehole, the latter being distinguished by the turbulent flow losses, the former by the laminar flow losses. Examination of the turbulent loss coefficient, c, can sometimes suggest high entry losses to the well due to poor design of the well screens or gravel packing. Good (i.e. low) figures for c in cQ^2, where Q is in m^3/min, are below 0.5. The laminar loss coefficient, b, describes the relative permeability of the aquifer(s) feeding the well.

Use of observation wells and monitoring

To determine aquifer characteristics (see Section 4.4) it is desirable to have two observation boreholes sited within the likely cone of depression of the water table about the borehole when test pumping takes place. The radial distance of these holes from the well to be tested should be such that:

(a) the water level drawdown they experience as a result of the test pumping is large enough to give a reasonably accurate measure of the drawdown;
(b) the drawdown amount is small relative to the depth of saturated aquifer which contributes flow to the pumped well.

The second requirement is needed to ensure that the flow through the aquifer at the observation bore site is essentially horizontal. Experience has shown that these requirements frequently result in observation holes being sited between 50 and 200 m of the pumped well. The larger this distance is, the greater is the 'slice' of aquifer brought under observation. Similarly more information will be gained if the two observation holes are on radials 90° apart. The radial distance of each bore from the well should be measured.

Water levels in these observation bores should be measured regularly before during and after test pumping, the frequency being increased after starting, stopping or altering the pump output, sufficient to log rapid change of water level in the observation bore. Preferably a water level recorder should be used. An existing well, such as a household well, can be used for observation purposes, provided the amount taken from it is too small to affect its standing water level. Usually, however, no conveniently placed existing well will be available and, if observation bores are required, they will have to be drilled. The expense of this often means that such observation boreholes are omitted, so reliance has then to be placed solely on results of the pumped borehole and the effect, if any, on existing wells and boreholes nearest to the pumped site.

It is also essential to monitor water levels in all wells and ponds, and the flows from springs and in streams, that could conceivably be affected. The area to be covered by such monitoring is a matter of judgement according to the particular circumstances, but would often be about 2 km radius about the test pumping site. But all important abstractions that could be sensitive to flow diminution, such as watercress beds and fisheries, should be monitored even if they appear to lie outside any possible range of influence of the proposed abstraction. The paramount need is to have proof they were not affected by the test pumping. The measurements of key indicator wells and springs may be started 6–9 months before the planned timing of the test pumping. Temporary weirs may be needed to measure springs and stream flows. On streams or rivers, measurements of the flow both upstream and downstream of the test pump site may be necessary. The aim is to get a sufficiently long record of water table levels and stream flows that their trend before and after the test pumping can show whether there is any interruption or change caused by the test pumping.

There is also a need to check that the underground water table 'catchment area' which contributes to the test pumping is the same as the topographical catchment. Increased abstraction from one underground catchment area can cause the water table divide to migrate outwards, reducing the area of an adjacent catchment feeding other sources which will, therefore, at least notionally suffer a reduction of potential yield.

Daily rainfall measurements must be taken also, if there are no existing raingauge stations to give the catchment rainfall. The work of monitoring is essential. In most countries, concerns about the need to preserve the natural aquatic environment and the rights of existing abstractors, mean that authorisation to abstract more underground water will only be obtained if the engineer is able to show, from records taken during test pumping, that no unacceptably harmful effect will be caused.

4.7   Geophysical and other investigation methods

Samples of strata will be collected during the sinking of a borehole, but the samples will inevitably be disturbed, and the groundwater quality will also be disrupted so that samples may not be properly representative of aquifer water quality. Any pumping from the bore will only give a mixture of all the flows entering the bore. To gain more precise information, a wide variety of subsurface 'down-hole' geophysical methods of exploration are valuable. Direct physical observation or measurement can be obtained by:

- use of a television camera to view the walls of a boring;
- using a calipering device to measure and record bore diameter with depth;
- use of recording instruments to obtain profiles of water temperature, pH and conductivity with depth;
- use of small sensitive current meters to detect differences of vertical flow rates in the boring, thus indicating changes in inflow or outflow rate from strata.

In addition electrical and nuclear instruments can provide information concerning the strata penetrated. Resistivity measurements taken down the hole, using an applied electrical potential through probes, can locate the boundaries of formations having different resistivities. This can aid identification of the type of strata penetrated and also permit fresh and saline waters in the formation to be distinguished. Measurement of the 'self' or 'spontaneous' electrical potential existing between strata at different depths of the formation can be of assistance in detecting permeable parts of the aquifer. Nuclear downhole tools used are mainly (in water) of three types.

- Natural gamma detectors which pick up natural radioactivity from potassium in clays (so correlating with clay content).
- Gamma-gamma tools which use a gamma radiation source to bombard the formation and measure back-scattered radiation produced, which is inversely proportional to the formation density and can be used to indicate degree of cementation or clay content also.
- Neutron tools which are used to give an indication of water content and porosity.

All these measures, co-ordinated and supplementing each other, can give extensive additional information about an aquifer. In particular they can indicate best locations for inserting well screens or gravel packs (see Section 4.8). More extensive information is available in IWES Manual No. 5.[16]

An additional method of investigation is to use packers down a boring to isolate individual aquifers contributing water to a borehole. Many aquifer systems comprise several separate aquifers. This permits measurement of the piezometric head in each aquifer and withdrawal of water samples for analysis. Also potential rates of inflow from different aquifers can be compared. Various multi-tube assemblies can be used to withdraw samples of water from the isolated aquifers and, if inserted in an observation borehole, they can remain in place for subsequent monitoring purposes. The production bore can, of course, be similarly investigated before a pump is installed. The method can reveal whether it is advisable to seal off inflows of undesirable quality (including signs of contamination), or which come from formations whose catchment outcrop is known to include potential sources of pollution best avoided.

4.8   Borehole linings, screens and gravel packs

The upper part of a boring is usually lined with solid casing which is concreted into the surrounding ground. This is to prevent surface water entering and contaminating the borehole and to seal off water in the upper part of the formation which may not be of such good quality as that obtained from the main aquifer to be penetrated.

Where a boring encounters weakly cemented or uncemented sands and gravels, it may be necessary to install perforated or slotted lining, designed to hold back such formations, permitting fine particles adjacent the screen to be washed out, to stabilise the formation and improve the yield of the borehole. There are many different types of screens; the simplest comprise slots cut in borehole casing, whilst the most sophisticated are stainless steel, wedge wire wound screens with accurately set apertures. In non-corrosive waters screens are usually made of steel; in corrosive waters, plastic coated steel, phosphor bronze, glass reinforced plastic, or rigid PVC slotted or perforated screens may be used. Figure 4.8 shows the aperture and percentage open area for some typical 300 mm diameter screens. The screen aperture required has to be decided according to the size of particles in the formation (see below). Because of unavoidable partial blocking of the screen by sand particles, the effective area of a screen is usually estimated at less than half its initial open area. The water entry velocity through the apertures needs to be limited to about 30 mm/s to avoid high turbulence losses. Working from the required yield, the open screen area and the amount of blockage that might occur over time, the length of screen necessary can be calculated.

If the apertures or slots of a well screen are suitably sized, withdrawal of the fines from the formation may result in the coarser material forming a 'naturally graded pack' against the screen which prevents further withdrawal of fines when the borehole is pumped. A formation is considered capable of forming a natural pack if the D10 size (aperture size through which 10% of the material by weight passes) is greater than 0.25 mm, and the D90 size (size through which 90% passes) is less than 1.0 mm, and the uniformity coefficient is between 3 and 10.

$$\frac{\text{Uniformity}}{\text{coefficient}} = \frac{\text{aperture size through which 60\% of material passes (by weight)}}{\text{aperture size through which 10\% of material passes}}$$

During test pumping and development of the boring, abstraction rates should be higher

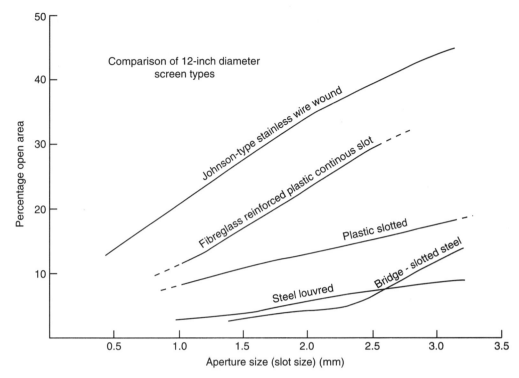

Fig. 4.8 Comparison of screen open areas.

than will occur during the normal permanent pumping so that no fines are withdrawn with normal rates of pumping.

Where the grading of the formation material is unlikely to form a 'natural pack', a gravel pack would be placed outside the screen. The screen is suspended centrally in the borehole by means of lugs fitted at intervals on the outside, and the gravel is tremied into the annular space. The gravel pack can be of uniform sized gravel or graded gravel. Uniform-sized packs are suitable when the Uniformity Coefficient of the formation material is less than 2.5, in which case the 50% size of the gravel pack should be four to five times the 50% size of the formation material.[17] Otherwise a graded gravel pack should be used, its grading being selected to parallel the grading curve of the formation material so the Uniformity Coefficients are similar, and the D15 size of the pack should be four times the D15 size of the coarsest sample from the formation, with the D85 size of the pack less than four times the D85 size of the finest formation material sampled.

The minimum thickness of the gravel packing should be 75 mm. The placing of a thinner pack becomes uncertain because of imperfect verticality of the boring, variations in its diameter, and the problem of placing the gravel evenly in the annulus at depth. Even with the use of a tremie pipe there may be zones where the pack is virtually absent, and others where the thickness is greater than necessary. Part of the development of a newly sunk borehole may involve the need to remove 'mud cake' from the formation face, and a thicker pack tends to reduce the ability to remove the cake. There are many methods for selecting an appropriate gravel pack[18,19] each usually for a particular type of aquifer formation.

In loose sandy formations, such as Greensand, the need for a gravel pack has to be taken into account when deciding on the starting diameter for a boring. The expected water pumping level is also an important factor to be taken into account because this determines where the pump is to be sited, and consequently the diameter required down to pump level to ensure the pump can be accommodated with reasonable flow conditions to its suction. Although it is possible to ream out a borehole to a larger diameter in some cases, it may not be practicable in a loose formation where solid or slotted lining has to be driven down closely following the boring tools to prevent collapse of the hole.

4.9   Construction of boreholes and wells

Borings for public water supply are usually of large diameter, so methods for sinking them differ from the smaller diameter holes sunk for oil drilling or for the water supply to a few houses. The percussive method is widely used. A heavy chisel (see Plate 1) is cable suspended and given a reciprocating motion on the bottom of the hole by means of a 'spudding beam' which alternately shortens the cable and releases it at a rate which varies with the size and weight of chisel used. Considerable skill is required by the operator to adjust the rate of reciprocation to synchronise with the motion of the heavy string of tools down the hole and avoid any violent snatch on the cable. This cable has to be paid out gradually and kept exactly at the right length for the chisel to give a sharp clean blow to the base of the hole. The wire cable is left hand lay and, under the weight of the string of tools tends to unwind clockwise, thus rotating the chisel fractionally with each blow. After a while the unwinding builds up a torque so that, when the weight is off the cable just after a blow on the base of the hole, the torque breaks a friction grip device in the rope socket attachment and the cable returns to its natural lay. Thus slow rotation of the chisel continues. Every so often the string of tools must be withdrawn and a bailer – a tube with a flap valve at the bottom – is lowered to clear the hole of slurried rock chippings (see Plate 2). Progress is therefore slow and, dependent upon the hardness of the formation, the drill chisel will require resharpening from time to time.

Another percussive method is use of a 'down the hole' hammer operated by compressed air. This is up to ten times faster in hard rock than a cable-operated chisel, but it is unsuitable for soft formations and cannot be used in substantial depths of water. The chippings have to be ejected to the surface by the exhaust air from the hammer passing through the annular space between the drill rods and the borehole. Unless the boring is of small diameter there may be difficulty in raising the chippings, in which case an open topped collector tube may be positioned immediately above the tool to collect the chippings. The down-the-hole hammer is predominantly used on small boreholes up to about 300 mm diameter (see Plate 1).

For rotary drilling of large holes in hard material a roller rock bit is used, equipped with toothed cutters of hard steel which rotate and break up the formation (see Plate 1). Water is fed to the cutters down the drill rods and, rising upwards through the annular space, brings the rock chippings with it. In the 'reverse circulation' method the water feed passes down through the annular space and up through hollow drill rods. This achieves a higher upward water velocity making it easier to bring chippings to the surface. Air is sometimes used instead of water, but the efficiency falls off with increase of depth below water. Heavy drill collars may have to be used to give sufficient weight on the rock cutters. Also a large quantity of water may be required for the drilling of a large diameter hole. Rotary rock bit drilling is widely used by the oil industry, usually for holes of substantially smaller diameter than those needed for public water supply. Clay or bentonite drilling fluids can be used to support unconsolidated formations and assist in raising chippings. However, the use of such suspensions is not always advisable in water well drilling in case they seal off water bearing formations. Instead polymer based, low solids, biodegradable muds are preferred since these are not so likely to seal off water bearing formations.

Diamond core drilling is seldom used for sinking water supply holes except for small diameter holes in hard formations or for trial holes where rock cores are needed.

An alternative method of rotary core drilling for large diameter water borings is to use chilled shot fed down to the bottom of a heavy core-cutting barrel (see Plate 2). The barrel has a thickened bottom edge under which some of the chilled shot becomes trapped and exerts a strong point loading on the formation causing it to crush and be fragmented. A small amount of water is added for lubrication and further chilled shot is added so that eventually, with the annular cut deepening, a core of rock enters the barrel. To break off the core and bring it to the surface, some sharp pea gravel or rock chippings are sent down, causing the core to be gripped by the barrel, rotation of which causes the core to be broken off. Progress in hard formations is slow and, if fissures are encountered, problems of loss of shot or water can occur; but large diameter holes can be cut in this manner.

When soft or loose material is encountered, a solid casing or slotted lining may have to be pushed down to the level required as the boring proceeds, or after withdrawal of the chisel. Once such lining is inserted, deepening of the hole may have to proceed in smaller diameter. Several such 'step-ins' of diameter may be required if the formation has several separated layers of loose material. With rotary drilling, biodegradable muds previously mentioned may be used to keep the hole open, provided the mud does not seal off water bearing formations or is not lost through fissures. In this case it may not be necessary to line the hole until it is completed to full depth. However rotary drill rigs may have difficulty in penetrating ground containing hard boulders and rotary drilling is considerably more expensive than percussion drilling.

Verticality of a boring is not of so much importance as are 'kinks' in the boring down to the level where the permanent pump will have to be sited. If any of these exist above pump level they may throw the pump siting to one side of the boring, making poor entry conditions to the pump suction and causing difficulty in lowering or removing the pump. A usual specification for verticality is that the hole should not be more than 100 mm off vertical in 30 m of depth, but it is not necessary to insist on this at depths below any possible siting for the pump. It is especially difficult to keep a boring vertical when steeply inclined hard strata are encountered. If correction is essential it may be necessary to ream out a hole to larger diameter to get the remainder of the boring on line.

The duties of the borehole driller are:

(a) not to lose drilling tools down the hole;
(b) to note and log down every change of drilling circumstances such as increased or decreased speed of drilling, change of tools necessary, quick or sluggish recovery of water level after bailing out, etc.; and
(c) to note and log down the nature and depth of all cores and chippings produced from the boring, keeping all cores and chipping samples.

Tools lost down a boring, such as a drill bit which comes off the rods, may cause days of delay in attempts to recover them and, if 'fishing' (with a wide variety of special tools) is unsuccessful it may be necessary to ream out a hole to larger diameter down to the tool to recover it. Sometimes tool recovery may defeat all efforts which means the hole has to be abandoned if it has not been driven to the required depth. When a water boring is being sunk it is particularly important to keep a watch for signs of encountering large fissures, such as a sudden drop of the chisel or drill, increased flow of water into the hole, or fast recovery of water level when bailing out. The engineer needs to see that logs of progress are meticulously kept. Core samples should be stored in wooden core boxes, their depths being clearly marked. Samples of chippings and soft material should be stored in plastic bags properly labelled. The samples need to be examined by an experienced geotechnical engineer or hydrogeologist following the guidelines set out in BS 5930 'Code of Practice for Site Investigations'.

Wells up to about 1.2 m diameter can be sunk in hard material by rotary drilling using a heavy core barrel and chilled shot as previously described. In less hard materials, such as chalk, a large percussion chisel may be used to chop up the formation, the excavated material being removed by suction pump or by grab. In very soft or loose material, wells may be sunk kentledge fashion, a concrete caisson of the required well diameter, with a lower cutting edge, is sunk into the ground by hand excavation of the core material below the cutting edge. As excavation deepens, further rings are added to the top of the caisson. The excavated material is removed by skips or by a crane grab. In extremely soft material, excavation by grab alone may be sufficient for the caisson to descend by its own weight until harder material is encountered. Early wells were hand dug with linings of brick; later wells had linings of cast iron segments bolted together. Either interlocking pre-cast concrete rings or concrete segments bolted together are now used. Except where a well is sunk in river bed deposits, its upper part would be sealed by cement grouting on the outside to prevent ingress of surface water to the well.

4.10   Development and refurbishing boreholes and wells

When a borehole or well has been completed it may be 'developed' to maximise its yield. The objective is to remove the clay or finer sand particles from the natural formation surrounding the slotted linings or well screens to improve the flow of water into the boring and to ensure that any gravel packing is properly compacted. Several different methods of development can be used. 'Surging' is the most common and consists of pumping from the well at maximum rate, and then suddenly stopping the pumping, causing flow to wash in and out of fissures in the formation thus dislodging silt and fine sand therein. The piston effect of using a bailer can also be used, but must not be too energetic or screens might be damaged or an uncased hole collapse. 'Swabbing' and 'surging' with a piston can be used, a swab having a valve in it which allows water upflow. Both tools promote energetic flow through a screen during their up and down movement, the surge plunger in particular causes strong agitation in a gravel pack, encouraging rearrangement of particles in the pack and removal of fines. But these tools cannot be used with wire wound screens which are supported by internal vertical bars; instead they have to operate in the casing above the screen where they may still be effective but considerably less so.

Air lifting, although an inefficient form of water pumping, can be useful for the development of a hole carrying sand-charged water which would be highly abrasive to normal centrifugal pumps. The conditions for air lift were shown in Fig. 4.6. There are several variations on the use of an air-lift for this purpose which depend on the vertical movement of the air pipe up or down inside the eductor pipe. It is also possible to seal the top of a boring and inject compressed air at the top, driving the water down into the formation, prior to release of the pressure and starting the air lift.

Jetting can be used. Depending on whether the gravel pack has to be agitated, or the screen or formation cleaned, either low or high pressure can be used. A jetting head with horizontal water jet nozzles is suspended on the end of a drill pipe and slowly rotated whilst jetting. Care has to be taken with non-metallic screens or where there is potential weakening of a screen through corrosion.

Other methods sometimes adopted comprise chemical treatment or acidisation, but it is strongly advisable that these should only be undertaken by an engineer experienced in the techniques. The most common chemical treatment uses a dispersant, such as sodium hexametaphosphate, e.g. Calgon, to assist in the removal of fine material. Acidisation is used in calcareous strata to enhance the size of fissures in chalk or limestone in the immediate vicinity of the borehole. Concentrated hydrochloric acid is usually used, together with an inhibitor to minimise corrosion of any mild steel casing. The acid is applied by pipe below the water level. If the top of the borehole is closed the pressure of the evolved carbon dioxide can enhance penetration of the acid into fissures. However, because of the danger of using acids, full safety measures must accompany the operation, and the production of carbon dioxide can be hazardous unless special attention is paid to the need for ventilation.

4.11 Pollution protective measure: monitoring and sampling

The dangers of groundwater contamination are widespread and varied. Industry, agriculture and urbanisation with sanitation generate large and varied suites of ground-water pollutants, many of which are of public supply significance. The characteristics of the various pollutants vary in time and location as Table 4.3 indicates: but all represent the background addition of some substances to groundwater which may be controllable in due course, but cannot be prevented. The significance for public supply is very high. Treatment becomes necessary initially, and the resources may then become difficult to manage to avoid inducing or accelerating further contamination. Groundwater pollution can sometimes be 'cleaned up'; many methods have been used in the last twenty years, mainly in the USA where funds have been provided for the purpose. But the technology for restoration of contaminated groundwater and aquifers remains costly and imprecise. Remedial action may be impracticable where predicted clean-up time is decades or longer. The aquifer may become completely unusable so that substitute resources, such as imported supplies or treated surface water, are needed and a valuable and inexpensive resource is lost.

The concept of evaluating the vulnerability of groundwater sources to pollution is based on consideration of the lithology and thickness of the strata above the aquifer and the surface soil leaching properties. From this, the size of protection zone required around a borehole or well may be derived, based on the estimated 'travel times' of potential pollutants within the saturated zone to the abstraction point.[20] Work initiated by the National Rivers Authority and later taken over by the Environment Agency, suggested three zones of protection should be investigated for protection of underground abstraction works.

Table 4.3 Major sources of potential groundwater pollution

Occurrence	Local/linear mode	Distributed mode
Seasonal/periodic	Road salting	Agricultural fertilisers herbicides and pesticides
	Rail and road verge herbicides	
	Silage	
Continuous	Road drainage	Industrial atmospheric discharges
	Cesspool overflows	
	Septic tank effluent disposal to land	
	Solid waste tip and landfill leachates	
	Contaminated abandoned industrial land	
	Influent polluted rivers	
Random	Road/rail tanker spills	Nuclear and industrial accident fallout
	Pipeline and sewer breakages	
	Fires	
	Defective storage of industrial or agricultural chemicals	

- *Inner Zone I* defined by a 50-day travel time from any point below the water table to the source, based principally on biological decay criteria.
- *Outer Zone II* defined by the 400-day travel time, based on the minimum time required to provide dilution and attenuation of slowly degrading pollutants.
- *Source catchment Zone III* defined as the area needed to support the protected yield from long-term groundwater recharge from effective rainfall.

To delineate such zones about any particular source a series of field investigations are necessary; initially to provide a conceptual model which later, as more data is obtained, is sufficient to produce a calibrated model of the catchment so that the zones can be more accurately defined. In the UK such zones are used as a guide for the control of catchment activities by water undertakers; but elsewhere, as in Germany, the protection zones can be statutory, being set up under local Federal legislation.

Continual vigilance over the surface catchment to an underground source should be exercised. All potential sources of pollution – cesspools, septic tanks, farm wastes, industrial waste and solid waste tips, farm or industrial storages of chemicals, etc. – should be discovered and logged on a map of the catchment area. All these should be regularly monitored to ensure they are properly controlled. Poorly functioning household sewage disposal works may have to be replaced by more efficient plant and assistance to farmers may be advisable in order to aid improvement of their waste disposal practices.

Monitoring of the quality of water will reveal signs of contamination and whether it is recent and local, or more likely to be from a distant part of the catchment. The chemical constituents of the water can show evidence of the types of pollution listed in Table 4.3. If a pumped borehole penetrates several water bearing strata, depth specific sampling devices can be used below the pump with isolating inflatable packers. This can be useful for ascertaining at what level a contaminant is entering the borehole. Gas operated ejection devices or small diameter piston pumps can be used to withdraw samples from sections of the borehole isolated by using packers. Otherwise samples from the pumped water will be a composite sample of the water entering the hole from the full thickness of the formation penetrated.

Groundwater sampling needs a slightly different approach to surface water sampling. The chemistry of groundwater tends to be relatively stable, but dissolved oxygen and redox potential may be more important than for surface waters. Quality changes during sampling have to be considered, mitigated as much as possible by taking certain measurements in the field such as dissolved oxygen, pH and conductivity. The groundwater temperature may also be a useful indicator, because surface percolation to the upper levels of an aquifer will show a substantially greater seasonal temperature variation than water drawn from depth. Long-term abstraction of groundwater may cause an increase in dissolved solids, which can imply the abstraction is drawing water from more distant parts of the aquifer, so the catchment area monitored may need to be extended.

4.12 River flow augmentation by groundwater pumping

Regulating river flow by pumping from groundwater is growing in importance.[21] Water is pumped from an aquifer and discharged to a river to augment its flow during a low flow period. Usually the abstraction points lie within the river catchment basin. Successful pilot schemes of this type were undertaken by the Severn-Trent and Anglian Water Authorities in the UK in the 1970s, and were later developed into major supply schemes by the National Rivers Authority.[22,23] In the USA at least one example has been reported.[24] To be successful river regulation by groundwater pumping has to meet certain basic criteria.

(1) The scheme should provide a satisfactory increase in yield above that which could be obtained by direct abstraction from the aquifer.
(2) Discharge of the pumped groundwater to the river should result in economies in pipeline costs.
(3) The abstraction pumping capacity must be not less than the desired gain in river flow during the design drought period, plus any reduction of springflows feeding the river caused by the pumping.
(4) If recirculation losses through the river bed occur, it must be practicable to adopt equivalent extra abstraction capacity.

Prolonged groundwater abstraction will almost certainly reduce some springflows contributing to the river flow. To avoid iterative calculations it may be better, in a simple approach, to assume that all spring flows within a given distance from the groundwater abstraction points will be reduced to zero by the pumping during a prolonged drought. Pilot well tests and a detailed knowledge of the catchment hydrogeology are necessary to assess which springflows will be so affected. A better approach is to adopt groundwater modelling by computer. This permits the cumulative effect of pumping to be traced for droughts of different severity and length. The method can reveal the effect on local wells used by householders, farming and other interests, and whether any adjacent catchment is affected; and can check if percolation is sufficient to restore the aquifer storage during later wet periods.

In general river regulation by groundwater pumping is most successful where the groundwater pumping has a delayed, minimal, short-lived effect on natural river flows. To achieve this, several carefully sited abstraction points, pumped according to a specific programme, may be necessary. Where a regulated flow substantially in excess of the natural low river flow is required, prolonged pumping tests at the proposed abstraction points during a dry period are essential. This can be an expensive operation. Also caution is required in estimating the possible yield during some extreme drought condition not yet experienced, because aquifer drawdown conditions during some future critical event cannot be known with any accuracy.

4.13   Artificial recharge

Artificial recharge of an aquifer is more often practised for water treatment reasons than it is to augment groundwater yields. The two basic methods adopted are the use of spreading areas or pits, and the use of boreholes. The former technique has dominated because of its simplicity and the ease with which clogging problems can be overcome. The relatively unknown performance of wells and the need to pass only pre-treated water down them, limits their usefulness except where land for recharge pits is at a premium. Untoward events can happen during recharge operations: at an Israeli site[25] the injection of water down a well caused unconsolidated sands around it to settle and, in less than two days after pumping ceased, the borehole tubes and surface pump house sank below ground level. In a British experiment[26] even the use of a city drinking water supply for recharge did not prevent the injection well lining slots from becoming constricted with growths of iron bacteria.

At times when existing surface water treatment works are working below capacity, e.g. when there is a seasonal fall in demand, it may be possible to make water available for well recharge. This is being practised in the Lee Valley[27] north of London. Complete efficiency cannot be expected because of the relatively uncontrollable and 'leaky' nature of aquifer storage. Because the recharged source may not be used for some months, there is a loss of resources as the 'recharge mound' decays outwards and down the hydraulic gradient. Even where abstraction facilities have been located specifically to minimise this, the losses are considerable. Special artificial recharge operations in coastal areas have been used to prevent saline intrusion from the sea. This has been done in coastal Israel and California, for example.[28,29]

Recharge as a way of improving wastewater quality has been used in the Netherlands, Germany and Scandinavia for many years.[30] Alluvial sand aquifers of the coastal Netherlands are recharged with heavily contaminated River Maas water after primary filtration. The abstracted water still has to be treated using activated carbon to remove heavy metals but can be used for public supply. The strata which provide the natural treatment during infiltration have to be 'rested' and allowed time to recover and re-oxygenate. At Atlantis in south-west Africa a successful artificial recharge scheme is part of the local water resources management system. Treated domestic sewage effluent and urban storm runoff from the 67 000 population is recharged through lagoon systems and re-abstracted from specially sited production wellfields.[31]

Recharge boreholes are vulnerable to various types of clogging due to accumulation of suspended solids, gas bubbles which come out of solution, and microbial growth filling interstices or screen apertures. Borehole recharge systems using a polluted water are normally not acceptable. Hence only recharge pits are considered below.

Although dimensions vary, a recharge pit bears resemblance to a slow sand filtration bed because replaceable filter media normally covers the base. The rating of such a pit depends upon the rate at which the raw water will pass through the filter media and the rate at which the underlying aquifer will accept water. The former depends upon raw water quality, pre-treatment of the water (if any), and the depth of water kept in the pit. Little, if anything, can done to improve the rate at which the aquifer accepts the filtrate, but a decline in the infiltration rate must be guarded against by tests to ensure the filter media is working satisfactorily. Published results for infiltration rates vary widely.

Pilot tests are always required at a new site, as are initial investigation bores to ensure the pit floor will be above the water table by a margin as big as possible to give the best opportunity for water quality improvement, and to allow for increased groundwater storage and its 'mounding' below the recharge pit. Iron-pan layers and other impedances to vertical flow should be avoided.

4.14 Groundwater mining

In many countries of the Middle East and elsewhere groundwater is being used at a rate greatly in excess of the rate at which the same resources are being replenished. This tends to occur because there is no institutional framework for controlling development, and because demands for water cannot be ignored while adequate resources appear to exist. Climatic patterns have changed during the last few tens of thousands of years, and areas which are now arid were once comparatively humid. In that earlier period, high rates of recharge applied where none now takes place, as in the Libyan Desert and much of Saudi Arabia. In both regions there are groundwater developments on a vast scale which 'mine' this water. Many thousands of cubic kilometres of groundwater can be developed, though the cost of doing this and taking the water to where it is needed are high. The Great Man-Made River Project in Libya (see Table 4.1) is a modern example, where the capacity now exists to pump many millions of cubic metres of groundwater through 4 m diameter concrete pipelines for over 600 km from inland desert wellfields to the Mediterranean coast. In Saudi Arabia 'fossil' groundwater is being mined for public supply, and here – as in some other countries – groundwater exhaustion problems have only been avoided through large scale adoption of seawater desalination.

The use of water resources without a knowledge of their sustainability, or where it is known that mining is taking place, may be looked upon as an irresponsible course of action in most cases. But often the use of technology to minimise demand or maximise efficiency in the use of water is too costly to be politically acceptable, especially in respect of agriculture which remains the greatest and least efficient user of water resources. It is estimated that a 10% improvement in the efficiency of agricultural use would double the resources available for public supply.

4.15   Island water supplies

Large numbers of islands abound in all the major oceans, particularly off South East Asia and in the Pacific. In the Philippines alone there are over 7000 islands and Indonesia comprises over 13 000 mainly small islands. The Republic of the Maldives in the Indian ocean is a nation comprised completely of coral atoll islands, about 1300 of them, of which about 200 are populated. Geologically the islands range from the Maldive type of atoll islands which are less than 2 m above high tide level, to larger islands with a rocky core surrounded by a rim of sediment and reef. Apart from isolation and lack of natural resources, such islands are short of water. Such freshwater as exists often occurs as an extremely fragile lens, floating on saline water below. Population densities are high. On Malé island, capital of the Maldives, there are over 70 000 population on 200 ha of land. The fresh groundwater therefore not only became polluted with sewage but eventually was virtually consumed by over abstraction so that dependence on desalinated water and external resources is now near total. Similar problems can occur in other places where tourism increases water demand beyond the ability of local resources to supply.

On the larger rocky islands, surface water resources can sometimes be developed from perennial streams or springs, but stream flow is ephemeral on all but the larger islands. Usually there is heavy dependence on groundwater which exists as a reserve of freshwater, usually described as forming 'a lens' of freshwater within the strata forming the island, below which lies denser saline water infiltrated below from the sea. The lens is not usually regular, but strongly distorted by complications of geology which influence fresh groundwater flow to the sea. A lens thickness of 20–30 m is common, with a transition zone between the fresh and saline water, the thickness of which is a function of the aquifer properties, the range of tidal fluctuation, and any fluctuation due to variation in the rate of rainfall recharge. The amount of water represented by a given lens depends on the specific yield of the strata, a coral sand may have a high specific yield (20% or more) whereas a limestone or volcanic rock may have a much lower capacity to store water of less than 1%.

The highly simplified system illustrated in Fig. 4.9 shows the main elements of water balance and attempts to show a transition zone between fresh and saline waters. The overall situation is that the dynamic balance between percolation from rainfall and outflow of freshwater to the sea determines the dimensions of the lens of freshwater,

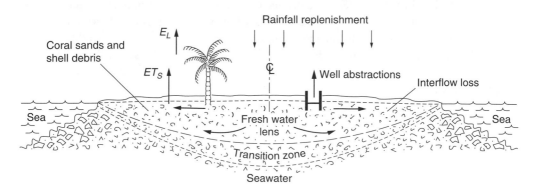

Fig. 4.9   Diagrammatic representation of lens of freshwater floating on seawater on a coral island.

mainly its depth because the width is limited. This is the 'classical view' of the lens configuration, in which the height of the lens surface above mean sea level is 1/40th of the depth of the lens below mean sea level because the ratio of the density of freshwater to that of seawater is 1:40. But observations from many islands show that a more realistic model incorporates vertical and horizontal flow due to the effect of a typical two-layered sand overlying a reef limestone system.[32-34] The transmission of the tidal fluctuations to the lens is not necessarily proportional to the distance from the shore, but rather to the depth of the sand aquifer and the transmissivity of the underlying reef limestone. Dispersion is also incorporated in the model to allow for the thickness of the transition zone between the fresh and saline water, which is typically of the order of 3–5 m.

The water balance equation for a simplified model is shown on Fig. 4.9. However putting a measure to all its components (other than rainfall) is very difficult. For instance the groundwater outflow is usually a significant proportion of the balance, but it cannot be directly measured; it can only be roughly estimated from the groundwater gradient and the hydraulic conductivity of the aquifer. Similarly evapotranspiration use by vegetation is difficult to estimate because of the unusual situation that such vegetation can find freshwater continuously available at shallow depth. Hence a water balance estimate can only reveal whether the likely replenishment of the lens is obviously less (or more) than the estimated losses from outflow and abstraction. Consequently the only safe way to evaluate the situation is to measure the thickness (and extent) of the lens and to keep it monitored.

The most sensitive monitoring of the freshwater reserves is obtained by measuring the electrical conductivity at the base of the freshwater lens and across the transition zone. Over this zone small changes in the elevation of the lens is reflected in large changes of electrical conductivity. Figure 4.10 shows the results of such probes at a number of depths made on Malé Island from 1983 to 1990, showing the virtual demise of the freshwater reserves during that period.

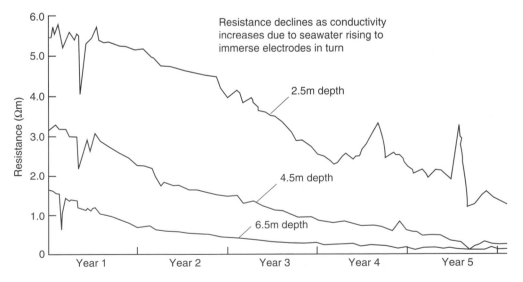

Fig. 4.10 Graph indicating rise of freshwater/seawater interface transition zone as lens of freshwater thins due to over-abstraction – Malé Island 1983–1988.

Controlled development of island freshwater is best achieved by shallow abstraction, using groups of shallow wells or collector systems, or even infiltration trenches. The use of boreholes is usually not the best approach because drawdown of the freshwater surface during pumping tends to cause upward movement of the saline water below. The rate of 'up-coning' of saline water depends on the geology, rate of pumping and depth of the borehole; but once contaminated with saline water the aquifer may take many weeks or longer to recover a usable quality of freshwater. The proportion of rainfall which recharges groundwater is subject to many factors. A relationship between rainfall and recharge derived by Falkland[34] for atolls and larger topographically low islands is shown in Fig. 4.11 and is a useful guide for preliminary use.

However the difficulty of preventing overdraw from an island aquifer has to be recognised. The local population will have been able to use house wells for their supply. But because the level of the water in such wells remains virtually constant (even though over-abstraction is causing the freshwater lens to thin), householders will see no evidence to suggest they should reduce their take. The technical solution required to preserve the lens is to close all the household wells, and install a surface abstraction scheme which takes only an amount equal to the average replenishment of the aquifer, supplying householders with a rationed supply from it via standpipes or metered connections. Any government or water authority would find such a policy difficult to implement because of the difficulty of explaining to householders the technical need, evidence for which is not visible to them.

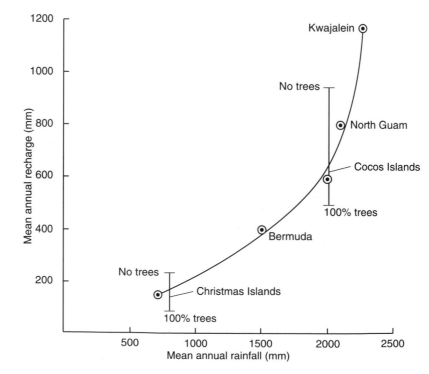

Fig. 4.11 Relationship between mean annual rainfall and estimated mean annual groundwater recharge.

4.16   Collector wells and other underground water developments

Collector wells sunk in river bed deposits are widely used on wadis where the dry weather flow of the river is underground. They take the form of large diameter concrete caissons sunk in the river bed, the caisson having to be of sturdy construction to withstand the force of the flood flow in wet weather. Access by bridge is necessary, the bridge and the top of the caisson being sited above maximum flood level. From the base of the caisson, collector pipes are laid out horizontally usually upstream, though sometimes across the stream bed or fanwise upstream. The collectors usually comprise porous, perforated or unjointed concrete pipes, 200–300 mm diameter, laid in gravel filled trenches cut in the river bed sediments, connected to the wet well. If ordinary centrifugal pumps (which are the cheapest) are used, they have to be sited above maximum flood level with their suctions dipping into the water, which limits the range of water levels from which water can be pumped. Use of submersible pumps can overcome this difficulty.

The *patented Ranney well* is an early form of collector well, usually made of cast iron and sunk in river bankside gravels or alluvial deposits fed from the river, with perforated collector tubes jacked out horizontally from the base in the most appropriate configuration, often parallel to the river. This type of well uses the bankside deposits to act as a filter. Consequently the well may have a short productive life if suspended sediment is drawn into the bankside deposits; but back-flushing of the collectors can prolong the useful life, sometimes very effectively. The advantage of a Ranney well is its relative cheapness of construction and avoidance of the need to construct a weir across the river and install intake screens.

Galleries are similar to the pipes of collector wells and are common in some areas of the world where seasonal watercourses cross thick and extensive alluvial strata which offer large storage potential. Galleries are comparatively large diameter perforated collector pipes buried 3–5 m deep or below the minimum dry season groundwater level, and surrounded by a filter medium. It is essential to locate galleries to avoid those parts of a watercourse where active erosion of the bed could occur when the river is in spate.

Qanats are believed to be of ancient Persian origin and are very widespread on the flanks of mountain ranges in Iran. They are also widespread in the Maghreb where they are known as 'foggaras', in the Arabian peninsula where they are known as 'aflaj' (singular 'falaj'), and are present in Afghanistan and China. The source or 'mother well' is dug to some depth below the water table relatively high on the flanks of a mountainous area (see Fig. 4.12). A series of additional wells is dug in a downhill line towards the area where the water is needed, and the wells are connected by means of tunnels below the water table so that the groundwater is drained down-gradient. The tunnel gradient is slightly less than the groundwater gradient, and much less than the topographical gradient, so the tunnel eventually emerges above the water table and subsequently to the surface where it acts as a canal to deliver the water where required.

The systems were hand dug and hence tunnels are comparatively large for ease of construction and to allow regular maintenance. The tunnel section from its point of emergence from the water table has to be clay-lined to minimise losses. In an increasing number of cases, competing large scale pumped groundwater abstraction is causing serious interference with these ancient gravity systems. As a result they are slowly going out of use through neglect as their discharge declines and the necessary maintenance is no longer done.

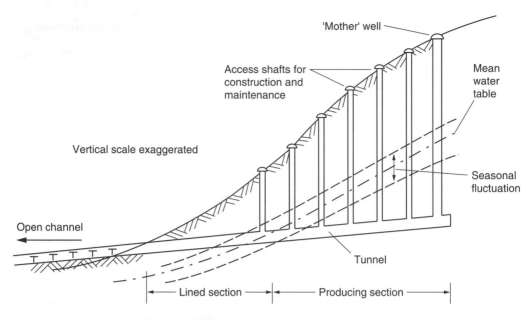

Fig. 4.12   Section through a typical Qanat system.

4.17   Borehole and well layouts

For most public supply purposes a borehole capable of giving upwards of 5 Ml/day is required and, for this, a boring of at least 300 mm diameter is necessary to accommodate the pump. In practice a more usual size of borehole would be 450–600 mm diameter because of the need for good flow characteristics in the boring and through the rising main, a possible need to reduce the diameter of the boring with depth or for installing screens, and to give allowance for any lack of verticality in the hole. If a boring is to be sunk on a site where no borings in the vicinity are available to give information concerning the likely nature of the strata to be penetrated it may be considered advisable to sink a pilot borehole first. Choice of the size of this boring can present a problem. If it is small, starting say 150 mm and reducing to 100 mm, it will be cheap and give the strata information required, but should it strike a good yield of water it may turn out too small for insertion of a pump to develop its yield. It can, perhaps, be reamed out to a larger diameter or a larger boring may be sunk nearby to accommodate the pumps. But there is no certainty in fissured formations such as chalk and sandstone, that a nearby hole will yield as much, or more. In one case where two 300 mm diameter borings were sunk in the chalk of southern England, the first hole gave just over 3 Ml/day which was considered not enough, so a second hole was sunk only 5 m away and gave more than 18 Ml/day.

A single boring of 450–600 mm diameter will permit only one pump to be installed and this may not adequately safeguard a public water supply. Even though the modern practice is to use submersible pumps, the time taken to remove a defective submersible pump from a boring and to substitute another may cause an interruption to the supply longer than can be tolerated. If the pumping level is not too far below ground level, say about 20 m, it may be economic to construct a well which can be sufficiently large to accommodate two pumps. The well can be sunk over a trial boring, if the latter gives a good yield; or a second boring can be sunk a few feet away and a connection between the two can be 'blown through' by using a directional charge (see Fig. 4.13). If, however, the trial borehole fails to give a satisfactory yield, but it is nevertheless thought the site should be capable of giving more, a well may be sunk elsewhere on the site, a second boring being sunk from its base. By use of a pump in this second boring, the well may be dewatered so that an adit can be driven towards the trial boring (also dewatered) connecting the two bores together. This makes it possible to utilise the combined output of the two borings.

Adits are seldom constructed now, but in the past many miles of adits were driven from wells sunk in chalk or sandstone formations. Their purpose was to increase the yield of a well by cutting more water bearing fissures: this was sometimes a successful policy, at other times not. The adits were generally unlined, about 2 m high by 1.2 m wide, driven with a slight upward grade from the well, and a grip on one side to drain incoming water to the well shaft when driving. The ever present possibility of meeting a large water-bearing fissure meant that quick, reliable escape means for the men cutting the adit had to be assured. Dual power supplies for the dewatering pumps and for the crane operating the access bucket were therefore essential; and adit cutting was slow because, for safety, the men had to come up with each bucket-load of muck. The consequent high cost of adit driving and an increased knowledge of developing yields by boreholes has therefore effectively ended their construction today save in rare circumstances.

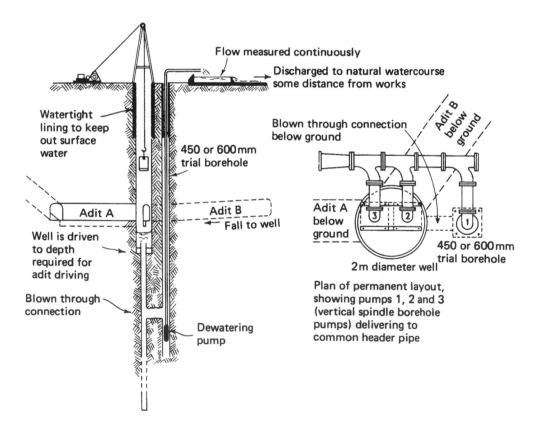

Fig. 4.13 Layout using trial borehole, deep well, and adits.

4.18   Choice of pumping plant for wells and boreholes

From the test pumping of the completed works the yield-drawdown characteristic curve for the source will be obtained, such as curve A shown in Fig. 4.14. It shows the pumping water level expected when the rest level is 'average'. However allowance has to be made for the fluctuation of the water table level during the wet and dry seasons of the year and the lowest rest level that might occur if an extreme dry weather period were experienced. There may also be a possible deterioration of the yield with time due to such unpredictable matters as silting of fissures, change of aquifer recharge due to changed conditions on the outcrop, or increased abstraction from the aquifer elsewhere.

The engineer needs to know what duty range to specify for the pump(s) to be inserted in the well, and what will be the range against which the pumps will most times be operating. The normal seasonal fluctuation of the water table level should be known from data collected before the pumping test takes place and, if historical records are available for existing wells in the same underground catchment, these may indicate the lowest water table level likely in an exceptionally prolonged dry season. The second problem is what to allow for deterioration of the yield with time. Experience of what has happened to boreholes in the same area, or in similar formations elsewhere can act as a guide; and where, as is often the case, such information is not available, the amount allowed for deterioration has to be arbitrary and regarded as a safety factor. A typical arbitrary

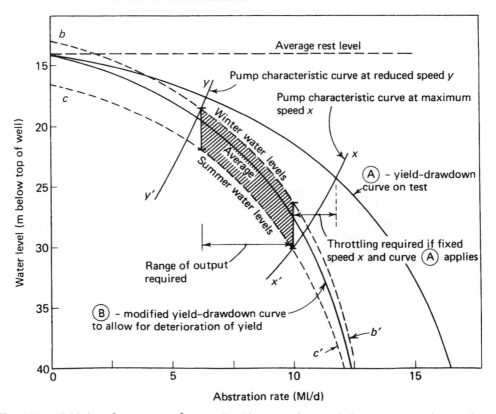

Fig. 4.14   Yield-drawdown curves for a well with pump characteristic, curves superimposed.

A good water-yielding fissure encountered in a chalk adit. (Contractors: George Stow & Co. Ltd., Slough, UK)

A 450 mm percussion chisel

A roller rock bit with reverse circulation water supply

A typical adit in the chalk at Falmer waterworks, Brighton, 1933

A 'down-the-hole' compressed air operated hammer drill

Plate 1

(Left) A large diameter sandstone core withdrawn from a 1.2 m diameter borehole drilled by the chilled shot method. (Contractors: George Stow & Co. Ltd., Slough, UK)

(Below) Bailing out a percussion drilled borehole. (Contractors: George Stow & Co. Ltd., UK)

Plate 2

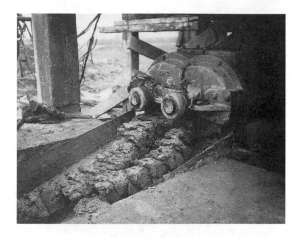

Puddled clay filling to a dam cutoff trench   A pugmill producing puddled clay

Taf Fechan reservoir, South Wales started 1913, completed 1925, showing drawoff tower and bellmouth. (Engineers: Sir Alex Binnie, Son & Deacon)

Plate 3

Tittesworth 33 m high earth dam of crushed sandstone under construction 1962, showing articulated concrete corewall. (Engineers: Binnie & Partners)

Brianne 91 m high rockfill dam, South Wales, showing side spillway. Completed 1973. (Engineers: Binnie & Partners)

Plate 4

allowance for future decline of yield is to assume there will be a 33% reduction in yield for a given drawdown level or drawdown will increase by 33% for a given pumping rate.

If, say, an arbitrary reduction of yield is assumed as shown by Curve B on Figure 4.14 curves bb' and cc' show its variation in the wet season and dry season respectively. The characteristic curve of a possible pump has also to be added onto the diagram, shown as curve x–x' for a given fixed speed pump. The curve x–x' applies to the head necessary to lift the water to the top of the well and to any additional lift to some storage tank. The pump characteristic has to be such that it is capable of giving the maximum output required (10 Ml/day in Fig. 4.14) if the well characteristics should decline to curve c–c'. However, if the well yield-drawdown characteristics remain as Curve A, the pump would have to be partially throttled by part closing of the delivery valve to keep the flow to 10 Ml/day. If the pump is driven by a variable speed motor, however, a reduced speed of rotation causes the pump output curve to shift. The curve y–y' represents the pump output at a lower speed of rotation, when only about 6 Ml/day, say, is the required output and the pumping water level in the well is higher.

The decision as to whether a fixed speed or a variable speed pump should be installed depends upon a costing of the alternatives. A fixed speed pump having to be throttled at high water levels in the well will cause some wasted energy costs. A pump with a variable speed motor will be more expensive than a fixed speed pump, and its average efficiency over the range of operation may not be so high as that of a fixed speed pump which can spend most of its time running at its design duty. The calculations will need to take into account that the yield characteristics of the well will appertain more closely to curve A for the first years of life of the well, and this can possibly be met by introducing a dummy stage into a fixed speed pump, the dummy stage being replaced by an additional impeller if the drawdown later increases. The best procedure for the engineer is to lay down all the expected circumstances of the case when approaching manufacturers as to what type of pump they consider best. The data provided should include:

- the well yield-drawdown characteristics;
- the additional lift required from the well head (and its characteristics);
- the expected time periods per annum for which the drawdown will be in the upper, middle, and lower range of values;
- what increase of drawdown should be assumed for the first 10 or 15 years;
- the average, maximum and minimum rates of pumping that will be required.

With respect to the last item, an important point that must not be missed, is that whereas the licensed abstraction is most often stated as an amount abstracted that must not be exceeded in any day of 24 hours, the actual pumping hours will be less, such as 22 hours to allow for shutdowns to attend to routine maintenance matters. Also, outputs less than the maximum would normally be met by pumping for reduced hours – which of course implies that some storage must be available to keep the supply going to consumers when the pump is not operating. Pumps most often now used for well and borehole pumping are fixed speed submersible pumps (see Section 12.7). Only for large outputs would vertical spindle centrifugal pumps be used because of their much greater expense. Variable speed drives also add to the expense and tend only to be chosen if use of a fixed speed pump would cause much extra power cost by having to be throttled. Given a fairly flat efficiency curve about the design duty, and a reasonable slope to the head-output curve, a mixed flow, fixed speed submersible pump (see Fig. 12.4(a) in Chapter 12) may be the most suitable for well pumping if the range of drawdown is not excessive.

References

1. Rodda J. C., Downing R. A. and Law F. M. *Systematic Hydrology*, Newnes-Butterworth, Oxford, 1976, see Table 3.10.
2. Singhai B. B. S. The Hydrogeological Characteristics of the Deccan Trap Formation in India, *Proc Rabat Symposium May 1997; Hard rock hydrosystems,* IASH Publications No. 241.
3. Wright E. A. and Burgess W. (eds). *The Hydrogeology of Crystalline Basement Aquifers in Africa*, Special Publication 66, Geological Society, London, 1992
4. Jones M. J. The Weathered Zone Aquifers of the Basement Complex Areas of Africa, *Q J Eng Geol* **18**(1), 1987 pp. 26–35.
5. Stanners D. and Bourdeau P.(eds). *Europe's Environment*, Chap. 5, Inland Waters, pp.57–108, European Environment Agency, Office for Official Publications, Luxembourg, Aug. 1995.
6. Sander P. Water Well Siting in Hard Rock Areas; Identifying Probable Targets Using a Probabilistic Approach. *Hydrogeology Jnl*, **5**(3), 1997, pp. 32–43.
7. Summers I. (ed.). *Estimation of Natural Groundwater Recharge*, Reidel, 1987, p. 449.
8. Headworth H. G. The Selection of Root Constants for the Calculation of Actual Evaporation and Infiltration for Chalk Catchments. *JIWE*, **24**, 1970, p. 431.
9. Theis C. V. The relation between lowering the piezometric surface and the rate and duration of discharge of a well using groundwater storage, *Trans American Geophysical Union*, **16**, 1935, pp. 519–614.
10. Walton W. C. *Groundwater Resource Evaluation*. McGraw-Hill, Maidenhead, 1970.
11. Jacob C.E. in: *Flow of Groundwater* (ed. H. Rouse). John Wiley & Sons, Chichester, 1950.
12. Kruesman G. P. and de Ridder N. A. *Analysis and Evaluation of Pumping Test Data*, 3rd edn. International Institute for Land Reclamation and Improvement, Wageningen, The Netherlands, 1983.
13. *Groundwater and Wells*, Edward E. Johnson Inc., Saint Paul, Minnesota. 1966, Table XXX.
14. Maidment D. R. (ed.). Computer Models for Sub-surface Water, Chapter 22, *Handbook of Hydrology*. McGraw-Hill, New York, 1992.
15. Bruin J. and Hudson H. E. *Selected Methods for Pumping Test Analysis*. Report 25, Illinois State Water Supply, 1955.
16. *Groundwater Occurrence, Development and Protection*, Water Practice Manual No. 5, IWES 1986, Chapter 9 on Downhole Geophysics.
17. Terzaghi K. *Theoretical Soil Mechanics*, John Wiley & Sons, Chichester, 1943.
18. Monkhouse R. A. The use of sand screens and filter packs for abstraction wells, *Water Services*, 78, 1974.
19. Campbell M. D. and Lehr J. H. *Water Well Technology*, McGraw-Hill, Maidenhead, 1973
20. Adams B. and Foster S. S. D. Land-surface zoning for groundwater protection. *JIWEM*, June 1992, pp. 312–320.

21. Downing R. A. and Wilkinson W. B. (eds). *Applied Groundwater Hydrology*, Oxford University Press, Oxford, 1991.
22. Skinner A. Groundwater Development as an Integral Part of River Basin Resource Systems, Groundwater in Water Resources Planning, *Proc Koblenz Conf*. Int. Assn. of Hydrogeologists, Vol. II, 1983, pp. 641–650.
23. Steering Committee. *Groundwater Pilot Scheme*, Final Report, Great Ouse River Authority, 1972.
24. Todd D. K. *Groundwater Resources of the Upper Great Miami River Basin and the Feasibility of their use for Streamflow Augmentation*, Report to Miami Conservancy District, Ohio, 1969.
25. Sternam R. Artificial Recharge of Water Through Wells: Experience and Techniques, *Proc Symposium on Artificial Recharge and Management of Aquifers*, IAHS, Haifa, 1967.
26. Marshall J. K., Saravanapavan A. and Spiegel Z. Operation of a Recharge Borehole, *Proc ICE*, **41**, 1968, pp. 447–473.
27. Edworthy K., Headworth H. G. and Hawnt R. Application of Artificial Recharge Techniques in UK, in: *A Survey of British Hydrogeology*, Royal Society, London, 1981.
28. Aberbach S. H. and Sellinger A. Review of Artificial Groundwater Recharge in the Coastal Plain of Israel. *Bull Int Assoc Sc Hydrology,* **12**(1), 1967, pp. 75–77.
29. Bruin A. E and Seares F. D. Operating a Seawater Barrier Project. *J Irrig & Drainage Div*, ASCE, 91, IR-1 pt 1, 1965, pp. 117–140.
30. IASH. International Survey on Existing Water Recharge Facilities, Publication No. 87, 1970.
31. Wright A. and du Toit I. Artificial recharge of urban wastewater; the primary component in the development of an industrial town on the arid west coast of South Africa, *Hydrogeol Jnl*, **4**(1), 1996, pp. 118–129.
32. Wheatcroft S. W. and Buddemeier R. W. Atoll Hydrology, *Ground Water* **19**(3), 1981, pp. 311–320.
33. Ayers J. F. and Vacher H. L. Hydrogeology of an atoll island; a conceptual model from a detailed study of a Micronesian example. *Ground Water* **24**(2), 1986, pp. 185–198.
34. Falkland A. Hydrology and Water Management on Small Tropical Islands. *Proc Yokohama Symp, Hydrology of Warm Humid Regions*, IASH Publication No. 216, 1993, pp. 263–303.

5

Dams, impounding reservoirs and river intakes

5.1 Introduction

The earliest dam Smith was able to report in his book, *A History of Dams*[1] was the 37 feet high Sadd el-Kafara dam, built between 2950 and 2750 BC the remains of which lie 20 miles south of Cairo. It had upstream and downstream walls of rubble masonry each 24 m thick at the base, with a 36 m wide, gravel filled space between, and it appeared to have had a short life because it suffered from the two principal defects that continued to plague many dams for the next 4500 years – it leaked and was probably overtopped.

There are many materials of which a dam can be made – earth, concrete, masonry or rockfill. The choice depends upon the geology of the dam site and what construction materials are nearest to hand. Concrete and masonry dams require hard rock foundations; rockfill dams are built on rock but have been built on alluvial deposits; earth dams can be built on rock and also on softer, weaker formations such as firm clays or shales. Masonry dams are still built in developing countries where labour costs are low. Where labour costs are high they have generally been replaced by mass concrete, which is compacted by using immersion vibrators or, in recent years by roller compaction. It was not until the middle of the nineteenth century that concrete and masonry dams began to be designed according to mathematical analysis of the internal and external forces coming upon them; nor until the first quarter of the twentieth century was the behaviour of earth dams understood enough for mathematical design procedures to be applied to them also. Since then advances in dam design have become so specialised that only the salient principles involved can be given here.

For a successful dam construction the following conditions need to be fulfilled.

(1) The valley sides of the proposed reservoir must be adequately watertight to the intended top water level of the reservoir, and they must be stable under the raised water level.
(2) Both the dam and its foundations must be sufficiently watertight to prevent dangerous or uneconomic leakage passing through or under the dam.
(3) The dam and its foundations must be strong enough to resist all forces coming upon them.
(4) The dam and all its appurtenant works must be constructed of durable materials.
(5) Provision must be made to pass all flood waters safely past the dam.
(6) Provision must be made to draw off water from the reservoir for supply and compensation purposes, and for lowering the reservoir water level in emergencies.

5.2   Essential reservoir conditions

The reservoir site formed by the dam must not leak to an unacceptable degree, and the hillsides forming the reservoir basin must be stable under all possible water level conditions. A proposed reservoir site must always be subjected to a detailed geological investigation to ensure that no fault zones, hidden valleys, or permeable strata exist through which unacceptable leakage would take place. In addition the nature and stability of the hillsides to the reservoir under waterlogged conditions must be investigated. Inundation of hillside materials increases their unit weight and decreases their cohesion, so that, if the water level in the reservoir should be rapidly lowered, the loosened wet material slides into the reservoir. This happened in the Vaiont Dam disaster of 9 October 1963 in Italy[2] when a landslide of gigantic proportions fell into the reservoir, causing a 100 m high flood wave to pass over the crest of the 206 m high arch dam causing the deaths of 3000 people in the valley below. The dam was not destroyed but the reservoir was afterwards abandoned.

5.3   Watertightness

All dams leak to a greater or lesser extent: the danger to be avoided is that the leakage does not carry with it material from the dam or its foundations. Hence as much as possible of all such leakage should be collected by an underdrainage system and delivered to a collecting basin where it can be continuously measured to reveal any signs of increase, and inspected to see that the drainage water is not carrying material with it. The Dolgarrog disaster[3] in Wales on 25 November 1925 – the last to cause any loss of life in UK – was caused by continuing leakage below a low concrete wall only 3 m high used to heighten the level of water in Lake Eigiau. At one point the wall had been taken only 0.5 m deep into clay foundations, and leakage at this point so widened its passage that there was ultimately a sudden breakthrough of the lake waters. The wave of water destroyed another small dam below and engulfed the village of Dolgarrog, causing 16 deaths.

To reduce seepage below the foundations of a dam two methods are in common use:

(1) the construction of a 'cutoff' trench across the valley below the dam;
(2) grouting the foundations beneath the dam to reduce their permeability.

A typical cutoff is illustrated in Fig. 5.1 and it is taken down sufficiently far to connect into sound rock or clay at the base. It is usually filled with concrete, the trench being about 2 m wide. The junction of the top of the cutoff with the corewall of the dam is a matter requiring the most careful design: Fig. 5.1 shows the concrete finished in the shape of a 'spearhead' so that the clay could be brought down on either side of it. A variety of designs have been adopted. It is also important to extend the cutoff wall into the abutments of the dam to reduce seepage around the ends of the dam. The dam shown in Fig. 5.1 is typical of early dams which had a puddle clay corewall (see Plate 3); in the 1950s rolled clay corewalls came into use and puddle clay was no longer used.

A wide, shallow cutoff is shown in Fig. 5.2 and this is used where a sound foundation material exists not far below ground surface. The corewall material, in this case 'rolled clay' which is a mixture of clay and coarser materials, is taken right down to the bottom of the cutoff.

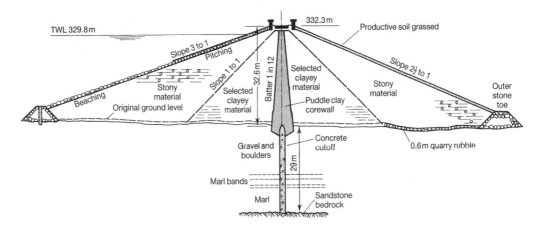

Fig. 5.1   The Taf Fechan dam; South Wales, 1928 (engineers: Binnie and Partners).

Some cutoffs have been extraordinarily deep and therefore difficult and expensive to construct. The classic case is the cutoff for the Silent Valley Dam for Belfast,[4] which reached a maximum depth of 84 m before sound rock was met. Cutoffs of up to 30 m deep have been common, but very deep cutoffs are seldom necessary today because of the development of techniques of grouting which are cheaper to adopt. As a 'rule of thumb', cut-offs are generally taken down to between one-half or equal to, the depth of retained water above. Two-thirds the impounded water depth is an often used figure, but may need to be adjusted locally for particular geology.

It is sometimes necessary to construct fill dams over alluvial deposits of high permeability. In such cases a cutoff can be formed by constructing two diaphragm walls about 6 m apart and grouting the intervening materials. The walls can be formed by sinking tangentially bored concrete piles 0.3–0.5 m diameter,[5] or by using smaller 0.2 m diameter overlapping 'secant' micropiles. Alternatively a trench of the required width is excavated in short lengths by clam-shell grab, the walls of the trench being prevented from collapsing by keeping the trench full of bentonite clay mixture. Concrete is tremied to the base of the trench and brought upwards, the displaced bentonite being piped away for re-use in the next contiguous section of trench excavated. Plate 8 shows use of this technique for construction of a diaphragm wall within the puddled clay corewall of an old dam to improve its watertightness.

Grouting consists of drilling vertical or inclined holes into the ground at intervals (see Fig. 5.3) and injecting into such holes mixtures of water and cement, or clay, or chemical mixtures. This mixture penetrates into the bedding and joint planes of any permeable strata which are cut and, when the grout sets, such fissures and joints are thereby sealed off. The drill holes have to be close enough for the grout to form a continuous impermeable barrier or 'curtain' from hole to hole. It is usual to grout a series of primary holes first, spaced perhaps 6 or 7 m apart, then to sink secondary holes between these, test them for permeability and grout them, and then if necessary to sink tertiary holes between them. The spacing of holes to form an effective grout curtain varies according to the type of strata penetrated, but is usually in the range 1.5–5.0 m. Sometimes several parallel rows of holes are grouted to form a wide band of grouted ground.

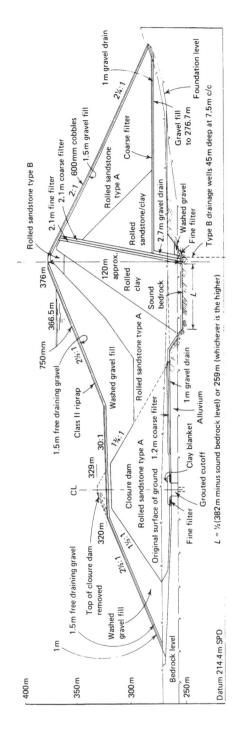

Fig. 5.2 The Mangla Dam for the Water and Power Development Authority, Pakistan, 1968 (engineers: Binnie and Partners (Binnie *et al.*, *Proc ICE*, 1967–1968)).

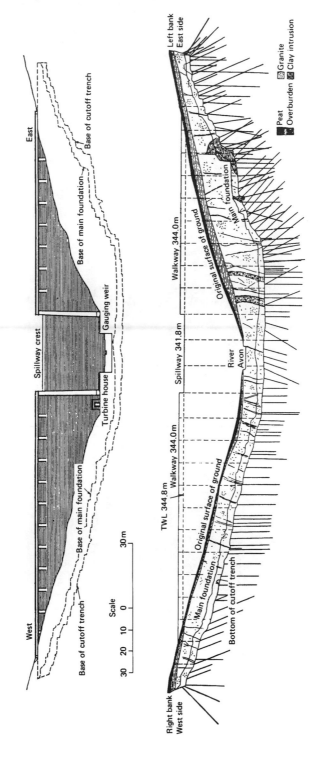

Fig. 5.3 The Avon Dam for South Devon Water Supply (engineers: Lemon and Blizzard (Bogle *et al.*, *Proc ICE*, 1959)).

Grouting is normally adopted to improve the watertightness of a rock which is already basically sound, reasonably impermeable, and not liable to decompose even with some leakage through it. In exceptional cases it may be economically justified to attempt to grout materials other than fissured rock, such as sands, gravel, silts, clay, and mixtures of these materials.[6] These are often difficult to grout despite use of a variety of methods, so that complete success may not come to hand and additional precautions will have to be adopted in the design of the dam to allow for unavoidable leakage.

'Stage grouting' is often used. In this process 'packers' comprising expandable rubber seals are fixed in the boring above and below a given length or 'stage' of borehole so that water tests and grout can be applied to this section only. It is usual to adopt 10 m stages working progressively upwards. When severely fractured rock, granular materials, or clay are encountered it is difficult or impossible to fix packers, but then the tube-a-manchette method can be used. The tube has slotted perforations at intervals down its length, each perforation being covered on the outside by a thin membrane. The tube is lowered into the borehole and the annular space outside it is grouted with cement grout. When the grout is set, packers are inserted inside the tube above and below a given perforation and, with a slightly raised 'breakout' pressure, the membrane is stretched and the annular grout in the vicinity of the perforation can be fractured, permitting grout to be injected into the formation at that level.

After completing foundation grouting it is usual to ensure that the contact zone between the dam material and the excavated ground surface is sealed by carrying out 'contact' or 'blanket' grouting, through a line of grout pipes with packers set above and below foundation level. After sufficient height of the dam has been constructed to provide weight, grout is injected into the pipes between the packers.

Grouting can be highly successful but tales are legion concerning the damage that can be done if the procedure does not come under careful control. Pushing too much grout under too high a pressure into the formation may do nothing to reduce the permeability of the ground, producing only lenses of grout. Sometimes the permeability of the ground can be increased by inappropriate grouting techniques and, in one case on record, the whole of a partially built dam was raised 150 mm or so by unnecessary grouting. It is important not to use too high a pressure to avoid 'hydrofracturing' or rupture of the formation – unless this process is intended for some special reason as advised by an experienced grouting specialist. Normally the maximum pressure applied at the point of injection should not exceed twice the overburden depth to that point. This holds for both grouting pressures and water test pressures. Tests for assessing the permeability of a formation should always be made on newly drilled, ungrouted holes: not on previously grouted holes. The water leakage from a boring under test (or of a stage of it) is usually expressed in 'lugeons', 1 lugeon being 1 litre per minute per metre of boring under 10 bar head. The diameter of the boring does not enter this definition as it has little significant effect on test results. Permeabilities are expressed either in cm/s units or m/s units. A frequent target for acceptable permeability of a formation after grouting is 1×10^{-4} cm/s.

In some places low dams have been constructed on alluvial or marine deposits extending to depths of several hundred metres making it impracticable to form an impervious cutoff. Steel sheet piles have proved effective in lengthening the leakage paths, thereby reducing the permeability below the dam and protecting its foundations.

In practice, especially under concrete dams, two means of foundation treatment are used together; an upstream grout curtain or cut-off is used to reduce seepage, and a downstream drainage system is used to reduce uplift. If the foundation has a high natural lugeon value then natural drainage exists and the focus should be on ensuring a cut-off. If the foundation is 'tight', say less than five lugeons, then seepage will be low and the focus should be on providing good drainage. In practice both systems are required over the length of a typical dam foundation.

5.4   Strength of the dam

Every dam must be proof against failure by sliding along its base. This applies whether the dam is of concrete, masonry, rock, or earth. Resistance to failure against shearing through any plane is another criterion to be applied. Usually the resistance of rock and earth dams against sliding or direct shear is ample on account of their large width. However masonry and concrete dams have failed in this manner. An early failure in 1844 was that of the Bouzey Dam in France,[7] a 19 m high masonry dam which moved 0.35 m downstream under the static force of the water. It was cemented back on its foundations, but 11 years later in 1895, it split horizontally about the middle. Another failure was that of the St. Francis Dam on 13 March 1928 which was a 60 m high concrete dam feeding Los Angeles.[8] It was placed on weak foundation material, so that a section of the dam broke out and a flood wave 40 m high travelled down the valley at 65 km/hour, causing 426 lives to be lost.

Concrete or masonry dams, being constructed to much steeper slopes than earth dams, have to be proof against overturning. Where concrete dams are arched, the abutments must be strong enough to take the end thrusts of the arch, and the foundations must be strong also as they may be highly stressed. The failure of the Malpasset Dam in France[9] on 2 December 1959 illustrates the catastrophic nature of an abutment failure. The arched concrete dam was 60 m high but only 6.5 m thick at the base. It was judged afterwards that, owing to rock joint pressurisation, the left abutment moved out ultimately as much as 2 m causing rupture of the arch in a few seconds, instantaneously releasing the whole contents of the reservoir which engulfed the Riviera town of Frejus 4 km downstream. The dam was completely swept away.

5.5   Durability of a dam

Durability is of the utmost importance because dams are one of the few structures built to last almost indefinitely. In Italy the Gleno Dam,[10] a multiple arch dam 43 m high, suddenly failed on 1 December 1923 when one of the buttresses cracked and burst in a matter of a few minutes. The cause was attributed to poor quality workmanship: the concrete in the arches was poor and inadequately reinforced with scrap netting used as hand grenade protection in the war and there was evidence of lack of bond with the foundations (see Plate 5).

The durability of an earth dam is primarily dependent upon the continuance of its ability to cope with drainage requirements. It has to contend with seepage, rainfall percolation, and changing reservoir levels. The movement of these waters must not dislodge or take away the materials of the dam, otherwise its condition and stability will deteriorate at a progressively increasing rate. It must perform this function despite long-term settlement of itself and its foundations (see Section 5.14).

Overflow and outlet works

5.6 Design flood estimation

The assessment of flood risk is a vital element in the safe design, maintenance and operation of impounding reservoirs. Earth dams are inherently erodible and uncontrolled overtopping can lead to catastrophic failure. Also overtopping of a rockfill, masonry or concrete dam needs to be avoided except where the design specifically provides for this. It is therefore necessary to specify a design flood, in combination with wave action, which the dam must be capable of withstanding. Greater security is required against dam failure where there is a major threat of loss of life and extensive damage and a lower security where the threat is less severe.

A wide range of methods are used for computing reservoir design floods. These methods include:

- empirical and regional formulae;
- envelope curves;
- flood frequency analysis;
- various types of rainfall-runoff and losses models, including the unit hydrograph, the US Soil Conservation Service, the transfer function, and the Gradex methods.

A publication entitled *The Design Flood* (*Guidelines*) produced by the International Commission on Large Dams (Committee on Design Flood) contains useful summaries of the more commonly used methods and their limitations. The report also contains general guidance and recommendations about reservoir design flood estimation, some of which are the following.

- The calculation of design flood is a complex design problem requiring the contributions of specialist engineers, hydrologists and meteorologists whose involvement must be sought through the whole process.
- The choice of the design flood involves too many and varied phenomena for a single method to be able to interpret all of them.
- All methods available are based on meteorological and hydrological records both for the river basin under investigation and often for other comparable areas. The results derived from the application of a particular method will essentially depend on the reliability and applicability of the data adopted.
- The exceptional circumstances causing extreme floods are often such that the records of historical moderate floods may provide only a very poor indication of conditions during a Probable Maximum Flood (PMF) or a 10 000 year event.

For dams in the UK the key advisory document is the third (1996) edition of *Floods and Reservoir Safety*[11] which recommends the levels of protection for different categories of dam. The recommended floods and waves protection standards in the UK are shown in Table 5.1. Chapter 3 of *Floods and Reservoir Safety* guides the engineer on the use of the 1975 Flood Studies Report[12] unit hydrograph rainfall-runoff and losses model to derive reservoir design flood inflows, while Chapter 5 contains a method for estimating wave surcharge allowances. The Flood Studies report has recently been superseded by the 1999 *Flood Estimation Handbook*[13] which presents the Flood Studies Report unit hydrograph and losses model in an updated and rationalised form. Volume 4 of the *Flood Estimation*

Table 5.1   Flood wind and wave standards by dam category according to *Floods and Reservoir Safety*, ICE 1996

| Dam category | Potential effect of a dam breach | Initial reservoir condition standard | Reservoir design flood inflow | | Concurrent wind speed and minimum wave surcharge allowance |
			General	Minimum standard if overtopping is tolerable	
A.	Where a breach could endanger lives in a community	Spilling long-term average inflow	Probable maximum flood (PMF)	10 000-year flood	Mean annual maximum hourly wind speed. Wave surcharge allowance not less than 0.6 m
B.	Where a breach (1) could endanger lives not in a community or (2) could result in extensive damage.	Just full (i.e. no spill)	10 000-year flood	1000-year flood	
C.	Where a breach would pose negligible risk to life and cause limited damage	Just full (i.e. no spill)	1000-year flood	150-year flood	Mean annual maximum hourly wind speed. Wave surcharge allowance not less than 0.4 m
D.	Special cases where no loss of life can be foreseen as a result of a breach and very limited additional flood damage would be caused	Spilling long-term average inflow	150-year flood	Not applicable	Mean annual maximum hourly wind speed. Wave surcharge allowance not less than 0.3 m

Note. Where reservoir control procedures require, and discharge capacities permit operation at or below specified levels defined throughout the year, these specified initial levels may be adopted providing they are stated in the statutory certificates and/or reports for the dam.

Handbook currently provides the most up to date advice on reservoir flood estimation in the UK.

The statistical analysis of flood events has a very limited role in reservoir design flood estimation in the UK. The reason for this is that extrapolation of statistical flood estimates to the high return periods relevant to freeboard and spillway design may lead to gross under- or over-design, given the relatively short period flood data typically possible. Flood estimates become increasingly unreliable when the return period of the design event is greater than about twice the length of the record. It is not possible here to provide a guide to each of the methods of flood estimation currently employed. The following paragraphs, however, provide some basic guidance on the unit hydrograph rainfall-runoff method which has been used for design flood estimation in many countries.

Unit hydrograph approach

There are numerous different forms of unit hydrograph and loss models, but all are similar in that they convert a rainfall input to a flow output using a deterministic model of catchment response. These deterministic models have three common elements:

- the unit hydrograph itself which often has a simple triangular shape defined by time to peak;
- a losses model which defines the amount of storm rainfall which directly contributes to the flow in the river; and
- baseflow (i.e. the flow in a river prior to the event).

Where possible the model parameters should be derived from observed rainfall runoff events. If no records exist the model parameters may be estimated from catchment characteristics.

A unit hydrograph graphs the flow from a given catchment resulting from unit effective rainfall in unit time on the catchment, e.g. the flow following, say, 10 mm effective rainfall falling in 1 hour. It assumes the rainfall is uniform over the catchment, and that runoff will increase linearly with effective rainfall. Thus the runoff from 20 mm of effective rainfall in one hour is taken as double that due to 10 mm and so on, and the ordinates of the hydrograph are doubled. Similarly if rainfall continues beyond the first hour, the principle of superposition can be used, the resulting hydrograph being the sum of the ordinates of the hydrographs for the first, second, third, and so on, hours.

Due to the complexities involved, the derivation of a catchment unit hydrograph from storm rainfall and runoff records is best carried out by an experienced hydrologist. Losses by evaporation, interception and infiltration have to be deducted from rainfall, and the baseflow has to be deducted from the measured streamflow. A complication is that rainfall events will seldom be of unit duration, so further analysis is required to produce a unit hydrograph for unit time and unit rainfall. It is usual to derive several unit hydrographs from separate rainfall events for comparative purposes and, where a catchment is too large for uniform rainfall intensity over the whole of it to be assumed, it is necessary to treat individual tributaries separately and assess their combined effect.

One relatively straightforward method of obtaining an adequately severe unit hydrograph is suggested by the US Bureau of Reclamation[14] as follows.

(1) Time T_c of concentration is given by

in Imperial units	in Metric units
$T_c = (11.9 \ L^3/H)^{0.385}$ hours	$T_c = (0.87 \ L^3/H)^{0.385}$ hours

where: L = length of longest tributary (miles–km); H = fall of this tributary (feet–metres); T_c = time elapsing between onset of storm and time when all parts of the catchment begin contributing to flow to measuring point.

(2) Time to peak T_p for the hydrograph for one inch of rainfall over the catchment is given by

$$T_p = 0.5(\text{rainfall duration}) + 0.6T_c$$

where T_p, T_c and rainfall duration are in hours.
(3) Peak runoff rate R_p

in Imperial units	in Metric units
$R_p = 484\ AQ/T_p\ \text{ft}^3/\text{s}$	$R_p = 0.2083\ AQ/T_p\ \text{m}^3/\text{s}$

where A is catchment area (sq. miles – km^2) and Q is the unit rainfall (inch – mm).

Use of this formula tends to over-estimate flood peaks in temperate, flat and permeable areas, for which a simple triangular unit hydrograph, with a base of 2.5 times the time to its peak should suffice. Its peak flow will equal $2.2/T_p$ m^3/s for each km^2 of catchment for every 10 mm of effective rainfall. The value of T_p often lies in the range $1.3(\text{area})^{0.25}$ to $2.2(\text{area}))^{0.25}$ where the area is in km^2.

A variety of alternative loss models can be incorporated into the unit hydrograph approach. The two most common are:

(i) to derive a single percentage runoff value applicable throughout the whole storm; and
(ii) to adopt an initial loss x mm at the start of the design storm followed by continuing losses of y mm/hour throughout the event.

Once each of the model elements has been defined for a catchment, the unit hydrograph method may be used to estimate the total runoff from any rainfall event. The design rainfall will be in the form of a hyetograph defined by a depth/duration characteristic of the area and arranged into a selected storm profile. A 'bell-shaped' profile is most commonly used for storm events up to about 24 hours duration, whereas the recorded profiles in severe historic events are used to define more realistic design storm profiles for longer duration events.

The design storm rainfall may be a statistically derived design event to produce a flood of specific return period (the T-year event), or may be a probable maximum precipitation (PMP) to produce a probable maximum flood (PMF). Rainfall depth–duration–frequency values, sometimes including estimates of the PMP, are available for many regions. Table 5.2 gives some examples of the PMP estimates adopted for the areas draining to a number of major damsites, and Fig. 3.3 of Chapter 3 also gives information on maximum precipitations that have been recorded.

5.7   Spillway flood routing

If a reservoir created by a dam is full to top water level, a flood inflow causes the reservoir water level to rise and this causes increasing rates of discharge over the spillway to the dam. This temporary ponding of water in the reservoir will result in the maximum overspill rate at the dam being less than the maximum flood inflow rate to the reservoir, and will delay its time of occurrence. The calculation of this maximum overspill rate is termed 'flood routing'.

The problem is that any inflow causing a rise in water level simultaneously increases the rate of discharge over the spillway. To solve this problem step by step calculations over short intervals of time are used.

For any short time period $t_1 - t_2 = t$

Table 5.2   Some probable maximum precipitation (PMP) estimates

Location	Country	PMP in mm				
		20 min	1 hour	6 hours	15 hours	24 hours
South Ontario	Canada	–	–	410	–	445
Guma	Sierra Leone	81	183	–	580	630
Selangor	Malaysia	–	160	300	–	460
Shek Pik	Hong Kong	101	220	–	915	1200
Garinono	Sabah	81	162	420	620	675
Brenig	Wales					
	May–Sept	74	109	183	–	254
	Oct–Apr	38	72	165	–	272
Tigris (50 000 km^2)	Iraq	–	–	60	–	167
Jhelum, Mangla (2500 km^2 sub-area)	West Pakistan					
	Dec–May	–	–	185	–	295
	Jan–Nov	–	–	365	–	575

Inflow = storage increase + outflow

Hence assuming, for small time intervals, the averages can be taken

$$\frac{(I_1 + I_2)}{2}.t = (S_2 - S_1) + \frac{(O_1 + O_2)}{2}.t \tag{5.1}$$

where I_1, S_1, O_1 and I_2, S_2, O_2 are the values of inflow, storage, and outflow at times t_1 and t_2 respectively.

Rearranging eqn (5.1) and dividing by t:

$$(S_2/t + O_2/2) = (S_1/t - O_1/2) + (I_1 + I_2)/2 \tag{5.2}$$

The terms on the right-hand side of this equation are all known because I_1 and I_2 are obtained from the flood inflow hydrograph, and S_1 and O_1 apply to the previous conditions (including the start values). This gives the value of the left hand side of the equation, from which it is possible to find O_2 because a graph can be drawn of O against $(S/t + O/2)$.

Data required before calculations can proceed are:

- the initial reservoir level and corresponding spillway overflow O_o;
- a graph of storage volume, S, above spillway crest for height, h, of water level above spillway crest;
- the overflow rate, O, for water level, h, above spillway crest;
- any major drawoff from (or effluent discharge, etc. into) the reservoir.

Calculations proceed as follows.

(1) A graph of $(S/t + O/2)$ against O for say $1/2h$ time intervals, is drawn as follows:
    (a) compute values for O and S for incremental values of the head, h, over the spillway;

Table 5.3 Flood routing calculation

(1)	(2)	(3)	(4)	(5)	
	Mean inflow		Outflow O_2	Reservoir level	
Period $1/2h$	$(I_1 + I_2)/2$	$S_2/t + O_2/2$	from graph	above spillway	
(h)	(m^3/s)	(m^3/s)	(m^3/s)	(m)	Notes
0–2	1.9	12.0	1.9	0.14	Initial steady state
2–1	1.9	12.0	1.9	0.14	
1–12	6.0[a]	16.1[b]	2.8	0.16	
12–2	18.1[a]	31.4[b]	6.7	0.31	

[a]Value from inflow hydrograph.
[b]Previous line Col (3) *minus* Col (4) + new Col (2).

    (b) sum $(S/t + O/2)$ for each increment of h and plot against O (where $t = 1.8 \times 10^3$ the number of seconds in $1/2$ h).

(2) Draw up a table as Table 5.3.

    An initial steady inflow/outflow is assumed so that $I_0 = I_1 = O_1$. Hence the starting value for $(S_1/t + O_1/2)$ can be known.

(3) For the next and subsequent lines of the table, new $S_2/t + O_2/2$ = previous $S_1/t + O_1/2 - O_1$ + new value $(I_1 + I_2)/2$ (as eqn 5.2 above). From the graph under (1) new O_2 can be obtained.

5.8   Diversion works and outlet culverts

When a concrete or masonry dam is built, the stream to be impounded may be taken through an opening left at the base of the dam in the valley bottom, this opening being finally sealed up when the dam is completed. Even if floods occur during construction of a concrete or masonry dam and it is overtopped before completion, no serious harm need come from the inundation if the construction is designed for this and the valley bottom is composed of rock.

It is not possible to follow this procedure with an earth dam which, if overtopped at any stage, would probably be destroyed. Also, except in the case where an earth dam is founded on sound rock (which is often not the case), it is undesirable to have a central culvert left through the body of an earth dam. Even if the culvert is later filled with concrete, or plugged with concrete at its upstream end, there is a danger that, unless it is on hard rock, it will settle more at the centre under the weight of the full height of the dam, than it does at the ends where it emerges from the dam. This differential settlement can cause the culvert to fracture, or open up a path for impounded water to leak through the junction where the culvert passes through the corewall of the dam.

The near failure of the Lluest Wen Dam in South Wales in 1970[15] illustrates the danger of a culvert through the centre of an earth dam even when the foundation is rock (see Fig. 5.4). The dam was built in 1896 in a coal mining area. A 'pillar' of unmined seams of coal was left below the dam but there was land settlement outside this pillar and this, combined with the weight of the upstream drawoff tower, caused the plug of concrete in the culvert to fracture within the zone of the puddled clay corewall. A small 150 mm drain pipe

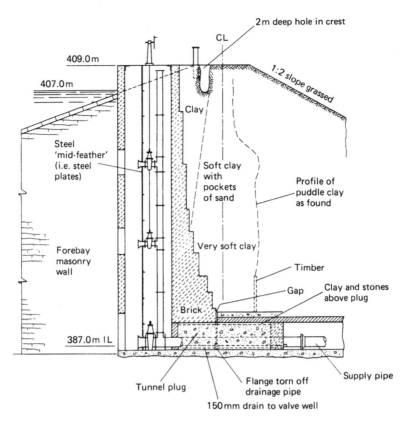

Fig. 5.4 Cause of the Lluest Wen Dam failure.[15]

through the plug was fractured also and, into this fracture seepage occurred bringing with it clay from the corewall. After an unknown length of time, probably of several years, a 2 m deep hole appeared in the crest of the dam 20 m above and, shortly afterwards clay slurry began emerging from the drain pipe. The dam had to be considered in danger of imminent failure because, once the puddle clay corewall of an earth dam is penetrated by impounded water, complete failure can follow rapidly. Emergency measures were taken to lower the reservoir water level and the dam was later restored by inserting a concrete cutoff wall through the clay corewall into bedrock below.

Instead of taking a culvert through an earth dam it is best to drive a tunnel through the abutment where differential settlement is unlikely. The same tunnel can then be used for housing the drawoff pipes from the reservoir. Alternatively the culvert should be trenched into the foundation rather than standing proud where it can become a focus for local soil rupture due to differential settlement.

The flood discharge capacity of the diversion works required during construction should reflect the consequences of damage should a larger flood occur. Early during construction the consequences of under-capacity may be slight. Later insufficient capacity may entail loss of the entire dam. It is not uncommon therefore to design the return period for diversion which increases as construction develops. Diversion works may initially be required to accommodate the one in 5-year flood, but eventually be required to take the one in 50- or one in 100-year event immediately prior to dam closure. Each case has, however, to be considered individually taking all the relevant circumstances into account, such as the nature of the formation, the type of dam and proposed method of construction, including the possibility of cofferdamming the dam foundations which could allow some ponding to alleviate the flood peak; or some ponding might be possible when the dam reaches a certain height – providing the reservoir can be kept empty in advance of a flood. Rainfall, runoff and flood routing studies need to be carried out to find the critical circumstances applying at each stage of the construction. The prudent engineer will additionally take all extra safety precautions which are possible at low cost, but some risk is unavoidable. In the case of the Oros dam in Brazil[16] a diversion tunnel of 450 m³/s (38 900 Ml/day) capacity had been built, but 600 mm of rainfall fell on the catchment in a week and, despite all efforts to raise the half built dam to beat the rising waters the dam was overtopped on 25 March 1960 and virtually destroyed, having to be rebuilt (see Plate 6).

5.9   Provision for passing flood waters

Spillway works can be classified into three types:

(1) overflows which permit water to flow over the crest of the dam;
(2) structures constructed to discharge via an opening or culvert through a dam, or a tunnel driven through an abutment;
(3) side channel spillways which convey overflow water through an open channel constructed on natural ground at a dam abutment or at a low point on the reservoir perimeter.

The crest of a dam has generally to be high enough above the maximum water level in the reservoir under flood conditions to prevent waves and spray passing onto the crest. For an earth dam the usual practice in the UK results in the top of the embankment being 2 m higher than the sill of the overflow weir, in addition a substantial wave wall capable of protecting the bank against spray from wave action must be provided. For dams in other parts of the world – where reservoirs and floods can be of much greater magnitude than in the UK – substantially higher freeboards may be necessary, as for instance in the case of Mangla Dam (Fig. 5.2) where 10 m is provided.

Direct discharges over a dam

Direct discharges are permissible over concrete or masonry dams and some rockfill dams, if provided with a suitably designed spillway. The practice is cheap because very little alteration to the profile of such a dam is necessary to accommodate the overflow section. Typical examples are shown in Fig. 5.3 and Plates 10 and 11. The water usually cascades down a smooth or stepped face of the dam. At the foot of the dam the water must be turned into a stilling basin to disperse some of its energy, because the main danger with this type of overflow is scour at the toe of the dam during a high flood. Sometimes the water is ejected off the face of the dam by a ski-jump, which throws the water some distance away from the toe of the dam, thereby lessening the danger to the toe. It is essential that all the construction is massive and soundly based upon solid rock. The design is not appropriate to a valley where good hard rock does not appear in the river bed at the dam and for some distance downstream.

Gleno buttress dam, Italy, before and after failure on December 1923

Failure of Baldwin Hills earth-embanked reservoir, Los Angeles, 14 December 1963. Excess flow was noticed in underdrains at noon; silty water emerging from the toe was observed at 1.00 pm; a hole appeared in the upstream face at 2.30 pm. The first photo shows the breach at 3.30 pm; the second, collapse of the crest road at 3.38 pm. Complete breaching followed at 3.40 pm. (Photo: *Los Angeles Times*)

Plate 5

Oros earth dam, Brazil, overtopped by flood during construction 25 March 1960. (Photo: Water & Water Engineering)

The failed left abutment of the Malpasset concrete arch dam, France, 2 December 1959 showing the crushed concrete due to, weakness and movement of the abutment. (Photo: Water Power)

Tarbela dam, Pakistan, showing erosion of concrete in the overflow tunnel, 1974, believed due to cavitation phenomenon. (Photo: A. Garrod)

Plate 6

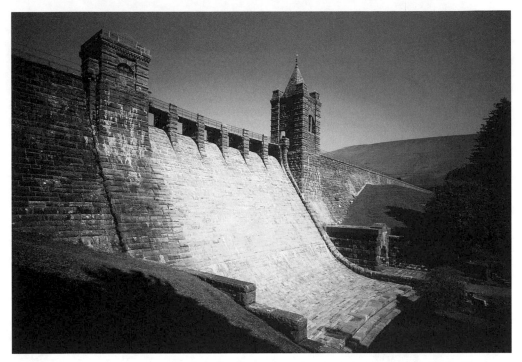

Upper Neuadd 24 m high masonry faced concrete dam, South Wales, completed 1902. (Engineer: Dr. G. F. Deacon)

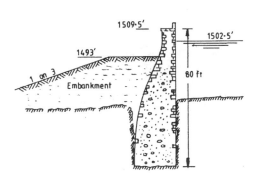

Profile of Upper Neuadd dam showing earth embankment support

Leakage through the masonry joints of the Upper Neuadd dam, 1963, due to acid moorland impounded water attacking the lime constituent of the mortar

Plate 7

Repairing the Lluest Wen 24 m high earth dam, South Wales, by insertion of a plastic concrete diaphragm wall within the old puddle clay corewall, 1971. (Engineers: Binnie & Partners)

Plate 8

Sometimes an additional feature of a concrete or masonry dam is the inclusion of automatic crest gates to the overflow weir. The gates permit water to be stored above normal top water level of the overflow sill but, after a certain level is reached, the gates are opened automatically and permit increasing discharge over the weir. This permits increased storage without increasing the flood overflow level. But, except in the case of some concrete dams on hydro-electric schemes in Scotland, gates have seldom been used on dams in the UK because storm conditions might interfere with the power supplies for operating the gates at the crucial time.

Another method is the use of a siphon hood over the weir crest. Upon a predetermined level being reached in the reservoir the siphon comes into action, and its discharge per unit length of siphon crest is more than would occur for the same water head over a fixed weir. However little increase in the discharge capacity occurs when the reservoir level rises above the siphon. The main use of siphons is to provide additional capacity to existing under-designed spillways. They are not suitable for use in cold climates where ice formation in winter might block or decrease the capacity of the siphon.

Spillway channels

Flood waters are frequently discharged via a spillway channel around the end of a dam, as shown in Plate 4. Such spillways should be constructed on natural ground and not on the dam itself where settlement will occur. The construction has frequently to be massive, of concrete or masonry, because of the need to prevent dislodgement of any part under flood conditions when the scouring force of the falling water may be very great.

At the upstream end of the channel there will be an overflow weir, usually with a rounded crest that will not be damaged by floating debris. The collecting basin immediately downstream of the weir must be designed to take the water away without any backing up of water on the weir. At the downstream end of this relatively level section of channel across the abutment, critical depth of flow will develop where the channel falls away more steeply acting. This critical depth of flow acts as a control point for the flow of water (see Sections 10.8–10.13). A design problem is that the floor of the collecting basin and channel upstream of the corewall of the dam will be subject to uplift when the reservoir is at top water level or overflowing, and the channel floor must bond to the corewall of the dam to prevent leakage below the floor. The overflow will pass at increasing speed down the steeper part of the channel, and sometimes attempts have been made to destroy some of its energy by constructing piers or steps in the channel, but they may not be very effective and can cause spray which erodes any soft material behind the walls of the channel. At the base of the spillway channel a 'stilling basin' is constructed to dissipate some of the energy of the fast-moving water. The most usual method is to induce the water to form a hydraulic jump within the stilling basin (see Section 10.12).

Bellmouth overflows

A bellmouth overflow (see Figure 5.5) may be adopted where the expense of cutting a spillway channel is great, or where a tunnel has to be constructed through one of the abutments for the diversion of the river during construction of the dam. The bellmouth must be constructed on firm ground, either clear or almost clear of the toe of the dam. If built within the body of the dam, its settlement might cause disruption of the fill of the dam, or the different levels of fill against it might cause it to tilt. The discharge over the

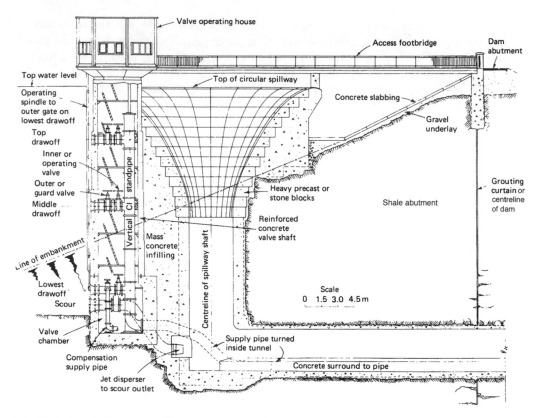

Fig. 5.5   The bellmouth spillway and drawoff tower of the Tittesworth Dam in Staffordshire (engineers: Binnie and Partners[20]).

bellmouth lip may be computed as if the lip were a straight weir of equivalent length, provided that the diameter of the vertical shaft and the dimensions of the tunnel are sufficient to take all the water away up to the designed maximum capacity. The bellmouth shaft may be vertical or sloping and it may join the tunnel with a smooth or sharp bend. The vertical profile of the bellmouth is usually designed so that no negative pressures, i.e. pressures below atmospheric, can occur on it.[17]

The junction of the base of the shaft with the tunnel requires careful design: the falling water brings with it appreciable quantities of air and the design should be such that this air does not accumulate at the soffit of the tunnel reducing its discharge capacity. Discharge characteristics vary with the head. Initially the discharge over the bellmouth lip will govern but, as the head increases, the bellmouth will become drowned and control will shift to orifice flow through the throat of the spillway shaft, or to flow through the tunnel dependent on their respective capacities. A disadvantage of the bellmouth spillway is that there will be little increase in discharge capacity after the discharge changes from weir to orifice or tunnel flow control.

Dangers arising from cavitation (see Section 12.20) must also be avoided. The phenomenon occurs when the high speed of the water reduces its pressure, so that air in solution and water vapour is released in minute bubbles. When swept to an area of higher pressure, such bubbles collapse with implosive force causing pitting of any adjacent surface followed by rapid erosion. The damage that can be done by cavitation is extraordinary. Plate 6 shows damage believed to be caused by cavitation[18] at the Tarbela dam in Pakistan in 1974. The fast flowing flood water eroded over 3 m thickness of concrete tunnel lining within about 24 hours.

An advantage of the bellmouth overflow and tunnel, apart from use of the tunnel for river diversion during construction, is that the supply pipe and compensation water pipe may be led through the tunnel. The scour valve may discharge directly to the base of the bellmouth shaft. By these means access to the pipes for maintenance is possible during periods when the reservoir is not overflowing. But a bellmouth overflow is not suitable in climates where substantial ice may form on the reservoir during winter. Also, even for small reservoirs the size of the bellmouth and its throat must be large enough to allow passage of any debris that might be brought down by the river. The bellmouth presents a danger to sailing and fishing boats when the reservoir water level is high, so guards must be fixed around its perimeter and these must also be designed not to hold back floating debris that could restrict the bellmouth discharge capacity.

5.10   Drawoff arrangements

Except when a reservoir is shallow, drawoff pipes are usually designed for withdrawing water from a reservoir at several different levels. A usual provision is three levels of drawoff: upper, middle and lower. The choice of levels is dependent upon the depth/ volume relationship of the reservoir and the expected variation of quality with depth, which may change seasonally. The upper drawoff must be a sufficient distance below top water level to avoid constant changing from it to mid-level drawoff with normal reservoir level fluctuations. The bottom drawoff level may need to allow for siltation. On each drawoff there are usually two valves, or else an outer sluice gate and inner valve. The inner valve is used for normal operation, the outer valve or sluice gate is used to permit the inner valve to be maintained. A typical arrangement is shown in Fig. 5.5.

A drawoff tower used with an earth or rockfill dam should not be sited in the body of the dam but near an abutment or elsewhere for the reasons given in Section 5.8. Where a reservoir is not impounding, but is a bunded storage reservoir for receiving pumped inflows, the tower may also be used for transfer of water into the reservoir. With a concrete or masonry dam there is no objection to making the drawoff structure part of the dam.

A scour pipe is required for a reservoir to ensure the approach channel to the lowest drawoff is kept free of silt, and to make it possible to lower the water level in the reservoir at a reasonably fast rate should any weakness become apparent in the condition of the dam. The scour capacity should preferably be enough to lower the water level by at least one-quarter of the full ponding height in a period of 14–21 days during a period of rainfall on the catchment equal to the average for the wettest month. This reduction usually means that both the volume of water stored and the stresses on the dam will be substantially reduced. Such a rate of lowering may not be feasible for a very large reservoir, but it is still important to size the scour so that, in conjunction with any other drawoff the water level can be dropped at a reasonable rate during the average wettest month.

Earth dams
5.11 Types of design

To avoid a misunderstanding it is necessary to mention that the term 'earth' does not mean that tillable soil can be used in an earth dam. All such 'soil' must be excluded because it contains vegetable matter which is weak and, in its decomposition, could leave passages for water percolation. The first operation when undertaking an earth dam construction is therefore to strip off all surface soil containing vegetable matter. The dam can be constructed, according to the decision of the designers, of any material or combination of materials such as clay, silt, sand, gravel, cobbles, and rock.

Earth dams of 'early' design had a central core of impermeable 'puddle clay', supported on either side by one or two zones of less watertight but stronger material. Examples are shown in Figs 5.1 and 5.6. The central puddle clay core is impermeable but its structural strength is low. This clay core must join to a cutoff in the ground below. The earliest dams had clay filled cutoff trenches, but use was soon made of concrete to fill the cutoff although the corewall continued to be made of clay. The inner zones of the shoulders of the dam contained a mixture of clay and stones, so was impermeable to some extent but the stony material added to its strength. The outer zones would contain less clay again and more stones, perhaps boulders and gravel. The main purpose of the shoulders was to hold the inner core of clay and add strength to the dam. Boulder clay (a glacier-deposited mixture of clay, silt, and stones with boulders) was a favourite material for this outer zone in the UK because it had the right characteristics and there were extensive deposits of it in those areas where impounding schemes developed.

Many such dams have been and continue to be successful. Their outer slopes were decided by accumulated experience and were often 3:1 upstream and 2:1 or 2.5:1

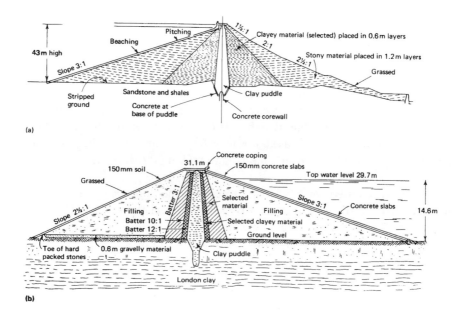

Fig. 5.6 Dams with puddle clay cores: (a) the Ladybower Dam (Hill, *JIWE*, 1949) and (b) The embankment of King George VI Pumped Storage Reservoir, London (*JIWE*, 1948).

downstream, with flatter slopes for the higher dams. The choice of material and design of the dam were decided by the engineer responsible. There were few measurement tests that could be applied. Although many new kinds of tests are now available, earth dam construction still involves judgement from experience and constant supervision as the work proceeds to ensure a sound construction. Only a minute proportion of the variable earth materials in a dam can be tested, so construction must always come under continuous detailed supervision by an experienced engineer.

Today a puddle clay core is no longer used. Instead the rolled clay core, made of a mixture of clay and coarser materials, 'rolled down' by compacting machinery, is used. It usually has a base width of 25–50% of the height of the dam to limit the hydraulic gradient across it. It is stronger than puddle clay and does not have to be placed at such a high moisture content so it is not so liable to erosion by seepage and, being spread over a wider area at the base, is not so susceptible to leakage at foundation level. A dam of this sort is shown in Fig. 5.2. Sometimes a dam may consist wholly of rolled clay with only an outer protective zone of coarser material on the waterface for protection against wave action.

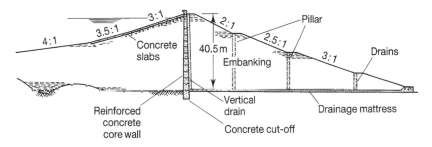

Fig. 5.7 The Gyobyu dam for Rangoon, Burma 1940 (engineers: Binnie and Partners).

An alternative used is the concrete corewall: Fig. 5.7 shows one of the earliest examples. Such a corewall may be more expensive than rolled clay, but it gives the advantage of being able to use a wide variety of materials for the shoulders, including quite permeable material such as sand. It results in a dam of good strength. The concrete corewall needs to be constructed in panels so that the joints between panels can be sealed with flexible sealers after the concrete has contracted upon cooling, and also to allow for small vertical movements due to compression of the foundation or lateral movements caused by the water level in the reservoir rising and falling. A concrete corewall is not, however, suitable where the foundations are soft.

With modern methods of analysis much more complicated designs for earth dams can be adopted than formerly. The principal aim is to use as much locally available material as possible because this keeps the cost of the dam as low as possible. The problem for the designer is how to incorporate the available materials in the dam in a manner which uses their different characteristics efficiently and safely.

5.12 Pore pressure and instrumentation in earth dams

An earth dam is built in layers of material, 375–450 mm thick, which are compacted with vibrating or heavy machinery. The material has to have an 'optimum moisture' content to achieve maximum *in situ* density. In crude terms, if the material is too wet, the voids between the grains of material become filled with water keeping the grains of material apart so that proper compaction cannot be achieved. If the material is too dry, there is insufficient water lubrication for the compaction machinery to cause the grains to move together. Hence in hot dry weather the layers of fill may have to be watered; but if the material is too wet it cannot be properly compacted. Often in the UK earth dam construction has to be suspended in winter. This however depends on the type of material being placed because different materials have different optimum moisture contents for good compaction. This optimum moisture content can be ascertained by a standard compaction test, known as the Proctor test. Often an earth dam construction specification will require the fill to achieve at least 95% Proctor optimum compaction.

As an earth dam is heightened the lower layers of material are compressed by the added load. This may cause moisture in the material to fill the pores between the material grains. Any further load is therefore resisted, not by grain-to-grain contact of the soil, but by the pore water, whose pressure therefore rises. This pressure can be measured by burying a piezometer – a small cell with a porous membrane – in the fill as shown in Fig. 5.8. The piezometer, its tubes and the pressure gauge are filled with de-aired water to avoid discrepancies caused by air coming out of solution in the water. The pore water pressure can be measured by the piezometer, and if the pressure equals the weight of the material above the instrument (termed 100% pore water pressure) an unstable condition exists. Any additional load from further heightening of the dam will therefore only be carried by the pore water, so the lack of grain-to-grain contact may cause the material above to slide out along a 'slip failure' line. Piezometers placed in the fill can therefore record the development of pore water pressure and, if this becomes too high, further placing of material in the dam may have to be stopped for a time to allow the pore water time to

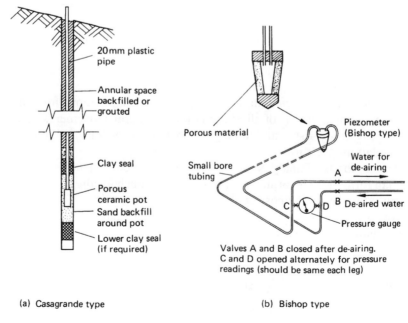

(a) Casagrande type   (b) Bishop type

Fig. 5.8   Pore water pressure sensing devices.

drain out. Drainage layers of suitably sized gravel are often incorporated at intervals in the material to speed up pore water drainage.

Piezometers placed in the upstream shoulder of an earth dam can register how the water level in the shoulder rises and falls with the reservoir water level. These can show how fast the water drains out of the upstream shoulder with a fall in reservoir water level. This is important in order to avoid a slip failure due to 'rapid drawdown'. Lowering the water level in a reservoir too rapidly may leave the shoulder material above undrained, so that its extra weight coupled with its high moisture content may induce a slip. This 'rapid drawdown condition' is particularly important during the early life of a dam when full consolidation and strength of the material may not have been reached. It is also important if faced with a need to pull down the reservoir level in some emergency; it must not be done too fast or the upstream shoulder may fail, adding to the emergency. Piezometers just downstream of the corewall to the dam at low level can show pressures which indicate leakage through the corewall – especially so if their pressure rises and falls with reservoir water level. It is thus possible to keep watch on the behaviour of an earth dam through use of piezometers.

Other instruments can measure the slow settlement of a dam; these consist of buried plates in the fill attached to non-metallic telescopic tubes. At each plate a magnet is attached so that a sensor lowered down the tube can measure their elevation. Internal horizontal movements are detected by lowering an inclinometer down a tube left in the fill, and markers permit surface movement and settlement to be detected by precise surveying, which must be based on benchmarks outside any land settlement caused by the dam and the weight of water in the reservoir.

5.13 Stability analysis in dam design

Figure 5.9 shows how a circular slip failure can occur in an earth dam if the shearing force along the slip plane is too great for the material to resist. To calculate the factor of safety against the material sliding out along the circular path. it is necessary to consider unit vertical slices of the material because the pressure on the slip plane varies from top to bottom.[19]

Taking moments about O for any slice of width δx

$$\text{Factor of safety} = F = \frac{\text{moment of weight about } O}{\text{resistance to sliding along base}} = \frac{r\delta l(c' + (N - u))\tan \phi'}{rW\sin \alpha}$$

(neglecting the force between adjacent slices). Where: r, δl, W and α are as shown in Fig. 5.9 and c' = the apparent cohesion of the material; u = average pore water pressure along δl; $N - u$ = the effective intergranular stress normal to δl; ϕ' = the angle of shearing resistance for the effective stress.

Substituting $W \cos \alpha/\delta l$ for N and simplifying this gives

$$F = \frac{\delta lc' + (W\cos \alpha - \delta lu)\tan \phi'}{W\sin \alpha}$$

If u is expressed as a percentage, p, of the weight of fill per unit area $u = pW/\delta x$, and noting $\delta l/\delta x = 1/\cos \alpha$

$$F = \frac{\delta lc' + (W\cos \alpha - pW/\cos \alpha)\tan \phi'}{W\sin \alpha}$$

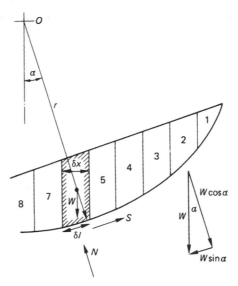

Fig. 5.9   Slip-circle stability analysis.

If c' and ϕ' can be taken as constant along the whole slip plane, the factor of safety for the whole slice

$$F \text{ (for whole slice)} = \frac{c'L + \tan \phi' \sum (W\cos\alpha - pW/\cos\alpha)}{\sum W\sin\alpha}$$

where L is the whole length of the slip.

To calculate this, the individual slice computations are tabulated and summed.

Where partly submerged slices occur, W has to be split into W_a for the weight of the portion above the water level and W_b the submerged weight below. In addition $u.\delta x/\cos\alpha$ has to be substituted for $pW/\cos\alpha$ where u is the *excess* of the pore water pressure above the water level in the fill in each case.

A large number of slip circles with different radii and centres have to be tested to find the one giving the lowest factor of safety. Some of those used in the design of the Mangla dam are shown in Fig. 5.10. Slip circle analysis applies only to unit width of the material, whereas in practice a slip can be three-dimensional, e.g. saucer-shaped. Wedge-type failures can occur as illustrated in Fig. 5.11.[20,21] These more complex modes of failure can be investigated by computer analysis which can also take account of a range of different materials in a dam, the various conditions applying during stages of construction, and fast drawdown of reservoir water level which can cause instability of the upstream embankment.

Because of uncertainties which always exist in material properties, standards of construction, confidence in predicted values of loading, scale effects and modelling inaccuracies, safety factors are always incorporated in designs. For initial construction, where pore water pressures may be temporarily high, or for transitory loadings such as earthquake, acceptable factors may be as low as 1.0–1.1. For rapid drawdown during the life of the embankment a higher figure of, say, 1.3 may be used. For normal longer term loading of the embankment, safety factors of around 1.5 are more usual.

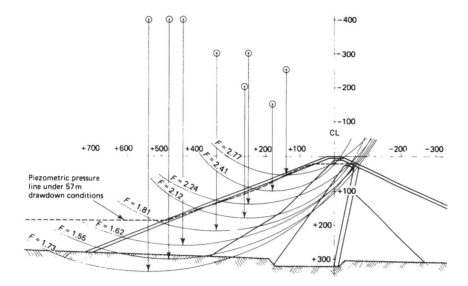

Fig. 5.10 Typical slip-circle stability analysis results for the upstream slope of a proposed design for the Mangla Dam. The results show a minimum factor of safety of 1.55 for the assumptions made. (This is Computer run no. 111 (150 circles) and is for static conditions with 57 m sudden drawdown.) (Engineers: Binnie and Partners.)

5.14   Drainage requirements for an earth dam

The preceding sections illustrate that an earth dam is subject to continuous water movements within it. Water levels in the upstream shoulder rise and fall with reservoir level; the downstream shoulder has a moisture content which rises and falls with rainfall, and it may have to accept some inflow from seepage through the corewall and from springs discharging to its base and abutments. The principal requirement is that all these water movements must not dislodge or carry away any material of the fill.

Drains are therefore necessary within a dam to accommodate the water movements that occur. However a drainage layer of coarse material, such as gravel, cannot be placed directly against a fine material such as clay, or water movement would carry clay particles into and through the drainage layer. The removal of the clay would progressively enlarge leakage paths until serious danger of disruption of the fill and breakthrough of water from reservoir would be reached. Every drainage layer of coarse material has therefore to be protected by a 'filter'. The filter material must have a particle size small enough to prevent the fine material moving into it, and coarse enough to prevent its own movement into the larger material of the drain. Sometimes two filters have to be placed when there is a large difference in size between the fine material to be drained and the coarse material of the drainage layer. All such drainage layers must convey their water individually to inspection traps where the flows should be regularly measured, and the water inspected to make sure it is not carrying suspended material from the dam. Any flow of cloudy drainage water from a dam should be tested for the amount and size of suspended solids it carries. An example of a drainage layout as adopted is shown in Fig. 5.12. The main drainage system comprises a 'chimney' drain against the downstream slope of the core for the transfer of seepage flows to a 'drainage mattress' laid at formation level, discharging to an outlet

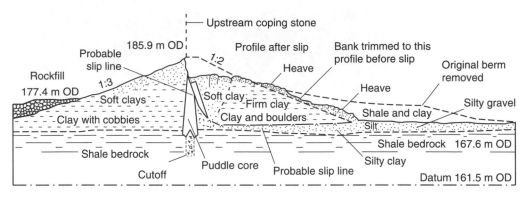

Fig. 5.11(a)   Failure of the Tittesworth dam 1960. Removal of a small berm to part of the downstream shoulder of this 100-year-old dam caused a wedge-type failure through its puddled clay corewall. Later investigations showed the dam had failed in the same area before and not been properly rebuilt.

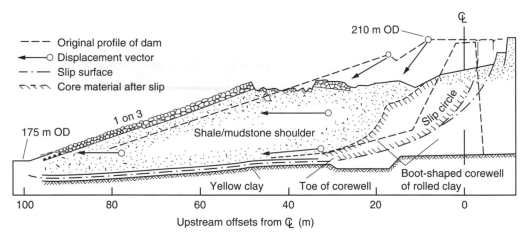

Fig. 5.11(b)   Failure of Carsington dam 1984. The mechanism of failure was attributed to overstressing of material near the toe of the corewall as construction neared full height. As each segment moved to a reduced residual strength under strain, more stress came on adjacent material leading to progressive slip circle failure.

chamber at the downstream toe of the dam. Drainage 'blankets' are laid at intervals in the upstream embankment to relieve pore pressures that build up during construction and for drainage of the upstream shoulder during rapid drawdown.

5.15   Surface protection of earth dams

The upstream face of an earth dam must be protected against erosion by water and wave action. Common forms of protection are: (1) stone pitching, (2) riprap, (3) concrete blocks, (4) concrete slabs, and (5) asphaltic concrete. Stone pitching, either grouted or open jointed, is generally found on older dams. Riprap is formed by random packing of pieces of rock; they must be large if wave action is expected to be strong and should

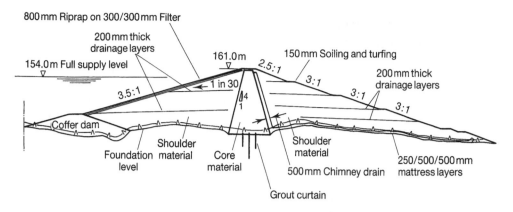

Fig. 5.12 Cross-section of 48 m high Upper Muar dam in Malaysia constructed of residual clay soil material on weathered rock overlying granite bedrock (engineers: Syed Muhammed, Hooi Binnie).

preferably be hand placed and locked together with small pieces of rock between the large pieces. Beneath the rock must lie layers of material which act as filters between the large rock and the finer material of the dam which must be prevented from being drawn through the rock surfacing. When concrete slabs are used for protection they need to be placed without open joints to avoid dislodgement by wave action, but drainage of the fill through them should be possible. They should be laid on one or two graded gravel filter layers.

For many years slope protection of dams has been designed empirically, with satis-factory results in most cases.[22] A conventional rule used by many engineers required the depth of riprap or concrete blockwork on dams in the UK to be not less than one-sixth of the 'significant wave height'.[11] The 'Rock Manual'[23] provides criteria for the grading of riprap, but further research into the behaviour of the surface protection material, granular underdrainage and embankment material is needed before detailed design procedures can be established.

Erosion of the downstream face of a dam by rainfall runoff must be prevented. It is usual to achieve this by turfing the surface of the dam or by soiling and seeding it with a short-bladed strongly rooting grass. If the dam is more than about 15 m high then, according to the type of rainfall climate, berms should be constructed on the downstream slope, each berm having a collector drain along it so that surface runoff is collected from the area above and is not discharged in large amounts to cause erosion of the slope below. Cobbles or gravel are sometimes used to surface the downstream slope. Surface drains on the downstream slope should discharge separately from the internal drains to the shoulder.

It is unwise to allow any tree growth on, or at the foot of, the downstream slope of an earth dam. Some tree roots can penetrate deeply in their search for water and drainage layers could be penetrated by such roots. Also the presence of trees may obscure proper observation of the condition of the slope. A damp patch, or development of a depression on the slope may indicate internal drainage systems not working properly or some leakage through the corewall. Any signs of this sort need to be investigated. It is best to treat the downstream slope uniformly with the same type of grass or turf throughout since this helps to disclose any discrepancies that might need investigation.

Concrete and masonry dams
5.16   Gravity dam design

It is primarily the weight of a gravity dam which prevents it from being overturned when subjected to the thrust of impounded water.[24] For the prevention of sliding, the dam must be sufficiently wide at the base, and adhere adequately to foundations of strong rock. In carrying out calculations to determine the stability of any proposed section for a dam, 'uplift' must be taken into account. Uplift is the vertical force exerted by seepage water which passes below a dam or which penetrates cracks in the body of the dam. Maximum or '100% uplift' on any section through a dam would be a triangular force as shown in Fig. 5.13 which assumes the upstream value equals the pressure exerted by the maximum water level during flood conditions above the section. Below the base of the dam, the force is trapezoidal if, at the downstream toe, there is potential uplift from any tailwater level.

Taking these three forces into account – uplift, water thrust, and weight of dam – the accepted rule is that the resultant of these forces should pass within the middle third of the section being analysed. This ensures that no tension is developed at the upstream face of the dam. At first sight the foregoing appears illogical since the assumption of 100% uplift assumes a crack exists, which the design should prevent. However, perfect construction everywhere is not likely to be achieved and cracks might progressively develop if there is any tension at the face. Even if no crack exists, there can be uplift pressure in the pores of concrete of the dam due to the passage of seepage. Figure 5.14 shows pore pressures measured inside the concrete of the Altnaheglish dam in Northern Ireland where concrete deterioration had occurred and before remedial work was undertaken. This pore pressure acts, of course, only on the proportion of the concrete which consists of voids so that the total uplift is only some proportion of the concrete area multiplied by the pore pressure.

In order to reduce uplift from seepage below a dam, a concrete-filled cutoff trench may be sunk at the upstream toe as shown in Fig. 5.14. Even if this cutoff does not make the foundation rock wholly watertight, it lengthens any seepage path and thus reduces uplift below the base. Relief wells or drainage layers can be inserted at intervals below the downstream half of the dam base; these also reduce the total uplift. With these provisions the amount of uplift assumed in the design can be less than '100% uplift'.

Ice thrust may have to be taken into account in the design of gravity dams in cold climates. Estimates[25] of the thrust force vary according to anticipated ice thickness, and range from 2.4×10^5 to 14.4×10^5 N/m^2 of contact with the vertical face of the dam. Seismic forces from earthquakes also have to be considered; it is now accepted that minor seismic damage can occur in areas, such as the UK, previously thought to be free of earthquake risk.

The design of gravity dams appears deceptively simple if only the overall principles described above are considered. In fact controversy raged for more than half a century over the subject of how stresses are distributed within a gravity dam, until computer programmes permitted computation of stresses using two- or three-dimensional finite element analysis. The method determines displacements at each node, and stresses within each element of the structure, the latter being considered for analysis as an assemblage of discrete elements connected at their corners.

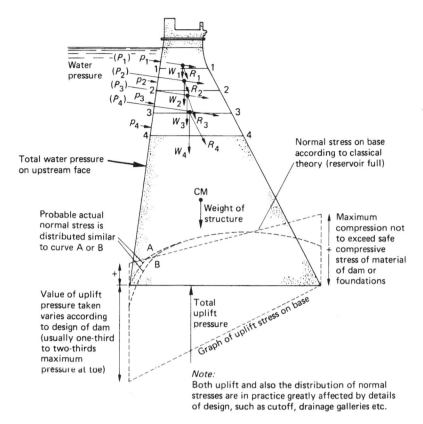

Fig. 5.13 Simple stability analysis of a gravity dam showing the effect of uplift. As well as taking into account uplift on the base of the dam, it may be necessary to take into account uplift acting below any portion of the dam above sections 1–1, 2–2, etc. (this is not shown on the diagram). If the resultants R1, R2, etc. of the forces *P* and *W*, together with uplift as deemed advisable, come within the middle third of sections 1–1, 2–2, etc. then theoretically no tension develops on the upstream face.

5.17 · Gravity dam construction

Concrete. The construction of concrete gravity dams is relatively simple. Most of the remarks concerning the preparation of foundations and cutoff trenches for earth dams apply also to gravity dams. The key problem with mass concrete dams is to reduce the amount of shrinkage which occurs when the large masses of concrete cool off. Heat is generated within the concrete as the cement sets and as this heat dissipates, which may take many months, the concrete cools and shrinks. In order to reduce this shrinkage the concrete is placed in isolated blocks and left to cool as long as possible before adjacent blocks are concreted. 'Low heat' producing cement is frequently used, or ice flakes may be added to water used in the concrete mix. In the larger installations, water cooling pipes may be laid within the concrete to draw off excess heat produced. Concrete dams are best constructed in areas where rock abounds which is suitable for the making of concrete aggregates.

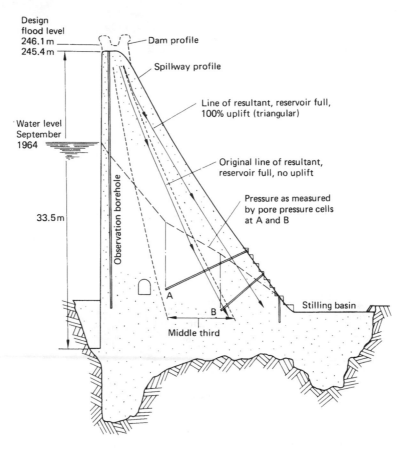

Fig. 5.14   The Altnaheglish dam.

At some locations concrete dams have been raised or strengthened by installing post-tensioned cables through the concrete mass near the upstream face, and grouting them into the foundation rock in order to provide additional vertical forces to counterbalance the overturning moments introduced by increased water loads.

Masonry. The construction of masonry dams is expensive because of the large amount of labour required to cut and trim the masonry blocks. In the UK masonry dams are no longer built, but elsewhere dams have had masonry facing and concrete hearting, primarily to improve the appearance of the dam and also to avoid the need to provide shuttering for the concrete. Overseas, particularly in India, construction of masonry gravity dams continued but seldom for large structures due to the unpredictability of masonry behaviour under seismic forces. Great care must be taken with masonry dams to fill all the joints and beds completely with a watertight mortar mix and to pay special attention to the quality and watertightness of the work on the upstream face. It was Dr G. F. Deacon's insistence on this when constructing the 44 m high masonry Vrynwy Dam for Liverpool in 1881 that has left the dam still in excellent condition nearly 120 years later. Up to then, he remarked – 'there was probably no high masonry dam in Europe so far watertight that an English engineer would take credit for it'.[26]

5.18 Roller-compacted concrete dams

Roller compacted concrete (RCC) has been developed as a time-saving technique for construction of a concrete dam since it was first used in the USA in 1982. It consists of a mixture of cement, water and rock aggregates that can support a roller and be compacted into a dense concrete material. A sound foundation similar to that required for a traditional concrete dam is needed, and then RCC can be placed in successive layers to form a monolithic gravity dam structure. Dam stability requirements are similar to those adopted for gravity dams. The major differences between RCC and ordinary concrete dams are in the design of the concrete mix, the method of placing it, and the design of spillways, outlet works, drainage and inspection galleries to facilitate RCC construction.

Initially different approaches were made to the type of concrete which would prove most economical for RCC dams.[27] One approach was to use a relatively lean and porous mix for economy for the body of the dam, relying on upstream waterproofing for watertightness, the latter usually comprising a thickness of richer, impermeable concrete on the water face. The other approach was to use a 'high paste' concrete containing more cement which was sufficiently impermeable not to need upstream waterproofing, and whose greater tensile strength permitted economy in the profile of the dam by use of steeper side slopes. Whilst both types have been adopted successfully, the lean mix dam depends on the upstream waterproofing for watertightness and can suffer seepage if this is not entirely effective; whilst the high paste RCC dam has a propensity to suffer shrinkage cracking if constructed without contraction joints. The more recent trend has therefore been towards use of a medium paste concrete, with formed or induced joints incorporating some form of water bar. In Japan a number of high RCC dams have been built incorporating formed joints with a water bar and using concrete with a cementitious content of $120–130$ kg/m^3 of which 30% may be PFA.

5.19 Arch dam design

The principle of design of an arch dam is greatly different from that of a gravity dam. The majority of the strength required to resist water thrust is obtained by arching the dam upstream and taking the load of the water upon the abutments. The abutments must therefore be completely sound; if this criteria does not apply failure will occur. The theory of design is complex: the dam resists the water partly by cantilever action from the base and partly by arching action from abutment to abutment. Hence an arch dam can be much thinner than a gravity dam (see Fig. 5.15). Early designs were based on the 'trial load' procedure. The dam was assumed to consist of unit width cantilevers one way and unit width arches the other. The water load at each point was then so divided between the 'cantilevers' and 'arches' that their deflections at every point matched. The modern method is to construct a three-dimensional finite element model of the dam to evaluate stresses under various loadings. Physical models are also often used to measure distortions likely to apply; these are useful, partly as a check to the mathematical calculations and partly as an aid to the calculations by giving a first approximation to the likely distribution of stresses.

There are many variations from the simple uniform arch shape, the most economic section being curved both vertically and horizontally, i.e. the horizontal arches vary in radii with level. A dam of this kind which curves 'both ways' is called a cupola dam and,

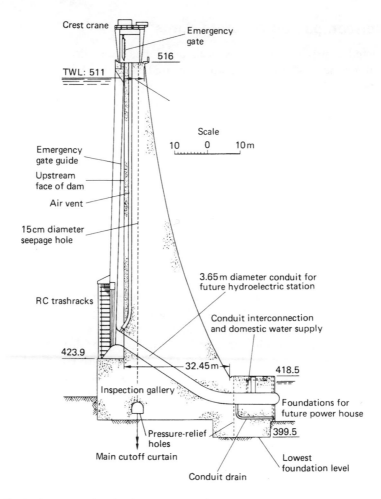

Fig. 5.15   The Dokan Dam for the Government of Iraq (engineers: Binnie and Partners) (Binnie *et al.*, *Proc ICE*, 1959).

although most economical in the use of concrete, it presents difficulty in the incorporation of drawoff and scour valves because these are usually operated from above by means of vertical spindles or shafts, but the upstream face of the dam is curved. River and flood diversions are usually taken in tunnels through the abutments; occasionally flood overspill may be passed over some central spillway. Arch concrete dams are among the highest in the world and are inherently stable when the foundations and abutments are solid and watertight. However the stresses in the concrete of the dam and in the foundations and abutments can be very high so the utmost care needs to be taken in the site investigations, and the design and construction. Whatever may be the results of the theoretical analyses of forces and stresses on the dam, the engineer responsible for the final choice of design has to ensure that a sufficient reserve strength is included in the design to meet unknown weaknesses.

5.20   Buttress or multiple arch dams

Where a valley in rock is too wide for a single arch dam, a multiple arch dam may be used which comprises a series of arches between buttresses, as in Fig. 5.16. Each section of the dam, consisting of a single arch and its buttresses, may be considered as achieving stability in the same manner as a gravity dam. There is a considerable saving of concrete as compared with a mass concrete dam: on the other hand this saving may be offset by the extra cost of the more complicated shuttering required and the more extensive surface areas requiring a good finish for appearance. A buttress dam has the advantage that uplift is negligible, because the space between the abutments allows uplift pressure to dissipate. Occasionally a buttress dam is advantageous when sound rock exists at a level which would make a mass concrete gravity dam more expensive.

Buttress dams are often used in connection with river control and irrigation works, where many gates have to be incorporated to control flow. In this case there are no arches: the buttresses are to gravity design and have to take the extra load from the gates when closed.

5.21   Rockfill and composite dams

Rockfill dams are appropriate for construction at locations where suitable rock can be quarried at or near the damsite, and foundations will not be subject to material settlement due to loading or to erosion from any seepage through or under the dam. The design must, of necessity, incorporate a watertight membrane which can comprise a concrete facing or an earth core. If concrete facing is used, the cutoff trench must necessarily follow the curve of the upstream toe of the dam in order to connect with it. The main difficulty with construction is that the watertight membrane must be flexible because rockfill dams are liable to continue to settle over long periods.

Early rockfill dams of dumped rockfill, with embankment slopes of about 1 on 1.5 and concrete faced, were satisfactory in service up to heights of about 75 m, but above that height, face cracks and excessive leakage occurred due to compressibility of the dumped rockfill. This problem was alleviated in the 1960s when vibrating rollers became standard equipment for the compaction of rockfill; this permitted the use of smaller sized and weaker rock. An earth core would be comprised of rolled clay, protected by zones of transitional material between the clay and the rockfill which act as filters and also take up progressive settlement of the rockfill. Usually two or more layers of transitional material are necessary, grading through from clayey-sand against the corewall, to a crushed rock with fines against the rockfill; the latter usually being graded with the coarsest material on the outside.

A variety of rocks can be used with modern methods of compaction; even relatively weak rocks such as sandstones, siltstones, schists, and argillites have been used. It is usual to examine all possible sources of available material and, by means of laboratory tests on samples and pilot construction fills, to base construction procedures and zoning of materials on the results of such tests. The specification for construction should set out the required layer thickness, normally 1–2 m for sound rock and 0.6–1.2 m for weak rock, and the number of passes required by a 10 tonne vibratory roller to achieve adequate rockfill breakdown and strength.[28] A concrete membrane on the water face would usually be laid on two graded filter layers, each 0.4–0.5 m thick, compacted in both the horizontal and

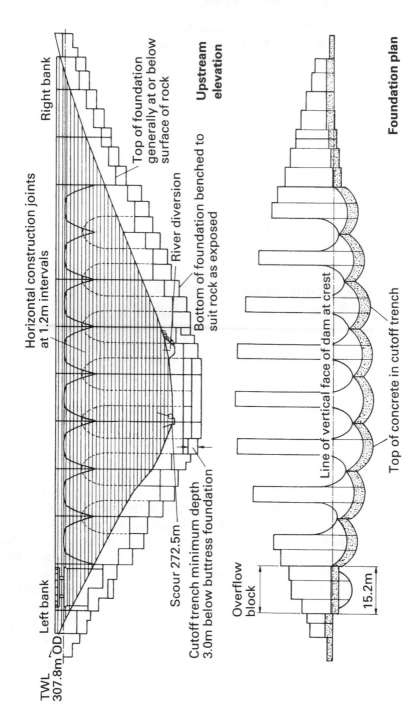

Fig. 5.16 Elevation and plan of the Lamaload Dam, Macclesfield UK (engineers: Hebert Lapworth and Partners).

sloping planes. Watertightness of the concrete slabbing is achieved by sealing vertical joints with waterbars. Horizontal joints are not required; construction joints being formed with reinforcing steel passing through them.

No concrete faced rockfill dams have experienced structural failure, hence a stability analysis on an assumed failure plane would be an academic exercise. However such an analysis is necessary when an earth core is incorporated to determine the extent to which the core strength and pore pressures affect the dam's stability.

A bituminous concrete facing can be used, and has an advantage over a concrete facing in that it can absorb larger deformations without cracking. However it can still be vulnerable to differential settlement at its connection with the cutoff. Cold weather affects the flexibility of bituminous materials, hot weather causes creep, and frequent variations in temperature can cause fatigue; and ultraviolet radiation can damage coating that is not submerged. Hence standard practice is to provide a protection layer and a membrane sealer. Modern practice is to use a single 100–200 mm thick impervious bituminous concrete layer spread by a vibratory screed and then rolled by a winched vibratory roller.

A centrally installed bituminous diaphragm wall avoids any problems with ultra violet radiation, is less vulnerable to damage, and permits the cutoff to be constructed along the centre-line of the dam. An example of an internal asphaltic core to a rockfill dam is shown in Fig. 5.17 which is one of two dams at High Island, Hong Kong,[29] 107 m high which are, unusually, founded below the sea bed across a sea inlet. The core consists of 1.2 m thick asphaltic concrete, made of 19 mm aggregate mixed with cement and water, pre-heated and mixed with about 6% of hot bitumen. Below a certain level the core is duplicated and above a certain level it is 0.8 m thick.

An unusual early design of a composite rockfill dam is the 103 m high Shing Mun Dam built in Hong Kong about 1937,[30] and shown in Fig. 5.18 and Plate 12. The upstream watertight membrane consists of heavy concrete blocks securely jointed together. This is supported by a concrete thrust block, which in turn is supported by the main rockfill. Between the concrete thrust block and the rockfill, a sand wedge is interposed which is designed to take up any movement of the rockfill. This bold and imaginative design proved successful.

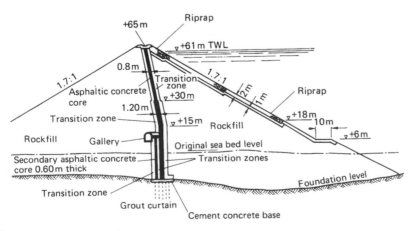

Fig. 5.17 High Island West Dam, Hong Kong, with asphaltic core membrane (engineers: Binnie and Partners[29])

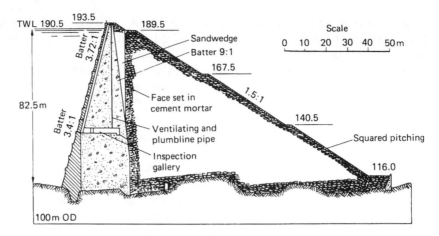

Fig. 5.18   The Shing Mun Dam forming the Jubilee Reservoir for Hong Kong (engineers: Binnie and Partners[30]).

5.22   Seismic considerations

Seismic loadings need to be considered in the design of new dams and the appraisal of old dams. In the UK the Department of the Environment has funded the preparation of a guide to the assessment of seismic risk to dams.[31] An application note for the use of the guide has also been prepared by the ICE.[32] These documents provide a structured framework for the consideration of seismic risk similar to that adopted for floods.

5.23   Statutory control over dam safety

Regulations are adopted in many countries to ensure that dams are regularly inspected, and are constructed or altered only under the charge of properly qualified engineers. In the UK the Reservoirs Act 1975[33] was not brought into force until 1986; in the meantime the earlier Reservoirs (Safety Provisions) Act 1930 remained in force. The 1975 Act followed the provisions of the previous Act but added some further requirements. The Act applies to all reservoirs having a capacity of 25 000 m³ or more above the level of any part of the land adjoining the reservoir and requires the Department of the Environment to establish panels of qualified civil engineers authorised to design, supervise the construction or alteration of, and inspect all such reservoirs. The main provisions are as follows.

(1) Initially the *county and metropolitan local authorities* were made responsible for keeping a register of all reservoirs in their areas to which the Act applied, and were required to act as enforcement authorities to see that the owners of such reservoirs complied with the Act. Some proposals have been made for the *Environment Agency* to take over the register of dams and to act as the enforcement authority, but have not yet been implemented.

(2) Where a new reservoir is proposed or an existing reservoir is to be altered, it must come under the control of a 'construction engineer' during design and construction, and afterwards, until this engineer issues a Final Certificate. He must issue a Preliminary Certificate permitting first filling of the reservoir, and his Final Certificate

must be issued not earlier than 3 years nor later than 5 years after his issue of the Preliminary Certificate.

(3) After the Final Certificate mentioned in (2) above, an independent 'inspecting engineer' must inspect the dam within 2 years and thereafter the same or another 'inspecting engineer' must inspect the reservoir at not less than 10-year intervals, or at such lesser interval as is stipulated in any previous Certificate issued. The inspecting engineer cannot be the same person who acted as 'construction engineer', nor may he be an employee of the owner of the reservoir.

(4) A panel (i.e. list) of engineers who are permitted to act as above is drawn up by the Secretary of State. Appointments of engineers to the panel are to be made for 5 years only, although an existing panel engineer whose term of appointment has expired can re-apply for appointment.

(5) Various prescribed formats for records, reports and certificates are set out in Regulations issued under the Act.

The provisions of this Act and the earlier Act of 1930 have worked well in that they have assisted in preventing any dam disaster in the UK for three-quarters of a century. An unusual character of the Act is that it imposes a personal responsibility on the engineer who issues a certificate. Such responsibility can only be effectively exercised by an engineer having adequate experience of dam design and construction, and who has access to the specialist services frequently required in making a proper inspection. Many dams in the UK are 100–150 years old and may require careful attention to ensure their continued safety.

5.24 Dam deterioration signs

Routine observations to ensure a dam remains in good condition are too numerous to list here. Reference can be made to publications by the BRE[22] and by CIRIA[24] which provide detailed check lists, but each dam requires its own specific programme of monitoring. Obviously all instrumentation of a dam, such as settlement or tilt gauges, pore water pressures and underdrain flows need regular monitoring plus checking for accuracy. Only some matters of importance, principally related to signs of leakage, can be mentioned here.

Good access to a dam is important and should be suitable for heavy constructional plant which might be needed for repairs. The situation arising if emergency work should be required has to be envisaged. Night work under heavy rainfall and high winds with a flood overflow may then be necessary. All gates, valves and other mechanical controls should be in easily operational condition, and access to shafts, galleries and inspection pits, etc. should be safe, properly ventilated and lit to safeguard against accidents to personnel adding to the troubles of an emergency.

The catchment needs monitoring for important changes of use, and for any evidence of hillside movement that could indicate instability.

For earth dams leakage may be evidenced by damp patches on the downstream embankment or areas with unusually luxuriant plant growth, increased underdrain flows or pore water pressures. But leakage may also cause settlement. When properly built an earth dam should have smoothly regular upstream and downstream slopes; and the crest should either be straight or to a pre-formed curve and show an even rise towards the centre or highest part of the dam because a settlement allowance should have been given, usually of about 1% of the dam height. Hence any irregularity of line and level of the crest, or any depression in the embankments may be evidence of leakage. The upstream facing of the dam, whether of riprap or concrete slabbing, also needs to be checked for settlement, sometimes easily evidenced by the reservoir waterline against the dam not following an even curve due to the settlement allowance, but showing a re-entrant curve. Concrete slabbing of the upstream face should be checked for signs of uneven settlement which may be caused by wave action pulling out the supporting material which is usually of gravel.

The ground at the toe of an earth dam is often wet, sometimes supporting a growth of rushes. The area is. of course, a natural collecting point for drainage of rainfall percolating the downstream bank and adjacent hillsides. But a wary approach is necessary, because undue wetness at the toe of an earth dam may be caused by a hidden defect within the body of the dam, such as cracking, erosion or settlement of the corewall. All underdrain inspection pits and drainage water should be examined for evidence of silt or clay being carried out of the dam. The puddled clay corewalls of old dams are particularly vulnerable to erosion through leakage, and thought must be given to the possibility the dam has suffered leakage or settlement in the past which has not been effectively dealt with.

Concrete dams can show signs of leakage by damp patches on the downstream face or by growth of moss and lichen at joints. Signs of stress or movement in a concrete dam are spalling of concrete at joints, opening up of joints and cracks, or displacement irregularities, both on the surface of the dam and in any internal gallery or shaft. Since displacement of shuttering when the dam was constructed, or initial settlement which has ceased, can cause irregularities, it is important to log them and the areas where no irregularities occur, so that any new signs of movement can be detected. In masonry or masonry faced dams, leakage will be through joints, and cracking or fallout of pointing needs investigation to see whether it is caused by weathering, mortar softening, increased stress, or possibly acid water seepage attack.

The flood discharge works to any dam need regular inspection for signs of settlement and for any other irregularities which could induce scour which might damage the works during a flood overflow.

A 'trained eye' is necessary for the inspection of dams, especially for old earth dams, where a small surface defect may eventually prove to be related to a dangerous condition of the dam developing internally. Hence the need to investigate the cause of anything that seems 'not quite right' or 'as it should be'.

5.25   Reservoir sedimentation

Figure 5.19 shows the range of suspended sediment loads experienced in rivers of different size as reported by Fleming.[34] To these must be added perhaps 10–20% to cover bed load which cannot normally be measured. Care must be taken to allow for unusually high loads following earthquake activity or landslides, or where loess or recent volcanic ash deposits are present. However, it is at reservoirs that sedimentation most concerns the water engineer. Dams in areas of active soil erosion have been known to fill rapidly with sediment in only a few years. Algeria has provided some of the best known cases of severe sedimentation. Of 13 basins, eight produced 60–700 t/km^2 and five 1100 t/km^2.[35] However, serious sedimentation need not be expected where there is good cover of vegetation. A survey of a few drained reservoirs in England showed a trapped sediment volume which can be expressed as between 40 and 90 parts per million of inflow (by volume) over a life of about 100 years.[36]

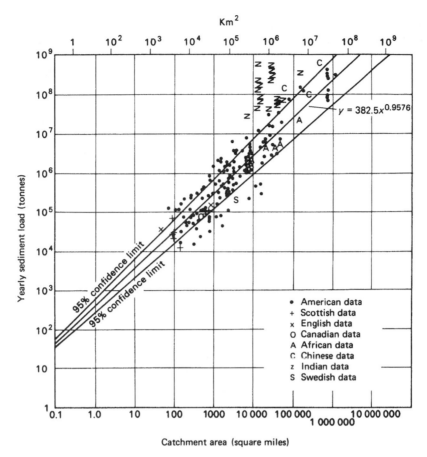

Fig. 5.19   Average suspended sediment concentration (after Fleming[34]).

A sediment rating curve giving a relationship between flow and solids carried may be obtained by sampling,[37] and then the average yearly amount carried into a reservoir may be obtained from a cumulative flow duration curve. However it is well known that flow on a rising flood frequently carries a larger proportion of sediment than on a falling flow, and also the bed load contribution of material rolled or shifted along the bed of a river will not be included. A severe flood may also move more material into a reservoir than is contributed by several years of moderate inflow. Sounding of the deposits in a reservoir at regular intervals of 5 years or so is a more direct method of measurement; and operation of the scour valve will show whether heavy sedimentation near the outlet works is taking place.

5.26   Fish passes

Where trout or salmon exist in a river it is necessary to provide a fish pass around a dam. The design of fish passes is a matter on which specialist advice should be taken. In 1942 the Institution of Civil Engineers published a report of the Committee on Fish passes which gave some suggested designs. A fish pass consists of a series of shallow pools arranged in steps, down which a continuous flow of water is maintained. It is essential to keep this flow running with an adequate amount of water because when trout and salmon are in migration upstream they will only be attracted to the pass by a good flow of water which should be turbulent. Examples of fish passes are shown in Fig. 5.20. An alternative to the fish pass is the automatic fish lift. This consists of an upper and lower chamber connected by a sloping conduit through which a continuous flow of water is maintained. Fish migrating upstream enter the lower chamber. In due course the major discharge from this lower chamber is automatically closed and the water level then floods up through the conduit to the upper chamber. The fish rise with the water and then proceed out into the reservoir at higher level. The cycle of operations is continuously repeated by automatic controls.

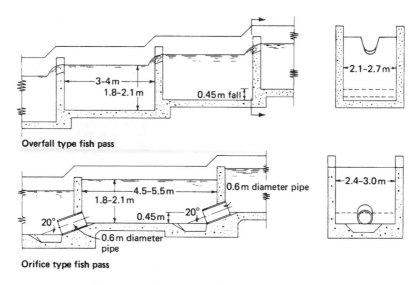

Fig. 5.20   Two types of fish pass used in Scotland (Fulton, *Proc ICE*, 1952).

5.27   River intakes

It is surprisingly difficult to design a satisfactory river intake. Figure 5.21 shows the conventional solution of a weir across a river to raise its level, with a gated side intake just upstream. This solution is feasible only for rivers of moderate size which do not carry cobbles, boulders or debris during a flood – and also those rivers which do not alter their course from time to time. On wide rivers, or rivers subject to high flood flows, the cost of a weir becomes prohibitive because of the massive works necessary to prevent the weir being undermined by scour and destroyed. On a rock foundation the weir will be less costly, but a frequent problem is how to cope with the silt, gravel, cobbles and boulders brought down by the river. In some rivers this bed load can be sufficient to fill the weir basin every time there is a flood. In Cyprus where such conditions are liable to occur, the 'Cyprus groyne intake' shown in Fig. 5.22 has been developed. The groyne weir, extending only partly into the river, is sufficiently low to be overswept in a flood. Many other designs have been tried for intakes on fast flowing rivers which carry much bed load. A round 'hump-backed' weir with slots in its crest feeding an intake conduit is sometimes used.

Illustration of wave spray overtopping the crest of a dam during storm conditions and wave discharge to spillway. Loch Quoich dam of the North of Scotland Hydro-Electric Board. (Photos: Richard Costain)

The 138 m high Mangla earth dam main spillway, Pakistan, under construction 1967, designed for 25 000 m³/s overflow rate. (Engineers: Binnie & Partners)

Plate 9

Upper Silent Valley 38 m high concrete gravity dam, with precast concrete block facing, built for Belfast 1957. (Engineers: Binnie & Partners)

Kalatuwawa masonry faced concrete dam, 21 m high, built for Colombo, Sri Lanka. Completed 1954. (Engineers: Binnie & Partners)

Plate 10

The Waterfall 55 m high masonry faced concrete gravity dam of the Tai Lam Chung water scheme for Hong Kong, completed 1957. (Engineers: Binnie & Partners)

Mudhiq concrete arch dam 73 m high, Saudi Arabia, 1980. (Engineers: Binnie & Partners)

Plate 11

The 85 m high Gorge dam of the Shing Mun water scheme for Hong Kong, comprising a concrete block diaphragm upstream, supported by a concrete thrust block and a downstream rockfill embankment, completed 1936. (Engineers: Binnie & Partners)

The 116 m high concrete arch dam at Dokan, Iraq, completed in 1959. (Engineers: Binnie & Partners)

Plate 12

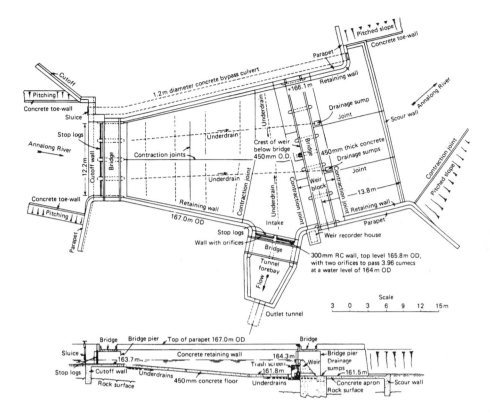

Fig. 5.21 The Annalong River intake for Belfast Water Supply (engineers: Binnie and Partners).

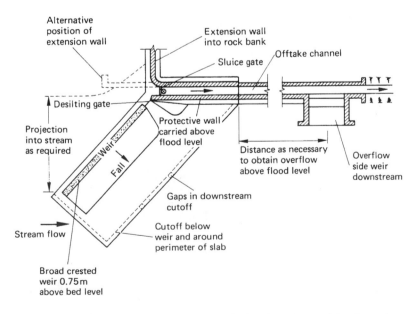

Fig. 5.22 Groyne intake for small abstraction on a flashy river (as used in Cyprus).

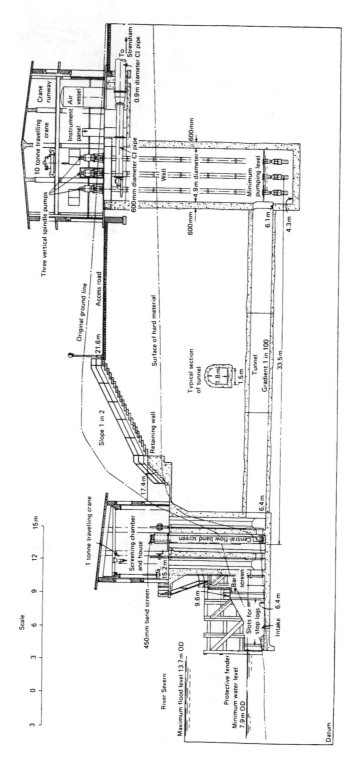

Fig. 5.23 The River Severn intake for Coventry Water Supply. Engineers: Binnie and Partners. (Hetherington and Roseveare, *Proc ICE*, 1954.)

With all such intakes the problem is how to obtain a self-cleansing offtake which is effective from high flows to low flows.

If the designer is fortunate he may be able to position the intake at a point where the river is consistently deep and constrained between stable banks. An intake of the type shown in Fig. 5.23 is then possible, usually sited on the outside of a bend in the river. Problems of avoiding siltation and fish entry have to be dealt with. Usually the intake is designed for about 1.0 m/s inflow rate, but this is a compromise; it will permit fish to swim out if they get in, it should keep siltation down to a reasonable amount, and it might avoid dragging into the intake the various floating and half submerged debris that a river often carries. In practice none of these things may be prevented entirely and, if bar screens of too fine an aperture are installed, the situation may be made rather worse by too much collecting on the screens which have to be cleaned daily – often by hand. A more costly solution is installation of an inclined heavy duty band screen if the water level is fairly constant, the perforated lifting plates being of mild steel and equipped with hooks, the band motive power being sufficient to pull out heavy debris such as bale of hay, or a tree branch. Economy can be achieved by the band screen only moving when the differential water head across it due to build up of debris exceeds a pre-set value. High pressure water

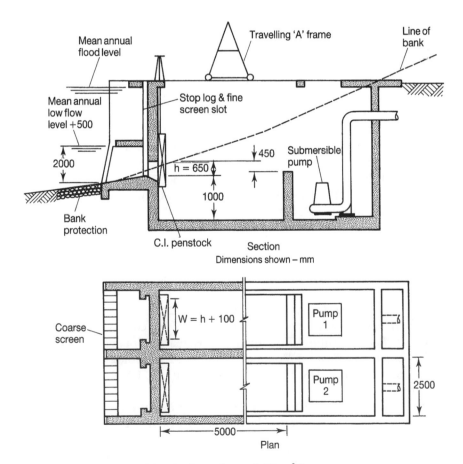

Fig. 5.24 A river side channel intake for a flow of 600 m³/h.

jets wash debris off the band screen into a hopper at the head of the bandscreen, so that manual labour is reduced.

A type of intake widely used on Malaysian rivers which carry heavy silt loads after rainfall, is shown in Fig. 5.24 for abstractions up to 0.5 m^3/s. For larger intake quantities the settlement bays are constructed with hopper bottoms for collection of silt which is removed at intervals by suction pipes after agitation with air delivered through pipes to the hoppers.

Infiltration galleries and bankside collector pipes (described in Section 4.16) are other means of abstracting water from a river which avoid the need for construction of a weir or direct intake. Sometimes a piled crib from which submersible pumps are suspended in the river can be used if the river flow is deep enough; otherwise suction pumps must be used with their suctions dipping into the water. Care must be taken to ensure pump strainers are of large size so that a pump can keep going with a partially blocked strainer, and reverse flushing of strainers should be possible. A crib intake is relatively cheap, but it may need to be of very sturdy construction if there is boat traffic on the river. Floating pontoon intakes have been used, but usually only for relatively small abstractions. They have the advantage of rising and falling with the river level; their disadvantage is the need for secure anchoring if the river is subject to a high flood flow.

References

1 Smith, N. *A History of Dams*. Peter Davies, London 1971

2. Jaeger C. The Vaiont Rock Slide. *Water Power*, 17 March 1965, pp. 110–111, and 18 April 1965, pp. 142–144.

3. *Engineering News Record*, 7 Jan. 1926 and 25 Nov. 1926.

4. McIldowie G. The Construction of the Silent Valley Reservoir. *Trans ICE*, **239**, 1934–1935, p. 465.

5. Carlyle W.J. Shek Pik Dam. *JIWE*, March 1965, p. 571.

6. Ischy E. and Glossop R. An Introduction to Alluvial Grouting. *Proc. ICE*, 21 March 1962, p. 449. See also Geddes W. E., Rocke S. and Scrimgeour J. The Backwater Dam. *Proc. ICE*, **51**, March 1972, p. 433.

7. The Failure of the Bouzey Dam (Abstract of Commission Report). *Proc. ICE*, **CXXV**, 1896, p. 461.

8. *Engineering News Record*, 29 March 1928, p. 517, and 12 April 1928, p. 596.

9. Jaeger C. The Malpasset Report. *Water Power*, 15 Feb. 1963, pp. 55–61.

10. *Engineering News Record*, 7 August 1924, pp. 213–215.

11. *Floods and Reservoir Safety: An Engineering Guide*, 3rd edn. ICE, 1996.

12. *Flood Studies Report*. Natural Environmental Research Council, London, 1975.

13. *Flood Estimation Handbook*, 5 Volumes. Institute of Hydrology, Wallingford, 1999.

14. *Design of Small Dams*, US Bureau of Reclamation, US Govt Printing Office, Washington, 1977.

15. Twort A. C. The Repair of Lluest Wen Dam. *JIWES* July 1977 p.269

16. The Breaching of the Oros Dam, North East Brazil. *Water & Water Engineering*, August 1960, p. 351.

17. *Design of Small Dams*. US Bureau of Reclamation, US Govt Printing Office, Washington, 1977, Section F.

18. Kenn M. J. and Garrod A. D. Cavitation Damage and the Tarbela Tunnel Collapse of 1974. *Proc. ICE*, Feb. 1981, p. 65.
19. Bishop A. W. The Use of the Slip Circle in the Stability Analysis of Slopes. *Geotechnique*, ICE, March 1955, p. 7.
20. Twort A. C. The New Tittesworth Dam. *JIWE*, March 1964, pp. 125–179.
21. Skempton A. W. and Vaughan P. R. The Failure of Carsington Dam, *Geotechnique*, ICE, March 1993, pp. 151–173 and Dec. 1995, pp. 719–739.
22. Johnston *et al. An Engineering Guide to the Safety of Embankment Dams in the United Kingdom*, Report 171. Building Research Establishment, 1990, new edn 1999.
23. CUR, *Manual on the Use of Rock in Hydraulic Engineering*, Report 169, Centre for Civil Engineering Research and Codes, Balkema (Rotterdam), 1995.
24. Kennard M. F., Owens C. C. and Reader R. A. *An Engineering Guide to the Safety of Concrete and Masonry Dams in the UK*. Report 148, CIRIA 1995.
25. Davis C. V. *Handbook of Applied Hydraulics*. McGraw-Hill, Basingstoke, 1969.
26. Deacon G. F. The Vyrnwy Works for the Water Supply to Liverpool. *ICE Trans*, **CXXVI** Pt IV, 1895–1896, pp. 24–125.
27. Mason P. J. The Evolving Dam. *Hydro Power & Dams*, Issue Five, 1997, pp. 69–73.
28. Cooke J. B. Rockfill and Rockfill Dams. *Stan Wilson Memorial Lecture*, University of Washington, 1990.
29. Vail A.J, The High Island Water Scheme for Hong Kong. *Water Power & Dam Construction*, Jan. 1975, p. 15.
30. Binnic W. J. E. and Gourley H. J. F. The Gorge Dam, *JICE*, March 1939, p. 174.
31. Charles J. A., Abbiss C.P., Gosschalk E.M. and Hinks J.L. *An Engineering Guide to Seismic Risk to Dams in UK*. ICE 1998.
32. *An Application Note to 'An Engineering Guide to Seismic Risk to Dams in UK'*. ICE, 1998.
33. The Reservoirs Act 1975 and Associated Regulations, 1985 and 1986.
34. Fleming G. Design curves for suspended load estimation. *Proc. ICE*, **43,** 1965, pp. 1–9.
35. Hydrotechnic Corporation, Report on *Resources en Eau de Surface et Possibilités de leur Amendagement*. Region d'Algerie Orientale, April 1970.
36. Rodda J., Downing R. A and Law F. M. *Systematic Hydrology*. Newnes-Butterworth, Oxford, 1976, Table 6.13.
37. United States Geological Survey, *Techniques of Water Resources Investigations of the USGS*, Book 3, Chapters C1–C3 on fluvial sediments, US Govt. Printing Office, Washington 1970.

6

Chemistry, microbiology and biology of water

6.1 Introduction

This chapter is divided into five parts.

Part I lists alphabetically some of the more usual physical and chemical characteristics of water and describes their significance. Reference is made to the following standards:

(1) the World Health Organisation *Guidelines for Drinking-Water Quality* 1993[1] and subsequent editions and amendments;
(2) the European Commission Directive on the quality of water intended for human consumption, 1998;[2]
(3) the Water Supply (Water Quality) Regulations 1989, and subsequent amendments, which apply to England and Wales,[3] with similar regulations applying to Scotland and Northern Ireland;
(4) the United States Environmental Protection Agency (USEPA) National Primary Drinking Water Regulations, as specified in the 1996 amendments to the Safe Drinking Water Act.[4]

Part II looks at the derivations of the aforementioned standards and discusses their application. It is important to appreciate that the WHO guideline values, which are of world-wide application, are advisory values. The EC Directive applies only to member states of the European Community. The Directive sets numerical standards for health related chemical parameters, along with standards for a number of indicator parameters. The latter are not health related but are set for monitoring purposes. Member States are required to adopt the Directive into national law within two years of it coming into force. Member States are allowed until December 2003 to achieve compliance with most of the standards specified in the Directive (i.e. 5 years after the Directive came into force), at which point the predecessor Directive of 1980[5] will be repealed. The current UK regulations incorporate the maximum admissible concentrations specified in the 1980 Directive. The standards for some parameters, such as lead, are more stringent and limits are also specified for a number of parameters not included in the 1980 Directive. These regulations are currently mandatory upon water undertakers in the UK. New UK regulations will be implemented, possibly by 1 January 2001, when the 1998 EC Directive has been transposed into national law. The USEPA regulations set 'maximum contaminant levels' (MCLs), which are mandatory, and secondary levels (SMCLs) or maximum contaminant level goals (MCLGs) which are not mandatory. It should be noted that the unit of measurement and the notation for a particular parameter can vary between the above standards. The WHO Guidelines and the European standards use a mixture of mg/l and μg/l, whilst the USEPA standards are mainly in mg/l.

Part II also looks at the levels of monitoring and analysis required for chemical parameters.

Part III looks at the microbiology of water and the most common waterborne diseases; the requirements for the bacteriological quality of drinking water as set out in the above standards; and the testing of water for pathogenic organisms.

Part IV looks at water biology in terms of the significance of macro organisms on water quality and Part V looks at new areas of concern in respect of drinking water quality.

Part I Significant chemical and physico-chemical parameters in water

6.2   Acidity

An acid water is one which has a pH value of less than 7.0 (see Section 6.33). The acidity of many raw waters is due to natural constituents, such as dissolved carbon dioxide or organic acids derived from peat or soil humus. These are unlikely to lead to pH values much below 5.5. Some apparently unpolluted moorland waters may have pH values below 4.5 due to 'acid rain' which is formed when atmospheric sulphur dioxide derived from the burning of fossil fuels combined with water vapour to form dilute sulphuric acid.

Surface waters can sometimes be contaminated with acidic industrial effluents. Acidic wastes from disused mines can also provide a significant source of acid contamination to waters in some parts of the world.

Whilst a pH value of 7.0 is termed 'neutral', a water with a pH value of 7.0 or above can still be corrosive depending on the amount of carbonate hardness and free carbon dioxide present (see Section 6.12). Certain treatment processes, notably coagulation with aluminium or iron sulphate, can lower the pH of the water and an alkali such as lime may need to be added to the water to make the final treated water non-corrosive. The various guidelines and standards specify slightly dissimilar ranges of pH value in drinking water, but the main criteria are to adjust the pH value to achieve a non-corrosive water in supply.

6.3   Algal toxins

Algae are discussed in detail in Part IV.

Blooms of blue-green algae can produce toxins which, if concentrated by reason of the algae collecting in large masses at the shallow margins of a reservoir, may prove fatal to fish or to animals drinking at the water's edge.[6] Some algae such as *Microcystis, Oscillatoria* and *Anabaena* produce hepatotoxins, the best studied of which is Microcystin LR. Other algae, including *Oscillatoria* and *Anabaena*, produce neurotoxins. The WHO Guidelines give a provisional guideline value of 1 μg/l for Microcystin LR.

Activated carbon has been used successfully to remove Microcystin from raw waters and ozone has also been shown to be effective at breaking down Microcystin into less toxic by-products.[7,8]

6.4   Alkalinity

In a general sense 'alkalinity' is taken to mean the opposite of 'acidity', i.e. as the pH value increases (see Section 6.33) alkalinity increases. In a more accurate sense, however, the alkalinity of a water principally comprises the sum of the bicarbonates, carbonates, and hydroxides of calcium, magnesium, sodium and potassium. Calcium and magnesium bicarbonates predominate in waters which are associated with chalk or limestone and comprise the temporary hardness of a water (see Section 6.24). Where the alkalinity is less than the total hardness, the excess hardness is termed permanent hardness. Conversely where the alkalinity is greater than the total hardness, the excess alkalinity is usually due to the presence of sodium bicarbonate which does not affect the hardness of the water. Because bicarbonate ions can exist at pH values below pH 7.0, a measurable alkalinity is still obtained with 'acidic' waters down to pH values of 4.5.

   Alkalinity provides a buffering effect on pH which is an important factor in many water treatment processes. It is also a key factor in determining the corrosive nature of a water. There is no maximum value for alkalinity in the WHO Guidelines. There is a minimum standard for alkalinity in a water which has been softened or desalinated in the current UK regulations.

6.5   Aluminium

Aluminium can occur in detectable amounts in many natural waters as a consequence of leaching from the sub strata. It is also found in the run-off from newly afforested areas. However the most usual source of aluminium in public water supplies comes from incomplete removal during the treatment process where an aluminium salt is used as a coagulant (see Section 7.21). It should be possible, by taking a holistic approach to optimising water treatment processes, to keep the residual aluminium in water going into supply after clarification and filtration to less than 0.05 mg/l (or 50 μg/l) Al. A percentage of this aluminium is still likely to settle out in the distribution system and will tend to accumulate as flocculant material, especially in areas where the flow is low. Inadequate treatment or deficiencies in process control can result in much higher concentrations of aluminium in the water leaving the treatment works. This is also likely to deposit as flocculant material within the distribution system and any disturbance of these sediments, either through flow reversals or changes in flow, may result in consumer complaints of 'dirty' water. The WHO Guidelines recommend that aluminium concentrations should not exceed 0.2 mg/l Al in the water in supply in order to avoid such complaints. This is the same level as the current UK standard of 200 μg/l and also the standard in the EC Directive, where it is an indicator parameter. Aluminium is likely to remain as a mandatory parameter in the new UK Regulations. The USEPA sets an advisory SMCL for aluminium in public water systems of 0.05–0.2 mg/l.

   Although the above limits set for aluminium are based on the aesthetic quality of the water supplied, concerns have been expressed about its possible association with neurotoxic effects. It has been established that the aluminium content of water used for renal dialysis should be no greater than 0.01 mg/l to avoid neurological problems in dialysis patients. Water companies in UK have agreed to notify health authorities of supplies where this level is exceeded, but such a low level can only be ensured by additional water treatment at the point of use. There has also been evidence thought to suggest a possible connection between aluminium in drinking water and the incidence of Alzheimer's disease. This was examined by the UK Department of Health CASW Committee (CASW – Committee on Medical Aspects of the Contamination of Air, Soil and Water) but the relationship was not found to be established. Official advice is that further research is necessary before any change in the use of aluminium compounds in water treatment could be justified.[9] A connection would appear unlikely in view of the small contribution of drinking water to the total daily intake of aluminium.

6.6   Ammoniacal compounds

Ammonia is one of the forms of nitrogen found in water (see also Section 6.30). It exists in water as ammonium hydroxide (NH_4OH) or as the ammonium ion (NH_4^+), depending on the pH value, and is usually expressed in terms of mg/l 'free' ammonia (or 'free and saline ammonia'). 'Albuminoid ammonia' relates to the additional fraction of ammonia liberated from any organic material present in the water by strong chemical oxidation. 'Kjeldahl nitrogen' is a measure of the total concentration of inorganic and organic nitrogen present in water.

Ammoniacal compounds are found in most natural waters; they originate from various sources, but the most important is decomposing plant and animal matter. Increased levels of free ammonia in surface waters may be an indicator of recent pollution by either sewage or industrial effluent; the ammonia level in a typical UK sewage effluent being of the order of 50 mg/l N. However some deep borehole waters which are of excellent organic quality may also contain high levels of ammonia as a result of the biological reduction of nitrates. The source of any substantial amount of ammonia in a raw water should always be investigated, especially if it is associated with excessive bacterial pollution.

The level of free ammonia in a raw water is of importance in determining the chlorine dose required for disinfection. Chlorine first combines with any ammonia present in the water to form chloramines. Free chlorine, which is a more effective disinfectant than chloramine, can only be formed on completion of this reaction (see Section 9.9). Ammonia removal may have to be considered if the concentration is sufficiently high to create a large chlorine demand (see Section 8.27).

The WHO considers that there is no health risk associated with the levels of ammonia found in drinking water and suggests a maximum level of 1.5 mg/l ammonia to avoid taste and odour problems. The current UK standard for ammonia in drinking water is 0.5 mg/l as NH_4, which is the same concentration in the EC Directive where ammonia is an indicator parameter.

6.7   Arsenic

Arsenic is toxic to humans and, if detected in water, the origins should always be investigated. It is unusual to find significant levels of arsenic in UK waters, but it is a natural water contaminant in many parts of the world particularly in areas of geothermal activity. Arsenic can also be found in surface runoff waters from areas where there are certain types of metalliferous ore or mining waste tips. Its presence may also be the result of pollution from weedkillers and pesticides containing arsenic.

High levels of arsenic in the drinking water supplies in parts of Bangladesh have recently caused significant health problems. The WHO has set a provisional guideline value for arsenic of 0.01 mg/l. The current UK standard is 50 μg/l which is the same as the USEPA interim standard for arsenic of 0.05 mg/l as a maximum contaminant level. The US standard will be included in a final National Primary Drinking Water Regulation to be finalised by 1 January 2001. The EC Directive sets a much lower maximum value of 10 μg/l and a number of source waters in Europe are likely to require treatment to achieve this level.

6.8   Asbestos

The widespread use of asbestos cement pipes for conveyance of water has raised the question whether the resulting fibres found in water are a danger to health. The Water Research Centre (WRc) found that some drinking waters in the UK can contain up to 1 million asbestos fibres per litre, over 95% of which are less than 2 microns in length.[10] The CASW Committee (see Section 6.5) reported in 1986 that there was a substantial body of evidence that asbestos, as found in drinking water, did not represent a hazard to health.[9] The WHO likewise considers that the concentrations of asbestos normally found in drinking water are not hazardous to health. Asbestos is not included as a parameter in the EC Directive and there is no UK standard. The USEPA sets a MCL of 7 million asbestos fibres of more than 10 μm in length per litre.

6.9   Biochemical oxygen demand (BOD)

The BOD test gives an indication of the oxygen required to degrade biochemically any organic matter in a water, as well as the oxygen needed to oxidise inorganic materials, such as sulphides. The test provides an empirical comparison of the relative oxygen requirements of surface waters, wastewaters, and effluents. For example, if a sewage effluent of high BOD is discharged into a stream, the oxygen required by biological organisms to break down the organic matter in the effluent is taken from the overall oxygen content of the receiving water. This depletion could potentially destroy fish and plant life.

The UK Royal Commission on Sewage Disposal in 1915 observed that unpolluted rivers rarely had BOD values of more than 2 mg/l and could accept added pollution up to a total BOD value of 4 mg/l with no apparent detriment. This gave rise to the maximum permissible BOD value of 20 mg/l in a sewage effluent entering a watercourse, with an 8:1 dilution with freshwater.

6.10   Bromide and iodide

Seawater contains 50–60 mg/l bromide so the presence of bromide in well or borehole sources near the coast could be evidence of seawater intrusion. There is no standard for bromide in drinking water; however its presence may lead to the formation of bromate as a disinfection by-product during water treatment (see Section 6.21).

Many natural waters contain trace amounts of iodide, usually at levels of less than 10 μg/l, although higher concentrations are found in brines and brackish water. The levels found in drinking water are unlikely to contribute significantly to dietary requirements.

6.11 Calcium

Calcium is found in most waters, the level depending on the type of rock through which the water has passed. It is usually present as calcium carbonate or bicarbonate, especially in waters that are associated with chalk or limestone, and as calcium sulphate. Calcium chloride and nitrate may also be found in waters of higher salinity. Calcium bicarbonate forms temporary hardness; the sulphates, chlorides and nitrates forming permanent hardness (see Section 6.24).

Calcium is an essential part of human diet, but the nutritional value from water is likely to be minimal when compared to the intake from food. There is no health objection to a high calcium content in a water; the main limitations are on the grounds of excessive scale formation.

6.12 Carbon dioxide

Free carbon dioxide in a water (as distinct from that existing as carbonate and bicarbonate) depends on the alkalinity and pH value of the water. It is of much importance regarding the corrosive properties of a water (see Section 8.40).

Surface waters usually contain less than 10 mg/l free CO_2, but some groundwaters from deep boreholes may contain more than 100 mg/l. A simple means of reducing free CO_2 in a water is by cascade aeration (see Section 8.18).

6.13 Chloride

Chlorides are present in nearly all waters, mostly in combination with sodium and to a lesser extent with potassium, calcium and magnesium. They are one of the most stable components in water, being unaffected by most physicochemical or biological processes. Chlorides are derived from a number of sources including natural mineral deposits; seawater intrusion or airborne sea spray; agricultural or irrigation discharges; urban runoff due to the use of de-icing salts; or from sewage and industrial effluents.

Most rivers and lakes have chloride concentrations of less than 50 mg/l Cl and any marked increase may be indicative of sewage pollution or, if the increase is seasonal, urban runoff. The chloride content of a sewage effluent under dry weather flow could increase the chloride content of the receiving water by as much as 70 mg/l.

The main problems caused by excessive chlorides in water relate to corrosion and taste. High concentrations enhance the corrosion rate of iron, steel, and plumbing metals. A sensitive palate can detect chlorides in drinking water at as low a level as 150 mg/l and concentrations above 250 mg/l can impart a distinctly salty taste (see Section 6.41). However waters containing more than 600 mg/l of chloride are drunk in some arid or semi-arid places where there is no alternative supply.

Conventional water treatment processes do not remove chlorides. If the chloride content of a water has to be reduced then some form of desalination has to be applied (see Section 8.43 *et seq*.). The WHO guide level for chloride of 250 mg/l is based on taste grounds and corrosion; a similar advisory level has been set by the USEPA. This is the same level required by the EC Directive, as an indicator parameter, with the caveat that the water should not be aggressive. The current UK Regulations set a standard of 400 mg/l Cl, based on an annual average.

6.14   Chlorinated solvents

The term chlorinated solvent covers a wide range of volatile organic chemicals used as solvents, metal cleaners, paint thinners, dry cleaning fluids, etc. They tend to be found as micro contaminants in groundwater sources that have been subjected to industrial pollution and can remain long after the original source of pollution has been removed.

The three most commonly detected chlorinated solvents in drinking water are tetrachloromethane (or carbon tetrachloride); trichloroethene (TCE); and tetrachloroethene (PCE). All three are potential carcinogens and the latter two compounds may degrade in anaerobic groundwaters to produce more toxic substances such as vinyl chloride. There is some variation in the recommended standards, especially for TCE and PCE (see Table 6.1(C)).

6.15   Chlorine residual

Chlorine remains the principal biocide and disinfectant used in water treatment in most countries (see Chapter 9). It is applied as a gas at major treatment works, or via powders, crystals or solutions containing chlorine at smaller works or at sites where the procurement of chlorine gas presents difficulties. The effectiveness of chlorine as a disinfectant arises from its high chemical reactivity. However it is recognised that the use of chlorine for disinfection can result in the formation of a wide range of undesirable chlorinated organic compounds, known collectively as chlorination by-products (see Section 6.21). Although found at very low concentrations, some of these compounds might be potentially hazardous to health; others produce objectionable tastes and odours. Notwithstanding these problems, which can be minimised by appropriate means (see Section 9.7), the use of chlorine is essential for ensuring a water is bacteriologically safe to drink unless other reliable means of disinfection (e.g. boiling) can be used.

There is no evidence that the levels of chlorine residual normally found in drinking water are harmful to health. Most consumers become familiar with the normal levels of chlorine residual in their local water supply. However sudden increases in chlorine residual are likely to be noticed and result in complaints, particularly from consumers with sensitive palates. Taste and odour complaints can also arise from the reaction of chlorine with other trace substances present in the water (see Section 9.11).

The WHO has set a guideline value of 5 mg/l free chlorine in drinking water, a level that most consumers would find objectionable on taste grounds. The WHO also recommends that for effective disinfection, there should be a free chlorine residual of 0.5 mg/l after 30 minutes contact time at a pH of <8.0. There is no maximum value for residual chlorine at the consumer's tap specified in the EC Directive or in the current UK Regulations; the UK Regulations merely require that all water leaving a water treatment works is disinfected. In the USA the Stage 1 Disinfectants/Disinfection By-products Rule, promulgated in December 1998, sets a maximum level of 4 mg/l as Cl_2 as a running annual arithmetic average of monthly averages for either residual chlorine or residual chloramines.

6.16   Colour

The colour of a water is usually expressed in Hazen units, which are the same as TCU (true colour units) or mg/l on the platinum-cobalt (Pt-Co) scale. True colour can only be determined after filtration, usually through a 0.45 micron filter, as a water often appears coloured because of material in suspension. The colour in unpolluted surface waters is caused by the presence of humic and fulvic acids which are derived from peat and soil humus. In some waters the brown colour is enhanced by the presence of iron and manganese. Waters subject to industrial pollution may also contain a wide variety of coloured materials.

The level at which colour becomes unacceptable depends largely upon consumer perception. The WHO Guidelines give a guide level of 15 TCU, above which the colour would be noticeable in a glass of water by most people. The USEPA advisory limit is the same. The current UK Regulations set a maximum value for colour of 20 mg/l Pt-Co scale. Colour is an indicator parameter in the EC Directive but with no numerical value. The level merely has to be acceptable to consumers and show no abnormal change.

6.17 Copper

Copper is rarely found in unpolluted waters, although trace amounts can sometimes found in very soft, acid moorland waters. The more usual source of copper in drinking water is the corrosion of copper and copper-containing alloys used in domestic plumbing systems. Water containing as little as 1 mg/l of copper can cause green stains on sanitary fittings, and even much lower concentrations can cause accelerated corrosion of other metals in the same system.[11] It is therefore inadvisable to use galvanised steel piping or storage tanks downstream of copper piping.

Copper is an essential element in the human diet, but acute gastric irritation can be caused in some individuals at concentrations above 3 mg/l. Excessive concentrations of copper can also impart an unpleasant and astringent taste to the water. This has resulted in two standards. One is on health grounds (the WHO provisional guideline value of 2 mg/l; the UK maximum concentration of 3000 μg/l and the USEPA maximum guide level of 1.3 mg/l). The other is set at a lower value to avoid problems of staining, etc. (WHO and USEPA at 1 mg/l). The EC Directive specifies a maximum value of 2.0 mg/l as sampled at the consumer's tap, but taken so as to be representative of a weekly average value ingested by the consumer.

6.18 Corrosive quality

There is no definition of the corrosive or aggressive quality of a water because of the many factors that determine whether or not corrosion will take place. These relate to the water itself and the materials with which it comes into contact. Nevertheless three specific characteristics can result in a water being corrosive or aggressive. These are:

● a low pH value, i.e. acidity;
● a high free carbon dioxide (CO_2) content; and
● an absence or low amount of alkalinity.

Waters which are frequently corrosive include soft moorland waters; shallow well waters of low pH with little temporary hardness but a lot of permanent hardness; water from iron-bearing formations; chalk and limestone waters having a high CO_2 content; waters from greensands and coal measures; and waters having a high chloride content. Desalinated water is also very corrosive unless suitably treated. The aggressiveness of a water can be tested by putting the water in contact with powdered chalk or marble for a defined period. The initial and final values for pH and alkalinity will disclose whether the water will readily dissolve the calcium carbonate, or whether calcium carbonate has been deposited (see Section 8.40). If the former, the water is likely to be corrosive to iron and steel as well as cement. If the latter, a protective layer is likely to form on the interior of a metal pipe.

Another aspect of corrosion potential which should always be checked is whether a water is plumbosolvent or cuprosolvent, i.e. whether it is aggressive to lead or copper (see Section 6.27). Such waters may require pH and alkalinity adjustment as well as the addition of orthophosphate or silicate, either individually, or in various combinations, to reduce the corrosion potential.

6.19 Cyanide

Cyanide and cyanide complexes are only found in waters polluted by effluents from industrial or mining processes involving use of cyanide. Most cyanides are biodegradable and can be removed by chemical treatment before any effluent is discharged into a receiving water. Chlorination to a free chlorine residual under neutral or alkaline conditions effectively decomposes any remaining cyanide present in a raw water.

The WHO guideline value for cyanide is 0.07 mg/l, whereas the USEPA MCL is set at 0.2 mg/l. The EC and current UK standards are lower at 50 μg/l.

6.20 Detergents

There are several substances that can cause foaming in water, the largest group being synthetic detergents or surfactants. In recent years the increased use of biodegradable detergents, which can be removed by normal sewage purification processes, has led to a reduction in the residual detergent discharged in sewage effluent. The main limitation on detergents is to prevent foaming in drinking water, although some components of anionic surfactants are toxic to aquatic life. Methylene blue active substances (MBAS) relate to the more common anionic surfactants found in detergents, which react with methylene blue. Many surface waters downstream of urban areas contain detergent.

The WHO does not set a limit for detergent. The parameter is not included in the EC Directive but the current UK standard is 200 μg/l for surfactants, measured as lauryl sulphates. The USEPA sets a SMCL of 0.5 mg/l for 'foaming agents'.

6.21 Disinfection by-products

Chemical oxidants, such as chlorine and ozone, are traditionally used as disinfectants to control pathogenic organisms present in raw waters (see Chapter 9). They are also used to assist in the oxidation of iron and manganese and to breakdown taste and odour forming compounds. Ozone is used extensively to breakdown pesticides. These side reactions with organic and inorganic constituents present in a raw water can give rise to low concentrations of chemicals, collectively known as disinfection by-products or DBPs. There is concern that some DBPs in water comprise minute quantities of chemicals which, in dosages vastly greater, would be potentially carcinogenic (see Sections 6.31 and 6.48). Hence keeping DBPs as low as possible in considered advisable to reduce any risk their long-term ingestion might pose to health. Other DBPs can cause objectionable tastes and odours. The level of DBP formation can be reduced by optimising the treatment processes prior to the final disinfection stage. However the WHO has emphasised that it is vital that the efficacy of the disinfection process is never compromised in order to meet the guidelines for DBPs. In 1991 there was a major cholera epidemic in Peru after water officials were put under pressure to reduce chlorination because of a perceived hypothetical cancer risk from chlorination by-products.

The most common DBPs and their standards, where applicable, are given below.

Bromate

Bromate and other brominated DBPs are formed by the oxidation of any bromide ion present in the water with ozone. Traces of bromide may be present in some raw waters (see Section 6.10); it is also found in commercially available sodium hypochlorite and can be present in the sodium hypochlorite produced on site by the electrolytic generation of chlorine from brine (see Section 9.15). The rate of bromate formation depends on the amount of natural organic matter present in the water, the alkalinity, and the ozone dose applied.[12] The WHO has set a provisional guide value for bromate of 25 μg/l, which is mirrored in the EC Directive as a maximum value in the water at consumers' taps. A more stringent standard of 10 μg/l has to be met by December 2008 and, where possible, Member States are expected to strive for a lower value without compromising disinfection. The USEPA has a proposed MCL of 0.010 mg/l.

Chloral hydrate

Chloral hydrate, or trichloroacetaldehyde, can be formed during chlorination when the necessary organic precursors are present. The WHO provisional guideline value is 10 μg/l, with a proposed USEPA MCL of 0.04 mg/l.

Chlorate and chlorite

Although not strictly DBPs, chlorate and chlorite can be found in water treated with chlorine dioxide (see Section 9.17). Chlorate, like bromate, may also be present in commercially available sodium hypochlorite and in sodium hypochlorite produced on site by the electrolytic generation of chlorine from brine. The WHO provisional guideline value is 200 μg/l for chlorite, which is much lower than the USEPA MCL of 0.8 mg/l proposed under the Stage 1 Disinfectants/Disinfection By-products Rule. The UK Regulations stipulate a combined concentration of chlorine dioxide, chlorite and chlorate of less than 0.5 mg/l as chlorine dioxide for water treated with chlorine dioxide as it leaves the treatment works. In the case of electrolytically produced sodium hypochlorite, the concentration of chlorate must not exceed 0.7 mg/l (as chlorate) in the treated water.

Chloroacetic acids

Monochloro-, dichloro- and trichloro-acetic acids can be formed during chlorination when the necessary organic precursors are present. The WHO provisional guideline values for dichloro- and trichloro-acetic acids are 50 and 100 μg/l, respectively. The USEPA standard of 0.060 mg/l for total haloacetic acids came into effect under the Stage 1

Disinfectants/Disinfection By-products Rule, with compliance for all public water systems by December 2003.

Trichloroacetic acid can be used as a herbicide and has been found, in this context, in detectable concentrations in some raw waters.

Chlorophenols

Chlorophenols can be formed from the chlorination of trace levels of phenolic compounds present in the raw water or as degradation products from the breakdown of phenoxy acid herbicides. They have very low organoleptic thresholds and, if present, are immediately noticed by consumers as an antiseptic taste. The lowest reported taste threshold for 2,4,6-trichlorophenol is 2 μg/l. The WHO guideline value for 2,4,6-trichlorophenol of 200 μg/l is based on a lifetime cancer risk.

Trihalomethanes (THMs)

Trihalomethanes can be formed when raw waters containing naturally occurring organic compounds, such as humic and fulvic acids, are chlorinated. They are also formed by the reaction of chlorine with some algal derivatives (see Section 9.7). They are best controlled by removing as much of the organic precursors as possible before chlorine is applied to the water. Chloroform (or trichloromethane), bromodichloromethane, dibromochloromethane and bromoform (or tribromomethane) are the most important members of the group in terms of monitoring for DBPs. The detection of THMs in the water at consumers' taps is indicative of presence of other chlorinated by-products.

The WHO has set guideline values of 200 μg/l for chloroform, 100 μg/l respectively for dibromochloromethane and bromoform and 60 μg/l for bromodichloromethane. The Guidelines also specify that the sum of the ratio of the concentration of each to its respective guideline value should not exceed one. The EC Directive sets a maximum value of 100 μg/l for the sum of detected concentrations of trichloromethane, bromodichloromethane, dibromochloromethane and tribromomethane, with the caveat that a lower value should be aimed for without compromising disinfection. The current UK Regulations require the three monthly average total concentration of the same four THMs not to exceed 100 μg/l. The USEPA standard of 0.080 mg/l for total THMs came into effect under the Stage 1 Disinfectants/Disinfection By-products Rule. Compliance, which has to be achieved by all public water systems by December 2003, is based on a running annual average of four quarterly samples.

Nitrites (see also Section 6.30)

Nitrites can be formed as a DBP where ultra-violet radiation is used to disinfect waters containing moderate to high concentrations of nitrate (see Section 9.24). Undesirable levels of nitrite can also be formed at the end of long and complex distribution systems where disinfection by chloramination is practised (see Section 9.8).

6.22   Electrical conductivity and dissolved solids

Conductivity is a measure of the ability of a solution to carry electrical current. As this ability is dependent upon the presence of ions in solution, the measurement of conductivity provides an indication of the total dissolved solids, or salts, in a water. For most waters a factor in the range 0.55–0.70 multiplied by the conductivity gives a close approximation to the dissolved solids in mg/l. The factor may be lower than 0.55 for waters containing free acid, and greater than 0.70 for highly saline waters. Conductivity is temperature dependent and a reference temperature (usually 20 or 25°C) is used when expressing the result. One of the advantages of conductivity determination is that it can be easily measured in the field or used for continuous monitoring of a supply.

The current UK Regulations have a maximum value of 1500 μS/cm at 20°C, based on an annual average. Conductivity is an indicator parameter in the EC Directive with a standard of 2500 μS/cm at 20°C and a caveat that the water must not be aggressive. There is no WHO guideline or USEPA standard for conductivity. Instead the WHO has set a guideline level of 1000 mg/l for total dissolved solids on the basis of acceptability to consumers. This is higher than the USEPA advisory SMCL of 500 mg/l.

High levels of dissolved solids can result in taste complaints as well as causing excessive scaling to domestic and industrial water systems. Water with low dissolved solids is desirable for many industrial processes but may be unacceptable to consumers, again on taste grounds, and also be corrosive to domestic plumbing.

6.23 Fluoride

Some waters used for drinking water supplies in the UK naturally contain several mg/l of fluoride. These are mainly deep well waters from chalk or limestone drawn through overlying clay formations, particularly the London Tertiaries and Oxford clays, where concentrations of 2.0–5.0 mg/l of fluoride may be experienced. In other parts of the world much higher concentrations occasionally occur, particularly in areas associated with fluoride-containing minerals. Specialised treatment has to be used to remove excess fluoride from a water, but it is costly to operate (see Section 8.15). Blending is always the preferred option provided there is sufficient low fluoride water available.

Low levels of fluoride may be added to public supplies in the UK,[13] the USA and elsewhere by decision of the relevant health authority as an effective means of reducing dental caries. The greatest reduction of dental decay occurs if fluoridated water is drunk in childhood during the period of tooth formation. Fluoride levels have to be closely controlled as excessive amounts can lead to fluorosis, with resultant mottling of the teeth and in extreme cases even skeletal damage. The addition of fluoride to public water supply is a principle which many people do not accept. Also other developments in dental care have greatly reduced the incidence of dental decay in UK.

The maximum fluoride concentration has to be related to climatic conditions and the amount of water likely to be consumed. The WHO guideline value is 1.5 mg/l, which is higher than the 1.0 mg/l generally adopted for fluoridated water supplies. The current UK standard is the same at 1.5 mg/l, as is the EC Directive maximum value. The USEPA MCL is appreciably higher at 4 mg/l.

6.24 Hardness

The term 'hardness' comes from the fact that a typical hard water reacts with ordinary soap to form a curd or scum. Some hardness is also precipitated by boiling and forms the scale found inside kettles and utensils. The precipitate formed by heating is the temporary, or carbonate, hardness and consists of the bicarbonates of calcium and magnesium.

Permanent, or non-carbonate, hardness (which is not precipitated by heating) is due to other salts of calcium and magnesium present in the water, usually in lesser quantities than the bicarbonates. Hardness, like alkalinity (see Section 6.4), is usually expressed in mg/l as $CaCO_3$ although other notations, such as French or German degrees, may be found in association with domestic appliances such as dishwashers.

The descriptive terms commonly applied are as follows:

Hardness description	Hardness as $CaCO_3$ mg/l
Soft	0–50
Moderately soft	50–100
Slightly hard	100–150
Moderately hard	150–200
Hard	Over 200
Very hard	Over 300

The problems caused by excessive hardness are mainly related to scale formation in boilers and hot water systems. Consumers in hard water areas also complain of scale deposition on kitchen utensils and increased soap usage, with associated scum formation. Conversely, waters containing less than 30–50 mg/l total hardness tend to be corrosive and may need additional treatment to reduce the risk of plumbo- and cupro-solvency (see Section 6.27). Desalinated water has virtually zero hardness and is highly corrosive, requiring treatment to render it non-aggressive to metallic plumbing materials.

A statistical inverse relationship between the hardness of a water and the incidence of cardiovascular disease has been found. The relationship is non-linear, being most evident when the water is very soft. Many other characteristics vary with the hardness of a water, and it is not clear whether the relationship observed might be due to a protective factor in hard water or a harmful factor in soft water. However on the advice of the Department of Health, few UK water undertakings continue to soften public supplies.[14]

The WHO Guidelines suggest a maximum desirable concentration for total hardness of 500 mg/l as $CaCO_3$, based on aesthetic grounds. The current UK Regulations do not set an upper limit for hardness but a lower limit of 60 mg/l Ca (equivalent to 150 mg/l as $CaCO_3$) for supplies which have been softened or demineralised.

6.25   Hydrocarbons

Hydrocarbons include petroleum, mineral oils, coal and coal tar products, and many of their derivatives such as benzene, styrene, etc. produced by the petro-chemical industry for industrial processes. In oil producing parts of the world, background concentrations of hydrocarbons may be present in surface and groundwaters from natural sources. However their presence is more usually the result of pollution.

On very rare occasions, traces of benzene and other solvents have been reported in domestic water supplies. This has usually occurred where there has been localised permeation of plastic water pipes from contaminated ground conditions or other external sources of hydrocarbons such as petroleum spillages.

The current UK standard is 10 μg/l for dissolved and emulsified hydrocarbons. The EC Directive, the WHO Guidelines and the USEPA have standards for benzene, although at varying levels of concentration (see Table 6.1(C)). The WHO Guidelines and the USEPA standards also specify values for other hydrocarbons.

6.26   Iron

Iron is found in most natural waters and can be present in several forms, namely in true solution; as a colloid or in suspension as visible particles; or as a complex with other mineral or organic substances. Iron in surface waters is usually in the ferric (Fe^{3+}) form, but the more soluble ferrous (Fe^{2+}) form is likely to be found in deoxygenated conditions that may occur in some groundwaters or in the bottom waters of lakes and reservoirs. On exposure to air, such waters rapidly become discoloured as the iron oxidises to the ferric form and precipitates out. Iron linings and iron rising mains and pumps can be readily corroded in boreholes with aggressive waters, especially in borings pumped for only short periods daily.

Iron salts are extensively used as coagulants in water treatment (see Section 7.23). With good process control, it should be possible to keep the residual concentration of iron in the water entering supply to less than 0.05 mg/l. However incomplete removal during treatment can lead over a period of time to significant deposits of iron in the associated distribution system. Corrosion products from unlined cast iron mains also contribute to the build up of ferruginous material in a distribution system. Regular flushing, providing it is adequately controlled, can help to control this build up of deposits which otherwise gives rise to discoloured water problems when disturbed. The deposits can also contribute to the growth of iron bacteria, which in turn can cause further water quality deterioration by producing slimes or objectionable odours.

Iron is as essential element in the human diet. The concentrations usually found in drinking water are not harmful but iron can impart a bitter taste when present above 1 mg/l. At lower concentrations it may cause 'dirty' water problems, with consumers rejecting the water on the grounds of appearance. It can also cause brown stains on laundry and plumbing fixtures. The WHO guideline value is 0.3 mg/l to avoid discolouration and staining, and the USEPA SMCL is the same. Iron is an indicator parameter in the EC Directive, with a value of 200 μg/l. This is the same as the standard in the current UK Regulations. Iron is likely to remain as a mandatory standard in the new UK Regulations because of the associations with discoloured water.

6.27   Lead

Lead is a cumulative poison and the hazards of exposure to lead in the environment have been well documented over a number of years. Recently there has been growing concern about the possibility that quite low levels of exposure can affect learning ability and behavioural problems in children. With the decrease in levels of lead in atmospheric emissions and also the risk of exposure from lead in paint, lead in drinking water is now considered to be the main source of controllable exposure.

Lead is rarely found in detectable concentrations in most natural waters except in areas where soft acidic waters come into contact with galena or other lead ores. The main source of lead in drinking water is the dissolution of lead service pipes and internal domestic plumbing which may still present in older properties in the UK and elsewhere in the world. Although plumbosolvency tends to be associated with soft, acidic waters from upland and moorland catchments, it can also occur in hard water areas, especially if the hardness is mainly non-carbonate. Thus the concentrations of lead in drinking water should always be monitored closely in areas where lead pipes are known to be in use.

Unacceptable levels of lead have also been found to occur with some new unplasticised PVC pipes, where lead compounds used as stabilisers in the pipe manufacture are leached out. The use of solder with a high lead content for copper pipe joints can also give rise to unacceptable levels of lead at the consumer's tap.

The WHO guideline for lead is 0.01 mg/l, based on a Provisional Tolerable Weekly Intake for children and expressed as an average value. The current UK standard is 50 μg/l in the water at the time of supply (i.e. the first sample taken from a consumer's tap, without flushing). The UK Regulations place a responsibility on the water undertaker to reduce lead concentrations at consumers' taps where there is a risk of the standard being exceeded. In addition, water undertakers are required to replace their part of a lead service pipe, in areas at risk, if the householder has replaced or is about to replace his or her part of the lead service pipe.

The EC Directive mirrors the WHO guideline with a lead standard of 10 μg/l maximum value based on a weekly average value ingested by consumers. Compliance with this standard is being phased in over a 15-year period (i.e. by the end of December 2013), with a requirement to meet an interim standard of 25 μg/l after 5 years (i.e. by the end of 2003). Member States are also required to take appropriate measures to reduce the concentration of lead in water intended for human consumption. The USEPA introduced the Lead and Copper Rule in 1991 which sets a zero MCLG for lead and an action level of 0.015 mg/l, based on the concentration of lead in more than 10% of tap water samples collected during any monitoring period. If this level is exceeded, treatment has to be installed to reduce plumbosolvency and, if treatment proves to be ineffective, the lead service pipes have to be replaced.

Treatment for plumbosolvency (see Section 8.14) by stabilising the final water pH value and/or orthophosphate dosing has been shown to be very effective in many parts of the UK and the USA. Since 1990 there has been a gradual reduction in the number of samples in England and Wales failing to meet the 50 μg/l standard as the longer term benefits of the treatment are realised.[15] Furthermore the percentage of samples with lead concentrations of less than 10 μg/l has increased from 81% of the annual number of samples taken in 1990 to nearly 84% in 1996. However it is recognised that in some areas the failures are due to particulate lead, which sloughs off the interior of the pipe. In such situations the only viable options are pipe replacement or, if the pipe is still structurally sound, lining with a thin plastic liner.

6.28   Magnesium

Magnesium is one of the earth's most common elements and forms highly soluble salts which contribute both carbonate and non-carbonate hardness to a water, usually at lesser concentration than the calcium component. Excessive concentrations of magnesium are undesirable in domestic water because of problems of scale formation. Magnesium also has a cathartic and diuretic effect, especially when associated with high levels of sulphate.

6.29 Manganese

Manganese can be found in detectable concentrations in both surface and groundwaters. The concentration of manganese in solution rarely exceeds 1.0 mg/l in a well aerated surface water, but much higher concentrations can occur in groundwaters subject to anaerobic conditions. Manganese can re-dissolve from the bottom sediments in impounding reservoirs if the bottom water becomes deoxygenated. This leads to an increase in the overall manganese content of the water when the reservoir 'turns over'.

Large concentrations of manganese are toxic, but treatment is required on grounds of taste and aesthetic quality long before such levels would be reached. Manganese is aesthetically undesirable in drinking water, even in small quantities, as it can precipitate from water in the presence of oxygen or after chlorination, coating the interior of distribution mains with a black slime. These slimes occasionally slough off, giving rise to justifiable consumer complaints. The tolerable concentrations of manganese in a distribution system are generally lower than those for iron for although deposition of manganese is slow, it is continuous. Thus the onset of serious trouble may not become apparent for some 10–15 years after putting a manganese rich supply into service.

The WHO has set a provisional guideline value for manganese of 0.5 mg/l on health grounds but recommends a much lower level of 0.1 mg/l on the basis of staining laundry and sanitary ware. Manganese is an indicator parameter in the EC Directive, with a value of 50 μg/l. This reflects the current UK standard, which is likely to remain as a mandatory standard under the new Regulations. The USEPA has a SMCL at the same concentration but expressed as 0.05 mg/l

6.30 Nitrate and nitrite

Nitrite is an intermediate oxidation state of nitrogen in the biochemical oxidation of ammonia to nitrate. It can also be formed by the reduction of nitrates under conditions where there is a deficit of oxygen. Surface waters, unless badly polluted with sewage effluent, seldom contain more than 0.1 mg/l nitrite as N. Thus the presence of nitrites in surface waters in conjunction with high ammonia levels indicates pollution from sewage or sewage effluent. The presence of nitrites in groundwater may be a sign of sewage pollution. On the other hand it may have no hygienic significance as nitrates in good quality groundwaters can be reduced to nitrite under anaerobic conditions, especially in areas of ferruginous sands. New brickwork in wells is known to have a similar effect.

Nitrate is the final stage of oxidation of ammonia and the mineralisation of nitrogen from organic matter. Most of this oxidation in soil and water is achieved by nitrifying bacteria and can only occur in a well oxygenated environment. The same bacteria are also active in percolating filters at sewage treatment works, resulting in large amounts of nitrate being discharged in sewage effluents from such works.

The use of nitrogenous fertilisers on the land can give rise to increased nitrate concentrations in both surface and underground waters. Nitrate levels in surface waters often show marked seasonal fluctuations, with higher concentrations occurring in winter when runoff increases due to winter rains at a time of reduced biological activity. During summer the nitrate levels are likely to be reduced by biochemical mechanisms and by algal assimilation in reservoirs. Bacterial denitrification and anaerobic reduction to nitrogen at the mud interface can, in addition, substantially reduce the nitrate levels in reservoirs. UK and EC measures to control farming practices liable to cause increased nitrates in surface and groundwaters are described in Section 2.6. By 1995/1996, 32 Nitrate Sensitive Areas had been introduced in the England and Wales and these measures are expected to reduce nitrate levels in the medium term in limestone and sandstone aquifers. The 68 Nitrate Vulnerable Zones, where there will be mandatory restrictions on the use of organic manures and inorganic fertilisers, become compulsory at some time before the end of 1999.

Waters containing high nitrate concentrations are thought to be potentially harmful to infants. At the neutral pH of the infant stomach, nitrate can undergo bacterial reduction to nitrite, which is then absorbed into the bloodstream and converts the oxygen-carrying haemoglobin into methaemoglobin. Whilst the methaemoglobin itself is not toxic, the effects of reduced oxygen-carrying capacity in the blood can obviously be serious, especially for infants having a high fluid intake relative to body weight. However it has become apparent that methaemoglobinaemia is a problem associated with rural shallow wells subject to microbial contamination, rather than public water supplies. There are many examples of public water supplies with relatively high nitrate levels in Europe and North America where no problems with methaemoglobinaemia have been reported. In 1997, an ecological epidemiological study carried out by Leeds University indicated that there was a small but statistically significant correlation between the incidence of childhood diabetes and nitrate in drinking water. The study was restricted to a very small geographical area and the correlation could have arisen by chance. A follow up study, commissioned by the Department of the Environment, Transport and the Regions examined the incidence of childhood diabetes over large areas of England and Wales and the results failed to demonstrate any correlation with nitrate concentrations in drinking water.[16]

Concern has also been expressed on the possible formation of nitrosamines, which are potentially carcinogenic, within the digestive tract. Ingested nitrites, some of which may be formed by bacterial reduction of nitrates, can react with secondary and tertiary amines found in certain foods to give nitrosamines. There is however no epidemiological evidence of an association between nitrite levels in water and cancer incidence.

There are no simple methods for treating a water to reduce its nitrate–nitrite concentration, other than by blending it with another supply with low or negligible concentrations of the same. Treatment processes currently used for nitrate removal include desalination by ion exchange or membranes, and biological removal under controlled conditions using denitrifying bacteria (see Sections 8.23 *et seq.*).

The WHO guideline values are 50 mg/l as NO_3 for nitrate and 3 mg/l as NO_2 for nitrite, with a further provisional value of 0.2 mg/l as NO_2 for nitrite based on chronic effects. There is a caveat that the sum of the ratio of the concentrations of nitrate and nitrite to their respective guideline values should not exceed one, i.e.

$$\frac{\text{Concentration of nitrite}}{\text{Guide value for nitrite}} + \frac{\text{Concentration of nitrate}}{\text{Guide value for nitrate}} \leq 1.$$

The current UK standards are 50 mg/l as NO_3 for nitrate and 0.1 mg/l as NO_2 for nitrite. The maximum value for nitrate remains the same under the EC Directive but that for nitrite is increased to 0.5 mg/l as NO_2 in the water at consumers' taps. A ratio of $NO_3/50 + NO_2/3 \leq 1$ has to be met. There is also a treatment standard of 0.1 mg/l as NO_2 for nitrite in the water leaving a treatment works. The USEPA MCL is slightly lower for nitrate at 10 mg/l as N. The MCL for nitrite is marginally higher than the WHO guideline at 1 mg/l as N and the USEPA also has a combined nitrate/nitrite standard of 10 mg/l as N.

6.31   Organic matter and chemical oxygen demand (COD)

The organic matter found in water can come from a variety of sources such as plant and animal material, including partially treated domestic waste and industrial effluents. The total organics present in a water can be estimated from the chemical oxygen demand (COD), the oxygen absorbed permanganate value (PV) or from the total organic carbon content (TOC). The oxygen demand tests are indirect measures of the total organic carbon by determining its oxidisability. All are gross measures of the total concentration of organic substances and are usually dominated by the concentration of natural organics, e.g. humic and fulvic acids. (Humic and fulvic acids are complex molecular structures in water comprising a major portion of the decay of humus, which is the product of biodegradation and decay of plant, animal, and microbial residues.) There are no numerical standards for TOC in drinking water.

Organic micropollutants

The application of gas chromatography and mass spectrometry to water analysis since the 1970s has revealed the presence of many hundreds of different organic compounds. These are derived from naturally occurring substances in the environment, as well as from materials produced, used or discarded by industry and agriculture, and, when present, are usually at very low concentrations of less than 1 μg/l. Some are known to be toxic or carcinogenic to animals at concentrations far higher than detected in drinking water. Some are known to be mutagenic (i.e. capable of making heritable changes to living cells) under laboratory testing. A number have yet to be identified fully.

Assessment of the risk these compounds might present to human health at the low levels found in water is difficult. Different methods of extrapolating data obtained from observing the effect of high concentrations on animals, in order to deduce the risk to human health of ingesting far smaller quantities, have given results in some cases varying by several orders of magnitude.[17] Epidemiological studies to discover some statistical relationship between cancer mortality and water type (e.g. comparing the effect of using surface waters likely to contain organic matter, with the effect of using groundwater likely to contain less) show 'some association' in some studies, whilst other studies have given inconsistent results. Thus any risk to health due to the presence of organic micropollutants in water is likely to be very small, difficult to measure, and to be manifest only after a long period of exposure. However the potential risk has led to the view that, where possible, the concentration of certain pollutants should be kept as low as possible in drinking water.

The WHO Guidelines list guide values for a number of organic compounds, as does the USEPA under its various organics rules (see Table 6.1(C)).

6.32   Pesticides

Pesticides cover a wide range of compounds used as insecticides, herbicides, fungicides, and algicides. The term can also refer to chemicals with other uses such as wood preservation, public hygiene, industrial pest control, soil sterilisation, plant growth regulation, masonry biocides, bird and animal repellents and anti-fouling paints.

Pesticides find their way into natural waters from direct application for aquatic plant and insect control, from percolation and runoff from agricultural land, from aerial drift in land application, and from industrial discharges. Organic pesticides are often toxic to aquatic life even in trace amounts and some, particularly some of the organo-chlorine compounds, are very resistant to chemical and biochemical degradation. Where pesticides and algicides are used for aquatic control they can cause deoxygenation as a result of the decomposition of the treated vegetation. This in turn can cause other problems such as the dissolution of iron and manganese and the production of tastes and odours.

Even with controlled application, traces of pesticides are still found in sewage effluents and in a number of underground waters and rivers used for public supply purposes. Pesticides used in non-agricultural situations, particularly on hard surfaces, pose a higher risk of contaminating water sources. Accidental discharges of pesticides in bulk to watercourses occasionally occur and can have serious implications, causing fish death and making it necessary for a temporary shutdown of any water intakes.

Some 450 individual substances are currently approved for use as pesticides in the UK. The DoE publication *Guidance on Safeguarding the Quality of Public Water Supplies* lists and provides information on the pesticides most widely used and likely to be detected in water supplies in the UK.[9] However the amounts of pesticides which may be consumed from drinking water are in most cases only a small fraction of those likely to be consumed from foodstuffs.

Many pesticides are insoluble in water or have limited solubility; others degrade readily, depending on the soil type. Many are rapidly adsorbed onto sediment or suspended material and this property can be utilised for removing pesticides in treatment processes involving coagulation followed by sedimentation and filtration. Oxidation using ozone, chlorine, chlorine dioxide, or potassium permanganate effectively breaks down organic pesticides, although in some cases the use of chlorine or ozone may result in degradation products which are more toxic than the original pesticide, or give rise to odorous compounds (see Section 6.21). Ultra-violet radiation, when used on good quality groundwaters, has also been shown to effectively degrade low concentrations of atrazine.[18] For more effective pesticide removal, adsorption onto activated carbon should be adopted either with or without a pre-oxidation stage (see Sections 8.35 and 8.36). A preferable policy is to restrict the levels of pesticides in water prior to treatment, rather than having to remove them in the treatment process. To this end atrazine and simazine, which were commonly used herbicides in the UK, have been banned for non-agricultural use.

The WHO has set guideline values or provisional guideline values for 41 individual pesticides, including TCA which may also be present as a disinfection by-product (see Section 6.21). The EC Directive sets a maximum value of 0.1 μg/l for most individual pesticides and 0.5 μg/l for total pesticides (i.e. the sum of the individual pesticide results for any one sample). The individual pesticide limit was introduced in the EC Directive of 1980. It was based on a surrogate zero, effectively derived from the analytical limits of detection available at the time, and does not take into account of the wide variation in the toxicity of different types of pesticide. The current Directive specifies lower standards for aldrin, dieldrin, heptachlor and heptachlor epoxide on toxicological grounds. However aldrin, dieldrin, heptachlor and heptachlor epoxide are no longer approved for use in the UK. The UK Regulations stipulate the same limits of 0.1 μg/l for individual pesticides and 0.5 μg/l for total pesticides but the *Guidance on Safeguarding the Quality of Public Water Supplies*[9] contains a list of 39 named pesticides with health related advisory values. The USEPA has MCLs for 29 individual pesticides or their derivatives, some of which have been restricted for use since the early 1980s.

6.33 pH value or hydrogen ion

The pH value, or log of the reciprocal of the hydrogen ion concentration, is a measurement of the acidity of a water. It is one of the most important determinations in water chemistry as many of the processes involved in water treatment are pH dependent. Pure water is very slightly ionised into positive hydrogen (H^+) ions and negative hydroxyl (OH^-) ions. In very general terms a solution is said to be neutral when the numbers of hydrogen ions and hydroxyl ions are equal, each corresponding to an approximate concentration of 10^{-7} moles/l. This neutral point is temperature dependent and occurs at pH 7.0 at 25°C. When the concentration of hydrogen ions exceeds that of the hydroxyl ions (i.e. at pH values less than 7.0) the water has acid characteristics. Conversely when there is an excess of hydroxyl ions (i.e. the pH value is greater than 7.0) the water has basic characteristics and is described as being on the alkaline side of neutrality.

The pH value of unpolluted water is mainly determined by the inter-relationship between free carbon dioxide and the amounts of carbonate and bicarbonate present (see Section 8.40). The pH values of most natural waters are in the range 4–9, with soft acidic waters from moorland areas generally having lower pH values and hard waters which have percolated through chalk or limestone generally having higher pH values.

Most water treatment processes, but particularly clarification and disinfection, require careful pH control to fully optimise the efficacy of the process. The pH of the water entering distribution must also be controlled to minimise the corrosion potential of the water. The use of cement mortar relining of mains in soft water areas can result in unacceptably high pH values over a long period of time until the system stabilises.

The WHO proposes a treatment pH value of <8.0 for effective disinfection with chlorine. The EC Directive sets a pH range of ≥6.5 to ≤9.5, whereas the current UK Regulations give a range of 5.5–9.5 for the pH value of water at consumer's taps. Likewise the USEPA sets a pH range of 6.5–8.5 as a SMCL.

6.34 Phenols

The phenolic compounds found in surface waters are usually a result of pollution from trade wastes such as petrochemicals, washings from tarmac roads, gas liquors, and creosoted surfaces. Natural phenols can be released in water from decaying algae or higher vegetation. Traces of phenols and other phenol like compounds may also be found in good quality groundwaters, especially in areas with coal or oil-bearing strata. Most phenols, even in minute concentrations, produce chlorophenols on chlorination (see Section 6.21). These are objectionable in both taste and odour. The taste thresholds vary depending on the chlorophenol formed, but all are at very low concentrations. Phenols can be effectively removed by superchlorination, whereby the excess chlorine chemically decomposes the phenol; by oxidation with ozone; or by adsorption onto activated carbon. However it is preferable to seek to eliminate the source wherever possible.

The WHO sets health based guidelines for trichloro- and pentachlorophenol, although the threshold taste concentration for trichlorophenol, at 2 μg/l, is one-hundredth that of the health level. The current UK Regulations standard for phenols is 0.5 μg/l.

6.35   Phosphates

Phosphates in surface waters mainly originate from sewage effluents which contain phosphate-based synthetic detergents, from industrial effluents, or from land runoff where inorganic fertilisers have been used in farming. Groundwaters usually contain insignificant concentrations of phosphates, unless they have become polluted. Phosphorous is one of the essential nutrients for algal growth and can contribute significantly to eutrophication of lakes and reservoirs (see Section 6.70).

Orthophosphates may be added during water treatment for plumbosolvency control (see Section 6.27). The applied dose is initially around 1 mg/l as P and is gradually decreased to around 0.6 mg/l as P as the treatment takes effect. The amount added is considered unlikely to be of environmental significance in most situations.

6.36   Polynuclear aromatic hydrocarbons (PAHs)

PAHs are a group of organic compounds comprising two or more benzene rings of carbon atoms and occur widely in the environment as the result of incomplete combustion of organic material. Trace amounts of PAHs have been found in industrial and domestic effluents. Their solubility in water is very low but can be enhanced by detergents and by other organic solvents which may be present.

Whilst PAHs are not very biodegradable, they tend to be taken out of solution by adsorption onto particulate matter. If present in raw waters, they are usually removed during coagulation, sedimentation and filtration. However, they can be re-introduced in the distribution system from mains that have been lined with coal tar pitch. Until the 1970s coal tar, which can contain up to 50% of PAHs, was used to line iron water mains to prevent rusting. In some situations this lining may eventually break down, releasing PAHs in solution and as particulates into the water. There is also evidence that chlorine dioxide, when used as a disinfectant or for taste and odour control, can result in elevated concentrations of PAHs at consumers' taps.[9] There tends to be a seasonal variation in the concentrations found, with solubility linked to water temperature.

Several PAHs are known to be carcinogenic at concentrations considerably higher than those found in drinking water, with exposure from food and cigarette smoke being far greater than that from drinking water. Monitoring is normally carried out for six indicator parameters, namely fluoranthene, benzo(b)fluoranthene, benzo(k)fluoranthene, benzo(a)-pyrene, benzo(ghi)perylene and indeno (1,2,3-cd) pyrene. The WHO Guidelines set a value of 0.7 μg/l for benzo(a)pyrene (or benzo-3,4-pyrene) only. Likewise the USEPA has a MCL of 0.0002 mg/l (or 0.2 μg/l) for benzo(a)pyrene only. The current UK standard is 0.2 μg/l for a total of the six indicator PAHs, with a separate standard of 10 ng/l (or 0.01 μg/l) for benzo(a)pyrene as an annual average. The EC Directive sets a maximum value of 0.1 μg/l for a total of five of the indicator PAHs, excluding fluoranthene, the most commonly detected PAH in drinking water. It also sets a maximum value of 0.01 μg/l for benzo(a)pyrene.

6.37   Potassium

Although potassium is one of the abundant elements, the concentration found in most natural waters rarely exceeds 20 mg/l. Much higher levels being found in spa waters.

6.38 Radioactive substances

People are exposed to a number of naturally occurring and man-made sources of environmental radiation. The average global human exposure from natural sources of radiation is estimated at 2.4 mSv/year.* This exposure from natural sources is estimated to contribute more than 98% of the total dose to the population, excluding medical exposure. There are large local variations in this exposure depending on the type of radionuclides in the soil, the height above sea level and the amount taken in from air, food and water. The contribution of drinking water to total exposure is generally extremely small and is mainly due to naturally occurring nuclides in the uranium and thorium decay series.

The concentrations of radionuclides in water sources used for drinking water supplies may be increased by a number of human activities. For example radionuclides from the nuclear fuel and power industry, from medical uses and from other uses of radioactive materials may contaminate water sources. Contributions from these uses are normally limited by regulatory control of discharges to the aqueous and other sectors of the environment. If there was concern about contamination of a water source or a drinking water supply these regulatory controls should be used to ensure that remedial action was taken.

The International Commission on Radiological Protection (ICRP) provides detailed advice and recommendations on the control of exposure to radiation.[19] The ICRP advice and recommendations has been used by the WHO in formulating its recommendations on the radiological aspects of drinking water supplies. The WHO recommends that:

- the reference level of committed effective dose is 0.1 mSv from 1 year's consumption of drinking water. This reference level of dose represents less than 5% of the average effective dose each year from natural background radiation; and
- below this reference level of dose, drinking water is acceptable for human consumption and action to reduce radioactivity is not necessary.

These recommendations apply to existing routine operational water supplies and to new supplies. They do not apply to drinking water supplies contaminated during an emergency involving the release of radionuclides into the environment – other advice is available in these circumstances. The Department of the Environment, Transport and the Regions would provide such guidance in the event of a civil emergency involving radioactive substances occurring in England and Wales.

For practical purposes the reference level of dose needs to be expressed as activity concentrations of radionuclides in drinking water and the WHO guidelines give concentrations of 0.1 Bq/l† for gross alpha activity and 1.0 Bq/l for gross beta activity for monitoring purposes. These concentrations are regarded as screening levels for drinking water and no further action is needed if the monitoring shows that activities are below these levels. It does not necessarily mean that the reference level of dose is exceeded if these gross alpha or gross beta screening levels are exceeded. In such circumstances the specific radionuclides should be identified and their individual activity concentrations measured. A dose estimate should then be made for each radionuclide using activity to dose conversion

---

*mSv (millisievert); sievert is the 'effective dose equivalent'. It is a measure of the effect produced by different types of radiation on a person, taking into account the nature of the radiation and the organs exposed.

†Bq (becquerel) corresponds to one nuclear transformation per second; 1 curie (Ci) $= 37 \times 10^9$ Bq.

factors (for 1 year's consumption of 2 litres of water per day) and the sum of the doses calculated to determine whether the reference level of dose is in danger of being exceeded. When this estimate of dose is made from a single sample, the reference level of dose would only be exceeded if exposure to the measured concentrations were to continue for a full year. Thus when a single sample indicates that the reference level of dose is in danger of being exceeded, it does not necessarily mean that the water supply is unsuitable for consumption, but it is a signal for further investigation of the supply. This should include further sampling and if this shows that the reference level of dose is likely to be exceeded, the options available to reduce the dose should be considered and implemented where they are justified. Further details of the application of the gross alpha and gross beta screening levels and the reference level of dose to drinking water supplies can be found in Chapter 4 of the 1993 WHO Guidelines for Drinking-Water Quality.[1]

The WHO has also recommended a provisional guideline value for uranium in drinking water of 2 μg/l based on its chemical toxicity and without taking into account any radiological effects. This is considerably lower than the proposed USEPA MCL of 20 μg/l. The USEPA also has proposed and interim standards for a number of radionuclides, including alpha and beta emitters and radium-226 and -228.

About 30 water sources (prior to treatment for drinking water supply) in England and Wales are monitored regularly for the Department of the Environment, Transport and the Regions for gross alpha and gross beta activities and for selected specific radionuclide concentrations. These sources supply about one-third of the population. The results are reported annually in the Department's Digest of Environmental Statistics. Practically all samples have been below the gross alpha and gross beta screening levels and any exceedences of these values have been marginal and transient. Many of the radionuclides detected are of natural origin, but traces of some artificial radionuclides are found; these are mainly from atmospheric fall out and sometimes from aqueous discharges, such as iodine-125 from medical use. Similar monitoring is carried out in Scotland and Northern Ireland.

6.39   Silica

Silica can be found in water in several forms, caused by the degradation of silica-containing rocks such as quartz and sandstone. Natural waters can contain between 1 mg/l of silica in the case of soft moorland waters and up to about 40 mg/l in some hard waters. Much higher levels are found in waters from volcanic or geothermal areas.

There is no evidence of silica in drinking water constituting a health hazard and no standards have been set. However it is a troublesome material in a number of industrial processes because it forms a very hard scale which is difficult to remove.

6.40   Silver

Trace amounts of silver are occasionally found in natural waters but it is rarely found in detectable concentrations in drinking water. However silver is sometimes used as a disinfectant in domestic water treatment units or point of use devices (e.g. ceramic filter candles or granular activated carbon impregnated with silver) and waters so treated may contain levels of 50 μg/l or more as silver chloride. The current UK standard is 10 μg/l, except where silver has been used in the treatment process, when up to 80 μg/l may be authorised. Silver vessels were prized in antiquity for storing water because of their disinfectant property.

6.41   Sodium

Sodium compounds are very abundant within the environment and are also very soluble. The element is present in most natural waters at levels ranging from less than 1 mg/l to several thousand mg/l in brines. The threshold taste for sodium in drinking water depends on several factors, such as the predominant anion present and the water temperature. The threshold taste as sodium chloride is around 150 mg/l as Na, whereas the threshold taste as sodium sulphate is much higher at around 220 mg/l as Na.[20]

The use of base exchange or lime-soda processes to soften hard waters can lead to a significant increase in the sodium concentration of the softened water. Apart from the possibility of imparting taste, such increases could be detrimental to consumers on sodium restricted diets. The WHO guideline value for sodium is 200 mg/l, based on taste. The current UK standard is a maximum of 150 mg/l, based on 80% compliance; however sodium is an indicator parameter in the EC Directive at the same value as the WHO guideline. There is no USEPA standard.

6.42   Sulphates

The concentration of sulphate in natural waters can vary over a wide range from a few mg/l to several thousand mg/l in brackish waters and brines. Sulphates come from several sources such as the dissolution of gypsum and other mineral deposits containing sulphates; from seawater intrusion; from the oxidation of sulphides, sulphites, and thiosulphates in well aerated surface waters; and from industrial effluents where sulphates or sulphuric acid have been used in processes such as tanning and pulp paper manufacturing. Sulphurous flue gases discharged to atmosphere in industrial areas often result in acid rain water containing appreciable levels of sulphate.

High levels of sulphate in water can impart taste and, when combined with magnesium or sodium, can have a laxative effect (e.g. Epsom salts). Consumers tend to become acclimatised to high sulphate waters and, in some parts of the world, waters with very high sulphate contents have to be used as there is no alternative.

Bacterial reduction of sulphates under anaerobic conditions can produce hydrogen sulphide, which is an objectionable gas smelling of bad eggs. This can occur in deep well waters and the odour rapidly disappears with efficient aeration. It can also occur if there is seawater intrusion to a shallow aquifer which is polluted by sewage.

The WHO guideline value is 250 mg/l SO_4, based on taste and corrosion potential. This is the same value as the current UK standard; the EC Directive standard, where sulphate is an indicator parameter; and the USEPA SMCL. The WHO recommends that health authorities be notified if sulphate levels exceed 500 mg/l in drinking water and this is the same level as the proposed USEPA MCL on health grounds.

6.43   Suspended solids

The suspended solids content or filter residue of a water quantifies the amount of particulate material present in the water. This includes both organic and inorganic matter such as plankton, clay and silt.

The suspended solids content of a surface water can vary widely depending on flow and season, with some rivers under flood conditions having several thousand mg/l in suspension. The measurement of suspended solids is usually on a weight–volume basis and gives no indication as to the type of material in suspension, the particle size distribution, or the settling characteristics. Adequate treatment should always be applied to remove suspended solids prior to final disinfection, otherwise the efficacy of the disinfection process is likely to be impaired.

6.44   Taste and odour

There are four basic taste sensations, namely sweet, sour, salt, and bitter. What is regarded as taste is in fact a combination of these sensations with the sensation of smell. In examining water samples, the odour rather than the taste of a slightly warm sample is often evaluated because it avoids putting a possibly suspect sample into the mouth. It is desirable however to sample the taste of a final treated water. A subjective or qualitative assessment of the taste or odour is often carried out at the time of sampling. This can then be supplemented by an quantitative measurement, reported as the Dilution Number (or the Threshold Number), which is carried out under controlled laboratory conditions. The quantitative test is based on the number of dilutions of the sample with taste and odour-free water necessary to eliminate the taste or odour.

Many tastes and odours are caused by natural contaminants such as extracellular and decomposition products of plants, algae and micro fungi. Certain of the blue-green algae and actinomycetes, when present in a raw water, can give rise to very distinctive earthy and musty tastes and odours (see Section 6.69). Raw waters contaminated by agricultural and industrial discharges may also give rise to severe taste and odour problems, which are often exacerbated by the use of chlorine for disinfection. Chlorine itself can often give rise to extensive taste and odour complaints, especially if the applied dose has been increased suddenly for operational reasons. Astringent tastes can arise from the dissolution of plumbing materials such as zinc or copper.

The taste and odour of drinking water can be very subjective. Consumers become familiar with a particular taste or odour associated with their local water supply and tend only to notice significant changes or differences, such as when there has been a change in treatment for operational reasons or when they move to a new water supply. The WHO Guidelines stipulate that the taste and odour of drinking water should be acceptable to consumers. The same approach is adopted by the EC Directive, with the additional caveat that there should be no abnormal change. However the current UK Regulations specify maximum threshold numbers at 25°C and the USEPA has a similar SMCL, without the temperature limitation.

6.45   Turbidity

The measurement of turbidity, although not quantitatively precise, is a simple useful indicator of the condition of a water. It is defined as the optical property that causes light to be scattered and absorbed rather than transmitted in straight lines through a sample. Although turbidity is caused by material in suspension, it is difficult to correlate it with the quantitative measurement of suspended solids in a sample, as the shape, size, and refractive indices of the particles in suspension all affect their light-scattering properties. For the same reason turbidity measurements can vary according to the type of instrument used. Nephelometers measure the intensity of light scattered in one particular direction and are highly sensitive for measuring low turbidities. Other instruments measure the amount of light absorbed by particles when light is passed through a water sample. Early measurements were made using the Jackson Candle Turbidimeter, with the results reported in Jackson units (JTU). However this is a fairly crude form of measurement, as are some of the other earlier instruments measuring in silica scale units, and is not really suitable for measuring low turbidities. The preferred methods now use a primary standard based on a chemical called formazin which, if used to calibrate nephelometric instruments, gives an equivalent turbidity measurement in FTU (i.e. FTU = NTU).

Raw water turbidities can vary over a very wide range, from virtually zero to several thousand NTU. Effective treatment should be able to produce final waters with turbidity levels of less than 1 NTU, which is the level recommended for efficient disinfection with chlorine. Turbidity meters are valuable in monitoring the various stages of treatment, such as the amount of flocculant material passing from clarifiers to filters and the performance of individual or groups of filters. However particle size counters may well eventually take over from turbidity meters as the optimum analytical tool for monitoring the performance of rapid gravity filters, particularly at the start of a filter run.

Turbidities above 5 NTU may be discernible to consumers. The WHO Guidelines set this as the maximum level acceptable to consumers but also set a treatment standard of less than or equal to 1 NTU for successful disinfection. The EC Directive has a similar requirement in that the water at consumers' taps must have a turbidity which is acceptable, without specifying a value, but goes on to set a treatment standard of less than 1 NTU for effective disinfection. The USEPA has likewise set turbidity standards of 5 NTU for surface waters used as drinking water without filtration; less than or equal to 1 NTU for slow sand or diatomaceous earth filtered water; and less than or equal to 0.5 NTU for conventional or direct filtration. The current UK regulations has a maximum standard for turbidity of 4 FTU at consumers' taps.

6.46 Zinc

Zinc tends to be found only in trace amounts in unpolluted surface waters and groundwaters. However, it is often found in the water at consumers' taps as a result of corrosion of galvanised iron piping or tanks and dezincification of brass fittings. The concentrations usually found in drinking water are unlikely to be detrimental to health.

Excess zinc can give rise to an undesirable astringent taste and also cause opalescence. The WHO guideline value of 3 mg/l is based on taste and appearance. The current UK standard is 5000 μg/l at the consumer's tap; the equivalent USEPA SMCL of 5 mg/l is based on taste.

Part II Water quality standards for chemical and physical parameters

6.47 Drinking water standards (chemical and physical)

Tables 6.1(A)–(D) list many of the chemical and physical standards given in the 1993 WHO *Guidelines for Drinking Water Quality*; the current UK *Water Supply (Water Quality) Regulations* of 1989; the EC Directive (98/83/EC) on the quality of water intended for human consumption which came into force on 25 December 1998; and the

Table 6.1(A) Physical characteristics and substances undesirable in excess

	WHO Guidelines 1993 guideline levels [a]	UK Water Supply (Water Quality) Regulations 1989	EC Directive 98/83/EC November 1998 [b]	USEPA Regulations under 1996 Safe Drinking Water Act amendments SMCL
Colour	15 true colour units	20 mg/l Pt/Co scale	Acceptable to consumers and no abnormal change	15 colour units
Turbidity	5 NTU <1 NTU (TS)	4 FTU – including suspended solids	Acceptable to consumers and no abnormal change < 1 NTU (TS)	5 NTU max 0.5–1.0 NTU (TS)
Taste/odour as threshold numbers	Acceptable	Dilution number of 3 @ 25°C	Acceptable to consumers and no abnormal change	3
Hydrogen sulphide	0.05 mg/l	Included in odour		
Temperature		25°C		
pH (Hydrogen ion)	<8.0 (TS)	5.5–9.5	≥6.5 and ≤ 9.5	6.5–8.5
Residual chlorine	600–1000 μg/l			
Dry residues	1000 mg/l	1500 mg/l[(c)]		500 mg/l
Conductivity		1500 μS/cm @ 20°C[(d)]	2500 μS/cm @ 20°C	
Alkalinity (HCO_3)		30 minimum[(e)] mg/l		
Total hardness (Ca)		60 mg/l minimum[(e)]		
Chloride (Cl)	250 mg/l	400 mg/l[(d)]	250 mg/l	250 mg/l
Fluoride		see Table 6.1(B)		2.0 mg/l – see also Table 6.1(B)
Sulphate (SO_4)	250 mg/l	250 mg/l	250 mg/l	250 mg/l – see also Table 6.1(B)
Ammonia and ammonium ion	1.5 mg/l as NH_3	0.5 mg/l as NH_4	0.5 mg/l as NH_4	
Calcium (Ca)		250 mg/l[(d)]		
Magnesium (Mg)		50 mg/l		
Phosphorus (P)		2200 μg/l		

Table 6.1(A) *(continued)*   Physical characteristics and substances undesirable in excess

	WHO Guidelines 1993 guideline levels (a)	UK Water Supply (Water Quality) Regulations 1989	EC Directive 98/83/EC November 1998 (b)	USEPA Regulations under 1996 Safe Drinking Water Act amendments SMCL
Potassium (K)		12 mg/l		
Sodium (Na)		150 mg/l(f)	200 mg/l	
Aluminium (Al)	0.2 mg/l	200 µg/l	200 µg/l	0.05–0.2 mg/l
Copper (Cu)	1.0 mg/l – see also Table 6.1(B)	3000 µg/l		1.0 mg/l – see also Table 6.1(B)
Iron (Fe)	0.3 mg/l	200 µg/l	200 µg/l	0.3 mg/l
Manganese (Mn)	0.10 mg/l – see also Table 6.1(B)	50 µg/l	50 µg/l	0.05 mg/l
Silver (Ag)		10µg/l(g)		
Zinc (Zn)	3.0 mg/l	5000 µg/l		5 mg/l
Dissolved and emulsified hydrocarbons		10µg/l		
Kjeldahl Nitrogen (N)		1 mg/l		
Oxidisability (O$_2$)		5 mg/l	5 mg/l	
Phenols (C$_6$H$_5$OH)		0.5 µg/l		
Surfactants or foaming agents		200 µg/l as lauryl sulphate		0.5 mg/l
Total organic carbon		No significant increase over normal	No abnormal change	

Notes.
SMCL – Secondary maximum contaminant level (USA, not mandatory). TS – Standard applies to treated water ex treatment works.
(a)Guideline Levels above which customer complaints may arise. (b)Indicator parameters with values set for monitoring purposes. Any exceedances must be investigated. (c)After drying at 180°C. (d)Annual average. (e)For softened water only. (f)80% compliance. (g)Exceptionally 80 µg/l if silver is used in treatment.

Table 6.1(B) Inorganic substances of health significance

	WHO Guidelines 1993 Guideline values	UK Water Supply (Water Quality) Regulations 1989	EC Directive 98/83/EC November 1998	USEPA Regulations 1996 Safe Drinking Water Act amendments MCL
Antimony (Sb)	0.005 mg/l (P)	10 μg/l	5 μg/l	0.006 mg/l
Arsenic (As)	0.01 mg/l (P)	50 μg/l	10 μg/l	0.05 mg/l[a]
Asbestos > 10 μm				7×10^6 fibres/l
Barium (Ba)	0.7 mg/l	1,000 μg/l[b]		2 mg/l
Beryllium (Be)				0.004 mg/l
Boron (B)	0.5 mg/l (P)	2,000 μg/l[b]	1.0 mg/l	
Bromate (BrO$_3$)	25 μg/l (P)		10 μg/l	0.010 mg/l
Cadmium (Cd)	0.003 mg/l	5 μg/l	5 μg/l	0.005 mg/l
Chlorate (ClO$_3$)		0.7 mg/l[c]		
Chlorine (Cl$_2$)	5 mg/l			4.0 mg/l
Chlorite (ClO$_2$)	200 μg/l (P)	0.5 mg/l[d]		0.8 mg/l
Chromium (Cr)	0.05 mg/l (P)	50 μg/l	50 μg/l	0.1 mg/l
Copper (Cu)	2 mg/l (P)	3,000 μg/l	2.0 mg/l	1.3 mg/l[e] – see also Table 6.1(A)
Cyanide (CN)	0.07 mg/l	50 μg/l	50 μg/l	0.2 mg/l
Fluoride (F)	1.5 mg/l[f]	1,500 μg/l	1.5 mg/l	4 mg/l – see also Table 6.1(A)
Lead (Pb)	0.01 mg/l	50 μg/l[g]	10 μg/l[h]	0.015 mg/l[i]
Manganese (Mn)	0.5 mg/l (P)			
Mercury (Hg)	0.001 mg/l	1.0 μg/l	1.0 μg/l	0.002 mg/l
Molybdenum (Mo)	0.07 mg/l			
Nickel (Ni)	20 μg/l (P)	50 μg/l	20 μg/l	0.1 mg/l
Nitrate	50 mg/l as NO$_3$[j]	50 mg/l as NO$_3$	50 mg/l as NO$_3$	10 mg/l as N

(Continued)

Table 6.1(B) (*continued*)

	WHO Guidelines 1993 Guideline values	UK Water Supply (Water Quality) Regulations 1989	EC Directive 98/83/EC November 1998	USEPA Regulations 1996 Safe Drinking Water Act amendments MCL
Nitrite	3 mg/l as NO_2 [i]	0.1 mg/l as NO_2	0.5 mg/l as NO_2 (0.1 mg/l ex works)	1 mg/l as N
Nitrate + nitrite	Sum of ratio of concentrations ≤1		$NO_3/50$ mg/l + $NO_2/3$ mg/l ≤1	10 mg/l as N
Selenium (Se)	0.01 mg/l	10 μg/l	10 μg/l	0.05 mg/l
Sulphate (SO_4)	————— See Table 6.1(A) —————			500 mg/l(proposed)
Thallium				0.002 mg/l
Tritium			100 Bq/l	
Total indicative dose for radioactivity			0.10 mSv/year [k]	
Uranium	2 μg/l (P)			20 μg/l (proposed)

Notes.

MCL – Maximum contaminant level (USA, mandatory). P – Provisional guideline value (WHO).

[a] Interim standard.
[b] Annual average.
[c] In the treated water where on-site electrolytic chlorine generation is used
[d] Where chlorine dioxide is used in treatment, the combined concentration of chlorine dioxide, chlorite and chlorate should not exceed 0.5 mg/l as chlorine dioxide in the water entering supply.
[e] Action level, exceeded if the concentration of copper in more than 10% of tap water samples collected during any monitoring period is greater than 1.3 mg/l.
[f] Value must depend on climate and local conditions.
[g] The standard applies to the first draw sample without flushing.
[h] The value applies to a sample of water intended for human consumption obtained by an adequate sampling method at the tap and taken so as to be representative of a weekly average value ingested by consumers. Interim standard of 25 μg/l.
[i] Action level, exceeded if the concentration of lead in more than 10% of tap water samples collected during any monitoring period is greater than 0.015 mg/l.
[j] Acute effects; provisional standard of 0.2 mg/l for nitrite based on chronic effects.
[k] Excluding tritium, potassium-40, radon and radon decay products.

Table 6.1(C) Organic substances of health significance (The EC Directive covers only a short list of mandatory standards for organic constituents and these are listed first)

	WHO Guidelines 1993 Guideline values	UK Water Supply (Water Quality) Regulations 1989	EC Directive 98/83/EC November 1998	USEPA Regulations under 1996 Safe Drinking Water Act amendments MCL
Acrylamide	0.5 µg/l		0.10 µg/l[(a)]	TT
Benzene	10 µg/l		1.0 µg/l	0.005 mg/l
1,2-dichloroethane	30 µg/l		3.0 µg/l	0.005 mg/l
Epichlorohydrin	0.4 µg/l (P)		0.10 µg/l[(a)]	TT
Pesticides – Total		0.50 µg/l[(b)]	0.50 µg/l[(b)]	
Pesticides individual substances, except for:		0.10 µg/l[(b)]	0.10 µg/l[(b)]	
Aldrin	0.03 µg/l		0.030 µg/l	
Dieldrin	0.03 µg/l		0.030 µg/l	
Heptachlor	0.03 µg/l		0.030 µg/l	0.0004 mg/l
Heptachlor epoxide	0.03 µg/l		0.030 µg/l	0.0002 mg/l
Polycyclic aromatic hydrocarbons		0.2 µg/l[(c)]	0.10 µg/l[(d)]	
Benzo(a)pyrene	0.7 µg/l	10 ng/l[(e)]	0.010 µg/l	0.0002 mg/l
Tetrachloromethane	2 µg/l	3 µg/l[(e)]		0.005 mg/l
Trichloroethene	70 µg/l (P)	30 µg/l[(e)]	10 µg/l[(f)]	0.005 mg/l
Tetrachloroethene	40 µg/l	10 µg/l[(e)]	(f)	0.005 mg/l
Trihalomethanes	(g)	100 µg/l[(h)]	100 µg/l[(i)]	0.08 mg/l[(i)]
Vinyl chloride	5 µg/l		0.50 µg/l[(a)]	0.002 mg/l

(continued)

Notes.

'Ethene' is the same as 'ethylene'.

MCL – Maximum contaminant level (US, mandatory).

TT – Treatment technique requirement.

(P) Provisional guideline value (WHO).

[a] Residual monomer concentration.

[b] 'Pesticides' means organic insecticides, herbicides, fungicides, nematocides, acaricides, algicides, rodenticides, slimicides and related products (*inter alia*, growth regulators) and their related metabolites, degradation and reaction products. 'Total pesticides' means the sum of all individual pesticides detected and quantified in the monitoring procedure.

[c] The sum of the detected concentrations of fluoranthene, benzo(b)fluoranthere, benzo(k) fluoranthene, benzo(a)pyrene, benzo(ghi)perylene and indeno (1,2,3-cd) pyrene.

[d] The sum of the detected concentrations of benzo(b)fluoranthene, benzo(k)fluoranthene, benzo(a)pyrene, benzo(ghi)perylene and indeno (1,2,3-cd) pyrene.

[e] Annual average.

[f] Sum of tetrachloroethene + trichloroethene.

[g] Standards for individual THMs of 200 μg/l chloroform (trichloromethane), 100 μg/l bromoform (tribromomethane), 100 μg/l dibromo-chloromethane and 60 μg/l bromodichloromethane, with the sum of the ratio of concentrations for each to their respective GV not exceeding one.

[h] Three-monthly rolling mean.

[i] The sum of the detected concentrations of trichloromethane, tribromomethane, dibromochloromethane and bromodichloromethane. A lower value should be aimed for, if possible, without compromising disinfection.

Table 6.1(D)   Other pesticides and associated substances (not listed specifically under the European Directive).

	WHO Guidelines 1993 Guideline values in μg/l	USEPA Regulations under 1996 Safe Drinking Water Act amendments MCL in mg/l	UK Advisory Values (see Section 6.33) in μg/l
Pesticides and associated substances			
Alachlor	20	0.002	
Aldicarb	10	0.001 – not final	
Atrazine	2	0.003	2
Bentazone	300		
Carbofuran	7	0.04	
Chlordane	0.2	0.002	0.1 – total isomers
Chlortoluron	30		80
Cyanazine	0.6		
DDT	2		7 – total isomers
1,2-dibromo-3-chloropropane	1	0.0002	
1,2-dibromoethane (EDB)	0.4–15 (P) (a)		
2,4-D	30	0.07	1000
2,4-DB	90		
2,4,5-T	9		
2,4,5-TP (Silvex)	9	0.05	
1,2-dichloropropane	40 (P)	0.005	
1,3-dichloropropene	20		
Dalapon		0.2 (a)	
Dichlorprop	100		40
Dinoseb		0.007	
Dioxin		3×10^{-8}	
Diquat	10	0.02	
Ethylene dibromide		0.00005	
Endothall		0.1	
Endrin		0.002	
Glyphosate	(b)	0.7	1,000
Hexachlorobenzene	1	0.001	0.2
Hexachlorobutadiene	0.6 (c)		
Hexachlorocyclopentadiene		0.05	
Isoproturon	9		4
Lindane	2	0.0002	
MCPA	2		0.5
MCPB			0.5
Mecoprop (MCPP)	10		10
Methoxychlor	20	0.04	30

(Continued)

Table 6.1(D) (*continued*)

	WHO Guidelines 1993 Guideline values in μg/l	USEPA Regulations under 1996 Safe Drinking Water Act amendments MCL in mg/l	UK Advisory Values (see Section 6.33) in μg/l
Metolachlor	10		
Molinate	6		
Oxamyl (Vydate)		0.2	
Pendimethalin	20		
Pentachlorophenol	9 (P)	0.001	
Permethrin	20		
Picloram		0.5	
Propanil	20		
Pyridate	100		
Simazine	2	0.004	10
TCA (Trichloroacetic acid)	100 (c) (P)	0.3 (c)	
Terbuthylazine	7		
Toxaphene		0.003	
Trifluralin	20		
Other organic substances			
Chlorinated alkanes:			
Dichloromethane	20	0.005	
1,2-dichloroethane	30	0.005	
1,1,1-trichlorethane	2000 (P)	0.2	
1,1,2-trichlorethane		0.005	
Chlorinated ethenes:			
1,1-dichloroethene	30	0.007	
1,2-dichloroethenes	50	*cis* 0.07 *trans* 0.1	
Aromatic hydrocarbons:			
Toluene	700	1	
Xylenes	500	Total 10	
Ethylbenzene	300	0.7	
Styrene	20	0.1	
Chlorinated benzenes:			
Monochlorobenzene	300	0.1	
1,2-dichlorobenzene	1000	0.6	
1,4-dichlorobenzene	300	0.075	
Trichlorobenzenes	Total – 20	1, 2, 4 – 0.07	

(Continued)

Table 6.1(D) (*continued*)

	WHO Guidelines 1993 Guideline values in $\mu g/l$	USEPA Regulations under 1996 Safe Drinking Water Act amendments MCL in mg/l	UK Advisory Values (see Section 6.33) in $\mu g/l$
Disinfection by-products:			
Chloroform	200		
2,4,6-trichlorophenol	200		
Dichloroacetic acid	50 (P)		
Trichloroacetic acid	100 (P)	0.060 total acids	
Chloral hydrate	10 (P)	0.04 proposed	
Dibromoacetonitrile	100 (P)		
Dichloroacetonitrile	90 (P)		
Trichloroacetonitrile	1 (P)		
Formaldehyde	900		
Cyanogen chloride as CN	70		
Other organic substances:			
Di(2-ethylhexyl)adipate	80	0.4	
Di(2-ethylhexyl)phthalate	8	0.006	
EDTA	600		
Microcystin LR	1 (P)		
Nitrilotriacetic acid	200		
Polychlorinated biphenyls (PCBs)		0.0005	
Tributyltin oxide	2		

Notes.
(P) – Provisional.
(a) For excess risk of 10^{-5}.
(b) No WHO guide value as the substance is not regarded as hazardous to health in drinking water.
(c) Can also be found in chlorinated water as a disinfection by-product.

USEPA requirements based on the National Primary Drinking Water Regulations as amended under the Safe Drinking Water Act of 1996.

The former WHO International Standards for Drinking Water of 1971 and earlier dates may still apply in many countries where they were adopted as national standards with modifications to allow for local conditions. The 1993 WHO Guidelines and subsequent addendums consider many potential contaminants, particularly organic contaminants, which were not included in the earlier editions. Supporting information is given to show how each guideline value has been derived. The need to adopt a risk-benefit approach is also emphasised, recognising that water must be made available 'even if the quality is not entirely satisfactory'. The Guidelines recognise that there is a different type of health risk from toxic chemicals in drinking water to that posed by microbiological contamination, with consumers being likely to reject a water because of unacceptable taste, odour or appearance. Standards for chemical parameters have been set on the potential to cause adverse health effects over long-term periods of exposure, either as cumulative toxins such as lead or as possible carcinogens such as some of the organic parameters. In some instances provisional guideline values have been set for potentially hazardous parameters, where available data on health effects are currently poorly defined or limited. The guideline values for aesthetic and organoleptic parameters have been set at levels that are most likely to be acceptable to consumers.

The Directive, although specifying many of the same values as the WHO Guidelines, is essentially a legal document. The standards are defined to enable equal water qualities and obligations to be achieved throughout the European Community. The frequency of monitoring and the levels of analysis are also defined in the Directive. There are provisions for special dispensations whereby individual public water suppliers can be allowed to supply water not complying with a particular standard. Such 'derogations' are conditional and are time limited to a maximum of three years. Under the current UK regulations similar authorised relaxations are permitted when there are 'exceptional meteorological conditions' or 'situations arising from the nature and structure of the ground' from which the supply emanates, or in emergencies. The Directive has 38 numerical standards for chemical and radiological parameters. Twenty-six of these standards are for chemical parameters which are considered to be potentially hazardous to health. The remaining 12 standards, along with five parameters that do not have numerical values, are termed indicator parameters and are included for monitoring purposes. These 17 parameters relate mainly to the aesthetic and organoleptic characteristics of water. Member States are allowed to set values for additional parameters not included in the Directive. They can also adopt more stringent standards than those specified, although the Commission has to be notified. The Directive requires any failure of a standard to be investigated and necessary remedial action must be taken as soon as possible to restore water quality if there is a health risk associated with the failure.

The Water Supply (Water Quality) Regulations 1989 and subsequent amendments are statutory instruments linked to the Water Industry Act 1991 and reflect the standards set out in the 1980 EC Directive.[5] The Regulations apply to water undertakers in England and Wales and are enforced by the Drinking Water Inspectorate. Similar legislation applies in Scotland and Northern Ireland. Some of the 52 chemical standards are more stringent than those set in the 1980 Directive and a legal obligation is placed on water undertakers to supply wholesome water as defined under the Regulations. New UK Regulations will be enacted as soon as the 1998 Directive is transposed into national law at the end of December 2000.

In the USA, drinking water quality regulations were mandated by law under the Safe Drinking Water Act of 1974. The most recent amendments were made in 1996, with Maximum Contaminant Levels specified for a wide range parameters. Many of the regulations have been promulgated as rules, for examples the Lead/Copper Rule and the Surface Water Treatment Rule. These rules define the practical application of the standards relating to them and the associated monitoring requirements. The standards are enforceable by the USEPA and are mandatory on all public water supply undertakings, except where a State authority grants a 'variance' to a water supply system that cannot comply because of the characteristics of its water source. Best available technology (BAT) has to be adopted however to treat the water. Good accounts of the EPA Regulations are given by Pontius.[21]

6.48   Comment on the application of standards

Standards for many of the substances listed in Tables 6.1(C) and (D) are derived from dietary experiments on animals (e.g. rats, mice, etc.). From these the 'no adverse effect level' (NOEL) of dosage is deduced, to which 'uncertainty factors' are applied which may vary from 10^{-2} to 10^{-4} to allow for species differences, nature and severity of adverse effects, and the quality of the data. Further factors are then applied to deduce the acceptable daily intake (ADI) for the average human life span (70 years) adjusted for the average human body weight (60–70 kg). The ADI is then divided into the estimated proportion (often as low as 1%) contributed by drinking water as compared to the intake from food, based on an average daily intake of 2 litres of water.

No acceptable daily intake is promulgated for carcinogenic or potentially carcinogenic substances; experimental data are derived from tests to ascertain the effect of such substances on laboratory animals. Mathematical models are also used to extrapolate the effect to dose levels low enough to reduce the risk of cancer development in human cells to an 'acceptable level', such as one case per lifetime per 100 000 individuals. There are significant problems with both procedures, for example, can limited term tests at high dosage rates be representative of life-duration ingestion of much smaller quantities; what is the extrapolation relationship that should be assumed; and what are the effects of combinations of substances or different forms of a substance.

The above techniques provide a simplified insight into the procedures used for setting drinking water quality standards; Fawell[22] gives a more scientific account. All the health related chemical standards listed in Tables 6.1(B)–(D) have large safety margins built in to ensure that the results achieved are no worse than those estimated on the basis of an 'acceptable degree of risk'. Failure to comply with a particular chemical standard does not therefore mean that the water is unsuitable for consumption. As the WHO comments 'The amount by which, and the period for which any guideline value can be exceeded without affecting public health depends on the specific substance involved'. The values adopted are precautionary even when, as in the UK and USEPA Regulations, compliance with them is a legal requirement.

The WHO involved over 200 experts from nearly 40 countries in the preparation of the 1993 Guidelines for drinking water quality. The Guidelines are subject to an ongoing rolling revision, with the highest priority being given to those substances with only provisional guideline values. The first set of amendments, including revised guide values for 20 substances, were issued in April 1998. The EC Directive includes a review of the standards at least every 5 years, 'in the light of scientific and technical progress'. The USEPA is likewise required to review and revise, as appropriate, each of the National Primary Drinking Water Regulations (NPDWR) at least every 6 years. The USEPA is also required to publish a list, and to update it every 5 years, of drinking water contaminants which are not subject to any proposed or promulgated NPDWR at the time of publication. This list, known as the Drinking Water Contaminant Candidate List, may include any chemical (or microbiological) contaminant which is known, or anticipated, to occur in public water systems. Such contaminants may then become regulated in the future.

6.49 Standards for raw water classification

An EC Directive of 1975[23] classified *surface* waters intended for use for the provision of drinking water supplies into three categories – A1, A2, and A3, each with a recommended minimum treatment requirement. The three categories were defined by mandatory maximum values for 21 physical and chemical parameters and by non-mandatory guide levels for total and faecal coliforms.

The faecal coliform guide level maximum numbers per 100 ml were:

- Category A1, 20;
- Category A2, 2000;
- Category A3, 20 000.

Surface water falling short of the mandatory values for Class A3 water was not to be used for abstraction of drinking water save in exceptional circumstances. These requirements were reflected in the UK Regulations of 1996 which adopted a similar classification in relation to only the mandatory physical and chemical parameters.[24]

The WHO issued a more detailed classification of raw waters in 1996[25] based on the degree of bacterial pollution of a water. It specified the required degree of treatment according to levels of faecal contamination. This also ensures that no viruses are present in drinking water on the basis that adequate disinfection must produce at least 99.99% reduction of enteric viruses. The classification is given in Table 6.2.

The raw water classification in Table 6.2 acts as a guide when deciding whether a source is suitable for public water supply. However in many parts of the world it may be necessary to take account of other circumstances when deciding what source of water to use and how to treat it. Consideration must be given to the unlisted or unquantifiable risks applying to a catchment, such as the desirability of choosing that source which is least likely to be affected by domestic and industrial wastes, or subject to dangerous accidental pollution. It is important to provide a palatable and aesthetically acceptable public supply of water, free of bacteria and objectionable taste and odour (especially chlorinous ones), to ensure that people use it, especially where there is a risk that otherwise an untreated local sources of doubtful quality may be used for drinking purposes. However allowances must still be made for physical and financial constraints which may make it impossible to provide a water which complies in all respects with the WHO Guidelines or any other standard. Account may need to be taken of the difficulty of maintaining consistent treatment and disinfection in the light of resources available and the skills of local labour.

6.50 Sampling for physical and chemical parameters

Sampling frequencies to WHO, EC, UK and USEPA requirements

The WHO Guidelines do not set out recommended sampling frequencies for physical and chemical parameters but emphasise the need to design a sampling programme for clearly defined objectives. This should cover both random and systematic variations in water quality and be representative of the water quality throughout a distribution system. The frequency of testing must be high enough 'to provide meaningful information while, at the same time conserving sampling and analytical effort'. The WHO points out that testing for parameters unlikely to vary in concentration during distribution need only be carried out on the water going into supply. Other parameters should be sampled at consumers' taps.

Table 6.2   WHO classification of water sources according to bacterial quality and the recommended level of treatment

Source	Level of contamination	Treatment
Ground waters (a)		
Protected deep wells	Free of faecal contamination; E. coli nil per 100 ml (b)	Disinfection (c) for distribution purposes only
	Evidence of faecal contamination; E. coli 20 per 100 ml	Disinfection (c)
Unprotected ground water, e.g. shallow wells	Faecal contamination; E. coli up to 2000 per 100 ml	Filtration (d) and disinfection (c)
	Gross faecal contamination; E. coli >2000 per 100 ml	Not recommended as a water supply source (e)
Surface waters (a)		
Protected impounded upland water	Essentially free of faecal contamination; E. coli <20 per 100 ml	Disinfection (c)
Unprotected impounded upland water or upland river	Faecal contamination; E. coli 20–2000 per 100 ml	Filtration (d) and disinfection (c)
Unprotected lowland river	Faecal contamination; E. coli 200–20 000 per 100 ml	Long-term storage or pre-disinfection; filtration; (d) additional treatment (f) and disinfection (c)

Notes to Table 6.2
(a) If the sources are contaminated with *Giardia* cysts or *Cryptosporidium* cysts, they must be treated by processes additional to disinfection (see Section 7.43).
(b) Water must comply with the WHO guideline criteria for pH, turbidity, bacteriological and parasitological quality.
(c) WHO conditions for final disinfection must be satisfied (see Section 9.6). ˙
(d) Filtration must be either rapid gravity (or pressure) preceded by coagulation–flocculation and where necessary clarification or slow-sand filtration. The degree of virus reduction must be >90%.
(e) Water from these sources should be used only if no higher quality sources is available. Drinking water from such sources carries a risk of inadequate virological quality.
(f) Additional treatment may consist of slow sand filtration, ozonation with granular activated carbon absorption or other processes demonstrated to achieve >99% virus reduction.

The EC Directive sets out a minimum frequency of sampling and analysis based on the volume of water distributed each day within a supply zone. Where water is supplied via a distribution system, the point of compliance is deemed to be the point within a building where drinking water is normally made available to consumers. Member States are required to establish appropriate monitoring programmes to meet these minimum requirements.

Under the current UK Water Supply (Water Quality) Regulations 1989, and subsequent amendments, the only chemical monitoring specified for water treatment works and service reservoir is residual disinfectant. In order to monitor the chemical quality of water in public supplies, water undertakings in England and Wales are required to divide their distribution systems for sampling purposes into water supply zones of not more than

50 000 population. A minimum sampling frequency is then specified for each parameter for which there is a standard. Samples have to be taken from point at which the water becomes available to the consumer, which is usually taken to be the kitchen tap. Most sample points are randomly selected although some samples may taken from fixed sample points which are predetermined and chosen in such a way as to be representative of the quality of water in the zone as a whole. All sampling points for copper, zinc and lead have to be selected at random, with the sample being taken as the first draw of water from the tap.

The USEPA sampling requirements are equally complex. They vary with the parameter to be monitored; the type of source used; and whether the source is defined as 'vulnerable' or 'non-vulnerable' to specific substances. Parameters with secondary maximum contaminant levels (SMCLs) are required to be monitored at intervals no less frequently than those for inorganic chemical contaminants. Monitoring is also required for many chemical contaminants for which final standards have yet to be set.

Minimum sampling requirements where no regulations apply

In countries where there are no legal requirements for sampling, the following regime would provide an adequate minimum level of monitoring, linked with the priorities suggested in Section 6.51:

Simple chemical tests should be carried out daily on the raw water and also the treated water leaving the water treatment works. Samples should also be taken at least weekly at consumers' taps. The tests should be for the more easily measurable but important parameters such as colour, taste, odour, turbidity, pH, conductivity, and chlorine residual in the case of treated waters. Other parameters might be included in respect of a particular source or situation. Among these might be chlorides to test for salt-water intrusion; nitrate and ammonia to indicate pollution; iron or lead in special cases; and residual coagulant and hardness for checking treatment performance.

Full chemical analyses should be carried out, including tests for toxic substances, on any raw water sources to be used for new supplies; whenever treatment processes are being altered; and when new sources of pollution are suspected. Routine samples for full chemical analysis of water in the distribution system should be taken quarterly, 6-monthly, or yearly, depending on the size of the population supplied. Checking for the presence of substances of health significance, for example trihalomethanes, pesticides, PAH, and the heavy metals, may need to be more frequent if they are a cause for concern.

Sampling techniques for physical and chemical parameters

It is of paramount importance that correct procedures are followed when taking water samples to ensure that the samples are representative. Whenever possible samples should be taken by trained and experienced personnel using dedicated sampling bottles and equipment.

Methods of sampling for chemical parameters are fully documented in a series of UK publications under the title of Methods for the Examination of Waters and Associated Materials.[26] Another useful publication is Standard Methods for the Physical and Chemical Examination of Water and Wastewaters,[27] jointly produced by the American Water Works Association, the American Public Health Association and the American

Water Pollution Control Federation. Most water companies in the UK have produced their own sampling manuals based on these publications.

On-site testing and field analysis

It is desirable that analysis for some parameters, such as temperature, pH value and residual disinfectant, is carried out at the time of sampling as significant changes may occur over even short time periods. Other parameters requiring on-site measurement include redox potential, dissolved oxygen and carbon dioxide. These are particularly important for groundwater samples where there can be a rapid change in concentration (or value) due to pressure changes as the sample is taken. The loss of carbon dioxide from a deep borehole sample will result in an increase of pH and, if present, soluble iron and manganese can precipitate out as the water comes into contact with atmospheric oxygen. Special sampling techniques are required in such situations.

Under some circumstances it may be necessary to carry out a more comprehensive analysis in the field, especially if the site is remote and without ready access to a laboratory. There are numerous test kits available for field analysis, covering a wide range of parameters, and varying greatly in complexity and accuracy (see Plate 18). The simplest are 'test strips' which are dipped into the water sample. The intensity of the subsequent colour development is then compared against a strip of standard colours for specific concentrations of the parameter under test. These provide a fairly crude but objective result. There is a wide range of pre-calibrated test discs also available for monitoring most of the common water quality parameters. Reagents, usually in tablet form, are added to a standard volume of sample contained in a glass sample cell. The resultant colour development is compared visually against a blank sample, using a test disc in the appropriate range of concentration. Most of these test kits are user friendly and can be used with minimal training. However the results should always be treated with a degree of caution, as they involve visual comparisons. It is also advisable to introduce an element of analytical quality control, as far as is practicable, to such systems, especially if the results are intended for regulatory reporting.

A broad range of parameters can be monitored in the field by means of electronic meters, for example for pH value, redox, dissolved oxygen, turbidity, and temperature. Multi-parameter meters are also available which can measure a number of parameters with a single instrument. Such instruments are available with an integral data logging system and are extremely useful for continuous monitoring surveys. Comprehensive 'field laboratories' are also commercially available. These typically provide reagents and apparatus for measuring single parameters, such as residual chlorine, or multiple parameters within a single unit. A pre-calibrated spectrophotometer is used for colorimetric tests, with electronic meters for pH and conductivity, and a digital titration system for parameters such as hardness and chloride. A higher level of skill is required to operate such instruments and a higher degree of analytical quality control should be applied to ensure the validity of the results.

Water quality monitoring at treatment works

On-line monitoring systems can be used to check the quality of the raw water and the final treated water as it leaves a water treatment works. Monitoring of the raw water is of first importance, especially if the source is of variable quality, although devising a reliable system can pose problems. It is impossible to monitor for all potentially harmful substances, so surrogates are used which can indicate changes in the concentration of the parameters of importance. One such surrogate is conductivity but this only indicates changes in the dissolved salts present in the water and is unlikely to reveal the presence of a toxic pollutant. A method with wider potential is a fish tank supplied continuously with the water being monitored, with the death of any of the fish in the tank indicating possible pollution. In a more sophisticated and reliable version of the fish tank, the gill movement or electrical responses of a number of fish are continuously monitored. Organic monitors that detect changes in the absorbance of the raw water at a specific ultra violet frequency provide another alternative. A range of raw water monitoring devices tend to be used at larger treatment works, especially if the raw water is abstracted directly from a river or canal source since these are most likely to experience sudden and dramatic changes in water quality or intermittent pollution (see Section 7.1). Bankside storage provides some guarantee of the raw water quality and helps to even out fluctuations. Ideally such storage should be divided into at least two compartments so that the raw water in each can be tested alternately before being drawn upon for treatment and supply. Distant monitoring of water quality upstream of an intake is also very frequently adopted.

On line monitoring systems are widely used to check the efficacy of the clarification and filtration processes at a treatment works and to ensure that the clarity and pH value of the filtered water meet the requirements for final disinfection (see Section 9.6). On line monitoring systems are also widely used for checking the residual disinfectant level and pH value of treated waters entering supply.

6.51 Priorities in water quality control

In some parts of the world, the availability of well equipped laboratories and resources for water quality testing can be very limited for many water undertakings. The level of testing in such circumstances must concentrate on the most essential parameters. The following is a suggested list of priorities in testing which, for the sake of completeness, includes bacteriological testing as defined in Sections 6.62 and 6.65 below.

Simple checks at source works

Twice daily checks should be carried out on chlorine dosage rate and the residual chlorine content of water entering supply.

Daily measurements on samples of raw and treated water should be carried out for turbidity; colour; odour; conductivity; and pH value.

Where coagulation, clarification and filtration are applied, daily checks should also be carried out on dosages of coagulants, and the pH and turbidity of the water ex clarifiers and ex filters.

Bacteriological testing (see Section 6.62)

Where possible, analysis for total and faecal coliforms should be carried out at least weekly on samples the treated water leaving a water treatment works. Samples of raw water and samples from the distribution system should be analysed monthly. Every effort should be made to meet the foregoing frequencies as a minimum, but, where they cannot be met, resources should be directed towards determining coliform counts in the treated water ex works. If no suitable laboratory for bacteriological analyses is available, consideration should be given to on-site testing by a trained operative visiting works on a routine basis. Portable test kits are available using filter membranes techniques, as illustrated in Plate 18, as are a number of test kits which determine whether an indicator organism is simply present or absent in 100 ml of sample.

Chemical testing

At all sources, chemical analyses covering the most important parameters of a given raw water should be conducted at least twice a year, at appropriate times, e.g. following the onset of heavy rainfall after prolonged dry weather. These should normally be accompanied by bacteriological analyses.

Source watch

A complete survey of the catchment to a new source should be undertaken, mapping and noting all potential sources of pollution, before the source is brought into use. Subsequent regular catchment surveys should note any changes. Sources of pollution closest to an intake, if they cannot be diverted, should be kept under frequent observation. Where chemical tests show toxic substances are present in a raw water, their source should be traced and if possible eliminated. If the latter is not possible, additional monitoring must be set up for such substances in both the raw and the treated water. The use of waters having consistently high coliform counts, or dangerously sited with respect to any waste discharge, should be avoided if at all possible.

In general a given raw water tends to have characteristics which normally range only between certain high and low values, often according to rainfall or other seasonal conditions. As soon as sufficient bacteriological and chemical data have been obtained to establish this usual range and the appropriate treatment has been adopted, routine testing can concentrate on monitoring the parameter or parameters which are most likely to indicate any abnormal change of quality. However routine bacteriological testing of the treated water leaving the treatment works should always be carried at an acceptable frequency to ensure that the treatment processes and, in particular, the disinfection process has not been compromised.

6.52   Methods of chemical analysis

It is important that methods of analysis should be standardised in order to achieve comparability of results. In 1972 in the UK, the then Department of the Environment established the Standing Committee of Analysts to set up working groups to produce suitable methods for water analysis. The committee, which now comes under aegis the Environment Agency, represents a wide range of interests in the water industry and has produced detailed guidance in a series of publications under the title of Methods for the Examination of Waters and Associated Materials.[26] Each publication looks at a single analytical method or linked group of methods. Another useful publication is Standard Methods for the Physical and Chemical Examination of Water and Wastewaters.[27] This comprehensive book, which is updated at regular intervals, provides a valuable reference and the methods described in it have had a wide influence on standards adopted in other countries. Field Testing of Water in Developing Countries by Hutton[28] provides further useful reference.

All analytical methods for chemical parameters should be fully evaluated and validated by the laboratory carrying out the tests before the method is adopted for routine use. The initial performance tests should demonstrate that the analytical system is capable of establishing, within acceptable limits of deviation and detection, whether any sample contains the parameter under analysis at concentrations likely to contravene the required standard. Such performance testing should cover the entire analytical procedure, including sample preparation and any concentration steps. The precision and accuracy of the test in terms of maximum tolerable values for total error and systematic error and limit of detection should be ascertained, along with checks for recovery and resilience against possible interferences. The required performance characteristics for chemical parameters are specified both in the EC Directive and the US National Primary Drinking Water Regulations. In the UK, guidance on analytical systems has been issued in the DoE publication on Guidance on Safeguarding the Quality of Public Water Supplies[9] and in subsequent information letters sent by the Drinking Water Inspectorate to water undertakers in England and Wales. It is recognised that the required level of performance can not always be met with the current methods of analysis available for some of the more obscure organic parameters.

The laboratory should also have established and documented procedures for routine analytical quality control as applied to each validated method. External quality control schemes or interlaboratory proficiency testing schemes, where available, provide further useful information on a laboratory's capabilities for carrying out an acceptable level of analysis.

Under the US National Primary Drinking Water Regulations all analyses, other than turbidity, chlorine residual, temperature and pH, have to be carried out by a certified laboratory. In England and Wales, water undertakings are expected to use suitably third party accredited laboratories or have their laboratories inspected annually by the Drinking Water Inspectorate.

6.53 UK standards applying to use of chemicals and materials in water supply

Under the Water Supply (Water Quality) Regulations 1989, all chemicals used in water treatment and materials which come in contact with water (such as pipe linings, etc.) are subject to approval by the Secretary of State. Similar provisions are contained within the Scottish and Northern Ireland Regulations. An annual list is published under the auspices of the Drinking Water Inspectorate, giving all substances, products and processes for which approval has been granted, refused, revoked or modified or for which use has been prohibited. The current list,[29] which is available on the Internet (www.dwi.detr.gov.uk), includes over 800 proprietary products and materials covering flocculants and coagulants, adsorbents; disinfectants; other chemicals; ion exchange resins; pipes; filters and membrane systems; linings and coatings; cementitious and associated products; products for emergency use and products based on traditional chemicals. All the products are identified by trade name, grade, and manufacturer. Maximum permissible dosages are stipulated for most proprietary products, together with other conditions of use as appropriate. The list also includes a section on traditional chemicals and filtration media, which are commonly used in water treatment and are considered to be unobjectionable on health grounds provided they are used in accordance with the manufacturers' instructions.

A number of European standards for water treatment chemicals have been produced as part of a harmonisation approach for European legislation. Chemicals meeting these standards may be used in the UK without approval from the Secretary of State, provided any national conditions of use are observed.

Part III Water microbiology

6.54   Diseases in humans which may be caused by water-borne bacteria and other organisms

Water-borne diseases in humans can be caused by the presence of pathogenic bacteria and other organisms such as protozoa and viruses in drinking water supplies, in water used for bathing or immersion sports, or via other routes. It has long been recognised that the ingestion of water contaminated with mammalian excrement can result in the spread of diseases such as cholera and typhoid and that adequate treatment and other measures are required to prevent outbreaks of such diseases. The WHO Guidelines state that 'The potential consequences of microbial contamination are such that its control must always be of paramount importance and must never be compromised'.

The following sections look at some of the classical intestinal diseases which are commonly, although not always invariably, water-borne. In all cases the associated organisms are present in large numbers in the excreta of an infected host and are relatively resistant to environmental decay. Many are likely to cause illness even when ingested in small numbers. Schistosomiasis and other parasitic diseases of the tropics are reviewed as part of water biology in Part IV.

6.55   Bacterial diseases

Cholera is caused by the bacterium *Vibrio cholerae* and its variant *Eltor vibrio*. Infection is usually contracted by ingestion of water contaminated by infected human faecal material, but contaminated food and person to person contact may also be sources. In recent years cholera has moved from the Far East to the Near East, Africa, and southern Europe, and could enter Britain and other European countries through carriers or by individuals who are in the incubation period of the disease. It is not likely to spread in communities with controlled water supplies and effective sewerage. However shellfish inhabiting polluted seawater and eaten uncooked or inadequately cooked, can carry the bacterium (and also the virus of hepatitis) and lead to serious outbreaks.

Typhoid fever is caused by the bacterium *Salmonella typhi*. Infection is usually contracted by ingestion of material contaminated by human faeces or urine, including water and food (e.g. milk, shellfish) *Salm. typhi* occasionally continues to proliferate in the gall bladder of a few patients who have recovered from the primary infection, and these carriers continue to excrete the organisms in their faeces or, occasionally, in their urine for long periods, even for life.

The largest water-borne outbreak of typhoid fever in Britain this century occurred in Croydon in 1937 and killed 43 people. It was investigated by Suckling[30] who found that it was caused by a combination of circumstances including a person who was a carrier of

Salm. typhi working down a well, which was pumping into supply, coincidental with the filtration and chlorinating plants being bypassed. There have been a number of more recent outbreaks, which are believed to have been due to water contamination coinciding with inadequate disinfection. *Paratyphoid fevers* are also caused by *Salmonella*, in this case *paratyphi A, B*, or *C*. Infection may exceptionally be via contaminated water.

Bacillary dysentery is caused by bacteria of the genus *Shigella* – *Sh. dysenteriae* 1, *Sh. flexneri, Sh. boydii* and *Sh. sonnei* are some of the several subspecies. Infection can occasionally be contracted via water contaminated by human faeces, but more commonly is due to ingestion of food contaminated by flies or by unhygienic food handlers who are carriers. The cause of *Traveller's diarrhoea* is not definitely known, but it may be some forms of pathogenic *Escherichia coli* or, rarely, *Shigella*. It is probably transmitted in the same way as bacillary dysentery and water may sometimes be the vehicle. The problems associated with enterovirulent *E. coli* are reviewed in Part V (see Section 6.76).

Leptospirosis is caused by very numerous serogroups of the motile, spiral organisms known as *Leptospira*, with symptoms ranging from mild fever to severe jaundice in the case of Weils' disease. The organisms are shed in the urine of infected rats, dogs, pigs, and other vertebrates, and are often present in ponds and slow-flowing streams haunted by such animals. People who bathe in, fish in, or sail on these waters are at risk, becoming infected via the mouth, nasal passages, conjunctiva, or abraded skin through which the organisms can enter. Normal water treatment eliminates these organisms but sewer workers remain at risk from rat infections.

Legionnaire's disease is caused by bacteria of the genus *Legionella*, of which some 20 species have been identified.[31] The bacterium *L. pneumophila* is regarded as the most dangerous, having been identified in all outbreaks of Legionnaire's disease. Outbreaks of the disease can be sudden and exhibit a high mortality. (The name comes from an outbreak of the disease causing a high mortality in war veterans attending a Legion Convention in Philadelphia, USA in 1976.) *Legionella* organisms are widely present in small numbers in surface waters and possibly also in groundwaters. They may survive conventional water treatment, including disinfection with chlorine, and retain an ability to colonise internal pipe surfaces. They are thermo-tolerant and ideally suited to grow in the warm water systems of buildings at 30–45°C. However they do not survive sustained temperatures above 60°C. Infection occurs from the transport of the bacterium by aerosols or air-borne water droplets, which are inhaled; there is no evidence of transmission by ingestion so infection is not attributable directly to drinking water supplies. Most outbreaks of Legionnaire's disease have been caused by droplets blown from exposed cooling towers of air-conditioning plants at hotels, hospitals, and similar large buildings. Other sources such as whirlpools and Jacuzzis, where recirculated warm water is sprayed, have been implicated in outbreaks of the disease.

Preventive measures include designing and maintaining water systems in buildings to minimise the risk of colonisation; minimising the accumulation of sediments and slimes; and maintaining hot water systems above 60°C and cold water systems below 20°C. Official advice is available concerning the measures to be taken.[32,33] It should be noted that biocides are generally less effective in controlling *Legionella* in water distribution systems within buildings but are effective in air conditioning systems using wet evaporative cooling towers.

Campylobacteriosis is caused by bacteria associated with the excrement from wild fowl. There are 14 species of the bacteria, some of which are pathogenic to man causing gastrointestinal illness. *Campylobacters* are frequently found in sewage and have been detected in surface waters, where they can survive for several weeks at cold temperatures. There have been a number of reported outbreaks of associated with unchlorinated or inadequately chlorinated surface water supplies or with contaminated storage facilities.

6.56   Other bacteria

The *Pseudomonas* group is commonly found throughout the environment. They may be present in human and animal excrement and can also multiple in water containing suitable nutrients. Their subsequent growth, if present in drinking water, may also result in an overall deterioration in the microbiological quality and lead to consumer complaints of taste and odour. *Ps. aeroginosa* is an opportunist pathogen, that is, it can cause infection in people whose natural defence mechanisms may be impaired, for example the very old, the very young or the immuno suppressed. Other pseudomonads may produce undesirable slimes within the distribution system.

The *Aeromonas* group is also naturally present in the aquatic environment. Their presence in drinking water does not necessarily indicate faecal pollution but highlights possible inadequacies in the treatment process or ingress within the distribution system. The number of such organisms likely to be found in a distribution system will depend on the residence time of the water, its organic content and the residual chlorine level.

6.57   Protozoal diseases

Amoebic dysentery is caused by the microscopic parasite *Entamoeba histolytica*. The parasite is distributed throughout the world and exists in two stages, only one of which, the cyst, is infective. The parasite infects mainly primates and, following infection, resides in the large intestine in humans where it produces further cysts, which are passed in the faeces. Infection takes place by ingestion of these cysts, which range in size from 10 to 20 μm. They can survive for several days in water at temperatures of up to 30°C and are resistant to chlorine.

Outbreaks can occur if water supplies are contaminated with domestic sewage containing viable cysts. More commonly the disease is transmitted by person-to-person contact or via food contaminated by carriers. Most infections tend to cause only minor symptoms, although liver abscess can be a serious *sequela*.

Cryptosporidiosis is an acute self-limiting diarrhoeal disease caused by the parasite *Cryptosporidium parvum*, which infects domestic and farm animals as well as humans. The normal routes of exposure are via direct contact with infected animals or humans or via contaminated water or food. Not all infected individuals necessarily develop the symptoms of the disease but the illness is likely to be serious or even life-threatening for patients who are immunologically compromised.[34]

The parasite has a complex life cycle, which takes place within the body of the host and can include repeated cycles of autoinfection. Infective oocysts of the parasite, which are 4–6 μm in diameter, are then shed in vast numbers in the faeces of infected animals and humans. These oocysts are often found in surface waters, particularly in areas associated with intensive animal grazing, and are also found occasionally in some groundwater sources. Treated sewage effluent can, on occasion, also contain large numbers. The oocysts can remain infective in water and moist environments for several months and are resistant to high concentrations of chlorine (see Section 9.6). The latter, coupled with the small size of the oocysts, can result in low numbers of oocysts occasionally penetrating conventional public water supply treatment processes. In recent years there has been an increasing number outbreaks of cryptosporidiosis in both the UK and North America where there has been a causal link with drinking water supplies. Following an outbreak in Swindon and Oxfordshire during the winter of 1988–89, an expert committee was set up by the Department of the Environment and the Department of Health, under the chairmanship of Sir John Badenoch, to examine the problem. Its report, known as the Badenoch Report[35] and published in 1990, made many recommendations concerning water treatment practices, monitoring and also the role of the various authorities in the event of a suspected waterborne outbreak. The Group of Experts produced further guidance in a second report issued in October 1995.[36] Following an outbreak of cryptosporidiosis in North London in 1997 in which groundwater was implicated, a new Expert Group was convened under Professor Ian Bouchier to review current concerns and practices. The report of their findings was published in November 1998.[37] Ongoing research is also being carried out into methods for assessing oocyst viability and infectivity; the life cycle of the organism; easier means of determining the presence of the oocyst in water; the infective dose; treatment processes to reduce the number of oocyst likely to get into supplies; and different disinfection processes.

Following the third report of the Group of Experts, new regulations on *Cryptosporidium* were introduced in England and Wales on 30 June 1999[38] (see Section 7.43). Under these regulations water companies are required to carry out a detail risk assessment at all water treatment sites. They are then required to install continuous sampling at those sites considered to be at risk and carry out daily analysis of the samples. The sample typically equates to one cubic metre of treated water filtered through a special filter over a 24-hour period. The treatment works so sampled will have to meet a treatment standard of an average of less than one oocyst in ten litres of water leaving the works. It will be an offence if the standard is breached and/or if the sampling and analysis requirements are not met.

The US National Primary Drinking Water Regulations specifies the level of monitoring required for *Cryptosporidium* for public water systems and, under the Interim Enhanced Surface Water Treatment rule, sets a maximum contaminant level goal of zero. The EPA is also considering requiring public water systems to monitor their source waters for *Cryptosporidium* and *Giardia* to determine appropriate levels of treatment.

Giardiasis is another diarrhoeal disease caused by a protozoal parasite, namely *Giardia lamblia* (or *intestinalis*). Like cryptosporidiosis, the disease is self-limiting and is caused by the ingestion of cysts by a susceptible host. The normal routes of exposure are via direct contact with infected animals or humans, or via contaminated water or food. Infected animals can contaminate surface waters and, in North America, beavers are frequently blamed for associated outbreaks. The cysts are larger than *Cryptosporidium* oocysts, being 7–10 μm wide and 8–12 μm long, and can survive for many days in a cool aqueous environment. Being larger, the cysts are effectively removed from drinking water by physical methods of treatment, such as filtration. They are also much more susceptible to disinfection than *Cryptosporidium*.

Giardiasis is a world-wide disease and there have been a number of reported waterborne outbreaks in the USA. Evidence of waterborne infection in the UK has been confined to situations where there has been direct faecal contamination of water used for drinking, such as holiday makers drinking untreated private supplies.

The USEPA has promulgated criteria under which filtration is required as part of the treatment for surface waters under the Enhanced Surface Water Treatment Rule, with an effective standard of zero for *Giardia* where such treatment is applied.

6.58   Viral diseases

Viruses differ from bacteria in that they are very much smaller and can multiply only within suitable host cells, in which they produce changes which give rise to a range of diseases. More than one hundred different types of virus have been identified in faeces and the main sources of human enteric viruses in the aquatic environment are sewage discharges. Little direct information is available on the removal of viruses by water treatment processes but information gained using cultured viruses indicates that the water treatment processes required for Category A2 and A3 waters (see Section 6.49) will, if applied properly, produce effectively virus-free drinking water. The main areas of risk then become the use of sewage polluted waters for recreational purposes or the recycling of wastewater for domestic use without adequate treatment and disinfection.

Human polioviruses which cause *Poliomyelitis* can be found in untreated sewage and even in the effluent from sewage disposal units. Like other viruses, they do not multiply in the absence of living cells and although there have been a few reports of water-borne infection there has been little confirmation.

The viruses associated with *Hepatitis A and E* have been detected in sewage and polluted rivers. Several large outbreaks of drinking water transmitted *Hepatitis* have been reported, usually where water treatment has broken down; where the distribution system has been disturbed; or where badly constructed wells have been contaminated from adjacent cesspits or as a result of heavy rainfall.

Other enteric viruses that are emerging as important risks in the transmission of waterborne viral diseases are reviewed in Part V.

6.59   Microbiological standards for drinking water

Pathogenic bacteria and other organisms are usually difficult to detect in an effectively treated water supply because, if present, their numbers are likely to be very small. Their presence even in sewage effluent or polluted river water may be only infrequent or at irregular intervals, depending on the level and source of contamination. Analysing directly for pathogenic bacteria is not therefore a practical safeguard for a water supply and, indeed, routine monitoring for process control purposes would be both impracticable and unnecessary. Instead, evidence of any pollution by the excreta of man or animals should always be sought and, if the evidence is positive, it should be assumed the water may also contain pathogenic bacteria and must therefore be regarded as unsuitable for supply purposes.

6.60   Use of coliforms as an indicator of bacteriological pollution

Coliform bacteria are widespread throughout the environment and have long been used as indicator organisms by water microbiologists because of their relatively simple analysis and detection at low numbers. The group contains species which can multiple in water but which are not of faecal origin. The group also contains species that are referred to as 'thermotolerant (faecal) coliforms' of which the bacterium *E. coli* is one. This is a natural inhabitant of the intestine and is present in large numbers in the faeces of man and other warm blooded animals, and also of fish and birds. Such bacteria can survive for a considerable time in water, longer than most pathogenic bacteria. Thus the detection of *E. coli* in drinking water supplies provides clear evidence of faecal pollution. If coliform organisms are detected, but no *E. coli*, the probability is that the pollution is via soil or vegetable contamination, or possibly a warning that more serious pollution could follow, especially after heavy rain. However the presence of any coliform bacteria in treated water indicates either deficiencies in the treatment process or some form of post treatment contamination and should be investigated immediately.

Although the absence of coliform organisms, and more particularly *E. coli*, implies that a water is unlikely to be polluted, it cannot be guaranteed absolutely that no other intestinal pathogens are present. This is because other pathogens such as viruses and protozoa, although less likely to be present, are likely to be more resistant to disinfection.

6.61 Standards of bacterial quality

Table 6.3 sets out the WHO guidelines, the current UK standards, the EC standards and the USEPA standards for bacterial quality. Coliforms should not be detected in the water leaving a water treatment works provided the treatment processes and, in particular, the disinfection process are adequate. The principal common requirements of all the standards for drinking water at the consumer's tap are:

- no faecal coliforms detected in 100 ml of sample; and
- 95% of 100 ml samples must not show the presence of coliform organisms.

Both these standards should be rigorously adhered to. Depending on the quality of the source water and the type of treatment adopted, biofilms may develop within the distribution system and give rise to occasional failures for total coliforms at service reservoirs and at consumer's taps. If a positive result is obtained, the number of organisms present should always be very low and shown to be non-faecal. All such failures should be investigated immediately and the need for corrective action, such as increasing the chlorine residual by booster chlorination, assessed. Immediate action should also be taken in the event of even a single faecal coliform (or *E. coli*) being detected in a 100 ml sample. Action should always be taken on the presumptive result, even if this does not subsequently confirm, and repeat samples should also be taken immediately both from the affected tap and from at least two other taps in the adjacent area.

Standards of bacteriological testing can vary in different parts of the world. Terminology is often inconsistent and confusing, with the same term being used to cover one test procedure in one country and a different test procedure in another. In hot climates many waters give total bacterial counts substantially higher than in temperate climates, and certain types of organisms may be more abundant. It is sometimes difficult for the engineer to interpret bacterial test results and the advice of an experienced water bacteriologist should always be sought in obtaining a definitive interpretation.

In addition to specific bacteriological standards the current UK Regulations contain a 'catch-all' clause that water supplied for drinking, washing or cooking, or for food production, must not contain any organism at a concentration which would be detrimental to public health.

Table 6.3 Bacterial standards

	Colonies/ml at 22°C	Colonies/ml at 37°C	Total coliforms	Faecal coliforms	Faecal Streptococci	Clostridia
WHO Guidelines 1993						
Water entering the distribution system			ND in 100 ml	ND in 100 ml (a)		
Water in the distribution system			ND in 100 ml (b)	ND in 100 ml (a)		
All water intended for drinking				Not detected		
Current UK Regulations (1989)						
Water leaving a water treatment works	No significant increase over that normally observed	No significant increase over that normally observed	0/100 ml	0/100 ml	0/100 ml	≤1/20 ml (d)
Water in service reservoirs	No significant increase over that normally observed	No significant increase over that normally observed	0/100 ml (c)	0/100 ml	0/100 ml	≤1/20 ml (d)
Water at consumers' taps	No significant increase over that normally observed	No significant increase over that normally observed	0/100 ml (c)	0/100 ml	0/100 ml	≤1/20 ml (d)
EC Directive (1998)						
Water at the point of supply				0/100 ml (e)	0/100 ml (f)	0/100 ml (g)
USEPA						
At sites representative of the water throughout the distribution system	Treatment requirement to be met	Treatment requirement to be met	0/100 ml (MCLG) 0/100 ml <5% of samples (MCL) (i)	(h)		

Notes.

ND – Not detected.

MCL – Maximum contaminant level (US, mandatory).

MCLG – Maximum contaminant level goal (US, not mandatory).

(a) *E. coli* or thermotolerant coliform bacteria.

(b) In the case of large supplies, where sufficient samples are examined, not present in 95% of samples taken over 12 months.

(c) 95% of the last 50 samples taken must meet the standard.

(d) Analysis by multiple tube method.

(e) *E. coli*.

(f) Entercocci.

(g) *Clostridium perfringens* in surface derived waters.

(h) Total coliforms includes *E. coli* and faecal coliforms.

(i) For a system, which collects at least 40 samples per month, total coliforms should not be present in more than 5% of samples collected in a month. Where fewer than 40 samples per month are collected, no more than one sample per month should contain total coliforms.

The EC Directive sets bacterial standards for water offered for sale in bottles and containers. These are more stringent than the standards for tap water and include numerical standards for colony counts and *Pseudomonas aeroginosa* as well as the above faecal indicators.

6.62   Routine tests for bacterial contamination of water

The rationale for routine bacteriological monitoring usually includes the following tests.

Colony counts at 20–22°C and 37°C

Large numbers of micro-organisms occur naturally in both ground and surface waters, many of which are associated with soil and vegetation and can survive for long periods in the environment. Counts of such organisms, grown as colonies on nutrient agar, provide a useful means of assessing the general bacterial content of a water. The colony count, or plate count, following incubation at 20–22°C gives an indication of the number of all types of bacteria present at normal environmental temperatures. Although the result does not have any direct health significance, it provides a useful means of assessing the efficacy of the various water treatment processes in terms of overall bacterial removal. It also gives an indication of the general bacteriological state of a given distribution system. Incubation at the higher 37°C temperature encourages the growth of bacteria that can thrive at body temperature and which, therefore, may be of animal origin. The main value of both tests is to provide a reference background level for a particular source water, treatment works or distribution system. A sudden marked increase, particularly in the 37°C count, could be indicative of treatment deficiencies or of a more serious problem developing. Marked changes over and above the normal seasonal trends for colony counts at both temperatures could indicate longer term changes in the bacteriological quality of the water. The counts are also of value where the water is used in the manufacture of food and drink as they could be an indication of a potential spoilage problem.

Total coliform count

As already discussed in Section 6.60, coliform organisms are easy to detect in water and also easy to enumerate. The term 'total coliforms' traditionally refers to bacteria capable of growing at 37°C in the presence of bile salts and of fermenting lactose at this temperature, producing acid and gas after 24–48 hours incubation. The bacteria are also oxidase-negative and non-spore forming. Under normal laboratory test conditions a 'presumptive' result is available after some 20 hours of incubation and further tests are then carried out to confirm the result. With the advent of a number of rapid and direct test methods, the traditional definition has been extended to include the possession of the β-galactosidase gene.

Faecal coliform count

Coliforms of faecal origin, including *E. coli*, and thermotolerant coliform organisms are capable of growth and of expressing their fermentation properties at the higher temperature of 44°C. As with total coliforms, a 'presumptive' result is available after some 20 hours of incubation, with the need for subsequent confirmation. Specific confirmation for *E. coli* includes the production of indole from tryptophan. A number of confirmatory test kits are commercially available.

The detection of *E. coli* provides a reliable indicator of recent faecal contamination. The presence of other thermotolerant coliforms, particularly when detected in warmer tropical or sub-tropical waters, provides a less reliable indication of such contamination. However for most routine monitoring purposes there is an acceptable correlation between *E. coli* and faecal coliforms, as defined by thermotolerance.

Other tests

Tests for faecal streptococci and sulphite reducing clostridia can be used as secondary indicators of faecal pollution. As such, they would not normally be included in routine monitoring but would be undertaken whenever there is a need to confirm a problem.

The test for faecal streptococci can be used to assess the significance of coliform organisms in the absence of confirmed *E. coli*, as the organisms, although rarely multiplying in polluted water, are more persistent than *E. coli*. The ratio between the numbers of faecal coliforms and faecal streptococci present may also provide some indication as to whether the source of the contamination is human or animal, although careful interpretation of any such ratios is necessary. Even then the outcome may be unreliable.

The presence of spore-forming, sulphite-reducing anaerobes, such as *Clostridium perfringens* (*C. welchii*) is also associated with faecal contamination. The presence of such organisms, especially in well or borehole supplies, can indicate remote or intermittent contamination. The presence of *C. perfringens* in a filtered water may indicate deficiencies in the filtration process and, as such, may indicate the breakthrough of protozoan cysts such as *Cryptosporidium*.

6.63   Frequency of sampling for bacteriological monitoring

The minimum UK requirements for testing samples of water for total and faecal coliforms and residual chlorine, as set out in the Water Supply (Water Quality) Regulations 1989 and subsequent amendments, are as follows.

Water leaving a water treatment works	52 samples per annum for $\leq$2000 m^3/day of water supplied for domestic purposes 104 samples for 2001–6000 m^3/day output 208 samples for 6001–12 000 m^3/day output 365 samples for > 12 000 m^3/day output, as a minimum standard number
Water leaving service reservoirs	Weekly sampling of each service reservoir when in supply
Water at consumers' taps (50% of sample points to be selected at random)	12 samples per annum for $\leq$5000 population supplied 24 samples for 5001–10 000 population 48 samples for 10 001–20 000 population plus 12 samples for every additional 5000 population supplied to a maximum of 50 000 population in a water supply zone (i.e. up to a maximum of 120 samples)

The above accord with the EC requirements and are also similar to the WHO recommendations and to the monitoring regime set out in the US National Primary Drinking Water Regulations.

Routine bacteriological samples should also be taken from raw water sources to ensure that the levels of treatment and applied disinfection are adequate. Where untreated water is supplied without disinfection, a minimum sampling frequency should be established based on local conditions.

Additional samples must always be taken for testing in the event of a bacteriological failure either in the water leaving a water treatment works or in the distribution system. Such samples should also be tested for faecal streptococci and clostridia. Routine samples are also advisable following any interruptions to supplies or any repair work that might have compromised the integrity of the distribution system.

Additional samples for a variety of biological tests should also be taken when there is a suspected water-borne outbreak of illness.

6.64   Sampling for routine bacteriological parameters

The sample should always be representative of the water in supply and care should always be taken to avoid accidental contamination either during or after sampling. Personnel taking bacteriological samples should be adequately trained and aware of the responsibilities of their role. They should be aware of the need to avoid cross-contamination between raw and treated water samples during transit and of the need to deliver the samples to the analysing laboratory without undue delay. Ideally bacteriological samples should be examined within 6 hours of sampling. Where this is not possible, samples should be stored in the dark at temperatures of between 2 and 10°C and analysed as soon as practicable and preferably within 24 hours.

Raw water samples should be taken at the inlet to the treatment works and preferably at a dedicated sampling point. Wherever possible dip samples should be avoided unless there is no alternative.

Water leaving a water treatment works should be sampled from a dedicated sample tap, which should be metal and of an approved design. The sample point should be located so as to be representative of the water entering supply. In some cases it may be necessary to have more than one sample point. Delivery pipework should be of a suitable material and as short as possible. The sample tap should be kept clean and be prominently labelled. It is not advisable to take bacteriological samples from constantly running taps as the action of turning the tap off to disinfect it and then turning it on again to run to waste could dislodge particulates or biofilm from the sample line.

Sample taps at service reservoirs should also be of metal and to an approved design. Delivery pipework should be as short as possible and the system designed to ensure that water sampled is representative of the reservoir as a whole.

Sampling from consumers' taps can often be difficult. The tap should be clean, in good repair and free of attachments. It should also supplied direct from the rising main to be representative of the water in supply. Many of the mixer taps currently available in developed countries are made of materials that are very difficult to disinfect. Therefore extra care is always needed to ensure that the tap is adequately cleaned and sterilised before being sampled.

Method of sampling

The order of sampling should always be physico-chemical samples and then samples for bacteriological analysis. This is because some chemical tests for samples taken at consumers' taps have to be carried out on first draw samples (e.g. copper, zinc and lead) and samples for PAH should be taken before the tap is disinfected. As soon as any chemical samples have been taken, the sample tap should then be disinfected either by using a blow torch in the case of a metal tap or by swabbing the outside and as much of the inside as possible with sodium hypochlorite solution, or an equivalent chemical disinfectant, and allowing a few minutes for the process to work. The tap should then be run to waste until the water is cool or until all the chemical disinfectant had been removed. The flow rate of the water should remain steady throughout this period and when the sample is being taken in order to reduce the risk of any biofilm being dislodged into the sample.

Only sterilised sample bottles should be used and, if the water being sampled contains residual chlorine, the bottle should contain a small amount of 2% w/v sodium thiosulphate to dechlorinate the sample. The sample bottle should not be rinsed but filled in one movement, with the cap or stopper being removed for the minimum time possible. The lip of the bottle should not be allowed to come into contact with the tap and the bottle should be filled without splashing, leaving a small air space below the cap or stopper. Care should always be taken not to contaminate the cap or stopper and if accidental contamination is suspected the sample should be discarded and taken again in a new bottle.

In some cases it may be necessary to take samples from hydrants or standpipes, particularly for new and repaired mains. The hydrant box should always be cleared of any accumulated water and the outlet should be dosed with sodium hypochlorite before the standpipe, which should be kept in a clean condition, is attached. The hydrant should then be cracked open to fill its outlet and the standpipe and allowed to stand for at least 5 minutes before flushing. Flushing should continue until the residual chlorine level is that of the mains supply and the bacteriological sample should be carefully taken without turning the water off.

Occasionally dip samples have to be taken for investigational purposes. Special wide mouthed sterile bottles should be used. These should have a sterilised wire attached for taking the sample and care must always be taken to avoid incurring any contamination of the water being sampled.

6.65 Methodology for bacteriological examination

The methods for the routine bacteriological examination of water are well documented in the Standing Committee of Analysts publication – The Microbiology of Water 1994; Part 1 – Drinking Water (still often referred to as 'Report 71' being first issued under that number by the Ministry of Health in the 1950s).[39] Equivalent test procedures are given in the American Standard Methods (see Section 6.52).

Established methods should always be used for routine analyses and good laboratory practice should be adopted at all times, with special precautions being taken to avoid accidental contamination of samples once they are in the laboratory. Appropriate quality control procedures should be used at all stages of the analysis and laboratories should partake in external quality control schemes, such as inter laboratory tests, where these are available.

The choice of method for coliform analysis depends to a certain extent on the number of organisms likely to be present. Membrane filtration is the best approach for treated waters but it is not suitable for highly turbid waters or for waters containing only a small number of indicator organisms in the presence of large numbers of other bacteria capable of growing on the media used. The multiple tube method is particularly suitable for very contaminated samples or samples containing a lot of sediment.

Present–absence tests for total coliforms

These are a modification of the multiple tube procedure, based on a single 100 ml volume of appropriate medium, such as minerals modified glutamate, instead of a series of tubes of different volumes. The tests provide a very good indication as to whether total coliforms, and therefore possibly faecal coliforms, are present in a sample incubated at 37°C for 18–24 hours. However the draw back of such tests is that a positive sample cannot be enumerated.

Other test kits

There are a number of other field test kits available on the market. Most consist of a small portable incubator, which can if necessary be plugged into a car battery, and a means of aseptically filtering a sample. The most simplistic form consists of a small sterile membrane on a frame and housed in a dedicated container containing an appropriate medium. The membrane is dipped in the water sample and after incubation the result is essentially the same as that obtained for the present–absence coliform test, although some estimate can be made of the number of coliforms present in a positive sample. However, since only a small volume of sample is used, the sensitivity of the test is reduced, thus such equipment is most suited to waters containing moderate rather than low levels of coliform organisms (e.g. more than 200 per 100 ml). The more sophisticated test kits provide the equivalent of a portable laboratory. Some test kits offer initial results after only 5–6 hours incubation.

Considerable care has to be taken with all field test kits to ensure that samples do not become contaminated during analysis, and that adequate sample dilutions have been prepared to cover the expected concentration ranges. Colony identification and counting must be carried out by an experienced person, familiar with techniques of identifying coliform bacteria, since non-coliform organisms may also be visible on the membrane. This is particularly so in tropical or sub-tropical climates. A basic level of analytical quality control should always be adopted with test kits, to include positive and negative control samples and a record of the incubator temperature.

6.66   Protozoal examination

Water companies in England and Wales are expected to implement a defined monitoring programme for all sources considered to be at risk of *Cryptosporidium*. Continuous sampling of the treated water entering supply is also required at treatment works considered to be at risk of breaching the new *Cryptosporidium* standard (see Section 6.57).

The US National Primary Drinking Water Regulations also specifies the level of monitoring required for *Cryptosporidium* and *Giardia lamblia* in public water systems, linked to the level of treatment and the potential for contamination of the raw water.

Sampling methods include continuous sampling, large volume grab samples, or composite samples. For general surveillance purposes, a large volume of water sampled through a filter over a long period of time is likely to give the best result. However this may not prove to be possible if the water is very turbid or contains a lot of algae. Grab samples provide a more immediate result for operational purposes.

Standard methods of analysis are available for both protozoa. The biggest problem associated with some of the methods for *Cryptosporidium* is the low percentage recovery. There is no guarantee that oocysts are not present simply because no oocysts have been detected. Research is ongoing to further improve methodology and performance criteria. Other research has indicated that *Clostridium perfringens* could provide a useful indicator when protozoal contamination of a treated water is suspected.

6.67 Virological examination

Human enteric viruses, although many times less numerous than *E. coli* bacteria in domestic sewage, occur widely in surface waters and are found in some well waters subject to contamination.

Routine testing of water for viruses is not recommended, as it would require trained virologists and adequate laboratory facilities. Large volume samples of 5–50 litres would also be required, depending on the type of water, and the analytical procedures tend to be time-consuming and complicated.

Bacteriophages (or coliphages), which are viruses that can infect bacterial cells, have been proposed as possible viral indicators in drinking water, since it has been found that bacteriophages and enteroviruses tend to be inactivated at similar rates during treatment. Furthermore the isolation of bacteriophages is relatively straightforward and rapid, thereby providing an effective means of monitoring treatment processes for the removal and inactivation of enteroviruses.

Neither the WHO nor the EC set standards for viruses in drinking water. The USEPA has set a maximum contaminant level goal of zero, with a treatment requirement of at least 99.99% (or 4-log) removal and/or inactivation of any viruses present in the raw water.

6.68 Nuisance organisms

Iron bacteria

There are several groups of iron bacteria, all of which are capable of abstracting and oxidising any ferrous and manganous ions present in a water. The process is continuous with a large accumulation of rust coloured or black deposits developing over time. These deposits tend to accumulate in storage tanks and on the walls of pipes in low flow areas of the distribution system. Pipes may become blocked or the flow seriously impaired and any disturbance of the deposits results in badly discoloured water.

The growth of iron bacteria also results in an increase in the organic content of the water that could in turn encourage the growth of other nuisance organisms. In combination with sulphur bacteria, they also contribute to the corrosion of iron and steel pipelines.

The development of the organisms can only occur where there is sufficient iron or manganese present in a water at the ideal oxidation–reduction potential. Thus iron bacteria are widely found in the bottom muds of raw water reservoirs where the depletion of oxygen provides adequate ferrous ion. They can also be associated with ferruginous groundwaters containing high levels of free carbon dioxide and low levels of oxygen.

Sulphur bacteria

There are two groups of sulphur bacteria with implications for the water industry. Sulphate reducing bacteria grow in anaerobic conditions and reduce any sulphate present in the water to hydrogen sulphide. They contribute to galvanic corrosion of water mains and can cause taste and odour problems.

Sulphur oxidising bacteria grow in aerobic conditions and produce sulphuric acid from any sulphides present, for example in sewers.

Actinomycetes

Actinomycetes and other micro fungi can give rise to earthy or musty tastes and odours, particularly in water derived from nutrient rich lowland sources. The two compounds most usually formed during actinomycete development are geosmin and 2-methylisoborneol, both of which have very low threshold taste/odour values.

Part IV Water biology

6.69   Introduction

Rivers and lakes can support a wide range of plants and animals. These living organisms (the biota) form ecosystems which are in balance with important variables such as climate, water quality and human uses. The biota usually contribute beneficially to water quality of surface waters but may, particularly when not in balance, also have deleterious effects. A large number of organisms have been reported as causing problems in rivers and reservoirs used for water supply, during treatment and within the distribution system. Most organisms can readily be removed during treatment but some create problems and special care has to be taken to ensure that they do not pass through to the distribution system and to the consumer.

Figure 6.1 summarises the most important issues raised by biota and the key interactions at the various stages of water supply are briefly described in the following paragraphs.

Consumers regard the presence of living organisms in potable water as aesthetically unacceptable, as they indicate an impure product. Apart from their mere presence, such biota can cause deteriorating water quality by introducing turbidity, tastes and odours as a result of their metabolism, and greatly increasing the chlorine demand. Algae – which are largely microscopic plants – are of considerable importance to the treatment process designer because of the problems they can cause with clarification and filtration.

Only one animal, whose adult life is free living, has been positively identified as a health hazard in drinking water. This is a water flea, which is the host for a human parasite, the guinea worm, which can infest man if ingested. There are other animals, which are parasitic in man and which produce life stages that are free living and could be transmitted if not killed or removed by treatment. Finally there are animals which if not removed would proliferate in the distribution systems where their activities could facilitate the survival of disease organisms such as bacteria or viruses.

6.70   Source water and storage reservoirs

There are two main groups of plants in rivers and standing waters, algae that are largely microscopic and the larger, readily visible, macrophytes – the plants commonly named water weeds. Algae are simple green plants, most of which are free floating (planktonic) forms ranging in size from single celled species of 2–5 microns diameter to larger colonial forms up to several millimetres. There are however some species of algae which grow to macroscopic size as attached fronds in rivers or as upright forms on lake beds. The number of algae and the species found depend upon the local environmental conditions, such as temperature and the concentrations of dissolved salts and particularly the available

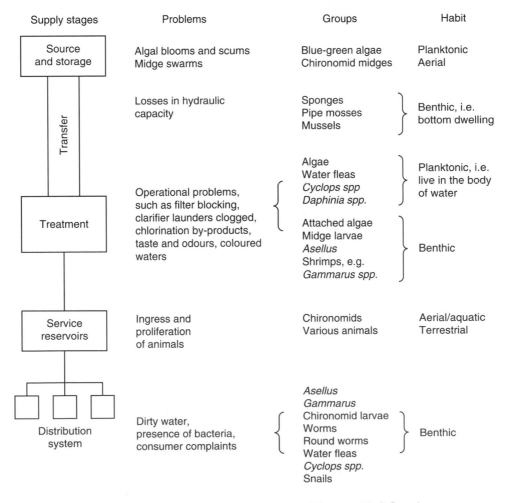

Fig. 6.1   Biological interactions with water supply (adapted from a WRC figure).

nutrients such as nitrogen and phosphorus which the plants, with energy from sunlight, build into their biomass.

Given the right conditions, plants will grow until the available nutrients are exhausted, and if not controlled may cause a variety of nuisances. For example water weeds or larger attached algae can cause major engineering problems such as mechanical breakages to control structures and even river channel blockages. Surface 'blooms' of floating algae can affect slow moving and standing waters, causing fish kills through deoxygenation of the water and the sliming of fish gills as the algae die and decay. In 1998 a major fish kill of over 150 tonnes of trout and coarse fish occurred in a fish farm fed from the Kennet and Avon canal in the UK due to this effect.

In recent years many rivers have shown an increase in nutrients, a feature which also occurs as rivers flow downstream. Starting as clean upland streams they often become polluted lowland waters from direct and indirect releases of wastes. This tendency to nutrient enrichment has become known as eutrophication and is outlined in Chapter 7. Whilst additional nutrients may improve plant and animal growth up to a certain point, beyond that point the excess nutrients can have a deleterious biological effect. This is because the growth of rapid growing biota is favoured and these species tend to be tolerant of most forms of pollution. They take over or dominate the other biota at the expense of the less tolerant forms. Many of these less tolerant forms are clean water biota and are highly valued for attracting animals such as salmon and trout, dragonflies and mayflies. By comparison the more tolerant species include nuisance organisms such as midges or some of the bloom-forming algae, few of which are regarded as attractive.

The macrophytes, the larger aquatic plants, may be floating plants or grow as emergent or submerged waterweeds. These plants maintain healthy ecosystems by providing food and shelter for numerous animals. They also create water quality improvements by allowing settlement of particulates in quiescent areas, by incorporation of nutrients from the water and by oxygenation. When in excess macrophytes can prove troublesome by reducing the carrying capacity of water courses, blocking intakes and water control structures, whilst sudden die off can lead to foul water.

There has been a great deal of research and development of methods to reduce the adverse effects of eutrophication, varying from control of the cause at the sewage works using a range of phosphorus and nitrogen removal technologies[40] to in-lake methods of lake management. These include direct intervention, such as the traditional use of copper sulphate[41] dosed direct to the water at times of blooms to kill the algae (a process now not acceptable to most authorities due to the toxicity of copper to other biota including humans), or control using barley straw which decays in the water releasing an algal toxin, as well as more subtle and indirect methods using introduced species or biological communities as a form of bio-manipulation. This book cannot cover all the literature on the control of algae and other biota in eutrophic waters but the reader may find two recent reviews useful.[42,43]

Although lakes and reservoirs vary considerably in size and other conditions, the range of species of algae is small and there is a characteristic make-up of species of algae that is found in similar water bodies anywhere in the world. As the waters leave the uplands and enter lowland plains there is an increase in dissolved solids, in temperature, and, in many developed countries, in nutrients from sewage discharges. There is commonly a transition in the algae from diatoms that tend to dominate in cold upland waters to green and to blue-green algae that thrive in warm, often shallow and nutrient rich lowland waters.

Nutrient poor upland waters are known as oligotrophic (poorly fed) whilst the richer nutrient waters are eutrophic (well-fed) waters. Oligotrophic waters tend to have a few species of flagellate algae such as the Chrysophyte genera *Synura, Uroglena* and in the spring there may be blooms of diatoms such as *Cyclotella, Tabellaria* or *Asterionella*. Green algae are common in many ponds and shallow water bodies, common species being *Chlamydamonas, Chlorella, Euglena, Chlorococcus, Coelastrum, Cosmarium, Oocystis, Pediastrum, Scenedesmus and Staurastrum*. Blue-green algae are a group of primitive algae more closely related to bacteria than algae whose current name, Cyanobacteria, is gradually increasing in use, although the older term is retained here for convenience. Many blue-green algae are colonial, usually with variable numbers of cells in the colonies and able to grow very rapidly in warm waters. A number of the commonest blue-green have flotation 'devices' such as gas vacuoles in the cells or mucus which binds the colonies into large rafts; examples are *Anabaena, Microcystis, Aphanizomenon,* and *Oscillatoria*. As already noted in Section 6.3 some of the blue-green algae also produce toxins present in the mucopolysaccharides that make up the mucus released by the cells. The toxins are not always present and their presence cannot currently be predicted except that high concentrations of the named blue-green species give rise to a higher risk. There have been numerous accounts of the ecology and biology of algae but the works of Palmer in the USA[44] and Bellinger in the UK provide a good general introduction. Illustrations of algae genera of significance in water supply taken from Bellinger[45] are shown in Fig. 6.2.

6.71 Transfer stages

The effects that plant growth can have on the carrying capacity of water courses has already been mentioned. Plants can serious reduce the cross-section of open channels and also increase the frictional resistance to flow. This is usually a minor problem in UK but can be an important issue in tropical countries where plant growth can be substantial. It is also more often a problem associated with bulk water transfers as in irrigation canals, in which velocities are commonly already low.

Plant growth in water courses, whilst maintaining a good aquatic habitat, can at the same time promote other problems for the water manager. In the Ely Ouse scheme in UK, the good habitat provided by plants has encouraged coarse fish. These then shoal and block the intake screens for this inter-river transfer scheme. In many tropical countries snails, carrying the intermediate vector for schistosomiasis (bilharzia or sleeping sickness), thrive in clean water canals and can create a major health hazard. Molluscicides are needed to control the snail, combined with adequate education of the local population with regard to sensible personal hygiene practice.

Plants do not in themselves cause problems in pipelines, as without light they cannot grow. There are however a number organisms that do give rise to problems in raw water transfer pipelines. These tend to be filter feeding animals such as sponges, moss animalcules (these are colonial forms resembling the common hydroid *Hydra*), mussels and, where sediments build up in major transfer systems, even cockles can be found within the sediments. These forms thrive upon algae and other fine plant debris carried into the pipelines and can result in increased headloss in the system, tastes and odours imparted to the water, and blockages when dead animals fall from the pipe walls. Water louse, shrimps and fish also occur in low velocity bulk water transfer systems.

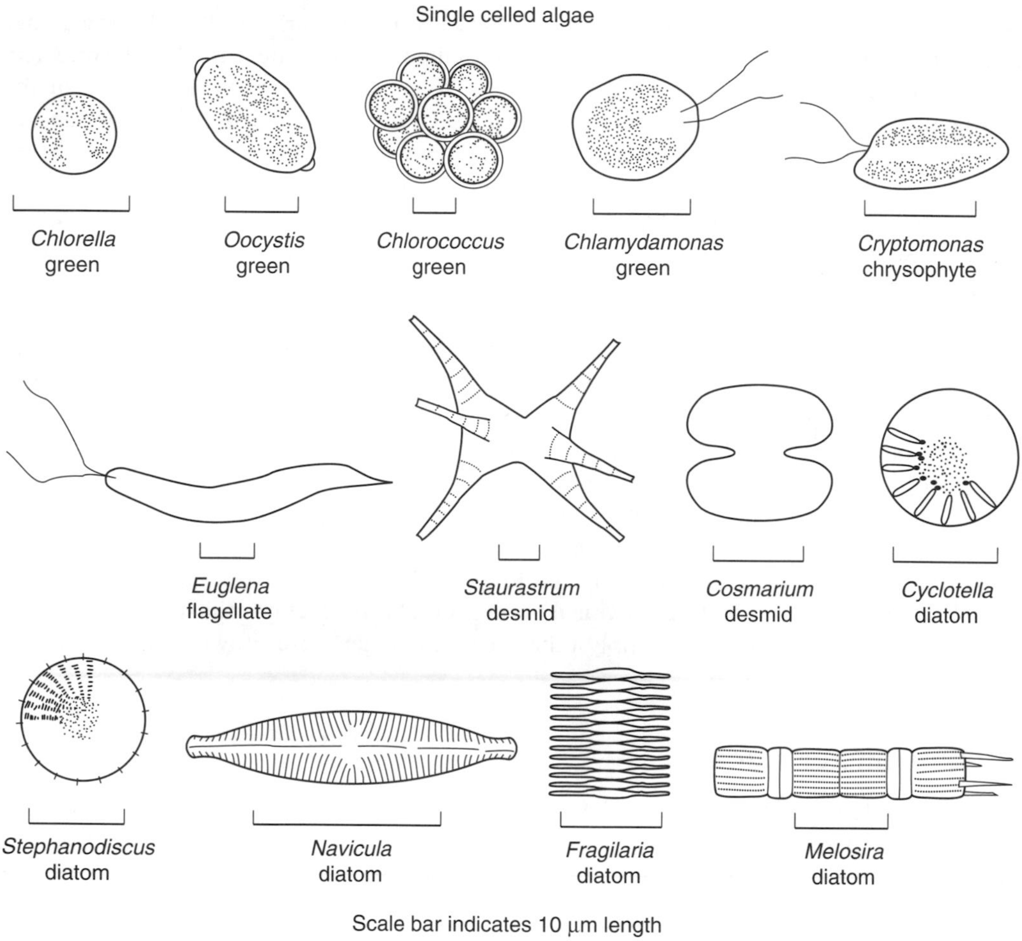

Single celled algae

| Chlorella green | Oocystis green | Chlorococcus green | Chlamydamonas green | Cryptomonas chrysophyte |

| Euglena flagellate | Staurastrum desmid | Cosmarium desmid | Cyclotella diatom |

| Stephanodiscus diatom | Navicula diatom | Fragilaria diatom | Melosira diatom |

Scale bar indicates 10 μm length

Fig. 6.2 Algae genera of significance in water supply.

In a recent, unusual instance in Hong Kong, animals have given water quality benefits. 'Moss animalcules' – which grow in a layer resembling a moss – grow naturally in a number of the extensive raw water tunnels. The presence of copious nitrifying bacteria in the matrix of fibres making up the colony of animals is able to strip any ammonia from the water in mains or tunnels that carry this biota. The improvement in water quality, and the reduced chlorine demand for pre-chlorination, has been sufficient to encourage management to promote this situation in other pipelines.

6.72 Treatment stages

The seasonality and speciation of algae has become an important issue in water supply as many of the algae have highly specific effects on the treatment process. In investigating problems with algae in the treatment process, it is important to be able to measure algae and the specific removal rates accurately. This has been difficult as the techniques of counting or measuring the plants have to vary to suit the organisms being counted. The

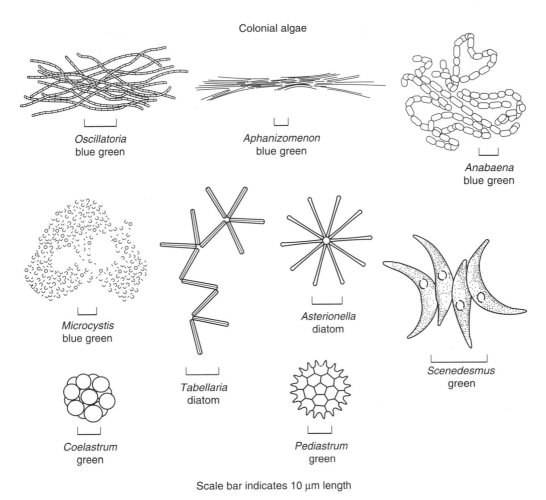

Colonial algae

Oscillatoria
blue green

Aphanizomenon
blue green

Anabaena
blue green

Microcystis
blue green

Asterionella
diatom

Scenedesmus
green

Coelastrum
green

Tabellaria
diatom

Pediastrum
green

Scale bar indicates 10 μm length

Fig. 6.2 (*Continued*). Algae genera of significance in water supply.

most standard measurement in the past has been direct counting under a light microscope with numbers being reported as cells or organisms or colonies per unit volume, with some agencies preferring to report the plan area or even the volume of the cells per unit volume.[28] The cost and subjectivity of counting has led to other methods being used such as the use of chemical analogues – the measurement of particulate organic carbon (POC) or the green pigment chlorophyll a – which is extracted and measured, or can now be measured fluorimetrically. In recent years particle counting using modified optical Coulter counter techniques has been use in special studies.

In Chapter 7 the treatment processes are discussed individually and in Table 7.5 of that chapter a list of a number of common algae species and their reported effects on the individual treatment processes is provided. For example, some species can penetrate and bypass filters whilst also blocking the same sand filters when present in large numbers; others may be damaged on the filter and release cell contents. These compounds may also be present in raw waters and interfere with flocculation; in addition some react with chlorine to produce chlorinated by-products, some of which have been shown to be carcinogenic.[25] As has been noted in Section 6.3, a number of the cell contents of some algae, notably the blue-green algae, may contain substances which are toxic to humans. The details of the specific effects of the algae on the unit processes are dealt with in Chapter 7.

As noted in the introductory section of this part there are no limits set for animals or plants in drinking waters though there is a presumption that they should be absent. In practice it is difficult to prevent some carry over of algae and small animal components from treatment processes though these are killed by the final chlorination before leaving the treatment works. Values for an algal content of between 100–1000 cells/ml, which is a range close to the lowest discernible limit for direct optical counting, have been used as a guide over the years. The lower number of 100 cells/ml has been advocated as a practical limit for the treated water prior to final chlorination.[46]

6.73   Service reservoirs and distribution systems

Water in service reservoirs and distribution systems will be chlorinated and, in the majority of cases, kept in the dark as most service reservoirs are covered to avoid photo-oxidation of the chlorine residual or aerial contamination of the water entering supply zones. It is therefore often surprising how many living organisms are found in such areas. Plants are seldom found in such systems as light levels are too low, but where mains have been laid too near the surface the roots of trees can penetrate brick or concrete conduits and have caused blockage and a pathway for other organisms to enter.

Service reservoirs should have midge- and mosquito-proof mesh fitted to all the essential air vents provided near the roof of the reservoir as such gaps have been shown to allow ready colonisation by the flying adults which lay their eggs on cool damp surfaces.

Apart from contamination at the surface contact points, the main source of animals found in the distribution system is the treatment process and, in an number of cases in the past, it has been found that the source may be not simply the by-passing of the clarification and filter stages by animals in the raw water, but rather that animal communities build up in the sedimentation tank and filters and provide a regular slippage of animals to the final water. These animals include a wide variety of freshwater forms including nematode worms, water fleas, the water louse and chironomid midge larvae. Correct control of treatment processes including, where there is a eutrophic water body providing the raw water, pre-treatment oxidation by chlorine or ozone, and where necessary occasional oxidation of process stages has been shown to be effective in controlling such nuisance animals.

Part V New and emerging issues

6.74   Introduction

The WHO has a continuing process of revision for the *Guidelines for Drinking-water Quality*.[1,25] Additional guidelines will be issued as new substances are evaluated or as new scientific information becomes available.

A similar process of review and revision over a specific period of time is built into the new EC Directive[21] and into the American National Primary Drinking Water

Regulations.[23] However a number of topics, both chemical and microbiological, have recently emerged which have caused concern for consumers. Some of the more major issues are discussed below.

6.75   Chemical issues

Methyl tertiary butyl ether (MTBE)

MTBE is the main fuel oxygenate for unleaded petrol. It is highly soluble, highly mobile in the aquatic environment, not significantly absorbed and not significantly attenuated or biodegraded. There is a possibility of it contaminating groundwaters via leaking storage tanks and pipes.

MTBE is included on the EPA Drinking Water Contaminant Candidate List because of its potential as a contaminant of drinking water, but will not be subject to monitoring until 2001. In the UK, water companies carry out limited monitoring.

To date there are no published data on concentrations in drinking water and data on health effects are not very comprehensive. However it would be detectable in drinking water by taste/odour at concentrations much lower than any perceived health risk.

Perchlorate

Perchlorates can be used in the manufacture of explosives but are more generally used as an additive to petroleum products to increase the octane number.

Perchlorates have been detected in some US drinking waters. However there are significant gaps in knowledge of health effects, levels in drinking water and effective removal during treatment. The USEPA has included the parameter on the Contaminant Candidate List to ensure that the required research is undertaken.

Endocrine disrupters

Since the early 1970s a large amount of research has been carried out world wide on substances that may disrupt human or animal endocrine systems. These substances can either be oestrogenic or androgenic and it has been claimed that they may be responsible for increases in breast and testicular cancer and for decreases in sperm counts.

Research in the UK has confirmed that steroids excreted by the human population, which are present in sewage effluents, are implicated in the feminisation of male fish. However the results applied to fish kept in cages in sewage effluent or to fish at points close to sewage effluent discharges. No equivalent effect has been reported for fish at points where water is abstracted for drinking water or in water storage reservoirs.

A recent study commissioned by the Department of the Environment, Transport and the Regions[47] investigated a number of approved water supply products as potential sources of exposure to endocrine disrupters. Transient and low levels of leaching were observed in some construction materials. Higher and more persistent levels were noted in some *in situ* glues and additives but exposure is likely to be very low. These products were all used within buildings rather than the public water supply system.

A method has been developed for the determination of oestrone, 17α-ethynl oestradiol and 17β-oestradiol in drinking water at levels of concern. Laboratory scale tests have shown that large concentrations of these substances can be effectively removed by aeration or coagulation and sedimentation, or powdered activated carbon. These oestrogens can also be removed by chlorination at concentrations typically used for final disinfection. Research is continuing on other substances such as nonyl phenol, bis phenol a and nonyl ethoxylates, which have been shown in bio-assays to possess endocrine disrupting properties.

Bovine spongiform encephalopathy (BSE)

BSE, or mad cow disease, is believed to be caused by the presence of an aberrant prion protein in the central nervous system. Other hypotheses put forward include damage to the central nervous system in cattle from the use of organophosphorus pesticides and autoimmunity triggered by exposure to bacteria of the *Acinetobacter* species. There is no evidence as yet to support these theories and no evidence that exposure to *Acinetobacter*, via contaminated water, has any role in human CJD.

The aberrant prion has the same chemical structure as the normal prion but a different molecular shape. The protein in the BSE agent is considered likely to be in an aggregated form and could be likened physically to a small virus. It may be assumed that the BSE agent is biologically and chemically stable. Furthermore, it is considered likely to be hydrophobic.

Assuming that the aberrant prion is the cause of BSE, it is unlikely that it would enter a water source by normal routes. Even if a water source were to be contaminated by BSE infected material, it is considered that the prion would be effectively removed by conventional treatment. However additional treatment may be needed at groundwater sites, as a precautionary measure, if risk assessment studies identify a potential risk.

6.76   Microbiological issues

Aeromonas

There are eleven named species of *Aeromonas*, of which there are a number of sub-species. The organisms are widely found in the aquatic environment and have been associated with biofilms in water distribution systems. Most treatment processes effectively remove Aeromonads and free chlorine residuals of 0.2–0.5 mg/l are considered sufficient to control the organisms in distribution systems.

The main health effects are the causation of wound infections and septicaemia. Aeromonads have also been associated with diarrhoeal illness. Various methods of detection are available.

Arcobacter

Four species of *Arcobacter* are generally recognised as gram-negative, spore forming rods. The organisms have been associated with diarrhoeal illness and there have been isolated incidents when water supplies might have been implicated in infections. However the epidemiology of such infections is not well understood.

No data are currently available on removal efficiency during treatment. *Arcobacter* species are similar to *Campylobacter*, so they are likely to be susceptible to inactivation by disinfectants such as chlorine and ozone.

Enterovirulent *Escherichia coli*

Certain strains of *E. coli* are an important cause of diarrhoeal illness and have been termed entrovirulent *E. coli*. Several classes have been defined based on the possession of distinct virulence factors. These are the enteropathogenic, enterotoxigenic, enteroinvasive and vereocytotoxigenic (VTEC) *E. coli*. The most data currently available relates to *E. coli* 0157, one of the VTEC strains that does not ferment sorbitol. The organisms can cause a wide range of symptoms including vomiting, fever, bloody and mucoid stools and, in severe cases, acute renal failure.

Depending upon the source of contamination, these organisms may be presence in water supplies. Most outbreaks have been associated with the consumption of contaminated food. However a number of minor outbreaks have been linked to small private water supplies, usually on farms and with no treatment. A more significant outbreak occurred in June 1998 and involved a municipal water supply in Wyoming, where there were 50 confirmed cases, almost half requiring hospitalisation.

Specific techniques exist to detect these organisms in water. Water treatment processes provide effective removal at the same level as for normal *E. coli*.

Adenoviruses

Human adenoviruses belong to the *Mastedenovirus* genus in the family *Adenovirus*. There are 47 serotypes, grouped into six sub-groups, which can cause upper respiratory infections, conjunctivitis and febrile illness. They are identified by electron microscopy.

They can be transmitted by water but should be inactivated by conventional water treatment processes.

Astroviruses

Seven different serotypes of human astroviruses have been described. They can be identified by electron microscopy, using ELISA techniques as well as molecular techniques. They cause gastroenteritis and can be transmitted by water.

It is anticipated that conventional water treatment should be able to produce virus free water.

Norwalk-type viruses

Molecular characterisation of the RNA genome has demonstrated the Norwalk-like virus to be part of the *Calcivirus* family, distinct from classic members. Different strains can be circulating in the community at any one time. These viruses can be detected by electron microscopy, using polymerase chain reaction and ELISA techniques. These viruses are responsible for nausea and diarrhoea and have been implicated in contaminated water.

Normal water treatment processes should remove viruses.

References

1. WHO. *Guidelines for Drinking-Water Quality*, 2nd edn, 1993, and subsequent amendments, including *Addendums to Volumes 1 and 2*, 1998. Vol. 1 – Recommendations; Vol. 2 – Health Criteria and Other Supporting Information, Vol. 3 –

Surveillance and Control of Community Supplies. HMSO, London, and WHO, Geneva.

2. Council of European Communities. *Directive 98/83,* 3 November 1998 on the Quality of Water Intended for Human Consumption. EC Official Journal L330/41, 1998.

3. Water Supply (Water Quality) Regulations 1989, and subsequent amendments. Statutory Instruments 1989 No. 1147; 1989 No. 1384; 1999 No. 1524.

4. *Public Law 93–523*, Safe Drinking Water Act (December 1974) and amendments under the Safe Drinking Water Act 1996, USA.

5. Council of European Communities, *Directive 80/778,* 15 July 1980 on the Quality of Water Intended for Human Consumption. EC Official Journal L229/11, 1980.

6. Carmichael W. W. Harmful Algal Blooms: A Global Phenomenon. *Proceedings of the 1st International Symposium on Detection Methods for Cyanobacterial (Blue-Green Algal) Toxins.* University of Bath Sept. 1993 (Royal Society of Chemistry).

7. Falconer I. R. *et al.* Using Activated Carbon to Remove Toxicity From Drinking Water Containing Cyanobacterial Blooms. *JAWWA*, August 1989, pp. 102–105.

8. Lahti K and Hissvirta L. Removal of cynaobacterial toxins in water treatment processes: review of studies conducted in Finland. *Water Supply* **7,** 1989, pp. 149–154.

9. *Guidance on Safeguarding the Quality of Public Water Supplies.* DoE 1990, HMSO, London.

10. *Asbestos in Drinking Water.* WRC Technical Report TR 202, 1984

11. Cambell H. S. Corrosion, water composition and water treatment, *JSWTE*, **19,** 1970, p. 277.

12. Siddiqui M. S. *et al.* Bromate Formation: A Critical Review. *JAWWA*, October 1995, pp. 58–69.

13. *Fluoridation Studies in UK and Results Achieved After Eleven Years.* Report of Research Committee of the MoH, No. 22, 1969, HMSO, London.

14. DHSS. *Circ.71/150*, 17 August 1971.

15. Reports by the Drinking Water Inspectorate on Nitrate, Pesticide and Lead, 1991 to 1994, and 1995 and 1996.

16. Study on Nitrate in Drinking Water and Childhood-Onset Insulin-Dependent Diabetes Mellitus in Scotland and Central England. Commissioned by the Department of the Environment, Transport and the Regions, March 1999.

17. Hunt S. M. and Fawell J. K. The Toxicology of organic Micropollutants in Drinking Water; Estimating the Risk, *JIWEM*, June 1987, pp. 276–284.

18. Bourgine F. P. *et al.* The Degradation of Atrazine and Other Pesticides by Photolysis. *JCIWEM*, **9,** August 1995, p. 417.

19. *Recommendations of the International Commission on Radiological Protection.* Annals of ICRP, 1990.

20. *Sodium, Chloride and Conductivity in Drinking Water.* WHO Regional Office for Europe, Copenhagen, 1979.

21. Pontius F. W. Complying with the New Drinking Water Regulations. *JAWWA*, Feb. 1990, pp. 32–52; A Current Look at Federal Drinking Water Regulations. *JAWWA*, March 1992, pp. 36–50; and New Horizons in Federal Regulation. *JAWWA*, March 1998, pp. 38–50.

22. Fawell J. K. Developments in health-related standards for chemicals in drinking water. *JIWEM*, Oct. 1991, pp. 562–565.

23. *EC Directive 75/440*. Concerning the Quality of Surface Water Intended for the Abstraction of Drinking Water. Official Journal L194.

24. The *Surface Waters (Abstraction for Drinking Water)(Classification) Regulations 1996*. Statutory Instrument 1996/3001. (Replacing the earlier Surface Waters (Classification) Regulations, 1989.)

25. WHO. *Guidelines for Drinking-Water Quality*, Vol. 2. WHO, Geneva, 1996.

26. Standing Committee of Analysts, *Methods for the Examination of Waters and Associated Materials* (a series of booklets). HMSO, London, 1976 onwards.

27. *Standard Methods for the Physical and Chemical Examination of Water and Wastewaters*. American Public Health Association, New York, 1971 and subsequent editions, latest edn 19th, 1995.

28. Hutton L. G. *Field Testing of Water in Developing Countries*. WRc, 1983.

29. *List of Substances, Products and Processes Approved Under Regulations 25 and 26* [of the Water Supply (Water Quality) Regulations 1989], Drinking Water Inspectorate, Department of Transport, Environment and the Regions, Dec. 1998.

30. Suckling E. V. *The Examination of Water and Water Supplies*, 5th edn. Churchill, Edinburgh, 1943.

31. Barrow G. I. Legionnaire's Disease and its Impact on Water Supply Management. *JIWEM*, Aug. 1987, p. 117.

32. *Legionnaire's Disease, Guidance Note EH48*, Health and Safety Executive, London, 1987.

33. *Code of Practice for Operation and Maintenance of Cooling Towers and Hot and Cold Water Services in Health Service Buildings*, Dept of Health and Social Security, HMSO, London, 1989.

34. Smith H. V. Cryptosporidium and Water: A Review. *JIWEM*, August 1992, p. 443.

35. *Cryptosporidium in Water Supplies: Report by the Group of Experts*. DoE, HMSO, London, July 1990.

36. *Cryptosporidium in Water Supplies: Second Report by the Group of Experts*. DoE, HMSO, London, Oct. 1995.

37. *Cryptosporidium in Water Supplies: Third Report by the Group of Experts*. DETR, London, Nov. 1998.

38. The Water Supply (Water Quality) (Amendment) Regulations, 1999.

39. Standing Committee of Analysts, *The Microbiology of Water; Part 1 – Drinking Water*. HMSO, London, 1994.

40. Anon. Biogrowth. *Water Quality International*, March/April 1997, p. 15.

41. *IWEM Manual of British Water Engineering Practice*, 4th edn. 1969, p. 168.

42. Sutcliffe D. W. and Jones J. G. *Eutrophication: Research and Application to Water Supply*. Freshwater Biological Association, Cumbria, UK, 1992.

43. Moss B. *et al. A Guide to the Restoration of Nutrient-Enriched Shallow Lakes*. Environment Agency and Broads Authority, 1996.

44. Palmer C. M. *Algae and Water Pollution Report*, EPA-600/9–77–036, 1977.

45. Bellinger E. G. *A Key to Common Algae: Freshwater, Estuarine and Some Coastal Species*, 4th edn. 1992.

46. Mouchet P. Potable Water Treatment in Tropical Countries: Recent Experiences and Some Technical Trends. *Aqua*, **3**, 1984, pp. 143–164.

47. Final Report to the Department of Environment, Transport and the Regions on Exposure to Endocrine Disrupters via Materials in Contact with Drinking Water, May 1999.

7
Storage, clarification and filtration of water

Part I Storage, screening, sedimentation and clarification

7.1   Raw water storage

This may often be regarded as a first stage in treatment as it may involve a complex combination of physical, chemical, and biological changes. Traditionally, raw water storage has been regarded as a major or almost essential 'first line of defence' against the transmission of water-borne diseases; this aspect is still of major importance if the unstored water is liable to excessive bacterial pollution from sewage, even though such pollution may only occur occasionally, e.g. if storm-water sewage overflows discharge into a river. A few days storage of a surface water will improve its physical and microbiological characteristics. This is the effect of a combination of actions including sedimentation, natural coagulation and chemical interactions, the bactericidal action of ultraviolet radiation near the water surface and numerous biotic pathways which help to reduce enteric micro-organisms.[1] When a water is stored in a reservoir for a period from one to several months, there is a substantial decrease in the numbers of bacteria of intestinal origin and the specific organisms of typhoid and cholera also disappear. The die-off rate for enteric coliforms, designated here as the time to achieve a 90% loss of bacteria or T_{90}, in lakes and other open waters varies from 2–3 hours in strong sunlight in clear waters, to 10 hours in more turbid waters.[2] A most valuable and practical management aspect of having short-term storage is that a river intake can be shut down to avoid or investigate any pollution which might, for example, be indicated by the death of fish or by other information[3] such as changes to physical chemical characteristics of the water.

Increasing risks to public water supplies from accidental spillages of industrial chemicals in transit on roads and on manufacturing sites led to the Department of the Environment (UK) recommending in circular No. 22/72[4] that water supplies downstream of effluent discharges should be protected by at least seven days' storage. The intended function of the storage was to allow closure of the intake until the pollution risk was over, to dilute any polluted water entering the intake with clean stored water and to allow further self purification to take place. By 1979 rather more than two thirds of river derived supplies in England and Wales had received some storage in a reservoir prior to further treatment.[5] Such buffer storage is still desirable but it is now possible to consider alternative strategies, particularly when economic or practical factors prevent the adoption of storage.

Alternative strategies might include catchment management, regulations to avoid or reduce the risk of industrial chemicals being spilt in locations where they might enter an aquifer or surface source, or protection of abstraction sites by continuous monitoring systems coupled with the provision of an alternative source of supply.[6]

Water quality parameters for which there are electronic methods of monitoring which gives stable measurements at appropriate concentrations, such as temperature, pH, conductivity, ammonia or dissolved oxygen have been available since the 1960s and have been in use by some undertakings from that time.[7] Incidents caused by accidental spillage of compounds, such as phenols which cause taste and odour problems at very low concentrations and were able to pass through conventional process works and enter the distribution system before being detected, have led to the development of increasingly sophisticated raw water monitoring.[8] Specific monitors for problem compounds such as phenols have been developed and successfully deployed, as at the Huntington Works on the River Dee (UK), where agricultural sources of phenol contamination require constant vigilance particularly in summer months.[9] Often there was a need for monitoring but no one compound could be generally said to be the main risk, as for example at an intake downstream of a road bridge at which accidental spillage might occur. In such instances fish monitors were traditionally proposed and have been in use for many years; in the past 20 years behavioural and metabolic activity monitoring of sensitive fish such as trout has been devised and improved by rigorous research and development.[10] There are several instrumental packages now capable of measuring soluble organic substances at very low concentrations such as UV absorption or total organic carbon (TOC) which can be used as general monitors to raise alarms when the background levels of such substances are suddenly raised. Most recently the manufacturers of liquid and gas chromatography have been able to produce reliable, stand-alone, instrumentation which together with facilities for alarms and computer-driven libraries of traces, allows notice to be given of sudden peaks and identification of the compounds within 10–20 minutes, which is usually sufficient time to close down an intake before a water supply is threatened by irretrievable contamination.[11]

Potential problems in raw water storage

There are potential disadvantages in the prolonged storage of raw waters which should be taken into account when considering storage and the management that may be needed. The most obvious of these is the likelihood of the growth of various forms of plants, either rooted aquatic types (macrophytes) which may choke shallow waters, and free floating or planktonic types such as algae (phytoplankton), which may increase the difficulties of treatment. The main issues for water supply raised by the presence of these plants and also animals have been discussed in Part IV of Chapter 6. Storage reservoirs that are less than about 10 m deep can allow light to reach the bottom and in this may encourage the growth of rooted plants unless the stored waters are sufficiently turbid to reduce light penetration. Shallow reservoirs are therefore generally avoided if there is any likelihood that plant growth could be high.

Waters which contain sufficient nutrient materials to support prolific growths of aquatic plants are usually described as eutrophic. Lakes and water bodies exhibit a range of concentrations from low nutrient conditions in upland lakes on igneous rocks (oligotrophic) through moderate or mesotrophic lakes where a balanced ecology with some shoreline macrophytes and a wide range of planktonic algae occur, through to lowland water bodies which tend to be eutrophic and even hypertrophic waters where prolific growths of plants commonly occur due to enrichment by sewage and where in temperate climates there are usually seasonal peaks. The boundaries of these 'trophic' classes have been defined for a number of parameters by the Organisation of Economic Co-operation and Development (OECD)[12] and are shown in Table 7.1.

In most reservoirs that are more than about 10 metres deep, and many are designed in this way to avoid excessive plant growth, an additional complication is that thermal layering or stratification may occur on a seasonal basis in temperate climates. As the water warms up in the spring, the water being warmed in the upper part of the reservoir tends to remain at the surface due to its lower density as it expands. In the absence of any strong wind induced circulation the colder and now denser water below remains and ceases to mix with the surface water. The upper and lower layers are known respectively as the epilimnion and hypolimnion; in between there is a zone known as the thermocline in which there is a relatively steep change with depth from the higher temperature of the epilimnion to the lower temperature of the hypolimnion. In reservoirs averaging less than 10 m deep thermal stratification in temperate lands is often only temporary, lasting for a few days at most or broken down by evening winds. In colder climates there may be a winter stratification due to ice formation which ensures the water body and its biological life is protected from lethal frosts.[13]

Thermal stratification can clearly affect retention time of an incoming water and is often of major importance with reference to water quality. In many large reservoirs, there are facilities for withdrawing the water for treatment at several different levels which can be chosen as circumstances dictate. Multiple drawoff facilities have been discussed in Section 5.10. In the case of eutrophic reservoirs the ability to avoid drawing from surface with high concentrations of algae is particularly useful, and this may be a further criterion in drawoff tower positioning. A number of schemes have sited the tower some way from the shore in order to avoid the build up of surface aggregations of algae that onshore winds can cause.

In some large reservoirs the water in the hypolimnion is of a high standard of purity, as well as being cool, e.g. in some places in Scandinavian countries, in the Lake District of England,[14] and in Lake Constance (Bodensee). However in eutrophic reservoirs organic impurities in the incoming water or released by leaching from bottom muds and inundated soils may accumulate in the bottom water. As a result of plant and animal respiration and bacterial activity, the concentrations of dissolved oxygen fall and may approach zero. Under such conditions major chemical changes take place in the transition from anoxic conditions (when only combined oxygen compounds such as sulphate (SO_4^{2-}) and nitrate

Table 7.1   Boundary values for trophic classes of OECD system

Trophic category	Average winter phosphorus (μg/l)	Average summer chlorophyll 'a' (μg/l)	Maximum chlorophyll 'a' (μg/l)	Secchi disc depth (m)
Ultra-oligotrophic	$\leq$4.0	$\leq$1.0	$\leq$2.5	$\geq$12.0
Oligotrophic	$\leq$10.0	$\leq$2.5	$\leq$8.0	$\geq$6.0
Mesotrophic	10–35	2.5–8	8–25	6–3
Eutrophic	35–100	8–25	25–75	3–1.5
Hypertrophic	$\geq$100	$\geq$25	$\geq$75	$\leq$1.5

(NO_3^-) are present), to full anaerobic conditions when even these compounds have been reduced to sulphide and nitrogen. This transition can be measured by redox potential* which falls from $+200$ mV when free oxygen is present to -200 mV in anaerobiosis. The important consequence for water quality is that under negative redox conditions numerous other chemical compounds can re-dissolve from the sediments into the overlying water. Iron, which is present as insoluble oxide or hydroxide, is usually the most prominent of these substances but if manganese is present (usually as insoluble oxides), it may also become soluble. Both iron and manganese are often in combination with organic colouring matter. The concentrations of plant nutrients, phosphates as well as ammonia commonly also increase in bottom water close to anaerobic sediment. Actinomycete fungi often proliferate in these conditions[15] and lead to tastes and odours in the raw water. Under such circumstances abstraction from the hypolimnion for water treatment should be avoided. Some authorities prefer to abstract this water for compensation flows to 'bleed' the higher dissolved contaminants from the stored water.

When the surface water cools down in the autumn and wind action becomes effective, the reservoir water mixes and, under normal circumstances, water from mid depth or deeper is of good quality. Rapid mixing caused by strong autumnal winds can cause sudden 'turnover' resulting in rapid deterioration in the quality of the water at the surface in respect of colour, iron and manganese. A more gradual effect of the resumed mixing is that the plant nutrients carried up from the bottom may increase the growth of algae in the water and cause an autumnal bloom of algae. This is not uncommon in eutrophic lakes and reservoirs and, often renders the surface water difficult to treat.

Reservoir management to control stratification of reservoirs has been established practice since the 1960s. Mechanical pumps have been used to induce mixing artificially although this method has met with varying degrees of success. The systems have usually been expensive to install and operate.[16] These systems are intended to have a dual purpose of oxygenating the bottom layers of a reservoir and controlling stratification. Pioneering studies in UK by the Metropolitan Water Board[17,18] (now part of Thames Water plc) showed that algal populations as well as water quality could be managed by water movement control. This approach led to arrangements for carefully controlled raising of cool low oxygen water from below the thermocline, using jetted inlets in those reservoirs which had pumped inflows, thus entraining the bottom waters (also low in algae) in the jet flow to the surface and increasing the amount of satisfactory water above it. This system was ideal in large flat-bottomed reservoirs where the bottom water volume is large relative to the surface epilimnion, but is not effective in narrow natural valley impoundments where the hypolimnion is small, and more vertically oriented mixing is needed. Air lift pumping systems such as the 'Bubble Gun'[19] and the 'Helixor'[20] which make use of rising

---

*Redox potential or oxidation–reduction potential (ORP) is the potential developed in a cell between a metal electrode (e.g. platinum) and a reference electrode during an oxidation–reduction reaction. It is reported with respect to the potential of the hydrogen electrode which is zero. The typical range is -1000 mV to $+1000$ mV. Oxidising agents (Cl_2, O_2) increase the ORP, whereas reducing agents (sulphites) lower the ORP. Therefore, it is used to determine the oxidising or reducing characteristics of a solution. At high ORP ions such as sulphate, nitrate or ferric predominate whilst the presence of sulphites, ammonia or ferrous ions in significant concentration are an indication of low ORP. In aquatic redox reactions ORP is related to bacterial activity (e.g. oxidation of iron in biological filters and reduction of feric hydroxides or oxides in lakes).

air bubbles in tubes to entrain bottom water and allow it to rise to the surface whilst being aerated by the bubbles have been shown to be more effective than mechanical pumps. A simpler and more economical method using compressed air was developed in the 1970s;[21] it is the use of perforated airlines laid along the bottom of the reservoir.[22] An alternative air injection system using a number of ceramic domes, to produce fine air bubbles has recently found favour.[23]

In some circumstances it may be beneficial to have an emergency bypass so that water may be taken directly from a river instead of from a reservoir. This could be used in the case of exceptional high algal growth in the water or of pollution having occurred, or being suspected in water in the reservoir.

Growth of plants can deplete nutrients from the water particularly if the growth is not limited by light or grazing. Management of stored water to make use of this characteristic has been carried out by Thames Water for many years.[24] Water from the River Thames in autumn and winter frequently has nitrate nitrogen concentrations which exceed the EC Directive value (see Section 6.30) of 50 mg/l as NO_3 due to pollution from agricultural runoff. Toms[25] showed that an empirical relationship could be used to predict the reduction of nitrate by storage, and long-term stored water with low nitrate could be used to reduce nitrate concentrations entering supply from other sources by blending. Thames Water use storage in one of their larger reservoirs for prolonged periods to reduce nitrates by the combination of algal growth and bacterial denitrification.

Screening
7.2   Bar screens

Practically all intakes are screened, even though the screens may be of the simplest type of bar grille. The bars must be quite substantial in size (of about 25 mm diameter) and are normally spaced at 75–100 mm centres. If the bars are inclined it is easier to clean them with a rake which may be a hand rake for occasional manual cleaning or mechanical rakes for continuous cleaning. Should a smaller mesh be necessary, it is best to group bars into frames so that each frame can be lifted out of the water, cleaned, and lowered back into position. To prevent unscreened water from passing through the intake when the screens are lifted, they should be provided in duplicate or provision should be made for stop log insertion upstream for temporary stoppage of the flow. At river intakes a great deal of trash may collect; this is often seasonal with a spring and autumn abundance of water weed and foliage. Screens may be provided with automatically operated raking mechanisms to assist the removal of captured debris. Bar screens generally having larger spacing will not arrest smaller size debris. In such cases it is advisable to install a robust band or cup screen downstream of the protective bar screen.

7.3   Band and drum screens

If fine screening is adopted, some means must be found of continuously cleaning the screens or they rapidly become clogged. For this reason fine screens are usually arranged as endless bands or rotating drums of material perforated with holes of about 6 mm diameter. Plate 13 shows a drum or cup screen and a band screen. The screening element is in continuous motion and having captured and lifted debris from the intake water passes over water jets which wash off the screened material into a trough. A pressure supply of clean water is needed for the washwater jets, and this may have to be pumped from the strained water. The total amount of water required for washing may be of the order of 1% of the throughput. Fine screening must always be preceded by a coarse screen.

7.4   Microstrainers

These are revolving drums mounted in open tanks with a straining medium which is usually a stainless steel wire fabric of a very fine mesh, fitted to the periphery of the drum. The drum is submerged for about 75% of its diameter (66% of area) and rotates at about 0.5–5 rpm (peripheral drum speeds of 3–50 m/min). Water to be treated enters the drum axially under gravity and flows out radially through the fabric, depositing particulate matter. Cleaning is accomplished by a row of water jets along the full length of the drum operating at about 2.5 bar maximum pressure. Particulate matter intercepted by the fabric rotates to the top of the drum where they are backwashed into a hopper running the full length of the drum and conveyed by a pipe which also acts as the axle for the drum assembly, to a point outside. Water jets use about 1–1.5% of the total quantity of water strained and this washwater should be filtered and chlorinated.

Total headloss through a microstrainer unit including inlets and outlets varies from about 150–200 mm. Single units have capacities of 10 m³/h to a maximum of 4000 m³/h for a 3.2 m diameter × 5 m wide drum. A typical microstrainer installation is shown in Plate 13.

Microstrainers bring about an improvement in the physical quality of a water but there is no change in the chemical characteristics of the water. The ideal water for microstraining is a lake or large reservoir supply which does not contain a large amount of suspended matter but which contains moderate quantities of zooplankton, algae and other microscopic-sized particles; total algae removal ranges from 50 to 75% depending on the size. When applied for the removal of zooplankton microstrainers are located either at the beginning or end of the treatment process. Fabrics commonly used with stored waters are made of woven stainless steel wires of 0.05 mm diameter with apertures of 23 and 35 μm. However a coarser mesh at 200 μm aperture is sometimes used after granular activated carbon (GAC) filters to remove eroding particles of the carbon and any bacterial flora that sometimes develops in GAC filters. Attempts to use plastic mesh materials have not been successful in water supply.

When designing a plant specifically for the removal of algae it is important to undertake pilot plant trials or, if this is not possible, laboratory tests to ensure that the microstrainers are effective; this is necessary because of the wide variation in the size of different species of algae or zooplankton and the difficulty in selecting the correct mesh size in advance. Microstrainer manufacturers produce their own test kits which may be available for extended testing on site as an alternative to pilot plant trials to select an appropriate fabric.

Microstrainer operation is fully automatic. A microstrainer screen can easily be damaged if too great a loading is placed upon it; hence head across the screen should be monitored and an alarm initiated if it approaches the maximum desirable value. The rotational speed of the drum assembly can be adjusted so that the optimum differential headloss across the fabric can be maintained to achieve maximum removal efficiency irrespective of the quality or quantity of raw water entering the unit. Most installations have an automatic fail-safe bypass weir which forwards unstrained water when the screens become overloaded. This causes deterioration of the treated water at times of peak loading and illustrates why microstrainers are not generally acceptable as an alternative to filtration by sand or other mixed media for potable waters. Their most extensive and successful use in water supply has been to lighten the loading upon rapid or slow sand filters such that the length of run of these filters between cleaning is extended thereby increasing their output by as much as 50%. Comparison with roughing filters, microstrainers (which have surface loading rates up to 80 $m^3/h.m^2$) require less space and produce lower headloss; their initial and running costs are lower. They therefore provide a valuable means of reducing the solids loading on to filters, when for example, installation or extension of filter plant is being contemplated. They may also be of particular advantage as the sole filtration technique for an industrial supply (e.g. 67 Ml/day Whitebull WTW, North West Water, UK) or for the artificial groundwater recharge.

Sedimentation and settling tanks
7.5   General design considerations

Sedimentation tanks are designed to reduce the velocity of water so as to permit suspended solids to settle out of the water by gravity. There are many different designs of tanks and most are empirical. No specific rules can be laid down and many contradictory results have been reported. A sedimentation tank which may be very successful on one kind of water may perform poorly when dealing with a different kind of water. The success of a tank may be judged on its ability to maintain the claimed throughput and the agreed effluent water quality under adverse raw water quality conditions. An effluent quality for suspended solids or turbidity of less than 5 mg/l and 5 NTU, respectively, would be acceptable to most designers. The amount of suspended solids in a water, the nature of these solids, their shape and relative density, the extent of clarification required, the temperature of the water, the rate of flow that must be handled – all these influence the performance of a tank. Therefore after performing laboratory tests using jar test apparatus on samples of raw water, the best guide to selecting a tank for a particular water is to research which type of tank has been successful before under similar conditions. This may lead to the conclusion that there is only one tank design for the treatment of a particular water, but this is far from the truth. The designer's task is to find the most economical and efficient solution for all conditions of water quality, although it would be foolish not to take due note of the type of tank which has been successful before under similar circumstances.

7.6   Plain settling

In plain settling (or sedimentation), suspended solids in a water are permitted to settle out by gravity alone: no chemicals are used. For this purpose the water can be left to stand in a tank, although with continuous supply at least two such tanks have to be used alternately. Such fill-and-draw tanks are seldom used in modern plants, except for filter washwater recovery (see Section 7.45). Instead plain sedimentation tanks are designed for continuous throughput, the velocity of flow through the tank being sufficiently low to permit gravitational settlement of a portion of the suspended solids to occur. In practice the application of plain sedimentation in waterworks is very restricted because impurities such as algae, aquatic plant debris and finely divided mineral matter, do not settle at a rate sufficient for a tank of reasonable size to be utilised. Plain sedimentation is most frequently used as a preliminary treatment for fast flowing river waters carrying much suspended solids as occur in the tropics or at intake works on water transfer schemes where it is desirable to minimise the amount of suspended material passing into the system. Under most circumstances chemically assisted sedimentation which is a more complex process is adopted, as described later.

The velocity with which a particle in water will fall under the action of gravity depends upon the horizontal flow velocity of the water, the size, relative density and shape of the particle and the temperature of the water. The theoretical velocity of falling spherical particles in slowly moving water V (mm/s), is given[26] by

$$V = \frac{g}{1.8 \times 10^4} (r - 1) \frac{d^2}{\gamma}$$

where $g = 9.81$ m/s^2, r is the relative density of the particles, d is the diameter of the particles in mm and γ is the kinematic viscosity of water in m^2/s, which varies with the temperature of the water as given in Table 7.2. (The coefficient of kinematic viscosity (m^2/s) = coefficient of absolute viscosity (Ns/m^2) divided by density (kg/m^3) where N (newton) is kg.m/s^2.) This is applicable for Reynolds numbers of less than 0.5.

A number of different (mainly empirical) formulae have been given for the settlement of sand and soil particles in still water; some of the values derived are given in Tables 7.3 and 7.4.

Table 7.2   Kinematic viscosity of water

Temperature (°C)	0	5	10	15	20	25
Value γ (m^2/s) $\times$ 10^{-6}	1.79	1.52	1.31	1.15	1.01	0.90

Table 7.3   For sand of relative density 2.65 in water at 10°C

Diameter of particle (mm)	Falling speed (mm/s)[a]	Falling speed (mm/s)[b]
1.0 Sand	100	140
0.6	63	–
0.5	–	70
0.4	42	–
0.2	21	22
0.1 Silt	8	6.7
0.06	3.8	–
0.05	–	1.7
0.04	2.1	–
0.02	0.62	–
0.01	0.15	0.08

[a]From *Water Treatment Plant Design*, AWWA, 1969.
[b]From *Disposal of Sewage* by K. Imhoff, Butterworth, 1971.

Table 7.4   For particles of relative density (*r*) values 1.5 and 1.2

	Settling speed (mm/s) in still water at 10°C	
Diameter of particle (mm)	Coal ($r = 1.5$)	Domestic sewage solids ($r = 1.2$)
1.0	40	30
0.5	20	17
0.2	7	5
0.1	2	1.3
0.05	0.4	0.3
0.01	0.02	0.008

Note: From *Disposal of Sewage* by K. Imhoff, published by Butterworth, 1971.

Aluminium and iron flocs have a specific gravity of about 1.002, particle size as large as 1 mm and a settling velocity (at 10°C) of about 0.8 mm/s (from Fair G. M., Geyer, J. C., Okun, D.A. *Water & Waste Engineering*, Volume 2, Chichester, John Wiley, 1986).

Clays generally have a grain diameter of 0.01 mm to less than 0.001 mm (1 μm) so that it is impracticable to remove them from a water by simple sedimentation, or even by filtration, without prior chemical coagulation treatment as described in Sections 7.11–7.24.

7.7   Maximum velocity to prevent bed uplift or scour

Apart from the settling rate in still water it is, of course, essential that once a particle has reached the base of the tank it shall not be resuspended by the velocity of flow of water over the bed. Camp[27] gives the channel velocity V_c (m/s) required to start motion of particles of diameter *d* (mm) as

$$V_c = \left(\frac{8\beta g}{10^3 f}(r-1)d \right)^{1/2}$$

where *r* is the relative density, *f* is the friction factor in ($4flv^2/2gd$), β is in the range 0.04–0.06 for sticky flocculent materials, and 0.10–0.25 for sand and $g = 9.81$ m/s^2.

7.8   Maximum horizontal velocity of flow

A third flow measure which must be taken into account is that the horizontal velocity of flow must not be so great as to prevent, by turbulence, the settling of particles under gravity. There is general agreement that this velocity should not be more than 0.3 m/s to allow sand grains to settle. This is, of course, too high a velocity for the settling of particles of light relative density (1.20 and less), but this is the figure normally used for sewerage grit chambers where the heavier material is to be deposited and the lighter material left to carry over. At 0.2 m/s faecal matter, i.e. organic matter, will begin to settle.

7.9   Theory of design of tanks

For a tank of length l, water depth d and width w, let the inflow rate be Q ($=$ outflow) (see Fig. 7.1).

Let a particle of silt entering the tank to have a vertical falling speed of V. Then

$$\text{speed of horizontal flow} = \frac{Q}{wd}$$

and

$$\text{time of horizontal flow} = \frac{l}{Q/wd} = \frac{lwd}{Q}$$

The time for falling distance d is d/V, and for the particle to reach the bottom before the water leaves the tank the time of fall must equal the time of horizontal flow, i.e.

$$\frac{d}{V} = \frac{lwd}{Q}$$

from which

$$V = \frac{Q}{lw} = \frac{Q}{A}$$

where A is the surface area of the tank. Q/A is known as the surface loading rate and is expressed as $m^3/h.m^2$, m/h or mm/s.

This is the limiting speed of fall to enable the particle to reach the bottom of the tank. All particles with a speed greater than Q/A will reach the bottom before the outlet end of the tank. Particles with a speed less than Q/A will be removed in the same proportion as their speed bears to Q/A (see Fig. 7.1), e.g. if the speed V is only half Q/A then only half the particles falling at this speed reach the bottom. Hence Q/A for any tanks is a measure of the effective removal of the particles in that tank. For example a tank of 300 m^2 surface area has a rate of inflow of 1.2 m^3/s, then Q/A = 14.4 m^3/h.m^2 (4 mm/s). Thus, theoretically, all particles having a falling speed of 4 mm/s or more will be removed, 50% of those having a falling speed of 2 mm/s, 25% of those with a falling speed of 1 mm/s, and so on. Thus on this basis the performance of a tank is independent of depth and retention time. This concept is the basis for the design of multi-tray horizontal flow tanks and inclined plate and lamella settlers (see Sections 7.14 and 7.17).

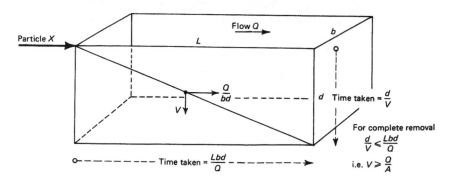

Fig. 7.1   Theoretical flow through a rectangular sedimentation tank.

The foregoing theory, however, assumes that the falling particles do not hinder each other, but McLaughlin[28] has shown in laboratory experiments with clay and aluminium sulphate that the faster particles, settling through the slower ones, gather some of the latter up and drag them out of suspension. This is the case with flocculated suspensions, where during settlement, agglomeration of the particles takes place with time and consequently particles settle much faster. Hence the performance of a tank in which this phenomenon occurs is related not so much to its surface loading rate as to the time of residence. These findings relate to still water. Thus, dependent upon the nature and size of the settling particles, the range of sizes, the degree of concentration of the suspension and the amount of turbulence, the performance of a sedimentation tank may relate to its surface loading rate or to its residence time, or partly to both.

7.10 Grit tanks

In waterworks, grit primarily constitutes sand, gravel and other abrasive material and is present mostly in water abstracted from rivers. If allowed to enter the works it could damage intake pumps, settle in raw water pipelines and inlet process units. Intakes should therefore be sited and designed to minimise the uptake of grit and tanks should be provided upstream of pumps to trap the grit. Grit tanks are also used as traps for sand, anthracite, granular activated carbon and other filter media carried over in used filter washwater to protect any pumping equipment downstream.

They operate as plain settlement tanks (unaided by coagulants) and are sized to capture particles with diameter larger than 0.1 mm. From Table 7.3 silt particle of 0.1 mm will fall in still water at 8 mm/s. The residence time in the tank should be at least 1.5 times the time taken for 0.1 mm particles to settle to the floor of the tank. Grit tanks for sand removal usually have surface loading rates of 10–25 $m^3/h.m^2$, a water depth of 3–4 m and length to width ratio of at least 4:1.[29] To prevent scour of settled material the horizontal velocity must be smaller than the scour velocity calculated from the formula in Section 7.7. This is usually about 0.3 m/s for sand.

Chemically assisted sedimentation or clarification
7.11 Chemically assisted sedimentation

This comprises several separate processes of treatment which together make up the complete system known as 'clarification'. This system is designed to remove from a water as much as is desirable, or possible, of suspended materials, colour and other soluble material such as those of organic origin and soluble metals such as iron, manganese and aluminium before it passes to rapid gravity filters. It is a delicate and chemically complex phenomenon having three stages: (1) the addition of measured quantities of chemicals to water and their thorough mixing, (2) coagulation and flocculation, or the formation of a precipitate which coalesces and forms a floc and, (3) sedimentation.

7.12 Chemical mixing

Most chemical reactions in water treatment applications are completed within 5 seconds and therefore the principal objective in chemical mixing is to obtain rapid and uniform dispersion of the chemical in the main flow of water to ensure that chemical reactions are completed in the shortest possible time. For example, inadequate mixing of a coagulant such as aluminium sulphate can impair the formation of a good floc in the clarifier and would result in poor plant performance which can only be corrected by using excess of the chemical. The addition and mixing of chemicals to the main flow of water is a continuous process and is frequently described as either rapid or flash mixing. The design of mixers is often based upon the concept of velocity gradient and its value is used to express the degree of mixing at any point in the liquid system. The velocity gradient G (s^{-1}) is defined in terms of power input by the following relationship developed by Camp and Stein[30] for flocculation:

$$G = \left(\frac{P}{\mu V}\right)^{1/2}$$

where P is the useful power input (W), V is the volume (m^3) and μ is the absolute viscosity (Ns/m^2 or kg/m.s) at the water temperature. V can be taken as the flow rate multiplied by the residence time in the mixer (see below). Mixing efficiency is directly related to the local flow turbulence created and should give a high degree of chemical-in-water homogeneity within a short time, with low absorption of power. Mixing designs should aim to achieve the optimum combination of maximum turbulence and low power input. The methods used for mixing can be hydraulic or mechanical. In hydraulic mixers all elements of liquid within the mixer will be subjected to the same retention time, akin to plug flow and they are suitable for many mixing applications. Mechanical mixers are mostly backmix devices and when applied to a continuous flow system the elements of liquid within them have a distribution of residence time. This is considered unsuitable for coagulant mixing. They require long residence time to make allowance for short-circuiting and the head loss across the mixing chamber is as much as that required for hydraulic mixing.

Hydraulic mixing makes use of the turbulence created due to the loss of head across an obstruction to flow such as from an orifice plate, pipe expansion, valve, or by a sudden drop in water level such as when water flows over a weir or hydraulic jump. The latter is usually formed at a flume[31] (see Plate 16) or constriction formed with the help of a chute along a channel designed to produce supercritical flow under all operating flows; the constriction is then brought back to the original channel width rapidly in a tapered section.[32] The ratio of the upstream depth d_1 (m) to the downstream depth d_2 (m) of such a device is given by the following equation:[33]

$$\frac{d_2}{d_1} = \frac{1}{2}\left[(1 + 8F^2)^{1/2} - 1\right]$$

where F (Froude number) $= V_1/(gd_1)^{1/2}$, V_1 is the velocity upstream of the jump (m/s) and g is the acceleration due to gravity (m/s^2). For the hydraulic jump to form the ratio (d_2/d_1) > 2.4 and therefore $F > 2$. For F between 4 and 9 about 40–70% of energy is available for mixing.[34] Hydraulic mixer efficiency is flow dependent. Mixers should be designed for the operating range of plant flow rates, if necessary with facilities for taking sections of the mixer out of service at low flow rates.

When pipelines are used for mixing, a pipe of equivalent hydraulic length of at least 20 pipe diameters should be allowed. Static or motionless mixers are pipeline hydraulic mixers with radial mixing employing stationary, shaped diverters which forces the liquids to mix themselves through a progression of divisions and recombinations until uniformity is achieved. These mixers comprise fixed mixing elements installed in a housing of diameter the same as the pipe diameter. They can easily become blocked and should be used with care on raw water particularly from river abstraction. They should be avoided for mixing lime or caustic soda as calcium carbonate formed due to localised softening could form scale on the elements. Chemicals are injected at about twice the velocity of pipe flow at a point within 1 m upstream of the mixer and samples of the mixed solution are taken at three pipe diameters downstream. Mixer unit should be designed for easy removal for cleaning.

The useful power input of hydraulic mixers is related to headloss by the equation:

$$P = Q\rho g h$$

where P is the useful power input (W), Q is the flow (m^3/s), h is the headloss (m), ρ is the density of water (kg/m^3), and g is the acceleration due to gravity (m/s^2).

In practice it is generally found that adequate mixing is obtainable with a headloss of between 300 and 400 mm, in any case it should not be less than 250 mm. Static mixers require a headloss of at least 600 mm. The residence time for hydraulic mixers (except when a pipeline is used for mixing) is typically 2–3 seconds and a G value of about 800–1000 s^{-1}. For static mixers the G value could be as high as 5000 s^{-1}.

Hydraulic mixers are usually simple and particularly suitable where some headloss can be tolerated. They have the advantages of having no moving parts or direct power consumption, maintenance is therefore negligible. A disadvantage is that the efficiency of mixing suffers if works throughput is lowered outside the operating range of the mixing device.

Mechanical mixing is achieved in purpose built chambers equipped with mechanical rotary impeller mixers such as radial flow turbines or axial flow propellers. Typical residence times are of the order of 20–30 seconds and the velocity gradient value G varies between about 300 and 600 s^{-1}. This gives a power input range equivalent of about 4–10 kW per m^3/s of water flow at 20°C and the value selected depends upon the raw water quality, chemical to be mixed and degree of short-circuiting in the chamber. In recirculation pump jet mixing about 2.5–5% of the plant flow is drawn upstream of the mixing chamber and returned in a pipe through an orifice plate onto a plate in the mixing chamber.[35] The resident time and the power input are the same as for the impeller type mixers. This concept has been extended to pipelines where the water is returned into the pipe against the direction of plant flow through an injection nozzle selected to give a full cone spray.[36] Due to a tendency for nozzle blockage in raw water mixing applications either clarified, filtered or plant service water should be used as the motive water for the mixer. Mixing time and velocity gradient are similar to those for hydraulic mixers. In both types of pumped jet mixers the chemical may be injected into the return pipe or at the point of turbulence. Mechanical mixers have the advantage that they are not affected by flow variations. To maintain a uniform velocity gradient at varying works throughputs the power input should be varied by fitting the mixer with a variable speed motor.

Pumps, in particular the centrifugal type, are good mixers. The lower the efficiency of the pump the better the mixing; for example, in a pump of 75% efficiency, a significant part of the 25% energy lost being in the generation of turbulence, can be used for mixing. However consideration should be given to possible volatilisation of the chemical due to low pressure on the suction side and hence corrosion of pump internals. For this reason gases such as chlorine in solution should not be mixed in pumps and even for other chemicals, materials of wetted parts should be carefully selected.

Mixing is rapid when viscosities of chemical and water are similar and therefore it is improved by applying the chemical in a diluted form. For lime, dilution should be limited to 2.5% w/v and dilution water should be low in alkalinity to prevent scaling of dosing lines. For a coagulant, it should be diluted to a level above where dilution does not shift the pH of the diluted solution into coagulation pH range. Polyelectrolyte should be diluted to about 0.05% w/v. Dilution is usually carried out downstream of the dosing pump.

The method of injecting chemicals also contributes to the performance of the mixer. For dosing chemicals into pipelines, single or multiple injection tubes or perforated diffusers should be used. The injection tubes when used for slurries or with chemicals likely to form precipitates should be of the withdrawable type. The nozzle velocity should be about 0.75 m/s or 50% of the velocity of flow in the large pipe, whichever is the greater. In all cases the injection tubes are mounted in a plane perpendicular to the direction of flow and in the case of horizontally laid pipes the tubes are inserted with their axes at 45° to the horizontal. For static mixers the injection criteria applied vary according to the mixer manufacturer.

Injection of chemicals at weirs or flumes is best achieved via perforated pipes (for clear solutions) or channels (for slurries) running the full width of the weir about 500 mm above the water surface and just upstream of the point of turbulence. Submerged perforated pipes similarly located are used for gases such as chlorine in solution.

7.13   Chemical coagulation and flocculation

Coagulation and flocculation are essential processes in the treatment of most surface waters; one exception being the application of slow sand filtration (see Section 7.38). The two processes operating in conjunction with solid–liquid separation processes, remove turbidity, colour, cysts and oocysts, bacteria, biological matter, viruses and many other organic substances of natural and industrial origin. Some are removed directly and some indirectly through attachment or adsorption on to particulate matter.

The terms coagulation and flocculation are sometimes used incorrectly with the same meaning, but they are in fact two separate processes. In coagulation the coagulant containing the aluminium or iron (ferric) salt is mixed thoroughly with the water and various species of the positively charged aluminium or iron hydroxide complexes are formed. These positively charged particles adsorb on to negatively charged colloids such as colour, clay, turbidity and other particles through a process of charge neutralisation. Flocculation is the process in which the destabilised particles are bound together by hydrogen bonding or Van der Waal's forces to form larger particle flocs during which further particulate removal takes place by entrapment into the flocs. Flocculation is usually achieved by a continuous but much slower process of gentle mixing of the floc with the water in one of the many possible types of plant. In the theory of flocculation the rate at which it takes place is directly proportional to the velocity gradient and the same equation as used in Section 7.12 for mixing is used for determining the velocity gradient G for flocculation. In fact the equation was first developed by Camp and Stein for flocculation applications and it is an inadequate parameter for design of mixers. However in the absence of a better design approach it is still being used for mixer design. If the residence time in a flocculation chamber is t seconds then the extent of flocculation which takes place, or the number of particle collisions which occur, is a function of the dimensionless expression Gt which is given by the equation:

$$Gt = \frac{1}{Q}\left(\frac{PV}{\mu}\right)^{1/2}$$

where the symbols have the same meaning as in Section 7.12.

For the common coagulants of aluminium and iron salts the value of G for flocculation is usually in the range 20–100 s^{-1} with the residence times in flocculation chambers varying from about 10–40 minutes. There are cases where flocculation times approaching two hours have been necessary for waters of extremely high colour and temperature close to 0°C. The value of Gt would be in the range from 20 000 to 200 000. The values of G and t depend on the raw water quality (e.g. colour, turbidity, algae) and downstream treatment process (e.g. clarifiers or filters), the type of clarifiers (e.g. dissolved air flotation) and water temperature. Therefore each application should be individually evaluated by pilot trials unless adequate information is available for almost identical conditions. In direct filtration where the intention is to form a micro-floc, G values of the order of 100 s^{-1} and a residence time of about 10 minutes are used. For dissolved air flotation the values used are, G about 50–70 s^{-1} and residence time about 15 minutes for algal laden water and 20 minutes for waters with colour. Flocculation for sedimentation would require G value in the range 30–70 s^{-1} and residence time between 20 and 40 minutes. For optimum flocculation the coagulated water should be subjected to a decreasing level of energy with time; the so-called 'tapered energy' flocculation. The G values quoted above would then normally be applied in the first stage reducing to any value between 10 and 30% in the last stage for two or three stage flocculation. The G and residence time values for the high rate clarifier 'Actiflo' are 150–300 s^{-1} and 6–8 minutes, respectively and is usually applied in one stage.

Types of flocculators

The agitation required for flocculation is usually provided by either hydraulic or mechanical means. The most common hydraulic flocculator is the baffled basin in which a sinuous channel is equipped with either around-the-end or over-and-under baffles. The flocculation energy is derived primarily from the 180° change in direction of flow at each baffle. The around-the-end type is preferred due to the ease of cleaning. For the around-the-end type the minimum water depth is 1 m and the head loss across the channel is in the range 500 mm to 1 m. The residence time is 20–25 minutes. The distance between baffles should be at least 500 mm and that between the end of each baffle and the wall should be about $1\frac{1}{2}$ times the distance between the baffles. The baffle spacing should be increased gradually with channel length to achieve tapered flocculation. The floor of the channel should slope towards the outlet. The advantage of the baffled basin is its simplicity because there is no mechanical or moving equipment, and near plug flow conditions occur with low short-circuiting. The disadvantages are that most headloss occurs at the 180° bends and therefore the value may be too high at the bends and inadequate in the straight channels for good flocculation, G value will vary as the flow Q varies (but this could be partly overcome by providing removable baffles) and settlement of suspended solids in the channel. When designing baffled sinuous channels some allowance must therefore be made for water quality and suggested velocities to minimise settlement are:

• for highly turbid waters   0.40 m/s

- for moderately turbid waters   0.30 m/s
- for water of low turbidity   0.25 m/s

Other hydraulic flocculators are helicoidal-flow, staircase-flow, gravel-bed and Alabama types.[34] Hydraulic flocculation is also used in sludge blanket clarifiers and is described in Section 7.16.

There are several types of mechanical devices for flocculation, the most common being the paddle type which is mounted either horizontally or vertically in the flocculating chamber. Axial flow turbines are also a suitable alternative[37] for vertical type and is frequently used for high energy flocculation. The power term in the velocity gradient expression for a paddle type is given by the equation:

$$P = F_D V = \frac{1}{2} C_D \rho A V^3$$

where P is the power input (W), F_D is the drag force (m kg/s^2), C_D is the drag coefficient, A is the submerged area of the paddles (m^2), and V is the relative velocity of the paddles (m/s) with respect to water; V may be approximated to 0.75 times the peripheral velocity of the paddle or equal to $1.5\pi r n$, where r is the effective radius of the paddle (m) and n is the number of revolutions (s^{-1}); ρ is the density of water (kg/m^3); C_D is about 1.8 for a paddle flocculator. The speed of rotation of the paddles varies from 2 to 15 revolutions per minute and the peripheral velocity of the paddles ranges from 0.2 to 0.8 m/s. For high energy flocculation the paddle tip speed could be as much as 3 m/s.

Flocculation chambers are usually made up of two or three equal size compartments in series to minimises short-circuiting and to permit tapered flocculation. Flocculators in each compartment should be fitted with variable speed motors with that in the last compartment having infinitely variable speed facilities. The compartments should be separated by either perforated baffle walls or over-and-under type which gives diagonal flow in the compartment; the headloss across the baffle walls will produce a G value of up to 20 s^{-1}. The velocity of flow should be limited to 0.25–0.4 m/s depending on the floc characteristics to minimise floc shear. The tank dimensions vary according to the type; square plan with depths up to 5 m for vertical shaft type and long and narrow (at least $L:W$ of 4:1 with a square cross-section perpendicular to the direction of flow) with depths of about 3 m for horizontal shaft type. For good performance horizontal shaft type with at least three paddles on each diametrical arm mounted in sinuous channels is considered better as short circuiting is reduced. When one or more flocculation tanks is dedicated to an individual clarifier, measures should be taken to prevent flocculator flow patterns being transmitted to the downstream clarifier.

Performance considerations

Impurities which require coagulation and flocculation before they can be removed in a waterworks by sedimentation and filtration can be of many types and origins, but broadly they can be classified as either inorganic or organic material. The more usual type of inorganic material encountered in water treatment settles easily, especially when it is of particulate size; when plain settling is used one or two hours retention will usually remove at least 40–50% of particulate matter and treatment with a coagulant such as aluminium or an iron salt will remove at least 98% of the remainder. On its own a primary coagulant, sometimes assisted by a coagulant aid, will usually remove 98% or more of the suspended particulate matter for lightly loaded water with suspended solids not exceeding about

300 mg/l, but such a removal rate may not be adequate for those river waters typical of the Middle East, India, the Far East, and parts of Africa where suspended solids are regularly measured in excess of 1000 mg/l. In such places it may be necessary to introduce a plain settlement stage before chemical sedimentation or, alternatively, to have two separate stages of chemical treatment and settlement in series because the quantity of solids in suspension, and incidentally the sludge formed by the addition of chemicals, is too great to remove in one stage of treatment.

Not all inorganic matter settles out quickly in plain settlement and in some waters, such as are found in Central Africa, it may be weeks or even months before any significant settlement of colloidal material takes place. With such a water there is no option but to use chemical coagulation at both the pre-settlement and the main clarification stages. Organic colour and finely divided mineral matter, including various forms of clay, comes in this category. Clay in water is a hydrophobic (water hating) colloidal suspension, or 'sol', in which the surfaces of the particles are considered to have a negative charge which contributes to the stability of the sol by helping to prevent the particles coalescing into larger particles which would have a relatively rapid rate of settling. It has often been thought that the excellent coagulating effect of aluminium or ferric salts is due to the triple positive charge on the trivalent aluminium or ferric ion neutralising the negative charges on the clay particles.

The literature on the coagulation of clay and similar materials has been reviewed by Packham,[38] who has also described his own investigations which have developed the theories in a somewhat different direction. His experiments have confirmed that it is not primarily the aluminium or ferric ions in solution which react with the clay particles, but it is the mass of rapidly precipitating hydroxides of aluminium or ferric which enmesh them. In subsequent work by other researchers this is identified as one of two mechanisms in coagulation and is called 'sweep coagulation'. It requires excess coagulant and occurs in about 1–7 seconds. The other mechanism is one of charge neutralisation of negatively charged colloids by positively charged hydrolysis products which are soluble hydrated hydroxide complexes of aluminium or ferric salts. The reaction is completed within a second.[39] The presence of some anions such as sulphate (SO_4^{2-}) helps the coagulation process.[33] Sweep coagulation uses more chemicals and produces more sludge than for charge-neutralisation, but removal of trace contaminants are better. The pH is an important factor in coagulation. The minimum pH is that necessary for hydrolysis of the coagulant and the optimum pH corresponds to minimum solubility of the hydrolysed products. Typically for ferric salts the coagulation pH is greater than 5 and for aluminium salts it is between 6.5 and 7.2. For coagulation of natural organic matter, e.g. colour, coagulation pH values for ferric and aluminium salts are about 4.5 and 5, respectively, whilst optimal turbidity removal typically occurs at pH 6–7.5.

The US EPA has included in the Disinfection By-Products Rule a requirement for 'enhanced coagulation' for greater removal of natural organic matter. It is defined as the addition of excess coagulant, a change in coagulant type, or a change in coagulation pH for improved total organic carbon (TOC) removal. US EPA proposes that enhanced coagulation is applied to surface waters treated by conventional treatment unless, the concentration of TOC in the treated water is <2 mg/l, the concentration of TOC in the raw water is <4 mg/l and the alkalinity is >60 mg/l as $CaCO_3$ or the concentration of chlorine by-products in the distribution system are below 50% of the maximum contaminant levels (MCLs) when chlorine is the disinfectant.

Polyelectrolytes (see Section 7.24) can be used as coagulants for waters containing high turbidities and sludge conditioning. For example in the 1365 Ml/day Karkh water treatment works for Baghdad a cationic polyacrylamide is used to assist in reducing the raw water suspended solids from 30 000 mg/l to about 500 mg/l in pre-sedimentation tanks. In the subsequent clarification stage the same polyelectrolyte is used as the coagulant aid to aluminium sulphate. For waters containing low turbidities they are ineffective as coagulants because, at the very low dosages applied, the residence time provided for flocculation is insufficient to produce large aggregates. They can however be made effective by increasing the number of particles by coagulation using aluminium or ferric salts. In this application polyelectrolytes are called coagulant aids and added after a time delay of zero to 60 seconds (warm water) or up to 5 minutes (cold water) following the mixing of the coagulant, but ahead of flocculation. This time delay is shown to be more important for ferric than aluminium coagulants[40] and found to be applicable in particular to polyacrylamide type polyelectrolytes. Coagulation by polyelectrolytes follows charge neutralisation or bridging or both. Therefore they would be effective even when the polyelectrolyte carries a charge of the same sign as that of the particles to be coagulated. Hence the best type of polyelectrolyte for an application should be determined by the laboratory tests.

Effect of organic content and algae

The variety of impurities present in raw waters and their varying concentrations explain the dictum that all waters are different and that water treatment is an 'art' as well as a science. This particularly applies to waters whose chief natural impurity is organic in origin, as these are very often the most difficult and obstinate to treat. Miscellaneous fragments of animal and vegetable matter contribute towards the organic content, as does the organic colouring matter derived from peat and similar sources consisting largely of humic and fulvic acids and of more complex compounds, partly in true solution and partly in colloidal form. The optimum chemical treatment for such waters is sometimes very difficult to achieve, particularly when it is necessary to remove dissolved iron or manganese. Laboratory jar tests are an essential tool in the formulation of treatment for all waters. Where possible this should be followed by testing at pilot scale.

Planktonic animals and plants, particularly the plants (phytoplankton) which are algae, can cause a variety of problems in treatment. Their removal in microstrainers has been mentioned in Section 7.4 and in Table 7.5 a short list of some commonly occurring algae and their implication for treatment in temperate climates is given. They often occur in large quantities in lowland and in eutrophic waters. Their removal by flocculation, sedimentation, and filtration is basically similar to, but more variable than that of other forms of suspended material because of the widely diverse types, densities, and shapes of plankton. Some algae, for example diatoms which have silica in the cell walls, are notably denser than water. These algae and many other algae with large or simple shaped cells are removed effectively by chemical treatment. On the whole however plankton do not settle very readily when growing rapidly as they are buoyant because of oxygen gas in their cells and the coagulant dose must be increased to be effective. Many small celled types and those which are active swimmers, due for example to their flagellae, are often not well retained in floc and are liable to pass through the subsequent rapid filters. Conversely problems often arise, because of the tendency of algae to form dense blooms in lake waters and cause a rapidly increasing loss of head in the filters. The removal of planktons by

Table 7.5 Organisms reported as having caused difficulties in British waterworks

Group	Genus	Difficulty experienced (✔)					
		Blocking of filters	Penetration of filters	Taste or odour	Growth in pipes, channels	Unsightly scums	Potentially toxic substances
Bacillario-phyceae (Diatoms)	*Asterionella*	✔					
	Cyclotella	✔					
	Fragilaria	✔					
	Synedra	✔		✔			
	Nitzschia	✔					
	Stephanodiscus	✔	✔				
	Melosira	✔					
	Diatoma	✔					
Chlorophyta	Green unicells		✔				
	Chlamydomonas		✔				
	Cladophora				✔	✔	
	Spirogyra	✔					
Cyanobacteria 'Blue-green algae'	Anabaena			✔		✔	✔
	Aphanizomenon	✔				✔	
	Lyngbya	✔	✔				
	Microcystis			✔		✔	✔
	Oscillatoria	✔	✔	✔	✔	✔	✔
Xanthophyceae	*Tribonema*	✔					
Chrysophyceae	*Synura*		✔	✔			
	Dinobryon	✔					
	Mallomonas	✔	✔				
Euglenophyta	*Euglena*		✔	✔			
Dinophyceae	*Peridinium*	✔					
Rotifers	*Diglena*		✔				
Crustacea	*Cyclops*	✔	✔				
	Diaptomus		✔				
	Daphnia	✔	✔				
	Asellus				✔		
	Bosmina	✔					
Nematodes					✔		
Oligochaetes	*Nais*				✔		
Insects	*Chironomus*		✔		✔		
	Tanypus		✔		✔		
Filamentous bacteria	*Sphaerotilus* (*Cladothrix*)				✔	✔	
Iron bacteria	*Crenothrix*				✔		
Polyzoa	*Plumatella*				✔		
	Cristatella				✔		
Porifera (sponges)	*Ephydatia*				✔		
	Spongilla				✔		
Molluscs	*Dreissena*				✔		

coagulation and settlement can be improved by the use of a polyelectrolyte as a coagulant aid and killing them or simply inactivating them using pre-treatment with ozone,[41] chlorine, or an algicide,[42] although both these latter methods have problems due to toxicity of the residues, or in the case of chlorine the formation of trihalomethanes (THMs), or other side effects such as taste and odour formation (see Section 8.32). Without killing them their removal by sedimentation is about 50–75%.[43] Some algae such as blue-green algae (*Cyanobacteria*) are especially adapted to live in the top layers of water, and the cells of these forms contain minute air vacuoles and are particularly buoyant and difficult to settle. The dissolved air flotation process is therefore particularly suited to blue-green algae removal because of the natural tendency for the algae to rise to the surface.[44,45] The process can also be effective with other less buoyant algae, such as some diatoms and is described in more detail in Section 7.18.

Micro-organisms such as cysts and oocysts are readily removed by coagulation and flocculation followed by filtration with or without clarification (see Section 7.43); a removal of 99.9% should be feasible for bacteria,[46] although it is reported that such removal efficiencies are achieved mostly in the summer months, reducing to 70% in the winter months.[47] Removal mechanisms include direct removal by charge neutralisation and entrapment in floc and through removal of particles to which bacteria are attached.

Of the other impurities which are encountered in raw waters, it is far less easy to be precise as to removal efficiency by chemical coagulation. In many lowland waters there are to be found small but appreciable concentrations of substances in the run-off from farmland, drainage from agricultural land and sewage and trade effluent discharges. Many of these are organic substances and are usually classified as volatile or non-volatile with many more subdivisions, as for example whether they are biologically degradable. Whilst it is known that chemical coagulation flocculation, clarification and rapid filtration are at least partially effective in their removal, in particular those proportions adsorbed onto particles, there is a need for a high degree of removal by different processes. These are described in Chapter 8.

Clarifiers

7.14   Horizontal flow clarifiers

Some simple types of settlement tank are in use for the clarification of flocculated waters. For large volumes of water containing a relatively heavy load of suspended solids a relatively dense floc is formed which settles easily, and in warm climates where the viscosity of the water is lower thereby permitting more rapid settlement of floc, the large horizontal flow sedimentation tank can be an economical solution for clarification. Although a large tank is necessary in order to keep velocities sufficiently low to permit settlement of floc to the base of the tank, its construction is simple because it need not be very deep and few internal walls are required. In the simplest design of horizontal flow tank, floc is allowed to accumulate on the floor of the tank until such time as the increasing velocity of water above the accumulated sludge stirs up some of the floc, thereby affecting the clarity of the effluent. When this occurs the tank should, at least in theory, be cleaned out. Moving scrapers operated continuously or intermittently add to the overall efficiency by pushing the settled floc to outlets in the base of the tank where it may be drawn off as sludge.

Raked bar screens at a river intake pumping station. (Hawker Siddeley Brackett Ltd., Colchester, UK)

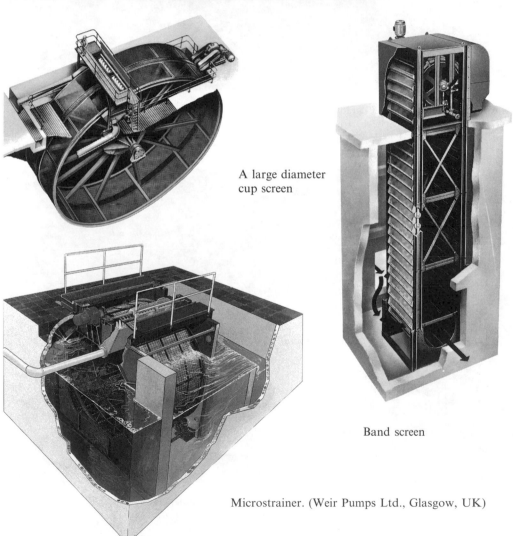

A large diameter cup screen

Band screen

Microstrainer. (Weir Pumps Ltd., Glasgow, UK)

Plate 13

Alton reservoir formed by 24 m high earth dam, with 46 Ml/day treatment works in foreground built 1987 for Anglian Water, UK. (Engineers: Binnie & Partners)

Layout of 82 Ml/day treatment works at Simley, Islamabad, Pakistan, completed 1995. (Contractor: Paterson Candy Ltd., Isleworth, UK. Photo courtesy Taisei Corporation)

Plate 14

Lamella plate clarifier. (Purac Ltd., Kidderminster, UK)

One of the DAF (dissolved air flotation) units, capacity 20 Ml/day, at the 450 Ml/day Frankley treatment works for Severn Trent Water, UK. (Engineers: Binnie & Partners)

Plate 15

Interior view of a CoCo-DAFF (counter-current dissolved air flotation filtration) clarifier unit. (Paterson Candy Ltd., Isleworth, UK)

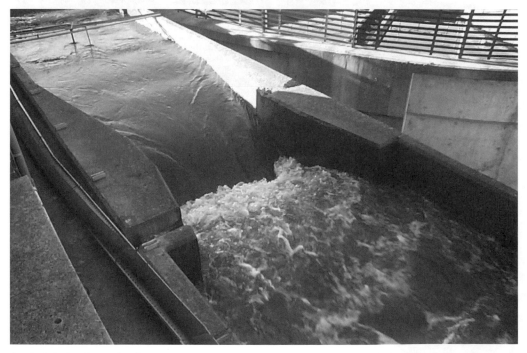

A 140 Ml/day flume for flow measurement and mixing of injected coagulation chemicals upstream in the downstream hydraulic jump, Lartington treatment works, Northumbrian Water, UK

Plate 16

Accepting their initial low rating and correspondingly high civil construction cost, horizontal tanks are nevertheless versatile clarifiers, and modifications to an original simple design to meet changing conditions of raw water can be fairly easily accommodated if provision has been made at the design stage. For example, rotating flocculators and sludge scrapers can be added, and some tanks have had their rate of flow increased by the addition of inclined tube or plate settlers (see Section 7.17). Since the depth of a tank does not influence its performance some horizontal flow tank designs use the principle of shallow depth sedimentation, with a sloping tray so that the direction of flow of the water reverses up and over the tray and exits near the inlet. Yet another variation is where the depth of the rectangular tank is divided by as many as four inclined trays and flow takes place in parallel streams between the trays. In this way the capacity of a single multiple tray tank can be made to equal the flow capacity of a more sophisticated design of sludge blanket or sludge recirculation tank. Flow in a horizontal direction is frequently maintained in circular tanks so that after chemical flocculation, which usually takes place in a central compartment, flow is radial and upwards to peripheral collecting launders or to a combination of peripheral and radial launders. This type of circular sedimentation tank is nearly always fitted with rotating sludge scrapers, either with the drive mechanism mounted on a central platform or bridge or, with the drive unit mounted on the outside wall and the scraper bridge pivoted on a central support. There are circular radial flow tanks in operation in many parts of the world, particularly for the treatment of heavily silted waters, and their use should always be considered as an alternative to rectangular horizontal flow tanks for the treatment of such waters.

In the design of rectangular or circular tanks for treating heavily silted water, the feed water should be conveyed in channels rather than pipes to permit easy access for cleaning of settled solids. The advantages of these tanks are: greater tolerance to hydraulic and quality changes, ideal for stop/start operation, infinite turn down, simplicity of operation, suitability for water containing high silt loads, and the influence of diurnal temperature change on the performance is small. The primary drawback is their low surface loading rate and hence the large area occupied and associated civil engineering costs. Also when compared to rectangular tanks, circular tanks do not lend themselves to a compact layout. One such circular tank design known as 'Centrifloc' is used for the treatment of R. Tigris water at the 1365 Ml/day Karkh water treatment works in Baghdad and is shown in Fig. 7.2. In this design water enters 18 No. 51 m diameter tanks through a 14.5 m diameter annular flocculating compartment of 16 minutes retention capacity and equipped with mechanical flocculators, and rises upwards and radial towards radial launders. A scraper with a central drive shaft moves the sludge to a central hopper for evacuation under hydrostatic head. The tank is operated at a surface loading rate of $1.7 \text{ m}^3/\text{h.m}^2$. The tanks receive water from a pre-settlement stage and are designed to treat a suspended solids load of up to 2500 mg/l and it has been successfully operated at values well in excess of 5000 mg/l.

7.15 Design criteria

In practice, a rectangular or circular tank does not work as efficiently as its theoretical capacity would indicate. Normally it will only work at about 33% efficiency although, in the case of chemically assisted sedimentation, the actual efficiency is generally much closer to the results obtained by laboratory jar tests. Hence, in general, the size of a plain

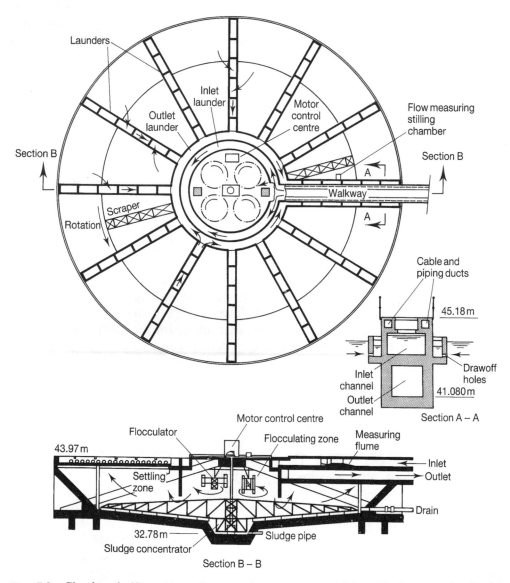

Fig. 7.2 Circular clarifiers 51 m diameter for treatment of River Tigris water for Baghdad. Engineers: Paterson Candy Ltd.

sedimentation tank must be three times the theoretical size to obtain the same results. This is due to the difficulty of achieving uniform flow through a body of water. However by careful design of the inlet and outlet to minimise short-circuiting the retention safety factor may be limited to about 1.5. These design features include multiple inlets at about 1.5 m centres sized to give an inlet velocity of about 0.5 m/s, perforated baffles at the inlet end across the cross section of the tank with orifices of diameters of 100–200 mm to give a head loss of less than 10 mm to minimise floc shear. Whilst these measures help, they may have little effect when the temperature of the incoming water differs from that in the tank; even if baffles do succeed in spreading out the inflow, the change of motion of the water up and over, or down and under such baffles creates turbulence which must subside before the particles can resume settling under quiescent conditions.

In horizontal flow tanks the horizontal flow of water should occur under laminar conditions (Reynolds number < 2000) and short-circuiting and instability of flow should be minimum (Froude number > 10^{-5}). For tanks of dimensions l, w and d (m), surface loading rate S (m³/h.m²) and flow Q (m³/s), Reynolds and Froude numbers are given by

$$Re = \frac{Q}{w + 2d}\frac{1}{\gamma} \text{ and } F = \frac{S^2}{g}\frac{l^2(w + 2d)}{1.3 \times 10^7 \, wd^3}$$

where γ is kinematic viscosity (m²/s) and $g = 9.81$ m/s²

Horizontal flow sedimentation tanks for the removal of organic matter and light flocculated particles using coagulants, have length to width ratios greater than or equal to 4:1 and, surface loading rates between 0.75 and 1.75 m³/h.m² which with a coagulant aid may increase to 2.5 m³/h.m². The mean horizontal flow velocity should be in the range 0.25–1.0 m/min. The flow over the outlet weir should be maintained below 50 m³/h per m of weir length and should preferably be about a quarter of this value. Alternatively, 90° V-notches about 50 mm deep and about 0.15 to 0.25 m apart may be used. Submerged orifice outlets are useful in minimising the passage of floating material to filters. Orifices should be sized to give a flow velocity of about 0.6–0.7 m/s with an orifice diameter greater than 30 mm and head loss 35–40 mm. They should be about 0.3 m below the water surface. To achieve the outlet flow, double sided launders are usually used. These are placed about 0.5 m from the end wall (rectangular tanks) or peripheral wall of circular tanks. Alternatively 'finger' launders of length <20% of the tank length (rectangular tanks) or radial launders (circular tanks) feeding a collector launder are used. The depth of the tank should be adequate for sludge deposits and storage; a minimum of 3 m is recommended for scraped tanks and 5 m recommended for manually cleaned tanks (see Section 7.19). Detention times for particles of low relative density, i.e. the theoretical time of travel of water in the tank, varies from a minimum of $1\frac{1}{2}$ hours to an average of 4 hours.

For large circular radial-flow clarifiers the detention time can be as high as 4 hours, but is more usually about $2\frac{1}{2}$ hours. These clarifiers generally operate with rates similar to those of rectangular tanks.

Overflow rates in clarifiers are often obtained by dividing the daily or hourly flow by the gross surface area of the tank. This can be misleading as it may not take into account the area occupied by the mixing and flocculation compartments (where these are provided), and the effluent launders, which usually account for at least 12% of the gross area. A better guide to the true overflow rate is to take the horizontal clarification area available at a depth of about 1 m below the surface.

7.16   Sludge blanket or solids contact clarifiers

Probably the earliest type of sludge blanket tank is that often referred to as the 'hopper bottomed tank'. It is believed that the first of these tanks were built in India during the 1930s, with hundreds of them having been built since, mostly in places where there has been a British influence. At the Johore River water works (Stages 1–3) in Malaysia serving Singapore there are 126 hopper bottomed tanks each of 10 × 10 m plan and 10 m deep and capacity 5.75 Ml/day in use. The sludge blanket effect is obtained simply by allowing the chemically treated water to flow upwards in an inverted pyramid type of tank with the angle of slope usually 60° to the horizontal. They are usually square in plan (occasionally circular), although rectangular multiple hopper tank variations have been constructed for large capacity works. As the water rises in the tank of increasing area, its velocity progressively decreases until, at a given level, the force of the upward flow on the particles is so reduced that it only just counterbalances the downward weight of the particles, which therefore hang suspended in the water. Their presence forms a kind of blanket in the water in which chemical and physiochemical reactions can be completed and in which a straining action to remove some of the finer particles may also take place. While the top of the blanket is very often well defined there is no distinguishable bottom to the blanket because of the varying density of the particles forming the blanket.

Flocculation takes place within the clarifier and is hydraulically induced as the coagulant dosed water enters downwards and flow reverses upwards at the base of the clarifier. Flocculation is completed within the sludge blanket through a process known as contact flocculation. The concentration of the sludge blanket is controlled by allowing it to bleed off from suitably placed 'concentrators' or hoppers. The hopper bottomed tank is not cheap to construct as the total water depth, governed by geometry, is usually about the same dimension as the side of the square in plan. During construction care must be taken to ensure that the inverted pyramid is reasonably accurate and the inlet pipe discharges at the geometric centre; if it does not, then streaming of the water to one side can occur and this can upset the stability of the sludge blanket. Some sludge blanket tanks have been criticised for the tendency of the blanket to 'boil' and this can happen particularly in hot climates (see below) or where the tank is in an exposed position and subject to high winds. However, the effect on water quality leaving a tank due to temperature boil is usually less severe than that due to wind effect. The surface loading of hopper bottomed tanks is within the range 2 to 3.5 m^3/m^2.h, the lower velocity may sometimes be needed for floc formed in the removal of colour from a soft reservoir water.[48]

In order to reduce construction costs there has been a continual striving towards simplification of the tank shape, and the most recent addition to the sludge blanket family is a flat-bottomed variety, usually rectangular in plan and illustrated in Fig. 7.3. In this type of design chemically treated water is directed downwards onto the base of the tank through suspended pipe 'tridents' before passing upwards through the blanket to the surface collecting launders. The sludge is collected in hoppers placed at the top level of the blanket and is removed under hydrostatic head.

Another special type of upward flow flat bottomed sludge blanket tank is the 'Pulsator' (see Fig. 7.4) where a proportion of the incoming chemically dosed water is lifted into a chamber built onto the main inlet channel by applying a vacuum of about 650 mm water gauge using a centrifugal fan and released into the tank. This creates a pulsing effect in the blanket. The frequency and duration of pulsation is varied according to the flow and water quality and typically would be of the order of 50–60 seconds and 8–10 seconds, respectively. These would normally be set to achieve a pulsation rate in the tank of about 7–8 $m^3/h.m^2$ for waters containing low settleable solids and 10–12 $m^3/h.m^2$ for waters containing high settleable solids. Flocculation in flat bottomed tanks is induced hydraulically as the water enters the tank through the distribution pipes and is completed within the blanket by contact flocculation.

Sludge blanket clarifiers are suitable for many types of waters, including turbid ones, provided that the particulate matter is of low density. A safe upper limit for turbidity is about 500 NTU and depending on the nature of the particulate matter, much higher (up

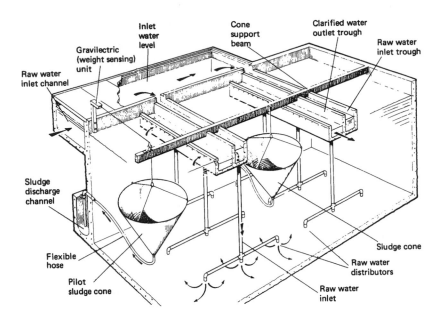

Fig. 7.3 A flat-bottomed upward flow sludge blanket clarifier (Paterson Candy Ltd).

to 1000 NTU) peaks could be accommodated. Heavy particulate matter tends to settle on the bottom of the clarifier. When used in such applications these clarifiers should be provided with scrapers or other methods for intermittent removal of bottom sludge. Alternatively they should be drained down and cleaned after the rainy season, or about once or twice a year. In this respect, hopper bottomed tanks are more appropriate for highly turbid waters. For heavily silt laden waters sludge blanket clarifiers should be preceded by simple rectangular or circular sedimentation basins as described in Section

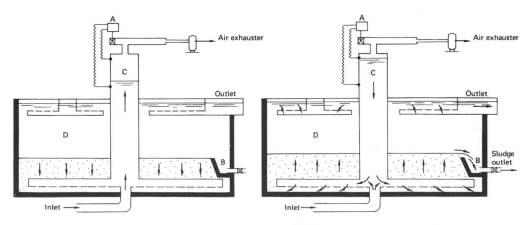

First half-cycle: Air valve A closed and water rises in vacuum chamber C. Water in clarifier D at rest and sludge settles.

Second half-cycle: Water in C rises to upper contact and air valve A opens. Water in C falls and enters D raising sludge which enters concentrator B. When water falls to lower contact, air valve A closes.

Fig. 7.4 The 'Pulsator' clarifier (Degremont UK Ltd).

7.14. The surface loading rate of flat bottomed sludge blanket clarifiers typically vary within the range 2–4 m^3/m^2.h. They have a side water depth of about 4.5–5 m which is typically made up of a bottom distribution zone 0.6–1.0 m, sludge blanket of 2.15–2.25 m and clarified water depth of 1.75 m. In some designs the clarified water depth could be as low as 1.0 m. In sludge blanket tanks, typically the blanket concentration is 20–25% v/v (after 10 minutes settlement in a 250 ml cylinder) and 0.1–0.2% w/v, although with some waters containing high settleable solids due to high colour and turbidity it could be as much as 30–35% v/v and 0.25–0.5% w/v.

The operation of sludge blanket tanks is somewhat sensitive to sudden changes in flow and raw water quality and requires greater operator skill. They are not suitable for stop/start operation. Restart following a lengthy shut down may take from 6 to 24 hours or more depending on the availability of sludge from a similar clarifier for seeding the blanket. Stoppages of 3–6 hours, can however be accommodated. There is also a constraint on flow turn-down, which is normally limited to about 60–75% of the maximum flow. The Pulsator clarifier because of its high intermittent flow can be operated down to about 30–40% of the maximum flow. The performance of sludge blanket clarifiers (and horizontal flow settling tanks) is known to be influenced by temperature.[49] The temperature effect is normally diurnal and is caused by the creation of thermal gradients within the clarifier due to the walls of the tank being heated by the sun or by warmer water entering the tanks, from an open raw water storage tank or an exposed raw water main, giving rise to density currents within the clarifier. The result is disturbances of the blanket and carry over of floc. Sometimes similar effects have been attributed to the release of gases due to bacterial activity in sludge. The measures taken to minimise carry over of floc are the use of polyelectrolyte as a coagulant aid or the inclusion of tube modules in the clarified water zone, or perhaps both techniques. Intermittent chlorination would help to overcome bacterial activity in sludge. So long as their sensitivity is appreciated and they are operated intelligently sludge blanket clarifiers will produce a good quality effluent (turbidity less than 2 NTU) and are very tolerant to changing conditions of raw water quality which would be detrimental to the operation of many other types of clarifiers.

Apart from the sludge blanket tanks already mentioned there are many other types of designs which endeavour to achieve the same high level of performance by providing extra mixing energy for flocculation or by the recirculation of sludge, or by a combination of both methods. Frequently the resultant design is something halfway between a wholly vertical flow sludge blanket tank and a radial flow tank. Nearly all such variations have circular configurations and are equipped with bottom sludge scrapers, but operate as solids contact clarifiers. The essential feature is that the settled floc is used to seed the incoming dosed water, thereby accelerating the flocculation process. This is achieved internally (i.e. Accentrifloc clarifier) by feeding the chemically dosed raw water into the flocculation zone where it is mixed with sludge recirculated at about twice the throughput by the aid of a rotor impeller or externally (i.e. Pre-treator clarifier) where a pump is used to recycle sludge (about 20% of tank throughput) to mix with dosed water. These clarifiers can be operated at surface loading rates about two to three times of those without recirculation, similar to sludge blanket clarifiers and performing in a similar manner. Optimisation of performance requires some skill as they have several operating variables such as the impeller speed, recirculation flow, blanket depth and sludge withdrawal rate. They have a high mechanical plant content and are comparatively high in capital and operating costs. At the 2015 Ml/day Bhandup treatment works in Bombay, which treats a stored water, solids recirculation tanks of 44 m square with pump recirculation are used for the clarification of water. The tanks are square in plan and have diametrical scrapers with pantograph extension (see Section 7.17).

7.17 High rate clarifiers

Commercial competition between treatment plant manufacturers has provided the main incentive for the development of new techniques for obtaining higher flow ratings in sedimentation tanks. The maximum rate for a sludge blanket tank using coagulants is about 2 $m^3/h.m^2$, the upper limit with polyelectrolytes is about 4 $m^3/h.m^2$. In proposals for clarification it is important to check the probable performance at the lowest water temperatures likely to apply when the viscosity of the water will be highest. This can be the limiting criterion for performance.

Tube settlers

Some success has been achieved to increase flow rates through existing clarifiers by the use of tube settlers, which make use of the principle established long ago by Hazen[50] that a settling tank should be as shallow as possible in order to shorten the falling distance for particles. The concept of shallow depth settlement by multiple trays is still used in horizontal flow tank designs in Europe (Section 7.14) and the use of all-plastic tube settler packs is a logical development which helps lessen the sludge removal problem of wide shallow trays. The 'tubes' are circular, hexagonal or square in cross-section, of hydraulic diameters in the range 50–80 mm and made of plastic usually polystyrene loaded with carbon black to protect against ultra violet light. The order of preference based on projected area is hexagonal, then square followed by circular.[51] The tubes have a large wetted perimeter relative to the wetted area and thereby provide laminar flow conditions which theoretically offer optimum conditions for sedimentation. Laminar flow is not achieved immediately when water enters the tubes, but after a transition length L_t (m) given by the Schiller formula,[52] $L_t = 0.0288\ Re.D$, where Re is the Reynolds number which should ideally be less than 280 and D is the hydraulic diameter (m) which is equal to $4s/p$, where s is the cross-sectional area (m^2) of the tube and p its wetted perimeter (m). In the design calculations this transition length must be deducted. Sedimentation takes place in the length following the transition length and it will retain all particles with a settling velocity less than V_s ($m^3/h.m^2$)[53] (see Fig. 7.5) which is given by

$$V_s = \frac{V.k}{\sin\theta + \frac{L_s}{d}\cos\theta}$$

where, V – average velocity of flow in the tubes ($m^3/h.m^2$) which is equal to $U\sin\theta$ where U is the average upward velocity ($m^3/h.m^2$) which is the surface loading rate of the settling tank (i.e. rate of flow ÷ plan area); L_s – settling length (m) which is equal to $L - L_t$ where L is the total tube length (m); k – a coefficient (1.0 for parallel plates, 1.33 for circular and hexagonal tubes and 1.38 for square tubes); θ – angle of inclination of tubes to the horizontal; and d – depth of water in a tube at right angle to the direction of flow (m).

When the angle of inclination of the tubes to the horizontal is between 55–60° the solids settled on the inclined wall of the tube leave the tube by sliding downwards under

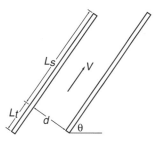

Fig. 7.5

gravitational forces along its lower side, the clarified water flowing in the opposite or counter-current direction. Most tube settlers have been used in circumstances where it was desired to increase the flow through or improve the performance of an existing clarifier, particularly horizontal and radial flow clarifiers and some sludge blanket clarifiers, so that the increased flow rates achieved were still in the range 2–5 m^3/h.m^2.[54] In sludge blanket clarifiers the clear distance between the tube pack and the top water level and the sludge layer below is usually maintained at about 500 mm in each direction. The vertical depth occupied by the tube pack is about 650–750 mm. An example of a sludge blanket clarifier with plates in the clarified water zone is illustrated in Fig. 7.6. This design is used at the 408 Ml/day Johore River water works (Stages 4 and 5) in Malaysia serving Singapore. The use of tube packs allow an increase of about 1.75 times on the surface loading rate without tubes; in essence the tube packs, which provide a much larger settling area than the clarifier plan area, traps the floc carried over from the blanket when subjected to such high rates. Major drawbacks with tube settlers particularly those in sludge blanket clarifiers is that the floc carried over to the filters tends to be fine and may not be retained well in the filters and that the clear water depth of only about 500 mm encourages algal growth on the tube which, along with slime growth if allowed to form, would partially clog them and affect performance. The use of the tubes helps to minimise the effect of thermal 'boiling' and wind on the clarifier performance. There are few examples of proprietary clarifiers where tube packs are incorporated into the design of new tanks.

Lamella plate systems

More recently the principle of shallow depth sedimentation has been extended to the design of parallel plate systems, sometimes referred to as lamella flow clarifiers. Clarifiers using the plate system are usually purpose built to take advantage of the high settlement rates which can be obtained and the greater density of sludge provided. They require an efficient flocculation stage which is critical for successful operation. The flocculated water enters at the base of lamella plates and travels upward between the lamellas counter current to settled sludge moving down (see Plate 15). Figure 7.7 illustrates the operation of a lamella clarifier. In some designs the uneven distribution to the inlet of the lamellas is corrected by introducing the flocculated water flow individually into each lamella space via slotted openings in the side walls of channels running on both sides of the plate pack along the length of the tank. Each space between the lamella plates therefore acts as an independent settling module. The lamella plates extend the full depth of the tank and rise about 125 mm above the top water level. Clarified water is collected by submerged orifices

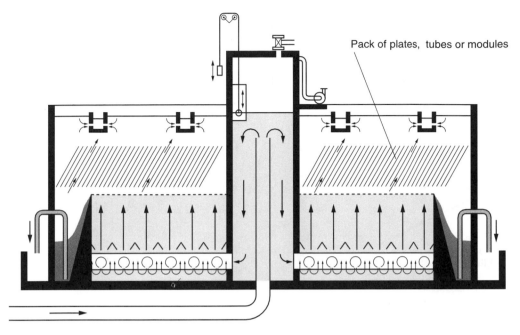

Fig. 7.6   Pulsatube Pulsator clarifier (Degremont UK Ltd).

or V-notches one between each pair of plates in decanting launders running along each side of the plate pack. About 1.5 m is allowed in the bottom of the tank for the collection of sludge which is removed by a scraper of the chain and flight or reciprocating type, to a series of small hoppers at one end of the tank. For small tanks, hoppers could be placed underneath the plate pack for collection of sludge. The plates are inclined at 55–60° to the horizontal.

With this arrangement the settling area available is equal to the sum of the projections of plates in a horizontal plane. Thus the settling area is very large on account of the overlapping of plates but occupies a relatively small plan area. The total settling area is equal to $(n-1)\,LW \cos \theta$ where n is the number of plates, L the plate length in water (m) (after deducting the transition length, W the plate width (m) and θ the angle of inclination of the plates to the horizontal. The value of n should be determined taking plate thickness and spacing between plates into consideration. The plates should be flat and not corrugated and they are usually made of stainless steel but, sometimes of plastic. Plate width is about 1.25 to 1.5 m and plate length is about 2.5 to 3.25 m including the length of 125 mm above the normal water surface; plate thickness is usually 0.7 mm for stainless steel. The horizontal spacing between plates is varied according to the application and is normally within the range 50–80 mm. Depending on the settling velocity of the particles the lamella clarifiers could be operated at surface loading rates of 20 $m^3/h.m^2$ or more and at about 40 $m^3/h.m^2$ under exceptional circumstances and can therefore give a much reduced surface area (up to 95%) compared with more conventional horizontal flow clarifiers. This also means that the retention time within the clarifier is low, sometimes 20 minutes or less, so that control of chemical treatment becomes more exacting. Johnson County water treatment plant (USA) of capacity 115 Ml/day uses lamella plates for pre-

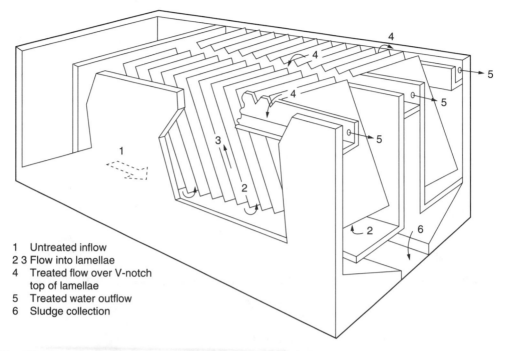

1   Untreated inflow
2 3 Flow into lamellae
4   Treated flow over V-notch
    top of lamellae
5   Treated water outflow
6   Sludge collection

Fig. 7.7   Lamellae sedimentation tank (Purac Ltd, UK).

settlement of Missouri river water prior to softening. The settling aided by cationic polyelectrolyte reduces turbidity from 2000 NTU to less than 10 NTU. The plate settlers are designed to a settling rate of 2.5 m³/h.m² and surface loading rate achieved is about 20 m³/h.m².

There is another type of lamella plate design which operates on the principle of co-current sedimentation in which the flocculated suspension passes downwards through the parallel plates, the sludge settling onto the plates and sliding down to the sludge collector. The angle of inclination of the plates is about 30–40° and the plates are about 35 mm apart. Clarified water is withdrawn at the bottom of the plates and is then made to pass upwards to outlet collecting launders on the surface. This design is used primarily in industrial effluent treatment.

An example of a sludge blanket clarifier which has made practical use of the plate system of sedimentation is the 'Super Pulsator' (see Fig. 7.8). The clarifier has many features in common with the Pulsator from which it has been developed and retains the principle features of the raw water feed and distribution, and formation of a sludge blanket. Parallel plates are located within the blanket about 300 mm apart, at an angle of 60° to the horizontal and perpendicular to the sludge collection hoppers; coagulated water passes upwards and sludge travels between the plates into the hoppers. The plates are fitted with deflectors to create internal sludge recirculation and to thicken the sludge; it is reported that the concentration achieved is twice that in a Pulsator operating at the same upward flow rate. By this arrangement of counter-current flow and plates and deflectors it is possible to obtain, with raw waters which respond to this type of treatment, clarification

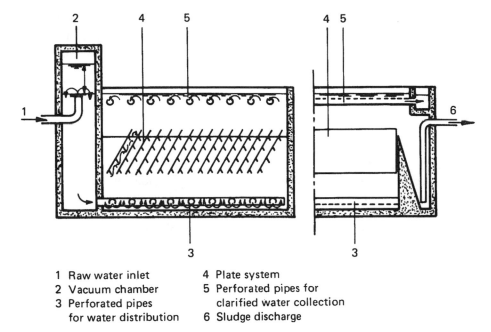

1 Raw water inlet
2 Vacuum chamber
3 Perforated pipes
  for water distribution

4 Plate system
5 Perforated pipes for
  clarified water collection
6 Sludge discharge

Fig. 7.8   The 'Super Pulsator' clarifier (Degremont UK Ltd).

rates in the range 5–10 m/h with enhanced sludge concentration. It is also claimed that this type of clarifier is capable of start-up in a matter of hours from a shutdown condition.

Other high rate clarifiers

In another high rate process a suspension of fine quartz sand (size range 50–135 μm and uniformity coefficient of 1.6) is used to form a weighted floc of density in excess of 2.5 kg/l resulting in very high settling velocities. In the process the coagulated water is mixed with sand and polyelectrolyte followed by high energy flocculation (see Section 7.13). Settlement takes place in flat bottomed or hopper bottomed tanks. The sludge containing the sand falls to the bottom and in the case of flat bottomed tanks is removed by scrapers to a series of hoppers located at one end. The underflow is then recycled at about 3–6% of works throughput via hydrocyclones to separate the sand from the sludge; the sludge discharge from the hydrocyclones is about 80% of the recycle rate. The recovered sand is then made up with fresh sand to account for losses in the sludge stream and recycled to 'seed' the incoming water again. The process requires about 3 g/l of sand which is equivalent to about 0.08% of the volume of the settling tank. The make up sand is of the order of 3 mg/l and is injected either continuously or in larger quantities intermittently (about once a week). Surface loading rates of 6 m^3/h.m^2 and over have been achieved with this type of clarifier. The process would remove *Cryptosporidium* oocysts and *Giardia* cysts like any other clarification process which follows coagulation and flocculation. There is however the potential risk of returning some oocysts and cysts back to the process along with the recycled sand. This aspect should be investigated by pilot trials. The proportion of water lost as sludge can be about 2.5% of works throughput consisting about 0.1% w/v solids. In the 'Actiflo' clarifier, by incorporating lamella plates in the clarifier rates in excess of 25 $m^3/h.m^2$ and sometimes as high as 60 $m^3/h.m^2$ have been achieved.[55] The energy consumption of the process can be in the range 0.01–0.02 kWh/m^3 of water treated.

In the 'Sirofloc' process magnetite (magnetic iron oxide of specific gravity 5.2) in fine particulate form (1–10 microns) is used as a ballasting agent to achieve high settling rates. The magnetite is first 'activated' by treating it with sodium hydroxide solution and is then introduced as a slurry of density 1.45–1.65 kg/l into the water at a dose of 0.5–4% w/v of the raw water flow whose pH may have to be lowered to 5.0–6.0 by the addition of sulphuric acid. A period of 5–20 minutes contact in a tank follows, during which the magnetite adsorbs destabilised colour and other fine colloidal matter in the water.[56] Aluminium sulphate or more usually a cationic polyelectrolyte is added to remove turbidity and bind it to the magnetite. The magnetite is then flocculated by passing it between the poles of a permanent magnet, and the floc thus formed settles readily to the base of subsequent radial flow clarifiers rated at about 7–10 $m^3/h.m^2$. Rapid gravity filtration must follow the Sirofloc process. By re-treating the magnetite sludge drawn from the clarifiers with sodium hydroxide at pH 11–12 and passing it through several washing stages, magnetite is recovered with minor losses for re-use. The make-up dose to counteract losses is about 2 mg/l. The effluent flow from the regeneration plant is about 3–5% of the works throughput and has a high colour (600–850° Hazen for raw water colour of 50–90° Hazen) and pH (about 12). It can either be discharged to a sewer or coagulated with ferric sulphate, settled and dewatered in filter presses as at the 45 Ml/day Littlehempston water treatment works. The process was developed in Australia and is primarily of use for a highly coloured, low alkalinity and low turbidity water. There are about ten water treatment plants of this type operating in UK and Australia.

The sirofloc process lacks a proper coagulation stage which is considered to be essential for the removal of *Cryptosporidium* oocysts and *Giardia* cysts in the downstream solid-liquid separation process. However there may be other mechanisms occurring which could also bring about their removal.

The principal advantage of high rate clarifiers is that they occupy a small area, but they need close attention and optimisation of chemical treatment. Most of them depend on polyelectrolyte dosing. The retention time in some of them is very short and therefore tolerance to changing water quality is significantly reduced and the sensitivity to optimum operating parameters is significantly increased.

7.18   Dissolved air flotation

Dissolved air flotation has been of increasing interest in its application as a clarification stage of treatment. It operates on the principal of the transfer of floc to the surface of water through attachment of air bubbles to the floc. The floc accumulated on the surface, known as the 'float', is skimmed off as sludge, as illustrated in Fig. 7.9 (see Section 7.19). The clarified water is removed from the bottom and is sometimes called the subnatant. Since rain, snow, wind, freezing could cause problems with the float, flotation tanks must be fully enclosed in a building; some users enclose the flocculation tanks as well. There are now several large plants operating in Scandinavia and other parts of Europe including the UK. The largest plant in the UK is at the 450 Ml/day Frankley water treatment works in Birmingham and treats stored water for the removal of colour, turbidity and iron. The

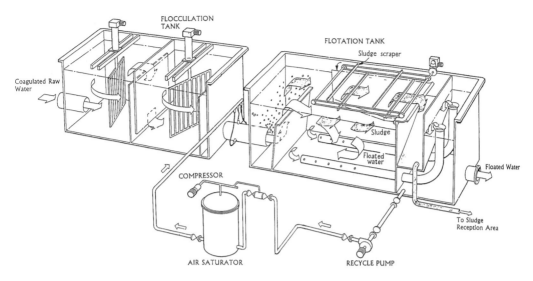

Fig. 7.9 Diagram of a dissolved air flotation plant (Purac Ltd, UK).

plant consists of two streams each of ten, 100 m^2 cells of unit capacity 26.4 Ml/day, rated at 8.5 m^3/h.m^2 and equipped with travelling bridge scrapers for float removal.[57] The process is particularly suited to treatment of eutrophic, stored lowland or otherwise algae laden waters and soft, low alkalinity upland coloured waters.[44,58,59] Although the process has been successfully used for some directly abstracted waters, other clarification methods tend to be more suitable for treatment of such waters especially when the turbidity consistently exceeds about 100 NTU.[60] Table 7.6 below shows some typical results when treating algal laden waters.

There is, however, some experience with eutrophic waters with very high counts of algae where dissolved air flotation has not been successful, so that caution is necessary when choosing the process. It should be noted that sedimentation can achieve degrees of removal comparable to flotation, if algae are first inactivated by chlorination. This would however result in the formation of THMs by the action of chlorine on algal metabolic products.

Flotation is preceded by a flocculation stage of the hydraulic or mechanical type usually dedicated to each flotation cell. The flocculation tank should have at least two

Table 7.6 Comparison of algal cells in the raw water and remaining after coagulation and sedimentation or flotation

Alga	Raw water	Sedimentation	Flotation
Aphanizomenon	179 000	23 000	2800
Microcystis	102 000	24 000	2000
Stephanodiscus	53 000	21 900	9100
Chlorella	23 000	3600	2200

Data from WRc.[42]

compartments in series (see Section 7.13). Flotation is normally carried out in rectangular tanks designed with surface loading rates between 8–12 $m^3/h.m^2$ but rates as low as 5 $m^3/h.m^2$ or as high as 15–20 $m^3/h.m^2$ have been used on some plants.[61,62] With such high rates there is a risk of air entrainment in the subnatant water causing problems, such as negative head (see Section 7.26) in downstream filtration processes. This can be overcome by installing lamellas in the subnatant section where, as the water flows down the lamellas, the entrained air is released and rises, counter-current to the water flow to the float.[63] The solids loading can vary in the range 4–15 kg dry solids/m^2.h. Typical tank depth is 2–3 m and the preferred length:width ratio is 2–2.5:1 with lengths up to 15 m using end-feed of air or 20 m with centre-feed of air. The retention time in the flotation tank is between 10–20 minutes.

For effective flotation the quantity of air required is about 6–10 g/m^3 of water treated and requires a recycle flow rate of about 6–15% which is dependent on water temperature.[64] Some designers determine the air dose from the weight ratio of air to suspended solids of about 0.1:1. Recycle water should be of at least clarified water quality and, in the case of a packed column absorber, preferably filtered water. Oil-free compressors are preferred but not essential for the air supply. Air is dissolved in recycle water under pressure either in pressure vessels equipped with an eductor on the inlet side for adding air or a packed column; the operating pressures of the two respective saturator systems are 6–7 bar and 3.5–5 bar. In packed columns a packing depth of 0.8 to 1.2 m of 25 mm polypropylene Pall or Rashig rings are used. The hydraulic loading rate of the air dissolving units lies in the range 50–80 $m^3/h.m^2$. Saturator efficiency* for packed column type is about 90–95% whilst that for unpacked type is about 65–75%.[65] Air saturated water is returned to the flotation tank through a series of nozzles or needle valves to give a sudden reduction in pressure and release of air bubbles in a white water curtain. Typically bubble size ranges from 10 to 100 μm with a mean diameter of 40 μm.[66] The outlets are usually spaced at 0.3–0.6 m for needle valves and 0.1 to 0.3 m for nozzles.[67]

In plants where there is a need for raw water ozonation and flotation, the two processes could be combined with air in the flotation process being replaced by an ozone–air or ozone–oxygen mixture.[68,69,70]

Since the clarified water is taken from the bottom of the tank in the flotation process it could be combined with rapid gravity filtration in one tank with the flotation section placed above the filters. Therefore the surface loading rates of the two processes need to be the same. 'COCO DAFF' (counter-current dissolved air flotation filtration) is an innovative combined flotation–filtration design in which air and water flow counter-current as against co-current in the conventional dissolved air flotation process (Fig. 7.10 and Plate 16). Air is introduced with recycle water across the total tank sectional area depth. It is claimed that this arrangement gives better particle–bubble interaction.[71] The process combines flotation and gravity filtration in one tank and uses a group of flocculation tanks common to all of the flotation cells.

---

*Saturator efficiency = $\dfrac{\text{amount of air measured in the recycle water}}{\text{amount of air that could be dissolved theoretically}} \times 100$

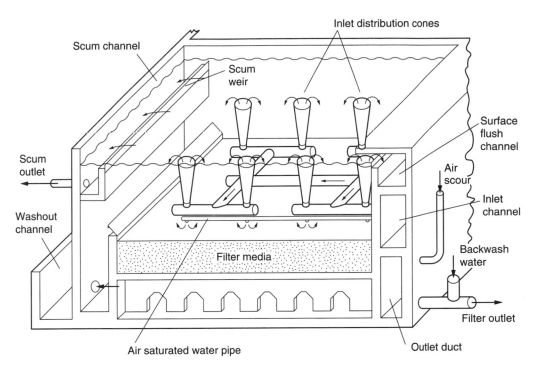

Fig. 7.10   Typical arrangement of COCO-DAFF unit (Paterson Candy Ltd).

Computational fluid dynamics is used to model the flotation process. It is used to simulate basin geometries and to study the effect of changes in process parameters and as is such becoming a valuable tool for design, optimisation and trouble shooting.

The flotation process is suitable for stop/start operation and has a flow turndown of about 30%. The former is one of its advantages when dealing with a water subject to high algal loadings; a plant can be 'switched in' as and when needed and will give a steady quality treated water within 45 minutes.[59] Apart from the drawbacks common to all high rate clarifiers, the flotation process has high energy requirements (about 0.05–0.075 kWh/ m^3 of water treated).

7.19   Sludge removal from clarifiers

Effective removal of sludge is very important for the efficient operation of clarifiers, but it is a subject frequently overlooked by designers. With a raw water having suspended solids not greater than about 250 mg/l – most waters used in the UK fall in this category – the sludge volume to be removed from the tank should not exceed about 2.5% of throughput. For raw waters having solids of about 1000 mg/l the sludge may be as high as 5–10% by volume of throughput; for very silty waters the sludge may have to be removed continuously in order to maintain the tanks in operation, even at a reduced output, and maintain an acceptable water quality. Under such conditions output may have to be reduced and the sludge volume can be as much as 20–25% of throughput. Special measures for sludge removal must obviously be provided for heavily silted waters and, for those having above 1000 mg/l solids or possibly less, depending on the nature of the solids, it is necessary to provide scraping equipment for all designs if throughput and quality are to be maintained. Scrapers move sludge to a series of hoppers located at the inlet end of rectangular tanks or in the centre or periphery of circular tanks. The floor of the tank should have a slope. Hoppers are of an inverted pyramid shape with an included angle of about 60°. Sludge is removed from hoppers individually under hydrostatic head using the full water depth in the tank. Scrapers used in rectangular tanks depend on the tank geometry and are travelling bridge with or without suction headers (for tanks up to 25 × 75 m), chain and flight or cable hauled type (for tanks up to 6 × 50 m). Bridge scrapers with suction headers (speeds varying from 1.0 to 2.0 m/min) and chain and flight type (speeds less than 0.5 m/min) are suitable for tanks treating heavily silted waters. The speed of bridge scrapers without suction headers are 0.5–1.0 m/min for scraping and about 2.5 m/min for the return.

Circular tanks have radial or diametrical scrapers with the bridge supported from the centre and driven with a central or peripheral drive unit. In larger tanks, support is also provided by a travelling wheel on the outside wall. The peripheral speed of the scraper is about 1.0–2.0 m/min. In some square tanks, corners are curved on the bottom and the scraper is then arranged to fold back on itself when traversing the four sides (e.g. solids recirculation type clarifiers at the Bhandup water treatment works, Bombay). Some circular tank designs include a suction header similar to the rectangular tanks. Tanks employing suction headers draw sludge from points just in front of the scraper blades or squeegees using pumps mounted on the bridge at an approximate rate of 2 l/s.m of tank width or radius; others are aided by submersible pumps or down pipes acting as ejectors. The floors of scraped circular tanks have slopes of about 1:10–1:20 and those of scraped rectangular tanks have slopes of about 1:300–1:500. Rectangular tanks equipped with suction headers do not require a slope except for drainage. Unscraped rectangular tanks usually have a cross fall of about 1:10 to a central channel running the length of the tank and a longitudinal fall of 1:200. By including high pressure water jets (at 3.5–4 bar) for cleaning the slope could be reduced to about 1:250. Valves for sludge removal are always better placed outside the walls of the tank. Both valves and pipework should be adequately sized to pass the maximum sludge withdrawal rate, which can be 400% or 500% greater than the average rate. The valves should be of the full bore type like plug valves.

Sludge removal from blanket tanks is generally easier than with other designs, although the same rules for valve and pipe sizing must be used. Certainly the positioning of sludge hoppers can be less critical than with other designs as a sludge blanket is in continual movement and will migrate towards the space left by evacuated sludge. Hoppers usually cover about 10–15% of the total settling area of the tank. The hydrostatic head available for sludge removal is limited to the clarified water depth (1–1.75 m). Many attempts have been made to improve the efficiency of sludge removal and concentration. Some sludge removal systems operate continuously, but this is usually wasteful except in the case of high suspended solids waters. Nowadays, even in what may be essentially a manually operated works, sludge removal is usually operated by an automatic system having adjustable timers for varying the duration of opening of sludge discharge valves at pre-selected but adjustable time intervals. Sludge should be withdrawn from hoppers individually; manifolding hopper outlets to allow for simultaneous withdrawal is not recommended.

Some success has been obtained using photoelectric cells for the detection of sludge build-up, but this method is unlikely to work for heavily silted waters. However, a method which has met with considerable success over the last few years is one patented and developed by Paterson Candy Ltd, and relies upon sensing the differential weight of concentrated sludge in water. The equipment is shown in Fig. 7.3 and consists of several flexible sludge cones suspended in water and one of which (called the pilot cone) is connected by a cable to a load cell. The load cell is sufficiently sensitive that when the weight of the sludge reaches a pre-set value (usually when the cone is about two-thirds full of sludge) then the load cell initiates the opening of the desludging valve on all the cone outlets.

In flotation tanks, sludge or the float collects on the water surface and is removed by mechanical or hydraulic means, or a combination of the two. Mechanical units are scrapers of the chain and flight, reciprocating or bridge type and the choice is made primarily on the tank dimensions. All types are known to cause 'knock down' of float solids; a process where subnatant water is contaminated with the sludge as a result of deaeration of the float due to the disturbances caused by the activities of the scrapers. In hydraulic desludging the subnatant water drawoff is restricted intermittently to raise the water level in the cell until the sludge layer overflows into a collection trough. This method produces sludge of low concentration compared to mechanical methods.

7.20 Chemical dosing equipment and treatment works layout

Chemical dosing plant comprises – storage facilities, solution or slurry preparation tanks, and chemical metering and conveying systems. It is usual for storage facilities to be sized for 30 days' demand at average dose and normal flow rate, or the size of one consignment plus the demand for the period between placing the order and receiving a delivery. Longer storage may be required for locations where access is affected by bad weather or chemicals have to be imported. Properties of some of the commonly used chemicals in water treatment are given in Table 7.7.

Most chemicals are made up into standard solutions or suspensions in batches; at least two batching tanks are required for each chemical in order to maintain continuity of dosing; additional tanks would allow maintenance and cleaning without interruptions to dosing. Each tank is normally sized so that one or two batches are prepared in a work shift. Accurate batching and dilution, with proper mixing, is required to maintain constant strengths for metered injection of chemicals. Solution strengths can be up to saturation concentration (e.g. aluminium sulphate); suspensions such as lime and powdered activated carbon (PAC) need to be maintained at a value of less than 10% w/v and must be continuously stirred. Concentration of batches must be checked for accuracy by using a hydrometer or chemical analysis.

The use of lime as a slurry for final pH correction increases the turbidity of the filtered water by up to 1 NTU depending on the proportion of impurities in lime and hence the lime dose. This can be overcome either by using caustic soda or a saturated solution of lime. Lime is usually prepared in saturators which are continuous upward flow hopper bottomed tanks with water fed from the bottom and lime water drawn from the top. Surface loading rates range from 1 to 1.2 $m^3/h.m^2$. In some designs, rates of 2.5–5.0 $m^3/h.m^2$ are achieved by using a turbine mixer to improve contact between lime and water, adding a polytrelectrolyte to improve settling rate, and lamella plates to increase settling

Table 7.7 Properties of some chemicals commonly used in water treatment (see Table 7.8 for coagulants and Chapter 9 for other chemicals)

Chemical	Function	Form	Density	Materials	Freezing point	Storage	Dosing concentration
Hydrated lime 96% w/w (a) $Ca(OH)_2$	pH correction	White fine powder	480 kg/m^3 (b) 400 kg/m^3 (c) 1.81 m^3/t (d)	Steels, thermoplastics (*Al, tin, Zn, brass, galvanised steel)	–	Bags (25 kg, 50 kg) on pallets, steel silos (see Figure 7.11)	>2.5% w/v <10% w/v (e)
Hydrated lime 17% w/w $Ca(OH)_2$	pH correction	Milky white liquid	1.11 g/ml	(As for hydrated lime)	0°C	Vertical steel or thermoplastic tanks with mixers or recirculation pumps	2.5% w/v to neat
Quicklime (f) 95% w/w CaO	pH correction	Hygroscopic powder	1230 kg/m^3 (b)	(As for hydrated lime)	–	Bags (25 kg, 50 kg) on pallets, steel silos	Slaked to form hydrated lime >2.5% w/v <10% w/v
Powdered activated carbon	Organics removal, dechlorination	Powder	410–600 kg/m^3 (depending on the grade) (b) 375–500 kg/m^3 (c)	Stainless steel (304, 306) mild steel (for slurry), thermoplastics	–	Bags (25 kg, 50 kg) on pallets, 450 kg bags, 1000 kg bins or steel silos (epoxy paint coated) (see Figure 7.11)	<10% w/v
Sodium carbonate (light grade) 95% w/w Na_2CO_3	pH correction	Anhydrous crystalline powder	550 kg/m^3 (b)	(As for hydrated lime)	–	Bags (25 kg, 50 kg) on pallets or steel silos	5% w/v (temperate) 20% w/v (tropics)
Potassium permanganate	Oxidation	Granular	1600 kg/m^3	Steels, thermoplastics (*Zn, Cu, Al, galvanised steel, rubber)	–	Kegs (50 kg), 150 kg drums	1.5 to 3% w/v; solubility 6% w/v at 20°C
Sulphuric acid (i) 98% w/w H_2SO_4 (g) (ii) 95% w/w H_2SO_4 (g)	pH correction	Corrosive liquid	1.84 g/ml at 20°C	Steels, PTFE (* most other metals)	(i) 3°C (ii) −10°C	Carboys (45 l), steel horizontal pressure vessels or steel vertical tanks (lagged for 98% w/w as applicable)	Neat or 10% w/w H_2SO_4

(Continued)

Table 7.7 (*continued*) Properties of some chemicals commonly used in water treatment (see Table 7.8 for coagulants and Chapter 9 for other chemicals)

Chemical	Function	Form	Density	Materials	Freezing point	Storage	Dosing concentration
Sulphuric acid 50% w/w H_2SO_4	pH correction	Corrosive liquid	1.4 g/ml at 15.5°C	Thermoplastics, rubber/steel	−37°C	Carboys (45 l), PVC/GRP or rubber/mild steel vertical tanks	10% w/w to neat
Caustic soda 47% w/w NaOH	pH correction	Corrosive liquid	1.497 g/ml at 20°C	Steels, thermoplastics, rubber, Ni and Ni alloys ($T <150°$) (*Al, Tin, Zn, galvanised steel, brass)	8°C (−25°C for 20% w/w NaOH)	Carboys (45 l) steel horizontal pressure vessel or steel or thermoplastic, PVC/GRP vertical tanks (heated and lagged as applicable)	Neat or 20% w/w NaOH
Hexafluorosilicic acid 20% w/w H_2SiF_6 (15.8% w/w F)	Fluoridation	Corrosive liquid. Highly toxic	1.18 g/ml at 20°C	Thermoplastics (PE, PP, uPVC), Neoprene rubber/steel (*Glass, stainless steel (304, 316), Al, brass, bronze, mild steel)	−11.6°C	Horizontal or vertical rubber/mild steel or thermoplastic vertical tanks	Neat or diluted to suit (i)
Sodium silicofluoride 98% w/w Na_2SiF_6 (59.4% w/w F)	Fluoridation	Crystalline powder. Highly toxic	1400 kg/m³	Thermoplastics, rubber lined mild steel	–	Bags (20 kg, 50 kg) on pallets, steel silos	0.2% w/v (h)
Orthophosphoric acid 75% w/w H_3PO_4 (24% w/w P)	Plumbosolvency control	Corrosive liquid	1.585 g/ml at 15.5°C	Stainless steel (316), thermoplastics (*mild steel, cast iron, Al, Al-alloys, brasses, tinned or galvanised)	−18°C	Lined steel drums (45 l, 200 l), horizontal or vertical stainless steel or rubber/steel or vertical HDPE or PVC/GRP tanks	Neat or diluted to suit (h)

(Continued)

Table 7.7 (continued)   Properties of some chemicals commonly used in water treatment (see Table 7.8 for coagulants and Chapter 9 for other chemicals)

Chemical	Function	Form	Density	Materials	Freezing point	Storage	Dosing concentration
Orthophosphates (i) mono sodium (20% w/w P) (ii) di sodium (17% w/w P) (iii) tri sodium (8% w/w P)	Plumbosolvency control	Crystalline powders	(i) 1200 kg/m³ (ii) 1200 kg/m³ (iii) 900 kg/m³	Thermoplastics, stainless steel (304, 316), rubber/ steel (*mild steel, Al)	–	Bags (50 kg)	(i) 40% w/v (h) (ii) 15% w/v (h) (iii) 20% w/v (h)
Sodium hypochlorite 15.5% w/w NaOCl (15% w/w Cl₂)	Disinfection oxidation	Hazardous liquid	1.27 g/ml at 20°C	Thermoplastics (PE, PVC, HDPE), rubber/steel (*mild steel, Al, Zn, Cu and their alloys, PP)	−17°C	Carboys (45 litres), horizontal or vertical rubber/steel or PVC/ GRP or HDPE tanks	Neat or diluted to suit (h)
Sodium bisulphite 32.5% w/w NaHSO₃ (20% w/w SO₂)	Dechlorination Deoxygenation	Hazardous liquid	1.28 g/ml at 15.5°C	Thermoplastics (PP, PVC) GRP, stainless steel (304, 316) (*mild steel)	10°C	Dums (45 litres, 210 litres), stainless steel or PVC/GRP or rubber/mild steel vertical tanks (heated and lagged as applicable)	Neat or diluted to suit
Hydrogen peroxide 35% w/w H₂O₂	Oxidation	Hazardous liquid	1.130 g/ml at 20°C	Aluminium (99.5%), Al-Mg alloys, stainless steel (304, 316), HDPE, PVC (*Fe, Cu, Ni, Cr, brass)	−33°C	PE carboys (50 kg), stainless steel or Al horizontal or vertical tanks or HDPE or PVC/GRP	Neat
Sodium chloride (Pure Dried Vacuum Grade) 100% w/w NaCl	Regeneration of ion exchange resin. On-site generation of sodium hypochlorite	Crystalline powder	1200– 1360 kg/m³	Thermoplastics, rubber/ mild steel, stainless steel (316), Aluminium alloy NS4 (*stainless steel, mild steel for moist or salt solutions)	–	PE bags (25 kg), 1 t containers, saturators of reinforced concrete of rich mix (1:1.5:3) with 40 mm cover or GRP	Saturated solution (26.5% w/w or 36%w/v at 20°C) or diluted to suit

(Continued)

Table 7.7 (*continued*)   Properties of some chemicals commonly used in water treatment (see Table 7.8 for coagulants and Chapter 9 for other chemicals)

Chemical	Function	Form	Density	Materials	Freezing point	Storage	Dosing concentration
Ammonium sulphate 25% w/w NH_3	Ammoniation	Crystalline powder	1120 kg/m^3	Thermoplastics, stainless steel (304, 316) (*iron, Cu, Zn, Tin and their alloys)	–	Bags (20 kg, 50 kg)	10% w/v (solubility at 10°C is 727 g/l)
Sodium chlorite 26% w/w NaClO$_2$	Chlorine dioxide generation	Hazardous liquid	1.27 g/ml at 20°C	Thermoplastics (PE, PVC) GRP (*Zn and combustibles)	−15°C	PE kegs (50 kg, 70 kg), HDPE vertical tanks	12.5–20% w/v

PTFE – Polytetrafluoroethylene; HDPE – High density polyethylene; PVC – Polyvinyl chloride; GRP – Glass reinforced plastic; PE – Polyethylene (Polythene); PP. – Polypropylene; PVC/GRP – PVC lined GRP; Rubber/steel – rubber lined mild steel; All stainless steel grades are to BS 970 or 1449.

*Unsuitable materials.

(a) x% w/w is x percent weight per weight = x grammes of the chemical in 100 g of the product.
(b) For calculating silo capacity.
(c) When aerated during bulk delivery.
(d) When stacked in bags.
(e) y% w/v is y percent weight per volume = y grammes of the substance in 100 ml of solution containing the substance.
(f) Quicklime gives off considerable amount of heat (1.14 × 10^6 J/kg) during slaking.
(g) Sulphuric acid gives off considerable amount of heat during dilution. Therefore when diluting, acid should be added to a large quantity of water.
(h) Softened water is preferred for solution preparation and dilution to prevent scaling.

area. The concentration of lime in saturated solution is temperature dependent; solubility is 1.76, 1.65 or 1.53 g/l at 10, 20 and 30°C, respectively. Saturators convert about 80% of the lime in the feed. They are also useful to produce a clear solution of lime when it contains grit. Unconverted lime and grit are removed by regular desludging.

Chemical dosing must be accurate and related to the flow of water to be treated. Positive displacement pumps of the reciprocating type with mechanical or hydraulic diaphragm heads are most frequently used for injection, but for viscous solutions such as polyelectrolytes, progressive cavity type positive displacement pumps are sometimes used. For lime and PAC suspensions, either diaphragm, progressive cavity or peristaltic pumps are used. Pumps should be provided with a calibration vessel on the suction side, a pressure relief valve, pulsation dampener, and a back pressure valve on the delivery side. All chemical dosing pumps must be of high quality design and materials, appropriate for the chemical handled, or their life will be short. The maximum stroking speed (spm) of reciprocating pumps should be about 100 spm, in particular for viscous or abrasive chemicals. The motor speed of progressive cavity and peristaltic pumps should be kept to less than 500 and 50 rpm, respectively. A typical layout for a lime or PAC dosing plant is illustrated in Fig. 7.11.

With the reciprocating pump, dosage adjustment is achieved by altering the pump stroke length. Where plant throughput is variable (greater than ±5%) the pump motor

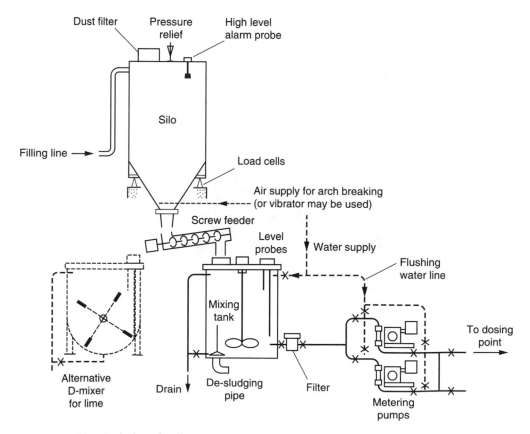

Fig. 7.11 Chemical plant for lime or PAC or soda ash mixing and dosing.

speed is automatically controlled in proportion to the flow rate measured near the chemical injection point. This type of control is called 'open-loop'; it has no feed-back or corrective action and the applied dose rate is strictly proportional to the flow. In a 'closed-loop' system, the pump output is corrected to maintain a given water quality value (such as pH) over a narrow pre-set band, measured downstream of the injection point after the chemical has been well mixed with the water ('feed-back' control). A process controller, working in conjunction with an appropriate water quality measuring instrument, sends a 4–20 mA signal back to the pump to adjust its stroke length. In some instances this water quality signal is combined with the rate of flow signal and used to control the pump motor speed. This latter method is used on pumps which have no stroke adjustment (e.g. progressive cavity and peristaltic types), or on chlorinators and similar equipment to control the orifice positioner so as to maintain a pre-set residual chlorine concentration in the water (see Section 9.12).

In some plants it is necessary to apply the same dose to two or more equal streams, e.g. water entering separate clarifiers; it is then vital to ensure accurate division of the metered chemical flow and this is often done by use of a splitter box with equally set V-notch weirs.

In remote parts of developing countries chemical dosing systems are kept simple, with manual preparation of solutions and slurries and the use of constant head solution feeders for dosing chemicals.[72]

Plant layouts

Figure 7.12 shows a typical layout for a conventional water treatment works using aluminium sulphate as the primary coagulant, polyelectrolyte as a coagulant aid, sulphuric acid for pH adjustment, chlorine for intermediate and final disinfection, plus partial dechlorination and lime for final pH correction, and the possibility of using powdered activated carbon to treat occasional taste problems (see Section 8.33). Among the most important considerations when planning a layout are the following.

(1) The flow through the works should be gravitational: it is inadvisable to re-pump water between clarifiers and filters or this would break up floc. Hence a site having a gentle gradient of 1 in 10 to 1 in 15 is most favourable.
(2) When siting works adjacent a river it is important to avoid siting any structure below highest flood level because of the difficulty and cost of countering uplift problems. The intake should ideally be sited on the inside bend of a river. No electrical or chemical plant should be put in a basement which could flood owing to a burst pipe.
(3) The works should be provided with means to safely evacuate overflow caused by fault or mal-operation. Typical locations are inlets to the works, filters and disinfection contact tanks.
(4) All structures conveying and retaining water downstream of filters should be sealed to prevent contamination. All water retaining structures should be provided with means for dewatering.
(5) Easy access, including turning-circles, should be provided for chemical delivery vehicles. Access for large equipment for plant repair or replacement should be possible to all buildings.
(6) It is preferable to provide chemical dosing lines in duplicate (one duty, one standby). Routing of chemical dosing lines should take account of the chemical handled. Corrosive chemical lines should not be laid in positions where any leakage could

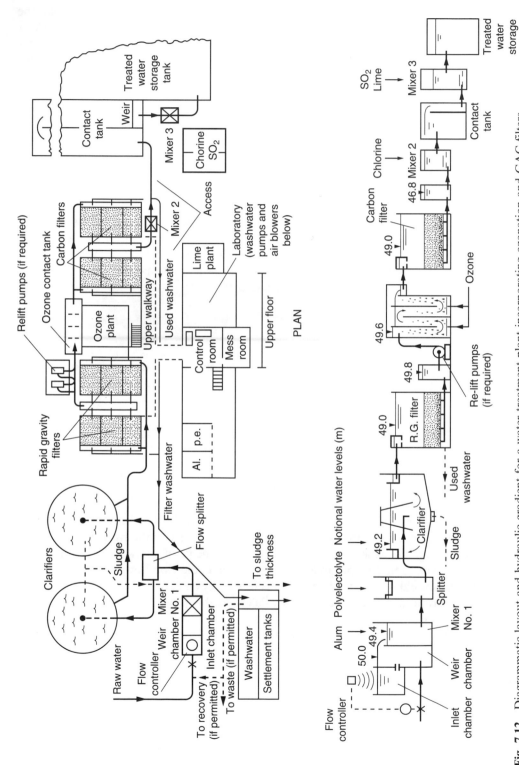

Fig. 7.12 Diagrammatic layout and hydraulic gradient for a water treatment plant incorporating ozonation and GAC filters.

damage other lines or cause injury to personnel. Toxic gas under pressure or solution lines should be laid outside buildings in separate ducting; when possible toxic gases should be conveyed under vacuum and mixed with water just before injection. Delivery lines for slurries, such as lime, are difficult to keep clean and should be of the flexible hose type and laid flat. An easy means of water-flushing of dosing lines should be provided. All chemicals can be diluted in-line after metering. Water for in-line dilution (and solution preparation) of those chemicals which react with calcium to form insoluble compounds (e.g. caustic soda) should be softened by base exchange (see Table 7.7). Lime should not be subjected to in-line dilution due to the tendency to scale dosing lines, unless the alkalinity of dilution water is less than 14 mg/l as $CaCO_3$. This is achieved by de-alkalisation (see Section 8.9) or by treating the dilution water with hydrochloric acid (HCl) to destroy the alkalinity followed by degassing to remove the carbon dioxide produced in the reaction; 1 mg/l of alkalinity as $CaCO_3$ requires 0.73 mg/l 100% HCl and produces 0.88 mg/l carbon dioxide. Alternatively dilution should be in tanks with at least 15 minutes residence time to allow any softening reactions to reach completion. Chemical pipes should be laid in trenches in the ground provided with removable covers for better access in preference to buried ducts.

(7) Instrumentation and electrical power lines will form a complex network of cabling inside buildings and on the site. In the early stages of planning the building layout, allowance should be made for chemical pipes, cable trays, ventilation ducts and other services such as site water supply. It is useful to include a major duct, connecting the main chemical building with other parts of the works, in which such cables and pipes can be laid. High tension electrical cables should be separately ducted. Ducts for chemical delivery lines can be adjacent, but must be separate. All ducts must have drainage outlets.

(8) Liquid chemical storage vessels will need bunding dedicated to each chemical. A bund should be designed to hold 110% of the contents of the largest tank. Dosing pumps and similar apparatus should also be surrounded by a low bund wall 150 mm high to contain leakage. Dust-nuisance chemicals, e.g. lime and PAC, should be fully segregated. PAC is an electrical conductor and should not be allowed to accumulate as dust on open electrical circuits. Toxic gas facilities (chlorine, sulphur dioxide and ammonia) should be located in fully segregated buildings, or rooms. A storage room should be physically separated from other rooms, with equipment handling gas under pressure confined to the storage room (see Section 9.12).

(9) Substances and products used in the Works which may come in contact with the water which is to be supplied for drinking, should not contain any matter which could impart taste, odour, colour or toxicity to the water or otherwise be objectionable on health grounds.

(10) All chemical drainage including that from hardstanding areas should be collected and disposed of separately and should not be allowed to contaminate water courses.

(11) Safety precautions for operational staff should receive careful attention. This should include the provision of safety showers, eye baths, first aid boxes, protective clothing and breathing apparatus.

(12) Any mess room for operatives should be positioned close to the principal control desks so that all alarm signals can be heard. In some cases floodlighting of works and closed circuit TV monitoring of outlying structures is advisable.

Coagulants and coagulant aids

7.21   Aluminium coagulants

Aluminium sulphate is the most widely used aluminium coagulant. It is available in a number of solid grades such as block, kibbled or ground and is also available as a solution. In waterworks practice aluminium sulphate is frequently but incorrectly referred to as 'alum'. The solid form has the composition $Al_2(SO_4)_3xH_2O$ where x may range from 14 to 21 containing 14–18% w/w Al_2O_3 (alumina) or 7.5–9% w/w Al (aluminium), depending on the number of molecules of water (x). The liquid form contains 8% w/w Al_2O_3 or 4.2% w/w Al. The amount of Al_2O_3 or Al in any solid grade of aluminium sulphate containing x moles of water is given by y% w/w $Al_2O_3 = [5.67 \div (19 + x)] \times 100$ and z% w/w Al = $[3 \div (19 + x)] \times 100$, respectively.

The aluminium sulphate dose is therefore normally expressed in mg/l as y% w/w Al_2O_3 depending on the grade of aluminium sulphate, or more usefully in mg/l as Al; 1 mg/l as y% w/w Al_2O_3 is equal to $5.29 y \times 10^{-3}$ mg/l as Al. Aqueous solutions of the solid grades are usually prepared in concrete tanks lined with acid resistant bricks, epoxy mortar, or fibre glass and equipped with a collector system of perforated pipe laterals in a bed of gravel. The tanks are usually built below ground level so that the material delivered in bulk can be tipped directly into the tanks. A saturated solution is prepared which contains about 660 g/l (10°C), 690 g/l (20°C) or 730 g/l (30°C) and is subsequently diluted about four- to six-fold in stock tanks before dosing. When the solid grade is delivered in bags, a 200–300 g/l solution is prepared in tanks containing two compartments separated by a timber grid to prevent solid in one compartment damaging the top-entry turbine mixers in the other. The liquid grade containing 8% w/w Al_2O_3 is stored in stainless steel, epoxy coated steel, rubber lined mild steel, or thermoplastic tanks and dosed by metering pump in the delivered form; after metering, it should preferably be diluted about five to six-fold, but to not less than 0.25% w/w Al_2O_3 to assist mixing at the point of application.

When dosed into water, the formation of an aluminium hydroxide floc is the result of the reaction between the acidic coagulant and the natural alkalinity of the water, which usually consists of calcium bicarbonate. A dose of 1 mg/l of aluminium sulphate as Al reacts with 5.55 mg/l of alkalinity expressed as $CaCO_3$ and increases the CO_2 content by 4.9 mg/l. Thus if no alkali is added the alkalinity will be reduced by this amount with a consequent reduction in pH. If a water has insufficient alkalinity or 'buffering' capacity, additional alkali such as hydrated lime, sodium hydroxide, or sodium carbonate must therefore be added; the alkalinity expressed as $CaCO_3$ produced by 1 mg/l of each chemical (100% purity) is 1.35, 2.5 and 0.94 mg/l, respectively. The aluminium hydroxide floc is insoluble over relatively narrow bands of pH, which may vary with the source of the raw water. Therefore pH control is important in coagulation, not only in the removal of turbidity and colour but also to maintain satisfactory minimum levels of dissolved residual aluminium in the clarified water. The optimum pH for coagulation of lowland surface waters is usually within the range of 6.5–7.2, whereas for more highly coloured upland waters a lower pH range, typically 5–6, is necessary. Lowland waters usually contain higher concentrations of dissolved salts, including alkalinity and may therefore require the addition of acid in excess of that provided by the coagulant. Under these circumstances it is usually more economic to add sulphuric acid rather than excess aluminium sulphate to obtain the optimum coagulation pH value.

Polyaluminium chloride (PACl) has the flocculating properties of aluminium sulphate. The principal advantages over aluminium sulphate are that it depresses the pH of the treated water less than aluminium sulphate, thereby reducing the alkali dose required for subsequent final pH correction; it forms a stronger and more readily settleable floc than aluminium sulphate; coagulation is less affected by low temperature and in many cases it performs as well as combined use of aluminium sulphate and polyelectrolyte, thus saving on the cost of the polyelectrolyte. In some waters it can be used in lower doses than aluminium sulphate and over a broader optimum pH range (6–9). There are several grades of PACl containing 10, 18 or 24% w/w Al_2O_3; the 10% w/w grade being the most commonly available. The other polymeric aluminium salts are polyaluminium chloro-sulphate and polyaluminium silicate sulphate. They behave in a similar manner to PACl. The properties of the most commonly used aluminium coagulants are summarised in Table 7.8.

7.22 Sodium aluminate

Sodium aluminate is prepared from aluminium oxide stabilised with caustic soda; it is used with aluminium sulphate to coagulate very cold waters which would not coagulate successfully with aluminium sulphate alone. It is also used in the 'double coagulation' of highly coloured waters; aluminium sulphate (with sulphuric acid) being added as the first stage to coagulate the colour at pH 4.5–5.0. The resulting soluble aluminium in the settled water from the first sedimentation stage is precipitated in the second sedimentation stage using the alkaline sodium aluminate at pH 6.5. Sodium aluminate is also used in lime-soda softening in which insoluble calcium aluminate is formed, and in turn flocculates the precipitated calcium carbonate and magnesium hydroxide.

In recent years a theory has been put forward that aluminium in drinking water may be associated with neurological disorders and Alzheimer's disease.[73] In addition its presence in filtered water can be harmful to users of renal dialysis (see Section 6.5). For these reasons some water undertakings have changed from aluminium to iron coagulants.[74] When making such a change care should be taken to clean process units free of all accumulated floc, which would otherwise dissolve and increase the aluminium concentration in the water if a ferric coagulant is used outside the optimum pH range for the aluminium coagulant.

7.23 Iron coagulants

Iron coagulants in the ferric form behave similar to aluminium sulphate and form ferric hydroxide floc in the presence of bicarbonate alkalinity. A dose of 1 mg/l of ferric sulphate or chloride as Fe neutralises 2.7 mg/l alkalinity expressed as $CaCO_3$ and increase the CO_2 content by 2.36 mg/l. The ferric hydroxide floc is insoluble over a much broader pH range (4–10) than aluminium sulphate. The lower end of the pH range (4–5.5) is useful for treating highly coloured moorland waters.

Iron coagulants are available as ferric sulphate, ferric chloride and ferrous sulphate. Ferric salts are very corrosive acidic liquids. All materials of construction suitable for aluminium sulphate are suitable for ferric salts, with the exception of stainless steel which is unsuitable for ferric chloride. Ferric sulphate is usually preferred to ferric chloride since the introduction of chloride ions may increase the corrosivity of a water. Ferrous sulphate,

Table 7.8 Physical properties of coagulants

	Sodium aluminate	Aluminium sulphate	Aluminium sulphate	Polyaluminium chloride	Polyaluminium chlorosulphate	Polyaluminium silicate sulphate	Ferrous sulphate	Ferric sulphate	Ferric chloride	Polymeric ferric sulphate
Physical form	Liquid	Solid	Liquid	Liquid	Liquid	Liquid	Solid	Liquid	Liquid	Liquid
Chemical formula	$NaAlO_2$	$Al_2(SO_4)_3 \cdot xH_2O$ $x = 14\text{--}21$	$Al_2(SO_4)_3$	$Al_x(OH)_y \cdot Cl_z$	$Al_2(SO_4)_x \cdot Cl_y \cdot (OH)_z$	$Al_w(OH)_x (SO_4)_y (SiO_2)_z$	$FeSO_4 \cdot 7H_2O$	$Fe_2(SO_4)_3$	$FeCl_3$	$Fe_2(SO_4)_3$
Typical commercial grade	25% w/w Al_2O_3	Blocks 14% w/w Al_2O_3	8% w/w Al_2O_3	10% w/w Al_2O_3	8.3% w/w Al_2O_3	8% w/w Al_2O_3	90% w/w $FeSO_4 \cdot 7H_2O$	40–42% w/w $Fe_2(SO_4)_3$	40–42% w/w $FeCl_2$	48–50% w/w $Fe_2(SO_4)_3$
Fe/Al content (% w/w) of the commercial product	10.5	$300/(19 + x)$	4.2	5.3	4.4	4.4	18	12	14–14.5	13.5–14
pH	12.5	1.5 for a saturated solution (670 g/l of water at 20°C)	1.3	2.3–2.9	2.8–3.0	3.6–3.8	1.7 for a saturated solution (390 g/l of water at 20°C)	<1.0	<1.0	1.0
Specific gravity	1.54 at 20°C	1 to 1.4 t/m³ (bulk density)	1.32 at 15°C	1.20 at 20°C	1.16 at 20°C	1.28 at 15°C	1.0 t/m³ (bulk density)	1.52 at 15°C	1.45 at 15°C	1.58–1.63 at 15°C

(Continued)

Table 7.8 (*continued*) Physical properties of coagulants

	Sodium aluminate	Aluminium sulphate	Aluminium sulphate	Polyaluminium chloride	Polyaluminium chlorosulphate	Polyaluminium silicate sulphate	Ferrous sulphate	Ferric sulphate	Ferric chloride	Polymeric ferric sulphate
Freezing point	−15°C	–	−15°C	−12°C	−12°C	0	–	−15°C	−2°C	−20°C
Viscosity at 20°C	470 m.Pa.s	–	20 m.Pa.s	3.5–4.5 m.Pa.s	4.5 m.Pa.s	11 m.Pa.s	–	30 m.Pa.s	7.5 m.Pa.s	55 m.Pa.s
Coagulation pH range	6.5–7.5 (10–11 for softening)	5.5–7.5	5.5–7.5	6–9	6.5–7.8	6.5–.8	4.0–10.0 (chlorinated)	4.0–9.0	4.0–9.0	>4.5

Notes.
All liquid coagulants are dosed neat and solid coagulants are dosed as saturated solutions. They can be diluted to suit (typically 20% w/v for metering), but any further in-line dilution following metering should be limited to a level that does not cause precipitation. Diluted solutions of some polymerised coagulants gradually hydrolyse with time with subsequent loss of effectiveness.
Suitable materials of construction are thermoplastic material such as polyvinylchloride (PVC), polyethylene, polypropylene, PVC lined glass reinforced plastic, rubber lined mild steel, stainless steel (316) except for those containing chlorides and concrete with suitable linings, e.g. acid resistant bricks, fibre glass or resin coated. Unsuitable materials are mild steel and most common metals such as Al, Zn, Cu and their alloys and concrete.

traditionally referred to in its hydrated form ($FeSO_4.7H_2O$) as 'copperas', is a powder and a solution for use is prepared in a manner similar to that described for the solid grade of aluminium sulphate. Sulphuric acid is added to a batch of a saturated solution of ferrous sulphate to eliminate the ferric hydroxide precipitate. It is used as a coagulant usually in conjunction with chlorine when it is oxidised *in situ* to ferric sulphate and ferric chloride. Chlorine and ferrous sulphate are metered separately and mixed together before dosing to the water. Theoretically for each part of ferrous sulphate as Fe, 0.65 parts of chlorine are required; in practice excess chlorine is used. A dose of 1 mg/l of chlorinated ferrous sulphate (excluding excess chlorine) as Fe neutralises 8.0 mg/l alkalinity expressed as calcium carbonate. A disadvantage with the use of chlorinated ferrous sulphate is the probability of the formation of THMs by the action of excess chlorine with THM precursors in the raw water. Ferrous sulphate on its own is used as a coagulant in processes utilising high pH values such as lime softening (pH 10–11) and manganese removal (pH 9). Iron coagulants have the advantage of producing a denser floc than that produced by aluminium sulphate thereby producing improved settlement characteristics but at the expense of about a 40% increase in the weight of hydroxide sludge when compared to aluminium coagulants. Water treatment using iron coagulants requires close process control because excessive residual iron will result in consumer complaints, whereas excess aluminium causes less noticeable effects.

Polymeric ferric sulphates are now available; they contain about 12.5% w/w Fe and are claimed to perform better and at lower doses than ferric sulphate.[75] There are ferric-aluminium sulphate coagulants; one such product contains approximately 8% w/w of the metal oxides made up of 6% w/w Al_2O_3 and 2% w/w Fe_2O_3. The properties of commonly used of iron coagulants are summarised in Table 7.8.

Many iron coagulants contain approximately 7 g of manganese per kg of iron as an impurity, and this will contribute to the manganese concentration in the water.

7.24   Coagulant aids and polyelectrolytes

Coagulant aids are used to improve the settling characteristics of floc produced by aluminium or iron coagulants. The coagulant aid most used for a number of years was activated silica; other aids included sodium alginates and some soluble starch products. These substances had the advantage of being well-known materials already used in connection with the food industry and were thus recognised as harmless in the treatment of water. Polyelectrolytes came later into use and were more effective. Originally of natural origin, for the most part they now comprise of numerous synthetic products: long chain organic chemicals which may be cationic, anionic, or non-ionic. The theory of their action has been reviewed by Packham.[76]

Polyacrylamides are the most effective of the synthetic group of polyelectrolytes, but for their safe use the toxic monomer residue (the raw materials used in their manufacture) which is not adsorbed by the floc, should be virtually absent from the product. In the UK the list of 'Approved Substances, Products and Processes' under the Water Supply (Water Quality) Regulations[77] states that 'no batch must contain more than 0.025% by weight of free acrylamide monomer based on the active polymer content'; the dose used must average no more than 0.25 mg/l and never exceed 0.5 mg/l. The US EPA allows a maximum dose of 1 mg/l on an assumed acrylamide content of 0.05% by weight.

Polyelectrolyte doses used are very small in relation to the dose of the primary coagulant. Natural polyelectrolyte doses vary between 0.5–2.5 mg/l whereas polyacrylamide doses vary between 0.05–0.25 mg/l. Polyelectrolytes are added as a coagulant for turbid waters, or after the primary coagulant as a coagulant aid (see Section 7.13). Sometimes they are added just prior to filtration in very small doses (less than 0.05 mg/l) to flocculate micro-floc particles carried over from the clarifiers and filter passing algae; care in control of the dose is necessary because excess polyelectrolyte could result in 'mud ball' formation and other problems in the filters.

Most polyelectrolytes are powders and a solution must be prepared for dosing. For successful preparation the powder must be wetted properly by using a high energy water spray before dissolving, and the solution should be allowed to age for about an hour in cold water or 30 minutes in warm water conditions before use. For polyacrylamide the solution should be prepared at about 2.5 g/l, whereas for natural polyelectrolytes the solution concentration could be as high as 25 g/l. Polyelectrolyte solutions are viscous and can be metered to the point of application using positive displacement pumps of the reciprocating or progressive cavity type; the stroking speed of the pump should be maintained under 100 spm. Following metering the solution should be diluted ten-fold to assist transfer in the pipe and dispersion at the point of application. Once a batch of stock solution is prepared it should be used within about 24 hours.

In some waterworks where activated silica is still used as a coagulant aid, it is prepared *in situ* by partial neutralisation of the alkaline sodium silicate to form a colloidal solution, known as silica sol. The neutralising agent can be sulphuric acid (which requires careful control to avoid the solution setting to a gel) or chlorine or aluminium sulphate if this is more convenient, or the more widely used sodium bicarbonate. The dosage of activated silica is commonly in the range 2–5 mg/l.

The practical effect of introducing polyelectrolytes in many existing waterworks has been to increase the settling rate and hence allow substantially greater output through the sedimentation tanks and filters; an additional advantage has been in the use of polyelectrolytes to assist in the used filter washwater recovery and thickening and dewatering of sludges.

Part II Water filtration methods and sludge disposal

Rapid gravity filtration
7.25 Mechanism of rapid filtration

Sedimentation with or without chemical coagulation is usually followed in a waterworks by solid-liquid separation processes which usually include rapid filtration, and the basic principles determining the removal of particles by such filtration will now be discussed. Usually rapid filtration is preceded by chemical treatment of the water; rapid filtration without chemical treatment is effective for relatively few waters, its main use being 'primary' filtration before slow sand filtration. The latter is discussed in Sections 7.38–7.41.

In rapid filtration the removal of particles is largely by physical action, although physicochemical processes may also occur. The size of grain of the filter media, usually sand, is normally within the range of 0.4–1.5 mm whereas particles which may be removed by simple filtration, e.g. mineral particles or diatoms, may be at least twenty times smaller. In fact a proportion of particles even several hundred times smaller than the size of the sand grains may be removed. In achieving effective removal of the smaller particles the addition of a coagulant to form a floc, containing aluminium or iron hydroxides, is usually necessary, but even the floc particles may be very small compared with the size of a grain of the filter media. It is therefore evident that filtration is quite different from a simple straining action such as that of microstrainers. In many instances there may be some straining action due to a coating on the surface of a filter, but in general filtration is a process in which some depth of the filter media is utilised. In filtration of water the flow within the filter bed is laminar or streamline; the loss of head through the media is proportional to the velocity of flow of water.

It is evident that it is necessary to consider by what mechanisms relatively small particles are removed by simple filtration, and a considerable amount of research has been carried out on this subject in recent years. Some of this work has been carried out and the concepts conveniently summarised by Ives *et al.*[78] The general conclusion has been reached that the principal mechanisms of simple filtration are physical and they may be considered under the headings of gravity (or sedimentation), interception, hydro-dynamic effects, and diffusion.

Research on filtration has also considered mechanisms of possible attraction (or repulsion) between particles and filter grains; in the absence of any attraction some particles would tend to become detached. Such considerations are, however, complex and any clear conclusions are of limited application. Van der Waals forces are well known as attractive forces between molecules and theoretically they apply to nearly all materials in water, but their range is usually limited to minute distances of less than 0.05 μm.

Finally it should be mentioned that in some water treatment processes physicochemical or chemical reactions occur in contact with filter grains. One example is the deposition of calcium carbonate from waters having a positive calcium carbonate saturation index. Another is the oxidation and deposition of compounds of iron and manganese. After such reactions have commenced the coated grains may provide active surfaces for their continuance. In a few instances the oxidation of ammonia to nitrate or other biochemical action may be an incidental effect of passing a water through a rapid filter, due to the development of the necessary bacterial flora in organic impurities on the filter grains.

The principle of biological reaction is now used in potable water treatment for reducing the concentration of nitrates, ammonia, iron and manganese to an acceptable level by the development of a biomass on the surface of filter sand (see Section 8.25).

7.26   Design and construction of rapid gravity filters

The working part of a rapid gravity sand filter or the part that removes the solids from the incoming water is the filtering medium which is usually sand. As already discussed in Section 7.25, there are many different mechanisms (sedimentation, interception, hydrodynamic diffusion, attraction and repulsion, etc.) which contribute towards the removal efficiency of a filter. Some have a greater role to play than others according to the nature of the water and the chemical treatment which has been used in previous treatment stages. Rapid gravity filters usually constitute the last solid–liquid separation stage in a treatment cycle for drinking water, and the objective with all designs of filters is to reduce the solids content measured as turbidity to less than 0.4 NTU with an upper limit of 1 NTU – these being the usual target values. More recently however, the need to ensure the removal of *Giardia* cysts and *Cryptosporidium* oocysts has led to turbidity targets of less than 0.1 NTU being applied to filtered waters.

To achieve these standards of purity there is some limited but nevertheless important scope for varying the basic design of a rapid gravity sand filter: for instance the sand size can either be (nominally) constant, i.e. monograde, or it can vary from fine to coarse; the depth of the sand can be shallow or deep; the direction of flow of water can be downflow, upflow, or can even be brought into the middle of the sand bed to flow both upwards and downwards. A further variation in design is to use three or more layers of sand and pebbles or, alternatively, to use other materials, such as anthracite or garnet, having differing grain size and relative density. All of these variations in filter design have been incorporated from time to time in different commercial (sometimes patented) filters. When a fine sand is used, the collection of solids during filtration, and hence the build-up of headloss, tends to be within the top layers of sand. In contrast, with coarser sands the solids penetrate to a greater depth and the lower layers of the sand bed are then called upon to do some of the work of solids removal. So long as there is an adequate factor of safety in bed depth against complete dirt penetration it makes good sense to utilise at least some of the bed depth for solids capture, but the proviso must be that the backwashing system can be relied upon to remove accumulated solids and achieve thorough cleansing of the sand before its next working cycle. In theory the ideal sand grading arrangement is to have a decreasing sand size in the direction of flow as this will bring about the greatest degree of solids capture. Attempts to do this are discussed in Section 7.31 dealing with the use of anthracite media.

Media data

The sand filter designs employed in Europe use either graded sand (fine to coarse or hetrogenous) or of coarse monograde sand (uniform size or homogenous). There is no single media specification (size and depth) that can applied universally for all waters. The choice will be dependent upon the water quality and upstream processes, filtered water quality objectives, cleaning method, filtration rate, length of filter runs, etc. In graded sand filter the bed depth typically comprises 0.7 m of 0.6–1.18 mm fine sand (effective size* 0.75 mm), 0.1 m of 1.18 to 2.8 mm coarse sand, 0.1 m of 2.36 to 4.75 mm fine gravel and 0.15 m of 6.7–13.2 mm coarse gravel. For applications requiring a finer sand the two upper layers are changed to 0.7 m of 0.5–1.0 mm sand (effective size 0.55 mm), 0.1 m of 1.0–2.0 mm coarse sand; the gravel layers remaining the same. Depending on the slot size of the nozzles the bottom gravel layer can be omitted and replaced by more of the adjoining media. The homogenous sand filter has a 0.9–1 m deep bed of 0.8–2 mm of sand (effective size 0.9 mm) placed on a 50 mm layer of 4–8 mm gravel. The stated size ranges for sand and gravel are generally 5 and 95 percentiles. For estimating the sand depth some employ the rule that the depth of sand should be $\geq$1000 times its effective size.[28] Some filter plant designers use the term 'hydraulic size' in place of effective size.[79] It is defined as the size particles would have to be, if all were the same size, in order to match the surface area of a sample covering a range of sizes.

---

*Effective size = size of aperture through which 10% by weight of sand passes.

The sand should be of the quartz grade with a specific gravity in the range 2.6–2.7. The uniformity coefficient* should be less than 1.6 and usually lie between 1.3 and 1.4. Loss in weight on ignition at 450°C should be <2% and the loss in weight on acid washing (20% v/v hydrochloric acid for 24 hours at 20°C) should be <2%. The sand should not be too friable to ensure that washing operations do not produce fines. It should therefore be tested for friability.[80]

Underdrain system

The filter media is placed on a collector system; its functions are to collect water from underside of the bed in an even manner and to spread air and water uniformly through the bed during cleaning. There is a choice of collector system design. One system comprises nozzles set in PVC pipe laterals, the spacing between laterals being infilled with concrete (see Fig. 7.13). The design can only be adapted to apply air and water separately during cleaning and therefore finds application in dual or triple media or granular activated carbon filters. In another design nozzles are set in a reinforced concrete false floor with a

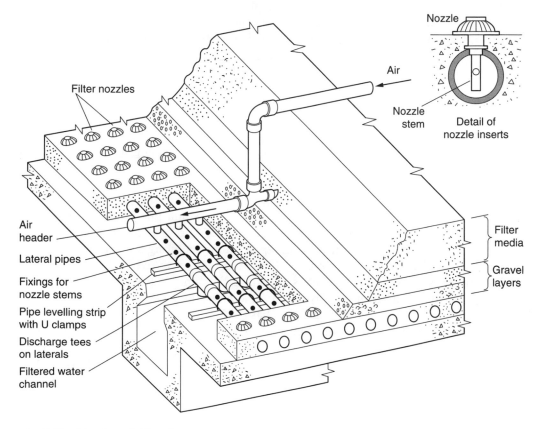

Fig. 7.13   Pipe lateral filter floor arrangement (Paterson Candy Ltd).

$$*\text{Uniformity coefficient} = \frac{\text{size of aperture through which 60\% sand passes}}{\text{size of aperture through which 10\% sand passes}} \text{ (by weight)}$$

space below (see Fig. 7.14 and Plate 17). The floor is either constructed *in situ* on plastic formwork or made up of pre-cast concrete slabs supported on concrete sills. The design allows water and air to be applied simultaneously or separately and the system provides for better distribution of air and water than the pipe lateral systems. In both these designs the gravel layers underneath the sand ensure uniform distribution of water and air with nozzles being in the gravel layer to minimise the risk of sand penetrating the nozzle which usually have slot sizes in the range 0.3–0.5 mm. Nozzle density depends on the type of nozzles and is about 40 nozzles/m^2. There are several other systems, mostly of proprietary designs, successfully used in many parts of the world.

Filter configuration

The overall number and size of filters vary. The number of filters is selected to minimise the effect of removing a filter from service for washing on remaining filters. Therefore the larger the number the better it is, as ideally it should be possible to take three filters out of service simultaneously (one draining down, one washing and one for maintenance). A minimum of six filters is desirable, although four filters may be used provided they are low rated or plant throughput can be reduced during maintenance of a filter. The limiting factors for size are the uniform collection of filtered water, even distribution of washwater and air, and the travel length of washwater to the collection channel during washing. Usually filter sizes vary from 25 to 100 m^2 with lengths in the range 8–20 m and widths 3–5 m. The washwater collection channel is located on one side along the length of the filter.

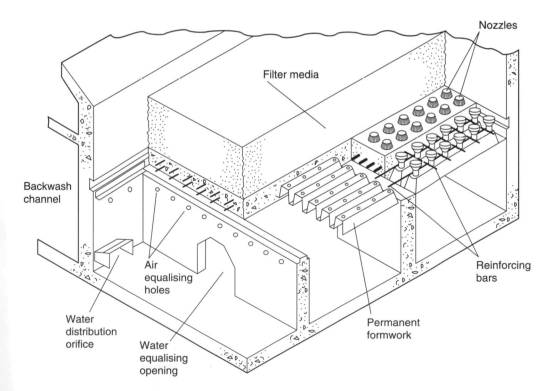

Fig. 7.14 Plenum filter floor (Paterson Candy Ltd).

Filter beds up to twice these sizes can be constructed by providing two identical beds separated by the washwater collection channel, thus limiting the travel length to 5 m. For some special applications, washwater channels above the filter bed (the normal US practice) are used (see Section 7.37).

Filtration rates

Filtration rates are selected on the basis of the application. Filters with deep bed coarse homogenous sand for iron removal are rated at 6–7.5 m^3/h.m^2 and for manganese removal at about 15–18 m^3/h.m^2. When used downstream of clarifiers coarse homogenous sand filters are rated at about 7.5–10 m^3/h.m^2 with the higher rate being used when water upstream of the clarifiers is treated by a combination of a coagulant and a polymer. At filtration rates above 15 m^3/h.m^2 the quality of filtrate tends to deteriorate and at rates in excess of about 20 m^3/h.m^2 the rate of headloss development becomes too rapid. Shallow bed, graded sand filters are usually rated at about 75% of the rates for deep bed, coarse homogenous sand filters. The rates achieved with multi-media filters are similar to those achieved with deep bed coarse homogenous sand filters (see Section 7.30). The recent concern over *Cryptosporidium* oocysts and *Giardia* cysts in raw water supplies has led to a reduction in filtration rates to about 6–7 m^3/h.m^2 to minimise the risk of particulate breakthrough.

Head losses, air binding and negative heads

In the downflow filter design with upflow washing it is usual for the filter to operate with about 1.5–2 m or more of water over the bed; however, there are some proprietary designs which operate with a much smaller depth of water (down to 0.5 m) and even with negative head conditions, but the latter is not regarded as good practice as difficulties can occur with cracking and mud balling in the filter bed and air binding, especially with high filtration rates. The pressure distribution in a filter bed is illustrated in Fig. 7.15. Negative head can occur when head loss (total head loss less the clean media loss) at any depth exceeds the static head (water depth) at that point. The point of negative head development varies with the filter media; nearer the media surface for graded sand, about one-third way down the media in coarse homogenous sand and just below the anthracite layer in dual media filters. Under negative head conditions dissolved gases in water are released into the space between sand grains restricting the water flow, increasing the head loss and prematurely terminating filter runs. It can also result in poor filtrate quality when air binding is restricted to part of the filter and due to channels formed by escaping gases. Because of greater solubility of gases in water at low temperature negative head is particularly a problem in filters treating cold waters, surface waters or well-aerated ground waters.

Negative head can be overcome by arranging the filter outlet to discharge at or above the top of the media level. The other option is to provide sufficient water depth above the top of the media or wash the filters at a head loss less than the static head down to the point of negative head development. The head loss in a clean filter is made up of clean media and underdrain losses which is usually less than 0.25 m. Therefore when a filter is returned to service after washing the loss of head through the bed and underdrain system should be less than 0.3 m. The rate of head loss development is a function of solids retention capacity of the filter and is lower for coarse homogenous sand filters than for

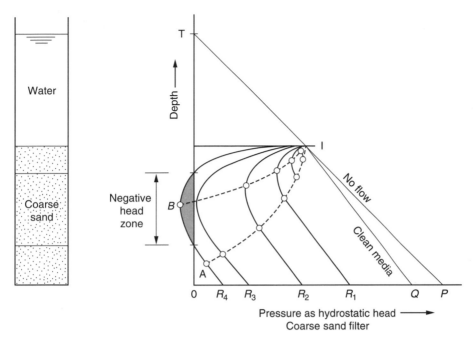

Fig. 7.15 Pressure distribution in a rapid gravity filter. Scales on X and Y axes are the same, hence pressure OP at floor level equals depth TO. Points A – slope of pressure curve is equal to the clean media curve. Points B – lowest water pressure during filtration. $IR_{1 \text{ to } 4}$ – pressure curves during clogging. PQ – initial head loss. PR_3 – maximum permissible operating head loss.

graded sand filters; in the latter it is improved by the use of an anthracite layer (see Section 7.30). The head loss allowed for retention of suspended solids (clogging head) is usually in the range 1.5–1.8 m. Therefore filters are allowed to run to 1.8–2 m head loss.

Solids retention

The maximum solids retention capacity of a filter is a function of the voids which is approximately equal to 45%. In practice only about a quarter of this space is available for solids removal. It is reported that the solids retention capacity of a gravity filter is limited to 10 and 35 g dry solids/litre of voids for light hydroxide floc and suspended solids in river water, respectively.[51] For example therefore, a filter of 0.9 m sand bed and filtration rate of 7.5 m^3/h.m^2, washing every 24 hours cannot accept more than 5 mg/l of light hydroxide floc or 17.5 mg/l of suspended turbidity in the influent over the run length. The respective equivalent filter loadings are 900 and 3150 g/m^2 and is in general agreement with the range of 550–5500 g/m^2 quoted by Cleasby.[60]

Flow control

As the bed becomes clogged the head loss through it increases resulting in the lowering of the filtration rate. For best filter performance it is desirable to have a constant filtration rate and any changes in filtration rate, such as when removing a filter for washing, should be as smooth as possible. Therefore control systems which divide flow equally between filters and allow filtration without fluctuations in rate are essential for good filtration results. Equal flow is achieved by using weirs to proportion the flow equally to all filters or it may be achieved by sizing the outlet pipework and valves of each filter to limit the maximum flow hydraulically. In such cases, after a filter is backwashed, the level of water in the filter box will rise to such a level above the outlet head on the filter so that it is sufficient to overcome the headloss through a clean filter and its underdrainage system. As the sand becomes progressively clogged during a filter run, increasing the head loss through the filter, so the water level in the filter box rises to the maximum possible. The simplicity of the rising level filter design is attractive and has many advantages,[81] but it suffers from the absence of any flow control on the outlet valve at start-up after a filter wash. At this time it is desirable to have a reduced rate of flow through the bed for a while because the full output through the newly cleansed sand can result in a temporary breakthrough of turbidity (see Section 7.27).

If an outlet controller is fitted, it modulates the outlet control valve so as to maintain a flow which does not exceed a permitted maximum. As the bed becomes clogged and the headloss through it increases, the controller opens the outlet valve to keep the output constant. There are also mechanical, electronic, electrical, or pneumatic linkages from this controller back to the filter, so arranged as to maintain a substantially constant water level during the course of a filter run. The principal reason for an outlet controller is to maintain a predetermined constant output from each filter under varying headloss conditions, the amount being such that the total inflow to the filtration plant is divided equally among all operating filters. A drawback with flow division on the inlet using weirs is that the free fall can break the fragile floc into small fragments which may not be effectively removed in the filters. In flow division based on outlet flow measurement this is overcome by having submerged inlets with common water level in all filters. A flowmeter in each filter outlet monitors the flow and modulates the outlet control valve to achieve the required flow which is derived by dividing the total flow by the number of filters in service. The constant level is maintained as before except that the level measurement could be made in the inlet channel because all filters have a common level. One of the drawbacks of constant rate filters is turbidity breakthrough occurs towards the end of the filter run. This is less likely in declining rate filters where the filtration rate decreases as the headloss develops.

Declining rate filters are based on a simple design.[49,82,83] The system is best suited for a group of six or more filters so that the additional flow to be shared when one filter is taken out of service for washing is not excessive. It is reasonable to design on the basis of a maximum flow range through each filter of ±35% of the average filtration rate, the average being taken as the total output divided by the total sand bed area provided. The inlet valve or penstock to each filter is usually submerged and, to restrict the filtration rate to the maximum, it is necessary to install some form of restricting orifice or valve on the outlet. Filters are washed in a fixed sequence and individual filter instrumentation for loss of head or quality of filtrate, i.e. turbidity, is used only to detect a filter whose behaviour is out of line with the rest for some reason. In terms of hardware the system is very simple and this is its chief merit; claims of lower head loss, longer filter runs, improved quality, and a less costly installation compared with other systems need to be critically examined as there may be doubt as to whether they are always obtained in practice. A good detailed analysis of rapid gravity filters and their hydraulics is given by Stevenson.[84]

7.27 Backwashing

Rapid gravity filters employing graded sand are washed by separate use of air and water through the bed by reverse flow. The first operation is to allow the filter to drain down until the water lies few centimetres above the top of the bed. Air is then introduced through the collector system at a rate of about 6.5–7.5 mm/s. The surface of the sand should show an even spreading of bursting air bubbles coming through the sand. The water over the bed quickly becomes very dirty as the air-agitated sand breaks up surface scum and dirt is loosened from the surface of the sand grains. This is followed by an upward flow of water at a carefully selected velocity to expand and fluidise the bed. Under this condition the voids between grains of sands are increased and resulting rotation of grains and consequent attrition between grains produces a scouring action to remove attached deposits. The wash rate should be just sufficient to achieve fluidisation velocity (incipient fluidisation) with little bed expansion. Increasing the backwash rate beyond this state will be counter productive because as the distance between sand grains increases, the scouring action will be reduced. High backwash rates may result in loss of sand and wastage of water and energy.

In the UK the practice is to use wash rates to produce 1–3% bed expansion. The rates are viscosity dependent and therefore are affected by water temperature, with higher rates used at warm water temperatures. Typical wash rates in mm/s used for sand at varying water temperature to give about 2.0% bed expansion are given in Table 7.9.

An empirical relationship has been developed to express bed expansion of graded media with temperature[85] which is the ratio of expansions at $T°C$ to $20°C$ is 1.57 exp $(-0.023T)$:1.

Applying this equation to the UK conditions it can be shown that if summer wash rates are used throughout the year, a 40% increase in the degree of expansion is shown in the winter. This would lead to wastage of water and could result in loss of sand. Therefore, in countries where water temperature varies, facilities for seasonal adjustment of wash rates are advisable.

The filters comprising deep bed coarse homogenous sand rely upon the application of air and water together in the wash phase, followed by a water rinse. In both phases the water rate is well below the fluidisation velocity and does not cause the bed to expand. This prevents hydraulic grading and maintains the homogeneity of the filter bed. The air rate is 16 mm/s of free air. The water rate in the wash phase is usually 2 mm/s and that in the rinse phase is 4 mm/s. Although some designers use a water rate of 4–5 mm/s in both the wash and rinse phases. In the combined air-water wash method, wash rates are not influenced by temperature. The duration of the wash phases depends on the method of wash and filter influent quality. For designs with air and water applied separately, air

Table 7.9 Wash rates (mm/s) for sand filters to give 2% bed expansion at varying water temperatures

Sand size range (mm)	Effective size (mm)	Water temperature (°C)					
		5	10	15	20	25	30
0.5–1.0	0.55	3.1	3.5	4	4.5	5	5.5
0.6–1.18	0.75	4.4	5.0	5.6	6.3	6.9	7.5

scour lasts about 3–4 minutes and the water wash lasts about 4–6 minutes; and for designs with the air and water applied concurrently, air is first introduced and after about 1.5–2 minutes to allow the air flow to become established, water is introduced and the combined air–water wash proceeds for about 6–8 minutes; the air flow is stopped while the water flow continues to rinse the bed for another 8–10 minutes. However the total period a filter remains off-line for washing is about 30–45 minutes which includes about 15–30 minutes for draining the filter down. The total water consumption per wash amounts to about 2.5 bed volumes.

It is sometimes very difficult to evacuate the whole of the dirty backwash water from a bed before the filter is refilled and put to use. In order to reduce the extent of this problem the filter influent (clarified water) is allowed into the filter and caused to flow across the top of the bed from the end remote from the wash water collection channel during the last stages of backwashing in order to flush the dirty water to waste from above the bed. This stage of the wash is called 'surface-flush' or 'cross-wash' and forms an essential part of the wash sequence and increases the washwater consumption to about 3 bed volumes. In one filter design the filter is washed as usual, but the used washwater is contained above the media by allowing the level to rise. At the end of the wash the used washwater is rapidly discharged to waste. An advantage of the design is high wash rates may be applied without fear of media loss. Typical free air and water rates are in the range 14–22 mm/s and 10–18 mm/s, respectively and can be applied either concurrently or separately. The water usage is about 1–2.5% of works throughput.

When the filter is returned to service after backwashing there is a short period, lasting about 15–60 minutes, when the filtrate turbidity is high. This is due to the displacement of residual washwater containing solids loosened from the bed during backwashing and also the lower solids removal efficiency of the freshly washed media. The options available for reducing the risk of this effect are either to return the first 15–60 minutes of filtrate to waste (or usually the works inlet), or to allow the filter to stand for up to about 30 minutes (so-called 'delayed start') or to allow time for the filter to 'ripen' by starting filtration at a slow rate (so-called 'slow start') or a combination of them. All of these features can be operated automatically as part of the wash sequence provided that the filters are appropriately designed.

Water used for backwashing should be filtered and preferably chlorinated. In works using aluminium coagulants it is best taken upstream of any final pH correction to minimise the risk of dissolving aluminium hydroxide floc retained in the filter. The total amount of washwater used has an important bearing upon the economy of a treatment works, especially in relation to the net yield of a source. The total washwater used should normally not exceed 2% of the treated water output and should preferably be less.

7.28   Operation of filters

Figure 7.16 illustrates typical relationships which exists between output, time, turbidity of effluent, and headloss during the course of a filter run for both constant flow and declining flow operation. The breakthrough of turbidity on start-up of a filter after backwashing is quite usual (see Fig. 7.16b), even though precautions may be taken to effect a 'slow start'. The length of a filter run between backwashings varies according to headloss build-up; Fig. 7.16(c) shows a typical headloss build-up for a sand–anthracite filter. The relatively

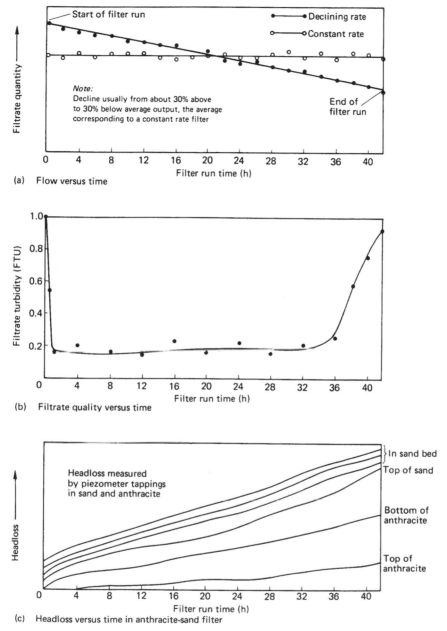

Fig. 7.16 Typical output characteristics of a rapid gravity filter.

high loss through the anthracite where the greater part of the dirt is retained and across the anthracite–sand interface should be noted.

When the maximum permitted headloss has been reached the filter run is terminated. It is usual to aim for a washing frequency not greater than once every 24 hours. The length of the run typically varies from about 24–60 hours. Long run lengths help to make savings on backwash water, but encourage bacterial growth in the filter bed, in particular if the raw water contains organic matter and is not treated with a disinfectant upstream. It is therefore desirable to restrict run lengths to 48 hours in warm water temperatures and to 60 hours in cold water temperatures. At the other end of the time scale, it can prove difficult to maintain full output from a works if each filter has to be washed more than about once every 6 hours. The sequence of backwashing including the draindown usually takes about 30–45 minutes, so that for a bank of filters requiring to be washed less than every 24 hours it could prove necessary for one filter to be draining whilst another is being backwashed. In difficult circumstances, when filter runs are as short as 6–12 hours, it is often the practice to discharge the total contents of the filter to waste instead of draining to supply in order to speed up the wash cycle: this is sometimes referred to as 'dumping' the filter.

Turbidity breakthrough is another criterion used in the termination of a filter run. The primary indications that a filter requires backwashing are duration of filter run, loss of head, or reduction in filtrate quality measured by turbidity. In most installations all three parameters are monitored and can be used to initiate the washing cycle automatically and eventually return the filters to service. More often than not, only the first two parameters are used for automating the start of a filter washing cycle. It is useful to monitor filtered water turbidity in total and from each filter and initiate an alarm on high value. The individual filter turbidity helps to identify filter breakthrough which may be used to give early warning of *Cryptosporidium* oocysts in the treated water, although particle counters on individual filter outlets would be more appropriate. In automatic filter plants manual control of washing should always be possible as an alternative so that special cleansing measures can be taken when necessary.

For air scouring, compressors working in conjunction with air storage cylinders or Roots type positive displacement blowers are necessary. Air is usually applied at about 0.35 bar pressure at the air inlet valve. For backwashing it is usually most economic to use gravity flow from a large elevated storage tank, since the rate of flow required is large and electrical demand charges are minimised by keeping such a tank topped up by a relatively small, continuously running pump drawing from the filtered water supply. The tank should have two compartments with each sized for at least one filter wash. A drawback with the system is that unless rate controllers are incorporated, the backwash rate decreases with the head in the tank. The pumps used to introduce water directly into filters are usually of the centrifugal type. The head required at the wash water inlet valve of the filter is about 5 m.

Flowmeters on backwash flow are important and these should have control facilities for setting the optimum rate, changing the rate between wash and rinse phases and, between winter and summer periods. On large filters or filters designed for automatic washing power assisted actuators to open and close a valve become necessary. Electric power is often used for operating the actuators but, pneumatically operated actuators are also successfully employed. In temperate and warm climates it is unnecessary to cover rapid gravity filters, but in very cold climates covering is required to prevent freezing. Once filtered, water should not be exposed to contamination and therefore all chambers and channels downstream should be closed. The water from a rapid gravity filter is not completely bacteriologically pure and, before the water passes into supply, it must be disinfected by the addition of chlorine or other disinfecting agent.

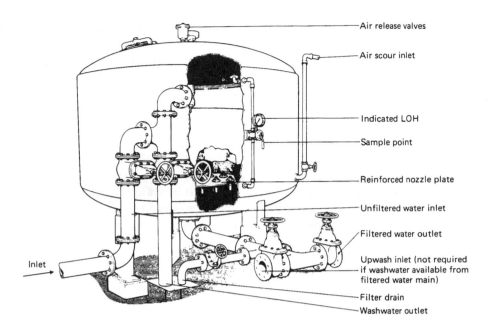

Air release valves

Air scour inlet

Indicated LOH

Sample point

Reinforced nozzle plate

Unfiltered water inlet

Filtered water outlet

Upwash inlet (not required if washwater available from filtered water main)

Filter drain

Washwater outlet

Inlet

Fig. 7.17   Sectional view of a pressure filter (Paterson Candy Ltd).

7.29   Construction and operation of pressure filters

Pressure filters are similar in bed construction to open rapid gravity filters, except that they are contained in a steel pressure vessel (see Fig. 7.17). Perforated pipes or a steel plate with nozzles are used for collecting the filtered water and for distribution of the washwater and air scour. The steel pressure vessel is cylindrical, arranged horizontally or vertically. With a pipe lateral underdrain system the bottom of the vessel is usually filled with concrete so as to obtain a flat base. In the plenum floor design a steel plate into which nozzles are screwed, is used. In a horizontal vessel, vertical plates are welded inside to give a rectangular shaped sand bed within the cylinder so that the bed may be washed evenly and there are no 'dead' areas beneath which air scour and water pipes cannot be placed. Ideally, the top of the sand in a horizontal vessel should coincide with the horizontal diameter to give the maximum sand surface area. The whole of the cylinder is kept filled with water under pressure and at the highest point an air release valve is inserted for the release of trapped air. For practical reasons the maximum diameter of vertical filters is limited to about 2.75 m; in the case of horizontal filters, to the same diameter and to a length of about 12 m.

The backwashing of such filters is very similar to that of an open rapid gravity filter. A bellmouth and pipe can be used for the removal of dirty washwater in a vertical filter; for most horizontal filters a single vertical plate located near to one of the dished ends will facilitate washwater removal, but for the larger filters, a central washout channel formed by two vertical plates is necessary.

The advantage of pressure filters is that the pressure of water in the mains is not lost when the filtration process takes place, as is the case with an open rapid gravity filter system. About 3 m head may be lost in friction through the sand bed and the inlet and outlet fittings which includes a head loss of about 1.5–1.8 m for retention of suspended solids. This combined head loss is between the common inlet and outlet bus mains serving a battery of filters. Pressure filters may be interposed on a pumping line or a gravitational line without a large loss of pressure on the supply. Air binding hence negative head rarely occurs in pressure filters. Pressure filters are suitable for direct filtration duties and are commonly used for iron and manganese removal in ground waters, and treatment of stored waters.

Pressure filters suffer from the disadvantage that the state of the bed under backwashing conditions and when the plant is working cannot be directly observed. It is of vital importance, therefore, that every pressure filter is fitted with an open box or dish in the front of it, into which the washwater is turned so that at least any washing out of the sand may be observed and the backwash rate immediately reduced.

When coagulation or other chemical treatment is required chemicals are injected and mixed under pressure and flocculation must be hydraulically carried out in pressure vessels fitted with baffles. The same applies when a contact tank is needed on the downstream side for disinfection.

In marked contrast to rapid gravity filters pressure filter installations are not normally provided with any form of flow control on each filter. It is not often appreciated that without this flow control each filter in a battery of pressure filters operates as a declining rate filter for at least part of its filtration cycle. Thus the only restriction to output through a clean, newly washed filter is the hydraulic constraint imposed by the size of pipework at the inlet and outlet. Sometimes an orifice plate is installed in the outlet. If an outlet controller is fitted it modulates the outlet control valve, operating in conjunction with a flowmeter on each filter to maintain an average flow equal to the total flow divided by the number of filters in service. As the head loss through the filters increases the outlet valves open to maintain the average flow. When employing flow controllers, head loss can be monitored on individual filters. The filter run can be terminated either on headloss, length of filter run or turbidity breakthrough.

The equipment required for air scouring pressure filters is similar to that used for rapid gravity filters, and the same applies to the type of equipment provided if power assisted opening and closing of filter valves is called for. For backwashing, as an alternative to direct use of pumps or overhead tank, a system is often devised to take advantage of the water pressure available from a battery of filters. For a filter washed by separate air and water the required rate of application of water is about four times as great as the rate of filtration which is about 5–6 m/h. Thus in a large filter battery of about 15 filters, groups of five filters can be taken out of service at a time and the combined filtrate from four of the filters can be used to wash the fifth filter, and so on, until all filters in the group are washed. In large pressure filter plants arrangements are usually made to wash filters in groups at a specific time each day. Monitoring of individual filters for loss of head is seldom done except when outlet flow control is used. In fact, because a large battery of filters is usually supplied by a common inlet bus main, and the outlets from filters also connect to a common outlet bus main, it is only meaningful therefore to measure the headloss across the battery of filters. Individual filtrates could usefully be monitored for turbidity.

The steel shells require careful maintenance so as to prevent both internal and external corrosion. Condensation upon the outside of the tanks is a continual nuisance as it will cause corrosion of the steel shell and staining of the floor below. The pressure applied to steel pressure filters is not usually in excess of 80 m head of water. This should normally be adequate for most distribution systems. Above this pressure the thickness of plates used for the steel shell may be so great that cost rises rapidly. Due to the limitations imposed on capacity and pretreatment, their application over rapid gravity filters is restricted to small plants which are required to be installed into a pumping system without breaking the hydraulic gradient and for iron and manganese removal, primarily in ground waters. More recently they are being used as GAC adsorbers for organics removal.

Multilayer and other methods of filtration
7.30 Use of anthracite media

As referred to previously in Section 7.26, the most efficient form of media grading for a rapid gravity downflow filter in order to obtain maximum capture of solids would be to have the sand decreasing in size in the direction of flow. Clearly this is not possible because hydraulic regrading takes place during backwash, so that the finer sand collects at the surface of the bed. This can be countered by using separate layers of different filter materials having different density and grain size, the denser materials being at the bottom of the bed and the less dense at the top. One type of multilayer filter bed in wide use at present is the two-layer filter using anthracite over sand. The specific gravity of anthracite (1.4–1.45) is lower than that of sand (2.6–2.7). It has been found that the filtrate quality from anthracite–sand can be as good as that from conventional sand only filtration, but because of in-depth clogging of anthracite and therefore its capacity to retain solids throughout its entire depth and the less rapid development of headloss, filter runs can be 1.5–3 times longer. A layer of anthracite is usually incorporated in a filter to extend filter runs. They are usually operated at filtration rates similar to those used on coarse homogenous sand filters although high filtration rates, even up to 15 m^3/h.m^2 can be achieved but, at the cost of short filter runs; a summary of the effects has been given by Miller.[86] The size of the anthracite used is usually 1.25–2.50 mm (effective size 1.3 mm) or 1.4–2.5 mm (effective size 1.6 mm) and uniformity coefficient <1.5. This is usually placed over 0.5–1.0 mm (effective size 0.55 m) but sometimes on 0.6–1.18 mm (effective size 0.75 mm) sand. Larger size anthracite has been tried, but variable results have been reported.[87] Typically the depth of the anthracite bed is of the order of 0.15–0.3 m with sufficient sand to give a combined depth of 0.75 m. Deeper anthracite layers (up to 0.5 m) can be used to give more capacity but with a corresponding increase in the depth of the sand layer (up to 0.6 m) to ensure good filtrate quality throughout the filter run. The sand should be supported on layers of gravel, the same as for graded sand filters (see Section 7.26). For plenum floor design a shallow gravel layer (100 mm of 4–8 mm) is normally used. Sources of anthracite (carbon content at least 90%) are limited to a few countries in the world. Therefore there is a tendency to use high grade bituminous coal (carbon content at least 80%) in place of anthracite. This is generally acceptable provided it is non-friable and can meet the standards for friability and acid solubility (see Section 7.26) and hardness.[80,88–90]

Beds of anthracite can be expanded or fluidised at about the same backwash rate as a sand bed when the size of the anthracite grains is about 1.5–3 times that of the sand grains. Backwash rates are used to give 10–15% expansion of the sand–anthracite bed with an occasional wash at higher rate to give about 30% bed expansion for regrading the media. The rates are viscosity dependent and therefore affected by the water temperature. Wash

Table 7.10   Wash rates (mm/s) for anthracite–sand filters at varying water temperatures

Bed expansion %	Water temperature (°C)					
	5	10	15	20	25	30
10	4.5	5.3	6.2	7.2	8.0	8.9
15	6.5	7.5	8.3	9.2	10.0	10.7
30	11.4	12.1	12.8	13.5	14.3	15.3

rates required for a bed of anthracite (effective size 1.3 mm) and sand (effective size 0.55 or 0.75 mm) are given in Table 7.10.

Filters are cleaned by air scour followed by water backwash. Air scour rate applied is about 8–12 mm/s. Simultaneous application of air and water in a conventional filter normally used for coarse homogenous sand filters is not suitable because the washwater will carry the lighter anthracite to waste. However this can be overcome by providing the filter with high level suspended washwater collection troughs in place of the conventional low level collection channel. The sequence consists of simultaneous application of air at 16 mm/s and water at about 2 mm/s as for coarse homogenous sand filters; air is turned off before the water level reaches the cill of the trough weir and the wash rate is then increased to about 16 mm/s for rinsing and regrading the media. The troughs are usually placed with invert level about 600 mm above the unexpanded media surface.

After backwashing, the two-layer bed settles down again with the anthracite on top (subject to a relatively small amount of mixing). Backwashing at fluidisation velocities results in the regrading each of the filtering layers by grain size, in particular with anthracite layer since it is heterogenous and fragile.

7.31   Use of anthracite to uprate filters

In recent years advantage has been taken of these higher filtration rates to uprate the output of existing filter plants by changing their media from sand to anthracite–sand, the previously conventional filtration rates of 4–6 m^3/h.m^2 for graded sand filters being increased to rates of 6–12 m^3/h.m^2. Increases of plant output capacity in the range 33–50%, and in some cases 100%, have been achieved. However, these increases have frequently been associated with the improvement of floc characteristics by the addition of a polyelectrolyte as a coagulant aid. It is important to pay attention to floc size when adopting anthracite filters: too fine a floc may pass through the anthracite layer and cause too large a load to reach sand below; too large a floc in relation to the anthracite size may place too large a load on to the anthracite, defeating the object of gaining filtration in depth by using anthracite and sand. Since anthracite is an expensive material compared to sand it is important to ensure a filter is adequately designed hydraulically before anthracite is used for uprating. This means checking that the inlet, outlet, and flow control pipes and valves can accept the higher flow rates; that the distribution of water at the inlet to the filter does not cause excessive scouring of the anthracite–sand bed because of the higher flow rate; that the higher backwash rate required does not result in loss of anthracite over the washwater weir; that the washwater discharge channel is of sufficient size and gradient to accept the increased backwash without backing up, and the design of the filter underdrain system can accept higher upthrusts.

With a water which produces a fragile floc at the sedimentation stage it may be necessary to add a dose of polyelectrolyte to the water before it passes on to the anthracite–sand beds in order to prevent excessive penetration of floc into the bed. The same precaution may be necessary to prevent the penetration of algae, as reported at the Iver treatment plant[91] where polyelectrolyte was used to arrest penetration through the anthracite and sand by small green algae, dominated by minute species *Nannochloris* and *Ankistrodesmus* with cell diameters of 4–8 microns.

7.32 Granular activated carbon adsorbers

Granular activated carbon (GAC) adsorbers employed for organics removal (see Chapter 8) are similar to sand filters of the rapid gravity or pressure type except that gravity type adsorbers are usually enclosed in a building or fully covered. Some GAC adsorbers contain two beds separated by the washout channel and operating in series; water flows up one and down the other. When the first bed is exhausted the partly spent carbon from the second bed is transferred to the first bed, and freshly reactivated carbon is added to the second bed. The flow rates, wash rates and wash methods are similar to conventional adsorbers. The principal advantage of the design is that the GAC is more efficiently utilised.

GAC is a good filter medium as well and therefore can be used on its own for filtration of turbidity. At the 160 Ml/day Iver treatment works in Buckinghamshire, UK, GAC is used as the primary filtration medium. GAC should be reactivated as and when it is exhausted with respect to organic compounds. Filtration rates vary from about 6–7.5 $m^3/$h.m^2 for filtration and up to 15 m^3/h.m^2 for adsorption. Media depth is a function of the empty bed contact time which could vary between 5–30 minutes; for filtration only duties the media depth is about 1 1.2 m while depths about 2 m (for gravity filters) and 3 m (for pressure filters) are used for adsorption. Effective size of GAC varies with the type and the application and usually lies in the range 0.6–1.1 mm. GAC should not be friable and should be tested for friability.[80] Water soluble ash should be less than 1% w/w. For GAC activated by phosphoric acid the phosphate content should not exceed 1% w/w.

GAC is placed in the adsorber on 50 mm of gravel or directly on appropriate filter nozzles. Since the adsorbers are washed by the sequential application of air and water most types of underdrain systems would be suitable. GAC adsorbers are washed using air at the rate of 14 mm/s applied at 0.35 bar followed by water at the rate of about 5–12 mm/s depending on the effective size, base material of the GAC (i.e. coal, wood, peat or coconut) and water temperature, to give 20–30% expansion of the carbon bed. Wash rates for different grades of a coal-based GAC at varying water temperatures to give 20 and 30% bed expansion are given in Table 7.11.

When reactivation of the carbon is required it is usually removed from the adsorber by means of a water operated eductor or by recessed impeller centrifugal pumps of rotational speed less than 1000 rpm. In some designs adsorbers are provided with a sloping floor or recessed drain in the floor discharging to a collector system. GAC should be removed as a 20%v/v (10% w/w) slurry and pipeline velocities should be maintained within 1.5–2.0 m/s. All pipework, in particular bends, should be in stainless steel. Straight lengths could be in ABS or uPVC. Bend radii should be 5–10 pipe diameters. The same equipment and design parameters should be used for carbon placement in adsorbers.

Table 7.11 Wash rates (mm/s) for GAC to give 20% and 30% bed expansions at varying water temperature (Source: Chemviron Carbon Ltd, UK)

Grade	Iodine No.[a]	BET surface area (m²/g)[b]	GAC size range (mm)	Effective size range (mm)	Water temperature (°C)					
					5	10	15	20	25	30
F200	850	900	0.425–1.70	0.6–0.7	4.5 (5.6)	5.0 (6.4)	5.8 (7.2)	6.3 (7.8)	6.7 (8.3)	6.9 (8.6)
F300	950	1000	0.600–2.36	0.8–1.0	6.4 (8.1)	6.9 (8.9)	7.8 (10.0)	8.5 (10.3)	9.0 (11.4)	9.3 (11.5)
F400	1050	1100	0.425–1.70	0.6–0.7	3.9 (5.0)	4.7 (6.0)	5.3 (6.7)	5.8 (7.2)	6.4 (8.1)	6.5 (8.2)
TL830	1050	1050	0.85–2.00	0.9–1.1	6.9 (8.9)	7.8 (9.7)	8.6 (10.3)	8.9 (10.8)	9.5 (11.7)	9.7 (11.8)

Values for 30% bed expansion shown in brackets.

[a] Iodine number: It indicates a GAC's ability to adsorb organic compounds and be regenerated. It should be greater than 500 mg/g of carbon (see AWWA Standard for GAC, AWWA B604–90).

[b] BET surface area: It indicates the surface area available for adsorbates in water. Measured by N_2-BET method (Brunauer S., Emmet P. H. and Teller E., Adsorption of Gases in Multimolecular Layers. *J Am Chem Soc*, **60**, 1938, pp. 309–319).

Virgin and reactivated GAC contain contaminants which would leach out into the filtrate when first placed in adsorbers. They include sulphides, sulphites and bisulphites (causing chlorine demand and odours), alkali (resulting in high pH), phosphates (if phosphoric acid is used in the activation process) and metals such as aluminium, iron, manganese and copper. Repeated backwashing with water followed by running to waste of the filtrate should be carried out until tests confirm that water is of acceptable quality for supply. The impact of reactivated GAC on water quality could be minimised by pre- and post-acid wash in the reactivation process.

7.33   Upward flow filtration

Upward flow filtration with upflow washing has been used for a few potable water treatment plants in the UK, but its use is more appropriate to industrial water applications, as roughing filters ahead of slow sand filters or to tertiary sewage filtration where a high standard of filtrate quality is not so important. The principle used in upflow filters is to have progressively finer sand in the direction of flow, which allows the filter to carry a greater load of impurity before backwashing because the larger particles tend to be held in the lower, coarser part of the filter, leaving the upper layers to deal with the smaller particles. However, unless the finer grades of sand are restrained they would be washed away at higher rates of filtration, as well as during the backwashing stage, and consequently designers have introduced a number of techniques to stop this occurring. One method is to use a filtrate collector pipe system just buried in the top layer of fine sand, with strainers located on the side of the pipes so that filtrate water flow has to change from a vertical to a horizontal direction, thus preventing expansion of the sand. During backwashing the filtrate collector is not used and dirty washwater escapes from an elevated trough. The design of upflow filter most commonly used in the UK is one which contains a grid, square in section, located about 0.1 m below the surface of the sand. During filtration the sand arches between individual members of the grid and prevents expansion of the sand, whilst during backwashing the arches are intentionally broken by successive applications of air and backwash water. Piped lateral floors with large orifice nozzles are used to distribute the incoming water. Screening of the raw water to remove leaves and other debris is essential to prevent blockages.

Claims have been made that upflow filters can be operated at filtration rates greatly in excess of conventional downflow rates: in practice the rates are not greatly or always in excess of what can be achieved with multilayer filtration (see Section 7.30).

7.34   Direct filtration

Some surface waters can be treated by coagulation, flocculation and rapid filtration (gravity or pressure), eliminating clarification. Such waters need to be carefully selected and pilot tested. In general a water source is considered to be suitable for direct filtration when turbidity and colour values are less than 10 NTU and 25° Hazen, respectively, with peaks of 40 NTU and 40° Hazen for periods less than 24 hours. The total organic carbon value should be less than about 2 mg/l as it influences the coagulant requirement. The coagulant dose should be no more than 1 mg/l as Al or 1.5 mg/l as Fe, although higher doses for short periods are acceptable. A polyelectrolyte may be used as a coagulant aid. Algae, both the filter clogging and passing types can cause problems such as shortened filter runs, if numbers are high;[92] an upper limit of 2000 asu/ml* for diatoms is reported.[93] The total flocculated solids load on to the filters should be limited to about 20–25 mg/l, with short-term peaks up to about 60 mg/l. Direct filtration operates well on micro-flocs; flocculation requirements are given in Section 7.13. The filters used are either monograde sand (ES 0.9 mm) deep bed (0.9–1.0 m) or anthracite–sand containing 0.3 to 0.4 m anthracite (ES 1.3 mm) and 0.6 m graded sand (ES 0.55 mm) (see Sections 7.26 and 7.30). Filtration rates should be maintained below about 7.5 m³/h.m². Plants provided with clarifiers treating water showing seasonal variations in quality may be provided with facilities to bypass the clarifiers when the water is suitable for direct filtration.

7.35   Filter problems

Filter problems are common in many of the older water treatment works. The problems are due to incorrect design (filtration and wash rates, media grading, flow control method, etc.) for the water to be treated, poor hydraulic design, use of unsuitable material, bad installation in particular of the underdrain system and mal-operation. These result in short filter runs, inefficient filter cleaning (dirty filters), very high starting head loss (up to 1 m) media loss, loss in capacity and shortfalls in filtered water quality and sometimes even ruptured filter underdrain systems. Installation problems include damaged or blocked nozzles, incorrect levelling of the floor or pipe laterals and nozzles (outside their tolerance limits), construction debris in the pipe laterals or the space below the floor and poor sealing of the floor slabs or pipe/duct joints. These flaws can be checked before placing the media by a hydraulic pressure test with all the nozzles plugged and by testing for uniform air distribution with about 150 mm of water in the filter sufficient to cover the nozzles. The observation of air scour pattern during backwash is a way of identifying underdrain problems in operating filters.

Blocked nozzles (usually the result of construction debris left in the underdrains) can result in high pressures in the underdrain system leading to its rupture. The risk of damage to the underdrain system can be minimised by incorporating a standpipe with free discharge in the washwater main. A pressure relief valve is not recommended. Damaged nozzles are a common problem with operating filters as this could occur when placing the media. Damaged nozzles allow sand into the underdrain system reducing its capacity and during washing they can cause channelling, sand boils, high localised velocities with sand ingress into support layers and gravel brought to the surface by jetting (see below) and loss of media. Sand boils and consequent upset of gravel layers can also be the result of sudden introduction of backwash water. Some channelling and sand leakage into the underdrain system could be attributed to incorrect sizing of sand and gravel media or use of nozzles with a slot size incompatible with the media size.

Deficient washing results in the build up of floc and organic matter (algal and detrital) ultimately leading to the formation of mud balls and jetting (build up of columns of support gravel through the media), cracks in the filter bed and bed shrinkage with media pulling away from the walls. Sometimes the presence of residual polyelectrolyte in the clarified water due to overuse can encourage mud ball formation. A good analysis of filter problems is given in References 49, 94 and 95. The problem of air binding is discussed in

---

*asu/ml = areal standard units/ml; 1 asu is 20 × 20 μm and for a medium-sized algae 1 asu/ml can be approximated to 0.1 μg/l of chlorophyll 'a'.

Section 7.26. It is important that filter washes are frequently witnessed to identify problems early to allow corrective measures to be taken. Observations to be made include uniformity of air distribution, undisturbed areas, sand boils, mud balls, and media carryover. The wash efficiency of a filter can be checked by taking core samples from several places in the bed after backwashing and analysing 250 ml of media from different depths in each sample for suspended solids by washing it thoroughly with water (up to 250 ml) and measuring the solids by volume using an Imhoff cone. For a good wash, it should be less than 2% v/v. An alternative criteria defined by Bauer[41] based on Thames Water experience is that the sand after backwashing should contain particulate organic carbon less than 0.4 g/l and suspended solids less than 2.4 g/l of filter medium.

7.36   Membrane filtration

Membrane filtration employs a semipermeable membrane to separate materials according to their physical and chemical properties when a pressure differential or electrical potential difference (electrodialysis – see Section 8.45) is applied across the membrane. Pressure driven processes can be broadly classified according to the membrane pore size and size of particles removed (see Fig. 7.18). They are microfiltration (MF), ultra-filtration (UF),

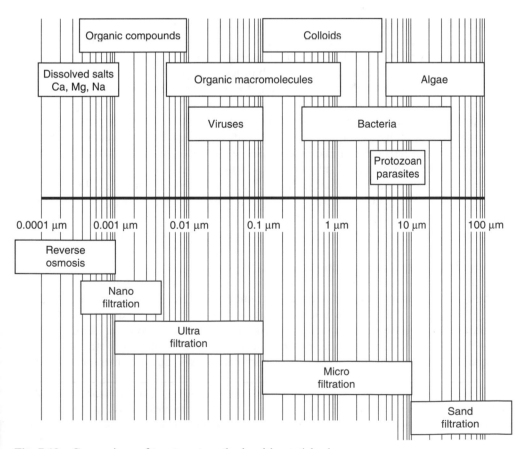

Fig. 7.18   Comparison of treatment methods with particle size.

nanofiltration (NF) and reverse osmosis (RO). NF, which is also called low-pressure RO, and RO are used to remove ions by osmosis and are discussed under desalination (see Section 8.46). MF and UF are micro-porous and remove particles by physical separation. They are therefore low pressure processes unlike RO and NF. The characteristics of membranes and therefore their performance is specific to the particular membrane and its pore size distribution. MF membranes have a nominal pore size 0.1 to about 5 μm with a typical cut-off value of 0.2 μm. UF membranes have a pore size 0.002–0.1 μm with a typical cut-off value of 100 kdalton* (equivalent to about 0.01 μm).

The membrane geometry most commonly used in MF and UF is hollow fine fibre; lumen (or internal bore) diameter range typically from 0.4 to 1.5 mm for MF and 0.35 to 1.0 mm for UF (see Plate 19).

MF and UF membranes remove particles which are an order of magnitude larger than their pore size on the membrane surface and particles less than the pore size in pores. They are used for particulate and micro-organism removal. Additionally, some UF membranes also remove large molecular weight organic compounds.

MF membranes provide a barrier to cysts and oocysts. Under laboratory conditions *E. coli* and *Cryptosporidium* oocyst numbers were reduced to values below their detection limits.[96] In practice MF membranes have an average removal of about 5-log† for *Giardia* and 4-log for *Cryptosporidium* and the average turbidity of the permeate (the product) is less than 0.2 NTU; values of 0.05 NTU or less have been reported.[97] Removal of viruses is comparatively low with values ranging from less than 0.5-log to 3.5-log.[98,99] UF membranes by virtue of their much smaller pore size, are very effective in retaining even smaller particles than MF membranes. UF consistently produces permeates with turbidity less than 0.1 NTU even from feed water turbidities as high as 300 NTU.[96] Removal of micro-organisms is reported to be greater than 5-log for *Giardia* and 4-log for *Cryptosporidium*, 8-log for *E. coli* and 4-log for viruses.[96] Therefore UF systems can provide total disinfection, and marginal chlorination of the permeate (the product) will be necessary for maintaining a residual in the distribution system. In practice, however, there is always the risk of particle and micro-organism leakage due to membrane imperfections and damage, and facilities for disinfection should be allowed.

MF and UF processes are operated either in cross-flow or dead-end mode. In the former only part of the inflow is filtered through the membrane; the remainder flows tangential to the membrane surface carrying with it the particulates removed on the surface and recirculated back to the feed water supply. The suspended solids concentration in the recycle is sometimes controlled by a continuous bleed (blow-down) to waste. In the dead-end mode all the water passes through the membrane and the retained particles build up as a cake on the surface; therefore there is total water recovery. This mode is usually used for UF when the raw water has an average turbidity <2 NTU and total organic carbon <2 mg/l;[100] higher values can be tolerated but, at the expense of frequent backwashing and fouling. Dead-end mode can only be used with hollow fibre systems. Both MF and UF systems can be designed to operate on both modes with changeover effected on feed water quality. The membrane flux for most commercial MF and UF membranes ranges

---

*UF membranes are normally classified based on molecular weight cut-off (MWCO) which is usually defined as the smallest molecular weight species for which the membrane has more than 90% rejection and is expressed in dalton, which is a unit which designates 1/16 mass of lightest and most abundant isotope of oxygen.

†x-log removal = 100 $(1-10^{-x})$%; therefore 4-log removal = 99.99%, 2.5-log removal = 99.68%.

between 0.08–0.2 $m^3/h.m^2$ at 20°C and varies with water temperature; flux is proportional to exp $(-0.0239)(T-20)$; 20°C being the reference temperature.[96] The feed pressure varies in the range from 0.5 to 3.5 bar with a drop in pressure across the membrane (trans-membrane pressure) in the range 0.1–1 bar. Feed water recovery varies between 85 and 98% depending on the operating mode and raw water quality having made an allowance for backwash water which can be about 1–10% of product flow.

The build up of particles on the membrane surface increases hydraulic resistance and reduces the membrane flux. This is called fouling. Surface fouling is usually reversible and the flux can be reinstated by backwashing. Membrane systems are operated either as constant rate (variable pressure) or declining rate (constant pressure). Once the membrane surface has a cake layer formed they are cleaned by water backwash using the permeate (one commercial MF design employs compressed air for dislodging particles from the membrane followed by crossflow flush using raw water on the shell side). Cleaning may be initiated by pressure, time or after a pre-set volume of permeate has been collected. In practice, depending on the feed water quality and operating mode, backwash frequency ranges between once every 30 minutes to 3 hours and duration ranges between 30 seconds and 3 minutes. The introduction of backwash water should be gradual to prevent membrane damage. The fouling in membrane pores is not reversible and cannot be removed by backwashing. This is mostly caused by adsorption of organic matter and precipitation of calcium carbonate and oxides of iron and manganese. Intermittent chemical cleaning is then employed. Typical cleaning agents are chlorine solutions, acids, alkali, detergents and oxidising solutions (e.g. hydrogen peroxide), depending on the type of fouling.

One of the most critical aspects of membrane plant operation is the risk of membrane failure. On-line particle counters are usually used for monitoring membrane integrity and their effectiveness depends on the particle load in the feed; turbidity monitors are not sensitive enough for the purpose. Other methods include routine microbiological testing, and membrane off-line tests such as sonic sensors for hydraulic noise analysis, air-bubble and air pressure holding tests[96,101] whereby fibre breakage can be measured against the differential pressure.

MF and UF membranes are commonly available as hollow fine fibre, hollow capillary fibre, tubular and spiral wound. For drinking water treatment hollow fibre types are mostly used. They have a high packing density (membrane area/m^3). The tubular type has a very low packing density. Spiral wound type cannot be backwashed. In the hollow fibre type flow can be either inside-out or outside-in. The former is better for cross-flow operating mode, has good hydrodynamic characteristics, gives a more efficient backwash but has a greater probability of fibres becoming blocked with particulates and comparatively higher head loss through the module. The latter gives greater filtration area:packing density ratio and lower head loss. When operating hollow fibre system in outside-in mode with waters containing high concentration of suspended solids, particulate can become trapped between fibres, eventually leading to fibre breakage. This would be the case with systems using high packing density. In inside-out systems particles tend to block the fibres and cause fibre failure. The material used for membrane making for drinking water application must have good tolerance to pH and oxidising agents. Most hollow fibre membranes are made from polyethersulphone (PES) (pH 1–13 and chlorine CT_{10} $>10^6$ mg.h/l*). Others include cellulose acetate (CA) (pH 4–8.5,

---

*Chlorine CT_{10} is the product chlorine concentration (mg/l) × the contact time (hours) resulting in 10% reduction in mechanical properties of fibres.

chlorine $CT_{10} > 10^5$ mg.h/l), polypropylene (PP) (pH 1–13 and $CT_{10} < 10^2$ mg.h/l) and polyvinylidene difluoride (PVDF) which has higher chlorine and pH tolerance. CA membranes are biodegradable and therefore bacteria need to be removed from its surface. In potable water treatment applications pH of the water is unlikely to fall outside the range for CA membranes and pH correction is rarely needed.

Both MF and UF systems can be applied directly on raw water with only a pre-filtration stage to reduce particle size to a range 50–200 μm depending on the internal diameter of the inlet ports and fibres. Membrane processes can be preceded by powdered activated carbon dosing or coagulation using aluminium or iron salts for organics removal, or used downstream of clarifiers as a filtration stage or downstream of rapid or slow sand filters as a polishing stage. When iron and manganese are present, they should be oxidised and precipitated prior to membrane filtration to prevent their precipitation within the pores or on the permeate side of the membrane during backwashing with chlorinated water. The quality of the permeate with respect to particle and micro-organism removal is not influenced by their numbers in the feed water.

A typical plant consists of raw water pumps feeding a bank or several banks of membrane modules,* recirculation pumps for cross-flow mode operation with capacity to give recirculation flow to raw water flow ratio in the range 3:1–6:1 depending on the raw water quality and backwash pumps. Membrane banks are backwashed sequentially. Backwash pressure varies in the range 1–2.5 bar. All pumps used on a membrane plant are usually of the centrifugal type. Both MF and UF systems can be staged with used backwash water and the bleed from the recirculation water treated in a second stage. A flow schematic diagram of a typical membrane plant is shown in Fig. 7.19. At the 65 Ml/day Homesford water treatment works, Severn Trent Water (see Plate 19), a groundwater is treated by MF using a hollow fibre membrane which utilises air to backwash; the product having particle numbers less than 10/ml (>2 μm) and turbidity less than 0.1 NTU. Used backwash water is treated in a second stage. Total water recovery is 99.9%. MF and UF membranes are usually pressure driven and are installed within pressure vessels. There is however an adaption of MF and UF membrane processes that consists of outside-in, hollow fibre membrane modules which are immersed vertically in the raw water in a tank. The permeate is removed from either one or both ends under a vacuum (125–325 mbar) produced by the use of a centrifugal pump. Air is diffused into the bottom of the membrane module to create turbulence which helps to clean the membranes, thus maintaining the flux rate. The concentrate containing suspended solids is periodically purged from the tank. Stored permeate is pumped in the reverse direction (inside-out) at regular intervals to control fouling and reduce the frequency of chemical cleaning.[102]

The operating costs for a membrane plant is primarily made up of energy, membrane replacement and labour components and in the case of an MF plant each contributes about 15–25% of the total cost with general maintenance, cleaning chemicals and waste disposal contributing about another 15–20%.[103]

---

*A module consists of several membrane elements in a pressure shell complete with feed inlet ports, distributors, outlets and permeate removal points. Several modules may be connected in series. In a bank several modules may be connected in parallel.

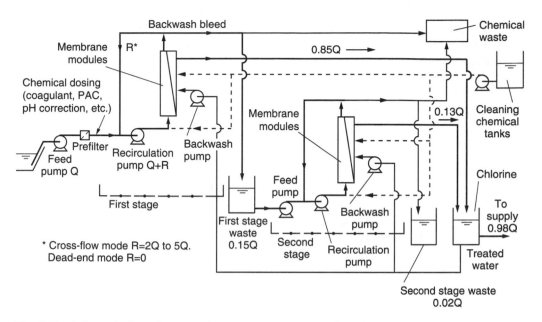

Fig. 7.19 Schematic flow diagram of a two-stage membrane filtration plant.

7.37 Miscellaneous filters

A design of filter mainly to be found in the USA known as the 'Green leaf' filter consists of a number of conventional, open type, rapid gravity filters, constructed in concrete, which surround a central control section, constructed in concrete or steel, which houses all the filter controls. The principal difference in operation is that siphons arc used for inlet flow control and backwash, instead of backwash pumps or an overhead tank. Water is maintained at a constant level by a weir in the effluent chamber and as one of the filter units becomes progressively dirty the level in the filter shell rises until the inlet flow is stopped, either manually or by automatic operation of a siphon break. Washing is usually performed by a separate air application, followed by a water wash using the filtered water from the remaining filters and the head created by the outlet weir in the effluent chamber. This head is about 1.5 m above the top of the sand bed which is low for backwashing a sand filter, particularly if it is in a dirty condition. A compact arrangement can be obtained using this design of filter, but it would be prudent not to use it for the treatment of difficult waters because of the backwashing constraints.

The 'ASF' automatic gravity valveless filter has three sections: backwash storage space, filter bed and filtered water storage space. Raw water flows downward through the sand bed whilst filtered water flows out via the backwash compartment. As the filter bed becomes dirty the level in the raw water pipe and backwash pipe (which contains a U-bend) rises and initiates the siphon action that backwashes the filter. When the level in the backwash compartment falls and exposes the end of the siphon breaker backwash stops and filtration re-starts.

There are some continuous flow filters in use. In one such design known as moving hood or moving bridge filter, the filter is made up of 10–50 individual cells each of plan dimensions up to 1 × 5 m and capacity up to 50 m³/h with a common filtered water compartment underneath; filter media depths range from 450 to 600 mm. Such filters are available in total lengths up to 50 m. Only one cell at a time is backwashed. A travelling bridge which spans the width of the filter carries beneath it a backwash hood of the same dimensions as a single cell, with a suction pump connected to it. During backwash, the hood is lowered and placed tightly over a cell, the pump starts and washes the filter by 'vacuum cleaner' action. The washwater is available from the remaining filter cells in service.

In another design which is made up of a vertical hopper bottomed cylindrical vessel filled with sand, raw water is introduced into the cylindrical section just above the hopper and flows upward through the sand bed counter-current to the sand moving continually downwards and leaves the filter at the top over an outlet weir, e.g. 'Dynasand'. The sand containing dirt is conveyed from the hopper section by an air-lift pump; the turbulent action of the pump cleans the sand with additional cleaning in a sand washer at the top where it is cleaned mechanically or hydraulically. The cleaned sand is returned to the top of the filter. Filtration rates can vary in the range 8–15 $m^3/h.m^2$. They are used as primary filters ahead of slow sand filters or in direct filtration mode with coagulation. In the latter case filtrate turbidity can be less than 1 NTU. Water loss through sand cleaning is in the range 5–8% of the inflow. Nowadays where filters are expected to meet stringent performance requirements with respect to washing and filtration, these specialist filters are no match to conventional rapid gravity filters.

Filtration of water and other fluids can be achieved with diatomaceous earth filters, or pre-coat filters, where the filtering medium is formed on septums or 'candles' of metal or other materials inside a type of pressure vessel. Excellent removal of suspended matter may be achieved. In the UK they are sometimes used for swimming baths and for small private supplies, but only occasionally for public water supplies. A good account of the process is given in *Water Quality and Treatment*.[60]

Cartridge filters of the woven fibre type rated at 5–10 microns are used in reverse osmosis plants preceding high pressure feed pumps. These filters remove particulate matter passing the anthracite–sand rapid gravity filters and are used as safety filters to protect the membrane and high pressure feed pumps.

There are so called 'depth filters' where the filtration efficiency increases with passage as a 'dynamic filtering' medium is deposited on a filtration surface, either from particles in the water being treated or by addition of a suitable fine material such as Kieselguhr (diatomaceous earth), Fuller's earth (a fine non-cohesive clay) or powdered activated carbon. The filtration surface can be fibres wound on to a cassette or a candle with grooves, fibres arranged in a bundle around a central core or a porous cloth woven to produce a series of filtration tubes. The fibrous filters are capable of filtering particles down to about 5 μm, and achieving *Cryptosporidium* oocysts removal up to 1-log consistently.[104] The porous cloth type is known to be capable of removing particles greater than 1 μm and therefore provide 2–3 log removal of oocysts.

Slow sand filtration
7.38   Introduction and history

The subject of slow sand filtration has not been discussed before in this chapter, principally for the reason that the considerable merits of the treatment can be better appreciated after a knowledge has been acquired of the complexities involved in treatment by coagulation and rapid gravity filtration. Slow sand filters were the first effective method devised for the purification in bulk of surface waters contaminated by pathogenic bacteria. They remain equally effective today and there is a growing number of circumstances which suggest that their use in new works should be encouraged.

'Slow' sand filters are so called because the rate of filtration through them may be only one-twentieth or less of the rate of filtration through rapid gravity or pressure filters. They were first constructed in the UK in the early nineteenth century. Their capacity for purification is well illustrated by the history of the cholera outbreak in Hamburg in 1892. Both Hamburg and Altona took water from the river Elbe which became contaminated with cholera, Hamburg suffered 8600 deaths from cholera but Altona which had slow sand filters had no cases of cholera. This vividly illustrates the ability of slow sand filters to purify a water bacteriologically, as well as physically, and many slow sand filters are in use today. Most of London's surface derived supplies are treated by slow sand filters, a 490 ml/day plant was added in 1972,[105] but since then a number of works have been closed down. On some of the remaining works beds have been taken out of service as a result of improvements made on pretreatment and filtration rate. The current total filter area is about 42 ha and they have also been adopted in works to treat water from Lough Neagh in Northern Ireland.[106] A summary of slow sand filter plants operated by Thames Water is given in Table 7.12.

Slow sand filters are an efficient method of producing water of good bacteriological and organic quality which requires marginal chlorination before distribution. They achieve 2–4 log removal of coliforms, *E. coli*, pathogenic organisms, cercariae of *Schistosoma*,[107] ova, cysts such as *Giardia* and oocysts such as *Cryptosporidium* and remove viruses, reduce colour, oxidise ammonia and organic matter and reduce turbidity to less than 1 NTU.

Table 7.12   Slow sand filter plants operated by Thames Water, UK

Works	Capacity (Ml/day)	Pre-treatment	Filtration area (m^2)
Ashford Common	690	Pre-ozone, anthracite–sand gravity filters, main ozone	32 × 3121
Coppermills	680	Sand gravity filters, main ozone	34 × 3400
Hampton	790	Sand gravity filters, main ozone	112 900 (25 filters)
Kempton	200	Sand gravity filters, main ozone	12 × 3640
Walton	140	Pre-zone, coagulation (iron salt) COCO-DAFF (with anthracite–sand gravity filters), main ozone, GAC filters	10 × 3320
Fobney	75	Pre-ozone, sand upflow filters, main ozone	12 × 900

Notes. All slow sand filters (except Walton) have a GAC sandwich of 100–150 mm in a total bed depth of 800–900 mm. In addition there is a gravel layer of 100 mm.
Filter underdrain systems are porous concrete floor type.
Filtration rate varies in the range 0.3 m/h (average) to 0.5 m/h (maximum).
For COCO-DAFF, see Section 7.18.
Post-treatment is chloramination.
All works except Fobney serves London.
Source: Thames Water, UK.

7.39 Mode of action of slow sand filters

The slow sand filter does not act by a complex straining process alone, but by a combination of both straining and microbiological action, of which the latter is the more important. The mode of operation is complex and is, from time to time, a subject of argument between experts. There is no doubt, however, that purification of the water takes place not only at the surface of the bed but also for some distance below. Van de Vloed[108] has given as clear an account as any of the details of the purification process. He distinguishes three zones of purification in the bed: (1) the surface coating, the 'schmutzdecke', (2) the 'autotrophic' zone existing a few millimetres below the schmutzdecke, and (3) the 'heterotrophic' zone which extends some 300 mm into the bed.

When a new filter is put into commission and raw water is passed through it, during the first two weeks the upper layers of sand grains become coated with a sticky reddish brown deposit of partly decomposed organic matter together with iron, manganese, aluminium, and silica. This coating tends to absorb organic matter existing in a colloidal state. After two or three weeks there exists in the uppermost layer of the sand a film of algae, bacteria, and protozoa, to which are added the finely divided suspended material, plankton, and other organic matter deposited by the raw water. This skin is called the schmutzdecke and it acts as an extremely fine-meshed straining mat.

A few millimetres below this schmutzdecke is the autotrophic zone, where the growing plant life breaks down organic matter, decomposes plankton, and uses up available nitrogen, phosphates, and carbon dioxide, providing oxygen in their place. The filtrate thus becomes oxidised at this stage.

Below this again a still more important action takes place in the heterotrophic zone which extends some 300 mm into the bed. Here the bacteria multiply to very large numbers so that the breakdown of organic matter is completed, resulting in the presence of only simple inorganic substances and unobjectionable salts. The bacteria act not only to break down organic matter but also to destroy each other and so tend to maintain a balance of life native to the filter so that the resulting filtrate is uniform.

The biological processes require oxygen and if it is absent anaerobic conditions would set in, resulting in the formation of hydrogen sulphide, ammonia, soluble iron and manganese and taste and odour producing substances. Therefore to ensure satisfactory operation the oxygen concentration in the filtrate should not be allowed to fall below 3 mg/l.[109] The efficiency of the process is also temperature dependent. For example the reduction in permanganate value (a measurement of the organic content (see Section 6.31) decreases by $(T + 11)/9$ where T is the water temperature in °C.[110] Below 6°C ammonia oxidation ceases. At low water temperatures the rate of biological reactions and activity of bacteria consuming micro-organisms reduces rapidly and the rate of reduction in *E. coli* falls sharply, thus requiring the chlorine dose to be increased. If water temperatures less than 2°C persist for prolonged periods, consideration should be given to covering the filters.

7.40 Construction and cleaning of slow sand filters

The bed of sand in a slow sand filter is 0.6–1.25 m thick and is laid over a supporting bed of fine gravel, beneath which a collector pipe system is constructed. The water passes downwards through the sand bed, through the gravel, and is collected by the pipes beneath, the whole arrangement being sited in a shallow watertight tank of large size. In some plants a porous concrete floor is used in place of pipes. It is important to note that the bed is 'drowned'; it is not a trickling filter as in sewage treatment works where air is permitted into the bed. Filtration rates used are typically 0.1–0.3 m/h. At the old Ashford Common Works of the Thames Water Authority the sand bed was 0.675 m thick (similarly at the 1972 Coppermills Works) and lies upon 75 mm of fine gravel which, in turn, rests upon a bed of porous concrete (see Fig. 7.20). Each filter is about 34 m × 90 m and the filtration rate is in excess of 0.2 m/h. Below the concrete, collector drains feed the filtered water to the main effluent pipes. The sand is ungraded; because they are not backwashed, hydraulic grading of the media does not occur. Size distribution in the bed is purely random. The sand has a size range 0.21–2.36 mm (effective size of 0.30 mm) and uniformity coefficient 1.5–3.5 (less than 2 is preferable). The stated size range is generally 5 and 90 percentiles. The sand should be hard and should not be liable to break down when subjected to skimming and washing processes. The suspended solids and particulate organic content of the sand should not exceed 0.5 g/l and 0.1 g/l, respectively.

The raw water is led gently on to the filter bed and percolates downwards. Directly after a bed has been cleaned a head of only 50–75 mm of water is required to maintain the design rate of flow through the bed. However, as suspended matter in the raw water is deposited on to the surface of the bed, so a mat (or 'schmutzdecke' as it is called) of organic and inorganic material builds up on the surface of the sand and increases the friction loss through the bed. To maintain the flow at a uniform rate as far as possible the head across the bed is gradually increased and when it reaches some predetermined value between 0.6 and 0.9 m the bed must be taken out of service and cleaned. The maximum permissible head loss should be kept to about 1 m. If fine grain sand is used then the depth

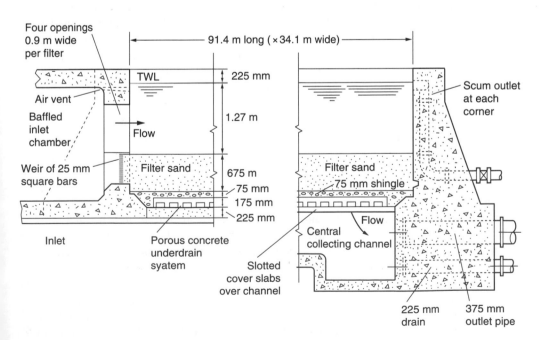

Fig. 7.20 Slow sand filters at Coppermills Works of Thames Water Authority, 1972.

should be reduced to minimise the resistance. To prevent negative head the maximum permissible headloss must be kept less than the depth of water on top of the sand. Normally the water depth above the sand bed is maintained at about 1–1.5 m with a maximum of 2 m. As a further preventative measure the outlet weir should be set above the media surface level.

When a sand bed requires cleaning it is drained of water and the top 12–25 mm of the sand surface are carefully scraped off. The filter is then returned to service by gradually increasing the flow over 24 hours, sometimes longer. When the sand bed again requires cleaning a further 12–25 mm of sand are scraped from the surface, and this process is repeated until the bed is thinned to the minimum practical thickness for efficient filtering which should not be less than 0.5 m.[110] When this stage is reached the bed is then topped up with clean new sand to its original level or the old sand may be replaced if it has been adequately washed and cleaned in a sand cleaning machine.

The interval between scraping may vary from several months during the winter when pre-filtration is installed, to 10 days where no pre-filtration occurs and algal growth is at a maximum. Resanding would only be necessary every 2–5 years depending on the scraping frequency. Originally, slow sand filters were invariably scraped by manual labour (typically for scraping: 5 person-hours/100 m^2 and resanding: 50 person-hours/100 m^2),[111] but the increased cost and decreased availability of this type of labour has led to the introduction of mechanical methods in particular for large plants thereby reducing labour, typically to about 2–4 person-hours/100 m^2 for scraping and 5–8 person-hours/100 m^2 for resanding. Resanding by wet slurry method is reported to reduce labour requirement further.[112] Some of the mechanical methods are fairly simple, e.g. using small vehicles with a skimmer bucket discharging to trucks. One of the difficulties of mechanising the cleaning of the older slow sand filters is that their sizes and geometry were often not uniform since these factors were unimportant in the days of manual cleaning. However, with new installations, provided that the filters are constructed with the correct dimensions, it is possible to span them with a standard sand lifting bridge which runs on tracks along the sides of the filters, thus greatly reducing the labour necessary for scraping.[113] Grey[106] reports details of such an installation at Lough Neagh. Other cleaning methods used are wet harrowing and flushing[114,115] which does not disturb the attached bacteria and the use of a suction dredge using laser depth control[116] which avoids the need to drain down the filter. Sand scrapings must be washed to remove dirt before being replaced in the filter. Washing must be carried out soon after removal. In small plants manual washing by high pressure water jet is used. In larger plants mechanised systems consisting screens, hydrocyclones and density separators are used. The standards aimed for cleaned sand are:[117,118] silt concentration in the range 0.5–1.0 g/l and particulate organic concentration 0.1 g/l or 0.1% w/w.

Each filter needs five valves; inlet, outlet, back-filling, waste and drain. The inlet valve is sized to give a velocity of about 0.1 m/s for good distribution of water and not to damage the schmutzdecke. It can also be used to drain the top water when the filter needs cleaning. The back-filling valve is used to refill the filter (after scraping) by reverse flow to a water depth of about 250 mm to remove air entrained in the voids. The waste valve is used to discharge the filtrate for a period until the filter is 'ripened' after it is brought on line following skimming or re-sanding; filtered water is discharged to waste until chemical and bacteriological tests prove it to be satisfactory. In the operation of filters it is important to minimise fluctuations in inflow and raw water quality. The plant should be designed to operate at constant rate with a constant level of water above the sand surface, or rising water level in the filter. With the latter the shallow water depth allows penetration of sunlight, encouraging algae growth and there is always the danger of disturbing the schmutzdecke by the incoming water. With the low filtration rate, manual control of filter inlet and outlet will be more than adequate and cheaper. Automatic control would however be useful for large plants. Water level in the filter can be kept constant by a level controlled inlet butterfly valve and a constant filtration rate is maintained by a flow controlled outlet butterfly valve.

7.41 Use of pre-treatment with slow sand filters

Slow sand filters may operate as the sole form of filtration for raw water turbidities of up to about 10 NTU. The object of pre-treatment such as storage or pre-filtration is to lighten the load of suspended matter on the slow sand filters and so permit longer intervals between cleanings or a faster rate of filtration, or both. Van Dijk and Oomen[119] recommend that if the average raw water turbidity exceeds 50 NTU for more than a few weeks, (or is in excess of 100 NTU for a day or so), then pre-treatment is indispensable. Pre-treatment can take the form of river bed filtration, Ranney wells, storage, plain settlement, filtration (horizontal, pebble matrix, rapid gravity or upflow)[120] or combination of these depending on the turbidity of the raw water. If the filtrate has a persistently low oxygen content (<3 mg/l) then aeration of the raw water will be necessary. Pre-treatment sometimes includes chemical coagulation and flocculation without clarification (Ivry plant near Paris) or with clarification (Orly plant near Paris) and rapid gravity filtration.[121] If raw water *E. coli* counts are in excess of about 10 000/100 ml, the filtered water should be subjected to disinfection as marginal chlorination would be inadequate. A summary of the effects of pre-filtration is given by Ridley,[122] who compared three London supply works – Hanworth Road with slow filtration only, Kempton Park with rapid gravity filters preceding slow sand filters, and Ashford Common with microstraining preceding slow sand filters. Figures for the period 1 April–30 September (when algal growths are usually most prominent) for the 5 years 1959–63 inclusive and the influence of more recent changes to the pre-treatment on operating criteria are shown in Table 7.13.

During the period 1956–63 Kempton Park rapid gravity filters used 1.15–1.62% washwater (as a percentage of the water filtered) while Ashford Common microstrainers used 1.64–2.03%. The extra washwater used and the capital and running costs of pre-filtration and ozonation are usually taken as economically justified in view of the very considerable increase in water filtered through the slow sand filters between cleanings.

7.42 Limitations and advantages of slow sand filters

Slow sand filters tended to be ignored when new plants were being considered because of the amount of land and labour they use and their relatively large capital cost. However, there is still a place in water treatment for slow sand filters, providing their advantages and limitations are carefully weighed in any particular case. They may also play a greater role in future water treatment technologies owing to pressures for less chemical usage in treatment.

Table 7.13 Performance of slow sand filters for London water supply (1998–99)

Station and type of pre-treatment currently adopted	Rate of filtration (m/h)	Quantity of water filtered per hectare of sand bed cleaned (Ml/ha)[a]
Hanworth Road[b] – no pre-treatment	(0.004–0.059)[c]	(315–705)[c]
Kempton Park; rapid gravity pre-filtration and ozone	0.3 (0.132–0.152)[c]	3300–4950 (1260–1600)[c]
Ashford Common; pre-ozone, rapid gravity pre-filtration and main ozone	0.3 (0.132–0.137)[c]	4140–6900 (1000–1570)[c]

Source: Thames Water plc, UK.
[a]1 Ml $= 10^3$ m^3. 1 ha $= 10^4$ m^2. So 1 Ml/ha $= 0.1$ m^3/m^2.
[b]Hanworth Road installation is now closed.
[c]Values in brackets show the information for the summer seasons in 1959–1963; since then ozone treatment and GAC sandwich have been introduced.

A review of the efficiency of slow sand filters by Lambert[123] gives degrees of removal for colour (36%), a series of organics groups (14–60%), and pesticides (e.g. 100% for Mecoprop and 2,4-D, but zero for Atrazine, Simazine and Bentazone). Slow sand filters do not significantly reduce the 'true colour' (see Section 6.16) of a water. They are thus only suitable for dealing with waters of relatively low colour. What the limiting colour should be depends upon the standards set for the final water, but generally one would say that raw waters of colour 10–15 Hazen units would, other things being equal, be quite suitable for slow sand filtration. With a colour of greater than 15 Hazen units, slow sand filters might still be used, but in this case it would probably be advisable to oxidise the colour in the feed or filtrate by ozonation (e.g. Invercannie water treatment plant, Grampian Regional Council, Scotland where ozone is used to oxidise colour, upstream of slow sand filters).[124] Another factor is that slow sand filters are not effective in removing iron and manganese in solution, and being a biological process there is a marked reduction in efficiency in the removal of contaminants at low temperatures (see Section 7.39).

A high concentration of algae in the raw water can cause treatment difficulties due to clogging of the filters, or if the cells die off in the filter, due to excessive respiratory demands causing anaerobiosis and tastes. The chlorophyll-a concentration of the feed should ideally be limited to 5 μg/l[111,125] with a peak of 15 μg/l.[125] Thus slow sand filters are not the treatment of choice for highly eutrophic waters in which sudden peaks of algae may appear in spring, summer or autumn. Methods to prevent or control algae include pre-treatment, covering of slow sand filters and chemical treatment. Low to moderate algal concentrations on the other hand are beneficial to the process.

Slow sand filters are also not very suitable for the removal of any substantial amount of finely divided inorganic suspended matter. With pre-filtration using primary filters, however, they may function successfully for years treating a water which is intermittently laden with fine silt.

Removal of organics by slow sand filters is not complete (see Section 7.38) and that of pesticides is type dependent as mentioned above. Therefore on some waters they are supplemented by additional treatment usually ozone and granular activated carbon (GAC) adsorbers used either upstream or downstream of slow sand filters. At Thames Water's Ashford Common Works a 0.135 m GAC sandwich layer is incorporated within the slow sand filters. Ozonation is carried out upstream of roughing and slow sand filters.[116] At Ivry and Orly plants near Paris, ozone and GAC are used downstream.[121]

Apart from the above, the other stated limitations of slow sand filters often appear of doubtful validity on closer inspection and four examples are given to illustrate this. Firstly, it is stated that slow sand filters occupy a greater area of land than do coagulation and rapid gravity plants. This is true so long as one ignores the question of sludge disposal, but if *adequate* sludge disposal facilities are included the land required for slow sand filters may well be no greater than that for a conventional treatment. Secondly, it is maintained that slow sand filters are expensive in capital costs. However, when the cost of chemicals is taken into account, slow sand filters may be cheaper in annual costs than conventional plants.[126] Thirdly, whereas it was usual to design slow sand filters for low rates shown in Table 7.13 for 1959–63, it has been found that with improvements to pre-treatment (e.g. pre-ozonation, GAC, redesigned primary filters, intermediate ozone) without covering (e.g. Ashford Common Works, Thames Water, UK)[116] or with covering (e.g. Barmby Works, Yorkshire Water, UK),[127] filtration rates have been almost doubled to about 0.3 m/h. Other techniques such as the use of replaceable non-woven synthetic fabric layer (geotextiles) on top of sand has helped to reduce the cleaning frequency;[128] at Hardhof plant, Zurich, a six-fold reduction in cleaning frequency is reported.[129] Fourthly, slow sand filters are said to be labour intensive, but with modern, properly designed, slow sand filters permitting the latest methods of mechanical cleaning, they may be no more labour intensive than conventional treatment plants, taking into account the work involved in sludge disposal.

In some European countries they are used as a final polishing stage of treatment, being preceded by as many as five or six stages of treatment including chemical coagulation, sedimentation,[130] ozone, GAC, softening, etc. (e.g. River-Lake plant, Loenderveen and Wesperkarspel, Holland[112]).

Cryptosporidium oocysts and *Giardia* cysts removal
7.43   *Cryptosporidium*

Removal of *Cryptosporidium* oocysts can be achieved by any process that removes particles down to the size of 4 μm or smaller. However, conventional water treatment processes in current use for drinking water supply cannot guarantee complete removal. Since the performance of disinfectants is poor (see Sections 6.57 and 9.6) the removal of oocysts in the upstream solid–liquid separation processes must be maximised. Well operated and maintained conventional treatment processes which include coagulation and flocculation, with or without clarification by dissolved air flotation (DAF) or settlement (horizontal flow, sludge blanket, etc.) followed by granular media filtration are expected to achieve 2-log to 3-log removal consistently.[131,132] The contribution by media filters to the overall removal can be between 1.5 and 2 log. To maximise their removal the water must be properly coagulated prior to clarification and filtration. Good mixing of the coagulant in water is paramount for effective coagulation of particles and the coagulation process should be operated at the optimum pH. A low soluble coagulant metal ion concentration (<0.05 mg/l as Al or Fe) is a good indication of effective coagulation. This should be followed by hydraulic or mechanical flocculation. The solid–liquid separation process can be by direct media filtration, or a combination of clarification followed by media filtration. Criteria for good clarification performance would be a turbidity of 1 NTU and total coagulant metal ion concentration less than 1 mg/l as Fe or 0.5 mg/l as Al depending on the coagulant used.[133] Filters should be operated at about 6 $m^3/h.m^2$ with all filters in service. Filtrate turbidity should be consistently less than 0.5 NTU, and ideally a filtered water turbidity from individual filters of less than 0.1 NTU should be aimed for.[134] Filtered water turbidities <0.1 NTU and particle counts <50/ml are indicators of good treatment for controlling *Cryptosporidium*.[135] Second stage filtration such as GAC adsorbers or manganese removal filters may provide a further barrier with removal <0.5 log but cannot be relied upon for effective removal because they are not used for residual floc removal.

Slow sand filters with an active schmutzdecke and at filtration rates up to 0.3 $m^3/h.m^2$ are expected to achieve up to 5-log removal. Microfiltration and ultrafiltration can achieve consistent removal rates better than 2-log and 4-log respectively (see Section 7.36); in some studies higher removals (>6-log)[98] have been reported. *Cryptosporidium* oocysts are resistant to even high concentrations of chlorine (see Section 9.6), but ozone when used in doses typically applied at water treatment works is reported to produce 1-log to 2-log inactivation of oocysts.[136,137] For ozone the '*ct*' value required to achieve 2-log inactivation is about 5 mg.min/l at 7°C and 1.7 mg.min/l at 22°C and that to achieve 3-log inactivation is about 8 mg.min/l at 7°C and 2.5 mg.min/l at 22°C.[138] Ozone if in use for other applications (e.g. oxidation, disinfection, etc.) would be beneficial for oocyst inactivation, although the installation of ozone solely for oocyst inactivation is not recommended until its effectiveness is confirmed in further studies and full scale plants. The process is used in the USA as a treatment method for oocysts. It is important to note that irrespective of the various degrees of removal achieved in different processes, the critical factor is the number of oocysts remaining in the water which is a function of the number in the raw water, and how it relates to the infectious dose.

In the UK, the Report of the Group of Experts which met under the chairmanship of Sir John Badenoch[136] made the following recommendations to minimise the risk of *Cryptosporidium* passing into water supplies:

(i)   the operation of rapid filters should avoid sudden surges of flow which may dislodge retained deposits;
(ii) rapid filters should not be restarted after shutdown without backwashing;
(iii) after cleaning, slow sand filters should not be brought back into use without an adequate 'ripening period';
(iv) bypassing part of the water treatment process should be avoided.

It also emphasised the importance of monitoring turbidity of individual filters to enable early detection of turbidity breakthrough in filters and questions the practice of recycling supernatant water from clarifier sludge and filter backwash water treatment facilities to the treatment works inlet.

The Second Report of the Group of Experts,[139] reiterated the earlier recommendations and proposed that rapid changes in flow through filters should be limited to between 1.5–5% per minute, depending on the fragility of the floc, to avoid penetration of particles through the filter. It also accepts the fact that turbidity breakthrough still occurs when using 'slow start' (see Section 7.27), albeit with lower turbidity values over a longer period, but the risk of oocysts released from the filter bed still remains. The effect of initial breakthrough with or without slow start on the filtrate can be overcome by diverting the initial filtrate to waste, or for later use as a source of backwash water or by recycling to the works inlet. Figure 7.21 illustrates the benefit of the first flush to waste over slow start in reducing the initial turbidity and particle breakthrough. Standing the filter for about 15–60 minutes (the so-called 'delayed start') following the backwash has also been shown to reduce the initial breakthrough of turbidity. The addition of polyelectrolyte to washwater for ripening the filter, which is practised mostly in the USA is reportedly not always successful.[140] Used backwash water should be treated by settlement with the aid of a polyelectrolyte before recycling; the target quality for the supernatant from the settling process being a turbidity of <5 NTU, suspended solids of <10 mg/l and total coagulant metal ion concentration of <5 mg/l as Al or Fe.[134] The recycle flow should be between 5–10% of raw water flow. Similar rules should be applied to the recycle of thickener supernatant. On the recycling of supernatant waters from clarifier sludge and filter backwash treatment system, the Second Report suggests that the efficiency of these facilities should be such that no more than 5% return of oocysts to the works inlet occurs to ensure minimal effect on the final water quality. This may require additional treatment of the supernatant such as membrane filtration, 'depth filtration' (see Sections 7.36 and

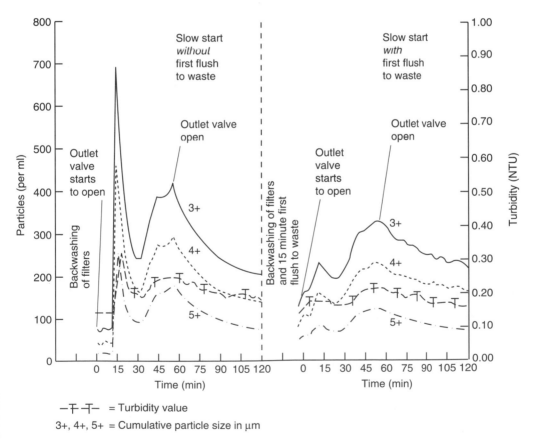

Fig. 7.21   Particle count and turbidity on filter start up – effect of first flush to waste (source: Portsmouth Water plc UK).

7.37) or ozonation. The use of particle counters which are more representative of oocyst breakthrough and sensitive to changes in filter performance than turbidity monitors on filters, is also emphasised in the Second Report. Turbidity value does not correlate well with particle counts, for example at a filtration works in the USA a 15% increase in flow through filters due to a filter being taken out of service for washing showed a turbidity increase from 0.03 to 0.05 NTU whereas particle count increased from 5 to 25 particles/ml.[141]

The importance of coagulation followed by solid–liquid separation as an effective barrier against *Cryptosporidium* is emphasised in the Third Report of the Group of Experts under the chairmanship of Professor Ian Bouchier.[142] The report supports the recommendations made in the previous reports and recommends that a risk assessment is carried out for each source, and water treatment and monitoring requirements should be reviewed against the level of risk. It makes the following additional recommendations on water treatment and monitoring:

Treatment

(i)   Works should be designed to treat typical worst raw water quality with respect to colour and turbidity and should not be operated under stress, nor in particular outside its design capacity. (It is therefore important to establish operating rules for a works based on its treatment limitations with respect to capacity and raw water quality.)

(ii)   Coagulation and flocculation processes should be regularly optimised, particularly when water quality changes.

(iii)   Works should be operated to minimise turbidity, colour and coagulant metal ion concentration in final water all the time.

(iv)   If the works are subjected to overloading or bypassing due to emergencies, strict monitoring regimen should be initiated to ensure turbidity increases are minimised.

(v)   For high risk sites, if slow start is ineffective, first flush (initial filtrate) should be run to waste or reused as washwater or recycled to the works inlet.

(vi)   Only dedicated washwater mains should be used to carry the returned washwater flow.

(vii)   Filters should be operated and maintained under optimum conditions with respect to quality, media depth and effectiveness of backwashing.

Monitoring

(viii) Process monitoring systems should be appropriate to the level of risk at each source and, for works treating high risk sources, monitoring should include continuous measurement of turbidity on each filter outlet and on the final water using instruments with a sensitivity better than 0.1 NTU. The use of particle count monitors to supplement turbidity monitors is encouraged.

(ix)   For individual treatment works, the value and duration that constitute a significant deviation in turbidity of final water should be defined, and monitored with alarms initiated outside the defined values. The turbidity should lead to action which might include shut down of filter(s) or the source, based on the level of risk of the source.

(x)   For sources with high risk, consideration be given to continuous sampling for *Cryptosporidium* with analysis times related to turbidity results (i.e. events of high turbidity), or sampling initiated by turbidity events.

7.44   *Giardia* cysts removal

Giardia cysts being larger organisms (8–12 μm) than *Cryptosporidium* oocysts (4–6 μm) are easier to remove by solids–liquid separation processes than *Cryptosporidium*.[143] In most cases *Giardia* removal is about 1-log to 2-log better than the corresponding value for *Cryptosporidium* removal. Because of their larger size they can be better correlated to turbidity than *Cryptosporidium* oocysts.

   Giardia cysts are less resistant to disinfection than *Cryptosporidium* oocysts, though chlorine is not very effective, a '*ct*' value of 100 mg.min/l (see Section 9.5) being required for 2-log inactivation at pH 7 and 5°C, improving to about 10 mg.min/l at 25°C.[144] Using ozone, the corresponding values for 2-log inactivation at pH 7 are 0.53 mg/l.min at 5°C, and 0.17 mg.min/l at 25°C.[145]

Disposal of sludge from treatment works
7.45   Types and quantities of sludges

Sludges may be classified according to the type of water treatment process adopted.

   Non-chemical sludges arise from microstrainers, clarifiers for pre-settlement unaided by chemicals, membrane filters except the effluent from chemical cleaning of membranes, and from slow sand filters and associated roughing filters which are used for the treatment of low turbidity waters, usually stored waters. These sludges are for the most part fairly innocuous and in some countries it may still be permissible to discharge them to a watercourse or water body without treatment; an exception occurs when the raw water has high populations of algae when treatment such as the use of drying beds may be required.

   Coagulant sludges derive from treatment plants using coagulation, whether by means of aluminium or iron salts. These are the most difficult to treat for disposal because of the relatively large volumes involved and the difficulties of dewatering due to their gelatinous nature. They are dealt with in detail below.

   Softening sludges are generated from softening by lime or lime-soda treatment, and the sludges, being of granular consistency, are comparatively easy to dewater.

Quantities

The quantities of dry solids in sludges produced in a treatment works employing coagulation, flocculation, clarification and filtration are a function of raw water quality and the chemical treatment applied. They are made up of coagulant hydroxide, suspended solids, precipitated colour, algae, iron, manganese and other chemicals contributing to solids, such as powdered activated carbon (PAC), impurities in lime and polyelectrolyte. Dry solids produced ahead of clarifiers or filters (in direct filtration) can be calculated as follows:

$$\textit{Sludge dry solids } (\text{mg/l}) = X + S + H + C + Fe + Mn + P + L + Y$$

where X – coagulant hydroxide (mg/l) $= f \times$ coagulant dose (mg/l as Al or Fe); $f = 2.9$ for Al and 1.9 for Fe.

S – suspended solids (mg/l); when suspended solids values are not available it can be approximated to $2 \times$ turbidity (NTU)

H – $0.2 \times$ colour Hazen

C – $0.2 \times$ chlorophyll 'a' in μg/l

Fe – $1.9 \times$ iron in water in mg/l as Fe

Mn – $1.6 \times$ manganese in water in mg/l as Mn

P – PAC dose (mg/l)

L – lime dose in mg/l as pure lime $\times$ $(\frac{1}{w} - 1)$ where w is the purity of lime expressed as a fraction

Y – polyelectrolyte dose (mg/l)

In a well-operated treatment works using clarification and filtration all but about 5 mg/l of the calculated dry solids will be removed in the clarifiers.

When softening follows clarification, the sludge dry solids produced in the softener can be calculated as follows (see Sections 8.2–8.4 for softening).

For lime softening to remove carbonate hardness:

$$Sludge\ dry\ solids\ \text{(mg/l)} = 2CaCH + 2.54MgCH + CO_2 + L = S(1)$$

where $CaCH$ – calcium carbonate hardness removed as $CaCO_3$ (mg/l)

$MgCH$ – magnesium carbonate hardness removed as $CaCO_3$ (mg/l)

CO_2 – carbon dioxide removed as $CaCO_3$ (mg/l)

L – as above

For lime-soda softening to remove both carbonate and non-carbonate hardness:

$$Sludge\ dry\ solids\ \text{(mg/l)} = S(1) + 2CaNCH + 2.54MgNCH$$

where $CaNCH$ – calcium non-carbonate hardness removed as $CaCO_3$ (mg/l)

$MgNCH$ – magnesium non-carbonate hardness removed as $CaCO_3$ (mg/l)

For caustic soda softening to remove carbonate hardness:

$$Sludge\ dry\ solids\ \text{(mg/l)} = CaCH + 0.54MgCH = S(c)$$

For caustic soda softening to remove both carbonate and non-carbonate hardness:

$$Sludge\ dry\ solids\ \text{(mg/l)} = S(c) + CaNCH + 0.54MgNCH$$

When softening and clarification processes are carried out together then the clarifier and softener dry solids should be combined.

Where coagulants are used, the water is usually filtered by rapid gravity or pressure filters. The coagulated water may pass directly to the filters, in which case there is only the sludge content of the filter backwash water to deal with, or there may be a clarification stage including the use of sedimentation or flotation tanks before filtration: the sludge drawn off from these tanks must be considered as well. Where there is no clarification before filtration (i.e. direct filtration with coagulation), the backwash water volume can range from 1 to 5% of plant throughput, with a quoted average of 3%. With clarification, a filter will use a washwater volume equivalent to $2\frac{1}{2}$ to 3 times the media bed volume per wash, which is about 1–2.5% of works throughput depending on the filter wash frequency. The actual values are dependent upon raw water quality, the dosage of coagulants and other chemicals used, and frequency of filter washing. Filter backwash water usually contains 0.01–0.10% w/v* dry solids with an average of 0.03% w/v and up to 0.2% w/v

---

*x% w/v is percent weight per volume = x grammes of dry solids/100 ml of liquid containing the solids either in soluble or insoluble form.

for direct filtration with coagulation. Sludge withdrawn from clarifiers may amount to about 1.5 to 2.5% of works throughput and contains 0.1–1.0% w/v solids with an average of 0.3% w/v. Dissolved air flotation tanks employing mechanical float removal methods will produce sludge flows of the order of 0.5–1% of works throughput, with dry solids concentrations ranging from 2 to 3% w/v. Hydraulic float removal methods will increase the sludge quantity to about 1–2% of works throughput with solids concentrations of less than 0.5% w/v and usually about 0.1% w/v.[146] For softeners the quantity discharged is usually in the range 0.5–2% of works throughput, although in some plants it can be as high as 5%, whilst the sludge is withdrawn at concentrations between 5 and 10% w/v. About 85–95% w/w of the solids is calcium carbonate depending on whether the softener is used for combined clarification and softening or for softening only.

For pre-settlement tanks and clarifiers which have to treat waters with heavy silt loads of over 1000 mg/l, the sludge volume removed can range from 5 to 10% of works throughput and may even reach as high as 30%. The concentration of such sludges can range from 2% to 5% w/v solids with a maximum of 10% w/v. These are exceptional cases in which special removal methods are necessary, but it is as well to be aware that they can occur with some raw waters, particularly in overseas countries. At the 1365 Ml/day Karkh water treatment works treating river Tigris water for Baghdad, when the raw water suspended solids loading was 30 000 mg/l, pre-settlement tank desludging accounted for 28% of works throughput and the concentration of the sludge was 10% w/v; the downstream clarifiers accounted for a further 7.5% of throughput at 2.5% w/v concentration. Losses due to filter backwashing remained below 1.5% of throughput. Total water losses were therefore 37% of works throughput.

Membrane filtration plants have a sludge stream and a chemical waste stream. The former is made up of the blow-down and used backwash water and has a composition similar to the feed, unless chemicals such as PAC and coagulants are used in the treatment of the feed. Chlorine may be present if used in the backwash water. The quantity varies between 2–15% of throughput depending on the feed water turbidity, and the suspended solids concentration is of the order of 250 mg/l, similar to that in rapid gravity filter used washwater. The quantity of chemical cleaning waste is a function of the cleaning frequency (which is usually once every few weeks). It is primarily made up of the chemicals used in cleaning and the rinse water.

In slow sand filter installations, the primary filter used washwater quantities and solids concentrations are similar to those of direct filtration plants. The sand washing plant produces up to 10 m^3 or possibly even 20 m^3 of water/m^3 sand, and contains solids primarily consisting of organic debris.

7.46 Methods of disposal

In a well operated clarification and filtration plant using a coagulant, a high proportion of the suspended matter would be removed in the clarifiers, leaving about 2–5 mg/l of suspended solids to be removed by the filters. Therefore used filter washwater is very dilute compared to clarifier sludge and it is preferable to dispose used washwater separately, or keep the two waste streams segregated until the used washwater has undergone settlement to concentrate its sludge solids. There are instances, however, where used washwater and clarifier sludge are combined and treated.

The simplest method of disposing filter washwater is to recycle it with the solids it contains to the treatment process at the works inlet, where it is mixed with raw water. The main benefit is that solids are subsequently removed in the coagulation and clarification process, and can be withdrawn as clarifier sludge at a steady rate and consistent solids concentration. Recycling also helps to maximise washwater recovery. Filter washwater is however generated intermittently in relatively large amounts over short periods, with the solids concentrations declining continually until it is almost clear. Averaged over a day the proportion of filter washwater used is usually within the range 1–2% of the works throughput and contains up to about 10% of the solids removed in the treatment plant. Uncontrolled return of used washwater as shock loads to the main treatment process may adversely effect the coagulation and downstream clarification processes. It should therefore be added uniformly at a steady flow of up to 5% of the works inflow, from a flow balancing tank.

Recycling can return undesirable material (which had already been removed) to the plant. The contaminants of primary concern are micro-animals (e.g. *rotifers, cyclopoids, copepods, chironomid midge larvae and nematode worms*) and protozoan organisms such as *Cryptosporidium* oocysts and *Giardia* cysts. If the numbers of these organisms in the raw water are high, then recycling of washwater (which would contain a large number removed in the filter) would exacerbate the problems. Holding tanks used for washwater can also provide breeding grounds for some of the organisms, and the used washwater returns can therefore potentially carry them in increased numbers to the treatment process. This seeding can, if it occurs, result in infestation of clarifiers and filters, increasing the risk of passing them to the treated water.

Treatment of the used washwater by settlement will reduce the risk of recycling micro-organisms. With 80% settlement efficiency, the increased loading of *Giardia* cysts and *Cryptosporidium* oocysts to the plant through recycling, will be only about 1.2 times the source loading; settlement will also help to reduce manganese, aluminium and iron present as particulates and THM precursor concentrations.[147] Settlement is usually preceded by a trap to remove any filter media carried over in used washwater. The settlement process is aided by a small dose of polyelectrolyte (0.02–0.2 mg/l of polyacrylamide) and can be by batch or continuous flow sedimentation. Batch tanks take the shape of hopper bottomed clarifiers (see Section 7.16) or horizontal flow sedimentation tanks (see Section 7.14) and usually consist of three or more tanks each sized for at least one filter wash and arranged to operate in rotation with one filling, one settling and one being emptied. Supernatant will be decanted using a floating arm drawoff arrangement, with sludge drawn-off from the bottom. Continuous flow sedimentation will be in lamella plate settlers (See Section 7.17). The settlement process will concentrate the solids to 0.3–1% w/v. The supernatant recovered will have a turbidity usually of less than 10 NTU and can be either discharged to a water course or returned to the works inlet at less than 5% of the works raw water flow.

The risk of returning *Cryptosporidium* or *Giardia* to the works can be further reduced by treating the supernatant by either membrane (see Section 7.36) or depth filtration (see Section 7.37) or ozonation (see Section 7.43). In works where pre-ozonation is practised, the high ozone dose required to inactivate *Cryptosporidium* oocysts and *Giardia* cysts can be achieved by injecting a proportion of ozone required for main stream ozonation into the supernatant return. An allowance should be made in the applied dose for the ozone demand of the supernatant. The minimum '*ct*' value required is typically 15 mg.min/l with about 2–3 minutes contact time.

The membrane filter sludge stream can be treated in a manner similar to rapid gravity filter washwater treatment. The chemical cleaning wastes from membrane plants can be either treated on site or removed for off-site disposal, depending on the quantity. On-site treatment usually consists of flow balancing followed by pH correction and oxidant residual removal (e.g. sodium bisulphite dosing to remove residual chlorine).

7.47 Sludge thickening and disposal

Clarifier sludge and settled sludge from used filter washwater settlement tanks are mixed and concentrated in a continuous flow thickener where the residence time of the supernatant and the sludge can be varied independently of each other. The thickeners must be preceded by flow balancing tanks to contain and mix intermittent sludge discharges, and feed the thickeners at a consistent concentration and uniform rate. In applications where used washwater (without settlement) is mixed with clarifier sludge, the sizing of the flow balancing tanks becomes critical because of the large surges of dilute, used washwater. The sludge concentration achieved in the thickener is independent of the feed concentration. However, the greater the feed volume, the larger the thickener will be. There are several thickener designs in use, most are of the settlement type developed for industrial applications which uses heavy duty scrapers with a picket-fence attachment. A design developed in UK by the Water Research Centre[148] specifically for waterworks sludge thickening applications, consists of a cylindrical tank of water depth 2 3.5 m with a shallow sloping floor (1 in 20). Sludge is introduced at the central feed well and the supernatant overflows a peripheral weir. The sludge is thickened by the action of a specially designed rake which also moves the thickened sludge to a central hopper, of included angle 60°, for intermittent discharge under hydrostatic head.[149]

With the aid of polyelectrolytes, coagulant sludges can be thickened to concentrations from 3 to 6% w/v solids. The polyelectrolyte dose can be in the range 0.1–1 g/kg of dry solids. The supernatant overflow will have a turbidity less than about 10 NTU. The hydraulic loading is usually about 1.5 $m^3/h.m^2$ and the dry solids loading is about 4 kg/$h.m^2$. Softener sludge is rarely thickened because it concentrates well in the clarifiers. If necessary it can be further thickened to 20–30% w/w or more at a dry solids loading rate of about 8 kg/$h.m^2$. To prevent the softener sludge becoming too thick underflow is sometimes recycled to the feed. Dissolved air flotation (see Section 7.18) is sometimes used in waterworks sludge thickening.[150] The hydraulic loading rate is in the range 4–5 $m^3/h.m^2$ and the solids loading rate is about 5 kg/$h.m^2$. The sludge is thickened to about 4% w/v. For optimum thickener performance polyelectrolyte dosing is required and the dose should be applied proportional to the feed solids concentration and flow. It should be well mixed using an in-line static mixer or similar.

Each thickener should have a dedicated feed pump of the progressive cavity type. Desludging can be initiated by sludge blanket level and terminated after a pre-set time, or when the solids concentration in the discharge measured in the outlet pipe falls to a pre-set value. In a well-operated thickener the supernatant is normally of an acceptable chemical and physical quality for recycling to the treatment process, but the potential risk of micro-organism and micro-animal return in clarifier sludges is far greater than in the case of recycling used washwater. Treatment for micro-organism removal similar to that for the supernatant from used washwater settling tanks will therefore be required. Chemical quality problems can also arise in thickeners when sludges containing oxides of iron and manganese and natural organic matter are allowed to age. Anaerobic conditions can then develop and release iron, manganese, colour and organics that would impart taste and odour into the supernatant, which if recycled, may consequently have an adverse effect on the treatment plant performance. Recycling can also lead to accumulation of toxic monomer (derived from polyelectrolytes used in thickening) in the treatment process, thereby affecting the treated water. This should be calculated by mass balance on the assumption that the monomer is fully dissolved and will be present in water.

Disposal of thickened sludge is always a problem. Discharging sludge back to a river from which the raw water has been abstracted is, of course, the simplest and most economical disposal method, but the practice is only possible where good dilution is available and there are no adverse environmental consequences. In some cases it is possible to discharge unthickened waterworks sludge to the public sewer, provided the proportion of sludge solids is less than 10% of the sewage sludge solids and adequate velocity (> 0.75 m/s) is maintained in the sewers at all times to prevent silting in pipes. Thickened sludge may be piped or tankered to a sewage works. Coagulant sludges may aid primary sedimentation, thus reducing the solids load to the subsequent biological stages and help in the removal of phosphorous. Thickened sludge may also be mixed with digested sludge for disposal. Iron sludges can prevent hydrogen sulphide gas formation and therefore corrosion of concrete sewers.[151]

7.48   Sludge dewatering

Dewatering of coagulant sludges is difficult because of their gelatinous nature. Materials such as lime, fly ash or diatomaceous earth gives body to the sludge and ease dewatering, but the quantity of dry solids is increased by 50–100% by weight. The dewaterability of softener sludge depends on the Ca:Mg ratio of the sludge; those with a Ca:Mg ratio less than 2:1 will be difficult to dewater while those with a Ca:Mg ratio greater than 5:1 dewater readily.

Disposal of thickened sludge by lagooning or discharging to drying beds is still widely used in some parts of the world where land is cheap and abundant. Lagoons are shallow structures (1.0–1.5 m deep) excavated from, or formed by, impoundment with earth embankments on porous ground above the water table. In the tropics they may be about 2.5 m deep. Once full, they are left to dry. The dried sludge is removed for disposal and the lagoons are returned to service for another fill. A complete installation should be provided with a large number of small lagoons, some being filled, some rested for drying and some being emptied. Dewatering is by percolation (although with coagulation sludges the base soon becomes impervious), settlement and decanting, and finally evaporation. The structure is provided with weir boards at several points to allow decanting of the supernatant and rainfall. In cold countries some dewatering through freezing and thawing can be accomplished, but it is unlikely that more than one application can be dewatered in this manner. In lagooning, sludge is usually placed in layers in several lagoons and allowed to dry before the next layer is placed. This is continued until the total depth is utilised. The coagulant sludges generally consolidate to 10–15% w/w dry solids* and softener sludge to greater than 50% w/w dry solids. The dried sludge is removed by using draglines or front-end loaders.

---

*x% w/w dry solids is x percent weight per weight $= x$ g of dry solids/100 g of wet solids (sludge).

Drying beds, unlike lagoons, have a permeable base of 150 to 250 mm of sand (effective size of 0.3–0.75 mm, uniformity coefficient less than 4), supported by about 300 mm of graded gravel laid over an underdrainage system of vitrified clay or plastic pipes laid with open joints and covered with coarser gravel. As an alternative porous concrete floors are sometimes used. Dewatering in drying beds is by drainage and evaporation, with the former accounting for about 40–50% of the dewatering. The average rate of evaporation of wet sludge is about 80% of that for free water. Rainfall delays the drying process, as only about 60% of the rainfall is lost by drainage, whereas rainfall during the later stages of drying when the bed is cracked, has little effect on the drying time. Sludge is usually placed to a depth of about 200–300 mm and coagulant sludges are removed at about 15–25% w/w dry solids. Drying beds are not extensively used for softener sludges because sludge penetrates the sand bed during drainage. This could be overcome by polyelectrolyte conditioning and sludge is dried to over 50% w/w solids. In drying beds, layering is not normally practised as it will retard the dewatering by drainage. The operating and emptying of drying beds is similar to those of lagoon. For lagooning and drying beds to succeed as dewatering processes the net evaporation rate must exceed rainfall for a considerable part of the year. In sizing an allowance should therefore be made for storage of sludge during winter and wet months when drying is minimal. The ability of a sludge to dewater by gravity varies a great deal and is very much dependent upon the characteristics of the individual sludge. Aluminium coagulant sludges usually drain more slowly than do iron coagulant sludges, although it has been suggested that the rate of draining for aluminium coagulant sludges can be increased by 50% or more by using a conditioning agent.[152]

Dewatering of sludge by plate pressing is becoming increasingly used for polymer thickened clarification sludges. In waterworks sludge dewatering, the recessed plate design is commonly used (the other being the plate-and-frame type). A press contains a set of a horizontal stack of rectangular or square plates covered with a filter cloth to provide a series of chambers clamped between two fixed end plates. The plates are suspended either from an overhead I-beam or on two side bars (see Plate 20). Plates are fabricated in steel, ductile iron or polypropylene, and cloths are of nylon, polypropylene or polyester fabric. Sludge is admitted from one or both ends through a central feed orifice in each plate. The cloth is secured centrally at the feed orifice by a sewn neck. Filtrate is collected through peripheral drainage ports. Initially the press is closed and the sludge is pumped using positive displacement pumps to fill the chambers. The separation of liquid and solids takes place in each chamber, with the solids being retained on the cloth. The filtrate passes through the cloth and emerges from the drainage ports. The filtrate flow rate is at its maximum at the start, remaining reasonably constant until the pressure begins to build up with the gradual formation of the cake. When the pumping pressure reaches the operation value the filtration rate begins to decline. The filtration time can vary from 60 to 90 minutes. There is then a period of constant pressure filtration (which can last several hours) with a continual decline in filtration rate which eventually ceases when the chambers are full of dry cake. The pump is then stopped, the pressure is released, the press ends are unclamped and the plates are separated from one end, one at a time, to release the cake. The presses are normally operated at pressures from about 7 to 8 bar or sometimes from 14 to 16 bar. For coagulant sludges a 25 mm cake thickness is used; thicker cakes (up to 35 mm) are feasible, but at the expense of increased filtration time.

Recessed plate presses can give a cake of 20–30% w/w dry solids. The volumetric capacity of a press is determined by the chamber depth (25–35 mm), plate dimensions, (0.5

× 0.5 m up to 2 × 2 m) and the number of chambers (up to about 160). For example the capacity of a 30 mm deep, 150 chamber, 2 × 2 m plate press is 14.25 m³ which is equal to the volume of the cake. If the sludge is to be dewatered to 20% w/w dry solids, then the weight of solids per pressing is about 3.25 t.* A drawback with the recessed plate press is that the filtration time can be up to 10 hours or even longer and the cycle time which includes downtime for filling, cake drop and cloth washing, can be 12 hours or longer. This drawback is overcome in the membrane plate press where one of the recessed plates in each chamber is replaced by an inflatable rubber or polypropylene ribbed membrane moulded round a steel insert plate, over which the filter cloth is laid. Filtration is carried out as in the recessed plate design at about 7–8 bar. The pump is then stopped, the membrane is inflated using compressed air or water and the remaining water in the sludge is squeezed out at about 15 bar. The cake thickness is reduced from say 32 mm to about 15 mm. These presses produce thinner and dryer cakes in a much shorter time. Cake-dry solids for coagulant sludges range from 25 to 45% w/w and those for softener sludges are usually greater than 60% w/w. The filtration and compression time can be about 2 hours (made up of 90 and 30 minutes, respectively) and the total cycle time is about 4 hours, thus giving six pressings a day, compared to two for the recessed plate press. In the example for the recessed plate press, the volume of cake after compression to 15 mm is 7.13 m³. If the sludge is dewatered to 30% w/w dry solids, then weight of solids per pressing is 2.6 t. The sludge processed per day by similar recessed and membrane plate presses will therefore be 6.5 and 15.6 t dry solids, respectively. In the Johore River water treatment works serving Singapore there are six membrane plate presses, each with 2 m × 2 m plates and 30 mm deep, with 134 chambers. The cake is compressed to 20 mm. The raw water is river derived and contains a high proportion of silt and therefore the sludge can be dewatered to 50% w/w dry solids. Press filling/filtration and compression times are 85 and 45 minutes, respectively. Total cycle time is 2.75 hours. Each press has a capacity of 6 t dry solids per pressing. The press feed pumps are usually piston ram, hydraulic diaphragm or progressive cavity type. In some application progressive cavity pumps are used only to fill the press. Polyelectrolyte is not always added to the press feed; sometimes polyelectrolyte conditioning is limited to thickening only. Filter pressing is a batch process. All operations in a cycle can be fully automated. The filtrate produced from a press initially contains about 100 mg/l suspended solids reducing to less than 10 mg/l as filtration proceeds. The overall solids capture is better than 98% (99–99.5%). The filter cake is usually discharged into a hopper located underneath the press and removed by screw conveyors for disposal.

Centrifuging of sludge has always had a place in the UK for the dewatering of softening sludges and, in recent years, has also been used successfully for coagulant sludges. Centrifuges of several designs are available. The solid bowl decanter type, also known as the scroll centrifuge is widely used on waterworks sludge. The bowl is a cylinder on a horizontal axis with a conveying conical section at one end, called the beach, and an inward facing flange or adjustable weir at the opposite end. A helical screw conveyor (scroll) is mounted coaxially inside the bowl with a very small radial clearance. The feed sludge is introduced into the revolving bowl of the centrifuge through a feed tube at the axis of rotation. The centrifugal action distributes the feed around the periphery of the

---

*1 t of wet sludge occupies $(1-0.6f)$ m³ where f is the fraction of dry solids by weight in the sludge. It is assumed that dry solids in waterworks sludge has a specific gravity of about 2.5.

A plenum floor for a rapid gravity filter under construction (see Fig. 7.14). Left shows dwarf walls constructed for supporting pre-formed plastic formwork for the underside of the filter floor. On the right, nozzle bushes (with sealing caps) are shown inserted into pre-formed holes in the floor formwork with a matrix of rebar reinforcement in place. The floor is then concreted flush with the sealing caps, and after curing, the sealing caps are removed and the nozzle stems and domes screwed in to a uniform level. (Paterson Candy Ltd., Isleworth, UK)

A rapid gravity filter under combined air and water backwash

Plate 17

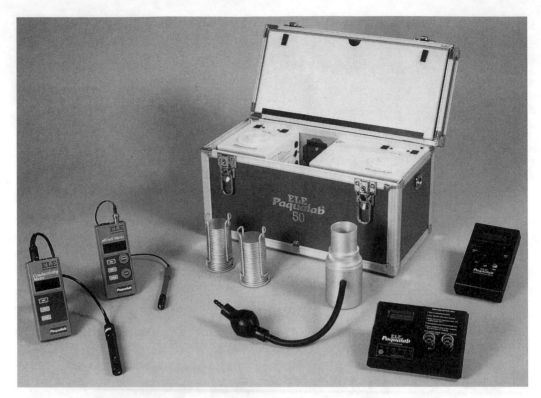

A typical portable test kit for carrying out comprehensive bacteriological and chemical analyses in the field. The kit shown includes equipment for membrane filtration, two battery powered incubators (permitting simultaneous faecal and total coliform analyses), meters for pH, conductivity, and turbidity, together with a photometer and a range of tablet reagents for conducting some 40 chemical tests of water quality. (ELE International Ltd., Hemel Hempstead, UK)

The portable test kit assembled for operation

Plate 18

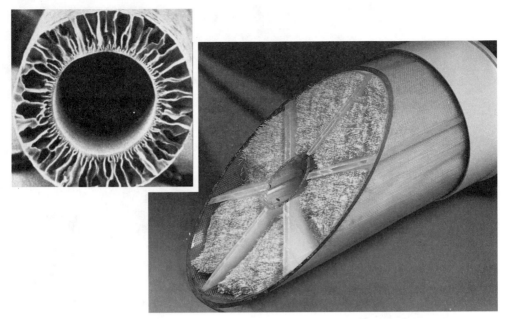

Sectional view of an ultrafiltration membrane cartridge containing about 10 000 fibres and with a treatment capacity of 4–7 m^3/h. The insert shows the cross-section of a hollow fibre membrane of internal diameter 0.8 mm, magnification ×30. (Manufacturer: Koch International UK Ltd, Stafford, UK)

A 65 Ml/day membrane filtration plant installed year 2000 at Homesford, for Severn Trent Water. (Engineers: Binnie, Black & Veatch)

Plate 19

Slow sand filters, Lough Neagh scheme, Belfast water supply 1972. (Engineers: Binnie & Partners)

Sludge press of handling capacity 35 t/week dry solids at Frankley treatment works, Severn Trent Water UK, 1993. (Engineers: Binnie & Partners. Manufacturers: Edwards & Jones, Stoke-on-Trent, UK)

Plate 20

bowl where sedimentation of solids takes place. The settled solids are continuously conveyed along the inside of the bowl, usually counter-current to the feed (the inlet is at the tapered end as opposed to co-current design where the inlet is at the cylindrical end), to the discharge end where the tapered reduction in the bowl diameter causes the solids to emerge clear of the water level. Further dewatering takes place during the movement of solids over this tapered section. The centrate builds up in the bowl and overflows the adjustable weir at the opposite end of the bowl.

The parameters that affect performance efficiency are bowl speed, scroll differential speed, pond depth, sludge feed rate and polyelectrolyte dose. To increase cake dryness, bowl speed and feed rate should be increased and scroll differential speed, pond depth and polyelectrolyte dose should be decreased. To increase solids recovery, bowl speed, pond depth and polyelectrolyte dose should be increased and scroll differential speed and feed rate should be decreased. Centrifuges are typically operated at about 1800–3500 rpm to give a centrifugal force at the wall of the bowl of about $1500-2000g$. The speed differential between the bowl and the scroll is in the range 2–40 rpm but, usually about 10 rpm. Unlike filter pressing, centrifuging is a continuous process. Its performance on coagulant sludges depends to a considerable degree on polyelectrolyte conditioning of the sludge, and polyelectrolyte usage is high and usually in the range 2–6 kg/t of dry solids.[153] The feed to the centrifuges should be maintained at <3% w/v to ensure good mixing of polyelectrolyte with the solids. Coagulant sludges can be dewatered to about 15–25% w/w depending on the nature of the raw water and softening sludges to about 40–50% w/w. Overall solids recovery is normally better than 95% (98–99%). Centrate (i.e. the effluent) water quality from a centrifuge is usually poor with a suspended solids concentration in the range 300–1000 mg/l. It may be spread on land or discharged to a sewage treatment works. High demand for energy (about 0.05 kWh/kg dry solids) is a disadvantage of the centrifuging process.

Freezing and thawing have also provided an effective method of dewatering clarification sludges. Natural freezing is more appropriate to countries with more severe winter conditions than in UK. There are a few plants where a mechanical freeze-thaw method is used for dewatering,[154] the oldest being at the Daer water treatment plant in Scotland.

Belt presses, where pressure is applied between moving endless belts, are being used successfully for coagulant sludge dewatering in limited numbers in USA[155] and France. They are very dependent on the polyelectrolyte dosing and the solids content obtained varies from 15 to 25% w/w. Vacuum drum filters are successfully used for dewatering softener sludges. With filter loadings in the range 300–450 kg dry solids/h.m^2, cake solids up to 65% w/w have been achieved.[156] Research will certainly continue on sludge dewatering because of the need in many countries to comply with increasingly stringent conditions for disposal.

7.49 Resources recovery from sludge

Economic and regulatory constraints, and environmental issues are driving water treatment plant operators to examine alternative disposal methods and in particular, resources recovery.[157] The uses of coagulant sludges include the manufacture of bricks[158] and cast iron[154] and, as an additive in the manufacture of cement,[159,160] also in the elimination of phosphorous in sewage treatment and agricultural use.[161] Coagulant sludges when used as soil conditioners are known to adversely affect plant growth due to the ability of aluminium and iron to fix phosphorous in the soil making it unavailable to plants. Coagulant sludges can be used in eutrophied lakes to fix phosphorus to minimise algal blooms.

Softening sludge cake which has a high purity calcium carbonate with little or no magnesium and is free from coagulant and suspended solids, is sold for agricultural use or to industry for products such as cosmetics. Such a pure calcium carbonate sludge is only likely to result from the softening of well water derived from a chalk aquifer.

The recovery of aluminium sulphate from sludge has received considerable attention in the past because it appeared to offer partial reuse of the coagulant and also reduce the volume of sludge to be handled. In theory every 1 g of aluminium hydroxide in the sludge could be recovered as 2.2 g of aluminium sulphate by treating the sludge with 1.9 g sulphuric acid at about pH 2. In practice the amount of acid can be much greater if other acid-soluble material is present in the sludge. Typically recovery rates greater than 75% are feasible.[162] The recovered aluminium sulphate is usually mixed with the commercial product before reuse in the water treatment process. Due to concerns over the possible accumulation of metals and other impurities such as organic material, the practice lost favour and has been discontinued in many installations. It is likely to be economical only when the cost of the sulphuric acid is lower than that of purchasing aluminium sulphate. There are also benefits from sludge volume reduction and improved dewatering characteristics of acidified sludge.

Lime recovery from softening sludges by recalcination is economically viable since the softening process causes calcium hardness to precipitate as calcium carbonate (see Section 8.3). Generally more lime is produced than is added for treatment. Recalcination is carried out in a furnace at about 900–1000°C. The available lime in the recalcined product may be only about 60–75% depending on the inerts present, such as magnesium, iron, silica and other compounds. Carbon dioxide which is a by-product is available for use in the recarbonation and pH correction of water on the plant.

References

1. Sykes G. and Skinner F. A. (eds). *Microbial Aspects of Pollution*. Academic Press, London, 1971, p. 160.
2. Kay D. and Hanbury R. *Recreational Water Quality Management*, Vol. 2, Ellis Horwood, Chichester, 1993.
3. Young E. F., Wallingford F. E. and Smith A. J. E. Raw Water Storage. *JSWTE*, **21**(2), 1972, pp. 127–152.
4. DoE. *First Annual Report of the Steering Committee on Water Quality*. Circular No.22/72, 1971.
5. DoE. *Second Biennial Report: February 1977–February 1979*, 1979.
6. DoE. UK Water Research Centre, *Bankside Storage and Infiltration Systems*. Report to the Department of the Environment, 1993.
7. Ward R. C. and Loftis J. C. Monitoring Systems for Water Quality. *Water Resources Bulletin* **19**, 1989, pp. 101–118.
8. Palmer D. *River Water Quality Monitoring Equipment Annual Handbook*. CIWEM, 1993.
9. Campbell A. T. *et al.* Identification of Odour Problems in the R. Dee. *JIWEM*, **8**(1), 1994, pp. 52–58.

10. Harman M. M. I. and Baldwin I. G. *Intake Protection Systems.* LR 45-M, WRc, 1988.
11. Fowlis I. A. On-line Analysis for Trace Levels of Phenolic Compounds. *Proc. Brighton Crop Protection Conference,* 1996, pp. 999–1006.
12. OECD, *Eutrophication of Waters: Monitoring, Assessment and Control.* Paris, 1982.
13. Moss B. *Ecology of Freshwaters.* Blackwell Scientific, Oxford, 1988.
14. Lund J. W. G., Mackereth F. J. H. and Mortimer C. H. *Phil Trans Roy Soc,* B246, 1963, p. 260.
15. Burman N. P. *Proc SWTE* **14,** 1965, p. 125.
16. Symons J. M. Quality Control in Reservoirs, A Committee Report, *JAWWA,* **63**(9), 1971, pp. 597–604.
17. Pastorok R. A. Review of Aeration/Circulation for Lake Management in Restoration of Lakes and Inland Waters. *Proc International Conference,* Portland, Maine, 1980.
18. Windle Taylor E., *Reports on the Bacteriological and Biological Examination of the London Waters,* Metropolitan Water Board, 1961–1968.
19. Henderson-Sellers B. *Engineers Limnology,* Pitman Advanced Publishing Program, Pitman, London, 1984.
20. Ridley J. E., Cooley P. and Steel J. A. P. Control of Thermal Stratification in Thames Valley Reservoirs. *JSWTE,* **15,** 1966, p. 225.
21. Davis J. M. Destratification of Reservoirs, A Design Approach. *Water Services,* August 1980, pp. 457–504.
22. Lorenzen M. and Fast A. *A Guide to Aeration/Circulation Techniques for Lake Management.* EPA-600/3–77–004.
23. Speece R.E. Lateral Thinking Solves Stratification Problems. *Water Quality International,* 3, 1994.
24. Steel J. A. P. 1975 Management of Thames Valley Reservoirs, in: *Effects of Storage on Water Quality,* WRc, 1975.
25. Toms I. P., 1981 Reservoir Management. *Proc Water Industry '81,* CEP Consultants, London, 1981.
26. Camp T. R. Sedimentation and the Design of Settling Tanks. *Trans ASCE,* **III,** 1946, pp. 895–936.
27. Camp T. R. Sedimentation and the Design of Settling Tanks. *Trans ASCE,* **III,** 1946, p. 913.
28. McLaughlin R. T. The Settling Properties of Suspensions. *JASCE,* HY12, Paper 2311, December 1959.
29. Kawamura S. *Integrated Design of Water Treatment Facilities.* John Wiley, New York, 1991.
30. Camp T. R. and Stein P. C. Velocity Gradients and Internal Work in Fluid Motion. *J. Boston Soc Civil Eng,* **30,** 1943, p. 219.
31. Chow V. T. *Open Channel Hydraulics.* McGraw Hill, New York, 1959.
32. Vigneswaran S. *et al.* Trends in Water Treatment Technologies. *Environmental Sanitation Reviews* No. 23/24, Asian Institute of Technology, Bangkok, 1987.
33. Fair G. M. *et al. Water and Wastewater Engineering,* Vol. 1. John Wiley, New York, 1968.
34. Schulz C. R. and Okun D. A. *Surface Water Treatment for Developing Countries.* John Wiley, New York, 1968.

35. Skeat W. O. (ed.). *Manual of British Water Engineering Practice*, Vol. III, 4th edn. W. Heffer & Sons, Cambridge, 1969, p. 204.
36. Amirtharajah A. (ed.). *Mixing in Coagulation and Flocculation*. AWWA, 1991, p. 385.
37. Fawcett N. S. J. and Sidall R. A Comparison of Gate and Turbine Impellers for Flocculation. *Proc IChemE Joint Subject Group Meeting on Mixing in the Water Industry*, London, Oct. 1992.
38. Packham R. F. The Theory of Coagulation Process. (1) The Stability of Colloids. (2) Coagulation as a Water Treatment Process. *JSWTE*, **11**(1), 1962, pp. 50 and 106, and **12**(1), 1963, p. 15.
39. Amirtharajah A. Design of Rapid Mix Units, in: *Water Treatment Plant Design* (ed. Sanks R. L.). Ann Arbor Science, Ann Arbor, Michigan, 1978, p. 132.
40. Brejchova D. and Wiesner M. R. Effect of Delaying the Addition of Polymeric Coagulant Aid on Settled Water Turbidity. *Water Science Technology*, **26**(9–11), 1992, pp. 2281–2284.
41. Bauer M. J. *et al.* Enhanced Rapid Gravity Filtration and Dissolved Air Pretreatment of River Thames Reservoir Water. *Proc IAWQ-IWSA Joint Specialist Conference on Reservoir Management and Water Supply*, Prague, May 1997.
42. Parr W. *et al. Method of Controlling Phytoplankton*. UM 1223, WRc, 1991.
43. Markham L. *et al.* Algal and Zooplankton Removal by Dissolved Air Flotation at Severn Trent Ltd. *Proc Int Conf on Dissolved Air Flotation*, London, April 1997.
44. Bare W. F. R. *et al.* Algal Removal Using Dissolved Air Flotation. *JWPCF*, **47**, 1975, pp. 153–169.
45. Edzwald J. K. and Wingler B. J. Chemical and Physical Aspects of Dissolved Air Flotation. *J. Water SRT-Aqua*, **39**(1), 1990, pp. 24–35.
46. WHO. *Guidelines for Drinking Water Quality*, Vol. 2, 2nd edn. WHO, Geneva, 1996.
47. O'Conner J. T. and Brazos B. J. Evaluation of Rapid Sand Filters for Control of Micro-Organisms in Drinking Water. *Public Works*, **128**(3), 1997, pp. 52–56.
48. Capacity and Loadings of Suspended Solids Contact Units, Committee Report. *JAWWA*, 1951, p. 263.
49. Hudson Jr H. E. *Water Clarification Processes*. Van Nostrand Reinhold, New York, 1981, p. 123.
50. Hazen A. On Sedimentation. *Trans ASCE*, **53**, 1904, p. 63.
51. Degremont *Water Treatment Handbook*, Vol. 1, 6th edn. Lavoisier Publishing, Paris, 1991.
52. Coulson J. M. *et al. Chemical Engineering*, Vol. 1, 3rd edn. Pergamon Press, Oxford, 1991, p. 319.
53. Yao K. M. Design of high rate settlers, *JASCE Env Eng Div*, **99**(EE5), 1973, pp. 621–637.
54. Galvin R. M. Lamella Clarification in Floc Blanket Decanters: Case Study. *J. Water SRT-Aqua*, **41**(1), 1992, pp. 28–32.
55. Cailleaux C. *et al.* Study of Weighted Flocculation in View of a New Type of Clarifier. *Aqua*, **41**(1), 1992, pp. 18–27.
56. Kolarik L. O. Magnetic Microparticles in Water Treatment. *Water Supply* **12**(3/4), 1994, pp. 253–262.
57. Schofield T. Birmingham Frankley Water Treatment Works Redevelopment. *Water Science Technology*, **31**(3–4), 1995, pp. 213–235.

58. Longhurst S. J. and Graham N. J. D. Dissolved Air Flotation for Potable Water Supply. *Public Health Engineer*, **14**(6), 1987, pp. 71–76.
59. Rees A. J. *et al*. Water Clarification by Flotation. *TR 114*, WRc, April 1979.
60. Pontius F. W. (ed.). *Water Quality and Treatment*, 4th edn. AWWA, McGraw Hill, New York, 1990.
61. Pfeifer B. J. *et al*. DAF and Catalytic Filtration for Treatment of a Challenging Water Supply in a Small Northern Community. *Proc AWWA Annual Conference*, Atlanta, Georgia, 1997.
62. Nickols D. and Crossley I. A. The Current State of Dissolved Air Flotation in the USA. *Proc International Conference on Dissolved Air Flotation*. CIWEM, London, April 1997.
63. Edzwald J. K. *et al*. Integration of High Rate Dissolved Air Flotation Treatment Technology Into Water Treatment Plant Design. *Proc AWWA Conference*, AWWA, Chicago, June 1999.
64. Edzwald J. K. *et al*. Flocculation and air requirement for dissolved air flotation. *JAWWA*, **84**(3), 1992, pp. 92–100.
65. Amato T. DAF: Its Place in an Integrated Plant Design, *Proc International Conference on Dissolved Air Flotation*, CIWEM, London, April 1997, pp. 373–379.
66. Zabel T. Flotation in Water Treatment, in: *Scientific Basis of Flotation* (ed. Ives K. J.). NATO ASI Series, Boston, MA, 1984, pp. 349–377.
67. Dahlquist J. State of DAF Development and Application to Water Treatment in Scandinavia. *Proc International Conference on Dissolved Air Flotation*, CIWEM, London, April 1997, pp. 201–213.
68. Boisdon V. *et al*. Combining of ozone and flotation to remove algae, *Water Supply*, **12**(3/4), 1994, pp. 209–220.
69. Baron J. *et al*. Combining Flotation and Ozonation: The Flattazme Process. *Proc International Conference on Dissolved Air Flotation*, CIWEM, London, April 1997, pp. 341–360.
70. Wilson D. *et al*. The Use of Ozaflotation for the Removal of Algae and Pesticides from a Stored Lowland Water. *Ozone Science and Engineering*, **15**(6), 1993, pp. 981–996.
71. Eades A. *et al*. (a) Counter-Current Dissolved Air Flotation/Filtration. *Water Science Technology*, **31**(3–4), 1995, pp. 173–178. (b) Counter-Current Dissolved Air Flotation Filtration, COCO-DAFF, *Proc International Conference on Dissolved Air Flotation*, CIWEM, London, April 1997, pp. 323–340.
72. WHO. *Practical Solutions in Drinking Water Supply and Waste Disposal for Developing Countries*, International Reference Centre for Community Supply, The Hague, 1977.
73. Packham R. F. and Ratnayaka D. D. Water Clarification with Aluminium Coagulants in the UK. *Proc International Workshop on Aluminium in Drinking Water*, IWSA, Hong Kong, January 1992.
74. Carroll B. A. and Hawkes J.M. Operational Experience of Converting From Aluminium to Iron Coagulants at a Water Supply Treatment Works. *Water Supply* **9**, 1991, pp. 553–558.
75. Jia-Qian Jiang *et al*. Coagulation of Upland Coloured Water With Polyferric Sulphate Compared Conventional Coagulants. *J. Water SRT-Aqua*, **45**(3), 1996, pp. 143–154.

76. Packham R. F. Polyelectrolytes in Water Clarification. *JSWTE*, **16**(2), 1967, pp. 88–102.
77. DoE. *List of Substances, Products and Processes Approved under Regulations 25 and 26 of the Water Supply (Water Quality) Regulations 1989*, Dec. 1998.
78. Ives K. J. and Gregory J. Basic concepts of filtration. *JSWTE*, **16**(2), 1967, p. 147. and Ives K. J. Theory of Filtration. *Proc IWSA Congress*, 1, p. K.1, 1969, also Filtration: The Significance of Theory. *JIWE*, **25**(1), February 1971, p. 13.
79. Stevenson D. G. The Specification of Filtering Materials for Rapid Gravity Filtration. *JIWEM*, **8**(5), 1994, pp. 527–533.
80. British Water. The Specification, Approval and Testing of Granular Filtering Material and Code of Practice for the Installation of Filtering Material. BW:P.18.96R, June 1996.
81. Cleasby J. L. Filter Rate Control Without Controllers. *JAWWA*, **61**(4), 1969, pp. 181–185.
82. Cleasby J. L. Declining Rate Filtration. *Water Science Technology*, **27**(7–8), 1993, pp. 11–18.
83. Cleasby J. L. Status of Declining Rate Filtration Design. *Water Science Technology*, **27**(10), 1993, pp. 151–164.
84. Stevenson D. G. *Water Treatment Unit Processes*. Imperial College Press, London, 1998.
85. Tebbutt T. H. Y. and Shackelton R. C. Temperature Effects in Filter Backwashing. *Public Health Engineering* **12**(3), July 1984, pp. 174–178.
86. Miller D. G. Rapid Filtration Following Coagulation Including the Use of Multi-Layer Beds. *JSWTE*, **16,** 1967, p. 192.
87. Miller D. G. Filtration: Two Experimental Developments. *JIWE*, **25**(1), 1971, p. 21, also remarks on p. 49 by Carr W. and p. 62 by Jeffery J.
88. *AWWA Standard for Filtering Material*. AWWA B100–96, 1996.
89. American Society for Testing Materials. *ASTM D3802*.
90. Humbly M. S. *et al*. Development of a Friability Test for Granular Filter Media. *JCIWEM*, **10**(3), 1996, pp. 87–91.
91. Crowley F. W. and Twort A. C. Current Strategies in Water Treatment Developments. *Proc ICE Conf Water Resources: A Changing Strategy*, 1979.
92. Chan Kin Man D. and Sinclair J. Commissioning and operation of Yau Kom Tau Water treatment works (Hong Kong) using direct filtration. *JIWEM*, **5**(2), 1991, pp. 105–115.
93. Hutchinson W. and Foley P. D. Operational and experimental results of direct filtration. *JAWWA*, **66**(2), 1974, pp. 79–87.
94. Lombard H. K. and Hanroff J. Filter Nozzle and Underdrain Systems Used in Rapid Gravity Filtration. *Water SA*, **21**(4), 1994, pp. 281–298.
95. AWWA. *Water Quality and Treatment*, 3rd edn. McGraw Hill, New York, 1971.
96. Mellevialle J. *et al*. (ed.). *Water Treatment Membrane Processes*. AWWA Research Foundation, McGraw Hill, New York, 1996.
97. Yoo S. R. *et al*. Microfiltration: A Case Study. *JAWWA*, **87**(3), 1995, pp. 38–49.
98. Jacangelo J. G. *et al*. Mechanisms of Cryptosporidium, Giardia and MS2 Virus Removal of MF and UF. *JAWWA*, **87**(9), 1995, pp. 107–121.
99. Adham S. S. *et al*. Membranes: A Barrier to Micro-organisms. *Water Supply*, **16**(1/2), Madrid, 1998.

100. Chow C. W. K. *et al.* A Study of Membrane Filtration for the Removal of Cyanobacterial Cells. *J Water SRT-Aqua*, **46**(6), 1997, pp. 324–334.
101. Anon. Predicting Log Removal Performance of Membrane System Using *In Situ* Integrity Testing. *Filtration & Separation* **35**(1), 1998,
102. Mourato D. and Best G. Application of Immersed Microfiltration Membranes, for Drinking Water Treatment. *AWWA Conference,* Florida, November 1998.
103. Gene A. R. Microfiltration Operating Costs. *JAWWA*, **89**(10), 1997, pp. 40–49.
104. O'Neill J. G. An evaluation Of Fibrous Depth Filters for Removal of *Cryptosporidium* Oocysts from Water in Protozoan Parasites and Water, in: *RSC Special Publication 168*. Royal Society of Chemistry, Cambridge, 1995.
105. Anon. The Coppermills Works of the Metropolitan Water Board. *WWE*, August 1972, p. 277.
106. Grey I. W. Lurgan and District Waterworks Joint Board's New Works at Castor Bay. *WWE*, July 1971, p. 297.
107. Pike P. G. *Engineering Against Schistosomiasis Bilharzia; Guidelines Towards Control of the Disease*. MacMillan, London, 1987.
108. Van de Vloed A. *Report to the 3rd Congress of the IWSA*, London, Subject No. 7, 1955.
109. Huisman L. and Wood, W. E. *Slow Sand Filtration*, WHO, Geneva, 1974.
110. Visscher J. T. *et al.* Slow Sand Filtration for Community Water Supply, Technical Paper No. 24, *IRC*, The Hague, June 1987.
111. Logsdon G. S. (ed.). *Slow Sand Filtration*. ASCE, New York, 1991.
112. Kors L. J. *et al.* Hydraulic and Bacteriological Performance Affected by Resanding, Filtration Rate and Pretreatment, in: *Advances in Slow Sand and Alternative Biological Filtration*, (eds Graham N. and Collins R.). John Wiley, Chichester, 1996, pp. 54–264.
113. Lewin J. Mechanisation of slow sand and secondary filter bed cleaning. *JIWE*, **15**(1), 1961, pp. 15–46.
114. Collins M. R. and Eighmy T. T. Modifications to the Slow Rate Filtration Process for Improved Trihalomethane Precursor Removal, in: *Slow Sand Filtration* (ed. Graham N.J.D.). Ellis Horwood, Chichester, 1988, pp. 281–305.
115. Joslin W.R. Slow Sand Filtration: A Case Study in the Adoption and Diffusion of a New Technology. *J New England WWA*, September 1997, pp. 294–303.
116. Glendinning D. J. and Mitchell J. Uprating Water-Treatment Works Supplying the Thames Water Ring Main. *JCIWEM*, **10**(1), 1996, pp. 17–25.
117. Toms I. P. and Bayley R. G. Slow Sand Filtration, Approach to Practical Issues, in: *Slow Sand Filtration* (ed. Graham N. J. D.). Ellis Horwood, Chichester, 1988, pp. 281–305.
118. Harrison N. Dunnore Point water treatment works – Phase 1, Extension. *Water & Sewage Journal*, Spring 1997, pp. 41–42.
119. Van Dijk J. C. and Oomen J. H. C. M. Slow Sand Filtration for Community Water Supply In Developing Countries. Tech. Paper 10, WHO, Geneva, 1978.
120. Galois G. *et al.* Comparative Study of Different Pre-Treatment Alternatives. *J. Water SRT-Aqua*, **42**(6), 1993, pp. 337–346.
121. Welte B. and Montiel A. Removal of BDOC by Slow Sand Filtration: Comparison with Granular Activated Carbon and Effect of Temperature, in: *Advances in Slow*

Sand and Alternative Biological Filtration (eds Graham, N. and Collins R.). John Wiley, Chichester, 1996, pp. 95–104.

122. Ridley J. E. Experiences in the Use of Slow Sand Filtration, Double Sand Filtration and Microstraining. *JSWTE* 16, 1967, p. 170.

123. Lambert S. D. and Graham N. J. D. A Comparative Evaluation of the Effectiveness of Potable Water Filtration. *J Water SRT-Aqua*, **44**(1), 1995, pp. 38–51.

124. Yordanov R. V. *et al.* Biomass Characteristics of Slow Sand Filters Receiving Ozonated Water, in: *Advances in Slow Sand and Alternative Biological Filtration* (eds Graham, N. and Collins R.). John Wiley, Chichester, 1996, pp. 107–118.

125. Bauer M. *et al.* The GAC/Slow Sand Filter Sandwich: From Concept to Commissioning. *Water supply*, **13**(3/4), Osaka, 1995, pp. 137–142.

126. English E. Water Treatment Problems at Lough Neagh. *JIWE*, **26**(4), 1972, pp. 201–210.

127. Wilson D. Uprating of Barmby Water Treatment Works, in: *Advances in Slow Sand and Alternative Biological Filtration*, John Wiley, Chichester, 1996, pp. 439–448.

128. Hendricks D. (ed.). *Manual of Design for Slow Sand Filtration*. AWWA Research Foundation, AWWA, Denver, Colorado, 1991.

129. Klein H.-P. and Berger C. Slow Sand Filters Covered By Geotextiles. *Water Supply* **12**(3/4), Zurich, 1994, pp. 221–230.

130. Schalekamp M. The Development of the Surface Water Treatment for Drinking Water in Switzerland. *Proc IWES Symposium: The Water Treatment Scene – The Next Decade*, 1979.

131. Standen G. *et al.* The Use of Particle Monitoring in the Performance Optimisation of Conventional Clarification and Filtration Processes. *Water Science Technology*, **36**(45), 1997, pp. 191–198.

132. Smith H.V. *et al. Cryptosporidiosis* and *Giardiosis*: The Impact of Waterborne Transmission. *J Water SRT-Aqua*, **44**(6), 1995, pp. 258–274.

133. Guidance Manual Supporting Water Treatment Recommendations from the Badenoch Group of Experts on Cryptosporidium (98/DW/06/5). UK Water Industry Research Limited, London, 1998.

134. Logston G. Removal of Micro-Organisms by Clarification and Filtration Processes, Special Contribution. *Water Supply*, **16**(1/2), Madrid, 1998, pp. 208–209.

135. Edzwald J. K. and Kelly M. B. Control of Cryptosporidium: From Reservoirs to Clarifiers to Filters. *Water Science Technology*, **37**(2), 1998, pp. 1–8.

136. *Cryptosporidium* in Water Suppliers. *Report of the Group of Experts*. HMSO, London, 1990.

137. Ransome M. E. *et al.* Effects of Disinfectants on the Viability of *Cryptosporidium Pavum* Oocysts. *Water Supply* 11, Amsterdam, 1993, pp. 103–117.

138. Finch G.R. and Black E. K. Inactivation of *Giardia* and *Cryptosporidium* Using Ozone. *Proc Eleventh Ozone World Congress*, Volume 2, S-19–1 to S-19–17, San Francisco, California, 1993.

139. *Cryptosporidium* in Water Supplies. *Second Report of the Group of Experts*. HMSO, London, 1994.

140. Kawamura S. Optimisation of Basic Water Treatment Process Design and Operation: Sedimentation and Filtration. *J Water SRT-Aqua*, **45**(3), 1996, pp. 130–142.

141. Rissel J. Particle Counting Augments Turbidity Measurements. *JAWWA*, **23**(1), 1997, pp. 6–7.

142. *Cryptosporidium* in Water Supplies. *Third Report of the Group of Experts*. HMSO, London, 1998.
143. Nieminski E. Effectiveness of Direct Filtration and Conventional Treatment in Removal of *Cryptosporidium* and *Giardia*. *Proc AWWA Conference*, June 1995, pp. 947–962.
144. Clark R. Analysis of Inactivation of *Giardia Lamblia* by Chlorine. *JASCE, Environ Eng Div*, **115**(1), 1989, pp. 80–90.
145. Wickramanayake G. B. *et al*. Inactivation of *Giardia Lamblia* Cysts With Ozone. *Applied Env Microbiol*, **48**(3), 1984, pp. 671–672.
146. Schofield T. Sludge Removal and Dewatering Processes for Dissolved Air Flotation System. *Proc International Conference on Dissolved Air Flotation*, CIWEM, London, April 1997, pp. 309–322.
147. Cornwell D. A. and Lee R. G. Waste Stream Recycling: Its Effect on Water Quality. *JAWWA*, **86**(11), 1994, pp. 50–63.
148. Warden J. H. *Sludge Treatment Plant for Waterworks*. Technical Report 189, WRc, March 1983.
149. Albertson O. E. Evaluating Scraper Designs. *Water Environment and Technology*, **4**(1), 1992, pp. 52–58.
150. Haubry A. and Fayoux C. La Flottation des Boues: Un Avenir Assure, L'Eau, l'Industrie. *Les Nuissances*, **79**(1), 1983, pp. 20–24.
151. McTique N. E. and Cornwell D. Impact of Water Plant Waste Discharge On Wastewater Plants. *Proc Residuals Management Conference*, AWWA/WPCF, San Diego, California, 1989.
152. Novak J. T. and Langford M. The Use of Polymers for Improving Chemical Sludge Dewatering on Sand Beds, *JAWWA*, **69**(2), 1977, p. 106.
153. Piggott G. A. *et al*. Waterworks Sludge Treatment and Disposal Options. Paper presented at *Water Malaysia '92' Conference*, Kuala Lumpur, October 1992.
154. Henke H. Application of Freeze-Thaw for Handling of Sludge from the Treatment of Great Dhunn Reservoir Water. *KIWA/AWWA Research Foundation Experts Meeting*, Nieuwegein, The Netherlands, 1989.
155. Migneault W. Santa Clara Valley District Sludge Concentration and Recycling Program, California Nevada Section. *AWWA Fall Conf*, October 1987.
156. Anon. Lime Softening Sludge Treatment And Disposal. *JAWWA*, **73**(11), 1981, pp. 600–608.
157. Sarfert F. *et al*. Treatment and Utilisation of Waterworks Sludge in Germany. *Water supply* **12**(1/2), Budapest, 1994, pp. 553–558.
158. Anon. *ENDS Report 250*. November 1995, pp. 13.
159. GWF-*Wasser/Abwasser* **137**(14), 1996.
160. Cornwell D. A. and Koppers H. M. M. Slib, Schlamm, Sludge. *AWWA Research Foundation Report*. AWWA, Denver, 1990.
161. Elliott H. A. and Dempsey B.A. Agronomic Effects of Land Application of Water Treatment Sludges. *JAWWA*, **83**(4), 1991, pp. 126–131.
162. Anon. Coagulant recovery system wins big award. *Water Engineering & Management*, **14**(7), 1994, pp. 12–13.

8

Specialised and advanced water treatment processes

Softening of water

8.1   Hardness compounds

A description of hardness was given in Section 6.24 and it was mentioned that softening of public supplies is not generally adopted in the UK in part due to unexplained statistical relationship between the softness of water and the incidence of cardiovascular disease. Modern detergents have also reduced the need to soften water for domestic purposes. But a hard water can form troublesome scale in hot water systems and if a water is excessively hard it may be desirable to soften it, but not below 150 mg/l as $CaCO_3$* the lower limit set by the EC Directive and UK Regulations for a softened water. For industry however, softening of a hard water is often required for process purposes and boiler feed waters.

The compounds producing temporary and permanent hardness in a water are shown in Table 8.1: the former being precipitated as scale when water is heated; the latter not. A large proportion of waters from underground sources are hard, particularly waters from chalk and limestone aquifers which often have a carbonate hardness of 200–300 mg/l as $CaCO_3$. The hardness compounds are taken into solution because the water acquires carbon dioxide from the soil formed by the oxidation of organic matter. A major source of non-carbonate hardness in surface waters is the calcium sulphate present in clays and other deposits. In contrast, many surface waters from the older geological formations in

Table 8.1   Compounds producing temporary and permanent hardness

Causing temporary hardness (carbonate hardness)	Causing permanent hardness (non-carbonate hardness)
Calcium bicarbonate $Ca(HCO_3)_2$	Calcium sulphate $CaSO_4$
Magnesium bicarbonate $Mg(HCO_3)_2$	Magnesium sulphate $MgSO_4$
	Calcium chloride $CaCl_2$
	Magnesium chloride $MgCl_2$

*1 mg/l as $CaCO_3$ = 0.4 mg/l as Ca.

the western and northern areas of the UK are soft or very soft, e.g. 15–50 mg/l as $CaCO_3$, because the rocks are largely impermeable and insoluble.

8.2   Principal methods of softening

There are three principal methods of softening a hard water. In the first, lime (calcium hydroxide) and soda ash (sodium carbonate) are added to the water to change the hardness compounds so they become insoluble and precipitate. The water then requires clarifying and filtering to remove the precipitate. In the second method the nature of the hardness compounds is changed by passing the water through a bed of 'ion-exchange' resin so that the changed compounds do not react with soap, and the water therefore appears 'soft'. In the third method, membrane processes such as reverse osmosis remove all dissolved salts from water at an efficiency of about 90–98%; nanofiltration removes bivalent ions (e.g. Ca^{2+}, Mg^{2+}, CO_3^{2-}, SO_4^{2-}) at an efficiency of about 85–95% and monovalent ions (e.g. Na^+, K^+, Cl^-) at an efficiency of about 40–70% (see Section 8.46). In the lime-soda process the hardness compounds are removed; in the base-exchange process the hardness compounds are changed; and in membrane processes a high proportion of all of the ions are removed. These differences are important because the chemical and membrane processes reduce the total dissolved solids in a water, a feature which is often desirable for industrial applications.

Softening in drinking water treatment is usually applied to a proportion of the flow (split-treatment) to soften the water to a hardness value below the required value and then blended with the unsoftened water – the so-called split-treatment.

8.3   The lime-soda process of softening

The aim of the lime-soda process is to convert calcium and magnesium compounds to the insoluble forms, calcium carbonate ($CaCO_3$) (partially soluble) and magnesium hydroxide ($Mg(OH)_2$).

Magnesium carbonate ($MgCO_3$) unlike calcium carbonate, does not precipitate in cold water. The stages of treatment involved are set out in Table 8.2. It will be observed that to remove calcium hardness, lime is added to remove the temporary hardness, and soda ash to remove the permanent hardness. Several applications of lime and soda ash are needed to remove magnesium hardness. In practice complete removal of hardness is undesirable because this renders a water highly aggressive. Since in most hard waters the calcium temporary hardness forms the major component it often suffices to remove only this by the addition of lime. Caustic soda (sodium hydroxide – NaOH) can be used in place of lime for carbonate and non-carbonate removal; alkalinity reduction is only 50% that of lime softening. Soda ash (sodium carbonate – Na_2CO_3) is formed in the reactions and sometimes it may be supplemented by soda ash addition to remove non-carbonate hardness. The advantages are that caustic soda is easier to handle than lime, only one chemical may be required and the quantity of calcium carbonate sludge produced is less (see Section 7.45). The drawbacks are that the concentration of sodium in the treated water is increased and the higher cost of caustic soda. Chemical softening, due to its high operating pH (9.5–10.5), also removes, by precipitation of many of the heavy metals (e.g. lead, copper, zinc, etc.), arsenic, iron and manganese.

Table 8.2   Lime-soda softening processes

To remove carbon dioxide in water add LIME (not a softening reaction) H_2CO_3 + **Ca(OH)₂** = $CaCO_3$ + $2H_2O$	8.1
To remove calcium temporary hardness add LIME. $Ca(HCO_3)_2$ + **Ca(OH)₂** = $2CaCO_3$ + H_2O	8.2
To remove calcium permanent hardness add SODA ASH. $CaSO_4$ + **Na₂CO₃** = $CaCO_3$ + Na_2SO_4 $CaCl_2$ + **Na₂CO₃** = $CaCO_3$ + $2NaCl$	8.3
To remove magnesium temporary hardness add LIME + more LIME Stage 1 $Mg(HCO_3)_2$ + **Ca(OH)₂** = $MgCO_3$ + $CaCO_3$ + $2H_2O$	8.4
The calcium carbonate precipitates, but the magnesium carbonate does not, so further LIME is added. Stage 2 $MgCO_3$ + **Ca(OH)₂** = $Mg(OH)_2$ + $CaCO_3$ The magnesuim hydroxide and calcium carbonate precipitate.	8.5
To remove magnesium permanent hardness add LIME and SODA ASH $MgCl_2$ + **Ca(OH)₂** = $Mg(OH)_2$ + $CaCl_2$ $MgSO_4$ + **Ca(OH)₂** = $Mg(OH)_2$ + $CaSO_4$ The addition of soda ash then converts the calcium chloride and calcium sulphate to calcium carbonate as in 8.3 above	8.6

Notes.
Compounds in bold are those being added and compounds in italics are those precipitating.
H_2CO_3 – carbonic acid (carbon dioxide in water); $Ca(HCO_3)_2$ – calcium bicarbonate; $Ca(OH)_2$ – calcium hydroxide (hydrated lime); $CaCO_3$ – calcium carbonate; $CaSO_4$ – calcium sulphate; $CaCl_2$ – calcium chloride; H_2O – water; Na_2CO_3 – sodium carbonate (soda ash); $Mg(HCO_3)_2$ – magnesium bicarbonate; $Mg(OH)_2$ – magnesium hydroxide; $MgCO_3$ – magnesium carbonate; $MgCl_2$ – magnesium chloride; $MgSO_4$ – magnesium sulphate.

8.4   Softening plant

Lime is used in the hydrated form, which is a dry powder. It is delivered by bulk tankers and blown into storage silos by compressed air. A dust filter and a pressure relief valve are therefore essential at the top of each silo. The base of the silo is conical and lime is usually drawn out by a screw conveyor which delivers a measured amount to a batch-mixer where a lime slurry of up to 10% w/v is prepared. The slurry must be kept in continuous suspension by an agitator. Compressed air is fed into aeration pads in the silo cone so that any 'arching' tendencies of the lime in the silo can be broken up by air injection. The air used must be dry. Alternatively mechanical arch breakers (usually vibrators) are used. A typical lime plant is shown in Figure 7.11. The lime suspension is injected into the water by positive displacement diaphragm pumps with mixing at the point of application, and delivered to the base of a hopper-bottom clarifier of the sludge-blanket type or to a solids recirculation clarifier (see Section 7.16). In sludge-blanket clarifiers, sludge is intermittently drawn off from the blanket, but some may be recirculated to assist in flocculation. The soda ash plant is similar, except that a solution is prepared instead of a slurry. The concentration of saturated solution varies with temperature; 90 g/l (5°C); 120 g/l (10°C),

160 g/l (15°C), 210 g/l (20°C) and 270 g/l (25°C). Usually a solution of about 60% of the saturated concentration at the lowest anticipated water temperature is prepared.

An excess of up to 10% lime and soda ash have to be added over that which is stoichiometrically required in order to achieve completion of the reactions in reasonable time. If magnesium is present, a portion of it may also cause precipitation but the reaction is slower than that of calcium. Softening reduces the hardness value to 35–50 mg/l as $CaCO_3$. If the water to be softened contains suspended solids and organic matter such as colour, these can be removed concurrently although coagulants and coagulant aids have to be added. The coagulants used are usually of the iron type because of their high coagulation pH values, but aluminium sulphates can also be used as they form the insoluble magnesium aluminate and not aluminium hydroxide. Sodium aluminate may also be used. These complications mean that the design of clarifiers and their rise rate has to be undertaken with care. Normally softeners are operated at surface loading rates in the range 3–4 $m^3/h.m^2$; when magnesium is to be removed the rates are about 2–2.5 $m^3/h.m^2$

The dosages of lime required for softening are high, being of the order of 100–200 mg/l. The process produces a large amount of liquid sludge due to the precipitation of hardness and coagulation of suspended solids and colour. Treatment and disposal of the sludge is discussed in Sections 7.46–7.49.

8.5   Water softening by crystallisation

Softening reactions can be accelerated by using sand grains for seeding the crystallisation of calcium carbonate. Softening takes place in a cylindrical reactor partially filled with filter sand (0.2–0.6 mm). The water to be treated is injected with the softening chemical (e.g. lime) and passed in an upward direction at a surface loading rate of 50–120 $m^3/h.m^2$ to fluidise the sand bed. Calcium carbonate deposits on the sand which grows in size to form pellets of about 1–2 mm in diameter which accumulate at the base of the reactor from where they can be periodically removed. Make up sand is added either from the top or at the base. The reaction tanks are called pellet reactors;[1] typically they are about 6 m deep and up to 4 m in diameter (see Fig. 8.1). The 'Spiractor' is an example of a pellet reactor (Fig. 8.2). The softening chemicals used are hydrated lime, sodium carbonate or soda ash. Lime is used when the ratio of carbonate hardness to total hardness is high, in the intermediate hardness range caustic soda is used, and when carbonate hardness is very low, soda ash is used. It is reported that crystal growth is adversely affected when the phosphate content of a water exceeds 0.5 mg/l as PO_4[1] (0.15 mg/l as P) and fluffy pellets are formed when iron is present in the water above about 1 mg/l.[2]

The advantages of the pellet reactor over the conventional softening process are its high surface loading rate, pellets are easier to handle than sludge and only a small excess of softening chemical is required. The disadvantages of the process are that it does not remove magnesium, the hardness after softening is in the range 50–100 mg/l as $CaCO_3$[3] (therefore unsuitable for most industrial uses), suspended solids in the product is high (up to 30 mg/l) and the removal of turbidity and colour in raw water by coagulants cannot be performed in the same reactor.

Lime is used as milk of lime (slurry of 10–100 g/l). The use of lime water (saturated lime) is not usually practical because of its low lime content (e.g. 1.76 g/l @ 10°C), the volume to be added would be over 10% of the volume of water to be softened. Nevertheless there are plants operating with lime water. Use of lime, in particular the slurry, leads to carry

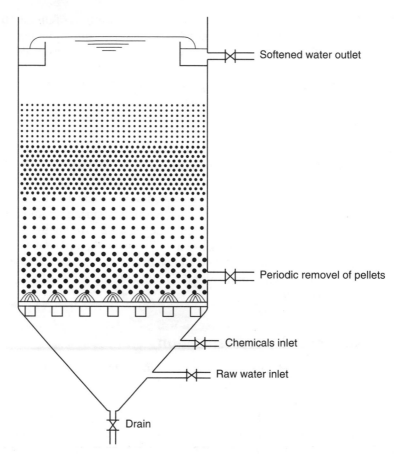

Fig. 8.1   Pellet reactor for softening by crystallisation.

over of calcium carbonate, un-dissolved lime and inert impurities in the lime amounting in total to about 20–30 mg/l as suspended solids. They are removed by dosing with acid and iron coagulants followed by filtration. By improving the quality of lime and its solubility, the carry over of suspended solids can be reduced by up to 80%.[4]

An alternative to chemical crystallisation is a physical process using magnetic, ultrasonic, electrolytic, electrostatic or electronic devices fitted to pipelines carrying hard water. The process does not change the chemical properties of the water but modifies the crystallisation of calcium carbonate giving an increase in particles in suspension and a decrease in the formation of scale. Tests carried out in water distribution networks in France have shown there was an increase in the precipitation of calcium carbonate in the water but a decrease in the formation of scale on metal surfaces.[5] Drawbacks were formation of sludge and inconsistency of performance. The devices are also used in waterworks on sample and lime dosing lines and on domestic hot water systems with some success.

8.6   Stabilisation after softening

The softening reactions of precipitation are not usually wholly completed in the clarification tanks and therefore the water leaving is usually supersaturated with calcium

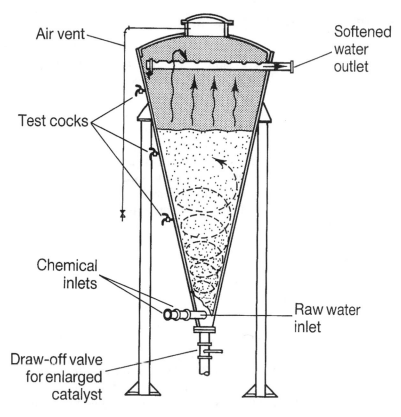

Fig. 8.2 The 'Spiractor' pellet softener.

carbonate and tends to form further deposits, mainly of calcium carbonate, in the later stages of the treatment plant. If calcium carbonate deposits only slowly on filter sand without cementing it together in lumps, this may be an acceptable method of stabilising the water. The sand will however require renewal every few years owing to the build up of deposits. A more acceptable method is to inject carbon dioxide into the water either as a gas using diffusers or as a solution (similar to chlorine injection using similar equipment) into a mixer to achieve an alkalinity of about 30–50 mg/l as $CaCO_3$* and a slightly positive Langelier Index. Carbon dioxide is stored as a liquid under pressure in a refrigerated vessel and vapourised before use. Sulphuric acid is used in some cases, but as it converts carbonate to sulphate, care must be taken to ensure that water is not rendered corrosive due to the low calcium carbonate concentration. The objective of adding carbon dioxide or sulphuric is to prevent after-precipitation while producing a non-aggressive water. Another method of avoiding after-precipitation is to add about 0.5 to 2 mg/l as P of a sequestrant such as a polyphosphate (e.g. sodium hexametaphosphate). This method is usually preferred for industrial applications. Softening is usually followed by anthracite–sand filters.

---

*1 mg/l as $CaCO_3$ = 1.22 mg/l as HCO_3 = 0.4 mg/l as Ca.

8.7   Ion (base) exchange softening

In the ion exchange process of softening, when water containing hardness salts is passed through a granular bed of strong acid cation exchange resin in the sodium form, calcium and magnesium are substituted by sodium. The hardness of the water is reduced to almost zero, but the total dissolved solids concentration undergoes little change; alkalinity and pH values are unaffected. When the resin's capacity to exchange hardness salts for sodium is exhausted the bed is regenerated by passing a strong sodium solution in the form of sodium chloride, through it. The reverse action then takes place, with the calcium and magnesium ions held in the bed being released in the effluent and the sodium from sodium chloride being substituted. The wastewater is very hard with a high concentration of dissolved salts and its disposal may present problems.

The ion exchange resins used are cross-linked polystyrene spherical particles or 'beads'. There are many different types of resins; typically they are classified according to their functional group: cation resins which are strong acid (sulphonated) or weak acid (carboxylic) and anion resins which are strong base (quaternary ammonium) and weak base (amines).

The total number of functional groups in the resin used determines its exchange capacity. The exchange capacity of a resin for softening is most frequently stated in terms of the hardness removed by a specific volume of resin, e.g. x g eq per litre* of resin and is specific to each resin; typical values vary in the range 1.5–2 g eq/l. A second measure of performance in base-exchange softening is the amount of salt that must be used in regeneration per unit of hardness removed. The theoretical figure for regeneration is 117 g of salt per 100 g of $CaCO_3$ (or its equivalent) removed, but in practice depending on the resin used up to 400 g of salt per 100 g of $CaCO_3$ removed, would be required. This can be reduced by 50% using the more efficient counter current regeneration techniques.

8.8   Plant for ion exchange softening

The plant required for ion exchange softening is relatively simple, Figure 8.3 shows a typical layout. The size of the steel vessel containing the base exchange media depends on the rate of softened water production required. The media bed would typically be about 1.2 m deep. A typical cycle comprises: in service (i.e. softening), backwashing with softened water, regeneration followed by rinsing to remove excess regenerant. Flow during both softening and regeneration is normally downwards and operation is automated. The surface loading rate for softening varies in the range 12–20 $m^3/h.m^2$ and that for regeneration is about 2.5 $m^3/h.m^2$ to give at least 30 minute contact time. A saturated solution of sodium chloride containing about 26% w/w NaCl (33% w/v) is initially prepared and diluted to about 5–10% w/v before use. Backwash water is applied at a rate to produce a bed expansion between 50 and 100%; rates vary with the water temperature and resin type. The plant is normally operated under pressure so that repumping after softening can be avoided; the loss of head is usually 4–5 m.

In general ion exchange plants are rarely used for softening public supplies; they are not suitable for softening turbid water (e.g. suspended solids should be <1 mg/l) or water

---

*g eq = gram equivalent. 1 g eq of hardness = 50 g as $CaCO_3$.

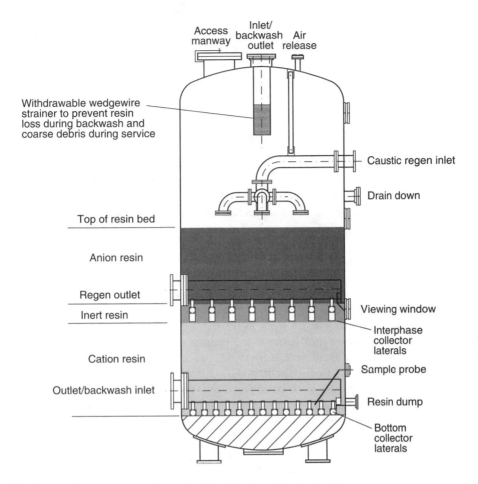

Access manway   **Inlet/ backwash outlet**   **Air release**

Withdrawable wedgewire strainer to prevent resin loss during backwash and coarse debris during service

Caustic regen inlet

Drain down

Top of resin bed

Anion resin

Regen outlet

Viewing window

Inert resin

Interphase collector laterals

Cation resin

Sample probe

Outlet/backwash inlet

Resin dump

Bottom collector laterals

Fig. 8.3   Internal arrangement of a mixed bed demineralisation unit.

which contains iron and organic matter. The process also adds an equivalent concentration of sodium to the water for the hardness removed. They are popular as domestic water softeners purchased by householders. In industry the tendency has been for ion exchange softening to be superseded by demineralisation plant which can produce a water more exactly tailored to the type of process water required, as Section 8.10 shows.

8.9   Removal of hardness and alkalinity by ion exchange

The process uses a weak acid hydrogen cation exchange resin and therefore hydrogen, instead of sodium, is exchanged for calcium and magnesium; sodium and potassium are not removed unless the alkalinity exceeds the hardness. Alkalinity is removed by converting bicarbonates to carbon dioxide which is removed in a degasser, leaving a residual carbon dioxide concentration of about 5–10 mg/l in the water depending on the efficiency of the degasser. Sulphuric or hydrochloric acid is used as the regenerant instead of sodium chloride. Alternatively alkalinity alone can be removed using chloride anion exchange resins (dealkalisation). Chloride is exchanged for bicarbonates and sulphates in the water. Sodium chloride is used as the regenerant.

Water low in alkalinity and carbon dioxide to give <14 mg/l of $CaCO_3$ with lime is useful in the preparation of lime slurry and as carrier water for lime slurry since the risk of scaling in tanks and pipework is minimised (see Section 7.20).

8.10 Demineralisation of water by ion exchange

The ion exchange process of softening is only a particular example of ion-exchange treatment and is more specifically an example of strong cation exchange with the resin in sodium form.

In demineralisation strong acid hydrogen cation exchange resin in which the calcium, magnesium, sodium, and potassium are all replaced by hydrogen ions to form carbon dioxide, is followed by a second stage of treatment using a strong base anion exchange process in which chloride, sulphate and nitrate are removed. The cation exchange resin is regenerated with sulphuric or hydrochloric acid and anion exchange resin is regenerated using sodium carbonate or caustic soda solution. Carbon dioxide formed in the first stage is often removed by 'degassing' or by aeration as an intermediate stage between the cation and anion exchange vessels. The product water has a pH between 7–9 and total dissolved solids concentration as low as, or lower than, ordinary distilled water; the conductivity is less than 20 μS/cm. Such a treatment is therefore called demineralisation and is now most frequently adopted in industry for the production of special quality process waters.

In 'mixed' bed ion exchange both cation and anion exchange resins as described above are mixed in one vessel. During backwashing the resins are hydraulically separated by virtue of their density difference. This allows separate regeneration of the two components. A mixed bed gives a water of neutral pH and conductivity of less than 0.2 μS/cm. Demineralisation can in theory be applied to brackish waters, but because of the expense, the process only finds application for waters having less than about 500 mg/l dissolved solids. For waters having dissolved solids higher than this the ion exchange process would not be economic. For public supply purposes other processes should be considered (see Section 8.43).

In potable water treatment the ion exchange process also finds application in the removal of arsenic (see Section 8.14), nitrate (see Section 8.24), ammonia (see Section 8.27) and radionuclides (see Section 8.49). One of the major drawbacks of ion exchange is difficulty in the disposal of wastewater which is highly saline and non-biodegradable. Options available include treatment or disposal to a water course, sea or sewer.

Removal of iron and manganese and other metals
8.11 Iron and manganese concentrations

Traces of iron and manganese are found in many waters. Occasionally concentrations may range up to 20 mg/l of iron and up to 5 mg/l of manganese, but at these high quantities it is usual to find that most of the metals, in particular iron, are in suspension so that they may be relatively easily removed by solid–liquid separation methods. It is the dissolved fractions of iron and manganese which can be troublesome and the disadvantages arising from their presence above certain levels are described in Sections 6.26 and 6.29 above. Removal is therefore often necessary.

8.12 Removal of iron and manganese from underground waters

In Chapter 7, in discussing the use of ferrous sulphate as a coagulant, it was stated that ferrous hydroxide is much more soluble than ferric hydroxide. The large difference is expressed quantitatively by the solubility products which are 1.6×10^{-14} and 1.1×10^{-36}, respectively. When iron occurs in underground waters it is usually in solution in the ferrous form in a water which is devoid of oxygen. Such waters are fairly common in water-bearing formations which are under an impermeable stratum, e.g. where greensand and other sand formations underlie a clay formation. Manganese in any appreciable amount occurs in only a minority of those raw waters which contain iron. Occasionally manganese is found without iron. Many, but not all waters from deep boreholes in sandstones contain iron and manganese in their lowest state of oxidation* as Fe (II) and Mn (II) and inorganic complexes which occur in solution. Since the organic quality of ground waters is generally good, iron and manganese are not commonly found complexed with organics. When a sample is first drawn it may appear perfectly clear, but after it has been exposed to air for short time it gradually acquires a turbid appearance, and after a further period a brown precipitate of ferric hydroxide is formed. Oxidation of Mn (II) is very slow under most conditions and does not precipitate out on long standing.

Removal of iron and manganese is effected by oxidation, followed by the separation of Fe (III) and Mn (IV) as ferric hydroxide ($Fe(OH)_3$) and manganese dioxide (MnO_2) precipitates by filtration; sometimes in the case of iron when its concentration is high being preceded by settling. Solids contact type clarifiers are suitable because the oxidation is catalysed by the oxides already present in the sludge blanket or recirculated sludge.[6]

Oxidation can be by oxygen in air or by the use of a strong oxidant such as chlorine, potassium permanganate, chlorine dioxide or ozone. In most cases the oxidation will be influenced by the pH. In the oxidation reactions hydrogen ions will be produced which in turn will destroy alkalinity. Table 8.3 below gives the stoichiometric quantities of the oxidant (in mg) required to oxidise 1 mg of iron or manganese and the corresponding reduction in the alkalinity. It also gives the optimum pH for the oxidation reaction.

Oxygen is added through aeration. In practice larger volumes of air will be required to accommodate the absorption inefficiencies of the aeration system used. Although the aeration requirements are small, the rate of reaction is slow and pH dependent, e.g. 90% oxidation of iron would require about 40 minutes reaction time at pH 6.9, but only 10 minutes at pH 7.2. In some waters aeration alone is adequate for complete iron oxidation because the removal of carbon dioxide by aeration raises the pH above 7.5. Manganese reaction is much slower and also requires elevated pH conditions for successful oxidation, e.g. at pH 9.5 manganese would require about 1 hour for 90% oxidation; for complete oxidation treatment by aeration, lime addition to elevate the pH to at least 10 followed by contact filtration is required.

Manganese removal by aeration at high pH values is in most instances a slow and inefficient process. Strong oxidants such as chlorine, potassium permanganate, chlorine dioxide or ozone are therefore usually necessary. These will also be effective in oxidising iron. The application of strong oxidants to water containing uncomplexed iron and

---

*Lowest oxidation state of iron and manganese is ferrous, expressed as Fe (II) and manganous expressed as Mn (II) and when oxidised these are converted to ferric, expressed as Fe (III) and manganic expressed as Mn (IV). Values within brackets denote the corresponding valencies.

Table 8.3   Oxidation of iron and manganese

Metal	Oxidant	Stoichiometric quantity of oxidant (mg/mg Fe or Mn)	Reduction in alkalinity (mg CaCO$_3$/mg Fe or Mn)	Optimum pH
Fe (II)	Oxygen	0.14	1.80	> 7.5
Mn (II)	Oxygen	0.29	1.80	> 10.0 (a)
Fe (II)	Chlorine	0.63	2.70	> 7.0
Mn (II)	Chlorine	1.29	3.64	> 9.0 (a)
Fe (II)	Potassium permanganate	0.94	1.49	> 7.0 (b)
Mn (II)	Potassium permanganate	1.92	1.21	> 7.0 (b)
Fe (II)	Chlorine dioxide	0.24[7]	1.96	> 7.0
Mn (II)	Chlorine dioxide	2.45[7] 0.49[8]	3.64 2.18	≤7.0 ≥7.5
Fe (II)	Ozone	0.43	1.80	(c)
Mn (II)	Ozone	0.87	1.80	(c)

(a) The use of a catalytic filter medium may reduce the pH to 7.5–8.5.
(b) Reaction is known to proceed at pH > 5.5.
(c) pH value at which the reaction occurs is less dependent than for other oxidants. Low pH values are preferred as ozone performs better under acidic conditions.

manganese results in rapid oxidation. In practice excess oxidant is used to satisfy the demands due to organic matter, hydrogen sulphide and, in the case of chlorine, the demand due to ammonia. Chlorine will be effective as both hypochlorous acid (HOCl) and hypochlorite ion (OCl$^-$), but the reaction with combined chlorine (i.e. chloroamines) is slow. The use of chlorine may be inadvisable when treating waters containing organic substances due to the possibility of trihalomethane (THM) formation.

Oxidation by potassium permanganate is more effective than by chlorine and it does not form THMs with organic substances. Manganese dioxide formed in the reaction adsorbs Mn (II) and catalyses its oxidation which brings about an improvement in Fe (II) and Mn (II) removal and a reduction in the amount of potassium permanganate required. The potassium permanganate dose applied must be carefully controlled to minimise any excess passing into supply which could give a pink colour to the water. Products of oxidation can 'muddy' the water and form 'mud balls' in filters.

Chlorine dioxide is particularly useful as an oxidant in the presence of a high ammonia concentration that would otherwise interfere with chlorine. Chlorine dioxide does not form THMs but excess chlorine used in the generation of ClO$_2$ will form some THMs. A disadvantage of chlorine dioxide is the possibility of adverse health effects from its disproportionation products (see Section 9.17) and the consequent UK limitation on the dose which may be applied which must not result in chlorite, chlorate and ClO$_2$ exceeding 0.5 mg/l as chlorine dioxide passing into supply (see Section 6.47).

Ozone reacts readily to oxidise soluble Fe (II) and Mn (II) in the absence of organic matter into the insoluble Fe (III) and Mn (IV). When both are present, iron is oxidised first, followed by manganese. An ozone overdose can oxidise Mn (II) to its higher oxidation state where it is present as permanganate and give a pink colouration to the water; stoichiometrically 1 mg/l of Mn (II) would require 2.2 mg/l of O_3. When organic matter is present, ozone oxidises organics before iron and manganese, thus the dose could be several times more than that required for oxidation in the absence of organics. Iron and manganese are more receptive to oxidation by ozone after organics have been removed by coagulation.

As shown above all the oxidation reactions are accompanied by release of hydrogen ions and hence a corresponding reduction in alkalinity. However since concentrations of iron and manganese present in waters are generally low, in all but exceptional cases there will be sufficient alkalinity in a water to buffer the effect of hydrogen ions and prevent a consequent reduction in pH which will otherwise reduce the reaction rates. A decrease in water temperature is also known to reduce the rate of oxidation.

Aeration of a water prior to oxidation is beneficial in not only oxidising iron and manganese, especially the former, and consequently limiting the consumption of costly oxidising chemicals for removal of any residual metal ion, but also in removing carbon dioxide and thus increasing the pH for the oxidation reaction and in the removal of volatile gases.

In the case of iron, oxidation is followed by settling and filtration, or filtration alone depending on the concentration of iron in the water. Plain settling would only be feasible if the Fe (II) concentration is greater than 20 mg/l which is rare. In the presence of turbidity (and colour) and when Fe (II) is greater than about 5 mg/l settling would be assisted by a coagulant and/or a coagulant aid. As an alternative to sedimentation, dissolved air flotation may be used.[9] Direct filtration is used when the iron concentration is less than about 5 mg/l. Sand (effective size 0.6 mm) or anthracite–sand filters with filtration rates between 5 (Fe <5 mg/l), 7.5 (Fe <3 mg/l) and 10 (Fe <2 mg/l) $m^3/h.m^2$ are most suitable for the application. Manganese is usually found in low concentrations compared to iron and, following oxidation, the water is subjected to direct filtration. In the absence of turbidity and upstream coagulation, filtration rates used are of the order of 10 $m^3/h.m^2$ or higher; an exception being oxidation using potassium permanganate when coagulation followed by settling and filtration are used. Filtration assists in the oxidation of Fe (II) and Mn (II) through the catalytic action of either previously deposited MnO_2 on a filter sand, or a media containing manganese oxide such as pyrolusite or a proprietary material such as 'Polarite'. The media usually contain at least 65% by weight MnO_2. Its specific gravity is in the range 3.5–4.0 and size range is usually 0.5–1.0 mm. It is used in the ratio sand to oxide media of 5:1. The retention capacity in filters is about 0.2–1.2 kg of Fe and 0.1–0.7 kg of Mn/m^2 of surface area. It is reported that the effective size and density of the media are not altered by the presence of the oxide coating.[10]

An alternative filter media is manganese greensand, formed by treating greensand (glauconite), which is a sodium zeolite, with manganous sulphate followed by potassium permanganate. Mn-greensand removes soluble iron and manganese by a process of ion exchange, frequently with the release of hydrogen ions. The process is therefore pH dependent, being virtually ineffective below pH 6.0 and very rapid at pH values above 7.5. When the Mn-greensand is saturated it is regenerated by soaking the filter bed with potassium permanganate. This procedure oxidises manganese on the surface of Mn-greensand to MnO_2 to become active exchange sites (intermittent regeneration – IR). It is reported that the exchange capacity is 1.45 g of Fe or Mn/l of Mn-greensand and 2.9 g of potassium permanganate (as a 1% w/v solution) per litre of Mn-greensand is required for regeneration.[11] Alternatively potassium permanganate is continuously supplied to the bed by dosing it at the filter inlet, which will oxidise some iron and manganese before the water reaches the filter (continuous regeneration – CR). Mn-greensand then acts primarily as a filter medium and oxidises any residual soluble manganese. Intermittent use of excess potassium permanganate will regenerate the bed. Mn-greensand has an effective size in the range 0.30–0.35 mm and a uniformity coefficient of 1.4–1.6 and is usually capped with a layer of anthracite to achieve longer filter runs. In the continuous dosing process chlorine can be used in place of potassium permanganate. The benefits are longer filter runs, the absence of risk of pink water and lower cost. It is reported that manganese dioxide coated filter media behaves in a similar manner to Mn-greensand with operation being carried out in either the IR or CR mode.[12] Chlorine is the most suitable oxidant for the CR mode of operation. Stronger oxidants such as ozone, chlorine dioxide and potassium permanganate tend to form colloidal precipitates which are not very well retained by the filters. The use of catalytic filtration media is usually limited to manganese which is otherwise difficult to remove. In most plants iron is oxidised ahead of filtration.

Organic substances such as humic, fulvic and tannic acids when found in groundwaters can form soluble complexes with iron and manganese which are not easily oxidised by oxygen (aeration). These and any soluble inorganic complexes such as silicates, sulphates and phosphates are removed by coagulation or oxidation using strong oxidants. However when iron is strongly complexed in the presence of significant concentrations of humic and fulvic acid, strong oxidants are sometimes ineffective. The oxidation of manganese is not similarly influenced by the presence of dissolved organics because it is not strongly complexed by organic matter.[7,13] In the presence of dissolved organic matter the oxidant dosage should be selected to satisfy the total oxidant demand of the water.

Fe (II) and Mn (II) can be removed biologically by utilising the ability of certain bacteria to produce enzymes and/or polymers which, by catalytic action, promote oxidation in the presence of oxygen in the water. Those which promote iron oxidation are generally considered to be autotrophs, but the physiology of those promoting manganese oxidation is poorly defined.[14] Those which are heterotrophs probably promote removal of dissolved iron and/or manganese as part of a process of detoxification of the ambient medium. The bacteria are usually present in the ground waters which contain these metals, e.g. *Gallionella ferruginea* (specific to iron) *Leptothrix* sp., *Crenothrix polyspora* and *Sphaeratilus natans*. If absent, they can be introduced from a suitable source; rapid gravity or pressure sand filters being used as biological reactors. When both metals are present together in water two filtration stages are usually necessary with manganese removed in the second stage because manganese removal bacteria require a completely aerobic environment.[15]

For iron removal aeration should be controlled, particularly at pH values greater than 7 to prevent the chemical process competing with the biological process, and therefore to minimise the risk of a chemically formed precipitate breaking through the coarse media used in high rate biological filters. Based on a number of plants operating in France it is reported that Fe (II) oxidation takes place preferably with a pH in the range 6.0–7.5 and a dissolved oxygen concentration of 0.25–1.5 mg/l. Mn (II) removal needs a pH greater than 7.5 and a dissolved oxygen concentration in excess of 5 mg/l.[15] Optimum water temperature is in the range 10–15°C depending on the predominant bacteria. Toxic elements and/or compounds such as heavy metals (e.g. zinc), hydrogen sulphide, hydrocarbons and chlorine must be absent. Ammonia interferes with the process and when present in excess of about 0.20 mg/l as N, the manganese removal stage must be designed for a simultaneous, slower nitrification process. When ammonia is present in excess of about 1.0 mg/l as N, a separate biological nitrification stage using a biological aerated filter (BAF) needs to be included between the iron and manganese removal stages (see Section 8.27). Due to the high oxidation kinetics, filters can be operated at rates ranging from 10 to 40 $m^3/h.m^2$ using coarse sand media (0.95–1.35 mm) with a solids retention capacity of 1–4 kg Fe or Mn/m^2 filter surface area.[15] Unlike the chemical process where a clarification stage is necessary when iron concentration exceeds 5 mg/l, biological iron removal process allows for direct filtration even when iron concentration is as high as 25 mg/l. Back washing is by combined air–water wash, using raw or treated but unchlorinated water. Filters can be of the open gravity or enclosed pressure type and the filtered water requires aeration (in iron removal only) and chlorination. Biological filters require a seeding period before they operate at their optimum removal efficiency; this is reported to vary from one week for iron to 3 months for manganese.[16] Sludge produced in biological treatment is well suited to thickening (thickened sludge concentrations of 3–8% w/v) and dewatering.[15]

Other merits of these biological treatments, compared to the conventional physical-chemical processes are: a better treated water quality (Fe-Mn residuals generally not detectable, no interference with dissolved silica which otherwise forms iron-silica complexes), longer filter runs, an easier operation, a dramatic decrease of capital cost (plants are much more compact) and operation cost (no chemicals for Fe-Mn oxidation or coagulation-flocculation, less washwater losses, less manpower).

8.13 Removal of iron and manganese from river and reservoir waters

Because river and reservoir waters frequently receive treatment which includes rapid sand filtration, often preceded by coagulation and sedimentation, the removal of iron and manganese is normally included, when necessary, in the same plant. Most river waters used as sources for water supplies are well oxygenated, if not saturated with oxygen. Usually, therefore, significant proportions of iron and manganese in such river waters are present in the suspended matter in their insoluble forms and are removed by the sedimentation and filtration treatment. However, this does not always apply, for instance when water is drawn from a reservoir in which the iron and manganese have been dissolved in the bottom water under conditions of deoxygenation, as described in Section 7.1. Generally the iron is fairly easily oxidised, but sometimes the iron and, more often, the manganese are combined with organic matter in a very stable form. These soluble organics are removed by coagulation. Strong oxidants may also be used to release Fe (II) and Mn(II) from complexed organics and to oxidise them to Fe (III) and Mn (IV) (see Section 8.12). One method by which both the iron and manganese may be removed is in the lime-soda softening process which operates at a pH value of 10.6. With a soft reservoir water, which does not require full coagulation treatment, it has sometimes been found that manganese can be removed by raising the pH above 9.0 with lime or caustic soda before rapid sand filtration;[17] the use of an oxidant and/or a catalytic filter medium would help to operate at a lower pH. A more general method for soft waters is to coagulate the water with ferrous sulphate (copperas) or with chlorinated copperas at a pH of about 9.5 before sedimentation and filtration. More often, however, when iron and manganese are found complexed with organics in a reservoir water a satisfactory treatment is oxidation using a strong oxidant followed by coagulation and clarification for colour, turbidity and Fe (II) removal, then pH adjustment to 9.0 for manganese removal in the downstream filters which may contain a catalytic medium. Pre-chlorination or the use of chlorinated copperas may not be desirable due to the possibility of THM formation.

In some cases where a trace of manganese (less than the WHO guide value of 0.05 mg/l as Mn), has passed through a filtration plant the subsequent chlorination has assisted in precipitating the metal to form objectionable deposits in the mains. Precipitation of iron and manganese in distribution systems can be controlled through sequestration (also called chelation). This process increases the solubility of the metal ion by forming a bond with it, thereby preventing it from forming a precipitate. Sequestering agents used are polyphosphates and sodium silicates. These agents do not remove iron or manganese; which will still be present as a soluble complex in the water.

8.14   Removal of other metals

There are several other metals that can be present in a raw water and some can be added in the distributed water due to the corrosion of water mains and plumbing systems. Metals present in dissolved form in raw water can usually be removed by precipitation as the metal hydroxide. This involves the correction of the raw water pH, usually by adding an alkali (caustic soda or hydrated lime), to a value at which the metal precipitates and can be removed by coagulation and filtration. In most cases the pH value required for precipitation can be achieved during coagulation by using aluminium or iron coagulants; the latter are more effective because they have a wider coagulation pH range. Metals which show good removal during coagulation are arsenic (pH 6–8), cadmium (pH > 8) chromium (pH 6–9), lead (pH 6–9) and mercury (pH 7–8).[18] Other metals which require high pH values and can be precipitated with lime or caustic soda include barium (pH 10–11), copper (pH 10) and zinc (pH 10).

Arsenic

Arsenic (As) occurs in the soluble form as As (III) (arsenite) under anaerobic conditions predominantly in groundwaters and as As (V) (arsenate) under aerobic conditions more commonly in surface waters. Both the forms can be effectively removed by coagulation followed by solid–liquid separation process. As (V) is removed with equal efficiency by aluminium and iron coagulants at pH < 7.5, however iron coagulants are more effective than aluminium coagulants in removing As (V) at pH > 7.5 and also As (III).[19] Alternatively As (III) could be more effectively removed by oxidising it to As (V) by pre-chlorination followed by coagulation using either of the two coagulants.[20a] It is reported that a drawback of the process is that arsenic is deposited on filter sand and requires intermittent backwashing with caustic soda.[21] A significant removal of As (V) can also be achieved during the oxidation of Fe (II).[19] Lime softening at pH 11 to ensure all magnesium is removed also removes arsenic. The removal methods applied to groundwater include activated alumina adsorption, ion exchange, reverse osmosis, nanofiltration and biological oxidation.

Activated alumina adsorption and ion exchange remove only As (V), although a new grade of activated alumina is known to remove As (III) equally well. Arsenic (V) adsorption on activated alumina is affected by silicates (at pH > 7) and phosphate (at pH > 7). The effect of silicates (at pH > 5) and phosphate (at pH > 6) is much adverse on As (III) removal.[22,23] Other ions which reduce the removal efficiency are sulphates and bicarbonates. The optimum pH range for adsorption is 5.5–6.5 although satisfactory performance has been achieved at a pH of 7.5.[23] Because of the effect of sulphate on arsenic removal, any pH correction should be by hydrochloric acid. The process design criteria for arsenic removal are given in Table 8.4. Ideally a plant should consist of an equal number of duty/standby vessels (see Section 8.24 for configurations) operating in the

Table 8.4 Process design criteria for fluoride and arsenic removal by activated alumina

Parameter	Fluoride	Arsenic
Media	Alcan Granular Activated Alumina	Alcan Granular Activated Alumina
Media type (a) (1)	AA400G (a)	AA400G (a)
Media size	0.6–1.2 mm	0.3–0.6 mm
Media capacity	1.4 g/100 g	1.5–1.7 g/100 g
Media type (b) (2)	AAFS50 (b)	AAFS50(b)
Media size	0.6–1.2 mm	0.3–0.6 mm
Media capacity	$\leq$ 2.8 g/100 g	3–8 g/100 g
Media depth	2 to 2.5 m	2 to 2.5 m
Flow rate	6–12 BV/h (c)	6–12 BV/h
EBCT(d)	6 minutes	6 minutes
pH of feed	6–7	5.5–6.5
Initial raw water concentration	3–8 mg/l	10–50 μg/l
Final raw water concentration	0–3 mg/l (e)	0–10 μg/l (e)
	Regeneration:	
Alkali	NaOH	NaOH
Concentration	1% w/w	1% w/w
Regenerant volume (f)	5–10 BV	5–10 BV
Rinse water volume (f)	2–5 BV	2–5 BV
Acid	H_2SO_4	HCl
Concentration	1–2% w/w	1–2% w/w
Acid volume (f)	2–5 BV	5 BV

(a) Specific gravity 2.6; bulk density 740 kg/m^3.
(b) New material. Removes both As(III) and As(V); patent pending.
(c) BV – Bed volume: bulk volume occupied by adsorbent in a vessel.
(d) EBCT – Empty bed contact time: time required for a volume of solution equal to the bed volume of adsorbent to pass through the column.
(e) Depending on requirements.
(f) Regenerant, rinse water and acid are applied at rates moderately lower than the raw water flow rate.
Source: Alcan Chemicals Ltd, UK.

386 Specialised and advanced water treatment processes

downflow mode. Following exhaustion the bed is backwashed to give about 50% bed expansion using raw water and then subjected to co-current regeneration using caustic soda and counter current reactivation using sulphuric acid with a co-current rinsing phase in between. Pilot trials have shown that regeneration recovers almost 80% of the bed capacity.[23]

The ion exchange process could be applied to waters containing dissolved solids of less than 500 mg/l and sulphates of less than 25 mg/l as SO_4. The process uses a strong base anion exchange resin of the chloride form at a pH between 8 and 9 and removes As (V); regeneration is by sodium chloride.[24] The exchange capacity of the resins is low but does not exhaust rapidly because arsenic concentrations in most ground waters are very low.

The spent regenerants from adsorption and ion exchange processes could be treated by aluminium or iron salts to form insoluble arsenates or by lowering the pH to 5–6.5 to form insoluble hydroxides.

Arsenic could also be removed by reverse osmosis or nanofiltration[25] or adsorption on to granular ferric hydroxide.[26] Other processes being developed include biological oxidation and photo-oxidation both in the presence of iron.

Lead

Lead is rarely a contaminant of any significance in natural water. In polluted waters the total concentration of lead could be as high as 10 mg/l with the dissolved fraction usually less than 0.01 mg/l[27] and it is effectively removed by coagulation using aluminium or iron coagulants or by lime softening at pH > 9. Lead in drinking water is mainly introduced through corrosion of the plumbing systems containing lead pipes and fittings, and lead compounds used in pipe jointing materials. Lead solubility increases with increasing alkalinity and decreasing pH.[28] The waters which are corrosive towards lead have a pH < 7 or a pH in the range of 7.5–8.5 and an alkalinity in excess of about 100 mg/l as $CaCO_3$[29] and the least corrosive have a pH greater than 8 and alkalinity of about 50 to 80 mg/l as $CaCO_3$. Corrective measures are therefore applied at the treatment works to minimise dissolution of lead, i.e. plumbosolvency. This is accomplished by appropriate control of pH and alkalinity of the water to assist in the formation of relatively insoluble lead compounds consisting of carbonates and hydroxide as a film on lead surfaces or by the dosing of orthophosphates to form sparingly soluble lead orthophosphate. For waters of low alkalinity (< 50 mg/l as $CaCO_3$), lead solubility can be greatly reduced by increasing the pH into the range 9–10[29] but this would contravene most national and international drinking water standards. In practice therefore elevation of pH value is restricted to about 8.5 and not less than 8.0 at the consumers' tap but it is generally unnecessary to increase alkalinity.[30] Raising the pH will generally increase the alkalinity sufficiently to provide the necessary protection against plumbosolvency.[29] Sometimes the alkalinity may have to be increased if the buffering capacity of the water is inadequate to maintain the pH unaltered in the distribution system. In a well-buffered water pH variation would be limited to about 0.5 units, whereas if buffering is inadequate the variation could be as much as 2.5 units. Soft and aggressive waters with alkalinity less than 50 mg/l as $CaCO_3$ could also be mineralised by lime dosing or passage through a limestone filter with or without the addition of carbon dioxide.[31] High alkalinity waters (hard waters) with pH > 7 may have to be softened or alkalinity removed to maximise the reduction in lead solubility and to prevent unwanted calcium carbonate scaling.[29] The objective in these treatment processes is to treat the water to a pH value greater than 8 and an alkalinity of about 50 mg/l as $CaCO_3$.

For hard waters elevated pH values are not possible because of carbonate precipitations and softening followed by elevation of pH is uneconomical, but orthophosphate dosing provides an appropriate treatment (see Plate 21); the most effective pH range is 7.2–7.8.[32] The treatment is usually limited to waters of hardness less than 300 mg/l as $CaCO_3$ due to the risk of calcium phosphate precipitation.[31] Sometimes low alkalinity waters are also treated with orthophosphate either as an alternative to pH correction or as a supplementary treatment after pH correction.[30] The orthophosphate dose applied is in the range 1–1.5 mg/l as P* and a reduction of plumbosolvency by about 80% with groundwaters and 70% with surface waters is reported after six months.[33] It is reported that the reduction in lead concentration is proportional to the orthophosphate concentration and several months of dosing may be necessary to provide total protection.[34] The process is reversible and interruptions to treatment should be kept to a minimum (<1 week); damage or repair to lead plumbing system would also affect the protection.[34] Orthophosphate is added as one of its sodium salts or as orthophosphoric acid. (See Table 7.7 for storage and Section 7.20 for dosing). A phosphate monitor could be used in closed loop control for phosphate dosing. Phosphate could be dosed with chorine used for final disinfection, but the injection point should be well separated from the lime dosing point for final pH correction. Polyphosphates have no effect in reducing plumbosolvency.

The application of plumbosolvent treatment could have adverse or beneficial effects in the following areas dependent on the characteristics of the distribution system – discoloration of water due to iron release, corrosion of other metals and cement lining of pipes, biofouling, scaling, THM formation, effects of chloramination, and effects of mixing waters in the system. Monitoring is critical to the success of plumbosolvent treatment. Parameters to be monitored should include pH, alkalinity and orthophosphate, and should be carried out both at the treatment works and in the distribution system. The above corrective treatment processes when correctly applied and controlled should help to reduce plumbosolvency and reduce concentrations to 10 $\mu g/l$ (WHO Guide Value).

Aluminium

Aluminium in surface waters is readily removed by coagulation using aluminium or iron coagulants in the pH range 6.5–7.2. In the presence of manganese, aluminium is first removed by coagulation and filtration (with or without clarification) followed by elevation of pH (> 10), oxidation and secondary filtration for manganese removal (see Section 8.12). A similar treatment regime is applied when treating raw water containing manganese, by aluminium coagulants. Aluminium is rarely found in ground waters in excess of 10 $\mu g/l$.

---

*1 mg/l as P $= 3$ mg/l as PO_4^{3-}.

Defluoridation and fluoridation

8.15 Defluoridation

Some groundwaters contain high levels of fluoride with concentrations well in excess of 1.0 mg/l as F (see Section 6.23). Levels in excess of 1.5 mg/l as F may cause dental fluorosis leading to mottling of the teeth; reduction of fluoride may therefore be necessary. Defluoridation can be achieved by chemical precipitation, adsorption, or by membrane separation processes. Hydrated lime precipitates fluoride as calcium fluoride, which has a solubility of about 7.7 mg/l. Therefore theoretically the lowest fluoride level achievable by this method is 8 mg/l. Aluminium sulphate coagulation is known to reduce fluoride levels by 10–60% but the pH must be kept in the optimum range for coagulation to ensure residual aluminium levels are low; the aluminium sulphate dose required could be as high as 750 mg/l. In the 'Nalgonda process'[35] the combined use of lime and aluminium sulphate has reduced fluoride levels to below 1.5 mg/l, but the chemical doses required were substantial.

Adsorption media successfully used are granular activated carbon (GAC),[36] bone char,[37] serpentinite,[38] activated bauxite[39] and activated alumina.[36–39] Of these all except activated alumina have limitations which would make the processes uneconomical; activated alumina is successfully used in several full-scale plants. Activated alumina is an excellent medium for fluoride removal. It is highly selective to fluoride in the presence of sulphates and chlorides when compared to synthetic ion exchange resins. In the presence of bicarbonates, although the fluoride level is reduced, the adsorption capacity shows a major decline. Silica is also known to interfere with the adsorption of fluoride. The adsorption process is best carried out under slightly acidic conditions (pH 5–7); the lower the pH the more effective the removal. Process design criteria for fluoride removal are given in Table 8.4. The plant and its operation are similar to that described for arsenic removal (see Section 8.14). The spent regenerant could be treated with lime or dried in evaporation ponds.

Reverse osmosis and electrodialysis (see Sections 8.45 and 8.46) which are membrane type desalination processes, can be used to remove fluorides along with other ions. Membranes exhibit anion and cation rejections in the range 95–99%. For these processes to be successful the feed water must be pre-treated to prevent membrane fouling and scaling.

8.16 Fluoridation

Many waters are deficient in natural fluoride and some regional or national health authorities consider that it should be added to reduce the incidence of dental caries. Fluoridation of water is carried out by using either hexafluorosilicic acid which is a solution usually containing 20% w/w H_2SiF_6 (15.8% w/w F) or disodium hexafluoro-silicate which is supplied as a powder usually containing at least 98% Na_2SiF_6 (59.4% w/w F). The acid is dosed into water either in the delivered form or after automatic pre-dilution with water. The powder is dissolved in water to form a 0.2% w/v (2 g/l) solution which is dosed into water. The water used for dissolving the powder should be soft or, where necessary, softened using a base exchange softener (see Section 8.7) to prevent precipitation and subsequent scaling. Alternatively a sequestering agent such as a polyphosphate could be used to minimise the problem of scaling. In the design of the dosing system care should be taken to ensure that the quantity of fluoride injected to supply over 24 hours does not exceed the maximum dosage allowable.

For systems dosing the acid drawn directly from storage, this is achieved by providing an intermediate storage tank (a day tank) which holds not more than 24 hours maximum usage. A pump should be used to transfer the acid from storage to the day tank. For the pre-dilution system, a day tank or a system to limit the number of batches per day is essential. For powder systems a hopper with 1 day's capacity should be used.

Dosing of fluoride should be by positive displacement pump of the reciprocating diaphragm type and should be of such capacity that it operates close to its maximum capacity most of the time. Where the works flow varies frequently by more than 5% the pump motor speed should be arranged to automatically vary with flow. Closed-loop control of the output based on a signal from a fluoride monitor to maintain a fixed fluoride value in the water is not recommended unless sufficient safeguard against over-dosing can be provided.

The point of injection of fluoride should be after any treatment. It is acceptable to dose fluoride alongside chlorine used for final disinfection, but the injection point should be well separated from any lime injection point for final pH correction. Following injection it should be well mixed in the water, and a sample representative of the water going to supply should be taken for monitoring of fluoride. Guidance on technical operation and safety requirements for fluoridation plants is published in a Code of Practice by a UK working group on fluoridation.[40]

Table 7.7 summarises the suitable materials for use with the acid and powder. Both the acid and powder are highly toxic; additionally the acid is very corrosive. Acid vapour and dust from the powder should be contained to prevent inhalation and ingestion by operators. Splashes of acid on skin should be washed with copious amounts of cold water.

Aeration

8.17   Purpose

Aeration has a large number of uses in water treatment. Listing the more usual, these are:

(1) to increase the dissolved oxygen content of the water;
(2) to reduce tastes and odours caused by dissolved gases in the water, such as hydrogen sulphide, which are then released; and also to oxidise and remove organic matter;
(3) to decrease the carbon dioxide content of a water and thereby reduce its corrosiveness and raise its pH value;
(4) to oxidise iron and manganese from their soluble states to their insoluble states and thereby cause them to precipitate so that they may be removed by clarification and filtration processes;
(5) to remove certain volatile organic compounds.

According to Henry's law, the equilibrium solubility of a gas in water is given by $C_s = kp$ where C_s is the saturation concentration of the gas in water (mg/l), p is the partial pressure in bar and k is the coefficient of absorption which is equal to $(55\ 600 \times M) \div H$ where M is molecular weight of the gas and H is Henry's constant. For oxygen–water and carbon dioxide–water systems k values at different temperatures are given in Table 8.5.

For example, air at 1 bar containing 20.8% v/v O_2 ($p = 0.208$ bar) and carbon dioxide 0.03% v/v CO_2 ($p = 0.0003$ bar) in contact with water at 15°C will, after prolonged

Table 8.5   Absorption coefficients for O_2 and CO_2

Temperature (°C)	0	5	10	15	20	25	30
kO_2[a] (mg/l bar)	70.5	61.5	54.8	49	44.8	40.4	38.2
kCO_2 (mg/l bar)	3380	2800	2350	2000	1740	1490	1310

[a]kO_2 decreases with increasing concentration of dissolved salts. In most waters the effect is insignificant; in brackish to sea water containing high chlorides, however, it can be significant and an approximate value can be calculated by the formula kO_2 (mg/l bar) $= (2284 - 12.74 m)/(33.5 + T°C)$ where m is g chloride ions/kg of water.

periods, reach an equilibrium, at which state the water will contain 10.2 mg/l O_2 and 0.6 mg/l CO_2.

The purpose of aeration is to speed up this process and there are two main types of aerators in general use: those in which water is allowed to fall through air, e.g. free-fall aerators, packed tower aerators and spray aerators, and those in which air is injected into water, e.g. injection aerators, diffused air-bubble aerators and surface aerators. For them to be effective it is essential they provide not only a large water–air interface (high area:volume ratio) but, at the same time, a high degree of mixing and rapid renewal of the gas–liquid interface to facilitate transfer of oxygen.

8.18   Cascade aerators

Cascade aerators are the simplest type of free-fall aerators. Figure 8.4 shows the design of a cascade aerator. Such aerators are widely used as water features: they will take large quantities of water in a comparatively small area at low head, they are simple to keep clean, and they can be made of robust and durable materials giving a long life. The plates can be made of cast iron, of reinforced concrete, timber or even of glass. The aerator should preferably be in the open air or, for protection against air-borne pollution, freezing and algal growth, in a small house which has plenty of louvred air inlets. They are efficient for raising the dissolved O_2 content but, not for CO_2 removal; reduction of CO_2 content is usually in the range of 60–70%. Weirs and waterfalls of any kind are, of course, cascade aerators. Where a river passes over artificial or naturally occurring obstacles a large contribution occurs to the self-purification of natural waters by the increase of the dissolved oxygen content which allows the process of breakdown of organic matter to proceed at a faster rate. From work carried out by the former UK Water Pollution Control Laboratories[41] an empirical relationship has been developed for the ratio, r, of the oxygen deficit just above a weir to that just below which is:

$$r = 1 + 0.38abh(1 - 0.11h)(1 + 0.046T)$$

where a is 1.25 in slightly polluted water, 1.00 in moderately polluted water, and 0.85 in sewage effluents; b is 1.00 for a free-fall weir and 1.30 for a stepped weir; h is the height of the fall in metres, and T is the water temperature in °C. The deficit ratio r is defined as

$$r = \frac{C_s - C_u}{C_s - C_d}$$

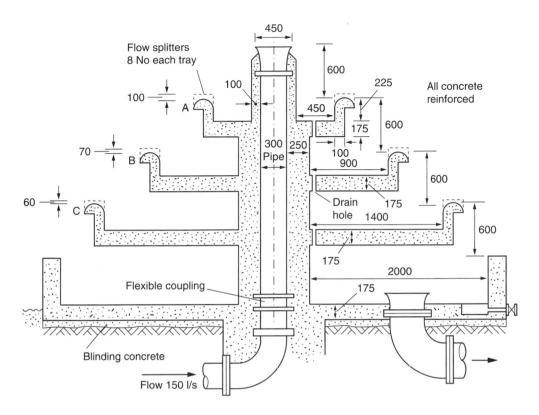

Fig. 8.4 Cascade aerator.

where C_u and C_d are oxygen concentrations upstream and downstream, and C_s is the saturation concentration at $T°C$.

Number of steps varies from 3–10 and fall in each step from 0.15 to 0.3 m. The rate of flow may vary between 20 and 100 $m^3/h.m$ weir length. To allow entrained air to mix in the water, each receiving basin should have a pool of water of depth 0.3 to 0.5 m. Weirs with serrated edges perform better,[42] as they help to break water flow into separate jets (see Plate 22). If the water is allowed to cling to the steps especially at low discharge rates, the efficiency is reduced. The space requirement is typically of the order of 0.5 m^2 per 1 m^3/h water treated. The oxygen transfer efficiency which is defined as the oxygen transferred (kg/h) per unit of power input could be as high as 2.5 kg O_2/kWh.[42]

8.19 Packed tower aerators

Packed tower aerators consist of a vertical steel, plastic or concrete shell filled with plastic shapes such as pall-rings, berls or saddles (see Fig. 8.5 and Plate 22). Water is applied at the top through a distributor and a counter current flow of air is usually blown (forced-draught) but, sometimes drawn (induced-draught) using a blower or a fan. Air must be filtered to prevent contamination of water; the air to water volumetric flow ratio is about 10:1. The water becomes distributed in a thin film over the packing thus providing the large water–air interface required for high mass transfer. Surface loading rates are of the

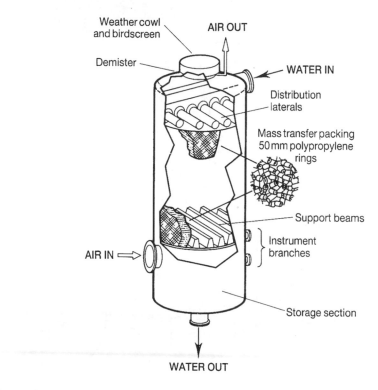

Fig. 8.5   Packed tower aerator.

order of 50–100 $m^3/h.m^2$. Packed towers constitute the most efficient form of aeration primarily for ground waters, as suspended solids in surface waters can rapidly clog the packing. They give over 90% oxygen transfer and 85% CO_2 removal efficiencies. They are also used for the removal of hydrogen sulphide, ammonia and volatile hydrocarbons; and can deal with varying throughputs without a decrease in the efficiency. This type of aerator is expensive in first cost and entails some additional head on the water supply because the top of the tower may be as much as 6 m above ground. The other drawbacks of the system are that CO_2 removal increases the tendency for scale deposition, whilst iron and manganese tend to foul the packing. A sequestering agent such as sodium hexametaphosphate can be used to eliminate scale deposition. After a number of years the packing will need replacing, dependent upon the deposits which have accumulated on the surfaces of the material. Initial cost, maintenance costs, and power costs all tend to make the packed tower aerator costly, although sometimes it is necessary when a high CO_2 content has to be dealt with, or ammonia (see Section 8.27) or volatile hydrocarbons must be removed (see Section 8.29).

8.20   Spray aerators

Spray aerators work on the principle of dividing the water flow into fine streams and small droplets which come into intimate contact with the air in their trajectory. Therefore they can be used on hard waters and those containing iron and manganese without the risk of scale deposition or fouling problems. About 70% CO_2 removal and 80% oxygen transfer efficiencies can be obtained with the best type of spray nozzles. There are a number of different types of nozzle on the market and the correct type of nozzle to be used must be chosen with care to prevent troubles caused by clogging or excessive pressure required to force the water through them. Up to 10 m head of water may be required and a large collecting area (typically 0.5–1 m^2 per 1 m^3/h water treated) is necessary because many hundreds of spray nozzles may be needed. At one installation 1000 nozzles are provided for an output of 9 Ml/day. Extra efficiency is obtained in some types of plant in which the spray is broken up by impinging on a plate. The sprays require protection from wind and freezing and should be enclosed in a well-ventilated housing with louvred air inlets. Because of the risk of blockages of nozzles with suspended solids they are normally used on ground waters.

8.21 Injection aerators

Injection aerators avoid the need to break the pressure of a water if this is particularly inconvenient or wasteful of energy. The water may be sprayed into a compressed air space at the top of a closed vessel under pressure, such as into the top of a pressure filter. The air has to be circulated by a compressor. Alternatively compressed air may be injected into the flowing water in a pipe upstream of a mixing device, such as an inline static mixer; or air at atmospheric pressure may be drawn into the pipe where a constriction, such as the throat of a venturi tube, reduces the water pressure below atmospheric. In the latter case the venturi tube is of special design, having a much narrower throat than usual and a much longer divergence cone downstream than in the case of a venturi designed for flow measurement. A defect of aeration under pressure is that it will not remove CO_2; its effect is to increase the oxygen content of the water and to saturate it with nitrogen at the operating pressure. The latter can be a disadvantage because air can appear as bubbles when the pressure is released and can lead to air binding in a downstream filtration process.

8.22 Other types of aerators

Mechanical surface aerators operate on the basic principle of entraining air by violent agitation of the water surface. There are several forms of surface aerators which are successfully used in sewage treatment works. Those where the axis of rotation of the impeller is vertical are used for aerating water in large open tanks. The oxygen transfer efficiency for a mechanical surface aerator is about 1.8 kgO$_2$/kWh.[43]

Diffused-air bubble aeration is usually carried out by passing air through some form of a diffuser placed at the bottom of a tank through which the water flows horizontally. The diffuser may be a perforated pipe or a porous ceramic plate. Aeration efficiency is a function of bubble size and tank depth; with smaller bubbles, efficiency increases with decreasing depth. Tank depth varies from 3 to 4.5 m and tank width should not exceed twice the depth. Residence time is about 10–20 minutes. Air (at atmospheric pressure) to water volumetric ratio is usually between 1:1.25 to 1:2.5. Air must be filtered to minimise contamination of water and clogging of the diffusers. The oxygen transfer efficiency is about 10%. This method is also widely used in sewage treatment. In water treatment, the perforated pipe arrangement finds application in the destratification of reservoirs[44] (see Section 7.1). When applied to a tank 4 m deep the oxygen transfer efficiency is of the order of 0.85 kgO$_2$/kWh.[43]

Nitrate removal
8.23   General

Nitrates are found in undesirable concentrations in some water supplies particularly those derived from groundwaters and some surface waters in areas where there is an increased agricultural use of nitrate fertilisers. In rivers its presence in appreciable concentrations can be seasonal with peaks during early rains due to agricultural run-off. Seasonal variation in groundwaters is much less.

Blending of sources is the most basic method for achieving low nitrate supplies. For surface waters long-term storage can be used to absorb the seasonal peaks and to bring about natural biological denitrification. Where such methods are not feasible, treatment of the water for nitrate reduction is necessary. The two nitrate reduction processes predominantly used are ion exchange and biological denitrification. Membrane processes such as electrodialysis reversal (EDR) and reverse osmosis can also be used for nitrate removal. All these processes bring about a nitrate removal in excess of 80%. Therefore treatment is applied only to a small part of the flow so that the desired concentration is achieved after blending with the remainder.

8.24   Ion-exchange process for nitrate removal

The ion-exchange process is similar to that used for demineralisation of water discussed in Section 8.10. In nitrate removal a strong base anion exchange resin in the chloride form is used. As the water passes through the bed of resin contained in a pressure vessel (see Plate 21), nitrates and other anions in water are exchanged with chloride in the resin, thus releasing chlorides into water. When the resin is saturated with respect to nitrate (indicated by the rise in nitrate level in the product water) the run is stopped and the resin is regenerated with sodium chloride solution (brine) and the bed returned to service. To ensure that a flow is maintained during regeneration ideally 100% standby units should be provided. Alternatively at least 3 × 50% flow for small works and 4 × 33⅓% for large works should be considered. During regeneration the exchange process is reversed; anions absorbed on the resin are replaced by chloride ions and discharged to waste with excess chloride ions. Either 'classical' anion exchange (see Section 8.10) or nitrate selective exchange resins can be used in the process. The former which has been used with success has the drawback that nitrate is less well absorbed than the sulphate, consequently the nitrate uptake falls off rapidly as raw water sulphate increases. Additionally, irrespective of sulphate level in the water, a bed which has run past its saturation point with respect to nitrate will begin to release nitrate (being replaced by sulphate) thus giving a higher level of nitrates in the treated water. This drawback can be overcome by terminating runs early. Other problems are that the presence of relatively high levels of sulphate reduces the resin capacity for nitrate and increases the regenerant (sodium chloride) consumption by reason of removal of anions other than nitrate, and as a result of sulphate removal, more chlorides are released from the resin to the water. Therefore for waters containing mass ratio of sulphate (as SO$_4$):nitrate (as N) in excess of 3.43:1* it is more economical to use nitrate selective exchange resins. These preferentially absorb nitrates and therefore do not show many of the drawbacks inherent to classical resins.

Surface loading rate for nitrate removal is about 30 $m^3/h.m^2$ and allowing about 20% of the time for regeneration, this corresponds to a mean throughput of 576 $m^3/day.m^2$. Volumetric flow rate should be less than 40 bed volumes (BV)/h; 20 BV/h is regarded as a suitable design value with all columns in service. The resin bed would be about 1.5 m deep and vessel diameters would usually be in the range 1–4 m. The length of a run (time between two consecutive regenerations) is a function of operating conditions and the resin and is limited to 8–12 hours. On completion of a run, the resin bed is first backwashed to remove suspended solids followed by counter-current regeneration using 6–10% w/w sodium chloride solution at 3–4 BV/h for 1.5 BVs; an upflow slow rinse follows at 3–4 BV/h for 2 BVs and a fast rinse in the direction of service flow at 6 BV/h for 4 BV, and the bed is then returned to service. Frequency of backwashing is about every 10–25 cycles depending on the raw water quality. It is normal to use raw water for backwash and treated water for rinse. The quantity of salt required is about 160 g NaCl/l of resin for classical resins and 125 g NaCl/l for nitrate selective resins. The nitrate removal capacity for the two types of resin is about 0.25 g eq/l of resin. When treating waters containing high alkalinity, provision for acid washing of the bed using hydrochloric acid is recommended to remove calcium carbonate deposits in the bed.

Drawbacks with the use of ion exchange for nitrate removal are that the process increases the chloride concentration and reduces alkalinity of the product water, and the need for disposal of the spent regenerant. Although an increase in chloride level of the water has no health implication, it can increase the chloride to alkalinity ratio in soft waters resulting in an increased potential for selectively dissolving zinc from duplex brass fittings; water with a chloride to alkalinity ratio above 0.5 is classed as corrosive to duplex brass[45] (see Section 8.42). The problem is less severe with nitrate selective resins. At the beginning of the run chloride ions increases, accompanied by a reduction in the alkalinity. The product water would therefore have a high chloride to alkalinity ratio initially; classical resins produce higher ratios than nitrate selective resins. The problem can be overcome by mixing the output of a run in a treated water reservoir before forwarding it to supply, or by dividing the flow between two or more parallel units and operating them out of phase; or using a combination of the two. A further but more costly option is to use a bicarbonate in place of the chloride solution in the last 10–15% of the regeneration phase, to replace the chloride ion in the resin with bicarbonate ion. Thus at the start of the run bicarbonate ions and not the chloride ions are released into water.

Waste from the plant contains high concentrations of nitrate, chloride and other anions and, depending on the nitrate removal required, the volume can amount to about 1.5–2.0% of the ion exchange plant throughput. This may be safely discharged to a sewer or for sites in coastal areas to the sea. In other cases removal by tanker may be necessary for its discharge to water courses may not be acceptable. For large plants tanker removal may need to be preceded by a volume reduction process such as electrodialysis reversal, or reverse osmosis; or in some countries solar evaporation may be used. The spent regenerant

---

*1 g of nitrate (as N) = 4.43 g of nitrate (as NO_3). The ratio of sulphate (as SO_4):nitrate (as NO_3) is 0.77:1.

could be treated by biological denitrification using a sequencing batch reactor; a 90% reduction in waste brine and 50% reduction in brine by reuse is reported.[46] However, this rarely, if ever, proves economic.

8.25   Biological process for nitrate removal

Biological denitrification of drinking water is based on the heterotrophic denitrification process that occurs in the anoxic zone of sewage treatment process. In the heterotrophic process an organic carbon source is used to sustain bacterial growth, using oxygen bound in nitrates for respiration, reducing it to nitrogen. Because most water supplies contain relatively low concentrations of organic carbon, these need to be added during treatment; the most common substrates (i.e. sources of organic carbon) are methyl alcohol, ethyl alcohol and acetic acid. Stoichiometric quantities of methyl alcohol, ethyl alcohol and acetic acid required for each mg of dissolved oxygen and nitrate (as N) are 2.57, 1.85 and 3.62 mg, respectively; the actual demand could be up to 1.5 times greater. The process has little effect on the alkalinity of the water. Trace concentrations of phosphates of less than 0.5 mg/l as P are also needed to assist bacterial growth, and should be added if this is not present in the water. The process is sensitive to temperature and reaction rates decrease markedly below about 8°C. The autotrophic denitrification process which uses hydrogen or sulphur compounds to sustain bacterial growth, is also used in some full-scale plants.

The biological process is usually carried out in fluidised bed (up flow) or fixed bed (up or down flow) reactors where the biological growth is physically supported on a medium. Fine sand is commonly used in fluidised bed reactors, whilst a porous medium such as expanded clay is used in fixed bed reactors. In a fluidised bed, water mixed with the carbonaceous substrate flows upward at 20–30 $m^3/h/m^2$ to provide 40–50% bed expansion and giving a detention time typically of 5–10 minutes. Before start up, the bed requires seeding with bacteria. This could take up to 1 month. As the biomass builds up, a proportion of the sand is periodically removed from the bed, the bacterial film is stripped in a sand cleaning plant, and the sand returned to the bed. Fixed bed reactors are usually based on conventional sand filter principles, the media used being a porous medium such as expanded clay, of coarse grain size. An empty bed contact time (EBCT) of 10–15 minutes is usually used. Backwashing regime is similar to that used in conventional sand filters. The treated water from the biological reaction is devoid of oxygen and contains dissolved organic carbon and bacterial floc carried over from the reactor. Hence the water needs to be reoxygenated and filtered through granular activated carbon and sand filters before passing to supply.

8.26   Membrane processes for nitrate removal

Membrane processes used are reverse osmosis (RO) and electrodialysis reversal (EDR) and are discussed in Sections 8.45 and 8.46. Both are demineralisation processes and are substantially non-selective. Nitrates are removed along with other dissolved salts. Membranes are prone to fouling, particularly by dissolved organic material, iron and manganese and suspended matter. Also, since it is a concentration process, sulphates and carbonates of calcium may increase to levels in the concentrate at which they begin to precipitate and scale the membranes. Thus a water that is to be demineralised must be pre-treated to remove a substantial proportion of the fouling contaminants. Additionally acid and sequestering agents such as sodium hexametaphosphate may be added to avoid scaling. Precipitation of sulphate is prevented by limiting the ratio of product water to raw supply which is normally in the range of 75–90%, depending on the sulphates and carbonates present in the water. The waste volume generated in membrane processes is consequently high and can vary between 10–25% of the feed supply; this is a major drawback of the process for inland installation. The process may also be uneconomic if extensive pre-treatment of the feed is required.

8.27 Removal of ammonia

Chemical and physical methods

Ammonia is present as saline ammonia in surface waters particularly those receiving municipal and industrial waste discharges. In ground waters, it is present commonly as free ammonia resulting from the bacterial reduction of nitrates (see Section 6.6). The most usual method used for ammonia removal is 'breakpoint' chlorination (see Section 9.9) where ammonia nitrogen is completely oxidised to nitrogen leaving a residual of free chlorine. Theoretically a chlorine (Cl_2) to ammonia (as N) ratio of 7.6:1* is required for complete oxidation; in practice the ratio required to reach breakpoint is about 10:1. The optimum pH range for the reaction is 7.0–7.5 which takes about 20 minutes to reach completion. To minimise THM formation 'breakpoint' chlorination should be practised after THM precursor removal. Organic nitrogen is not destroyed by chlorine. Ozone oxidises ammonia at pH greater than 8.5, but the reaction is slow.[47] Neither chlorine dioxide nor potassium permanganate affect ammonia.

When present as free ammonia or ammonium ion, ammonia can be removed in packed tower aerators. In the latter case the pH of the water needs to be raised to 10.5–11.5 to convert all ammonium ions to free ammonia. The type of tower design and the packing used are similar to those used for aeration (see Section 8.19) and for removal of volatile organic compounds (see Section 8.29). The desorber operates on the counter-current flow principle with air being blown upwards. An air-to-water volumetric flow ratio as high as 3000:1 is used because of the high solubility of ammonia in water.[48] To prevent 'flooding', i.e. inundation of the packing with water, the air velocity should be maintained at about 3 m/s.

Deposition of scale on the tower packing due to the desorption of dissolved carbon dioxide may reduce the tower efficiency. This can be minimised by using a sequestering agent such as a polyphosphate. Another problem is ice formation in the tower due to evaporative cooling when air temperatures reach freezing. Air temperature also influences the removal efficiency. The drop in temperature between summer and winter can reduce the removal efficiency by about 25%. This may be partly offset by increasing the air-to-water flow ratio.

Biological methods

Biological treatment methods are used for ammonia removal in sewage treatment and their application can be extended to drinking water treatment. The oxidation of saline

---

*1 g of ammonia (as N) = 1.21 g of ammonia (as NH_3) = 1.29 g of ammonia (as NH_4^+). Ratio of chlorine to ammonia (as NH_3) = 6.26:1 and chlorine:ammonia (as NH_4^+) = 5.92:1.

ammonia to nitrate, known as nitrification, takes place in two steps:[49] initially the formation of nitrite by *Nitrosomonas* bacteria, followed by change to nitrate by *Nitrobacter* bacteria. Both steps require oxygen. Carbon dioxide is the carbon source; 1 mg/l ammonia (as N) consumes about 7.2 mg/l alkalinity (as $CaCO_3$). When treating some soft waters therefore alkalinity may have to be added. The process also requires up to 0.2 mg/l phosphates (as P) to allow nitrifying bacteria to develop. The optimum pH range for the reaction is between 7.2 and 8.2. Nitrification is influenced by temperature; low temperatures retard the reaction. Water temperature should be greater than 10°C; there is no biological activity below 4°C. Optimum temperatures for bacterial growth are in the range 25–30°C. The process requires oxygen at the rate of about 4.57 mg per each mg of ammonia (as N).[49] At high ammonia concentrations simple saturation of the water with oxygen by aeration may therefore prove inadequate and oxygen must be continually added.

In biological removal of ammonia in drinking water treatment, nitrification is usually carried out in filters. A maximum ammonia concentration of 1.5 mg/l as N can be removed in conventional rapid gravity filters, depending on the temperature and the dissolved oxygen concentration of the influent water.[49-51] Filtration rates could be in the range 5–10 $m^3/h.m^2$ and to remove 1 mg/l ammonia (as N), the necessary EBCT is about 20, 10 and 5 minutes at 5, 10 and 30°C, respectively. After nitrification the water is devoid of oxygen and needs to be reaerated. For higher ammonia concentrations biological aerated filters (BAF) have been used.[49] These are similar in principal to the trickling and/ or aerated filters used in sewage treatment where there is a continuous flow of air through the media of the filter bed. The ammonia loading of a BAF filter is typically 0.25–0.6 kg of ammonia (as N)/day.m^3 (of media), depending on the type and effective size (ES) of media.

In drinking water treatment the aerated filter consists of a layer of coarse media (ES of between 1.5 and 3.0 mm); filtration is either downflow counter current or upflow co-current with air flow injected continuously into the bottom of the filter bed using either an independent pipe lateral system or special nozzles in a plenum floor design. The volumetric ratio of air:water is in the range 0.5–1.0.[20,49,50] Upflow filters are generally 15–25% more efficient than the downflow filters; to remove 1 mg/l of ammonia as NH_4^+ (0.78 mg/l as N) at pH 7.2 and water temperature of 10°C, EBCT required for upflow and downflow filters using 2 mm ES 'Biolite' is about 3 and 4 minutes, respectively.[52] The depth of the filter medium would be a function of EBCT and filtration rate, and is dependent on the ammonia concentration in the water; e.g. a water containing 2.5 mg/l ammonia (as N) would require a depth of about 2.5 m for an upflow filter when compared to 3 m for a downflow filter. Upflow BAFs would have a surface loading rate of 10–12 $m^3/h.m^2$ and a coarse medium (ES 1.5–2.0 mm) while downflow BAFs operate at surface loading rates of 8–10 $m^3/h.m^2$ (depending on the suspended solids loading) and a coarser medium (ES 2.5–2.85 mm), provided that the water is then filtered. Such high surface loading rates are feasible with expanded mineral filter media mostly of proprietary makes (e.g. 'Biolite'). Filters using naturally occurring media such as pozzolana or carbon (2–5 mm or even larger to give high specific surface per unit volume) operate at about 5 $m^3/h.m^2$ with EBCT of about 20–30 minutes.[53]

BAFs are washed by concurrent application of air and water at about 16 and 7 mm/s, respectively. Washwater should be free of chlorine. When treating surface water BAFs are best used after the clarification stage and downflow filters are generally preferred. Upflow filters are generally used for ground waters particularly those containing high ammonia concentrations. BAFs are best followed by a conventional rapid gravity sand or anthracite–sand filtration in order to produce a water free of suspended matter. The biological process is adversely affected by chlorine, hydrogen sulphide, heavy metals and precipitates from iron and manganese oxidation and other suspended solids in the water. When ammonia is present together with iron and manganese, the order of biological removal is iron, ammonia followed by manganese.[15] Iron is usually removed separately. If the ammonia concentration is high manganese will not be removed in the same filter unless adequate EBCT is provided. BAF beds also effectively remove organic carbon. Manganese dioxide coated sand filters have been successfully used to oxidise low concentrations of ammonia biologically and manganese by catalytic oxidation.[54]

The biological reaction principle can also be applied to sedimentation tanks of the sludge blanket type. The sludge acts as the medium for the growth of nitrification bacteria. Oxygen for the nitrification reaction is limited to that which can be contained in the feed water and the ammonia removal is limited to about 0.5 mg/l as N.

Removal of volatile organic compounds from groundwater
8.28   General

Some groundwaters are contaminated by volatile organic compounds (VOC) which are used as solvents in industry (see Section 6.14). The WHO Guidelines values and EC maximum admissible concentrations are given in Table 6.1(c) of Section 6.47. Although the compounds are normally present in trace concentrations, they can be significant and it may be necessary to reduce them.

They are not removed by conventional water treatment processes, being stable towards most oxidants including ozone, and they are not biodegradable. The most cost effective treatment is considered to be packed tower aeration; but adsorption onto granular activated carbon may be more economical in hard waters[55] (see Section 8.30).

8.29   Packed tower aerators

Packed tower aerators, also known as air stripping towers, are discussed in Sections 8.19 and 8.27. In these the contaminated water flow is downwards through a packing, while a counter-current air flow strips the VOCs into the gas phase and discharges them through the top of the tower. The treated water is collected at the bottom of the tower. The process removes over 99.99% of VOCs.[56]

The design parameters are air-to-water volumetric flow ratio, surface loading rate, type and size of packing, and the depth of packing. These are influenced by temperature, chemistry of the water and the mass transfer characteristics of the packing.[57] Towers are usually constructed in polyethylene, glass reinforced plastic or rubber lined mild steel, and need to be provided with a good water distribution system and mist eliminators on the air discharge. Packing types are pall rings or Rashig rings or saddles and materials are usually of plastic (see Section 8.19); packing height should be limited to about 6 metres. For greater packing heights two or more towers in series should be considered. Surface loading rate of the tower needs to be selected to prevent flooding of the packing and is usually about 60–75 $m^3/h.m^2$. Tower diameter should be limited to 3–4 metres. High water flow rates should be divided between two or more towers in parallel. The air-to-water volumetric ratio is usually about 25:1 to 30:1 to limit the air pressure drop across the packing to 10–40 mm of water per metre of packing.

In packed towers while the VOCs are stripped from the water, dissolved oxygen is increased and carbon dioxide is removed. The addition of oxygen can cause iron to precipitate if present in water which therefore fouls the packing. A problem of more concern is, where water alkalinity is high, the CO_2 removal can cause scale formation on the packing or further downstream due to the precipitation of calcium carbonate and is expected to occur at a Langelier Index greater than $+0.6$.[55] Sequestering agents such as polyphosphates may therefore be needed to alleviate scaling in the packings. Other methods include dosing of sulphuric acid to shift the CO_2/carbonate equilibrium, or base exchange softening to reduce hardness of the water. The air stripping results in release of VOCs to atmosphere. Although the quantities are small, this may be of some concern for plants located in urban areas.

8.30 GAC (granular activated carbon) adsorbers

Adsorption of organic compounds by GAC is well known and the technique can be used to remove volatile organic compounds effectively (see Section 8.35). The adsorbers are of the fixed bed type and can be in conventional rapid gravity or more commonly pressure filter form. The principal design parameters are the empty bed contact time (EBCT), defined as the bulk volume of adsorbent bed (m^3) divided by the water flow rate (m^3/min); the type of GAC used, the bed depth, and the hydraulic loading. These parameters should be evaluated by pilot scale tests. For pressure filters typical values are EBCT in the range 10–20 minutes, surface loading rates between 10–20 m^3/h.m^2 and bed depths of 2.5–3 m. The adsorption capacity of GAC for VOCs is generally small and the media will need frequent regeneration if VOCs are present in concentrations in excess of 100 μg/l.[58] Experience in the USA shows 12–18 month regeneration frequency.[59]

8.31 Chemical oxidation

VOCs can be oxidised by ozone when used in combination with hydrogen peroxide (H_2O_2) or UV radiation[60] (See Section 8.36). Ozone (O_3) is first injected into a reaction chamber consisting of two or three compartments to satisfy the ozone demand; hydrogen peroxide being injected into the second or the third compartment together with further addition of ozone. Tests must be carried out to optimise the ratio H_2O_2:O_3 and the contact time which depends on the concentration and the nature of the organic matter. The ratio can vary between 0.3 and 1. It is desirable to follow the oxidation stage of treatment with GAC absorbers to remove the by-products of the oxidation and excess hydrogen peroxide. The latter dechlorinates water.

Taste and odour removal

8.32 Causes of tastes and odours

The source of a taste or odour in a water is often difficult to identify, but the following list includes the most likely causes.

(1) Decaying vegetation, such as algae, may give rise to grassy, fishy, or musty 'pharmaceutical', cucumber, etc. type odours. Algae mostly cause offensive odours as they die off, but some living algae, e.g. blue-green (*Cyanophyceae*) cause taste and odour problems.

(2) Moulds and actinomycetes may give rise to earthy, musty, or mouldy tastes and odours in a water which may be wrongly attributed to algal growths. In stagnant waters and especially water in long lengths of pipeline left standing in warm surroundings, such as the plumbing system of a large building, the moulds and actinomycetes have favourable conditions for growth and the first water drawn in the morning may have an unpleasant taste or odour of the kind mentioned.

(3) Iron and sulphur bacteria produce deposits which, on decomposition, release an offensive smell. The sulphur bacteria in particular give rise to the hydrogen sulphide (rotten egg) smell. Hydrogen sulphide is also naturally present in some ground waters in concentrations up to 10 mg/l. The taste and odour thresholds for hydrogen sulphide in water are 0.06 mg/l and 0.01–0.001 μg/l, respectively.

(4) Iron above a concentration of about 0.3 mg/l will impart a bitter taste to a water.

(5) Excessive chlorides and sulphates will impart a brackish taste to a water; taste thresholds for sodium and calcium chlorides are 200 and 300 mg/l as chloride, respectively, and those for sodium, calcium and magnesium sulphates are from 250, 250 and 400 mg/l as sulphate, respectively.

(6) Industrial wastes are a prolific source of taste and odour troubles of all kinds, of which those produced by phenols are the most frequently experienced. In the presence of free residual chlorine the phenols form a 'chlorophenol' medicinal taste which is quite pronounced; a quantity as small as 0.001 mg/l phenol may react with chlorine to form an objectionable taste.

(7) Chlorine will not, by itself, produce a pronounced taste except in large doses, but many taste troubles accompany the injection of chlorine into a water because of subsequent reactions between chlorine and a number of organic substances. These tastes are usually described as 'chlorinous'. Reaction of chlorine with ammonia will produce tastes and odours; the limiting concentrations for monochloramine, dichloramine and nitrogen trichloride being 5, 0.8 and 0.02 mg/l, respectively.[61] This is a common problem where waters containing free chlorine and combined chlorine are mixed in a distribution system.

8.33 Methods for the removal of tastes and odours

Tastes and odours of biological origin (see Section 8.32(1) and (2)) are due to the presence of organic compounds such as geosmin, methyl isoborneol (MIB), phenols, and many others. Those of industrial origin are also invariably of organic nature. Oxidation and adsorption onto activated carbon are considered to be the most effective methods available for reduction of tastes and odours associated with organic compounds. Of the oxidants, ozone is very effective in destroying some of the taste and odour producing compounds. In theory saturated compounds such as geosmin and MIB, which are responsible for musty and earthy tastes, and some of the chlorinated hydrocarbons are not oxidised by ozone. In practice however ozone is known to be effective in removing geosmin and MIB.[62] Ozone when coupled with hydrogen peroxide however, increases the oxidation efficiency of ozone (see Section 8.36) and therefore can be more effective in reducing tastes and odour,[63] including those due to geosmin and MIB.[64]

Other oxidants such as potassium permanganate (see Section 8.12) and chlorine dioxide have been successfully used for taste and odour removal. Chlorine dioxide is useful when phenols are present as it does not form chlorophenols (see Section 9.17).

Activated carbon is the most effective method of removing taste and odour compounds of organic nature. It can be used in either the powdered activated carbon (PAC) or granular activated carbon (GAC) form. GAC is normally used when taste and odour removal is required continuously for long periods; for seasonal occurrence lasting several days at a time or dealing with pollution incidents it may be economical to use PAC. When used as PAC the dosages can vary between 10–30 mg/l divided between the works inlet ahead of clarifiers and the filter inlets, with dosages to the latter maintained below about 5 mg/l. When applied at the works inlet the PAC dose may need to be increased to accomodate those organics which otherwise have been removed by coagulation. Addition of PAC with the coagulant can reduce the adsorption efficiency because PAC is incorporated into the floc particles[65] and organics must therefore diffuse through the floc. When PAC adsorption is preceded by coagulation, organic removal is improved[66] because PAC adheres to the outer surface of the floc. Good mixing and sufficient contact time are the important design parameters. When dosed at the works inlet, a contact time of 15 minutes is sufficient.[67] Sludge blanket and solids recirculation tanks provide adequate contact time; however organics contained in sludge are known to utilise some of the adsorption capacity of PAC. For filter inlet dosing the filters provide the necessary contact time. PAC is stored as a powder and dosed as a slurry (see Section 7.20). GAC is used in adsorbers of the rapid gravity or pressure filter type (see Section 7.32). It is sometimes used in place of sand as a media in rapid gravity filters for turbidity removal, but organics removal is more effective when GAC adsorbers are used after rapid gravity filtration. For taste and odour removal only, an EBCT of 10 minutes is adequate and the life of a GAC bed between regenerations may be 2 to 3 years or longer if ozone or ozone-hydrogen peroxide oxidation is used with GAC (see Section 8.35).

Aeration can sometimes improve the palatability of water which is made poor due to stagnation, such as the bottom waters of reservoirs. It is also effective for taste and odour caused by compounds with high vapour pressures such as chlorinated solvents, some hydrocarbons and hydrogen sulphide. Taste complaints resulting from iron, etc. in water that has been left to stagnate in the ends of mains can be overcome by flushing.

Hydrogen sulphide removal

In natural waters hydrogen sulphide (H_2S) is in equilibrium with hydrosulphide (HS^-) and sulphide (S^{2-}); H_2S is predominant up to pH 7, between pH 7 and 13 more than 50% is HS^- and above pH 13, S^{2-} predominates. Aeration is effective in removing H_2S at pH <6 when over 90% is free H_2S; a packed tower aerator (see Section 8.19) removes >95%. Aeration removes CO_2 in preference to H_2S thus raising the pH. Acid dosing may therefore be necessary to maintain the pH < 6.

H_2S could also be removed by chemical oxidation using chlorine, hydrogen peroxide or ozone. In the oxidation reactions elemental sulphur is formed initially and then further oxidised to sulphuric acid. The latter reaction takes place at low pH. At pH values between 6 and 9, elemental sulphur reacts with residual sulphides even in the presence of an oxidant to form obnoxious polysulphides and a milky blue suspension of colloidal sulphur.[68]

Treatment consists of converting colloidal sulphur and polysulphides formed in the oxidation reaction to thiosulphate by the addition sulphur dioxide (or sodium bisulphite) and then converting the thiosulphate formed to sulphate by chlorination.[61,68] Polysulphides are not formed at pH > 9. Potassium permanganate oxidises H_2S only to elemental sulphur. It is also known to introduce soluble manganese into a water. The stoichiometric quantities of oxidant required for the oxidation of 1 mg of H_2S to elemental sulphur are 2.1 mg of chlorine, 1 mg of hydrogen peroxide, 1.41 mg of ozone and 6.2 mg of potassium permanganese. Due to high cost, chemical oxidation is used after aeration to remove the residual H_2S remaining in the water.

Micropollutants removal
8.34   General

Micropollutants were described in Section 6.31 and include pesticides, volatile organic carbons (VOC) such as chlorinated organic solvents, trihalomethanes (THM), polynuclear aromatic hydrocarbons (PAH) and polychlorinated biphenyls (PCB). THMs are sometimes present in surface waters which have received industrial discharges, but are commonly formed during the chlorination of water containing THM precursors, which are the natural organic substances such as humic and fulvic acids that impart colour to the water, and algal metabolic products and cell debris (see Section 9.7). Micropollutants occur in concentrations of the order of micro or sub-microgram per litre and in most instances their removal is necessary to comply with standards set for drinking water quality (see Section 6.47).

Conventional chemical coagulation, clarification and filtration and also slow sand filtration achieve partial removal depending on the type and the nature of the micropollutant. In most instances a significant reduction in the concentration of THM precursors is achieved by coagulation. The subsequent formation of THM depends on the concentration of precursors remaining, the chlorine dose applied, pH and temperature of the water and the contact time. For effective removal of micropollutants, conventional treatment processes alone are usually insufficient to meet the recognised standards for drinking water quality and need to be supplemented with advanced treatment processes including advanced oxidation processes.

8.35   Advanced treatment processes

Granular activated carbon adsorbers

It is well known that activated carbon is effective in removing organic compounds from water. It has been used in the past primarily in the form of powdered activated carbon (PAC) for the removal of taste and odour producing organic compounds. As PAC is normally dosed at the inlet to a treatment works, a high organic loading in the raw water will decrease the adsorption capacity of PAC in relation to any organic micropollutants present. PAC is used for the removal of pesticides from a raw water subject to seasonal changes. The removal efficiency increases with dose and contact time. An allowance should be made in the PAC dose to satisfy other organic demands particularly when the concentration of natural organic matter exceeds that of the pesticides. Complete removal of several pesticides has been reported in direct filtration plants treating ground waters using PAC doses of 5–10 mg/l supplemented by a small coagulant dose; PAC was available for adsorption throughout the duration of filter run.[69]

With the development of regenerable GAC and reactivation facilities, many water undertakings have switched from using PAC, which is costly and thrown away after using once, to GAC which although it is about two to three times the cost of PAC can be reused after reactivation. Since the introduction of more rigorous standards for drinking water quality the use of GAC has become widely used for the removal of micropollutants. For best results GAC adsorbers should be installed downstream of rapid gravity filters used for turbidity removal.

Activated carbon can be made from wood, coal, coconut shells or peat. The material is first carbonised by heating and then is 'activated' by heating to a high temperature whilst providing it with oxygen in the form of a stream of air or steam. Sometimes chemical activation by phosphoric acid is used. It is then ground to a granular or powdered form. It is a relatively pure form of carbon with a fine capillary structure which gives it a very high surface area per unit of volume. The adsorption capacity of GAC is described by various parameters including Iodine Number and BET surface area (see Table 7.11). GAC can be reactivated by heating to 800°C in steam or CO_2, or chemically (either on or off-site) but in the process a substantial proportion, up to 25%, of the carbon may be lost. Reactivation restores GAC to almost its original adsorption capacity.[70] However, it is reported that metals which are adsorbed by GAC are not effectively removed by reactivation and result in a reduction of adsorptive capacity (indicated by low BET surface area value) and, when GAC is returned to service, these metals may be released into the filtered water. Both problems can be overcome by including an acid wash of the GAC; superior results are obtained by a pre-reactivation wash over a post-reactivation wash[71] (see Section 7.32).

GAC adsorbers are of conventional rapid gravity or pressure filter design and the basic design parameters are the bed depth or hydraulic loading (flow per unit area; $m^3/h.m^2$) and the empty bed contact time (EBCT). When used as filters they are washed by water only or by air followed by water; combined application of air and water is considered unsuitable due to the high losses caused by attrition. When GAC is used for adsorption only, a water only wash is considered adequate, although at least an intermittent air–water wash may be beneficial to maintain a well mixed bed and to minimise the development of the adsorption front formed when in service. GAC is less dense than sand, and requires 20–30% bed expansion during backwashing, depending on the type of carbon used. The frequency of backwashing of GAC adsorbers at ground water sites could vary from once every 2–8 weeks depending on the raw water quality. At surface water sites, in order to maintain low bacterial counts in the filtered water, the backwashing should include an air scour and the frequency should be 2–3 days. This can also help to control the growth of micro-animals (nematodes, chironomid midge larvae) because the wash frequency is shorter than their reproductive cycle. The problem of micro-animals could also be overcome by chlorinating the backwash water or taking a filter out of service for a period sufficient to produce anaerobic conditions in the filter, to kill the micro-animals. This should be carefully controlled not to loose biological activity in the filter and to prevent the formation of ammonia and nitrite in the filter. Details of GAC adsorber design are given in Section 7.32.

GAC characteristics will vary according to the base material used. For example the absorptive capacity for the pesticide atrazine varies in the order wood > coconut shell > peat > coal.[72] However coal based GAC finds wide use for most water treatment applications. Pilot plant work or laboratory accelerated column tests (ACT) should be used to optimise the GAC type and other design parameters, such as adsorption capacity (by Freundlich adsorption isotherm) and life of carbon between reactivation (by ACT). Empty bed contact time will vary for different micropollutants and is usually in the range 5–30 minutes. For pesticide removal an EBCT in the range 15–30 minutes is used. An EBCT of about 10 minutes is considered adequate for THMs and VOCs.

Although GAC removes most micropollutants efficiently, the adsorption capacity towards some is low, so that frequent reactivation may be necessary which makes the adsorption process costly. For example using an EBCT of 10–30 minutes most pesticides may show breakthrough in 6–12 months; THMs in 6–12 months and VOCs in 6–18 months. But if only taste and odour removal is required, breakthrough normally occurs in 2–3 years when using an EBCT of about 10 minutes. In total organic carbon (TOC) removal breakthrough occurs in about 3 months. In one UK works it has been shown that the TOC removal efficiency reduced from 90% to 10% in 14 weeks, but this did not have any adverse effect on the THM concentration in the final water which was significantly less than that before the installation of GAC.[73] For TOC removal a criterion used for estimating GAC life is 50 m^3 water treated per kg of GAC.[74]

Biological activated carbon reactors

Dissolved organic carbon (DOC) present in water is primarily in the form of refractory or poorly biodegradable compounds with some biodegradable dissolved organic carbon (BDOC) which is also called assimilable organic carbon (AOC). In the GAC, all organics are adsorbed irrespective of their biodegradability but the presence of BDOC will lead to limited biological activity occurring after a few weeks. Although a high proportion of the organics will be removed by adsorption, the GAC life is ultimately determined by the refractory organic breakthrough.

Ozonation of the water upstream of GAC adsorbers can, however, transform a large proportion of the refractory organics into BDOCs, thus promoting biological activity in the adsorber which will help to remove organic matter which is otherwise unaffected or only marginally affected directly by ozone oxidation. VOCs and some pesticides are examples of such compounds. In the adsorbers the GAC acts more as a biomass substrate and less as an adsorber, hence GAC beds used in this fashion are called biological activated carbon (BAC) reactors. In the biological reaction the BDOCs are converted to carbon dioxide, thus more of the adsorption capacity of GAC is available to deal with a smaller proportion of refractory organics. The life of the GAC between reactivation is therefore increased many-fold. For example, for pesticide removal an increase in running time to 1.5 years is reported.[75] THMs once formed cannot be removed with ozone or BAC treatment. BAC reactors are of similar design to conventional rapid gravity sand or GAC adsorbers (see Sections 7.26 and 7.32). For best results they should be installed after rapid gravity filters which are used for turbidity removal. Zooplankton can grow in BAC reactors and bacterial counts in filtered water tend to be high. Therefore reactor washing should be efficient and frequent. Filter wash is similar to that described for GAC adsorbers except that unchlorinated water is preferred for washing filters. Filter washing should be by the application of air followed by water to prevent accumulation of mud balls formed as a result of slime growth in the filter and zooplankton and bacteria control. In some cases the filtered water may be devoid of oxygen due to the oxygen consumption during biological activity, therefore oxygenation of the filtered water may be necessary.[75]

The frequency of washing the reactors should be similar to GAC adsorbers. For most applications an EBCT of 10–15 minutes is generally considered adequate. Larger EBCTs will help to increase the service life of carbon between regeneration. In BAC reactors TOC is more effectively removed than in GAC adsorbers; following a very high initial removal it stabilises at about 30–40% thereafter.[76,77] A high proportion of the remaining TOC could be non-biodegradable refractory organic matter, therefore the use of ozone with GAC produces a biologically stable water thus reducing the risk of aftergrowths in the distribution system and minimise THM formation in the subsequent chlorination for disinfection.

In the ozonation stage pathogenic micro-organisms are significantly reduced, but the water from the BAC reactors will contain a higher level of non-pathogenic bacteria, thus requiring an effective disinfection stage such as chlorination for their removal.[78]

8.36   Advanced oxidation processes

The effectiveness of oxidation of organic matter by ozone can be increased by using it in conjunction with hydrogen peroxide.[79] This advanced oxidation process is sometimes called 'peroxone' and produces the so called hydroxyl free radicals which break down organics including those resistant to ozone such as some pesticides and VOCs. For example this process is known to improve atrazine removal, from 15 to 40% using ozone alone, to 70–85%.[80] Removal is better at low alkalinities and pH in the range 7–7.5. The hydrogen peroxide: ozone ratio used depends on the application and varies from less than 0.3:1 for controlling taste and odour compounds[63] to 0.5:1 for controlling pesticides and VOCs.[81] In the 'peroxone' process ozone is added to the first stage of a two or three stage contact chamber in which ozone rapidly oxidises reactive organic substances, followed by the second stage where the ozonation efficiency towards less reactive organic substances is substantially increased by the addition of hydrogen peroxide. Alternatively the second chamber is used for disinfection with the final chamber being used for hydrogen peroxide-ozone reaction. Again the process is best used downstream of rapid gravity filters after a significant proportion of the organic load has been removed. Like ozone the 'peroxone' process converts refractory organic substances to readily biodegradable compounds which should be removed in biological activated carbon reactors. These reactors will also act as safety filters in removing the by-products of oxidation and any unreacted hydrogen peroxide which would otherwise dechlorinate the water when final residual chlorination is adopted.

The high oxidative power of hydrogen peroxide and the large volume of oxygen generated and the heat evolved from its decomposition make it a hazardous chemical to handle. Correct material selection and good facility design can eliminate hazards. For waterworks application the commercial grade of solution strength 35% w/w would normally be used (see Table 7.7).

Ozonation coupled with UV-radiation instead of the hydrogen peroxide is also known to increase the oxidation efficiency of ozone.[82] but work on this process has only been carried out at laboratory and pilot plant level. UV irradiation alone is reported to give 70% removal of acid pesticides such as mecoprop, MCPA and bromoxynil at a radiation dose of 200 Wh/m^3. However the removal of atrazine was limited to 30%;[82] increasing the energy input to 700 Wh/m^3 increased the removal to about 70%.[83]

8.37 Colour removal

Natural colour is primarily due to humic and fulvic acids. It is colloidal, therefore it can be very effectively removed by coagulation using aluminium and iron salts.[84] Removal is most effective at pH values of about 5–5.5; iron salts at low coagulation pH values are considered to be most suitable. Nevertheless aluminium salts are successfully used with some waters for colour removal particularly lowland waters in the pH range 6.5–7.2 and upland waters in the pH range 5.5–6. Since colour is organic in nature it is susceptible to oxidation by chlorine, chlorine dioxide and ozone. Chlorine, although very effective in breaking down colour, is not recommended due to the formation of THMs.[85] Therefore chlorinated ferrous sulphate, although effective in colour removal, is not recommended due to the potential for THM formation with excess chlorine. The ozone dose required to breakdown colour can be high and it is therefore beneficial to use it in combination with chemical coagulation, sedimentation and filtration, the ozone being used to oxidise the residual colour. Grampian Water (UK) is using ozone (up to 5.5 mg/l with 15 minutes contact) ahead of slow sand filters to reduce colour from 70 to less than 20° Hazen at the Invercannie slow sand filtration plant.

Colour can be removed by membrane processes. There are several plants in Scotland using spirally wound cellulose acetate nanofiltration membranes to reduce true colour. It is reported that colour removal of 100° Hazen to less than 1° Hazen and total iron removal of 1.35 to less than 0.01 mg/l as Fe have been achieved.[86] Membranes were operated in the cross-flow mode (see Section 7.36) with recycle. Fouling of the membranes was high. Frequent cleaning by chemicals and detergent was necessary, with membrane replacement every three years. Tubular membranes also of cellulose acetate were effective without pretreatment and required less frequent cleaning. Performance was slightly inferior reducing the colour to less than 10° Hazen.

Corrosion causes and prevention

8.38 Physical and electrochemical corrosion

High velocities of water flow can cause deep pitting and erosion of surfaces due to the phenomenon known as cavitation. Section 12.20 describes how cavitation occurs, and Plate 6 shows a case where it is thought to have caused major destruction of concrete. The same kind of corrosion causes pitting and erosion of metal surfaces, such as on the blades of centrifugal pumps, on valve gates where valves are kept almost closed, at nozzles where high water velocities are created, and at bends in pipes, tunnels and conduits where high velocity flow, at a pressure below atmospheric, moves to an area of higher pressure. Cavitation corrosion can therefore be avoided by hydraulic design measures which avoid high velocities of flow impinging on surfaces where a sudden increase of pressure occurs, or where pulsating pressures occur.

Electrochemical corrosion occurs where two different metals having an electropotential between them are immersed in a common body of water. All natural waters can act as an electrolyte, but the degree to which they do so depends on the dissolved chemicals present. The metal which loses ions to the electrolyte and therefore corrodes is termed the 'anode'; the other is the 'cathode' which gains ions. Of the metals commonly used in water supply systems, zinc is anodic to iron and steel, which are themselves anodic to copper. Thus zinc-galvanised pipes or tanks may corrode if fed by water which has passed through copper pipes; and corrosion of lead joints can occur in lead-soldered copper pipework, because lead is anodic to copper. Electrochemical corrosion can also occur because the metal of a pipe may vary locally from place to place, creating 'pockets' of differing material which have an electropotential between them. Corrosion processes are complicated by the further chemical reactions that can occur. Thus anodic corrosion of iron piping can produce tubercles of rust, which alter the rate of corrosion, and beneath which other forms of corrosion (e.g. anaerobic) can occur. Ions deposited on the cathode may reduce its cathodic effect, so the electrochemical reaction dies away; or the ions may be constantly removed by flowing water or combine with other substances in the water so the anodic corrosion continues or may even increase.

Unprotected metal pipes when buried in wet soil which acts as an electrolyte, corrode externally. The corrosion can be increased if the pipes become anodic to other buried metallic structures in the soil, particularly so if the other buried structures are acting as 'earths' to electrical apparatus which introduce stray currents into the soil. Soil surveys can locate conditions likely to cause external corrosion which are said to include:[87]

- soils with a moisture content above 20%;
- acid soils with a pH of 4 or below;
- poorly aerated soils containing soluble sulphates (see below);
- soils with a resistivity less than 2000 ohm cm.

Cathodic protection and sleeving of pipelines are measures widely adopted to prevent external pipeline corrosion as described in Sections 13.6 and 13.12.

8.39   Bacterial corrosion

Sulphate-reducing bacteria (i.e. *Desulfovibrio desulfuricans*) existing in anaerobic conditions, i.e. in the absence of oxygen, can cause iron to be corroded. Many of these bacteria are capable of living on a mineral diet and, as a result of their metabolism, they produce hydrogen sulphide which attacks iron and steel and forms the end product ferrous sulphide. Steel thus becomes pitted and cast iron becomes 'graphitised',[88] in which state it becomes soft. Sulphate-reducing bacteria may be most numerous in waterlogged clay soils where no oxygen is present and where sulphur in the form of calcium sulphate is very likely to occur. In such clays the sulphate reducing bacteria are one of the most virulent forms of attack on iron and steel. Backfilling pipe trenches with chalk, gravel or sand to prevent anaerobic conditions arising, in addition to cathodic protection and sleeving, are used to prevent this form of corrosion.

Iron bacteria (e.g. *Crenothrix, Leptothrix* and *Gallionella* types) may be present in a water which lacks oxygen (as evidenced by the presence of sulphates and hydrogen sulphide in the water), and may cause internal corrosion. These are bacteria which have the power of absorbing oxygen and then oxidising iron from the water or iron pipes and storing it. They are aerobic and, under favourable conditions, can form large deposits of slime which are objectionable, giving rise to odours and staining. Their growth is an ever-present possibility when a water has a high iron content.

Tuberculation inside an iron main is sometimes set off by sulphate reducing bacteria, or more commonly by organic substances in the water, a low pH, or high oxygen content of the water. The external surface of a tubercle or nodule consists of a hard crust of ferric hydroxide, often strengthened by calcium carbonate and magnesium dioxide. Below this exterior crust, conditions tend to be anaerobic so that sulphate-reducing bacteria can flourish creating further products of corrosion. Hence two remedies for the prevention of internal tuberculation are the aeration of a sulphate or hydrogen sulphide containing water, and lining the interior of the pipeline with an epoxy resin or with cement mortar.

8.40 Corrosion caused by water qualities

Water is a ubiquitous solvent so that it readily takes into solution a wide variety of gases and substances. Hence distilled, demineralised or desalinated water which contains almost no dissolved substances is always highly aggressive to metals and concrete. Some waters have a tendency to corrode metals due to a high content of chloride or other saline constituents. In some cases free residual chlorine will tend to cause corrosion, due to its oxidation potential. However, the extent to which corrosion occurs depends in most cases largely upon the acidity or carbon dioxide content of the water and the extent to which this is countered by the presence of calcium carbonate alkalinity in the water and a sufficiently high pH. The relationship between alkalinity and dissolved carbon dioxide at equilibrium is given in Table 8.6 which is based on a graph by Cox.[89]

The primary consideration is whether or not the carbon dioxide in the water is sufficient to keep its dissolved calcium bicarbonate in equilibrium with solid calcium carbonate. If the amount of carbon dioxide is more than sufficient to achieve this balance, the excess is termed 'aggressive' free carbon dioxide, and the water will tend to dissolve any protective coating of solid calcium carbonate with which it comes into contact. For example, from Table 8.6, a water having an alkalinity of 125 mg/l as $CaCO_3$ at pH 7.5. is aggressive if free CO_2 concentration exceeds 7.5 mg/l. Conversely, a water with less free carbon dioxide will deposit some of the calcium bicarbonate as a protective coating of calcium carbonate on metals.

The 'chalk test' is therefore a useful way of determining whether or not a water is in equilibrium with calcium carbonate. A sample of water is placed in contact with powdered chalk and the pH is determined at which it no longer deposits or dissolves calcium carbonate. This pH value, denoted as pHs, is termed the 'saturation pH value' of the water. Langelier's Index, I, is then determined as the numerical difference between the pH of the water and its pHs, i.e.

$$I = pH - pHs$$

When I is negative the water is liable to be corrosive. If I is positive the water will tend to deposit calcium carbonate. If I is zero, then the water is in balance, being neither aggressive to, nor tending to deposit a protective coating of calcium carbonate. Consequently the main form of treatment for an aggressive water is to add chemicals to the water, sufficient to produce a thin protective coating of solid calcium carbonate on otherwise corrodible surfaces. The treatment must be continuous and usually consists of

Table 8.6 Relationship between alkalinity, equilibrium pH and dissolved carbon dioxide

Alkalinity as $CaCO_3$ (mg/l)	25	50	75	100	125	150	175	200	250	300
pH at equilibrium, i.e. pH$_s$	8.8	8.1	7.7	7.6	7.5	7.4	7.35	7.3	7.2	7.0
Free CO_2 at equilibrium (mg/l)	0.0	1	2.5	4.5	7.5	12	18.5	27	32	60

dosing of an alkali such as hydrated lime to ensure that the pH of the treated water is maintained only slightly above the pHs value (i.e. Langelier's Index of about $+0.2$ to 0.3), the resulting deposition then being in the form of a smooth hard scale. If the water is over-saturated with calcium bicarbonate, i.e. the pH value is significantly greater than pHs, then precipitation of calcium carbonate takes place rapidly, producing an excessive amount of the amorphous soft form of calcium carbonate, namely 'chalk'. Table 8.6 is a useful practical guide for determining the saturation pH (pHs) of a water.

8.41 Corrosiveness of various waters

There are a wide variety of naturally occurring waters with different characteristics which determine their aggressiveness to metals and concrete. The principal types of water occurring are dealt with below.

Hard surface waters, with fairly high carbonate hardness

A large proportion of lowland river waters associated with sedimentary geological formations have a substantially positive Langelier Index. Generally, therefore, such waters are very satisfactory for the avoidance of corrosion of metals and they are not usually plumbosolvent.

Hard underground waters with substantial carbonate hardness

The mineral composition of many underground waters from chalk, limestone and other sedimentary formations may often be similar to that of many surface waters; but the carbon dioxide content of the water is usually much higher, being near to, or above the amount which is in equilibrium with the solid calcium carbonate. The excess carbon dioxide has therefore to be removed, usually by aeration, to give a positive Langelier Index. But although a water with a positive Index should, theoretically, deposit calcium carbonate, in practice it will probably not do so to any appreciable extent in water mains or metal tanks. Its anti-corrosion effect, however, probably depends in part upon the rapid deposition of calcium carbonate on cathodic areas where electrochemical corrosion is initiated, the effect being to stifle this type of incipient localised corrosion. Waters with fairly high carbonate hardness due to calcium bicarbonate are plumbosolvent (see Section 8.14).

Underground waters with low carbonate hardness but high free carbon dioxide

Many small sources from some of the older geological formations, and some from gravel wells and springs comprise waters of this type; although many of the gravel sources can have a high non-carbonate hardness. Aeration can be adopted to reduce the carbon dioxide content, and lime can be added to achieve a further reduction. If the water is already hard it may be preferable to neutralise the free carbon dioxide by adding caustic soda. This reagent is very soluble and thus can be administered as a small volume of a strong solution. For the correction of the corrosive characteristics of very soft waters, passage through a filter containing granular limestone may be useful.

Soft waters from surface sources

Certain lake waters such as some in Scotland and north-east England give very pure waters, with a low organic quality approaching that of an underground water, but not containing any appreciable content of carbon dioxide. They are therefore almost neutral in reaction, but many may also have a very low carbonate hardness, e.g. 10 mg/l as $CaCO_3$. However from the point of view of corrosion protection, and of preventing plumbosolvent action in particular, they need careful treatment (see Section 8.14). With some of these waters an economical form of treatment is simply to add a small dose of lime which can prove satisfactory. In cases where a greater amount of calcium carbonate hardness must be added, dosing with carbon dioxide followed by dosing with lime or passing the water through calcium carbonate filters will be adequate. Alternatively use of calcium carbonate with soda ash or sodium bicarbonate may be convenient.

The majority of surface waters derived from upland moorland catchments, however, are coloured brown by peaty matter which contains organic humic and fulvic acids which give the water an acid reaction with a pH of 6 or lower. To counter the general corrosiveness of this type of water and its plumbosolvent action, the organic acids do not merely require neutralisation, but must be removed[90] and this is usually achieved by chemical coagulation followed by filtration. But if the coagulant aluminium sulphate is used, this actually removes carbonate hardness, the carbonate or bicarbonate being converted to free carbon dioxide. Hence lime should be added after filtration, and the carbon dioxide can then usually be largely converted again to calcium bicarbonate (see Section 7.21).

8.42 Dezincification

Galvanised iron or steel is susceptible to corrosion by excess of free carbon dioxide which dissolves the zinc, and to any waters in which the pH is excessively high, especially above 10 which may occur if the final treatment stage of a water is unsatisfactory. However, the term 'dezincification' usually refers to the effect on alloys containing copper and zinc, i.e. brass, when the zinc is dissolved out. If the corroded metal is cut and polished, the dezincified part will show the typical colour of copper instead of the original brass. However a particular form of this corrosion is called 'meringue' dezincification because of the voluminous white layer of corrosion product which appears, the effect of which is to cause failure of fittings, mainly hot water fittings when they are constructed of hot-pressed brass. The effect, investigated by Turner[91] was found to occur with waters having a high pH of over 8.2 and having a ratio of chloride to carbonate (temporary) hardness greater than is shown in Table 8.7. These findings reinforce the need for the addition of carbonate hardness to any very soft waters to reduce their aggressiveness to metals. If a water contains an appreciable amount of chloride, then simply raising its pH without increasing

Table 8.7 Limiting values of chloride-to-alkalinity ratio (at pH values greater than 8.2) for dezincification

Chloride (mg/l as Cl)	10	15	20	30	40	60	100
Alkalinity (mg/l as $CaCO_3$)	10	15	35	90	120	150	180
Chloride:alkalinity ratio	1:1	1:1	1:1.75	1:3	1:3	1:1.5	1:1.8

its carbonate hardness may increase its liability to cause dezincification and may also fail to control its corrosiveness towards other metals.

Desalination

8.43   Introduction

Desalination is a term used to describe processes used for the reduction of dissolved solids in water, usually referred to as total dissolved solids (TDS) and measured in mg/l. Sometimes conductivity is used as the measurement of total dissolved solids as it is easy to measure in the field with a conductivity meter and is expressed as microSiemens per cm (μS/cm). The SI units are milliSiemens per m. For most purposes the TDS value of a water can be obtained by multiplying the conductivity value by 0.66, except in certain special cases, e.g. mine waters. Natural waters may be classified in broad terms according to their TDS values as follows.

Type of water	TDS value (mg/l)
Sweet waters	0–1000
Brackish waters	1000–5000
Moderately saline waters	5000–10 000
Severely saline waters	10 000–30 000
Seawater	Above 30 000

Several methods have been commercially developed for the desalination of high TDS waters and the selection of the correct process, especially in the 500–5000 mg/l range, requires careful evaluation of process efficiency, plant capital and running costs. As a general guide, the most frequent application of the various desalination processes has been in the following categories of TDS waters.

Process	TDS value (mg/l)
Ion exchange	up to 500
Electrodialysis	500–3000
Low-pressure reverse osmosis including nanofiltration	1000–10 000
High-pressure reverse osmosis	10 000 and above
Distillation	Above 30 000

In terms of installed capacity, distillation accounts for about 75% of all desalination installations, with the largest concentration of plant in the Middle East and North Africa. In recent years, however, reverse osmosis systems have been developed for desalting seawater and the installed capacity of such plants is steadily increasing.

The use of freezing is theoretically capable of producing water free from saline constituents, but it has not been adopted on a commercial scale and so it will not be dealt with here. A review of desalination processes has been given by Wade *et al.*[92]

Ion exchange denitrification plant, capacity 4.5 Ml/day, with high lift pumps to the right, and process pumps in the background, 1990. (Paterson Candy Ltd., Isleworth UK)

Orthophosphoric acid dosing plant for plumbo solvency control, Alton treatment works, Anglian Water 1987. (Engineers: Binnie & Partners. Contractors: Degremont UK, Dunstable, UK)

Plate 21

A 4-step cascade aerator (each fall 0.5 m) for 2.75 m³/s flow rate at Izmit treatment works, Turkey. (Contractor: Paterson Candy Ltd. UK. Photo courtesy Thames Water, UK)

Inlet aerator disacharging 200 Ml/day to Grafham Reservoir of Anglian Water, UK

Packed tower aerator of capacity 8 Ml/day. (Purac Ltd., Kidderminster, UK)

Plate 22

A combined 150 Ml/day five-stage seawater flash distillation plant with 608 MW output, Ras Abu Fontas, Qatar, 1999. (Engineers: Mott MacDonald, Brighton UK. Contractors: Weir Westgarth and ABB, UK)

An 11 Ml/day reverse osmosis desalination plant at Kill Devil Hills, North Carolina USA, 1989. (Engineers: Black & Veatch USA)

Plate 23

A bank of wall-mounted chlorinators. (Manufacturers: Portacel, Winchester, UK)

A bank of triple validation monitors, comprising residual chlorine analysers, buffer solution containers and indicators. (Manufacturers: Portacel, Winchester, UK)

Plate 24

8.44   Ion exchange

The use of ion-exchange processes for reducing mineral or saline constituents in water is discussed in Section 8.10 where it is referred to as demineralisation instead of desalination. Whilst the two terms are synonymous, the word demineralisation is used most often in industry to describe the particular application of ion-exchange resins to remove some or all of the mineral and saline salts in water. The use of ion exchange rarely finds application for waters containing more than 1500 mg/l of dissolved solids.

8.45   Electrodialysis

In this process a potential gradient is applied between electrodes and the ionic constituents in the water are thus caused to migrate through semi-permeable membranes which are selective to cations and anions, and demineralisation occurs in a series of cells separated by alternate pairs of cation and anion membranes (see Fig. 8.6). A number of membrane pairs can be installed parallel between a pair of electrodes to form a stack and several stacks can be connected hydraulically together either in parallel or in series to increase output. The reduction in dissolved solids obtained by electrodialysis is related to the electrical energy input and therefore in practice to the cost of electricity. Electrodialysis membranes, like reverse osmosis membranes, are prone to fouling and feed waters require pre-treatment.

In the electrodialysis process the current flow is uni-directional and the demineralised and concentrate compartments remain unchanged. Fouling and scaling material which build up on the membranes are cleaned by taking units out of service. In electrodialysis reversal (EDR), the electrical polarity is reversed periodically which results in a reversal of

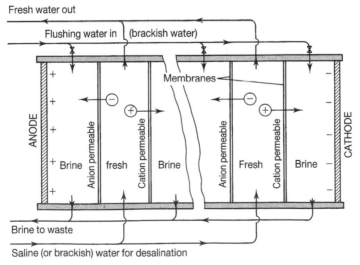

A d.c. voltage is applied across a battery of cells. The anions in the saline (or brackish) water, induced by the e.m.f. applied, pass through the anion-permeable membrane into the brine cells, but cannot pass the cation-permeable membrane. Similarly the cations are also collected in the concentrate or brine compartments which discharge to waste. The saline (or brackish) water in alternate chambers loses ions and reduces in salinity. A proportion of the saline feed is directed to the concentrate (brine) compartments to flush out the brine.

Fig. 8.6   Principle of electrodialysis plant.

ion movement and provides 'electrical flushing' of scale forming ions and fouling matter. This self cleaning can help to reduce pre-treatment needs and allow operation of units at higher levels of supersaturation of sparingly soluble salts thereby achieving higher water recoveries.

8.46   Reverse osmosis and nanofiltration

In osmosis when a salt solution is separated from pure water by a semipermeable membrane, the pure water flows across the membrane until the pressure on the pure water side is equal to the osmotic pressure of the salt solution. In reverse osmosis (RO) by applying a pressure greater than the osmotic pressure to the salt side, pure water can be forced from the salt side across the semipermeable membrane. The osmotic pressure is proportional to the TDS of the water and a pressure of at least twice the osmotic pressure is required to achieve an economically feasible flow. In RO, separation is by diffusion through a dense membrane. Nanofiltration (NF) follows the same principle but utilises a micropore membrane. In both the processes a further mechanism of separation is by size exclusion where larger sized solutes are physically prevented from passing through the membrane. NF is a variation of the RO membrane process.

RO rejects particles of size as small as 0.0001 μm whereas NF rejects particles of size greater than 0.001 μm (1 nanometre) (see Fig. 7.18). As a consequence in NF rejection of monovalent ions (Na^+, K^+, Cl^-) is poor (40–70%) whereas rejection of multivalent ions (Ca^{2+}, Mg^{2+}, SO_4^{2-}) is comparatively high (85–95%)[92] allowing the process to be used for softening of hard waters. RO rejects all dissolved salts at an efficiency of 90–98%. A comparison of nanofiltration and RO processes performance is illustrated by an example in Table 8.8. The other applications of the process include the reduction of organic compounds such as natural colour (see Section 8.37) disinfection by-products and pesticides, turbidity, bacteria and virus. RO would provide the same treatment as NF but more effectively and additionally give greater TDS removal. However because of the greater tendency for fouling the use of RO is limited to desalination. Even with NF, pre-treatment may be necessary to remove turbidity to prevent plugging of the membrane. In NF the high passage of the monovalent ions gives rise to an osmotic back pressure (hence lower $\Delta\pi$ – see below) with a resulting operating pressure (typically 5–15 bar) much lower than that for RO membranes (which may be up to 30 bar for brackish water and 70 bar for sea water).

The following equations give an approximate representation of the flow of water and dissolved solids (solutes) through membranes:

$$Q_w = W_p(\Delta p - \Delta\pi)$$

and

$$Q_s = K_p(\Delta C)$$

where Q_w is the water flux through the membrane (m^3/s.m^2), Q_s is the salt flux through the membrane (kg/s.m^2) which is independent of pressure, W_p is the membrane permeability coefficient for water (m^3/N.s), K_p is the membrane permeability coefficient for salt (m/s), Δp is the pressure differential across the membrane (N/m^2), $\Delta\pi$ is the osmotic pressure differential across the membrane (N/m^2), and ΔC is the concentration differential across the membrane (kg/m^3). Cellulose acetate membranes have water flux rates in the range

Table 8.8 A Comparison of NF and RO processes

Permeate flow:	700 m^3/day
Type:	Single pass design
Water temperature:	10°C
Recovery:	80%
Fouling allowance:	15%
Inlet pressure:	9.6 bar (NF); 10.5 bar (RO)
Concentrate pressure:	3.9 bar (NF); 4.8 bar (RO)
Type of membrane:	Thin film composite (TFC) softening membranes for NF and ultra low pressure membranes for RO

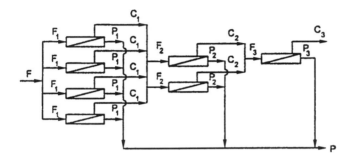

F=Feed C=Concentrate P=Permeate

System arrangement:

Bank	Tubes/Bank	Elements/Tube	Elements/Bank
1	4	6	24
2	2	6	12
3	1	6	6

System flows:

Bank	Total feed (m^3/h)	Tube feed (m^3/h)	Total concentrate (m^3/h)	Tube concentrate (m^3/h)	Average flux (l/m^2.h)	Inlet pressure (bar) NF	Inlet pressure (bar) RO	Final element (Beta)[a]
1	36.5	9.1	17.2	4.3	26.2	8.4	9.1	1.108
2	17.2	8.6	9.8	4.9	20.2	7.0	7.9	1.072
3	9.8	9.8	7.3	7.3	13.6	5.7	6.6	1.030

(Continued overleaf)

Water quality (concentrations in mg/l):

Parameter	Membrane feed	Nanofiltration (NF)		Reverse osmosis (RO)	
		Concentrate	Permeate	Concentrate	Permeate
Calcium as Ca	178	751.3	34.7	879.5	2.6
Magnesium as Mg	9.4	39.7	1.8	46.4	0.1
Sodium as Na	23.7	75.3	10.8	110.8	1.9
Potassium as K	2.2	6.9	1.0	10.1	0.2
Bicarbonate as HCO_3	459.4	1855.5	110.4	2248.7	12.3
Sulphate as SO_4	84.8	400.7	5.8	423.2	0.2
Chlorides as CI	80.0	336.6	15.9	395.8	1.1
Silica as SiO_2	22.0	93.8	4.1	107.1	0.7
Carbon dioxide as CO_2	96.9	96.9	96.9	96.9	96.7
Total dissolved solids	626.0	2617.1	128.3	3079.0	12.9
Hardness as $CaCO_3$	483.2	2039.6	94.1	2387.7	7.1
pH	7.0	7.6	6.4	7.7	5.4
Langelier index	−0.04	1.9	−2.1	2.1	−5.2
Osmotic pressure (bar)	0.34	1.37	0.08	1.64	0.01

(a) Beta defines the degree of concentration polarisation which is equal to the ratio of the concentration of salts at the membrane surface to that in the main body of the water and should ideally be less than 1.13.
Source: Koch Membrane Systems, Fluid Systems, UK.

5×10^{-6} to 1×10^{-5} $m^3/s.m^2$ and for polyamide membrane rates are about one-tenth of this range.

The equations assist in understanding the basic criteria controlling the reverse osmosis process. It can be seen that as the feed water pressure is increased water flux increases but salt flux remains constant, i.e. both the quality and the quantity of the product, known as the permeate, increase with a higher driving force. It is also evident that at constant applied pressure the water flux decreases as the solute concentration of the feed is increased. This is due to the reduction in the driving force caused by the higher feed osmotic pressure. As more water is extracted (called water recovery) from a given feed, the salt concentration in the bulk solution increases and, at constant applied pressure, the water flux falls due to high local osmotic pressure. Higher water recovery also raises the salt flux due to higher salt concentration in the bulk solution.

The two most commonly used membrane configurations are hollow fine fibre and spiral wound types (see Figs 8.7 and 8.8). The membranes are made from cellulose acetate and their derivatives, polyamides and thin film composites. Hollow fine fibre membranes are made from polyamides whilst the other materials are used in spiral wound membranes. Cellulose acetate membranes can tolerate continuous exposure to residual chlorine up to about 1.0 mg/l, but are sensitive to pH outside the range 4–6.5 and water temperatures > 30°C. They are also susceptible to biodegradation. Polyamide membranes have no tolerance towards chlorine or other oxidants, have a wider pH operating range (pH 4–9) and are not degraded by bacteria. Thin film composites are also sensitive to chlorine.

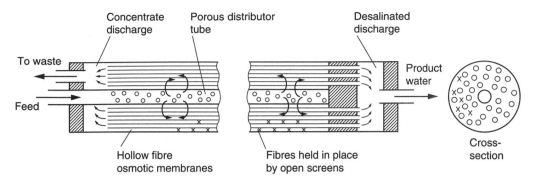

Fig. 8.7   Hollow fibre type of reverse osmosis membrane.

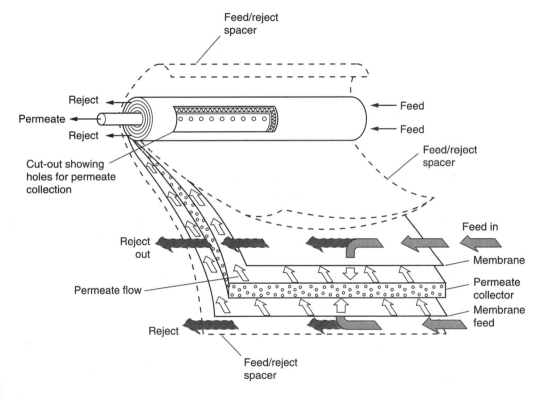

Fig. 8.8   Spiral wound reverse osmosis membrane.

The design bases for an RO plant are feed water quality and the required permeate water quality and output. In brackish water desalination, the high salt rejection allows high blending ratios and therefore smaller RO plants. In seawater desalination the high TDS of the feed precludes blending. The temperature of the feed water has a significant influence on the capacity of the membrane; for example 1°C rise in temperature can increase the flux by about 2.5–3%. However feed water temperatures consistently higher than the rated temperature for the membrane can have a detrimental effect on the membrane. A high temperature combined with high operating pressure may result in membrane compaction which would give a superior salt rejection but reduced output. Another parameter that affects the output is concentration polarisation. This is the effect of increased concentration of salts along the surface of the membrane in a stagnant boundary layer. This results in a reduction of output and an increase in the risk of precipitation of sparingly soluble salts on the membrane surface. This phenomenon can be reduced by maintaining a turbulent flow across the membrane surface at all feed flows.

The salt-rejection properties of a membrane can be varied by a membrane manufacturer so as to produce high salt-rejection and low flux rates or, alternatively, low salt-rejection and high flux rates. In practice, salt rejections of up to 90% can be achieved for brackish waters with most commercial membranes. Seawater desalination is carried out by using high-pressure reverse osmosis membranes operating at pressures between 55–70 bar with salt removal efficiencies greater than 99%. The recovery ratio or water conversion factor (i.e. permeate flow divided by membrane feed flow) can vary from 35 to 45%. Such higher salt rejections can be achieved by using a multi-pass system where the permeate from the preceding pass is the feed to the subsequent pass.

In seawater applications the high pressure and flow of the reject stream allow energy recovery. Typical energy consumption of a seawater reverse osmosis plant can be reduced by about 30% by connecting a recovery turbine to the reject stream.

Modern membranes are continually improving with longer life and robustness, nevertheless their performance can be significantly affected by fouling due to the presence of materials such as turbidity, organic compounds, silica, iron, manganese, and sparingly soluble salts in the feed water. Some degree of pre-treatment upstream of a membrane process is therefore necessary and depends on the contaminants present in the water. Most ground waters do not require pre-treatment except when iron, manganese, silica or turbidity is present. Surface and sea waters often require chlorination, coagulation, clarification, filtration (usually anthracite–sand) and GAC adsorbers. Sea water is sometimes drawn from beach wells, seabed filtration systems or horizontal gallery beach collection systems and therefore have lower turbidity which reduces the pre-treatment requirements. Pre-treatment requirements for NF need not be so extensive. In some instances microfiltration or ultrafiltration are used in the pre-treatment of RO feed. The objective of the pre-treatment is to eliminate the risk of membrane fouling which is measured by several indices of which the Silt Density Index is the most widely used. Typical SDI values for NF and RO are less than 3 and 2, respectively.[93]

RO and NF are concentration processes and therefore water recovery is affected by the presence of sparingly soluble salts such as the carbonates and sulphates of calcium. With increasing water recovery the concentrations of sulphates and carbonates of calcium in the bulk solution also increase. When the products of their molar concentrations exceed the solubility product, they precipitate and plug the membranes. Therefore salt precipitation may be prevented by limiting the recoveries to that allowable, just before the limiting salt precipitates. In sea water desalination calcium sulphate precipitation is prevented by limiting the water recovery to less than 45%. It is not economical to control calcium carbonate scaling by limiting water recovery. This is usually controlled by sulphuric acid treatment. Alternatively antiscalents such as sodium hexametaphosphate or other commercially available antiscalents are used to complex the metal ions (usually calcium) and prevent precipitation. They may be used in conjunction with acid. Usually, for brackish waters with pre-treatment of the feed, it is possible to recover about 75% of feed as permeate, although recoveries of up to 90% can be achieved with multistage systems in which the concentrate from each stage feeds successive stages. Any residual chlorine in the feed to the membranes should also be partially or totally dechlorinated by sodium bisulphite depending on the membrane material, and the feed is finally passed through 5 μm cartridge filters. Sometimes total dechlorination is achieved by passing through GAC contactors. A flow schematic of a typical sea water RO plant is given in Fig. 8.9.

Salt rejection and flux of membranes will deteriorate with time and manufacturers' warranty of the membranes for 3–5 years may be qualified by various operating criteria and cleaning requirements. Cleaning is by chemical agents which are effective for the removal of calcium carbonate scale and organic solutes. Iron, manganese, and turbidity are impossible to clean from the membrane surface. NF and RO membranes cannot be backwashed. Chemical cleaning is usually recommended by the membrane manufacturer as a standard procedure, but full recovery is not possible if the membranes are subjected to continual abuse.

The permeate from membrane processes are devoid of alkalinity, usually acidic (pH 5.5–6.9) and is very aggressive (see Table 8.8). It should be remineralised to contain a minimum alkalinity of 75 mg/l as $CaCO_3$. A positive Langelier Index value should also be achieved. The treatment methods include – hydrated lime and carbon dioxide dosing; carbon dioxide dosing followed by filtration through limestone (note: saturation pH cannot be exceeded; therefore should be followed by lime dosing); sodium bicarbonate and calcium chloride dosing (note: high chloride to alkalinity ratio is unsuitable for duplex

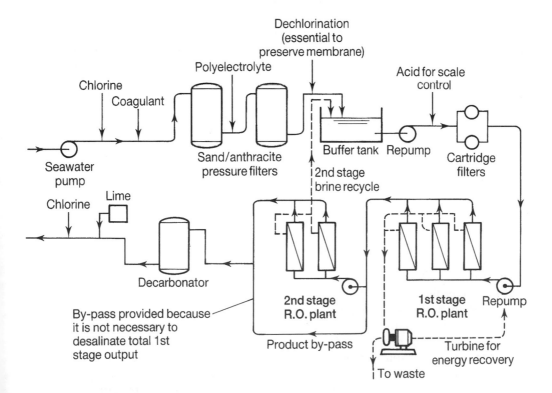

Fig. 8.9 Reverse osmosis desalination plant for seawater.

brass fittings); or blending with a suitable brackish water source.[94] Sources of carbon dioxide include liquid CO_2, on site production by combustion of propane or distillation plant by recovery from the air ejectors. Seawater itself cannot be used for blending with distilled water as it does not contain a sufficiently high ratio of alkalinity to salt to permit this.

Although membrane processes are very effective in the removal of bacteria and virus, the water should be chlorinated before distribution.

The reject stream, known as the concentrate, contains a high concentration of salts and can amount to a large flow. Disposal is a significant problem for both inland and seaboard installations, as mentioned in the next section.

8.47   Thermal processes

Distillation works on the principle that the vapour produced by evaporating seawater is free from salt and the condensation of the vapour yields pure water. The majority of plants use the multistage flash thermal process (MSF), but other processes used are horizontal tube multiple effect plants (HTME), multiple effect distillation using vertical, enhanced surface heat exchange tubing (VTE), and vapour compressing evaporators (VCE). Most large thermal distillation plants are constructed as dual purpose stations for both desalination and the generation of power. The trend towards higher operating temperatures means that greater attention has to be paid to the reduction of corrosion and the use of cost-effective materials and chemicals to combat corrosion.

Effluents from desalination processes have a higher density than seawater from the concentration of total dissolved solids to about double that of seawater. In addition the effluents also contain corrosion products (distillation processes), e.g. copper, zinc, iron, nickel and aluminium and any additives, e.g. corrosion inhibitors (distillation process) and sequestering agents. Distillation process effluents also have elevated temperatures. Disposal of such hot, hypersaline waters may give rise to adverse ecological effects on the environment. The use of a long sea outfall provided with discharge ports designed to achieve designated initial dilution of the effluent, followed by subsequent tidal dispersion makes it possible to meet a predetermined water quality criteria. It is usually necessary to carry out studies at each site to enable prediction of dispersal. For land installations deepwell injection, evaporation ponds, and discharge to sewer are options which have been adopted.

One of the major design parameters for all thermal desalters is the performance ratio which is, in effect, a measure of the efficiency of energy utilisation. The amount of energy required to desalt a given brine concentration varies according to the degree of sophistication of the plant installed, i.e. annual energy costs reduce as capital costs increase. Other factors to be taken into account include size of units, load factor, growth of demand, interest rate on capital and technical matters concerning the auxiliary services, repairs, and maintenance. For detailed design something in the order of 70 design parameters have to be settled. Many of these are concerned with the safe or most economic limits for the temperatures, velocities and concentrations of the coolants, brines, brine vapour, steam, steam condensate and boiler feedwater, with particular reference to the prevention of scaling, corrosion, erosion, the purity of the distillate, the efficiency of heat exchangers and the nature and cost of the auxiliary plant involved.

Distilled water approaches zero hardness with an alkalinity not likely to exceed 2 mg/l. It is aggressive to metal and asbestos cement pipes and will take up calcium from mortar-lined pipes. It is unsuitable for distribution and is unpalatable, being flat and insipid. To remedy these deficiencies the distilled water should be remineralised (see Section 8.46).

Major problems which can occur with seawater evaporators are: scale formation on heat transfer surfaces due to the presence of carbonates and sulphates of calcium; internal corrosion due to hot sodium chloride and the presence of dissolved gases such as oxygen, ammonia, and hydrogen sulphide; plant start-up problems and running at low capacity.

8.48   The costs of desalination

The subject of desalination costs can be discussed only cursorily because of its diversity and complexity. The purpose of this brief review is therefore to indicate the principal items contributing to the production costs of desalinated water and to point out how these are likely to be influenced by local conditions. The major factors to be considered in any desalting application include – type and characteristics of the saline feed; type of desalination process to be used; local costs of primary energy source (e.g. oil or gas or electricity); availability and costs of chemicals needed for feed pre-treatment, product conditioning and plant cleaning. Also important, in the case of distillation plants, is the choice of performance ratio and whether dual purpose operation is proposed (Section 8.47). Except for the smallest plants, staffing costs are not usually a major item. Table 8.9 summarises the principal cost contributory factors of the three main desalination processes currently popular (1999) as applied to single purpose plants (i.e. producers of water only). In the case of brackish waters (as defined in Section 8.43) the primary energy and power demands are usually less than half those for seawater desalination, whether electrodialysis or reverse osmosis is used. Thermal processes, including vapour compression, are not usually adopted for brackish water desalting because of their unfavourable energy demands. The feed water and chemical consumption of brackish water desalination are also only about one-third of those for seawater reverse osmosis.

The costs of desalted water are very variable and site specific. Amortisation of the initial capital investment, together with energy costs, usually accounts for up to 80% of the total water cost. The energy components can be calculated from the data in Table 8.9 if local primary energy unit costs are known. Only the briefest indication of capital costs can be given in the present context for the larger sizes of the four major types of desalination plant listed in Table 8.9. Plant costs, erected and commissioned, including typical costs of civil works, local product storage, and all other site-specific costs (such as cost of intake, discharge of effluents, fuels storage and handling, etc.) in early 1999 were as given in Table 8.10 below.[95] Typical water costs are for large desalination plant operating at base load and assuming an energy cost of £10 per barrel of fuel oil.

In the case of reverse osmosis and mechanical vapour compression plants, it is assumed that power can be taken from the grid supply or some other source. For brackish water the capital costs of desalination plants are much lower due to the smaller number and cost of membranes and other equipment.

Table 8.9 Cost contributory factors for major types of seawater desalination processes

Process	Usual maximum size of unit to date (m³/day)	Total primary fuel energy demand (MJ/m³) (a)	Components of energy demand		Chemical consumption (g/m³)	Feedwater consumption (m³/m³)	Annual cost of spares and replacements as % of initial capital cost of desalter
			Power (kWh/m³)	Heat (MJ/m³)			
Multistage flash evaporation (MSF)	32 000	120–400	3–4	185–300	4–6	5–10	2%
Mechanical vapour compression (MVC)	1500	88–130	8–12	0–60 (b)	5–7	2.0–2.5	2–4%
Seawater reverse osmosis (SWRO)	6000–30 000	45–65	4–6	Nil	10–20	3–4	4–6%
Brackish water reverse osmosis (BWRO)	3000–20 000	22–33	2–3	Nil	6–12	1.2–1.5	4–6%

Notes.
Data refer to stand-alone plants. Heat and power assumed generated on site.
Thermal efficiencies assumed: 80% for steam boilers for MSF plant; 33% for diesel generators for RO and MVC plant.
Higher energy value of diesel fuel taken as 37.85 MJ per litre.
1 MJ = 0.278 kW h.
(a) Lower value for dual purpose plant.
(b) Dependent upon heat recovered from jacket of diesel plant.
Table updated by Mott MacDonald.

Table 8.10 Capital and operating costs for large desalination plants

Type of plant	Plant capital cost ($£$ per m^3/day output)	Water cost ($£$ per m^3 produced)
Multistage flash evaporation plant	1000–1500	0.70–0.80
Mechanical vapour compression plant	1200–1600	0.50–0.60
Reverse osmosis plant (seawater)	400–500	0.30–0.40
Reverse osmosis plant (brackish water)	900–1200	0.50–0.60

Radionuclides

8.49   Removal of radionuclides

An introduction to radionuclides in water was given in Section 6.38. Radon is a water soluble gas primarily found in ground waters. It is effectively removed by aeration. Test work carried out on various aeration methods have shown packed tower aerators (see Section 8.19) to be the most efficient with removal efficiencies of the order of 98%, closely followed by diffused air system (see Section 8.22). The removal in a spray aerator (see Section 8.20) was limited to less than 75%.[96] For a packed tower the packing height is about 3 m with a surface loading rate of about 75 m^3/h.m^2. The packed tower performance was found not to be influenced by air to water flow ratio.[96] Packed towers also remove carbon dioxide thus affecting the carbonate-bicarbonate equilibrium leading to scaling of the packing. GAC adsorbers are also known to remove radon but not as effectively as aeration.

The removal of particulate or soluble radionuclides in water can be achieved by coagulation using iron or aluminium salts followed by solid–liquid separation (uranium, radium and β-particles), lime softening (uranium and radium), anion exchange (uranium), cation exchange (radium), or mixed-bed ion exchange (β-particles).[51,97] Strong base anion exchange resin of the chloride form has a high exchange capacity for uranium particularly in low sulphate waters[98] and strong acid cation exchange resins are used for radium removal.[99] Both the resins are regenerated with sodium chloride.

Table 8.11 summarises the treatment methods that can be effectively applied for the removal of radionuclides.[100]

Table 8.11   Treatment methods for radionuclides removal from drinking water and their efficiencies

Radon		Radium		Uranium	
Method	% Efficiency	Method	% Efficiency	Method	% Efficiency
Aeration:		Ion exchange	81–99	Ion exchange	90–100
– packed tower	up to 99+	Reverse osmosis	90–95+	Lime softening	85–99
– diffused air	up to 99+	Lime softening	80–92	Reverse osmosis	90–99
– spray	70–95+	Electrodialysis	90	Coagulation-filtration	80–98
GAC contactor	62–99+	Ra complexing	90–99	Activated alumina	90
		Greensand	25–50		
		MnO_2 filter	90		

Note. Highest efficiencies of some treatment methods are for small capacity domestic units.

In the treatment plant the radionuclides become concentrated in waste streams and need careful disposal. US EPA have suggested disposal alternatives based on the concentration of radionuclides in the waste stream. These include disposal of liquid wastes to watercourses, sewer, or deep injection wells, and treatment by evaporation or precipitation with disposal of solid wastes (dewatered sludges) to various types of landfill facilities.[101]

References

1. van Dijk, J. C. and Wilms D. A. Water Treatment without Waste Material – Fundamentals and State of the Art of Pellet Softening. *J Water SRT-Aqua*, **40**(5), 1991, pp. 263–280.
2. van der Veen C. and Graveland A. Central Softening by Crystalisation in a Fluidised-Bed Process. *JAWWA*, **80**(6), 1988, pp. 51–58.
3. van Honwelingen G. A. and Nooijen W. F. J. M. Water Softening by Crystallization. *European Water Pollution Control*, **3**(4), 1994, pp. 33–35.
4. van Eekeren M. W. M. *et al.* Improved Milk-of-Lime for Softening of Drinking Water – the Answer to the Carry-over Problem. *J Water SRT-Aqua* **43**(1), 1994, pp. 1–10.
5. Envrad M. *et al.* Effects and Consequences of Electric Treatment in Preventing Scaling of Drinking Water Systems. *J Water SRT-Aqua*, **46**(2), 1997, pp. 71–83.
6. Curry M. D. and Reynolds M. Using By-Products of an Iron Removal Process to Improve Water Treatment. *JAWWA*, **75**(5), 1983, pp. 246.
7. Knocke W. R. *et al.* Kinetics of Manganese and Iron Oxidation by Potasssum Permanganate and Chlorine Dioxide. *JAWWA*, **83**(6), 1991, pp. 80–87.
8. Faust S. D. and Aly O. M. *Chemistry of Water Treatment*, 2nd edn. Ann Arbor Press, Michigan, 1998.
9. Barnhoorn D. *et al.* The Treatment of Ferruginous Groundwaters for River 33 Augmentation in the Waller's Haven, East Sussex. *JIWES*, **38**(3) 1984, pp. 217–230.
10. Knocke W. R. Removal of Soluble Manganese by Oxide-Coated Filter Media: Sorption Rate and Removal Mechanism Issues. *JAWWA*, **83**(8), 1991, pp. 64–69.
11. Benfield L. D. *et al.* *Process Chemistry for Water and Wastewater Treatment*. Prentice-Hall, New Jersey, 1982, p. 476
12. Merkle R. B. *et al.* Dynamic Model for Soluble Mn^{2+} Removal by Oxide-Coated Filter Media. *J Environmental Engineering, ASCE*, **123**(7), July 1997, pp. 650–658
13. Knocke W. R. *et al.* Alternative Oxidants for the Removal of Soluble Iron and Manganese. AWWA Research Foundation, Denver, Colorado, 1990.
14. Rittmann B. E. *et al.* Biological Treatment of Public Water Supplies. *CRC Critical Reviews*, **19**(2), 1989.
15. Mouchet P. From Conventional to Biological Removal of Iron and Manganese in France. *JAWWA*, **84**(4), 1992, pp. 158–167.
16. Bourgine F. P. *et al.* Biological Processes at Saints Hill Water Treatment Plant, Kent. *JIWEM*, **8**(4), 1994, pp. 379–392.
17. Waterton T. Manganese Deposits in Distribution Systems. *Proc SWTE*, **3**, 1954, p. 117.
18. Committee Report on: A Review of Solid Solution Interactions and Implications for Controlling Trace Inorganic Materials. *JAWWA*, **80**(10), 1988, pp. 56–64.

19. Edwards M. Chemistry of Arsenic: Removal During Coagulation and Fe-Mn Oxidation. *JAWWA*, **86**(9), 1994, pp. 65–78.
20. Degremont. *Water Treatment Handbook*, 6th edn. Lavoisier, Paris, 1991, (a) p. 1331; (b) p. 1219.
21. Mouchet P. Potable Water Treatment in Tropical Countries: Recent Experience and Some Technological Trends. *Aqua* **3,** 1984, pp. 143–164.
22. Azizian F. Private Communication, 1999.
23. Simms J. and Azizian F. Pilot-Plant Trials on the Removal of Arsenic from Potable Water Using Activated Alumina. *AWWA Water Technology Conference*, Denver, Colorado, November 1998.
24. Pontius, F. W. (ed.) *Water Quality and Treatment.* AWWA, McGraw-Hill, New York, 1990, p. 617.
25. Waypa J. J. *et al.* Arsenic Removal by RO and NF membranes. *JAWWA*, **89**(10), 1997, pp. 102–114.
26. Driehans W. *et al.* Granular Ferric Hydroxide – A New Adsorbent for the Removal of Arsenic from Natural Water. *J Water SRT-Aqua*, **47**(1), 1998, pp. 30–35.
27. Galvin R. M. Occurrence of Metals in Waters: An Overview. *Water SA*, **22**(1), 1996, pp. 7–18.
28. Leroy P. Lead in Drinking Water – Origins; Solubility; Treatment. *J Water SRT-Aqua*, **42**(4), 1993, pp. 223–238.
29. Schock M. and Oliphant R. J. *The Corrosion and Solubility of Lead in Drinking Water, Internal Corrosion of Water Distribution Systems*, 2nd edn. AWWA Research Foundation, Denver, Colorado, 1996, Chapter 4.
30. Sheiham I. and Jackson P. J. The Scientific Basis for Control of Lead in Drinking Water by Water Treatment. *JIWES*, **35**(6), 1981, pp. 491–515.
31. Boireau J. *et al.* Lead in Water Treatment and Advances in Lead Pipe Replacement and Relining Techniques. *Water Supply*, **16**(1/2), Madrid, 1998, pp. 599–609.
32. Schock M. R. and Clement J. A. Lead and Copper Control with Non-Zinc Orthophosphate. *JNEWWA*, (3), 1998, pp. 20–41.
33. Hayes C. R. *et al.* Meeting Standards for Lead in Drinking Water. *JCIWEM*, **11**(4), 1997, pp. 257–264.
34. Colling J. H. *et al.* Plumbosolvency Effects and Control in Hard Waters. *JIWEM*, **6**(3), 1992, pp. 259–268.
35. Nawlakhe W. G. Defluoridation of Water by Nalgonda Technique. *Indian J Environ Health*, **17,** 26, 1975.
36. Shoeman J. J. *et al.* An Evaluation of the Activated Alumina Process for Fluoride Removal from Drinking Water and Some Factors Affecting Its Performance. *Water SA*, **11**(1), 1985, pp. 25–32.
37. Maier F. J. Defluoridation of Water Supplies. *JAWWA*, **45**(8), 1953, pp. 879–888.
38. Adikari A. I. T. *et al.* A Low Cost Defluoridation Method for Rural Water Supplies. *J Geological Soc Sri Lanka*, 1, 1988.
39. Commins B. T. Controlling Fluoride Levels. WHO Report PEP/85.12.
40. Code of Practice on Technical Aspects of Fluoridation of Water Supplies. HMSO, London 1987.
41. Notes on Water Pollution, Department of the Environment (UK), Note No. 61; June 1973.

42. Van der Kroon G. M. and Schram A. H. Weir Aeration – Part 1: Single Free Fall, H$_2$O. **22**(2), 1969, pp. 528–537.
43. Irving, S. J. In-stream Aeration of Polluted Waters. *Water and Waste Treatment*, **17**, 1974.
44. David J. M. Destratification of Reservoirs – A Design Approach for Perforated-Pipe Compressed Air Systems. *Water Services*, August 1980.
45. Croll B. T. *Nitrate Removal Using Ion-Exchange: Brass Corrosion Considerations*. IWEM Yearbook, London, 1991.
46. Clifford D. and Liu X. Biological Denitrification of Spent Regenerant Brine Using a Sequencing Batch Reactor. *Water Research*, **29**(9), 1993, pp. 1477–1484.
47. Rice R. P. and Netzer A. (eds). *Handbook Ozone Technology and Applications*, Vol. II. Butterworth, Oxford, 1984, p. 99.
48. Short C. S. Removal of Ammonia from River Water. TP101, The Water Research Centre (UK), July 1973.
49. Richard Y. *et al*. Study of the Nitrification of Surface Water. *Prog Wat Tech*, **10**(5/6), 1978, pp. 17–32.
50. Nauleau F. *et al*. Ammonia Removal by Biological Nitrification on Conventional Sand Filters as a Specific Stage of Treatment. IWSA Conference on Inorganic Nitrogen Compounds and Water Supply, Hamburg, 27–29 Nov. 1991.
51. Bablon G. *et al*. Developing a Sand-GAC Filter to Achieve High-Rate Biological Filtration. *JAWWA* **80**(12), 1988.
52. Mouchet P. Private Communication, 1995.
53. Lacamp B. *et al*. Advanced Nitrogen Removal Processes for Drinking and Wastewater Treatment. *Water Science and Technology*, **22**(3), 1990.
54. Janda V. and Rudovsky J. Removal of Ammonia in Drinking Water by Biological Nitrification. *J Water SRT-Aqua*, **43**(3), 1994, pp. 120–125.
55. Booker N. A. *et al*. Removal of Volatile Organics in Groundwater. Water Research Centre, UK June 1988.
56. Hess A. F. *et al*. *Control Strategy – Aeration Treatment Technique – Occurrence and Removal of VOCs from Drinking Water*. AWWA Research Foundation, Denver, Colorado, 1983.
57. Kavanaugh M. C. and Russell R. R. Design of Air Stripping Towers to Remove Volatile Contaminants from Drinking Water. *Aqua*, **6**, 1980, pp. 118–125.
58. Foster D. M. *et al*. New Treatment Process for Pesticides and Chlorinated Organics Control in Drinking Water. *JIWEM*, **5**(4), 1991, pp. 466–477.
59. Dyksen J. *et al*. Operating Experience at VOC Treatment Facilities, Part 1: GAC Opflow; *AWWA*, **25**(1), 1999, pp. 11–12.
60. Glaze W. H. Drinking Water Treatment with Ozone. *Environmental Science and Technology*, **21**(3), 1987, pp. 224–230.
61. White G. C. *Handbook of Chlorination*, 3rd edn. Van Nostrand Reinhold, New York, 1992.
62. Terashima K. Reduction of Musty Odour Substances in Drinking Water – A Pilot Plant Study. *Water Science and Technology*, **20**(8/9), 1988, pp. 275–279.
63. Ferguson D. W. *et al*. Comparing 'Peroxone' and Ozone for Controlling Taste and Odour Compounds. Disinfection By-products, and Micro-organisms. *JAWWA* **80**(4), 1990, pp. 181–191.

64. Duquet J. P. *et al.* New Advances in Oxidation Processes: The Use of Ozone/ Hydrogen Peroxide Combination for Micropollutant Removal in Drinking Water. *Water Supply* 7(4), 1989, pp. 115–124.
65. Gauntlet R. B. and Packham R. F. The Use of Powdered Activated Carbon in Water Treatment. *Proc Conf on Activated Carbon in Water Treatment*, Reading, Water Research Association, Medmenham, England, 1973.
66. Huang W. J. Powdered Activated Carbon for Organic Removal from Polluted Raw Water in Southern Taiwan. *J Water SRT-Aqua*, 44(6), 1995, pp. 275–283.
67. Najm I. M. *et al.* Using Powdered Activated Carbon: A Critical Review. *JAWWA*, 81(1), 1991, pp. 65–76.
68. Monscwitz J. T. and Ainsworth L. D. Treatment of Hydrogen Sulphide. *JAWWA*, 66(9), 1974, pp. 537–539.
69. Haist-Gulde B. and Baldauf G. Removal of Pesticides by Powdered Activated Carbon – Practical Aspects. *Water Supply*, 14(2), Amsterdam, 1996, pp. 201–208.
70. Dussert B. W. *et al.* Impact of Preozonation on Granular Activated Carbon Properties. *Water Supply*, 13(3/4), Osaka, 1995, pp. 7–11.
71. Pilard G. *et al.* Impact of Mineral Elements on the Regeneration Process of Activated Carbon Used in Potable Water Treatment. *Water Supply*, 13(3/4), Osaka, 1995, pp. 1–6.
72. Paillard H. *et al.* Technologies Available to Upgrade Potable Waterworks for Triazines Removal. IWEM Scientific Section, Pesticide Symposium, London, 11–12 April 1990.
73. Smith D. J. *et al.* Activated Carbon in Water Treatment. *Water Supply*, 14(2), Amsterdam, 1996, pp. 85–98.
74. Langlais B. *et al.* (ed.). Ozone in Water Treatment – Application and Engineering. AWWA Research Foundation, Lewis Publishers, Colorado, 1991, p. 290.
75. Graveland A. Application of Biological Activated Carbon Filtration at Amsterdam Water Supply. *Water Supply*, 14(2), Amsterdam, 1996, pp. 233–241.
76. Armenter J. L. L. and Canto J. Filtration by GAC and On-site Regeneration in the Treatment of Water for the City of Barcelona. Review of 15 Years in Operation. *Water Supply*, 14(2), Amsterdam, 1996, pp. 119–127.
77. Bauer M. *et al.* The GAC/Slow Sand Filter Sandwich – From Concept to Commissioning. *Water Supply*, 14(2), Amsterdam, 1996, pp. 159–175.
78. Rice R. P. and Netzer A. (eds). *Handbook of Ozone Technology and Applications*, Vol. II. Butterworth Publishers (1984), p. 151.
79. Dugnet J. P. *et al.* Improvement in the Effectiveness of Ozonation of Drinking Water through the use of Hydrogen Peroxide. *Ozone: Science and Engineering*, 7, 1985.
80. Mouchet P. and Capon B. Recent Evolution in Drinking Water Treatment Technology. Extract Revue Travaux, Degremont, July/August 1991.
81. Paillard H. *et al.* Removal of Nitrogonous Pesticides by Direct and Radical Type Ozonation. *EC Annual Conference on Micropollution*, May 1990, Lisbon.
82. Paillard H. *et al.* Application of Oxidation by a Combined Ozone/Ultraviolet Radiation System to the Treatment of Natural Water. *Ozone: Science and Engineering*, 4, 1987.
83. Bourgine F. *et al.* Treatment of Pesticides by UV photolysis. *Techniques, Sciences, Methodes*, 92(7/8), 1992, pp. 23–29.

84. Hall S. *et al.* Coagulation of Organic Colour with Hydrolysing Coagulants. *JAWWA*, **57**(9), 1965, p. 1149.
85. Rook J. J. Haloforms in Drinking Water. *JAWWA*, **68**(3), 1976, p. 168.
86. Merry A. *et al.* Membrane Treatment of Coloured Water. *Proc of International Symposium of Wastewater Treatment* and *7th Workshop on Drinking Water*, Montreal, Canada, November 1995.
86. Jarvis M. G. and Hedges M. R. Use of soil maps to predict the incidence of corrosion and the need for mains renewal. *JIWEM*, Feb. 1994, pp. 68–75.
87. Tiller A. K. Corrosion induced by bacteria. *The Public Health Engineer,* **12**(3), 1984.
88. Cox C. R. *Operation and Control of Water Treatment Processes.* WHO Geneva, 1964 p. 203.
89. Miles G. D. *J Soc Chem Indust*, **67**, 1948, p. 10.
90. Turner M. E. D. The influence of water composition on dezincification of Duplex brass fittings. *JSWTE*, **10**, 1961, p. 162 and **14**, 1965, p. 81.
92. Wade N. and Calister K. Desalination: The State of the Art. *JCIWEM*, **11**(4), 1997, pp. 87–97.
93. Mallevialle J. (Ed.), Water Treatment – Membrane Processes. AWWA Research Foundation, McGraw Hill, New York, 1996, pp. 94–96.
94. Applegate I. Y. Posttreatment of reverse osmosis product waters. *JAWWA*, **78**(5) 1986, pp. 59–65.
95. Private communications from N. M. Wade, Mott Ewbank Preece.
96. Dixon K. L. *et al.* Evaluating Aeration Technology for Radon Removal. JAWWA, **83**(4), 1991, pp. 141–148.
97. US Environmental Protection Agency, National Primary Drinking Water Regulations; Radionuclides; Advanced Notice of Proposed Rulemaking, Fed. Reg. 56: 138(40) CFR Parts 141 and 142, July 1991.
98. Zhang Z. and Clifford D. A. Exhausting and Regenerating Resins for Uranium Removal. *JAWWA*, **86**(4), 1994, pp. 228–241.
99. Snoeyink V. L. *et al.* Strong Acid Ion Exchange for Removing Barium, Radium and Hardness. *JAWWA*, **79**(8), 1987, pp. 66–72.
100. Parrotta M. J. Radioactivity in Water Treatment Wastes: A USEPA Perspective. *JAWWA*, **83**(4), 1991, pp. 134–140.
101. Committee Report: Research Needs for Inorganic Contaminants, *JAWWA*, **85**(5), 1993, pp. 106–112.

9

Disinfection of water

9.1 Disinfectants available

The term 'disinfection' is used to mean the destruction of infective organisms in water to such low levels that no infection of disease results when the water is used for domestic purposes including drinking. The term 'sterilisation' is not strictly applicable because it implies the destruction of all organisms within a water and this may be neither achievable nor necessary. Nevertheless the word is often loosely used, as in 'domestic water sterilisers'.

On a plant scale the following disinfectants are in common use:

(1) chlorine (Cl_2);
(2) chloramines (NH_2Cl, $NHCl_2$);
(3) chlorine dioxide (ClO_2);
(4) ozone (O_3).

For small plants or special circumstances the following may be used:

(5) ultra-violet radiation;
(6) products releasing chlorine, e.g. calcium hypochlorite (chloride of lime) ($Ca(OCl)_2$), sodium hypochlorite (NaOCl), chlorine tablets.

The organisms in water which it may be necessary to kill by disinfection include bacteria, bacterial spores, viruses, protozoa and protozoa cysts, worms and larvae. The efficacy of disinfection depends on numerous factors – the type of disinfectant used, the amount applied and the time for which it is applied; the type and numbers of organisms present; and the physical and chemical characteristics of the water.

Chlorine and chloramine processes of disinfection

9.2 Action of chlorine

The precise action by which chlorine kills bacteria in water is uncertain but it is believed that the chlorine compounds formed when chlorine is added to water rupture bacterial membranes and inhibit vital enzymic activities resulting in bacterial death. Chlorine is also a strong oxidising agent which will break up organic matter in a water; but, in so doing, because it is a highly reactive chemical it can form a wide range of chlorinated compounds with the organic matter present. Among these are the trihalomethanes (THM) for which limits have been set for health reasons (see Section 9.7). Chlorine can also restrain algal growth, react with ammonia and convert iron and manganese in the water to their oxidised forms which may precipitate. Hence there are a number of factors to be taken into consideration when using chlorine as a disinfectant.

9.3   Chlorine compounds produced

When chlorine Cl_2 is added to water which is free from organic matter or ammonia, hypochlorous acid HOCl is formed.

$$Cl_2 + H_2O = HOCl + HCl$$

The very weak HOCl is further dissociated into H^+ and OCl^-, the extent of the dissociation being dependent upon the pH and temperature of the water. The proportion of active HOCl at various pH values and temperatures are given in Table 9.1. The hypochlorous acid HOCl and hypochlorite ion OCl^- are together known as 'free chlorine' and are the most effective forms of chlorine for achieving disinfection. Free chlorine acts more rapidly in an acid or neutral water. Therefore when final pH correction is practised, the alkali should be added after the disinfection process has been completed. Of the free chlorine, hypochlorous acid HOCl is a more powerful bactericide than the hypochlorite ion OCl^-.

If ammonia is present in the water, or if ammonia is added to the water, chloramines will be formed in a stepwise manner: monochloramine NH_2Cl, dichloramine $NHCl_2$, and trichloramine NCl_3 (nitrogen chloride). Of these compounds, the monochloramine and dichloramine together in total are known as the 'combined chlorine'; total chlorine is the sum of combined chlorine and free chlorine.

Table 9.1   Variation of HOCl as percentage of free chlorine with pH and temperature values of water

pH	Percent HOCl						
	0°C	5°C	10°C	15°C	20°C	25°C	30°C
6.0	98.5	98.3	98.0	97.7	97.4	97.2	96.9
6.25	97.4	97.0	96.5	96.0	95.5	95.1	94.6
6.5	95.5	94.7	94.0	93.2	92.4	91.6	91.0
6.75	92.3	91.0	89.7	88.4	87.1	86.0	84.8
7.0	87.0	85.1	83.1	81.2	79.3	77.5	75.9
7.25	79.1	76.2	73.4	70.8	68.2	66.0	63.9
7.5	68.0	64.3	60.9	57.7	54.8	52.2	49.9
7.75	54.6	50.5	46.8	43.5	40.6	38.2	36.0
8.0	40.2	36.3	33.0	30.1	27.7	25.6	23.9
8.25	27.4	24.3	21.7	19.5	17.6	16.2	15.0
8.5	17.5	15.3	13.5	12.0	10.8	9.8	9.1
8.75	10.7	9.2	8.0	7.1	6.3	5.8	5.3
9.0	6.3	5.4	4.7	4.1	3.7	3.3	3.0

Note. Percent OCl^- = 100 − percent HOCl.

The free chlorine is many times more powerful as a bactericide than combined chlorine (i.e. mono and dichloramines). Of the chloramines, the dichloramine is more powerful than monochloramine requiring only about 15% of the monochloroamine dose for inactivation of *E. coli*. Whitlock,[1] quoting from the work of Butterfield *et al.*[2] estimated that 25 times as much combined chlorine is needed to achieve the same degree of kill of bacteria as free chlorine in the same time. Since ammonia is often naturally present in a water, it is usual to add sufficient chlorine to react with all the ammonia present and produce an excess of free chlorine sufficient to achieve speedy disinfection. As a consequence the efficacy of chlorine as a disinfectant is influenced by a number of conditions.

9.4 Factors relating to the disinfection efficiency of chlorine

The following factors have to be taken into account when treating water with chlorine.

The stage at which chlorine is applied. Chlorine is often applied at more than one stage in the treatment of a water. 'Pre-chlorination' comprises the application of chlorine to a water (often a raw water) before it is processed through treatment works, e.g. before clarification and filtration. 'Intermediate-stage chlorination' is chlorine added between stages of treatment. 'Final chlorination' means the final disinfection of a water before it is put into supply. The purposes of pre-chlorination and intermediate-stage chlorination are often partly biological such as to reduce bacterial content, prevent bacterial multiplication, and restrain algal growth; and partly chemical, such as to assist in the precipitation of iron and manganese and achieve other oxidation benefits. Final chlorination is always for the purpose of disinfecting the water and to maintain a residual in the distribution system so that it is safe for drinking.

Effect of turbidity. The effect of turbidity in a water is to make it difficult for the penetration of chlorine and therefore the destruction of bacteria in particles of suspended matter. It is always necessary, therefore, that final disinfection by chlorine is applied as a final stage in water which contains low turbidity. For effective disinfection the World Health Organization[3] (WHO) suggests a guide level value for turbidity of less than 1 NTU.

Consumption of chlorine by metallic compounds. A substantial amount of chlorine may be used to convert iron and manganese in solution in the water into products which are insoluble in water (see Section 8.12). Reduction of these parameters by upstream processes is therefore desirable. Typically iron and manganese should be less than 0.1 mg/l as Fe and 0.05 mg/l Mn, respectively. If at the point of chlorine application their levels are too low to justify removal, the dose must take their demand into account.

Reaction of chlorine with ammonia compounds and organic matter. The ammonia compounds may exist in organic matter or, alternatively, ammonia may exist separately from organic matter (see Section 6.6); in either case they will form combined chlorine which, as mentioned in the previous section, is not so effective a bactericide as free chlorine. Chlorine may be used in the oxidation of some organic matter. Ammonia should not exceed 0.015 mg/l as N. When this value is exceeded or when organic matter is present, an allowance should be made both in the chlorine dose and contact time to satisfy the chlorine demand prior to disinfection.

Therefore the substances that are causing a chlorine demand must be removed prior to disinfection by upstream treatment or an allowance for them must be made in the chlorine dose, otherwise disinfection could be compromised.

Low temperature causes delay in disinfection. A very substantial decrease in killing power takes place with lowering of temperature. The difference in kill rate of bacteria between the temperatures of 20 and 2°C is noticeable both with free and combined chlorine. This must be borne in mind when fixing the contact period (see below).

Increasing pH reduces effectiveness of chlorine. In free chlorine and in combined chlorine the more effective disinfectants in each case, i.e. hypochlorous acid and dichloramine respectively, are formed in greater quantities at low pH values than at high values. Thus disinfection is more effective at low pH values; the guide level value suggested by WHO[3] is less than 8.

The number of coliforms presented for disinfection. This has an influence on the disinfection efficiency. To be confident of achieving 100% compliance with the requirement of zero coliforms leaving the disinfection stage, the water subjected to disinfection ideally should not contain more than 100 coliforms/100 ml. Most ground waters satisfy this criteria. In surface waters, coagulation followed by solid–liquid separation processes, achieves up to 99.9% bacteria removal (see Section 7.13). Consquently pre-disinfection in addition to conventional treatment is only required for heavily polluted surface waters.

Time of contact is important. The disinfecting effect of chlorine is not instantaneous. Time must be allowed for the chlorine to kill organisms. This important factor is dealt with in the next section.

9.5   Chlorine residual concentration and contact time

The principal factors influencing the disinfection efficiency of chlorine are free residual concentration, contact time, pH and water temperature. The term 'free residual' means the amount of free chlorine remaining after the disinfection process has taken place. Given adequate chlorine concentration and contact time, all bacterial organisms and most viruses can be inactivated. Thus a useful design criterion for the disinfection process is the product of contact time (t in minutes) and the chlorine-free residual concentration (C in mg/l) at the end of that contact time. This is known as the 'Ct value'[4] or 'exposure value'.[5] On this basis the WHO guide level of 0.5 mg/l free residual concentration after 30 minutes contact would have a Ct value of 15 mg min/l (see Section 9.6). This is shown to provide a high factor of safety (12.5-fold) so that a degree of inefficiency in the contact tank performance can be tolerated.[6]

The WHO Ct criterion is for faecally polluted water and therefore may be varied according to the quality of the source water. For example a groundwater free of *E. coli* but containing no more than 10 coliforms/100 ml could have a Ct value of 10 mg.min/l with t not less than 15 minutes and for groundwaters where coliforms and *E. coli* are absent, marginal chlorination sufficient to maintain a residual in the distribution system with no contact at the treatment works could be acceptable. On the other hand, for a heavily polluted surface water a higher Ct value would be used, e.g. 30 mg min/l, with C and t being not less than 1.0 mg/l and 30 minutes, respectively.

In the USA the Surface Water Treatment Rule (SWTR)[7] for disinfection specifies the water treatment required to achieve removal by solid–liquid separation processes, and inactivation by disinfection processes, of 3-log (99.9%) for *Giardia lamblia* cysts and 4-log (99.99%) for enteric viruses, these two being chosen for their resistance to disinfection when compared to *Legionella*, heterotrophic bacteria and coliforms. According to the

SWTR, direct filtration treatment is given 2-log credit for *Giardia* removal and 1-log for virus removal, whilst conventional treatment of clarification and filtration is given 2.5-log credit for *Giardia* and 2-log for viruses. This leaves 1 to 0.5-log inactivation of *Giardia* and 2- to 3-log inactivation of viruses to be achieved by disinfection.

Tables 9.2 and 9.3 give the *Ct* values stated in the SWTR to achieve 1-log inactivation of *Giardia* and 2 and 3-log inactivation of viruses.

The WHO *Ct* criterion is applicable to the inactivation of bacteria and most viruses and therefore cannot be compared with the *Ct* values for inactivation of cysts given in Table 9.2. Comparison with *Ct* values in Table 9.3 confirms that the WHO criterion has a high factor of safety which is desirable to ensure complete inactivation.

In the expression *Ct*, the contact time, *t*, is the time the water remains in the contact tank. Due to eddies and short-circuiting, the total volume of the tank may not be available for contact, some water may pass through the tank in less time than the theoretical residence time (t_T) which is the volume of water in the tank divided by the rate of flow. It is therefore necessary to design the contact tank so that at least 90% of water passing through the tank remains in the tank for more than the required contact time at the design

Table 9.2 *Ct* values for achieving 1-log inactivation of *Giardia lamblia*

Disinfectant	pH	Ct value at water temperature					
		0.5°C	5°C	10°C	15°C	20°C	25°C
Free residual chlorine of 2 mg/l	6	49	39	29	19	15	10
	7	70	55	41	28	21	14
	8	101	81	61	41	30	20
	9	146	118	88	59	44	29
Ozone	6–9	0.97	0.63	0.48	0.32	0.24	0.16
Chlorine dioxide	6–9	21	8.7	7.7	6.3	5	3.7
Chloramines (pre-formed)	6–9	1270	735	615	500	370	250

Table 9.3 *Ct* values for achieving 2- and 3-log inactivation of enteric viruses at pH values 6–9

Disinfectant	Log inactivation	Ct values at water temperature					
		0.5°C	5°C	10°C	15°C	20°C	25°C
Free residual chlorine	2	6	4	3	2	1	1
	3	9	6	4	3	2	1
Ozone	2	0.9	0.6	0.5	0.3	0.25	0.15
	3	1.4	0.9	0.8	0.5	0.4	0.25
Chlorine dioxide	2	8.4	5.6	4.2	2.8	2.1	1.4
	3	25.6	17.1	12.8	8.6	6.4	4.3
Chloramines (pre-formed)	2	1243	857	643	428	321	214
	3	2063	1423	1067	712	534	356

flow rate. This is also called the t_{10} time, i.e. the time it takes 10% of water to pass through the tank. The ratio of t_{10}/t_T is a measure of short-circuiting in the contact tank and varies in the range 0–1. A value of one is an indication of total plug flow, but in practice depending on the design it varies between 0.5–0.7.

A good design can be achieved either by physical modelling[8] or mathematical models, e.g. CONTANK[9] or DISINFEX[10] or computational fluid dynamics (CFD) models. In the absence of these tools the following basic design parameters are suggested: the inlet jet should be baffled, the tank should be divided into long straight flow and return channels of length:width ratio greater than 10 and depth:width ratio less than 1.5 and a weir outlet should be provided to maintain the required volume of water under all flow conditions. With these provisions, a value of t_{10}/t_T of 0.5–0.7 may be achieved and the volume of the tank will have to be increased accordingly. The actual time of contact provided by a newly designed or existing tanks can be checked by timing the passage of a slug dose of a tracer chemical such as lithium chloride at a concentration of <0.1 mg/l through the tank[8] or by using models.

Pipelines can also be used to provide contact time, provided no consumers are supplied on route and control and monitoring of the residual chlorine at its downstream end is possible and convenient.

The free residual chlorine concentration of the water leaving the contact tank, after the requisite time of contact, can be reduced if desired by partial dechlorination (see Section 9.13) to suit the needs of the distribution system.

9.6   Efficiency of chlorine in relation to bacteria, enteric viruses, and protozoa

Bacterial kill. The work of Butterfield *et al.*[2] has shown that under nearly all conditions the typhoid bacillus and other enteric pathogenic bacteria are at least as susceptible to chlorination as *E. coli*. Due to the far greater concentrations of *E. coli* present in pollution of human or animal origin (see Section 6.60) it is therefore practicable to assume that, if *E. coli* are absent in a 100 ml sample of disinfected water, then the water is also free of pathogenic bacteria. The spores of bacteria are, however, more resistant to the action of chlorine than are the bacteria; fortunately the bacteria causing most waterborne diseases are not spore formers. The spore-forming *Clostridium perfringens* (*Cl. welchii*) used as an indicator of pollution (see Section 6.62) is not considered significant for health.

Virus kill. The pathogenic enteric viruses, described in Section 6.58, occur in far less numbers than *E. coli* in a polluted water. However they can survive for long periods in water and the minimum infective dose causing human infection is believed to be very low. Test methods available for detecting the presence of viruses (see Section 6.67) cannot be used for routine monitoring. The enteric viruses have also been shown to be more resistant to chlorine than *E. coli*. Poynter *et al.*[11] in 1972 reported Russian experiments had indicated that higher levels of residual chlorine and longer periods of contact were required to free water from viruses than were required to destroy *E. coli*, and similar results were obtained by Scarpino *et al.*[12]

The consequence of these difficulties is that tests showing the absence of *E. coli* in 100 ml samples of disinfected water do not give the same level of confidence that viruses are absent as they do for the absence of pathogenic bacteria. However the *E. coli* test remains the only practicable one at present for routine monitoring, and the WHO[3] considers: 'it has been demonstrated that a virus-free water can be obtained from faecally polluted source waters' if the following chlorine disinfection conditions are met:

- the water has a turbidity of 1 NTU or less;
- its pH is below 8.0;
- a contact period of at least 30 minutes is given; and
- the chlorine dose applied is sufficient to achieve at least 0.5 mg/l free residual chlorine during the whole contact period.

Protozoa resistance to chlorine. The principal protozoa pathogenic to humans – *Entamoeba histolytica*, *Giardia lamblia*, and *Cryptosporidium parvum* – were described in Section 6.57 where it was mentioned that human infection takes place through ingestion of them in their cyst stage. Like viruses the cysts can remain viable in the environment for long periods; the ingestion of a very few may be sufficient to cause human infection; and detection of their presence by routine testing is impracticable.

The cysts are more resistant to chlorine than viruses. The WHO describes the cysts of *E. histolytica* as 'among the most chlorine-resistant pathogens known', and the resistance of *Giardia lamblia* to chlorine as falling 'somewhere between *E. histolytica* and enteric viruses'.[13] The cysts of *Cryptosporidium* (called 'oocysts') were also found to be very resistant to chlorine, concentrations of 16 000 mg/l chlorine or more having been found necessary to reduce the viability of the oocysts to zero.[14] The small size of cysts – *E. histolytica* 10–20 μm; *Giardia* spp. 8–12 μm; *Cryptosporidium* 4–6 μm – means their complete removal by conventional processes of coagulation, clarification and filtration cannot be relied upon although these processes can be designed and operated to give between 2- and 3-log removal for *Cryptosporidium* (see Section 7.43) and higher log removal for others.

No completely satisfactory disinfection procedure to eliminate these protozoan cysts and oocysts has yet been found. The use of ozone in lieu of chlorine as a disinfectant appears likely to be more effective and research is continuing. Slow sand and membrane filtration offer the prospect of removing a high proportion of the cysts and oocysts. Apart from these possibilities the only recommendations that can presently be made relate to eradicating sources of pollution likely to give rise to the presence of such pathogenic protozoa in a water. If their presence is detected in a water an intensive search for the source and its elimination is necessary. When they are detected or suspected in the treated water, it is necessary to advise consumers to boil water used for drinking until evidence of their elimination from the supply is obtained (see Section 9.25).

9.7   Chlorination and the production of trihalomethanes

When chlorine is applied to water containing precursors which arise from the presence of natural colour and algal metabolic products, THMs are formed (see Sections 6.21 and 8.34). The reaction rate is favoured by the increase in precursor and free chlorine residual concentration, pH, temperature and contact time. Limits to their concentration in drinking water have been set because of their possible health effect. Emphasis is now placed on avoiding, or limiting the pre-chlorination of a raw water containing organic matter in order to minimise the formation of THMs. If a pre-disinfection stage is necessary, chlorine dioxide or ozone may be used instead of chlorine (see Sections 9.17 and 9.22 below). Alternatively, since chloramine does not react with organic matter to produce

THMs to the same extent as free chlorine, the chlorine dose may be kept low enough to produce only chloramines by making use of ammonia when present in the water, or by using chloramination of raw water (see Section 9.8). If relatively high levels of THMs are expected at the treatment works outlet, the degree of removal of organic substances prior to chlorination must be maximised. In some instances, it has been sufficient to move the pre-chlorination dosing point further downstream in the treatment process, e.g. after coagulation and clarification. If that is insufficient, advanced treatment processes, described in Sections 8.34–36, can be used to prevent THM formation by the removal of precursors before final chlorination is applied.

The maintenance of residual chlorine in the treated water may cause THM formation to continue within the distribution system by the reaction with residual precursors, particularly at high pH. THM formation reaction is initially fast with up to 50% formation in the first hour or so, but takes several hours or even days to complete and therefore its concentration at the consumers' tap could be much higher than in the treated water leaving the works. The position can be exacerbated in nutrient-rich waters when 'booster chlorination', i.e. addition of further chlorine at some key point(s) in the distribution system, has to be adopted to limit biological aftergrowths within the system. The formation of THMs in the distribution system can be minimised by controlled dosing of ammonia to convert free chlorine residual to combined chlorine residual (see next section)

Whatever measures are adopted, the WHO Guidelines emphasise that the disinfection process must not be compromised, and that 'inadequate disinfection in order not to elevate the THM level is not acceptable' (see Section 6.21).

9.8   The ammonia–chlorine or chloramination process

Combined chlorine, i.e. chloramine resulting from the reaction of chlorine with ammonia in water, is not commonly used as a primary disinfectant because of the acknowledged greater efficiency of free chlorine (see Section 9.3). In some instances a theoretical contact period of several hours would be required for chloramine to achieve adequate disinfection of certain difficult waters.[15] However ammonia is sometimes deliberately added after final chlorination to produce a chloramine residual in the final water passing into the distribution system. The primary reasons for using chloramines rather than chlorine are – the residual is longer lasting than chlorine, reduction in THM formation, superior control of biofilm growth (aftergrowth of bacteria or low forms of animal life) in the distribution system, and high doses (about 2 mg/l) can be applied with less risk of producing chlorinous taste. The weight ratio of chlorine to ammonia (as N) is usually in the range 3:1 to 4:1, and the ammonia is added after final chlorination when the free chlorine has acted for the requisite contact time. The reaction is fast taking only a few seconds to form a combined residual. The most effective pH range for chloramination is 7–8.5 when monochloroamine is predominant. Also excess ammonia ensures monochloramine is predominant (see disadvantages below).

An advantage of maintaining residual chloramine in water passing into the distribution system is that, if routine examinations show chloramine is present at the ends of the distribution system, this should indicate that no serious pollution has entered the pipes en route. The residual cannot be large enough to disinfect pollution entering the system, but it is useful for monitoring the state of the distribution system. The principal disadvantage of chloramination is the nitrification of any excess ammonia or free ammonia released by the breakdown of chloramines, to nitrite in the distribution system by nitrifying bacteria. Nitrites are toxic (see Section 6.30). Control methods include optimising the chlorine-to-ammonia ratio,[16] flushing out the affected sections, decreasing residence time in service reservoir, removing excess ammonia locally by break-point chlorination (see Section 9.9), or re-chloramination of the affected sections to eliminate bacterial growth.

Blending of chloraminated water with water containing free residual chlorine in distribution systems should be avoided as it could result in the formation of dichloramine and nitrogen trichloride which are well known for causing taste problems.

9.9 Breakpoint chlorination

Where a water already contains ammonia the production of chloramine is unavoidable when chlorine is added. To ensure the production of free chlorine to enhance bacterial kill, substantially more chlorine may have to be added because the additional chlorine at first only causes a reduction of the chloramines by oxidation. Only when this reaction is completed does the addition of further chlorine produce free chlorine. Stoichiometrically the breakdown of ammonia to nitrogen commences at a chlorine:ammonia (as N) ratio of 5:1 and completes at a ratio of 7.6:1.* In practice the ratio for complete breakdown could be as much as 10:1 and is pH dependent. The point at which the free chlorine begins to form is called the 'breakpoint' for the water, and adding enough chlorine to exceed this is called 'breakpoint chlorination'. This is illustrated in Fig. 9.1. In laboratory experiments Palin[17] observed that for neutral to slightly alkaline pH when the ratio of chlorine to ammonia (as N) is less than 5:1 (by weight) the residual was mainly NH_2Cl; breakpoint occurred at 9.5:1 for pH 6; between 8.2:1 and 8.4:1 for pH 7 to 8; and 8.5:1 for pH 9. As the ratio increased to 10:1 and above there was a decrease in combined chlorine accompanied by increases in NCl_3 and free Cl_2. Apart from the advantage of producing free chlorine, breakpoint chlorination can sometimes reduce taste and odour problems. The foregoing reactions are complex, being dependent on numerous factors such as temperature, pH and contact time. The breakpoint reaction could take about 20 minutes to complete and depends on the water quality. In some waters the ammonia content may be so high (0.5–1.0 mg/l) that the amount of chlorine to be applied to achieve breakpoint is uneconomic and other means to reduce the ammonia first should be adopted (see Section 8.27).

9.10 Superchlorination

By 'superchlorination' is meant the dosing of a water with a high dose of chlorine, often much larger than the usual condition of the water demands. The method is most often used on a borehole or well water which, though normally free of pollution, may be subject to the onset of pollution to an unknown degree following heavy rainfall or some other circumstances. The normally unpolluted water may only require a small protective dose of chlorine of the order of 0.2 mg/l. To wait for the pollution to occur, detect it, and then increase the dose is impracticable since action could not take place in time to prevent some

---

*Ratio of chlorine to ammonia as NH_3 = 6.26:1, and of chlorine to ammonia as NH_4^+ = 5.92:1.

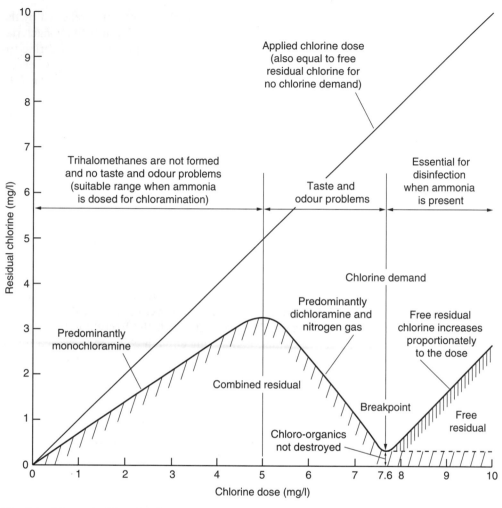

Fig. 9.1   Theoretical breakpoint curve for a water containing 1.0 mg/l ammonia (as N) (adapted from White[4]).

of the pollution passing into supply. It is also used on heavily polluted river waters, where the pre-treatment process does not include a pre-disinfection stage. Therefore a continuous high dose and adequate contact time is given sufficient to counter the worst conditions likely. After the contact period the water can be partially de-chlorinated by the injection of sulphur dioxide or sodium bisulphite (see Section 9.13), and the dose may be so controlled that only a part is removed, leaving a residual to go into supply. These processes of chlorination and partial dechlorination to achieve some given level of residual chlorine can be controlled automatically. Super chlorination can also sometimes be found helpful in the case of waters liable to give rise to taste and odours with lesser doses of chlorine.

9.11   Typical chlorine doses; taste problems.

Typical chlorine dosages to final treated waters are frequently in the range 0.2–2.0 mg/l of free chlorine. The lower doses tend to be those used on clear well waters not subject to pollution; the higher doses relating to treated surface waters or to well or borehole supplies which are liable to experience sudden pollution, where superchlorination followed by partial dechlorination after adequate contact time may be advised. Some waters, particularly surface waters, can have a high chlorine demand of 6–8 mg/l.[15] For a very clear water with no history of pollution, marginal chlorination at 0.2–0.5 mg/l might be adopted.

Application of chlorine can cause taste and odours, but these are principally caused by the reaction of chlorine with some of the many trace compounds in the water (see Section 8.32).

Production and application of chlorine

9.12   Use of chlorine gas

Chlorine is contained as a liquid under pressure in drums or cylinders. The liquid occupies about 95% of the volume, the remaining 5% is occupied by the gas. Most cylinders are designed to draw gas only, but the drum design allows either gas or liquid to be withdrawn. The cylinders are available in 33, 65 and 71 kg capacities whilst drums are available in 864, 966 and 1000 kg capacities. Gas withdrawal rate, at 15°C and 2 bar back pressure, is 1.0 and 1.5 kg/h for 33 and 71 kg cylinders, respectively and that for a drum is 7 kg/h. Higher rates can be achieved by connecting more than one container in parallel. A practical limit to the number of containers connected to a header is about six for cylinders and four for drums. The larger the number of containers connected the greater the number of connecting joints and hence the risk of a leak. When higher rates are required the contents should be drawn as liquid and vaporised in evaporators before use. The liquid should not be drawn from more than one container. In a few large works where chlorine usage is high, chlorine is stored as a liquid between 4 and 8 bar pressure at 5–30°C in bulk storage vessels which are filled from road tankers. The bulk storage facilities must be designed to very high standards of safety.[18]

Chlorine gas is metered at sub-atmospheric pressure and made into a chlorine–water solution containing up to 3500 mg/l Cl_2 for application to the water supply by means of an auxiliary water flow passing through an ejector which, at the same time, creates the operating vacuum (Fig. 9.2). Thus interruption or failure of the operating water supply will immediately shut down the flow of gas.

The gas from the containers enters the system via a vacuum regulating valve which is positioned either as an integral part of the chlorinator or as a separate unit located as close as possible to the containers. The valve reduces the gas supply pressure from up to 10 bar down to a vacuum of about 170 mb. In the unlikely event that positive pressure should reach beyond the vacuum regulating valve a pressure relief valve will vent a small quantity of gas to outside and operate an internal check valve until such time as normal vacuum conditions are restored.

The flow of gas is metered by maintaining a constant differential pressure drop across a manually or automatically adjusted variable area orifice through which the gas passes, the rate of flow being indicated by a glass tube flowmeter. Should an excess vacuum condition

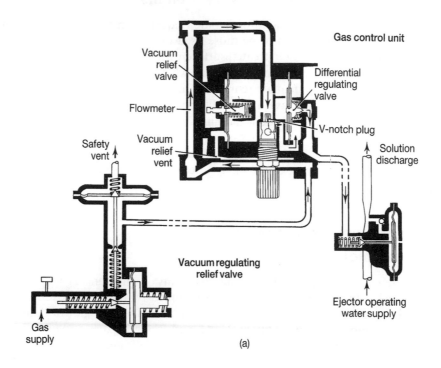

Gas control unit

Vacuum relief valve

Differential regulating valve

Flowmeter

V-notch plug

Safety vent

Vacuum relief vent

Solution discharge

Vacuum regulating relief valve

Ejector operating water supply

Gas supply

(a)

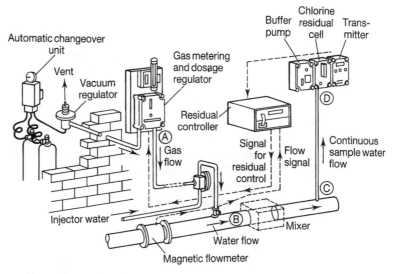

Automatic changeover unit

Vent

Vacuum regulator

Gas metering and dosage regulator

Buffer pump

Chlorine residual cell

Trans-mitter

Residual controller

Gas flow

Signal for residual control

Flow signal

Continuous sample water flow

Injector water

Water flow

Magnetic flowmeter

Mixer

Notes: Distance B to C should be about 10 pipe diameters.
Process time A(BC) to D should be about 4 minutes at average flow.

(b)

Fig. 9.2   Vacuum-operated gas metering unit for automatic dosage injection, controlled by residual recorder. Suitable for recorder, sulphur dioxide, or ammonia (as USF Wallace and Tiernan Ltd).

occur within the system, as when the gas supply is exhausted, a relief valve operates downstream of the metering section and a no-flow condition will be indicated on the flowmeter. Depending upon the type and capacity of the equipment the ejector may form an integral part of the unit, or be separately mounted either adjacent to the chlorinator or near the point of application. The chlorine dose rate may be automatically adjusted to accommodate variations in water flow or quality, or both, by means of an electric positioner fitted to the variable area orifice (see Fig. 9.2) to accept a control signal from a chlorine residual monitor to maintain a pre-set level of residual chlorine.

Chlorine is known to contain up to 0.1% w/w of bromine. This if allowed to react with ozone could form up to 1.6 μg/l of bromate for each 1 mg/l of chlorine dosed upstream of ozonation.

Safety precautions. Chlorine gas reacts explosively with many organic compounds. When mixed with water or moisture it is extremely corrosive. Pipework carrying liquid or gaseous chlorine should be of carbon steel and must be designed to exclude water or moist air; uPVC pipework can be used for chlorine gas under vacuum and for chlorine solution applications.

Chlorine is very toxic with an 8-hour exposure limit of 0.5 ml/m^3 and a 15-minute exposure limit of 1.0 ml/m^3.[19] Liquid chlorine leaks are far more dangerous than chlorine gas leaks since, upon evaporation, 1 kg of liquid chlorine yields about 335 litres of gas at 15°C and 345 litres at 25°C. Chlorine gas is heavier than air. A chlorine solution leak gives chlorine fumes laden with moisture and is very dangerous as it seems more tolerable to the respiratory tract than inhalation of dry chlorine.[4]

Safety precautions must be taken in the design and layout of chlorine installations to safeguard the operators and the public.[18,20] For sites in the UK, the Control of Major Accident Hazards Regulations[21] must also be complied with.

Chlorine facilities should be designed to minimise leaks and to contain them if they should occur. A chlorine building should house storage and dosing equipment and should be separated from any ventilation intakes of other buildings by at least 25 m and from the site boundary by at least 20 m (for cylinder installations), 40 m (for drum installation using gas) and 60 m (for drum installations using liquid). Chlorine containers should always be installed in a separate 'gas tight' store constructed in substantially fire-resistant material and should be provided with access from open air. All doors should open outwards and emergency exit doors should have a push bar operated panic bolt. External windows should be avoided, artificial illumination being used throughout. A useful safety measure at chlorine installations is to fit the containers connected to headers with a remote shut-off device for effective isolation of the chlorine supply at the container in the event of a downstream leak. Risk of chlorine leaks should be minimised by conveying chlorine under vacuum, e.g. vacuum regulating valves should be located in the container store so that pipelines conveying chlorine liquid or gas under pressure are confined to the store and ejectors should be located remote from the chlorinators, local to the point of application so that the risk of a chlorine solution leak is confined to a short length of pipe at the point of application. Heaters (non-radiant type) may need to be installed in the container store to maintain a temperature greater than 10°C. The ventilation system in the container store should have high level fresh air inlets and low level extraction of foul air discharging to outside at high level. Ventilation systems should be designed to give not less than 10 air changes per hour and should be arranged to start at a low leak level and to shut down at a high leak level to contain major leaks. Leak detectors with a 1 ml/m^3 low detection limit and 3 ml/m^3 high detection limit should be installed in enclosed areas where chlorine is handled.

At chlorine sites in urban areas or near housing developments or with inventories of 10 or more tonnes, the use of a chlorine absorption plant for neutralising a leak should be considered. The absorption plant should be designed to treat the contents of a container (e.g. 1000 kg for a drum installation). The leakage rate could vary from 1.5 kg/min to about 35 kg/min. A typical system consists of a packed tower (see Section 8.19) where the foul air flows counter-current to a flow of dilute caustic soda (10–20% w/w NaOH) which is continually recycled back to the top of the tower. Stoichiometrically 1000 kg of chlorine (about 340 m^3 as gas) requires 1127 kg of caustic soda (100% w/w NaOH); in practice about 20% more caustic soda would be required.

9.13   Dechlorination

Dechlorination can be achieved with sulphur dioxide or sodium bisulphite and should be carried out after chlorine has had adequate contact time for disinfection. In disinfection treatment (see superchlorination, Section 9.10) only partial dechlorination is required. In pre-treatment for reverse osmosis, complete dechlorination is the objective (see Section 8.46). The dechlorination reaction is fast taking less than 1 minute to complete.

Sulphur dioxide, similar to chlorine, is contained in cylinders (30 and 65 kg) or drums (865 and 1016 kg) and is drawn for use as a gas or liquid. The gas withdrawal rates, at 15°C and 2 bar back pressure, are 0.3 and 0.45 kg/h from 30 and 65 kg cylinders, respectively, and 2.3 kg/h from a drum. Higher withdrawal rates can be achieved by connecting several containers in parallel, or withdrawing sulphur dioxide as a liquid and vaporising in evaporators. The equipment used for sulphur dioxide is very similar to that used for chlorine. The dose rate may be automatically controlled proportional to water flow, or to maintain a pre-set level of chlorine residual, or both. Stoichiometrically 0.9 mg of sulphur dioxide removes 1.0 mg of chlorine; in practice at least 15% more sulphur dioxide would be required.

Safety precautions. Sulphur dioxide is corrosive; materials of construction used for chlorine can be used for sulphur dioxide. Sulphur dioxide is a toxic gas with an 8-hour exposure limit of 2 ml/m^3 and a 15-minute exposure limit of 5 ml/m^3.[19] Liquid sulphur dioxide leaks are far more dangerous than a corresponding gas leak since, on evaporation 1 kg of liquid sulphur dioxide yields about 370 litres of gas at 15°C and 380 litres at 25°C. It is heavier than air and is compatible with chlorine and therefore can be stored in the same room as chlorine.

Sodium bisulphite is available as a solution containing 20% w/w SO$_2$ (for properties see Table 7.7). Stoichiometrically 4.5 mg of sodium bisulphite (20% w/w SO$_2$) removes 1.0 mg of chlorine; in practice at least 15% excess sodium bisulphite would be required. Any spillages should be neutralised with soda ash to prevent sulphur dioxide emission, and then be oxidised to neutral sulphate with sodium hypochlorite. The chemical is dosed using positive displacement metering pumps. Automatic dose control requires control of the pump motor speed proportional to water flow and adjustment of the stroke length to maintain a pre-set chlorine residual in the water.

Other dechlorination methods include filtration through granular activated carbon (GAC) or ammonia (see Section 9.9) or hydrogen peroxide treatment. GAC is sometimes used to protect chlorine sensitive reverse osmosis membranes; flow rates of

10–15 $m^3/h.m^2$ are typical with EBCT between 10 to 15 minutes. GAC is consumed by chlorine and cannot therefore be regenerated for reuse.

Sulphur dioxide leaks are treated in absorbers of a similar design to those described for chlorine, using caustic soda as the absorbent.

9.14 Ammoniation

Ammonia or ammonium sulphate can be used for ammoniation and should be added after disinfection and partial dechlorination (if applicable). Final pH correction should follow ammoniation. Ammonia is available as a liquid in cylinders (49 and 65 kg) or drums (500 kg). It is withdrawn as a gas at 0.5 kg/h from a cylinder or 2 kg/h from a drum at 15°C, or as a liquid to evaporators. Apparatus used is very similar to that used for chlorine. Ammonia dosing may be automatically controlled proportional to water flow or to a pre-set ratio of free residual chlorine:ammonia, or both. Ammonia is very soluble in water and is corrosive. Steel piping is suitable for conveyance of ammonia liquid and dry gas. Iron, copper and zinc are attacked by ammonia solution, but uPVC is suitable. Water for use in ammonia dosing solution should be softened to a hardness value of less than 25 mg/l as $CaCO_3$.

Ammonia is toxic with an 8-hour exposure limit of 25 ml/m^3 and a 15-minute exposure limit of 35 ml/m^3,[19] and is lighter than air. It is flammable and flammable limits are 15 and 28% by volume. Electrical apparatus in areas where ammonia is handled should therefore be suitably protected. Ammonia gas forms explosive mixtures with chlorine and sulphur dioxide gases and should therefore be stored away from them. Storage room design requirements are similar to those for chlorine, except for the ventilation system which requires low level fresh air inlets and air extraction and disposal at high level.

Ammonium sulphate is a crystalline powder (for properties see Table 7.7). Dosing of ammonium sulphate solution is by positive displacement metering pump. The speed of the pump may be automatically controlled proportional to water flow, or the stroke length of the pump may be automatically adjusted to maintain a pre-set ratio of free chlorine residual:ammonia, or both.

Absorbers for treating ammonia leaks are of a similar design to those used for chlorine. Absorbent used for ammonia is sulphuric acid (10%w/w H_2SO_4).

9.15 Hypochlorite production on site by electrolysis

Sodium hypochlorite can be produced by the electrolysis of brine (a solution of sodium chloride). A direct current passed through a solution of sodium chloride (common salt) containing Na^+ and Cl^-, produces chlorine at the anode, and hydrogen at the cathode. With mixing of the catholyte, anolyte and sodium ions in the solution, sodium hypochlorite (Na^+OCl^-) is produced. The principal reactions are as follows:

At the anode	$2Cl^- - 2e = Cl_2$
At the cathode	$2H_2O + 2e = 2OH^- + H_2$
On mixing	$2Na^+ + 2OH^- + Cl_2 = Na^+OCl^- + Na^+Cl + H_2O$

The process is favoured in UK because the hypochlorite solution can be manufactured on site, thus avoiding the risks of transporting and handling liquid and gas chlorine and the difficulty of meeting all the associated safety measures required. The on-site generation gives a hypochlorite solution easy to handle, with favourable running costs as compared with the use of purchased liquified chlorine. The process can be used on large or small supplies.

A diagram of the plant used is shown in Fig. 9.3, and Plate 25 shows a 320 kg Cl_2/day electrolyser. The brine for electrolysis is prepared by withdrawing saturated brine from a salt saturator at a concentration 358 g/l (10°C), 360 g/l (20°C) or 363 g/l (30°C) and diluting this to 20–30 g/l. About 3.5 kg of sodium chloride is required to form 1 kg of chlorine; approximately 50% of sodium chloride is converted to hypochlorite. The salt should be of high purity grade, free from calcium, magnesium and heavy metals. Pure dried vacuum salt containing at least 99.7% w/w NaCl is most suitable (for properties see Table 7.7). Lower quality salt should be treated by chemical softening to remove hardness. Water for saturation and dilution of salt must have a hardness less than 25 mg/l as $CaCO_3$ and may therefore need to be softened; base exchange softening being used. The process water requirements are about 125–150 litres per kg of chlorine. The feed water temperature should be maintained above 5–10°C to maximise anode coating life; the product solution temperature should be below 40°C to inhibit chlorate formation (see by-products below). The temperature rise in the electrolyser is limited to about 20°C. In cold climates it may be necessary to warm the feed water, and in hot climates to cool it.

The chlorine content of the hypochlorite produced is in the range 6–9 g Cl_2/l. The electrical consumption is 4–5 kWh per kg of chlorine produced; a low voltage direct current (e.g. 40 V) is used. Hydrogen gas is produced at the cathode at the rate of 330 litres

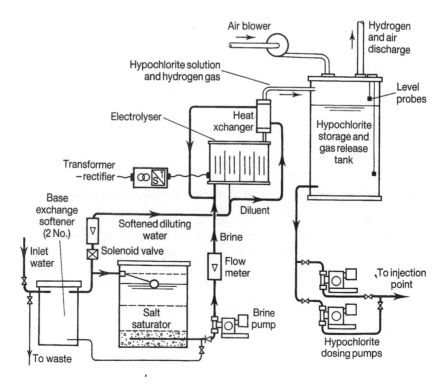

Fig. 9.3   On-site hypochlorite production by electrolysis of brine.

at 15°C and 1 bar pressure, per kg of chlorine produced. The sodium hypochlorite solution with the hydrogen is fed to an enclosed hydrogen disentrainment tank which can also act as a solution storage tank. The hydrogen degasses very rapidly to an airspace at the top of the tank, which is force ventilated with air at a rate 100 times the maximum hydrogen production rate. This reduces the hydrogen concentration to 1%, i.e. one-quarter of the 'lower explosion limit' which is 4% of hydrogen in air by volume. The diluted gas is then vented to the atmosphere. The hypochlorite solution, having checked for chlorine content, is injected by a positive displacement metering pump into the water to be treated. Hypochlorite solution tanks can be constructed in high density polyethylene or GRP with a PVC lining.

At coastal installations hypochlorite can be produced from seawater. Due to contaminants present in seawater, such as hardness, heavy metals and suspended matter, a special cell design with appropriate anodes and cathodes is used, and the chlorine content of the hypochlorite solution is maintained at about 1 g/l. The feed water should be free of pollution and have a relatively constant salinity and must be passed through a strainer to remove coarse suspended particles. Hypochlorite produced from seawater is used primarily for biological growth control in raw water, or cooling water treatment at desalination, or coastal power generation plants.

Safety measures. Electrolytic plant presents a potential fire and explosion hazard due to the hydrogen produced. Precautions should therefore be taken in the siting, layout, and design of the plant.[22] The design of electrical apparatus needs particular attention: as much as possible of such apparatus should be sited in a non-hazardous area or else specially protected.

By-products. Several side reactions take place during the electrolysis of brine. One such produces chlorate, ClO_3^-, which is considered potentially harmful to health and consequently a limit of 0.7 mg/l chlorate in the treated water has been set under the UK Water Regulations (see Section 6.53). Commercial units are known to produce less than 0.1 mg of chlorate as ClO_3^- per 1.0 mg of chlorine and this acts as a limiting factor to the amount of hypochlorite that can be applied. The process also produces bromates from bromides present as an impurity in the salt. The use of high purity salt containing 0.015% w/w of bromide restricts the bromate concentration to less than 1.0 μg per 1.0 mg of chlorine.

All the sodium chloride used in the feed to electrolyser is injected into the water being treated as sodium and chloride. Account therefore has to be taken of the limits set for the sodium content of drinking water (see Section 6.41) and the design of the disinfection process has to take this factor into account. Also chloride coupled with low alkalinity in a water can contribute to corrosion of metals (see Section 8.42). The quantities of sodium and chloride produced are 1.38 and 2.12 mg, respectively per mg of Cl_2 produced.

9.16 Testing for chlorine

On-line measurement of the chlorine content of a water is achieved by passing a continuous sample of the chlorinated water to a cell containing electrodes of dissimilar metals, e.g. platinum and copper, or platinum and silver. The galvanic current induced between the electrodes reduces because of polarisation of the electrodes, but the presence of chlorine acts as a de-polarising agent, so the current produced by the cell varies in proportion to the concentration of chlorine in the cell water. The resulting variations of the micro-ampere current are used to measure the chlorine concentration. The flow rate of sample water through the cell must be held constant so that the chlorine content of the sample is always proportional to the chlorine concentration in the main flow of water. The electrodes have to be constantly cleansed, and the pH of the sample water may have to be automatically adjusted if it is outside the range for which the apparatus is designed. Additional equipment can provide numerical display and print-out records of chlorine measurements.

Modern treatment works using chlorine for final disinfection are usually provided with automatic chlorine residual recorders and controllers. Most engineers consider it desirable that the chlorine content of the final water leaving the sourceworks is also checked manually once or twice a day. Manual testing of the chlorine content of water is also necessary on samples drawn from the distribution system.

Manual testing for chlorine comprises the addition of reagents, usually in the form of tablets, to a prescribed sample quantity of water and comparing the colour developed in the sample with a range of standard colour discs. A wide range of discs and reagents are available so that both free chlorine and combined chlorine can be measured for concentrations from 0.01 mg/l upwards. The reagents used for measuring chlorine content were developed by Palin in the 1960s and comprise diethyl-p-phenylene diamine (DPD) to produce a pink colouration, the intensity of which is proportional to the free chlorine present. The further addition of potassium iodide to the same sample gives a colour intensity proportional to the total chlorine concentration (i.e. free and total chlorine), the difference between the two readings gives the amount of combined chlorine present.

As an alternative to visual comparison, the colour can be measured photometrically, with display and print-out of recorded values. Such apparatus is helpful in removing the subjective element involved in judging colour intensity by eye. In the use of all such apparatus a practical requirement is the need to keep sample tubes clean and to have available a stock of new tubes for use when those in use can no longer be cleaned satisfactorily.

9.17   Use of chlorine dioxide

Chlorine dioxide is produced on-site due to its relatively short half-life, commonly by the reaction between a solution of chlorine (in water) and sodium chlorite ($NaClO_2$) in a glass reaction chamber packed with porcelain Rashig rings. It may also be generated by the reaction between sodium chlorite and hydrochloric acid (HCl). Sodium chlorite is supplied as a liquid containing 26% w/w of $NaClO_2$ in small containers or in bulk. For properties of sodium chlorite see Table 7.7. Spillages of sodium chlorite should be washed quickly as evaporation leads to deposits of high flammable sodium chlorite powder.

The proportion of chlorine to sodium chlorite (100% w/w $NaClO_2$) will vary from the stoichiometric ratio of 0.39:1 to as much as 1:1, dependent upon the alkalinity of the water. Excess chlorine, above the stoichiometric requirement, should be limited to that required to neutralise the alkalinity, otherwise any further excess chlorine will promote chlorate (ClO_3^-) production and cause THMs to form if precursors are present. Some of the alkalinity also reacts with the hydrochloric acid produced by the action of chlorine on water which otherwise would have reduced the chlorine dioxide yield by about 20%. The pH value should be about 4 and therefore for most waters the chlorine concentration needs to be over 500 mg/l. In practice chlorine dioxide solution concentrations are maintained at less than 1 g/l in open systems, and 10 g/l in fully enclosed pressurised systems.

The acid process uses more sodium chlorite than the chlorine process; about 1.25 times to produce the same amount of chlorine dioxide. Stoichiometrically 1.67 g of sodium chlorite (100% w/w $NaClO_2$) and 0.54 g of HCl are required to produce 1 g of chlorine dioxide. In practice about 50% more sodium chlorite is required. Furthermore between 300–350% of the stoichiometric quantity of acid is required to lower the pH (to ≤ 0.5) and neutralise alkalinity and maximise the yield.

Chlorine dioxide (ClO_2) is most commonly used as a disinfectant in cases where problems of taste and odour arise with chlorine, particularly those due to the presence of phenols.[23] It is a powerful oxidant, but at the limited dose levels it can be used (because of by products formed), its oxidation potential is not fully utilised. It is known to oxidise iron (II) and manganese (II)[24] (see Section 8.12), and remove colour[25] and certain types of tastes and odours. It does not produce THMs, nor oxidise THM precursors; nor does it react with ammonia or phenols in water. In the USA it is primarily used as a substitute pre-oxidant for chlorine at the inlet to the works for taste and odour control, colour removal, pre-disinfection and iron and manganese oxidation because unlike chlorine it does not produce THMs.[26] Its bactericidal efficiency is comparable with that of free chlorine in the neutral pH range[27] but, unlike chlorine, its efficiency increases with increasing pH.[25] Chlorine dioxide therefore has particular advantages for disinfecting waters liable to produce chlorophenol tastes, or which have a high pH or which contain substantial concentrations of ammonia.

The principal drawback with chlorine dioxide use is the formation of chlorate (ClO_3^-) and chlorite (ClO_2^-) in the generation process and in the water. Chlorates can be produced by generating chlorine dioxide at too low pH and with high excess chlorine. At concentrations less than 10 g/l chlorine dioxide disproportionates* under alkaline and acidic conditions to form chlorate and chlorite. In practice the disproportionate products are kept to a minimum by maintaining the pH of the solution in the range 3.5–7.5. Chlorates are also known to form by the exposure of water dosed with chlorine dioxide to sunlight, by increased pH (such as in softening) or by the action of chlorite ion with free residual chlorine in the contact tank and distribution system. Chlorite originates from the reactants in the generation process and from the disproportionation of chlorine dioxide.

As mentioned in connection with the production of hypochlorite solution by electrolysis (see Section 9.15), chlorate is considered to be of health significance.[28,29] In the UK, therefore, under the Water Supply Regulations 1989 (see Section 6.47), a limit has been set for health reasons for the combined residual levels of chlorine dioxide, chlorite and chlorate of 0.5 mg/l as ClO_2 in the water entering supply. The corresponding USEPA value is 1.0 mg/l as ClO_2. These values would restrict the chlorine dioxide dosage to about 0.75 mg/l (UK) and 1.5 mg/l (USA). The WHO have a provisional Guideline Value of 0.2 mg/l for chlorite.[3] These relatively low limits have curtailed the use of chlorine dioxide in practice.

---

*Disproportionation is the transformation of a compound into two dissimilar compounds by a process involving a simultaneous oxidation and reduction.

When used as a pre-oxidant it is seldom applied at a dose greater than 1.0–1.5 mg/l. At a dose of 1.0 mg/l, chlorate ion in the treated water would be in the range 0.2–0.4 mg/l. Chlorates can be removed by ferrous chloride and it is reported that 6–7 mg/l ferrous chlorides as Fe per 2.5 mg/l of chlorine dioxide dose was effective in reducing the combined species concentration to 0.2 mg/l.[26,30] Sodium bisulphite or sulphur dioxide is effective in removing chlorite ions; 95% removal efficiency in the pH range 5–6.5. Due to restrictions on by-products in the finished water chlorine dioxide is rarely used as the primary disinfectant.

9.18   Calcium hypochlorite powder ('chloride of lime')

Calcium hypochlorite in powder form, commonly known as 'bleaching powder' or chloride of lime is widely used in the less developed countries of the world for disinfection of water supplies.

The bleaching powder contains 30–35% w/w of releasable chlorine and excess lime. The powder has a bulk density of about 400 kg/m^3 and has the advantage that sealed drums of it can be held in store for long periods without serious loss of chlorine. It is best to make up a solution in batches. A measured quantity of bleaching powder is mixed in a tank of water and allowed to settle. The supernatant containing chlorine is drawn off and diluted to the dosing concentration in a storage tank. The supernatant is injected into the water to be treated by means of a positive displacement reciprocating pump of the diaphragm type or constant level feeders (see Ref. 72 in Chapter 7). In large plants, saturators may be used (see Section 7.20). The strong chlorine solutions resulting rapidly lose their chlorine content if exposed to air or sunlight; hence they need to be made up daily or every second day. Likewise a drum of bleaching powder will begin to lose its chlorine content once opened.

The strength of the solution used depends on the quantity of water to be treated daily and the capacity of the injection apparatus. Taking the powder to contain 33% by weight of chlorine, a 1% w/v Cl_2 (10 g Cl_2/l) solution would be made up by mixing 30 kg of powder in 1000 litres of water. A 100 litres of this 'batch' would be sufficient to give 1.0 mg/l in 1000 m^3 of water. The solution strength needs checking regularly because the chlorine content of the powder may not be known, the batching procedures used may not be very exact, and solutions left standing may lose substantial chlorine. A sample needs to be diluted by a factor of 5000–10 000 to determine the chlorine content of the solution using a chlorine comparator.

9.19   Calcium hypochlorite granules

Calcium hypochlorite granules contain 65–70% w/w chlorine and can be supplied in 45/50 kg drums with plastic liners. It has a bulk density of about 800 kg/m^3. The granules are readily soluble and a standard strength solution can be made up for dosing. A 1% w/v Cl_2 (10 g Cl_2/l) solution is prepared by mixing 15 kg of granules containing 65% w/w Cl_2 in 1000 litres of water. They give a clearer solution than chloride of lime powder but, after initial standing, the supernatant of a 0.5% w/v chlorine solution will still contain up to 1% suspended matter. It is preferable to use water with low alkalinity, such as rainwater, for making up the solution as the use of a hard water results in some precipitation of its hardness. As with solutions made from chloride of lime, the chlorine content will reduce substantially in a few days if left exposed to the air.

9.20   Sodium hypochlorite solution

Sodium hypochlorite solution is available as a household disinfectant ('bleach') under many brand names. It can be produced by electrolysis of brine (see Section 9.15) and, as used for waterworks purposes, is a clear solution containing 14–15% w/w of available chlorine. It can be supplied in small containers or in bulk, but loses its chlorine strength when exposed to atmosphere or sunlight (for properties and storage see Table 7.7). The rate of decomposition increases with increased temperature as shown in Table 9.4.

Dosing of hypochlorite can be by positive displacement reciprocating diaphragm pumps. The use of sodium hypochlorite solution on a plant scale is expensive because of the costs of packaging and freight for the relatively large volumes of solution required. The cost rules out its use in developing countries, but elsewhere it is used in place of chlorine gas for safety reasons.

Attention has to be paid to limiting the introduction of chlorate and bromate to the treated water, as mentioned in connection with the on-site production of sodium hypochlorite by electrolysis (see Section 9.15). Commercial hypochlorite of 15% w/w chlorine, produced using the membrane process contains chlorates (as ClO_3^-) and bromates (as BrO_3^-) of about 0.25 and 0.035%, respectively, by weight of the product.

9.21   Chlorine tablets

Chlorine tablets are available in a range of sizes and chlorine content. They are normally too expensive to use on a large scale, but are useful in certain circumstances for the chlorination of small supplies. The smallest tablets contain 2.5 mg of chlorine and are used by travellers for the disinfection of drinking water; they are designed for rapid release of chlorine. The larger tablets, containing 10 g or more of chlorine, are frequently used for disinfection of mains and tanks. They are designed for slow release of chlorine over several hours to disinfect mains and tanks.

Table 9.4   Decomposition of chlorine in sodium hypochlorite of 14.5% w/w Cl_2 due to age and temperature

	Chlorine content % w/w at solution temperatures				
Time (weeks)	10°C	15°C	20°C	25°C	30°C
1	14.43	14.39	14.08	13.57	12.41
2	14.39	14.08	13.57	12.41	10.88
4	14.17	13.58	12.41	10.52	8.70
26	11.72	9.54	7.00	4.5	2.59

Note. Tests were carried out in the dark, in the absence of metallic contaminants except traces of iron, copper and nickel present in the product.

Ozone process of disinfection

9.22   Action of ozone

Ozone gas, O_3, is a powerful oxidising agent widely used for disinfection and oxidation. The bactericidal effect of ozone is rapid, the usual contact time being between 4 and 10 minutes with dosages of the order of 2–3 mg/l. It is also known to be more effective than chlorine in killing viruses, cysts and oocysts (see Sections 9.5 and 7.43). The WHO Guidelines[3] state: 'Ozone has been shown to be an effective viral disinfectant, preferably for clean water, if residuals of 0.2–0.4 mg/litre are maintained for 4 minutes'. Therefore the criterion adopted by European designers for disinfection is to maintain a free ozone residual of 0.4 mg/l for 4 minutes (i.e. $Ct = 1.6$ mg.min/l); this residual being the required level after the initial ozone demand is satisfied. The US practice is to design for the Ct values based on the inactivation of cysts and viruses given in the SWTR (see Section 9.5).

WHO Guidelines do not define the t value whereas the European practice has been to define t as the 'hydraulic' residence time (volume ÷ mean flow). Ideally t should be the t_{10} value which for ozone contactors could vary in the range 0.5–0.65 of the t_T (see Section 9.5). According to USEPA[31] since ozone uses more than one point of application in the contactor, inactivation credits of 1-log for viruses and 0.5-log for *Giardia* are given if ozone residuals of 0.1 mg/l and 0.3 mg/l, respectively, are maintained at the outlet of the first cell of the contactor. Therefore following from Section 9.5 the maintenance of an ozone residual of 0.3 mg/l or greater at the outlet of the first cell of an ozone contactor appreciably reduces the Ct required in the subsequent stages. For the method for calculation of Ct values of the subsequent cells the reader is referred to the USEPA design criteria.[31,32]

The contact chambers for ozonation for disinfection comprise a minimum of two stages (each with two cells) in series. In the first stage, water is retained to satisfy the initial ozone demand, sufficient ozone gas being introduced to ensure adequate free ozone residual for disinfection in the water leaving this stage. The retention time in this first stage and the ozone dose applied will vary according to the ozone demand. In the second stage, disinfection is achieved, the stage being sized to give a Ct value of 1.6 mg.min/l where the contact time t is equal to 4 minutes and the ozone dose being controlled to maintain $C = 0.4$ mg/l free ozone residual throughout or to suit any other criteria (e.g. USEPA). In the third and final stage, ozone is allowed to decay, no ozone being normally dosed in this stage. Most of the excess ozone after treatment may be removed by cascading the outlet water. The remaining residual ozone could be removed by dosing sodium bisulphite which is an ozone scavenger.

The number of stages to be provided in a contactor is governed by the need to achieve uniform flow and minimise short-circuiting. The greater the number of stages, the closer is the flow regime to plug flow. For economic reasons the number of stages (each with two cells) is usually limited to three with a maximum of four. The preferred flow configuration in each stage is one in which ozone gas, injected by diffusers in one cell, flows upwards counter-current to water flowing downwards. In the second cell of each stage the flow is reversed using underflow baffles to enable counter-current flow to be re-established in the first cell of the following stage (see Fig. 9.4). This provides good liquid–gas contact and helps to minimise short-circuiting. In some contactors each stage is provided with one cell, and counter-current and co-current flow configurations are used in alternate stages to minimise costs. When large water flows have to be treated it is necessary to divide the flow

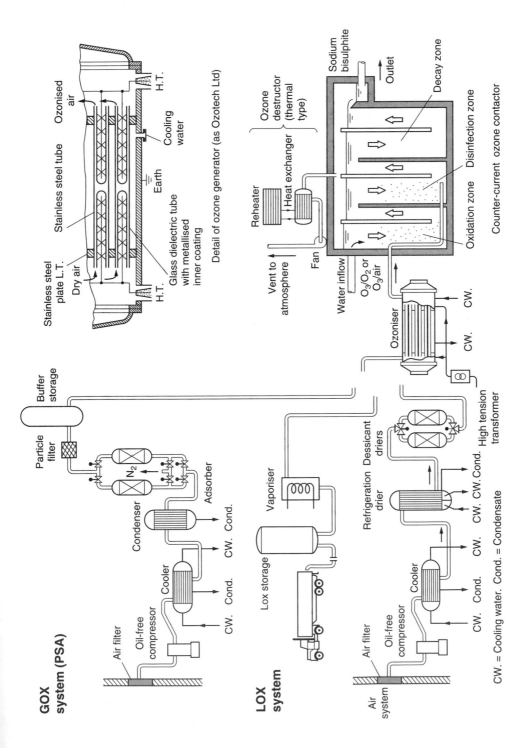

GOX system (PSA)

Buffer storage

Particle filter

N₂

Condenser

Adsorber

CW. Cond.

Cooler

CW. Cond.

Air filter

Oil-free compressor

LOX system

Lox storage

Vaporiser

Refrigeration drier

Dessicant driers

High tension transformer

CW. CW. Cond.

CW. Cond.

Air system

Air filter

Oil-free compressor

Cooler

CW. Cond.

Ozoniser

CW.

CW.

Water inflow

O₃/O₂ or O₃/air

Fan

Vent to atmosphere

Reheater

Heat exchanger

Ozone destructor (thermal type)

Sodium bisulphite

Outlet

Decay zone

Disinfection zone

Oxidation zone

Counter-current ozone contactor

Stainless steel plate L.T.

Dry air

H.T.

Stainless steel tube

Ozonised air

Cooling water

Earth

H.T.

Glass dielectric tube with metallised inner coating

Detail of ozone generator (as Ozotech Ltd)

CW. = Cooling water. Cond. = Condensate

Fig. 9.4 Ozone plant.

into several equal parallel streams. When hydrogen peroxide dosing is practised (peroxone process, see Section 8.36), it is added to the inlet of the first cell in the second stage. However, since the use of hydrogen peroxide leaves the water without any ozone residual, when disinfection is also necessary the hydrogen peroxide is added after disinfection, to the inlet of the third stage, with ozone being dosed into its first cell.

Ozone oxidises many organic substances to less complex compounds (see Sections 8.33–8.36). It oxidises colour and some organic substances responsible for taste and odour. It also oxidises iron(II) and manganese(II) to form precipitates. It is useful for the disinfection of a water containing a high amount of ammonia which would otherwise require a heavy dose of chlorine. Ozone in conjunction with GAC filtration has been found useful for the reduction of some pesticides in water.[33] Pre-ozonation results in improved particle and turbidity removal in a coagulation process[34] and in sand and GAC filtration.[35] To enhance coagulation, the ozone dose required is about 0.4 mg/mg total organic carbon.[32]

It was thought that use of ozone might limit the formation of THMs by oxidation of organic precursors in a water, but experiments have shown that ozone does not remove organic matter but breaks it down into smaller, more bio-degradable compounds. Some of these compounds react more readily with chlorine used for providing a residual, leading to an increase rather than a decrease in the formation of disinfection by products such as chloroacetic acid and chloro aldehydes and ketones during final chlorination.[33,36] These organic by-products, being more biodegradable than their precursors, can provide nutrients for biological growth and thus promote aftergrowth in the distribution system. The potential health significance of the many by-products of ozone is still not well understood. Among the identified by-products are formaldehyde, organic peroxides, unsaturated aldehydes, epoxides, haloacetic acid, and the inorganic by-product bromate. However the addition of biological activated carbon reactors (see Section 8.35) after ozonation can remove a high proportion of the organic by-products, but not the inorganic by-products such as bromates.

Bromates are formed during ozonation of waters containing bromides and can be minimised by optimising the ozone dose and the contact time to the minimum required, leaving little or no ozone residual and improving the contactor design and operation. Depression of pH to less than 7 minimises the bromate formation but would not be economical for high alkalinity waters.[37] Addition of ammonia before ozonation reduces bromate concentration through monobromoamine formation.[38] Hydrogen peroxide (H_2O_2) dosed after ozonation decreases the bromate formation with the increase of its dose. For maximum reduction, the weight ratio of H_2O_2:O_3 would be greater than 2:1.[39] Hydrogen peroxide is usually added immediately after the oxidation stage in the post-ozone contactor and would therefore prevent disinfection because it reacts with the residual ozone. Bromides are usually found in very small concentrations (<0.2 mg/l) in water and their removal (by ion exchange or reverse osmosis), is not cost effective. There are no proven methods in full scale operation for the removal of bromate after its formation.

A disadvantage of ozone is that its half life in water is too short for it to be effective as a residual disinfectant in the distribution system. Consequently to provide a residual, chlorination, chloramination or chlorine dioxide dosing has to be used as a final stage of treatment.

9.23 Production of ozone

Ozone is produced on a commercial scale by passing air or oxygen through a silent electrical discharge. A high voltage AC current is applied between two electrodes separated by a dielectric and a narrow gap through which the air or oxygen to be ozonised is passed. The gap width is a function of the feed gas, the dialectric material, and the power supply's characteristics. Typical gap widths are between 0.5 and 3 mm. Electrical power options commonly available for electrolysis are low frequency (50–60 Hz at 14–19 kV); medium frequency (60–1000 Hz at 9–14 kV); and high frequency (greater than 1000 Hz typically 5000–7000 Hz at 10 kV). Medium frequency generators are used primarily for oxygen feed systems, they have a smaller discharge gap and allow higher ozone production per tube. Low frequency systems have the simplest power supply. High frequency systems have the advantage of low voltage and therefore few dielectric failures.

In practice, the two electrodes are concentric tubes but may also be arranged as plates. In concentric tube designs, the outer electrode is a stainless steel tube. The dielectric is a glass or ceramic material, which can be plated onto the outside tube or the inner electrode. Historically a glass tube with an inner metallised coating forms the inner electrode. A typical ozone generator consists of several hundred of such tubes assembled in a large vessel (see Fig. 9.4 and Plate 25). The high voltage is applied to the inner tube; the low voltage is connected to the stainless steel outer tube. Approximately 90–95% of the energy input appears as heat and must be removed by applying cooling water. Manufacturers recommend that cooling water has a chloride content <30 mg/l to minimise corrosion of stainless steel. Depending on the temperature and the chloride content the cooling water could be once-through type using filtered plant water or closed-circuit type which uses water-cooled heat exchanger or refrigerated water chillers. The system capacity should be adequate to limit the increase in gas stream temperature to less than 5°C. The feed (air or oxygen) to the ozoniser must be completely oil-free to prevent detonation (hydrocarbon content <10 ml/m^3), dust free to a level better than 99% at 0.5 micron to reduce electrode fouling and very dry; the normal requirement is that the feed should have a dew point below -80°C to prevent sparking and formation of nitric acid. In the air feed to achieve this dryness it is usual to both refrigerate the feed and pass it through a desiccant. The feed should be cool with a temperature below 25°C, and compressors or blowers used for the feed must be of the oil-free type. The diagram of an ozone plant is given in Fig. 9.4.

With air as the feed, typical production is up to 4% w/w ozone in air (52.3 g O$_3$/m^3 of air at 0°C, 1.013 bar). Air feed systems can be either low pressure (<2 bar g) or high pressure (>4 bar g). In a high pressure system most of the moisture can be removed in the aftercooler and eliminates the need for a refrigerant dryer. High pressure systems are commonly used in packaged ozone generators.

With oxygen as the feed typical production is about 10% w/w ozone in oxygen (149 g O$_3$/m^3 of oxygen at 0°C, 1.013 bar). Methods of supplying oxygen include use of liquid oxygen (LOX), delivered in bulk to site by tanker or pipeline, on-site production of LOX, or on-site generation of gaseous oxygen (GOX). LOX delivered in bulk is almost 100% pure oxygen with a dew point lower than -80°C and therefore does not normally require further treatment. It is stored on site in a vacuum insulated storage tank. Vaporisers which use ambient air, warm water, steam or electrical energy convert LOX to gas for use in the ozone generators.

LOX is produced on-site in pre-engineered packaged plants with capacities >20 t/day by cryogenic air separation where liquefaction of air is followed by fractional distillation to separate oxygen and nitrogen. This method is commonly used in commercial gas production plants where both oxygen and nitrogen are required as products. Oxygen concentration in the product gas is usually greater than 95%.

More commonly GOX is produced on site by pressure swing adsorption (PSA) or vacuum swing adsorption (VSA) processes. These comprise of one to three vessels containing a synthetic zeolite material which selectively adsorbs nitrogen from the air at elevated pressures allowing oxygen to pass through. Adsorbent also removes moisture, CO_2 and hydrocarbons. Once the bed is saturated, it is regenerated by subjecting the bed to a lower pressure to release nitrogen. A continuous stream of oxygen is maintained by switching the beds periodically at 30 second to 2 minute intervals. In PSA, adsorption is at 1–3 bar and regeneration is at atmospheric pressure whilst in VSA adsorption is at 0.2–1 bar and regeneration is under vacuum. The processes produce oxygen of purity 90–94% with low hydrocarbon content (<10 ml/m^3) and dew point lower than $-60°C$. If the system has less than three adsorbers an oxygen buffer storage vessel is usually provided. For VSA systems an oxygen booster compressor would be required for feeding the ozone generator. In general bulk LOX is the best economic choice for oxygen capacities up to 20 t/day. Absorptive processes are suitable for small to intermediate capacities (PSA: 5–20 t/day and VSA 20–60 t/day) and cryogenic plants are best for capacities of more than 60 t/day.

The output of ozone generators increases with the increase in oxygen concentration, improved quality of the feed gas (i.e. dryness, hydrocarbon content and dust content), decrease in the gas discharge gap between the electrodes and the increase in the frequency of the current applied to the dielectric.

The specific energy consumption for ozone generation is dependent upon a number of factors and is in the range 20–27 kWh per kg O_3 produced for low pressure air feed plant, and about 15 kWh per kg O_3 for oxygen feed plants. Specific energy consumption for on-site oxygen production processes are about 0.3–0.35, 0.3–0.4 and 0.4–0.45 kWh/kg O_2 for cryogenic, VSA and PSA, respectively. Ozone production plants require cooling water between 1.5 and 3 m^3/kg O_3 produced.

Ozone must be transferred from the gas phase to water efficiently to minimise losses. Therefore the design of the ozone-water contacting units is critical to the transfer process. In ozone contactors for disinfection fine bubble diffusers are commonly used to transfer ozone into water. They are either porous rod, dome or disc type and require submergence depths between 5.5–7.5 m. Transfer efficiencies achieved are usually greater than 93%. They are operated under the ozone generator discharge pressure. Clogging of the diffuser pores due to the precipitation of iron and manganese is a disadvantage and tendency to foul precludes their use in pre-ozonation applications if raw water suspended solids concentrations are high.

Alternatively, transfer is either by turbine mixers, static radial diffusers, venturis, or in-line static mixers.[40] The turbine mixers aspirate ozone-feed gas into the water and eject the resulting mixture into the contactor in a manner which encourages mixing with the bulk liquid.[32] In some designs the mixer motor is submerged. They require up to ten times the energy required for bubble diffusers. In static radial diffusers about 10% of the water flow is pumped to a submerged radial diffuser where it meets the ozone gas stream and divides this up into very fine bubbles when injected into the contactor through the diffuser head.[41] If a pressurised water flow is available, a vacuum can be created through the throat of a venturi to draw ozone gas into solution, resulting in excellent ozone transfer at the cost of pressure decrease for the treated flow. In-line static mixers are similar to those described in Section 7.12. If the flow into the contactor has a turn down of less than 2:1 a single in-line static mixer in the main pipe is adequate. If the turn down in flow is greater than 2:1, the main line static mixer needs to be supported by a side stream static mixer in a pumped recirculation system with ozone injected into the side stream.[42] The side stream flow is about 10% of the main stream flow. The main line static mixer requires about 2 m headloss and it should be followed by a contactor for oxidation and disinfection reactions to complete.

About 5–10% of the ozone introduced into the water remains in the spent gas at the top of the contactors. A destructor is required to convert unused ozone to oxygen for safe discharge. This conversion can be achieved by heating the gas to 350°C at which temperature decomposition takes place in 5 seconds, or by heating the gas 5–10°C above the inlet temperature (to decrease relative humidity) and passing it through a catalyst to accelerate decomposition. The higher temperature thermal system usually incorporates a heat recovery system; its main drawbacks are a longer start-up time and increased energy use. The catalytic system on the other hand can be prone to fouling.

Ozone is toxic and a dangerous gas to handle; its odour perception threshold is less than 0.02 ml/m^3. The 8-hour exposure limit is 0.1 ml/m^3.[19] Ozone leak detectors must be installed with facilities to shut down generators in the event of a leak. Other precautions must be taken in the design and layout to minimise hazards to health.[43] In addition ozonised air is highly corrosive in the presence of moisture; hence piping and other equipment must be of special materials, mostly stainless steel grade 316 S13. (British Standard BS 970 Part 1 or BS 1449 Part 2). Electrical plant and insulations may also need special protection against the possibility of ozone leakage.

Other disinfection practices
9.24   Ultra-violet radiation

Ultra-violet (UV) radiation is suitable for the disinfection of waters which are free of suspended matter, turbidity and colour. It is more effective against bacteria than viruses and bacterial spores.[13] UV radiation lies between 15 and 400 nm* wavelength, but UV-C radiation required for disinfection is achieved in the range 200–280 nm peaking between 240 and 280 nm with a maximum of 265 nm,[44] falling to zero at 320 nm. The mechanism of bacterial kill is not fully understood but it is believed that the UV radiation causes irreversible inactivation of the micro-organism's DNA (deoxyribonucleic acid) molecules rendering them unable to replicate. The degree of kill depends *inter alia* upon wavelength, the incident intensity (mW/cm^2) and contact time. UV dose (mWs/cm^2) is the product of UV-irradiance (incident intensity) (mW/cm^2) and the time of exposure (seconds) of the water to the radiation. The UV penetration depends upon the transmittance of the water and is highest in waters with very low natural colour; the concentration of other organic matter that respond to UV adsorption at 253.7 nm and the turbidity (<1.0 NTU) which scatters UV. Iron and manganese should be absent in raw water because they form a coating on the quartz tube; iron may also originate from the pipework. If a coating is present wipers could be fitted to clean the quartz. UV is best suited to groundwaters which require disinfection only. The exposure time for disinfection varies between 0.5–5 seconds

*1 nm = 10^{-9} m.

depending on the level of micro-organisms present and the optical transmissivity of the water. The minimum UV dosage level for effective bacterial disinfection (> 3-log removal) is generally agreed as 16 mWs/cm^2 at the reference wavelength of 253.7 nm which is the dosage specified by the US Department of Health. When viricidal action is also required, dosage levels of 30–40 mWs/cm^2 are considered more appropriate. At dosage levels greater than or equal to 40 mWs/cm^2 a 4-log reduction of bacteria and viruses is achieved.[45] The germicidal output of UV lamps slowly declines in use due to lamp ageing and fouling of the lamp sleeves. Typically lamp life is about 4000 hours for medium pressure lamps with a loss of initial intensity of about 30%. Low pressure lamps usually have an effective life of about 8000 hours with 60% loss of initial intensity. In sizing units this loss must be taken into account.

Recent work[46] in the USA has shown that UV is very effective at inactivating *Cryptosporidium* oocysts and 4-log inactivation at doses not exceeding 40 mWs/cm^2 using medium pressure lamps have been demonstrated. Other UV technology has been developed which provides much higher UV exposure than conventional systems. One such device which subjects *Cryptosporidium* oocysts to a total UV dosage of 8000 mWs/cm^2 in two passes using low pressure lamps is known to achieve greater than 4-log inactivation.[47]

UV radiation is produced by an electrical discharge (arc) in a low or medium pressure ionised, mercury vapour atmosphere. Lamps are made from fused silica quartz and are physically and thermally isolated from the water by an additional, UV transparent, quartz sleeve. The complete assembly is mounted in a stainless steel cylindrical chamber. The water to be disinfected is passed through the annular space between the quartz and the stainless steel tubes. A diagram of a UV chamber is given in Fig. 9.5 (see Plate 25).

Low pressure lamps are used in low flow or domestic applications. They produce a monochromatic output at a wave length of 253.7 nm and deliver about 85–90% of the energy input at this value. For medium pressure lamps only about a third of the energy input is delivered between 240 and 280 nm but the energy intensity is about 50 times greater than that for low pressure lamps.[45] Low pressure lamps require considerably less energy than medium pressure lamps and lasts up to 2.5 times longer.[48] The energy consumption of UV radiation using low pressure lamps is typically of the order of 10–20 Wh/m^3 of water treated and up to twice that for medium pressure lamps.

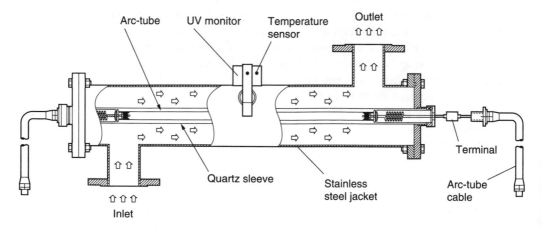

Fig. 9.5   UV treatment chamber (as Hanovia Ltd).

A typical system consists of one or several lamps per chamber with flow more commonly parallel, but in some cases perpendicular, to the axes of the lamps. A plant can consist of several parallel trains, each train containing one or more chambers in series. The number connected in series is determined by the dose required and the retention time; the number in parallel is governed by the water flow to be treated and includes standby facilities. Each chamber is provided with a UV monitor to ensure the correct level of UV-irradiance (mW/cm^2) is applied and to compensate automatically for deterioration in water quality or build up of deposits on the lamp. In contrast to chemical disinfection UV disinfection at the applied low doses in the effective wavelength range is not known to have any effect on the chemical composition of the water.[49] At wavelengths of less than 240 nm however it can enhance the formation of by-products by photolysis of organic and inorganic contaminants in the water.[44] An example is the formation of nitrite when nitrate is present in water. This is a particular risk at high disinfection doses and when flow through UV chamber is intermittent. The problem could be eliminated in the manufacture of UV sources by using quartz that suppress these short wavelengths. Phosphates used for plumbosolvency control should be dosed after UV application. Degradation of organic compounds could be achieved with high energy inputs (see Section 8.36). UV does not produce a disinfectant residual and if this is required chlorination or chloramination must be practised.

9.25 Boiling water

Boiling water is an extremely useful process for disinfecting water because boiling kills bacteria, viruses, ova and cysts present in polluted water. In an emergency it may be necessary for a water undertaking to advise consumers to boil all water used for drinking or cooking or brushing teeth. It is reported to be equally effective whether the water is clear or cloudy, relatively pure or highly contaminated (though obviously contaminated or cloudy water should be used only as a 'last resort'). When a boiled water notice is issued by a water undertaking because a normally safe supply is contaminated with micro-organisms, it is sufficient to bring the water to boil and the use of electric kettles with automatic switch-off is acceptable.[50] In the UK, immunocompromised individuals are advised to boil all drinking water from any source.[50] For the complete sterilisation of a polluted water the WHO recommends bringing the water to 'rolling boil' (large bubbles continuously coming to the surface) and then maintaining this for at least 1 minute for a clear water. At high elevations 1 minute extra time should be given for every 1000 m above sea level because of the lower temperature at which boiling takes place.[51] Turbid water should preferably be filtered through a clean cloth before boiling. Alternatively the water should be boiled for up to 5 minutes. Boiled water can become recontaminated once it has cooled and therefore should be stored in a clean closed container.

Disinfection of waterworks facilities
9.26 Disinfection of water mains and tanks

The UK recommended practices for disinfecting mains have been set out in Technical Guidance Notes published by UK Water.[52]

Before a new or renovated main is put into service it should first be swabbed clear of dirt and debris with a foam swab and flushed with water. It should then be filled with water containing about 20 mg/l of free chlorine and allowed to stand for 16 hours, after which it should be flushed and recharged with mains water and allowed to stand for a further 24 hours. Samples should then be taken from a number of points along the main and at its extremities and, if samples are found to be free of coliform organisms and give satisfactory results for residual chlorine, taste, odour and appearance, the main can be brought into service. All service pipes connected to the main should be flushed out. Alternatively a renovated main may be disinfected for a minimum period of 30 minutes with 50 mg/l of free chlorine followed by flushing, filling with mains water and sampling. A similar treatment is advised for an operation which involves cutting the main, and where there is a risk of contamination from water in the trench or other foul water (e.g. from sewer). For repairs to a live main which has to be cut with minor soiling at opening, all surfaces which will come in contact with the drinking water are to be cleaned down with water containing 1000 mg/l free chlorine. In all cut main repairs, following disinfection and flushing the main must only be returned to service when bacteriological and other qualitative tests have proved satisfactory.

Tanks and service reservoirs are usually hosed down with strong jets of clean water and the walls and floor are then brushed down with a chlorine solution containing not less than 20 mg/l chlorine. Initially a reservoir is half filled with water containing at least 0.5 mg/l chlorine and, after standing for 24 hours, a sample of the water is tested for coliform organisms. If the sample fails, the chlorine dose of the inlet water is increased to 1.0 mg/l and fill the reservoir, the water being retested after a further 24 hours. If coliforms are absent, the reservoir can be filled with water having 0.5 mg/l chlorine, which is resampled and tested when the reservoir is full. Personnel entering a service reservoir which has been emptied for inspection, etc. should scrub their footware in a tray containing 1000 mg/l chlorine immediately before entering the reservoir. Normally chlorine tablets, bleaching powder, calcium hypochlorite granules or sodium hypochlorite are used as chlorine sources, but for large new transmission mains and reservoirs chlorine water from a gas chlorinator might be injected into the filling water. All test water containing chlorine should be dechlorinated before disposal; chemicals available include sodium bisulphite, sulphur dioxide, sodium sulphite and sodium thiosulphate.

9.27   Control of aftergrowth in distribution mains

Considerable aftergrowth may occur in distribution mains especially with nutrient rich waters. Slime, iron and manganese deposits, algae, and corrosion products may foster growth of bacteria and other forms of life within mains as biofilms which are resistant to control measures.[53] In such systems it is often difficult to retain any residual chlorine in the water. No practicable level of residual chlorine will prevent problems of aftergrowth if algae, residual iron or aluminium floc, suspended solids or substantial amounts of dissolved organic matter (see Section 8.35) are allowed to remain in a treated water. Phosphates used in plumbosolvency control can also contribute to aftergrowth. Swabbing followed by slug dosing with a heavy dose of chlorine passed slowly along the mains affected is a short-term remedy, but the permanent cure lies in optimising the treatment process and thus improving the quality of the water distributed.[54] In extensive distribution systems carrying nutrient-rich waters it may be necessary to adopt 'booster chlorination', i.e. the addition of further chlorine or chloramine to the water at some key distribution point or points in the system, usually at the inlet or outlet of some service reservoir.

Self-contained packaged plant for electrolytic sodium hypochlorite generation. Output up to 2 kg/day chlorine. (Manufacturers: USF Wallace & Tiernan, Tonbridge, UK)

Ozoniser with end cap opened, showing nest of interior electrode tubes. (Manufacturers: Ozotech Ltd, Burgess Hill, UK)

On-site hypochlorite generating plant at Heigham treatment works, Anglian Water UK. Output 320 kg/day chlorine

A 20 kW ultra-violet disinfection plant treating up to 40 Ml/day of water. (Manufacturers: Hanovia Limited, Slough, UK)

Plate 25

(Above left) A 2270 m^3 capacity concrete water tower used on Malaysian rural water supplies project, 1992. (Engineers: Syed Muhammad, Hooi dan Binnie)

(Above right) A 95 m^3 capacity elevated concrete tank for the roof rainwater collection scheme for Male, Maldives, 1988. (Engineers: Binnie & Partners)

(Left) A 1010 m^3 29 m high elevated concrete tank for a 1350 Ml/day water supply in the Middle East 1989. (Engineers: Binnie & Partners)

Plate 26

Chloramination has also become more common, where there is a need to control biofilms and avoid exceeding chlorination by-product (THMs) concentrations. This has become essential in some utilities.[55] As noted in Section 6.63 weekly bacteriological testing of water in service reservoirs is required under the UK Water Regulations.

9.28   Disinfestation of distribution mains, wells and boreholes

Some distribution systems can be become infested with small aquatic animals, especially systems with old mains carrying treated lowland surface waters which contain considerable organic matter. The most commonly experienced animals are – *Asellus* the 'water louse'; *Gammarus* the 'freshwater shrimp'; nais worms and nematode worms. Occasionally the larvae of midges and flies may be found, having passed through filter beds or gained entry to a service reservoir. Many other small aquatic organisms can occasionally be present. The flushing of mains and chlorination is largely ineffective where animal growth is prevalent. The ability of the animals to leave reproductive spores or to reproduce from fragments means that reinfestation tends to be rapid: hence disinfestation is necessary. Formerly pyrethin was used for this purpose, but nowadays permethrin, a synthetic pyrethroid, is used. The application has to be most carefully controlled; the average and maximum concentration must not exceed 10 and 20 μg/l, respectively, under the UK Water Regulations[56] and must not be applied for longer than 7 days. The Secretary of State must be notified prior to the use; and dialysis users, aquarium owners, beekeepers and hospitals must be forewarned of the intention and, if necessary, assisted to obtain temporary alternative supplies whilst the disinfestation is proceeding.[57]

In cases where corrosion of the main is the cause of high turbidity or coloured water flushing and pigging with polyurethane foam swabs have been shown to be effective.[58]

In mains iron bacteria may develop, and occasionally other types of bacteria may multiply into substantial colonies, of which one of the aeromonas strains, Aeromonas hydrophila, is believed capable of causing gastroenteritis.[59] The bacterial infestation of mains can usually be dealt with by flushing and swabbing mains, and then chlorinating them. This has usually to be done in sections when, as is likely, a whole system is affected. Maintenance of an enhanced chlorine residual can assist in preventing regrowth, but this practice may be inhibited by the need to keep THMs below the permitted maximum (see Section 6.21). In the case of iron bacteria in mains, the proper remedy is to revise the treatment of the water to remove the iron or, if the bacteria grow because of deterioration of old iron mains, relining them is effective.

Iron bacteria are particularly prone to develop in wells and boreholes drawing water from ferruginous formations. The bacteria are sessile i.e. attach themselves to a surface, and the majority produce large masses of extra-cellular covering material in the form of slime which clings to well screens and borehole linings. Often the presence of large iron bacterial growths remains unevidenced until part of the slime detaches and is discovered in the water, or the yield of the well or borehole falls off because of clogging of screens. The slimes protect the bacteria against any biocide, hence physical removal of the slimes is necessary. Pumps and rising mains in wells and boreholes can be withdrawn and cleaned. But to clean well screens, surging, jetting, chemical applications, or steam injection may have to be adopted. Prevention of regrowth in well screens and borehole linings after physical cleaning may have to comprise slug-dosing with chlorine and then flushing. Chlorine should never be applied direct to a well or borehole; this prevents knowledge of the degree of pollution of the water and may encourage corrosion of the well or borehole lining and of the pump inserted. The continuous application of chlorine may also be inadvisable since it will tend to precipitate the iron in solution.

References

1. Whitlock E. A. The Application of Chlorine in the Treatment of Water. *JWWE*, Jan. 1953, p. 12.
2. Butterfield C. H. *et al*. *US Public Health Reports 58, 59, 61*, 1943–46; also *JAWWA*, **40,** 1948, p. 1305.
3. WHO, *Guidelines for Drinking-Water Quality*, Vol. 1, Recommendations, 2nd edn. WHO, Geneva, 1993.
4. White G. C. *Handbook of Chlorination*, 2nd edn. Van Nostrand Reinhold, New York, 1992.
5. WHO, *Disinfection of Rural and Small-Community Water Supplies*, WRc, Medmenham, England, 1989.
6. Stevenson D. G. *Water Treatment Unit Processes*. Imperial College Press, London, 1998, p. 96.
7. USEPA, National Primary Drinking Water Regulations: Filtration, Disinfection, Turbidity, *Giardia lamblia*, Viruses, *Legionella* and Heterotrophic Bacteria. Federal Regulation 54:124:27486, 29 June, 1989, and see Reference 31.
8. Falconer R. A. and Tebbutt T. H. Y. A Theoretical and Hydraulic Model Study of a Chlorine Contact Tank. *Proc ICE*, Part 2, **81**(6), 1986, pp. 255–276.
9. Stevenson D. G. The Design of Disinfection Contact Tanks. *JCIWEM*, **9**(2), 1995, pp. 146–152.
10. Dawson M. Cleaning Up with Disinfection Design Software. *Water Treatment and Supply*, **11,** 1998, pp. 14–15.
11. Poynter S. F. B., Slade J. S. and Jones H. H. The Disinfection of Water with Special Reference to Viruses. *JSWTE*, **22,** 1973, pp. 194–206.
12. Scarpino P. V. *et al*. A Comparative Study of the Inactivation of Viruses in Water by Chlorine. *Water Research*, August 1972, p. 959.
13. WHO. *Guidelines for Drinking Water Quality*, Vol. 2, *Health Criteria and Other Supporting Information*, 2nd edn, Geneva, 1996.
14. Smith H. V. *et al*. The Effect of Free Chlorine on the Viability of *Cryptosporidium* Oocysts. WRC report PRU 2023-M, 1989.
15. Smith D. J. The Evolution of an Ozone Process at Littleton Water Treatment Works. *JIWEM*, **4**(4), 1990, pp. 361–370.
16. Lieu N. I. *et al*. Optimising Chloramine Disinfection for the Control of Nitrification. *JAWWA*, **85**(2), 1993, p. 84.
17. Palin A. T. Chemical Aspects of Chlorine. *JIWE*, **IV**(7), 1950, pp. 565–581.
18. Health & Safety Executive. *Safety Advice for Bulk Chlorine Installations*, HS/G28, HMSO, London, 1999.
19. Health & Safety Executive. *Occupational Exposure Limits*, Guidance Note EH40/92, HMSO, London, March 1998 (published annually).
20. Health & Safety Executive. *Safe Handling of Chlorine from Drums and Cylinders*. Guidance Note HS/G40 (rev.), HMSO, London, 1999.

21. Control of Major-Accident Hazards Involving Dangerous Substances (COMAH), *Council Directive 96/82/EC*, Official Journal 14.1.97.
22. Health & Safety Executive. *Fire and Explosion Hazards at Electrochlorination Plant*. Document No. HSE 490/11, Nov. 1987.
23. Walker, G. S. *et al*. Chlorine Dioxide for Taste and Odour Control. *JAWWA*, **78**(3), 1986.
24. Knocke W. R. *et al*. Kinetics of Manganese and Iron Oxidation by Potassium Permanganate and Chlorine Dioxide. *JAWWA*, **83**(6), 1991, pp. 80–87.
25. Aieta E. C. and Berg J. A Review of Chlorine Dioxide in Water Treatment. *JAWWA*, **78**(6), 1986, pp. 62–72.
26. Rittmann D. D. Can You Have Your Cake And Eat It Too With Chlorine Dioxide? *Water Engineering & Management*, April 1997, pp. 30–35.
27. Bernarde M. A. *et al*. Efficiency of Chlorine Dioxide as a Bactericide. *Applied Microbiology*, **13**(9), 1965, p. 776.
28. Lykins B. S. *et al*. Concerns With Using Chlorine Dioxide Disinfection in the USA. *J Water SRT, Aqua*, **39**(6) 1990, pp. 376–386.
29. Condie L. W. Toxicological Problems Associated With Chlorine Dioxide. *JAWWA*, **78**(6), 1986, pp. 73–78.
30. Hurst G. H. and Knocke W. R. Evaluating Ferrous Ion for Chlorite Removal. *JAWWA*, **89**(8), 1997, pp. 98–105.
31. USEPA, *Guidance Manual for Compliance with Filtration and Disinfection Requirements for Public Water Systems Using Surface Water Sources*, Appendix O, Guidance to predict performance of ozone disinfection system. Oct. 1990.
32. Langlais B. *et al*. (eds). *Ozone in Water Treatment, Application and Engineering*. AWWA Research Foundation, Lewis Publication, Michigan, 1991, p. 414.
33. Foster D. M., Rachwal A. J. and White S. L. New Treatment Processes for Pesticides and Chlorinated Organics Control in Drinking Water. *JIWEM*, **5**(4), 1991, pp. 466–477.
34. Tobiason J. E. *et al*. Effects of Ozonation on Optimal Coagulant Dosing in Drinking Water Treatment. *J Water SRT-Aqua*, **44**(3), 1995, pp. 142–150.
35. Bourgine F. P. *et al*. The Effect of Ozonation on Particle Removal in Drinking Water. *JCIWEM*, **12**(3), 1998, pp. 170–174.
36. *Proceedings, Seminar on Ozone in UK Water Treatment Practice*, Papers by Hyde and Zabel, and Greene *et al*. IWES Sept 1984.
37. Siddigui M. S. *et al*. Bromate Ion Formation: A Critical Review. *JAWWA*, **87**(10), 1995, pp. 58–70.
38. Ozekin K. *et al*. Molecular Ozone and Radical Pathways of Bromate Formation During Ozonation. *J Environmental Eng ASCE*, **124**(5), 1998, pp. 456–462.
39. Kruithof J. C. and Meijers R. T. Bromate Formation by Ozonation and Advanced Oxidation and Potential Options in Drinking Water Treatment. *Water Supply*, **13**(2), 1995, pp. 93–103.
40. Davidson C. *et al*. New Concepts in Ozone Systems at North Surrey Water – Design, Construction and Early Operating Experience. Paper to SE Branch of CIWEM, Dec. 1997.
41. Anon. Ozonia Radial Diffuser. *Ozonia Product Information*, Zurich, 1994.

42. Boisdon V. *et al*. The Effect of Static Mixer on Mass Transfer in a New Design of Ozone Reactor. *12th World Congress*, International Ozone Association, 15–18 May 1995, Lille, France.

43. Health and Safety Executive, *Ozone Health Hazards and Precautionary Measures*, Guidance Note EH 38 (revised), HMSO, London, 1996.

44. Von Sonntag C. and Schuchmann H. P. UV Disinfection of Drinking Water and By-Product Formation: Some Basic Considerations. *J Water SRT-Aqua,* **41**(2), 1992, pp. 67–74.

45. Hoyer O. Testing Performance and Monitoring of UV Systems for Drinking Water Disinfection. *Water Supply*, **16**(1/2), 1998, Madrid, pp. 419–442.

46. Bukhari Z. *et al*. Medium Pressure UV for Oocyst Inactivation. *JAWWA*, **91**(3), 1999, pp. 86–94.

47. Dyksen J. E. *et al*. Cost of Advanced UV for Inactivating *Crypto*. *JAWWA*, **90**(9), 1998, pp. 103–111.

48. Degremont, *Water Treatment Handbook*, 6th edn, Lavoisier, Paris, 1991, p. 911.

49. Cairns W. L. UV Technology for Water Supply Treatment. *Water Supply*, **13**(3/4), Osaka, 1995, pp. 211–214.

50. *Cryptosporidium in Water Supplies*. Third Report of the Group of Experts, HMSO, London, 1998.

51. WHO, *Guidelines for Drinking Water Quality*, Vol. 3, *Surveillance and Control of Community Supplies*. WHO, Geneva, 1997.

52. UK Water, *Technical Guidance Notes 1 to 7*. UK Water, October 1998.

53. Fleming H. C. Biofilms in Drinking Water System. *GWF Wasser/Abwasser*, **139**(13), 1998, pp. 65–72.

54. Mouchet P. *et al*. Physico-Chemical Degradative Changes in Water While in the Distribution System. *Techniques, Sciences, Methods*, **6**, 1992, pp. 299–306.

55. Norton C. D. and Le Chevallier M. W. Chloramination, Its Effect on Distribution Water Quality. *JAWWA*, **89**(7), 1997, pp. 66–77.

56. *List of Substances, Products and Processes Approved under Regulations 25 and 26 of the Water Supply (Water Quality) Regulations, 1989*, HMSO, London.

57. WRc. Permethrin for the Control of Animals in Water, *Technical Report TP 145*, 1980.

58. Smith G. Aruba. A Case Study in Pipeline Maintenance. *Water & Wastewater Int.*, **1**(3), 1986, pp. 4–5 and 7.

59. Edge J. C. and Finch P. E. Observations on Bacterial Aftergrowth in Water Supply Distribution Systems. *JIWEM*, Aug. 1987, p. 104.

10
Hydraulics

10.1 The energy equation of fluid flow

A fluid moves, in accordance with Newton's laws, under the action of external forces. If there is a net force acting on an element of the fluid, then that element will either accelerate or decelerate depending on the direction of that force; or, if the forces are in balance, then the element will remain at rest or at the same velocity. There is a resistance to motion, however, in the form of drag on that element of fluid and, in moving, energy is expended in overcoming that drag. This expenditure of energy appears in the form of turbulence in the water created by the drag of the surfaces of the conduit and by any obstructions or changes to the shape and direction of the conduit. As the eddies of turbulence decay, their kinetic energy is transmitted to the motion of individual water molecules, so the temperature of the water increases by a very slight amount, scarcely detectable and too small to be of any practical use. The balance of energy remaining is that in which the engineer is interested, because it is this which determines the subsequent level, pressure, and kinetic energy of flow.

The energy of a unit mass ρ of water can be expressed as

$$E = \underset{\substack{\text{kinetic} \\ \text{energy}}}{\rho u^2/2} + \underset{\substack{\text{pressure} \\ \text{energy}}}{p} + \underset{\substack{\text{potential} \\ \text{energy}}}{\rho g z} \tag{10.1}$$

where u is the velocity of the water, p its pressure and z its height above some given datum. For convenience this expression is divided by ρg so that the energy and pressure are expressed in terms of a height, or 'head' of water to give the expression in the form:

$$\frac{E}{\rho g} = H = \frac{u^2}{2g} + h + z \tag{10.2}$$

For flow between two points, A and B, as shown in Fig. 10.1

$$H_A = H_B + H_L$$

where H_L is the energy head lost by the water flowing from A to B.
Hence

$$\frac{u_A^2}{2g} + h_A + z_A = \frac{u_B^2}{2g} + h_B + z_B + H_L \tag{10.3}$$

This is the general form of the energy equation and is fundamental to almost all hydraulic calculations. It will be necessary to return to it many times.

This general energy equation is also referred to as the modified Bernoulli equation. Bernoulli's equation itself, which is one of the most widely quoted equations in fluid

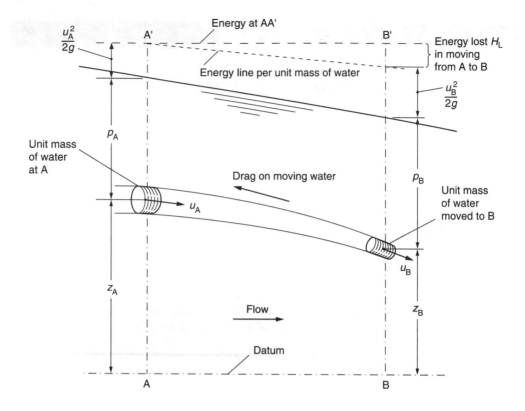

Figure 10.1   Variation in energy as unit mass of water moves from position A to B.

dynamics, is a particular case of this general energy equation, in which there is no energy loss between points 1 and 2.

Bernoulli's equation is:

$$\frac{u^2}{2g} + h + z = \text{constant}$$

The limitations of Bernoulli's equation must be carefully noted. It applies only to steady flow within a streamline and to flow where no energy is lost through turbulence. In practice this means that it can only be applied as an approximation over short distances; in all civil engineering applications, there is a loss of energy as flow moves from point A to point B. The problem faced is how much energy is lost and how accurately can it be calculated.

10.2   Boundary layers

The concept of streamlines is most relevant to the idealised condition of flow moving uniformly in a large body of water. Away from the influence of any solid boundary, a particle of water will follow its streamline. This condition is known as potential flow. However, when a boundary is introduced into the flow the situation is different. The water in immediate contact with the solid surface must be stationary. Moving away from the surface the velocity increases until at some distance from the surface it is essentially back to its undisturbed value. Thus there is a region adjacent to the surface in which there is a velocity gradient, with adjacent streamlines at different velocities. This region is known as the boundary layer. The thickness of the boundary layer depends on a number of factors but the boundary layer can be envisaged as growing from the point of contact of the flow with the solid boundary. In a conduit of finite size the boundary layer will increase in the direction of flow until it fills the full depth of flow in an open channel or the full cross-section of a pipe.

The velocity profile within a boundary layer is primarily a function of the size of the conduit, the velocity of flow and the density and viscosity of the fluid. These parameters are combined in a single dimensionless grouping known as the Reynold's number *Re* where:

$$Re = \frac{VD}{\nu}$$

V being the average velocity of flow, *D* some representative dimension of the flow (for example, the depth of flow in an open channel or the diameter of a pipe) and ν is the kinematic viscosity of the fluid, defined as μ/ρ where μ is the viscosity and ρ the density.

At low Reynold's numbers the flow is described as *laminar*. In laminar flow viscosity is dominant and the momentum and inertia of the flow have little effect. The boundary layer is small and the velocity gradient within it may be high. A good example is treacle flowing over an object. Laminar flow occurs rarely in water supply systems. To put this in perspective, laminar flow occurs when the Reynold's number is below about 2000. Since the value of the kinematic viscosity of water is about 1.1×10^{-6} m²/s at 15°C, this requires that $V \times D$ be less than about 0.002 m. Thus, for example, it would apply to a case where the flow velocity was less than 0.02 m/s in a pipe of 100 mm diameter.

As the Reynold's number increases so turbulence becomes increasingly important. When the Reynold's number increases above about 2000 there is a sudden transition in the flow as turbulence begins to affect the motion, and laminar conditions no longer apply. Resistance to motion then increases markedly. Viscosity still remains a major influence at Reynold's numbers beyond this transition and the flow is described as 'smooth turbulent'. As the Reynold's number increases further so the effect of viscosity gradually reduces and the influence of turbulence increases. At a higher Reynold's number, depending on the roughness of the boundary, the effect of viscosity becomes negligible and the flow is said to be '*rough turbulent*'. The region between smooth and rough turbulent flow is described as the '*intermediate zone*' where both viscosity and turbulence have an influence.

This relationship between resistance to flow and Reynold's number is best illustrated by the diagram developed by Moody,[1] which is shown in Fig. 10.2 for pipe flow. The flow resistance is characterised by the friction factor, *f*, which will be explained later. In the laminar flow region, *f* is a function solely of the Reynold's number but above the transition region, *f* is a function of both Reynold's number and the roughness of the pipe surface, which is expressed in terms of the relative roughness, k_s/d, where k_s is a linear measurement of the surface roughness and *d* the pipe diameter.

Turbulence is generated by water moving at a different velocity from that in a streamline adjacent to it. This generates a shear force between the streamlines tending to retard the faster flow and speed up the slower flow. Thus, in a boundary layer, the faster flow is dragged down towards the boundary and the slower flow pulled away from the boundary,

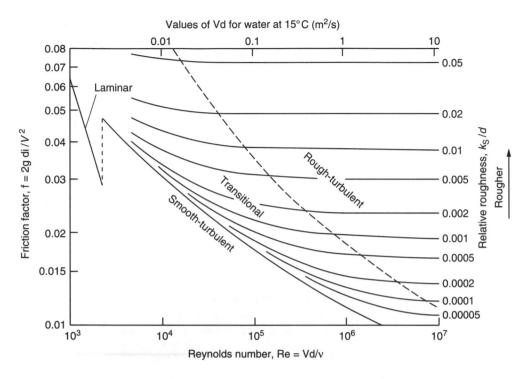

Figure 10.2 Moody diagram illustrating the variation of hydraulic resistance with Reynold's number (*i* is the gradient of the energy line, the 'hydraulic gradient').

creating a lateral velocity component and the formation of turbulent eddies. There is a transfer of momentum across the streamlines and the velocity profile is no longer linear but approximately logarithmic with zero value at the boundary, as noted above.

10.3   Pipe flow

The engineer is concerned with two types of flow – open-surface channel flow and closed conduit flow. In the former the depth of flow can vary. In the latter the area of flow is fixed and for a known flow in a given size of conduit the velocity can be calculated directly. Pipe flow will be considered first as it is more straightforward.

At the entry of water into a pipe from a large tank or reservoir, the flow, as it accelerates into the inlet, approximates to the idealised condition of potential flow. However a boundary layer is generated from the lip of the inlet and, within a relatively short distance downstream of the entry, this has expanded to fill the whole pipe. At Reynold's numbers just above the laminar transition the velocity profile will be logarithmic from the boundary wall to the centreline of the pipe. At higher Reynold's numbers the turbulence is such that momentum is transferred across the streamlines. Across much of the section the time-averaged velocity will be constant but there will be steep velocity gradients near the pipe wall.

Figure 10.3(a) illustrates the velocity profile across a pipe operating in the smooth turbulent zone. At the centre of the pipe the velocity is greatest and is about 1.2 times the

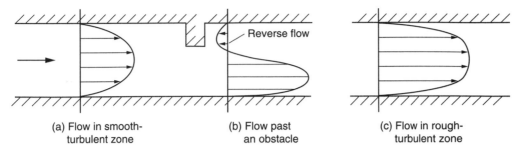

(a) Flow in smooth-   (b) Flow past   (c) Flow in rough-
    turbulent zone   an obstacle   turbulent zone

Figure 10.3   Velocity profiles in a pipe flowing full.

average value, although the ratio of the maximum speed of the flow to the average varies as a function of the Reynold's number and the roughness of the walls. Of course, it may vary very greatly if an obstacle in the pipe affects the flow as illustrated in Fig. 10.3(b). Figure 10.3(c) shows the velocity profile across a pipe operating in the rough-turbulent zone, with a much more uniform velocity across much of the cross-section, but with steep velocity gradients close to the walls.

The average velocity in a pipe is simply the rate of discharge divided by the cross-sectional area of the pipe, i.e. Q/a, where Q is the discharge and a is the cross-sectional area. This average velocity is denoted as V to distinguish it from the velocities in individual streamlines, u_1, u_2, u_3, etc. The energy equation applies to any streamline flow but to apply the same equation to the whole flow within a pipe the energy equations for the separate streamlines must be summed. Thus the total kinetic energy of the flow is:

$$\sum(u_1^2/2g + u_2^2/2g + u_3^2/2g + ... + u_n^2/2g)$$

However this sum does not equal $V^2/2g$. The ratio of $(\Sigma u^2/2g)/(V^2/2g)$ is therefore given the symbol of α and the total kinetic energy of the flow is $\alpha V^2/2g$. In a similar fashion the summation of the pressure energies is βP, where P is the average pressure across the cross-section of the pipe. As it is a linear function, the average potential energy in a circular pipe is given by the level of the centre of the pipe, provided the pipe is flowing full. Thus for the whole pipe flow the energy equation is correctly:

$$H = \alpha V^2/2g + \beta h + z$$

where V and P are the average velocities and pressures across the section, respectively, and Z is the level of the centre of the pipe above some datum. In practice, the values of α and β are close to unity and in most practical applications the errors involved in ignoring the difference are so small as to be insignificant. Hence this assumption permits the flow at any section of a pipeline to be related to that at any other point of the same line.

Units used

As indicated in Section 10.1, for water supply systems the common practice is to express the terms in the energy equation in '*head of water*'. In metric units this is normally expressed in metres. This is convenient as levels are expressed in the same form. Thus, for example, the difference in level between two reservoirs connected by a pipeline can be considered as a direct measure of the available potential energy between the two locations. In metric units, if the energy, H is expressed in 'metres head of water' then z must also be expressed in metres and the kinetic energy in the same units. This requires that the velocity is in m/s and the gravitational acceleration, g, is 9.81 m/s^2.

10.4   Headlosses in pipes (1) – the Colebrook–White formula

The energy equation for the total flow in a pipe can now be expressed as:

$$V_1^2/2g + h_1 + z_1 = V_2^2/2g + h_2 + z_2 + H_L \qquad (10.4)$$

where H_L is the energy loss in the pipeline between the two sections 1 and 2. Knowing the total energy level at location 1 (for example, the level of a reservoir) the energy level at any other point can be calculated if the energy loss H_L can be estimated.

The energy lost through turbulence is caused by two mechanisms – (i) the drag of the pipe walls on the flow, and (ii) turbulence generated whenever there is a change to the direction or area of the flow. A flowing fluid has momentum and does not want to change direction. Thus any change to the angle of the boundary walls, particularly if the boundary turns away from the direction of flow, may lead to the flow 'breaking away' from the surface leaving an area of turbulence, the 'wake'. The more abrupt the boundary change the greater is the potential for energy loss. The former mechanism for energy loss is known as the hydraulic resistance or 'friction' loss; the latter is the 'form' loss due to the geometry of the change of cross-section or obstruction in the flow. The friction losses are continuous over the length of a pipeline; the form losses are localised in the immediate vicinity of the element causing the energy loss and are also referred to as 'local losses'.

The loss of energy due to the hydraulic resistance of a pipe is a function of the velocity of the flow, V, the internal diameter of the pipe, d, the length of the pipe, L, the roughness of the surface of the pipe and the characteristics of the flowing fluid , expressed in terms of the kinematic viscosity, ν.

Darcy–Weisbach formula

A dimensionally correct formula for the head loss is the Darcy–Weisbach equation which gives the head loss in a length of pipe as:

$$H_L = fLV^2/2gd \qquad (10.5)$$

where f is a non-dimensional coefficient, known as the *friction factor*, which includes the effects of pipe wall roughness and the fluid viscosity. (Note that the friction factor f is also commonly designated by λ.) Unfortunately f is not constant but varies with the size of pipe and the degree of turbulence of the flow.

Colebrook–White formula

Colebrook and White showed that f in the Darcy–Wiesbach formula is a function of the relative roughness of the pipe surface, the viscosity of the flow, and the Reynold's number, R_e. From a combination of theoretical analysis and empirical data they showed that

$$\sqrt{(1/f)} = -2\log_{10}\{k_s/3.71d + 2.51/Re.f^{0.5}\}$$

which can also be written as:

$$\sqrt{(1/f)} = -2\log_{10}\{k_s/3.71d + 2.51\nu/d(2gdi)^{0.5}\}$$

where i is the *hydraulic gradient*, H/L; k_s is the roughness of the internal surface of the pipe; ν is the kinematic viscosity of the water.

This variation of f with Reynold's number and relative roughness was plotted by Moody and forms the basis of the diagram illustrated in Fig. 10.2.

The Colebrook–White equation is a formulation of the Darcy–Weisbach equation and is usually written as:

$$V = -2(2gdi)^{0.5}.\log_{10}\{k_s/3.71d + 2.51\nu/d(2gdi)^{0.5}\} \tag{10.6}$$

This equation can be solved directly only for V (and hence Q), knowing d and i. More commonly it is required to find i, knowing Q and d. The equation is then not explicit but must be solved via an iterative technique of successive approximations. Whilst this used to be a major objection to its widespread use, now with programmable calculators and PC-based spreadsheets, it is no longer a significant drawback. The Colebrook–White equation, or one of the approximations given below allowing a direct solution, is now widely used and is recommended as the formula that should be used to estimate pipeline head losses. One of the advantages of this equation over the empirical formulae discussed below is that the roughness coefficient, k_s is a function only of the surface roughness of the pipe and does not change with the size of the pipe or velocity of the flow. The factor k_s is sometimes referred to as the *equivalent sand roughness* because the original experiments carried out by Nikuradse (the data from which Colebrook and White used in the development of their formula) utilised sand grains stuck to the inside of the pipes. The value of k_s is meant to represent the equivalent diameter of the sand particles giving that degree of roughness. Although this is only a notional concept, it does provide a physical meaning to the roughness measurement which does not apply to the coefficients in any of the empirical formulae given in Section 10.5 below.

A second advantage of the Colebrook–White formula is that it applies over the full range of turbulent flow from the smooth turbulent condition at Reynold's numbers as low as 3×10^3 to the rough turbulent flow condition at Reynold's numbers in excess of 1×10^7. The first term in the logarithmic function, $k_s/3.71d$, represents the effect of the pipe roughness and dominates at high Reynold's numbers, whilst the second term includes the dynamic viscosity and dominates at low Reynold's numbers.

There are approximations to the Colebrook–White formula that allow direct calculation of d or i knowing the flow, Q, and the other of the two parameters. These provide an accuracy within 0.5% – well within the accuracy with which the roughness is known. Two such equations which can be recommended are those of Barr, to solve for i, and Pham, to solve for D. Thus the three explicit equations can be written in the following similar format.

The Colebrook–White equation, to solve for Q:

$$\frac{V}{\sqrt{(2gdi)}} = \frac{0.9003Q}{d^2\sqrt{(gdi)}} = -2\log_{10}\left\{\frac{k_s}{3.71d} + \frac{2.51\nu}{d\sqrt{(2gdi)}}\right\} \tag{10.7}$$

The Barr approximation, to solve for i:

$$\frac{0.9003Q}{d^2\sqrt{(gdi)}} = -1.9\log_{10}\{(k_s/3.71d)^{1.053} + (4.932\nu d/Q)^{0.937}\} \tag{10.8}$$

The Pham approximation, to solve for d:

$$\frac{0.9003Q}{d^2\sqrt{(gdi)}} = -1.884\log_{10}\{0.365(g.i)^{0.2}k_s/Q^{0.4} + 3.55\nu/Q^{0.6}(g.i)^{0.2}\} \qquad (10.9)$$

Units must be consistent, e.g. d (m); Q (m^3/s); g (9.81 m/s^2) and k_s must be in m (although quoted below in mm). The kinematic viscosity, ν, of clean water is 1.310×10^{-6} at 10°C, and 1.011×10^{-6} at 20°C.

For *water mains* there is considerable guidance available on the choice of k_s values to adopt for design.[2] Typical values for new clean pipes are as follows:

	For new pipes
Steel or ductile iron pipes:	
with spun bitumen or enamel finish	0.025–0.05 mm
with cement mortar lining	0.03–0.1 mm
Concrete pipes	0.03–0.3 mm
Plastic pipes	0.003–0.06 mm

For design purposes to allow for deterioration of interior condition the following values are indicative of what might be used:

	For design
Raw water mains	1.5–3 mm
Treated water trunk mains	0.3–1.0 mm
Distribution systems	0.5–1.5 mm

For *service pipes* typical values of k_s are

	For new pipes
Galvanised steel	0.06–0.3 mm
Copper	0.002–0.005 mm
MDPE	0.003–0.006 mm
uPVC	0.003–0.06 mm

However (i) k_s values below 0.01 mm show no significant change in V or i from that at 0.01 mm; (ii) loss at joints, elbows, tees, etc. on the line to the consumer's tap may add 50–70% to pipe losses (see Table 15.5, Chapter 15); and (iii) it is not possible to quote k_s values for old service pipes due to the wide range of interior conditions that can apply.

In raw water mains there may be a tendency for organic slimes to develop on the walls. This slime tends to be ripped off as the velocity increases and there is evidence that the roughness reduces at higher flows. The designer of a new pipeline must make a judgement as to whether this effect can allow a reduction in the design head losses. It may also be necessary in a raw water pipeline to allow for some increase in roughness due to presence of sediment in the invert and possibly fresh water organic growth such as mussels. The latter may greatly increase the roughness. Also if the water quality is such that tuberculation is likely over the life of the pipeline, then much higher values may be appropriate.

The development of computer programs for the analysis of water distribution system flows allows a much more detailed approach. Such programs allow calibration of network models by the adjustment of roughnesses in each pipe. Once a model using the Colebrook–White formula is calibrated then the prediction of the system's performance under other flow regimes or the design of improvements can be undertaken with some confidence, without the need for adjustment of the roughness values. However it is important to remember that measurements are rarely more accurate than $\pm 5\%$ in practice, with errors often greater. Some network programs still cater for the use of the Hazen–Williams equation, which is discussed further below. Given the uncertainties involved in network modelling this does not necessarily lead to greater inaccuracies but it does mean that a model calibrated for one set of flow conditions must be used with greater caution for other flow conditions.

10.5 Headlosses in pipes (2) – empirical formulae

There are several other formulae for the calculation of headloss in pipes which have been and are still used by water supply engineers. They have to be used with caution as each applies over a limited range of Reynold's numbers and in different areas of the Moody diagram (Fig. 10.2). Thus, for example, the Blasius formula applies at low Reynold's numbers in the smooth-turbulent zone, where viscosity dominates, the Hazen–Williams formula applies in the intermediate zone and Manning's formula applies in the rough-turbulent zone where the pipe roughness dominates. The Blasius equation has relatively limited applicability in the civil engineering context but the latter two are discussed in more detail below.

The *Hazen–Williams formula* has been used for many years in water supply. It is well documented and, until the advent of programmable calculators and computers, was considerably easier to use than the Colebrook–White equation. The equation can be expressed in metric units as

$$H \text{ (m)} = \frac{6.78L}{d^{1.165}}(V/C)^{1.85} \tag{10.10a}$$

or

$$V \text{ (m/s)} = 0.355C.d^{0.63}.i^{0.54} \tag{10.10b}$$

where C is a coefficient and $i (= H/L)$, d, V, H and L are as defined earlier.

It will be noticed that the coefficient, C, is not dimensionless. It has units and is therefore a function of the other parameters. As noted above, the Hazen–Williams equation is most accurate for typical pipe sizes and velocities found in water supply practice. The flow in a pipe of 0.6 m diameter with a velocity of 1.0 m/s has a Reynold's number of about 5×10^5. This is in the intermediate zone in the Moody diagram and the Hazen–Williams formula can be applied with reasonable accuracy provided that velocity and pipe size do not vary greatly from these values. The formula becomes increasingly inaccurate as the Reynold's number varies further away from this mean value. Thus different values of C apply for different pipe sizes and even for the same pipe at different flows.

The value of C can, of course, be adjusted to provide a more accurate answer if the parameters do vary significantly. Figure 10.4 shows how the coefficient varies with pipe diameter for a range of pipe roughnesses, and indicates the approximate adjustment

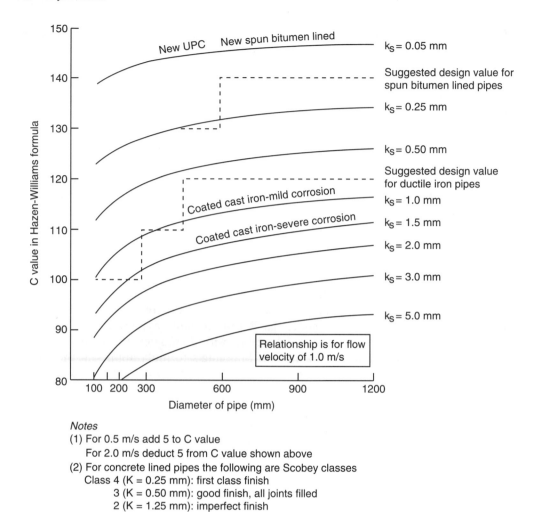

Figure 10.4 *C* values in the Hazen–Williams formula as a function of pipe size and pipe roughness for a velocity of 1.0 m/s.

needed for velocities varying from 1.0 m/s. The Hazen–Williams flow diagram in Fig. 10.5 is useful in giving an approximation of the size of pipe required for a given flow, etc. for subsequent checking by the Colebrook–White formula.

Manning's equation is appropriate for use when the flow is in the fully-turbulent range, either at high Reynold's numbers or when the conduit is particularly rough. It is widely used in open channel flow, for which there are extensive data, and will be referred to again in that context. It is not generally recommended for pipeline systems, except possibly in large, rough conduits such as unlined tunnels. Manning's equation, in metric units, is

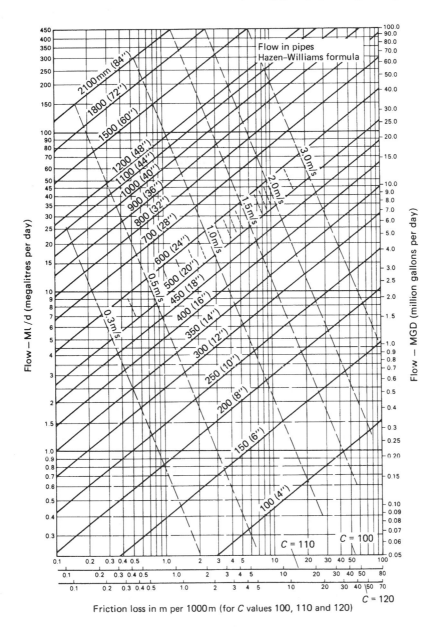

Flow in pipes
Hazen–Williams formula

Flow – Ml /d (megalitres per day)

Flow – MGD (million gallons per day)

$C = 110$ $C = 100$

$C = 120$

Friction loss in m per 1000 m (for C values 100, 110 and 120)

Figure 10.5   Hazen–Williams flow diagram.

normally written in the form:

$$V = \frac{R^{2/3}.i^{1/2}}{n}$$ (10.11)

where R is the hydraulic mean depth (also known as the 'hydraulic radius') – the area of flow divided by the wetted perimeter, i is the hydraulic gradient ($= H/L$), and n is a roughness coefficient known as the Manning coefficient. For a circular pipe,

$R = \pi d^2/4\pi d = d/4$, so for a pipe the equation can be written

$$V \text{ (m/s)} = \frac{0.397 d^{2/3} . i^{1/2}}{n} \tag{10.12}$$

Again, it should be noted that n has dimensions and is a function of the size of the conduit. However it is relatively insensitive to the diameter of a pipe and for many calculations can be assumed to be constant. Appropriate values for n are given in Table 10.2 in Section 10.11.

While both Hazen–Williams equation and Manning's equation have been used in the past for pipe flow and, in the case of the former equation, quite widely used, the simplicity of their formulations is no longer a significant advantage over the Colebrook–White equation, which is applicable over the full range of conditions likely to be found in water supply. The use of the Colebrook–White equation, or the direct solution approximations of it, is therefore strongly recommended for all pipeline calculations.

10.6   Local head losses at fittings

As discussed earlier, headlosses occur at every location where there is a geometric change to the conduit, such as at a bend in the pipeline, a change of section, an obstruction to flow such as a valve, or simply the entry to a pipe or exit from a pipe into a tank. In such cases the headloss is usually expressed as a proportion of the velocity head, or kinetic energy, of the flow, i.e. ΔH (or H_L) $= KV^2/2g$, where K is a coefficient depending primarily on the type of fitting or fixture in the pipeline. V is normally taken as the velocity in the upstream pipeline: not, in the case of valves or other obstructions, the velocity through the opening in the fitting itself.

Values of K are almost entirely empirical but there have been extensive experimental measurements on standard fittings on which estimates can be based. Table 10.1 gives values of the headloss coefficients for some standard fittings and suggested values for design. Some standard data is available for valves but in most cases it will be prudent to obtain values from manufacturers. The suggested design values in Table 10.1 are generally conservative, and it is prudent to use them when estimating losses in short conduits containing several fittings or changes of direction, etc. in close proximity. Short conduits conveying water from one tank to another often occur in water treatment works, and it is important not to under-estimate losses when designing weir and overflow levels.* Also allowance for imperfection in lining up fittings and for increase of roughness with age may need to be allowed for. If accurate analysis is being carried out on an existing installation then reference to more accurate data may be required[3] and it should be noted that the losses at some fittings, particularly bends, are a function of the pipeline roughness as well as the geometric shape. For complicated arrangements a hydraulic model may be advisable.

10.7   Open channel flow

The flow in an open channel follows the same principles that have been developed for pipe flow. Thus the energy equation, eqn (10.2), is still valid and energy is lost in the same way

---

For some calculations such as surge and transient flow analyses and assessing the range of duties on a pump, it may be necessary to take minimum losses possible.

Table 10.1 Loss coefficients through pipeline fittings

	k value in $kV^2/2g$	
	Laboratory test values	Suggested field value[a]
Entrances: V = velocity through pipe or gate		
Standard bellmouth pipe	0.05	0.10
Pipe flush with entrance	0.50	1.00
Pipe protruding	0.80	1.50
Sluice-gated or square entrance	–	1.50
Bends – 90° (45° half values given)		
Medium radius (R/D = 2 or 3)	0.40	0.50
Medium radius – mitred	0.50	0.80
Elbow or sharp angled	1.25	1.50
Tees – 90°: Assumes equal diameters		
In-line flow	0.35	0.40
Branch to line, or reverse	1.20	1.50
Exits		
Sudden enlargement: ratio 1:2	0.60	1.00
Gradual (well tapered) exit	0.20	0.50
Sudden contractions: Loss on contraction and subsequent expansion; V = velocity through contraction		
Contraction area ratio:		
1:2	1.00	1.50
2:3	0.65	1.00
3:4	0.40	1.00
Expansion only	–	1.00
Gate valve fully open	0.12	0.25
Butterfly valve fully open	0.25	0.5

[a]The k values in the second column are recommended for assessing loss through short conduits containing several fittings, bends, etc. in close proximity.

through the resistance of the channel surfaces and locally where the geometry of the channel changes. However, in open channel flow there is the added complication that the depth and hence area of flow can change. Whereas in a given pipe the area of flow is fixed and hence the velocity is a function only of the flow, in a channel of known dimensions, the velocity is not only a function of the flow but also of the channel depth and width.

Returning to the energy equation (10.2) in Section 10.1

$$H = \frac{V^2}{2g} + h + z$$

the pressure head, h, is now the depth of water, y; and z is the level of the channel invert above some datum. For a case where the energy is constant any increase in velocity, and hence the kinetic energy must be accompanied by a fall in the potential energy and hence a drop in water surface. Similarly a retardation of the flow and reduction of the kinetic energy must be accompanied by a corresponding rise in the water surface if there is no energy loss. In practice, whilst there may be very little energy loss in a case where the velocity is increased and potential energy is converted to kinetic energy, it is always the case that there is some loss of energy in the reverse process. The latter process, known as *recovery of head*, is never 100% efficient.

The *specific energy head*, H_S, is defined as the energy head above the channel invert and eqn (10.2) can be written as

$$H_S = \frac{V^2}{2g} + y$$

where y is the depth of flow.

Since $V = Q/A$ where Q is the flow and A the area of the flow,

$$H_S = \frac{Q^2}{(A^2 2g)} + y \tag{10.13}$$

Since A is a function of the depth y; eqn (10.13) relates the specific energy of flow H_s to a cubic function of y. This is made clearer if a rectangular channel is considered, for which $A = y.b$ where b is the width of the channel. Substituting for A gives:

$$H_S = \frac{Q^2}{y^2 b^2 2g} + y \tag{10.14}$$

Substituting q for Q/b, where q is known as the unit discharge, i.e. the flow per unit width in a rectangular channel, *gives*

$$H_S = \frac{q^2}{y^2 2g} + y \tag{10.15}$$

For any value of H_s (above a certain minimum value to which reference is made later) there are two possible depths y which satisfy eqn (10.15) for the same flow q, i.e. H_s can be made

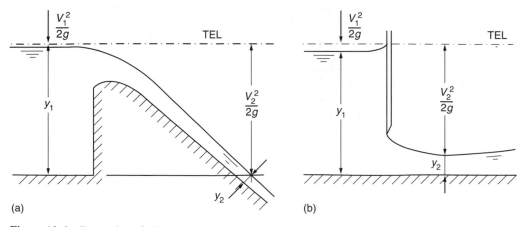

(a)　　　　　　　　　　　　　　　　　　　　(b)

Figure 10.6　Examples of alternative flow depth.

up of two different proportions of the velocity energy and the depth energy. There are a number of common examples which demonstrate this, two of which are shown in Fig. 10.6.

The first is the flow over a dam spillway or weir. Upstream the flow is deep and the velocity very low. Downstream the same flow has a much higher velocity and the depth is much reduced. The second case illustrates the flow under a freely-discharging sluice gate. Again the upstream flow is deep and slow; the downstream flow is shallow and fast. In both cases it is assumed that there is negligible energy loss across the structure. For reasons which will become clearer below, the slow deep flow is known as sub-critical flow, whilst the fast shallow flow is known as super-critical flow.

Equation (10.15) can be plotted for y as a function of the specific energy head, H_S, with a constant unit discharge, as illustrated in Fig. 10.7. This shows how the equation provides two answers for a particular value of the specific energy. It also shows that there is a minimum value of the specific energy at which there is only a single value of depth which solves the equation; this is known as the *critical depth*. The critical depth is an important concept in open-channel flow and is discussed in more detail below. It marks the boundary between sub- and super-critical flow. If the depth is less than the critical depth then the flow is supercritical; if the depth is greater than the critical depth the flow is sub-critical.

10.8 Critical depth of flow

The graph in Fig. 10.7 shows that, for unit discharge, the minimum energy use and critical depth of flow occur when dH/dy equals zero. Differentiating eqn (10.15) and putting the

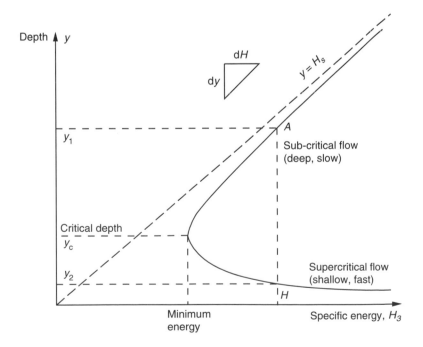

Figure 10.7 Relationship between specific energy and depth for a given unit discharge.

differential equal to zero for the minimum value gives

$$\frac{dH}{dy} = \frac{-2q^2}{y^3 2g} + 1 = 0$$

i.e. $y^3 = q^2/g$, from which the critical depth,

$$y_C = \sqrt[3]{(q^2/g)} \tag{10.16}$$

This equation is important because it demonstrates that the critical depth is a function only of the discharge per unit width. Furthermore, substituting y_C^3 for q^2/g in eqn (10.15), and simplifying, the following simple relationship between critical depth and the minimum specific energy is obtained.

$$H_{S(min)} = 1.5y_C = 1.5(q^2/g)^{1/3} \tag{10.17}$$

Thus the minimum energy is also a function of the unit discharge alone. Hence, given the unit discharge, the critical depth and minimum energy head at any location can be calculated.

Consider the case of flow in a rectangular channel with unit discharge, q, and assume the flow is sub-critical. As shown in Fig. 10.8, the specific energy level (the energy head above the invert) is denoted by the chain dotted line and comprises the depth plus the kinetic energy of the velocity. If part of the bed is raised, as indicated in the diagram, then the specific energy reduces. The energy level cannot increase so the depth must decrease and the velocity must therefore increase. The velocity head therefore increases and there must be a drop in water level over the raised section of the bed. If the bed is raised further, then eventually the specific energy will reach the minimum value and the depth of flow over the raised section will drop to the critical depth. Any further raising of the bed will reduce the energy below the required minimum value for that unit flow, and the only possible result is a reduction in flow over the raised sill. If the flow is to be maintained then there must be a rise in upstream water level to provide more energy to enable that flow to pass over the sill.

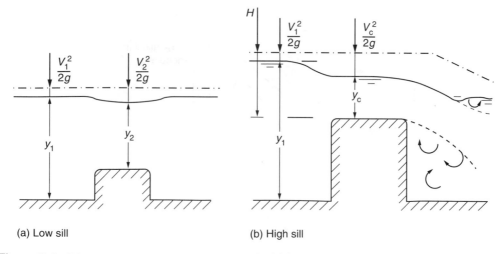

(a) Low sill   (b) High sill

Figure 10.8   Flow over a raised sill: (a) low sill, (b) high sill.

Critical depth of flow thus occurs when there is free discharge over a weir or gate. The flow utilises minimum energy to pass the maximum discharge possible for the available energy head. The expressions *free* discharge, or *modular* discharge, are used to denote that the downstream conditions do not affect the flow. Clearly, if there were some control downstream, such as a gate, which was progressively closed then eventually the downstream level would back up to the extent that the free discharge over the weir would be drowned out. Critical depth would not occur and the upstream water level would rise accordingly.

Critical depth conditions can also occur because of an increase in the unit discharge. If instead of raising the bed, the sides of the channel are brought in, squeezing the flow, the specific energy remains constant but the discharge per unit width must increase. Again, this can continue until the unit discharge reaches the value defined in eqn (10.15). If the channel is narrowed beyond this point then the specific energy must increase and the upstream water level must rise to provide this additional head. This effect is utilised in measurement flumes, which are discussed in Section 10.15. The entrance to a steep culvert is another example where critical depth may occur because the flow is squeezed through a narrower conduit. There are many instances where the flow passes through critical depth due to both a narrowing of the channel and the raising of the bed. The spillway of a dam is one such example. Another is illustrated in Fig. 10.9 which illustrates the flow from a reservoir into a channel. The channel is narrower than the reservoir and clearly the bed is raised, so critical depth would be expected at the channel entrance, provided the channel is steep and that the downstream water level is not sufficient to drown out the entrance to the channel. The water surface would also be seen to drop as the flow accelerates into the channel.

The proviso above that the water level downstream of the entrance must not be so high as to drown out the entrance is important. This depends on the slope of the downstream channel. If the slope on the channel is too flat to maintain the flow, the water level will rise and the entrance becomes drowned out with sub-critical flow throughout. At one particular slope the depth will remain at critical depth and the control at the entrance will remain. At steeper slopes than this *critical slope* the flow will accelerate away from the inlet with super-critical flow conditions.

The occurrence of critical depth in a system is described as a hydraulic control. As noted above it provides a location where the *rating curve*, i.e. the relationship between depth (or

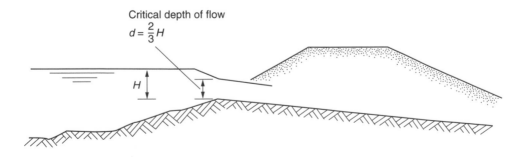

Figure 10.9 Critical depth of flow at a channel entrance with the condition that the slope is greater than the critical slope.

level) and flow, is fixed and a function only of the local geometry. If such locations can be identified in a system they form a starting point for assessing water levels throughout the system.

10.9   Weirs, flumes and gates

The above discussion considered the case of flow passing from sub-critical to super-critical through the critical, minimum energy condition. This can be achieved smoothly and without energy loss over a weir, or through a flume or gate. In each of these cases the flow is accelerating and this tends to dampen down turbulence. The reverse, passing from super-critical flow to sub-critical flow, involves the deceleration of the flow. The flow is less stable in such conditions and it is almost impossible to achieve the transition smoothly without energy loss. Generally, a region of high turbulence, known as a *hydraulic jump* will form. This is discussed further in Section 10.12.

Before dealing with the subject of energy losses in open channel flow, it is useful to consider further the hydraulics of weirs and gates which are widely used in water supply systems.

Weirs

Returning to eqn (10.17), it can be re-arranged in the form $(2/3H_S)^3 = q^2/g$
Hence

$$\text{flow per unit width } q = 2/3\sqrt{(2g/3)}H^{1.5} \tag{10.18}$$

and, for a rectangular channel of width, b

$$Q = 2/3\sqrt{(2g/3)}bH^{1.5} \tag{10.19}$$

This is now in the form of a *weir* equation, which can be generalised as

$$Q = C_d bH^{1.5} \tag{10.20}$$

where C_d is a *discharge coefficient*.

In eqn (10.19) the discharge coefficient $2/3\sqrt{(2g/3)}$ equals 1.705 in metric units. If the flow passes through critical depth over a weir crest then it might appear that C_d would always take that value. However, it is important to appreciate the inherent assumptions lying behind the theory that led to eqn (10.2) and all the subsequent equations derived from it.

(i)   There is a uniform velocity distribution across the section;
(ii) The streamlines are straight and parallel (i.e. there is no lateral pressure set up by curvature of the streamlines);
(iii) The vertical pressure distribution is hydrostatic (i.e. the pressure is a linear function of depth);
(iv) The effect of the longitudinal slope is negligible.

Provided these conditions are met, as is very nearly the case with a broad-crested weir, such as illustrated in Fig. 10.8(b), then the discharge coefficient is indeed about 1.705. However, many other crest profiles are used, ranging from simple plate *or sharp-edged* crests, to rounded tops of walls to triangular and ogee-profile crests. They each have advantages and disadvantages and their use depends on whether accurate flow measurement is required or whether they act merely as simple overflow hydraulic controls. Weir shapes used for flow measurement are discussed further in Section 10.14. The discharge coefficient can vary significantly from the basic broad-crested value of 1.71 and depends largely on the geometry of the crest, but it is also a function of the depth and velocity of the approach flow. The subject is wide-ranging and cannot be dealt with here. As a rule of thumb, however, the discharge coefficient is likely to be greater than 1.71 if there is strong curvature to the flow, for example over a half-round crest or an ogee crest. In the latter case, the profile of the crest is that of the underside of a free-falling jet of water, so, in theory, there should be little if any pressure on the solid surface. The pressure distribution through the depth of flow cannot, therefore, be hydrostatic and one way of considering the problem is that the back pressure on the flow over the crest is reduced, thus allowing an increased discharge and a corresponding increase in the discharge coefficient. Increases in C_d of 30–40% above the broad-crested weir value are possible.

Similarly, the discharge coefficient may be reduced if the weir crest is long, (in the direction of flow), or very rough – as might be case of flow over a grassed embankment. A value of 1.71 is, however, a good starting point for initial design or in the absence of more details of the weir shape.

One final point in this general discussion of weirs needs noting; namely, that a weir only acts as a hydraulic control if it has free or 'modular' discharge, i.e. the downstream water level is low enough to allow the flow to pass through critical depth. If the tailwater depth is high enough to affect the flow over the weir, then the weir is said to be drowned. This condition is usually catered for in the weir equation by introducing a drowning factor, f_d, which is a function of the height of the tailwater level above the crest. Referring to Fig. 10.8(b) it would appear that provided the tailwater level is not more than the critical depth, i.e. two-thirds of the upstream head above the crest, then critical depth flow will occur. Strictly, the tailwater level has a small effect even when it is at the level of the crest, particularly if lowered pressures can be generated below the nappe of the falling water, but this two-thirds criterion is a useful rule of thumb. Although the crest shape does affect the drowning factor, in many instances where a weir is not being used for measurement it is reasonable to assume that the hydraulic control remains at the weir crest until the tailwater level rises to a level greater than two-thirds of the upstream head above the crest.

Flumes

The foregoing comments also apply to flumes. Figure 10.16 (in Section 10.15) shows a typical flume with a narrow throat forcing the flow through critical depth. The weir equation can be applied to the throat and, provided the underlying assumptions as mentioned for weirs above apply, the discharge coefficient will again be 1.705. Thus a flume, with parallel sides and level invert to the throat, will have a discharge coefficient close to this base value, i.e.

$$Q = 1.71 C_v b h^{1.5}$$

in metric units where C_v is the approach velocity in the channel of width b, and h is the depth of the water in the channel. As a general guideline, a flume will also be drowned out when the downstream water level rises to a level greater than two-thirds of the upstream head. In practice it is often possible with a well designed exit transition from the throat to recover some of the velocity head, and it may even be possible to design the flume such the downstream water level can rise above this level and the overall head loss is less than $H/3$.

Gates

Water control gates, as distinct from other types of gates as used in navigation locks, etc., may be either undershot or overshot. The latter case is akin to the flow over a weir and only the former type is discussed here. There are many designs of gates within these broad categories. The hydraulic principles are similar and will be illustrated in the following discussion with reference to vertical-lift gates, also called *sluice gates* (and, in the UK, *penstocks*), which are by far the most common in the water supply systems. This is another case of flow passing from sub- to super-critical flow, but with a gate the flow does not pass through the critical depth.

Figure 10.10(a) illustrates free discharge through a vertical penstock gate in the wall of a tank. The gate opening forms an orifice. The flow passing under the gate has a vertical contraction and, in the plane of the gate itself, there will be vertical components of flow. Thus the minimum depth occurs downstream of the gate at a point referred to as the *vena contracta*. Assuming negligible energy loss between the upstream section and the vena contracta, Bernoulli's equation can be applied along a streamline flow between points 1 and 2.

$$H_1 + z = V_1^2/2g + y_1 + z = V_2^2/2g + y_2 + z$$

where y_1 and y_2 are the upstream and downstream depths from the water surface to the centre of the opening. If y_2 is small, then

$$V_2^2/2g = H_1$$

and, since $Q = VA$

$$Q = A_2\sqrt{(2gH_1)}$$

If the area of the gate opening is A_0 and the area of the vena contracta A_2 is $C_C A_0$, where C_C is the contraction coefficient, then

$$Q = C_C A_0 \sqrt{(2gH_1)}$$

In practice the assumption that energy losses are negligible is not strictly true and the

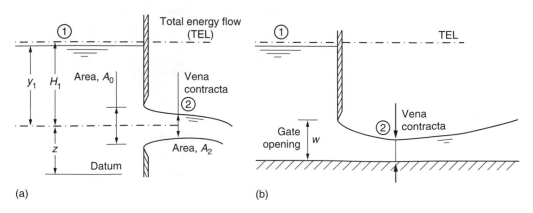

(a)   (b)

Figure 10.10   Orifice and undershot gate flow: (a) orifice flow, (b) undershot gate.

equation needs to be written in the more general form

$$Q = C_D A_0 \sqrt{(2gy_1)} \tag{10.21}$$

where C_D is a discharge coefficient, and the flow is related to the upstream depth rather than the energy head. Typically, for a sharp-edged opening, C_D is about 0.6. If the edges of the opening are more rounded, then the value approaches closer to 1.0.

For the case where the jet is not freely discharging but the opening is drowned, y_2 becomes significant and, referring back to Bernoulli's equation above and going through the same procedure

$$Q = C_D A_0 \sqrt{2g(y_1 - y_2)} = C_D A_0 \sqrt{(2g\Delta y)} \tag{10.22}$$

where Δy is the difference in water levels across the gate.

The same equations apply to the case of a sluice gate in a channel, as illustrated in Fig. 10.10(b), but now the depth is the total water depth to the channel invert rather than to the centre of the opening. For a full-width, freely-discharging gate with sharp-edged lip, the discharge coefficient varies between about 0.5 for low depths of submergence ($y_1/w = 2.0$) up to about 0.58 for submergence ratios of 10 and more, where w is the height of the gate opening as shown in Fig. 10.10(b).

10.10 Froude numbers

Before dealing with headloss in open-channels, one further useful concept is worth introducing. The flow can be characterised by its Froude number. This is a non-dimensional grouping of the flow parameters, which can be shown to represent the relative magnitudes of inertial and gravitational forces acting on the fluid. The Froude number, F, is defined as

$$F = \frac{V}{\sqrt{(gy)}} \tag{10.23}$$

where y is depth of flow (m) and g in metric terms is 9.81 m/s^2.

Its usefulness lies in the fact that at the minimum energy condition, the critical depth,

$$y_C = \frac{2H_S}{3}$$

as shown in eqn (10.17). Hence the kinetic energy of the critical velocity, V_C, equals the balance of H_S available, i.e.

$$\frac{V_C^2}{2g} = \frac{H_C}{3} = \frac{y_C}{2}$$

Thus

$$V_C = \sqrt{(gy_C)} \tag{10.24}$$

So, when the flow passes through critical conditions the Froude number, $V/\sqrt{(gy)}$, has a value of 1.0. For sub-critical flow the depth is greater and the velocity lower, therefore the Froude number is always less than 1.0; for supercritical flow the opposite is true and the Froude number is always greater than 1.0. Calculation of the Froude number thus provides an immediate check on the type of flow and how near the flow conditions are to those at critical depth. The closer the Froude number is to 1.0 the more unstable the water surface becomes with waves and surface disturbances. As a general rule it is preferable to design channels with Froude numbers outside the range of 0.6 to about 1.5.

10.11 Head losses in channels

The concepts of head losses in channels are exactly the same as those in pipes. There are both 'friction' losses and local losses and the same equations can be applied. However, as noted before, there is the complication that the water depth can vary so that the velocity is not a function only of the flow.

Normal depth

In a long straight channel of constant cross-section the flow will reach an equilibrium depth when the rate of loss of energy through the turbulence generated by the drag of the boundaries equals the rate at which potential energy is given up by the fall in elevation. When the flow reaches this condition the depth remains constant and the slope of the channel bed, the water surface and the energy line are all equal. This equilibrium depth of flow is known as *the normal depth*. It should be noted that it applies to both sub- and super-critical flow. Strictly, the resistance equations apply to this condition of normal depth. The pipe equations can be used for open-channels but they need to be modified for the different cross-sectional parameters. Instead of the pipe diameter, the relevant dimension is the hydraulic mean depth, R, also commonly referred to as the *hydraulic radius*. This is defined as

$$R = A/P$$

where A is the area of the cross-section of the flow and P is the wetted perimeter, i.e. the portion of the perimeter of the channel surface in contact with the flow.

Manning's formula

Although it is possible to use the full Colebrook–White equations for open-channel flow, by far the most widely used equation for calculation of normal depth in open-channel flow is Manning's equation:

$$V = \frac{R^{2/3} i^{1/2}}{n} \tag{10.25}$$

where n is a coefficient relating to the roughness of the channel. This form of the equation is in metric units. In some countries, particularly in Europe, it is also known as Strickler's equation. It should be noted that this equation is not dimensionally balanced and hence n has dimensions of length to the power of one-sixth. Some typical values of Manning's coefficient are given in Table 10.2. For more detailed information on channel roughness, see Reference 4.

Despite the dimensional nature of the Manning's equation, it is almost universally used for open-channel head loss calculations and there is a great deal of guidance on suitable values for the coefficient. Its use is strongly recommended.

Table 10.2 Values of Manning's roughness coefficient n

Surface	Manning's coefficient n
Smooth metallic	0.010
Large welded steel pipes with coal-tar lining	0.011
Smooth concrete or small steel pipes	0.012
Riveted steel or flush-jointed brickwork	0.015–0.017
Rough concrete	0.017
Rubble (fairly regular)	0.020
Old rough or tuberculated pipes	0.025–0.035
Cut earth (gravelly bottom)	0.025–0.030
Natural watercourse in earth	0.030–0.040
Natural watercourse in earth but with bank growths	0.050–0.070

Chezy formula

The equivalent equation to the Darcy–Weisbach equation is the Chezy equation which is written as

$$V = C(Ri)^{0.5} \tag{10.26}$$

where C is a coefficient related to the roughness of the channel and i is the hydraulic gradient $= H/L$ (the slope of the energy line). Like the Darcy equation this equation is dimensionally correct so that C has no units.

Comparing equation 10.26 to the Darcy–Weisbach formula, eqn (10.5), and noting that for a pipe running full $R = \pi d^2/4\pi d = d/4$

$$C = \sqrt{(8g/f)}$$

So it can be seen that, strictly, C will vary, in a similar manner to the friction factor, as a function of relative roughness and Reynold's number. In practice the Reynold's numbers related to most open-channel conditions are such as to put the flow regime into the rough turbulent region of the Moody diagram, where the variation of friction factor and hence of C is small. Thus the assumption that C is constant is reasonable in most cases.

The value of C can be derived from values of Manning's n because, for given values of i and V:

$$R^{2/3}/n = CR^{1/2}$$

i.e. $C = R^{1/6} n$.

Local head losses

As with pipeline fittings, head losses occur in open channels at any change of channel geometry. The calculation of losses is made in a similar manner, i.e. as a function of the velocity head. The complication with open channel losses is that it is often not possible to use a single, standard velocity as in the case of a pipe of constant diameter. It is common, therefore, to express the headloss at a change of section in terms of the difference in velocity heads, thus

$$\Delta H = K\left(\frac{V_1^2}{2g} - \frac{V_2^2}{2g}\right) \tag{10.27}$$

where V_1 and V_2 are the upstream and downstream velocities respectively and it is the positive value of the difference to which the empirical coefficient is applied. Care must be taken in interpreting data on the coefficients, however, as this form of the expression is not universal. Some data may be presented in terms of a single reference velocity as is done for pipes. For example, losses at a channel bend of constant cross-section can only relate to a single velocity. Table 10.3 provides some values of K.

10.12 Hydraulic jump

One other form of local energy loss should be introduced at this point. In the discussion of critical depth and its development over a weir crest or in a flume, there was the inherent assumption that the flow could smoothly accelerate from sub-critical conditions through critical depth to super-critical flow. Practical observation confirms that that is the case. Accelerating flow tends to damp down any turbulence. The reverse is not true. To pass from super-critical to sub-critical flow is much more difficult without a significant energy loss. The flow is decelerating and expanding and almost always this leads to a region of turbulence and energy loss. This is known as *a hydraulic jump* in which the water surface rises abruptly from the fast shallow flow to the deeper sub-critical flow downstream. Figure 10.11 illustrates a hydraulic jump downstream of a sluice gate.

Table 10.3   Local channel headloss coefficients

Feature		Loss coefficient
Bends[a]		
$r/B > 3$ (B is surface width)		0.15
$r/B < 3$ but > 1		0.15 – 0.6
90° single mitre		1.2
45° single mitre		0.3
Transitions		
square ended	– inlet	0.5
	– outlet	1.0
cylinder quadrant	– inlet	0.4
	– outlet	0.6
smooth tapered	– inlet	0.1
	– outlet	0.2

Loss coefficients for other elements can be taken as the same as for pipe elements provided the Froude no. of the flow is less than about 0.3.
[a]Based on average channel velocity.

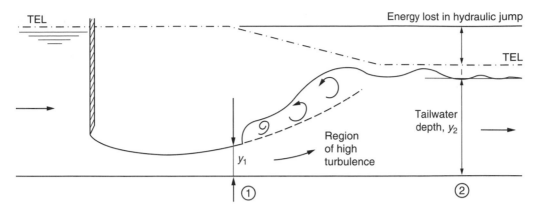

Figure 10.11 Hydraulic jump downstream of a gate.

A hydraulic jump is a good way of 'destroying' surplus energy and is often deliberately introduced for that purpose, such as at the foot of a dam spillway or downstream of a control gate. It enables the potentially erosive power of the high velocity flow to be reduced in a controlled fashion before the flow is released into the river downstream. The region of high turbulence is contained locally within the structure.

Because there is a high energy loss across a hydraulic jump, it is not possible to use the energy equation to analyse the phenomenon. Instead, the theory of a hydraulic jump is based on the theory of momentum. With reference to Fig. 10.11 it can be seen that the only external force on the jump is the friction on the floor of the channel. Within the jump itself this is relatively small and can be ignored, so that the assumption made is that the net pressure force acting on the flow – the difference in the downstream and upstream hydrostatic pressures – is equal to the net momentum loss of the flow. This is treated more fully in other hydraulic texts[4,5] but it can be shown that for a jump to form, the downstream or *tailwater* depth, y_2, must be equal to or greater than that given by:

$$y_2 = y_1\{(1 + 8F_1^2)^{0.5} - 1\} \tag{10.28}$$

where F_1 is the Froude number of the upstream, super-critical flow of depth y_1. The value of y_2 given by this equation is known as the *sequent depth*.

If the tailwater level is less than the sequent depth then the super-critical flow will continue further downstream until it has lost enough energy or until there is sufficient tailwater depth for the jump to form. If, on the other hand, the tailwater depth is greater than the sequent depth, then the jump will be forced upstream until the balance is re-established. Thus, for example, considering the jump downstream of the sluice gate shown in Fig. 10.11, if there were another gate downstream controlling the tailwater level, then closing this second gate would increase the tailwater depth and force the hydraulic jump to move upstream. Ultimately it would drown out the first gate, which would then operate with a submerged, drowned discharge as discussed earlier.

10.13 Non-uniform, gradually varied flow

So far only flow in a long straight channel where the equilibrium of normal depth can be reached has been considered. In practice long straight channels are something of a rarity;

in many instances the channel shape will be changing or the slope varying. Even if the channel is straight and uniform, it may not be long enough for the flow to reach normal depth. Thus although normal depth and uniform flow are useful concepts, in many cases this condition will not exist and neither the water surface nor the hydraulic gradient will be at a uniform slope.

One obvious example of conditions where the longitudinal profile of the water surface is curved, is the local drawdown of the flow as it speeds up to pass over a weir. This is a local effect and, as discussed earlier, the simplifying assumptions underlying the theory no longer hold true. It also the case that the depth of flow upstream of a weir, dictated by the hydraulic control of the weir, is unlikely to be the normal depth in the channel, but moving further upstream the depth will gradually change towards normal depth.

This can be more clearly visualised by considering the flow in a river entering a reservoir. The flow is slowing down and getting deeper, with the water surface changing from a slope approximately parallel to the average river bed slope, to a horizontal level at the reservoir, as illustrated in Fig. 10.12. On the other hand, downstream of the crest of the dam spillway, the flow will accelerate away from the critical depth and will get shallower and faster, as shown in Fig. 10.12. Another good example of the curvature of the water surface is that of flow under a sluice gate into a horizontal channel downstream as illustrated in Fig. 10.11. Energy must be lost through the resistance of the channel bed and the flow must slow down. This means that the depth must increase and the water surface will curve upwards.

The problem is how to calculate the water surface profile in any of these cases or other situations, such as that of a river with varying cross-section and with occasional obstructions such as bridges or weirs? Before examining this problem it needs to be stressed that, in all the following cases, steady flow is assumed. The depth and velocity of flow may vary with position, but the total flow at any point remains constant. The introduction of changes to the flow with time adds another level of complexity to the problem; such conditions are then described as *dynamic* or *transient* and are almost always analysed using computer programs, and are beyond the scope of this book.

Returning to the energy equation, consider the flow at two points, distance L apart, in a non-uniform channel, as shown in Fig. 10.13.

Between the two points, ΔL apart, the head loss, ΔH is given by

$$\Delta H = H_1 + \Delta z - H_2 = (V_1^2/2g + y_1) + \Delta z - (V_2^2/2g + y_2)$$

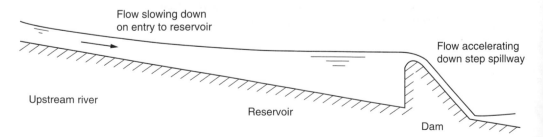

Figure 10.12 Examples of gradually-varied flow profiles.

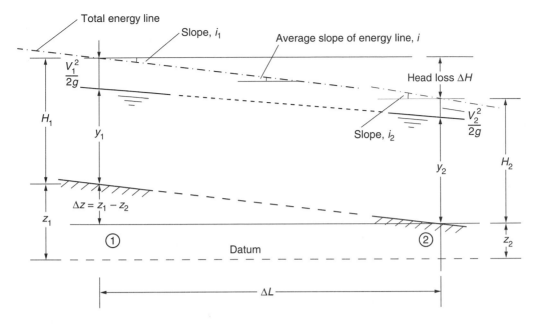

Figure 10.13 Energy loss in non-uniform flow.

Substituting Q/A for V

$$\Delta H = Q^2/(A_1^2 2g) + y_1 + \Delta z - Q^2/(A_2^2 2g) - y_2 \qquad (10.29)$$

If the length ΔL is short, the average slope, i, of the energy line can be taken as approximating to the mean of i_1 and i_2 so that

$$\Delta H = \Delta L(i_1 + i_2)/2 \qquad (10.30)$$

where i_1 and i_2 are the slopes of the energy line at Sections 1 and 2.

There are thus two equations for ΔH. From the known geometry of the channel and starting from a known depth of flow y_1 at Section 1 for flow Q, the hydraulic gradient i_1 can be calculated from the Manning or Chezy formula. Using $i_1 \Delta L$ as the loss of energy ΔH to Section 2, a first approximation of the depth of flow y_2 can be obtained (because $H - \Delta H = Q^2/(A_2^2/2g) + y_2$). From this i_2 can be calculated, so that a revised estimate of the hydraulic energy loss $\Delta H' = \Delta L(i_1 + i_2)/2$ can be used to calculate a more accurate value of y_2. By further repetitions, y_2 can be calculated to the desired degree of accuracy. Thus the computation can be progressively carried out for further points downstream giving the profile of the water level. For reasonable accuracy, the distance steps ΔL should be reasonably short, having regard to the change of slope of the hydraulic gradient.

Such calculations can be carried out by hand, (though for non-rectangular channels they are very tedious), but are now normally undertaken with a spreadsheet or more probably with a computer program developed for the purpose. Many organisations have developed their own program and there are a number of programs commercially available, the most widely known probably being HEC 2 (now marketed as HEC2-RAS), developed by the Hydrological Engineering Centre in the USA.

There are three further points to make about this type of calculation.

(i)   The basic theory includes the inherent assumptions as listed in Section 10.9 for weirs. Thus, the approach cannot be used accurately where there are rapid changes to the water surface, for example close to a weir crest or through a flume.

(ii)   The starting point for any calculation must be a known flow and depth. In many cases this will be a hydraulic control, such as flow over a weir crest or through a gate where the water level can be directly calculated. It should be noted that such a control defines the water profiles both upstream and downstream. Thus for the sub-critical flow, the calculation proceeds upstream from the control, whilst for super-critical flow the calculation can proceed in the downstream direction, i.e. the profile in sub-critical conditions is affected by what happens downstream, and the super-critical flow profile is affected by what happens upstream.

(iii)   In some cases there may be no hydraulic control which defines a specific depth at any point; for example, a long river reach with no control structures. In such a case it may be adequate to assume a normal depth control in the reach downstream of the length of interest (assuming sub-critical flow) and to check the sensitivity of the water levels to that assumption.

The most common calculation relates to finding the water profile of sub-critical flow, flowing towards some hydraulic control. This type of calculation is known as a *backwater calculation* and the water–surface profile is known as a *backwater profile* as the calculations proceed upstream from the control point.

10.14 Measurement weirs

In Section 10.9 it was shown that, provided there is free, undrowned discharge over the crest of a weir, then the flow will pass through critical depth and the upstream water level will be uniquely controlled by the flow rate and the geometry of the crest. In theory, therefore, any weir can be used for measurement but, in practice, a limited number of standard crest shapes tend to be used, partly because experience has shown that these are the most accurate and practical structures for a particular application and partly because the discharge coefficients and the limiting conditions for their accuracy are well known. British and other international standards are available for such weirs, covering the standard geometries, measurement requirements and other features. Reference 6 provides a definitive discussion of the weirs ands flumes used for open-channel flow measurement. For non-standard shapes, either detailed model tests or site calibration must be carried out.

For conditions, which meet the underlying assumptions laid out in Section 10.9, the upstream total head, H, is given by eqn (10.19), i.e.

$$Q = 2/3\sqrt{(2g/3)}bH^{1.5} = 1.705bH^{1.5} \tag{10.31}$$

with $g = 9.81$ m/s^2 in metric terms, and the general equation is therefore

$$Q = C_d bH^{1.5} \tag{10.32}$$

For *measurement weirs*, the general weir equation is (by tradition) normally written as

$$Q = C'_d 2/3\sqrt{(2g)}bH^{1.5} \tag{10.33}$$

where, for eqn (10.33) to be equivalent to equation 10.31, C'_d has the value $1/\sqrt{3}$.

In practice C'_d varies from the value $1/\sqrt{3}$ according to the shape and type of weir crest.

Broad-crested weir

For the broad-crested weir, illustrated in Fig. 10.8(b), the discharge coefficient, C'_d is indeed about $1/\sqrt{3} = 0.577$. However, it is easier to measure the actual water level, h, above the weir crest, than total energy head H (including the velocity of approach head), and hence the equation is normally written as

$$Q = C_v 0.577 \ (2/3)\sqrt{(2g)}bh^{1.5} = C_v 1.705 bh^{1.5} \qquad (10.34)$$

where h is measured away from the local draw-down of the water surface as it passes over the crest, and C_v is a *velocity coefficient* to account for the approach velocity head. If the approach velocity is small then C_v can be taken as 1.0 and for other cases the effect of approach velocity head can be calculated and included in C_v.

Sharp-crested, or thin-plate weirs

Sharp-crested weirs, usually formed from a metal plate, are used in a variety of situations. They are simple, and because the plate can be machined to a high degree of accuracy, they can be very accurate flow measurement devices. However, because the plate becomes heavy and unwieldy for large flows, they are used particularly for small and medium flows, and not generally for river flow measurement. There are a number of standard shapes adopted, which are described by the shape cut out of the plate – rectangular of full channel width; or rectangular with side contractions as illustrated in Fig. 10.14; or V-notch with a central angle, θ ; or less commonly, trapezoidal.

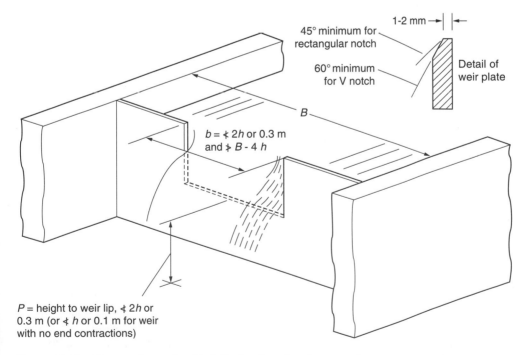

Figure 10.14   Rectangular weir with fully developed end contractions.

For a rectangular weir with no end contractions, i.e. across the full width of the channel, the discharge coefficient, for use in eqn (10.33), can be taken from the empirically-derived Rehbock formula:

$$C_d' = 0.602 + 0.083h/P \qquad (10.35)$$

where h is the water depth over the weir crest, and P is the height of the weir crest above the floor of the channel.

For a rectangular weir with fully developed end contractions as illustrated in Fig. 10.14

$$C_d' = 0.616(1 - 0.1h/P) \qquad (10.36)$$

For a 90° V-notch weir, i.e. with the central angle = 90°, C'_d is given by the following table:

h (m)	0.050	0.075	0.100	0.125	0.150	0.200	0.300
C_d'	0.608	0.598	0.592	0.588	0.586	0.585	0.585

In this latter case correction factors must be applied if the base of the channel is less than $2.5h$ below the base of the notch, or if the width of the approach channel is less than $5h$, where h is the depth of water measured above the vertex of the notch.

In all cases for thin plate weirs there are a number of requirements for accurate measurements:

(i)   the water depth should be measured at distance between $3h$ and $4h$ upstream;
(ii) if the waterway of the approach channel is less than 10 times the area of the waterway over the weir crest then a correction factor must be included to take into account the velocity of approach;
(iii) the downstream water level must be below the level of the crest, i.e. no submergence can be allowed; and
(iv) most importantly, the nappe (or underside) of the falling water must be well aerated to ensure atmospheric pressure there. This means that, for a full width rectangular weir in a channel whose walls extend downstream, an air pipe must be provided to allow aeration of the pocket beneath the nappe.

A limitation of sharp-edged weirs is that they can be damaged by debris brought down by the flow, and the shape of the machined edge to the plate is important for accurate measurement.

Crump weirs

Where new measurement weirs are constructed on rivers and stream in the UK by far the most widely used weir is the Crump weir, so-called after the engineer who developed it as a measuring device. The profile is triangular with an upstream slope of 1:2 and a downstream slope of 1:5 as shown in Fig. 10.15.

There are several advantages to the Crump weir, not least of which is that the discharge coefficient is very nearly constant over a wide range of discharges. C_d for use in eqn (10.20) is almost exactly 2.0 (C_d' in eqn (10.33) thus equals 0.677). Other advantages are – (i) the weir allows movement of sediment past the weir (deposition of silt against the vertical upstream face of most other types of weirs being a problem); (ii) the weir is easy to build accurately, particularly with a pre-formed bronze crest angle built into the concrete

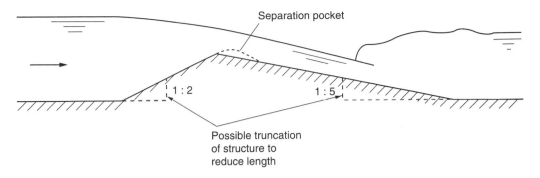

Figure 10.15 Crump weir profile.

structure; and (iii) the high modular limit which the weir can tolerate before the discharge is significantly affected – the limit for the ratio of tailwater depth above the crest to the upstream depth being about 0.8.

Furthermore, the Crump weir can provide reasonably accurate flow measurement even if it is operating beyond the modular limit. The flow over a weir, once it is drowned, becomes a function of both the upstream and downstream water levels. For most weirs, this requires measuring the water level downstream in the area of high turbulence where the jet expands. This can be difficult and inaccurate, particularly as the critical difference between the tailwater and upstream depth is small. With a Crump profile weir, instead of the direct measurement of the tailwater level, the pressure in the separation pocket immediately downstream of the weir crest is measured (see Fig. 10.15). The pressure in this pocket reflects the downstream tailwater level when the weir is drowned. It is a much more stable parameter than the turbulent water surface level downstream and can be measured more accurately. It thus enables the Crump weir to be used relatively accurately for measuring flow even when drowned. A rating curve can be developed for drowned flow using this parameter.

The one disadvantage of a Crump weir is that it can be a fairly long structure with consequent extra costs but it has been extensively model tested and criteria have been developed for truncation of the shape, i.e. neither the upstream nor downstream slopes need necessarily extend to the floor of the channel. Also investigated thoroughly has been the calibration of Crump weirs constructed with the crest sloping towards the centre such that the crest, in elevation, appears as a shallow Vee. This has the advantage of concentrating low flows towards the centre of the weir and giving greater accuracy of measurement over a wide range of flows.

10.15 Measurement flumes

The principles behind the use of a flume as a measurement device are identical to those for a weir. The basic concepts of a flume were discussed in Section 10.9. In essence a flume is a constriction in the channel such that the width is reduced, forcing the flow through critical depth. Provided the requirements behind the simple theory, set out in Section 10.9 are met, then again eqn (10.31) applies to undrowned flow through a flume, i.e.

$$Q = 1.705bH^{1.5}$$

Rectangular-throated flume

The flume equivalent to the broad-crested weir is the rectangular-throated flume as illustrated in Fig. 10.16. A flume can also have a raised floor as well as a narrowed throat but since one of the advantages of a flume over weir is that it can more readily pass sediment, any raising of the floor probably needs to be limited and with as smooth an inlet transition as practical. Downstream of the throat the flow will be super-critical and a hydraulic jump will form in the downstream channel. It is important that the tailwater level downstream of the jump does not drown out the critical depth in the flume and if necessary a fall must be introduced into the downstream channel to ensure that the water flowing away is at a low enough level.

For a rectangular-throated flume as in Fig. 10.16, the length of the throat must be not less than 1.5 times the maximum total head upstream. It is also necessary that the surfaces of the flume are smooth, whether of concrete or constructed with a pre-formed steel or fibre-glass insert, and also that the divergence downstream must not exceed 1:6 and that the approach flow must be sub-critical. The upstream level measurement should be between $3h$ and $4h$ upstream of the flume inlet, where h is the upstream depth of water above the invert of the throat.

As with weirs, the discharge equation is usually modified to relate to the upstream depth of flow in the channel, h:

$$Q = C_v 1.705 b h^{1.5} (\text{or } Q = C_d b h^{1.5}) \tag{10.37}$$

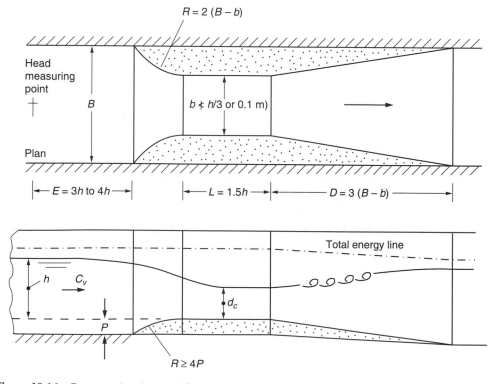

Figure 10.16 Rectangular-throated flume.

where C_v is again a coefficient to account for the approach velocity. More generally the equation is written as

$$Q = C_v C_s 1.705 b h^{1.5} \qquad\qquad (10.38)$$

where C_s is a shape coefficient to take into the particular geometry of the flume. For the smooth-entry flume illustrated in Fig. 10.16, the values for C_s depend on the ratios, L/b and h/L as shown in the table below, where L is length and b the width of the throat.

h/L	L/b				
	0.4–1.0	2.0	3.0	4.0	5.0
	C_s				
0.1	0.95	0.94	0.94	0.93	0.93
0.2	0.97	0.97	0.96	0.95	0.95
0.4	0.98	0.98	0.97	0.97	0.96
0.6	0.99	0.98	0.97	0.97	0.96

Other standard flumes

There are, of course, many other designs of flumes and the geometry determines the particular discharge coefficient C_d for use in eqn (10.37), or the shape coefficient for use in eqn (10.38). In the USA, the Parshall flume is widely used but it has a relatively complicated geometry which varies depending on the range of flows to be measured and is rarely used in the UK. One disadvantage of the rectangular-throated flume is that it is relatively long and the cut-throat flume has been developed as a shorter alternative. This has tapering inlet and outlet sections which meet at an angle at the throat. The shape coefficient is reduced from that for a rectangular-throated flume.

One final point regarding the use of flume and weirs should be noted. For free or modular discharge, the tailwater level can, as a general approximation, be no higher than the level of the critical depth flow at the throat or crest. (For thin-plate weirs a much lower limit must be set.) Generally, therefore, at least one third of the upstream specific energy head must be lost at the structure. Because flumes force critical depth by increasing the unit discharge rather than reducing the specific energy as do weirs, a flume will have a greater critical depth in the throat than a weir in the same location. Thus a flume will cause a greater head loss than a weir. (For a flume the drop in water level is one-third the depth of flow in the approach channel; for a weir it is one-third the height of water above the weir cill – which is less – see Figs 10.8b and 10.16.) Hence, if head loss is a critical issue, then this must be a factor in the choice of measurement structure.

10.16 Venturi and orifice flow meters

These types of flow meter apply to flow in closed conduits and work on the principle demonstrated by the Bernoulli equation: if the velocity of flow is increased then the pressure must drop. In both Venturi and orifice meters the flow is passed through a constriction in the pipeline causing the velocity to increase. The two types of meter are shown in Fig. 10.17 and the measurement principle is illustrated in Fig. 10.18.

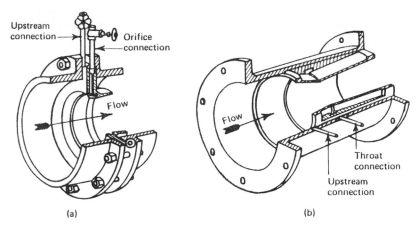

Figure 10.17   (a) Orifice meter, (b) Dall-type Venturi meter.

Returning to the energy equation and assuming the pipeline is level and that head losses between sections A and B are negligible, the head difference between the two sections, A and B, is given by

$$\Delta h = h_A - h_B = \frac{(V_B^2 - V_A^2)}{2g} \tag{10.39}$$

Since

$$Q = \pi D^2/4.V_A = \pi d^2/4.V_B$$

where D and d and A and a are the pipeline and constriction diameters and areas respectively, eqn (10.39) can be re-written as

$$Q = \frac{\pi d_2 (2g\Delta h)^{1/2}}{4[1-(d/D)^4]^{1/2}}$$

or

$$Q = ak\Delta h^{1/2}$$

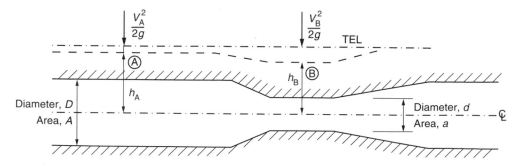

Figure 10.18

where

$$k = \frac{(2g)^{1/2}}{[1 - (d/D)^4]^{1/2}}$$ (10.40)

 In practice there are head losses between A and B and it is not necessarily the case that the stream lines at the constriction will be parallel even for the venturi. Thus a further coefficient C must be introduced, so that

$$Q = Cak\Delta h^{1/2}$$ (10.41)

For a venturi meter C varies from about 0.95 to 0.99 depending on the exact geometry. Typically a well-designed meter with parallel-sided throat will have a value of about 0.98. The Dall tube illustrated in Fig. 10.17 is a shortened version of the venturi and the value of C will be less.

The theory behind an orifice meter is exactly the same, and eqn (10.40) equally applies. But C will vary much more, depending on the velocity of flow and the ratio d/D. If the latter is in the typical range of 0.4–0.6 and for pipe diameters greater than 200 mm, the value of C will be in the range 0.60 to 0.61 for the usual velocities experienced in a pipeline.

The accuracy of measurement of both venturi and orifice meters depends on the lateral flow distribution through the device and can be severely affected by flow disturbances created by fittings in a pipe system. Detailed conditions for accurate measurement are laid down in BS 1042 Part 1. These can generally be met by ensuring that for d/D ratios not exceeding 0.6, there are at least 20 diameters of straight pipe without a fitting upstream of the meter and seven diameters of straight pipe downstream. The venturi is designed to minimise the head loss and this can be made very small with a well-designed expansion downstream of the throat providing good recovery of pressure head. The head loss through an orifice is substantially higher because of the sudden expansion of the diameter downstream. If head loss is an important consideration, then a venturi meter such as the illustrated Dall tube, should be considered, but venturis are more expensive and require greater space than an orifice meter

10.17 Other flow meters

There are a number of other types of flow meters that are widely used in the water industry. The principles on which they operate are not the same as those outlined in the foregoing material.

Propeller or turbine meters

These use a freely-rotating, bladed rotor positioned in the flow, the speed of rotation being measured, usually by a pick-off coil mounted in the housing, which senses the passage of the rotor blades. Pulses are thus generated which can be counted over a known time. There are many variations on the basic concept and each manufacturer has his own designs for use in different applications. The speed of rotation is related to the fluid speed so this type of meter is essentially a velocity meter and it is necessary to know the area through which the total flow is passing and the velocity distribution across that area. For large conduits it may be necessary to carry out velocity traverses across the section to establish the relationship between velocity at a particular location and the total flow rate.

Electromagnetic flowmeters

The operation of this type of flow meter is based on Faraday's Law of electro-magnetic induction from which it can be shown that an induced voltage in a conducting fluid as it moves through a magnetic field is proportional to the mean velocity of the fluid, the strength of the magnetic flux, and the size of the conduit. For any particular installation a relationship can be developed relating the induced voltage to the total flow through the conduit. They are thus integrating meters and do not depend on the velocity profile across the section. To induce a measurable potential in the electrodes of the instrument placed either side of the conduit, a magnetic field is generated by means of a coil or coils usually arranged as saddles wrapped around the pipe or, in the case of a river measurement, buried under the river bed. These types of meter are increasingly being used in the water industry, particularly for measuring flows in pipes. They provides zero headloss and no obstruction to the flow and are available in a wide range of sizes. Further information is given in Section 15.15.

Ultrasonic flowmeters

The usual type of ultrasonic meter operates on the principle of the measurement of the time difference between sound waves transmitted through the flow in opposite directions. If the beams are transmitted diagonally across the flow then the wave travelling downstream will have a higher velocity than the one travelling back upstream. The time difference is a function of the flow velocity. There is a second type of ultrasonic meter in which a beam of known frequency is reflected back from particles in the flow and the Doppler shift in frequency of the reflected wave is indicative of the velocity.

As with propeller or turbine meters, this type of flow meter measures average velocity along the path of the beam. The measurement of flow is therefore dependent on the velocity profile. In large systems, particularly in channels and rivers, multi-path systems are commonly installed, in which several pairs of transducers are located at different levels to provide better accuracy. Also, in a open-channel situation, measurement is needed of the water level, or depth, and this can be provided with an ultrasonic water level probe located above the water surface such that a sound wave is transmitted vertically downwards and reflected from the surface.

Ultrasonic flowmeters of the transit time type are reasonably accurate if installed and operated correctly but have a tendency to long-term instability. Doppler meters are less accurate and can only be used on dirty water. As with electro-magnetic systems, ultrasonic meters need provide no obstruction to the flow and provide no head loss. Clamp-on meters, requiring no intrusion into a pipeline, are becoming increasingly reliable and accurate.

References

1. Moody L. F. Friction factors for pipes. *Trans ASME*, **66**, 1944.
2. Hydraulics Research Station and Ball D. I. H. *Tables for the Hydraulic Design of Pipes, Sewers and Channels*, 7th edn. Thomas Telford, London.
3. Miller D. S. *Internal Flow Systems*, 2nd Edn. BHRA, Bedford UK, 1990.
4. Ven te Chow, *Open Channel Hydraulics*, McGraw-Hill, New York, 1959.
5. Henderson F. M. *Open Channel Flow*, Macmillan, Basingstoke, 1966.
6. Ackers P. *et al. Weirs and Flumes for Flow Measurement*, John Wiley, Chichester, 1978.

11

Service reservoirs

11.1 Functions

A service reservoir has four main functions.

(1) To balance the fluctuating demand from the distribution system, permitting the source to give a steady or differently phased output.
(2) To give a suitable pressure for the distribution system and reduce pressure fluctuations therein.
(3) To provide a supply during a failure or shutdown of treatment plant, pumps or trunk main leading to the reservoir.
(4) To provide a reserve of water to meet fire and other emergency demands.

It is seldom possible or economic for a source to give a fluctuating output in step with demand. Treatment processes need to be run 24 hours a day with only infrequent, carefully controlled changes of output. Pumps need to be run near their design point for maximum efficiency, whilst electricity tariffs may influence their running times; it is not economic for a long supply main to have an overlarge capacity simply to meet the peak demand of a few hours duration. A technical and economic study of the capital and operating costs of the various options available should take place, including deciding where the reservoir should be sited (see Section 14.9). The reservoir minimum water level should generally be just high enough to maintain the required minimum pressures in the distribution system at peak flows during the planned life of the scheme. A balance has to be struck between choosing an unnecessarily high elevation which would raise leakage levels from the distribution system and increase pumping costs; and the need to have some reserve elevation to meet increased demand or extension of the system later.

11.2 Reservoir capacity required

Minimum storage to even out hourly demand

A typical graph of the hourly variation of demand for a town of about 75 000 population is shown in Fig. 11.1(a). The demand reaches a peak typically between 0700 and 0900 and remains high until after midday. It slackens off in the afternoon but rises again to a second peak about 1800–1900 in the evening when people, home from work, prepare an evening meal. During the summer months the evening peak may be higher and more prolonged due to garden watering. There will be a different pattern of demand at week-ends and on holidays with peaks occurring later and lasting longer. In the winter there will also be a higher rate throughout if a severe frost causes many pipe bursts on consumers' premises. On Fig. 11.1(a) the average demand for the day is shown and, assuming this is the steady

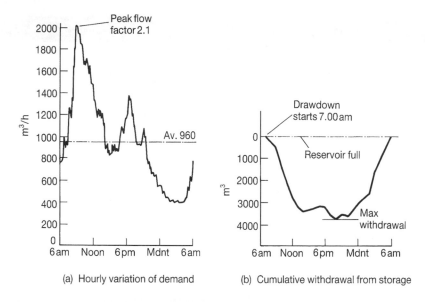

(a) Hourly variation of demand   (b) Cumulative withdrawal from storage

Fig. 11.1   Typical demand variation of 23 ml/day supply.

input from some source into the reservoir, Fig. 11.1(b) shows the consequent withdrawal from storage starting at 0700 when demand starts to rise above input. A total of 3750 m^3 of stored water is used for an average demand of 960 m^3/h, i.e. approaching 4 hours' storage is required to support the hourly variation of demand. This is commonly found to be the case for towns in temperate climates.

In practice a greater storage volume will be required for the following reasons:

(1) there will be variations from one day to the next both in daily peak demands and due to seasonal changes in the daily demand pattern;
(2) it will not always be convenient, or possible, to have the reservoir full every morning at about 0600;
(3) provision must be made for contingencies (see below) and some storage volume must be retained in the bottom of the reservoir for settling sediments.

As a guide much greater flexibility will be provided if the capacity of the reservoir is increased by about 50%, i.e. to about 6 hours' supply or 25% of the average daily demand. This is about the minimum practicable storage required solely for the purpose of levelling out the hourly fluctuations of demand from a distribution system. It must not, however, be regarded as sufficient to safeguard the continuance of the supply against all contingencies.

Contingency storage

Contingency storage is required to meet breakdowns at sources, loss of supplies from major mains bursts and the time to repair them, and loss of water used to meet major fire demands. The storage required to meet these contingencies cannot be arbitrarily fixed: it depends upon the nature of the source, the layout of mains, and what safety precautions are possible.

Borehole pumping stations or boosting stations can have 4–5 hours interruption of supply due to electricity supply failure, but in any case such stations should be designed for not more than 22 hours working out of 24, or routine maintenance is difficult. Water treatment plants also need the ability to shut down for 4–5 hours as a minimum in order to effect any necessary repairs or alterations. River intake sources present particular hazards due to sudden pollution, but this is best met by providing raw water storage at the intake since the polluted water should not pass to the treatment works (see Section 7.1). For the repair of major trunk mains the following figures are indicative only of what may occur.

Time elapsed before burst reported	$\frac{1}{2}$ hour (say)
Mobilising repair gang and closing valves to isolate burst pipe	2 hours
Repairing main:	
– 300–600 mm diameter	6–8 hours
– over 600 mm	8–12 hours
Refilling mains including disinfecting	2–4 hours

In practice, a risk assessment should be carried out for mains deemed to be critical if they burst. Emergency action plans must be prepared to assess the total time supplies to consumers would be interrupted following a burst, and what possibilities exist of re-routing supplies to the affected area. The loss of water following a major burst will also need to be considered and how such loss would be regained. A major fire can use 5000–15 000 m^3 of water, but usually notice will be received of this demand and measures should be possible to increase the output from sources accordingly. The breakdown times quoted above are what might be termed 'normal': if things go wrong, double the time may elapse before supply can be restored and every water engineer can think of an instance when this has happened. Allowance has also to be made for the fact that a major mains burst or fire demand may occur when the reservoir is already substantially drawn down, such as during the maximum demand period when the overdraw may rise to 25% of the day's supply.

Hence, storage to meet contingencies as well as hourly and daily variations of demand needs to be provided; but where it should be provided depends on the layout of the distribution system and the output capacity of sourceworks. A study of this is also required, since contingency storage needs to be located where, through interconnection of mains, water can be re-routed to areas cut off by a burst main. Taking all these considerations into account, the overall desirable storage for a system is 24 hours' supply to cope with variations of demand, major mains bursts and sourceworks servicing and repair times. This is a widespread provision, albeit not all undertakings achieve it because of financial constraints. Some undertakings, where reservoir storage levels are continually monitored from a control room, may find 18 hours' storage adequate provided they can accept the risk of having to cut or curtail supplies temporarily in an exceptional emergency. Each water undertaking will need to devise its own strategy for sizing service reservoirs, involving:

(1) use of a water balance/strategic transfer model especially where a reservoir forms one component of a complex system;
(2) use of a formula to take account of peak hour and peak week demand factors applying to the area served;

(3) the provision of additional capacity to meet the extra seasonal demand in holiday areas where visitors can cause local demands to increase by as much as 50% at peak holiday times;

(4) the provision of additional capacity in service reservoirs receiving pumped inflows to permit maximum use of cheap-rate off-peak electricity supplies and keep the maximum demand charge to a minimum.

11.3   Reservoir shape and depth

Service reservoirs are generally built with at least two compartments so that one can be drained for maintenance without putting the whole reservoir out of service. Reservoirs which are circular in plan are less suitable for subdivision, especially if construction of the two compartments is to be phased, when a rectangular shape is usually preferred. Nevertheless circular tanks permit the use of pre-stressed concrete walls which may offer an advantage in cost in certain circumstances. For a two-compartment rectangular reservoir the most economic plan shape is usually obtained when its length (measured perpendicular to the division wall) is 1.5 times its breadth (measured parallel to the division wall). These proportions may need alteration in the light of the shape and slope of the site, the cut and fill balance, depth of inlet/outlet mains and pipework configuration in relation to circulation, and any future extension likely or amenity requirements. If significant or abnormal soil settlements are expected, there may be advantages in providing two adjacent single-compartment reservoirs, which are structurally independent of one another, instead of one two-compartment reservoir with a common dividing wall.

There is an economic depth for any service reservoir of a given storage capacity. The greater the depth the less length of wall and area of roof and floor is needed, though the unit cost of the wall increases with increased water depth. There can, however, be other constraints on the depth such as the character of the available site or the desirable range of distribution pressure required. Depths most usually used for rectangular concrete reservoirs are

Size (m^3)	Depth of water (m)
Up to 3500	2.5–3.5
3500–15 000	3.5–5.0
Over 15 000	5.0–7.0

The following data then needs to be determined for the design to proceed:

(a) top water level (TWL); usually the level at which the supply into the reservoir is to be shut off;

(b) the overflow weir level (OWL); giving a small margin above (a);

(c) the maximum water level (MWL) needed to discharge the maximum possible inflow over the overflow weir;

(d) bottom water level (BWL), being the lowest level to which the water should be allowed to fall for the purposes of supply;

(e) the fall to be given to the roof and the allowance to be given between the roof soffit and the MWL.

The freeboard between highest water level and the roof soffit is required for ventilation and should not be less than 150 mm above MWL or less than 300 mm above TWL. Settlement of suspended solids may occur in reservoir compartments and will concentrate on or near the floor. To prevent turbid water being drawn into supply, the operating BWL should be not less than 150 mm above the highest level of the floor. It may need to be higher, depending on the outlet arrangements (see Section 11.10).

11.4   Covering and protecting service reservoirs

Flat-roofed concrete reservoirs are usually covered over with earth and grass for appearance and temperature insulation. This involves maintenance and grass cutting, but an uncovered reservoir is likely to meet amenity objections. The earth cover to the roof should comprise grassed top soil 150 mm thick, over a fabric filter membrane laid over 100 mm of single size 20 mm round gravel forming a drainage layer. If there are no amenity objections, it is possible to use only gravel for the roof cover which should then be 150 mm thick. The earth banks against the external reservoir walls must be designed to stable slopes and, for ease of grass cutting, should not be steeper than 1 on 2.5. Topsoil cover to banks should be not less than 150 mm vertical thickness.

Special attention has now to be paid to ensuring service reservoirs are secure against vandalism, acts of terrorism and theft. An impact and vulnerability assessment should be undertaken to determine the level of risk and hence the security measures necessary. Secure perimeter fencing can minimise ordinary vandalism but is not sufficient protection by itself. Access manholes to the reservoir can be screwed and locked down; but, if possible, concealment is better, their location being known. Roof air vents are a problem because they are a potential source of pollution and access (see Section 11.13). Valve and instrument houses need to be of strong construction and should have no windows. Sampling points are necessary on both inlet and outlet mains, and they too should be protected. All reservoirs should be visited frequently to ensure that none of the protective measures have been tampered with.

11.5   Statutory consents and requirements

Reservoirs and associated structures will require the consent of the national and local Planning Authority. Regulations will vary from country to country but, in the UK reference will have to be made under the Town and Country Planning Act 1990 for consent, and this action will normally attract comment from various statutory bodies, local organisations and groups depending on the sensitivity of the location. Early consultation with these bodies is necessary to resolve any difficulties which could otherwise result in delay or refusal of the planning application. In the UK, where nationally or internationally designated nature conservation sites could be affected, consultation with English Nature, the Countryside Commission for Wales, or Scottish Natural Heritage is necessary as set out in the Wildlife and Countryside Act 1981. Where the proposed works may affect the apparatus of other statutory undertakers, they must be consulted, and Notices will need to be served under the New Roads and Street Works Act 1991. A discharge consent from the Environment Agency will be required for overflow and drain-down discharge into a sewer, soakaway or watercourse. Where public roads are affected, the Highway Authority should be consulted at an early stage. A range of non-statutory consultation may also be required (e.g. The Countryside Agency for landscape, and English Heritage for archaeology).

In the UK, if a service reservoir will store more than 25 000 m^3 above natural ground level of any part of the land adjoining the reservoir, its design and construction have to be supervised by a Construction Engineer appointed by the water undertaking and who is one of the civil engineers on a panel of engineers appointed under the Reservoirs Act 1975 Act.[1] The Construction Engineer is required to issue certificates during and after completion of the reservoir, specifying the water level(s) to which the reservoir may be filled. During the life of the reservoir, it has to be inspected at intervals not exceeding 10 years by an Inspecting Engineer appointed also from the panel (see Section 5.23).

11.6   Structural design of concrete reservoirs

The type of design adopted depends on the reservoir capacity required, the topography of the site and nature of its subsoil, and the unit costs of the principal materials required – concrete, reinforcement and formwork. Two main types of design adopted are: (a) jointed concrete reservoirs and (b) monolithic concrete reservoirs.

(a) Jointed concrete reservoirs

Jointed concrete reservoirs normally have joints: (i) between lengths of wall, (ii) between the top of walls and the roof, (iii) between floor panels and their junction with wall bases and columns, and (iv) in roofs dependent on their area. The floor and walls may be of plain or reinforced concrete; the roof and its supporting columns are always reinforced.

The *wall design* probably most frequently used is the reinforced concrete free-standing vertical wall cantilevered from a substantial base resting on firm foundation material (see Fig. 11.2). It has to be designed for active loading of the external soil with the reservoir empty, and for maximum water pressure less the passive resistance of the soil. The wall base must be proof against horizontal movement. An unreinforced mass concrete wall can be used alternatively if cheaper than a reinforced concrete wall and the foundation material is satisfactory. This resists water pressure by its weight and the passive resistance of the soil backing. A sliding joint must be provided between top of the reservoir wall and the roof to prevent roof expansion or contraction imposing a lateral force on the wall. Contraction joints equipped with waterstops and sealing grooves must be provided at intervals in the wall, usually at intervals equal to twice the column spacing. Expansion joints with waterstops must be provided at intervals not exceeding 45 m.

Columns of reinforced concrete are normally arranged on a rectangular (usually square) grid pattern. A column spacing of 5 m results in a flat-slab roof of economic thickness without the need for dropped panels. This distance is reduced where an expansion joint is necessary in the roof and between a wall and the nearest row of columns. Columns may be square or round with mushroom heads (pyramidal or conical). The side dimension or diameter should be not less than 300 or 350 mm, respectively and not less than one-twentieth of the height from reservoir floor to bottom of column head. Columns need to be of uniform height for economy in cost of construction, this being obtained by making the floor and roof parallel.

The *floor* may be of plain or reinforced concrete, cast in square panels having a side length equal to the column spacing, and may comprise one of the following designs.

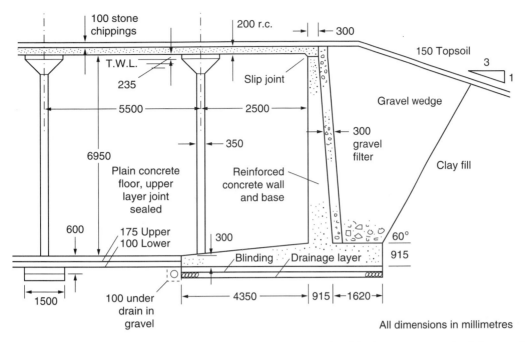

Fig. 11.2 Design for a jointed service reservoir where only chippings were required for roof covering.

(a) An upper layer, with thickness dependent on the water depth but typically 175 mm, below which is a lower layer, 100–125 mm thick. A membrane between the layers permits sliding of the upper layer. This design is suitable for a clay subsoil.
(b) A single layer, typically 175 mm thick, below which is a 75 mm layer of blinding concrete. This design is suitable for founding on a firm, non-compressible material, e.g. solid rock, unweathered chalk, firmly compacted and free-draining sand or gravel.

In the two-layer floor (a), usually only the top layer is reinforced, the reinforcement being discontinuous through the contraction joints. Where a single-layer floor (b) is used, it is laid on a membrane of low frictional resistance, and it may be unnecessary to provide reinforcement if the subsoil is firm and of uniform bearing capacity. It is, however, recommended that reinforcement is provided where the floor is resting on a clay subsoil. With these types of jointed floors, uplift pressures must be prevented by provision of an underdrainage system which has a free discharge to a lower level.

Joints separating floor slabs should be of the 'complete contraction' type, incorporating a joint sealant at the water face (see Fig. 11.3). Only the upper layer of a two-layered floor need be so jointed. Centrally-placed waterstops are often used, particularly at floor-to-wall and the floor-to-column base junctions, but great care is needed to ensure proper compaction of the concrete under the waterstop which is difficult to place. This problem is avoided if 'rearguard' waterstops at the base of the floor are used instead; but these tend to be less effective in preventing outward leakage although they will prevent contamination from the ingress of ground water. In a two-layer floor, the joints in each layer should be staggered to avoid vertical alignment. The joints in the lower slab should be arranged midway between the columns; the joints in the upper slab may then lie on the column grid

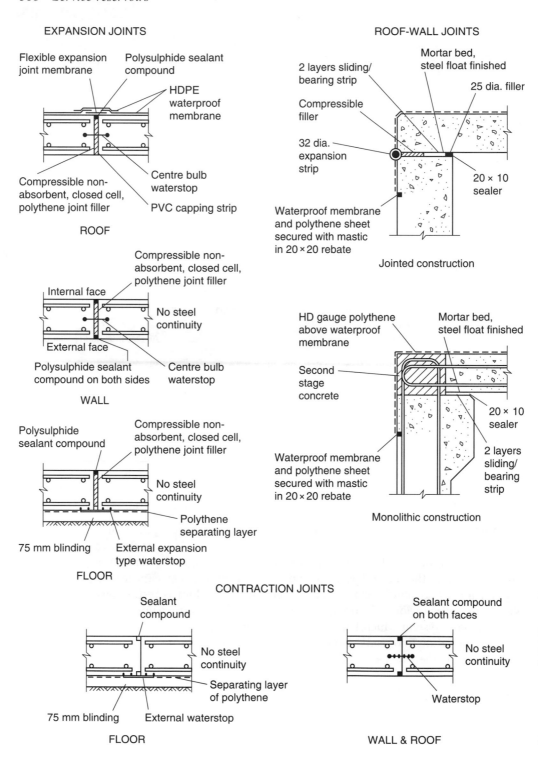

Fig. 11.3   Typical joints for concrete reservoirs.

lines between the column bases. Where possible, the upper floor slab should not be cast until the reservoir, including the roof, is substantially completed. This helps to avoid excessive shrinkage and temperature movements and (an important point) will assist in preventing the joints from becoming damaged and fouled before the joint sealant is applied. In a single-layer floor, the slab joints may be arranged in a manner similar to that described above for the top layer of a two-layer floor. The blinding concrete beneath a single-layer floor may be cast continuously, construction joints being provided as necessary.

The *roof* is a reinforced concrete slab of uniform thickness, minimum 200 mm and is monolithic with the column heads. Contraction joints are not needed (because the columns are flexible enough to permit roof contraction), but expansion joints with waterstops must be provided at intervals not exceeding 45 m. The roof design must allow for the impact loading of constructional plant placing gravel and soil on the roof, and for any live loading that may possibly occur.

(b) Monolithic concrete reservoirs

A monolithic concrete reservoir has reinforced concrete walls, floors, columns and roof, in which there are few (if any) permanent movement joints. In some cases the walls and floor are monolithic but there are sliding joints between roof and top of walls. This type of design has been found to be structurally economical in most situations where the underlying ground (after improvement if necessary) will support the load without risk of appreciable differential settlement. The reservoirs are normally rectangular in plan, see Fig. 11.4, but circular and other shapes are feasible.

External walls are usually vertical or near vertical on the inner face but battered on the outer face to give the tapered section appropriate to the form of loading (see Fig. 11.5). Depending on the height of the walls and the length of the roof slab, monolithic connections with floor and roof slabs can result in lower bending moments and shear forces (especially in the vertical plane) than is the case with jointed structures. Within the walls, joints are usually restricted to partial contraction joints (discontinuities in the concrete with a proportion of the reinforcement passing through) and with a sealing groove on the water face. The maximum spacing of partial contraction joints should be 7.5 m to avoid unacceptable cracking. For operational reasons, the division wall is usually full-height and can therefore assist in supporting the roof. The columns are arranged on a square grid, the span to external walls being typically reduced to three-quarters of the normal spacing, and the roof should be a reinforced concrete slab of uniform thickness.

An economical form of floor is a reinforced concrete slab of uniform thickness except at the perimeter, where it should be thickened to cater for moments transferred from the walls or resulting from differential vertical movements as between perimeter and centre (see Fig. 11.5). Local thickening of the floor below columns should be avoided as it can be awkward and costly to construct; instead additional reinforcement under the columns can be used to increase the shear strength. Local thickening is usually required at drainage channels and sumps where these are included. Joints are normally restricted to construction joints but partial contraction joints can be used as described for walls.

(c) Other types of reservoir

Whilst service reservoirs of the types described above are by far the most common type adopted, other types of construction are possible and have been used – mostly where

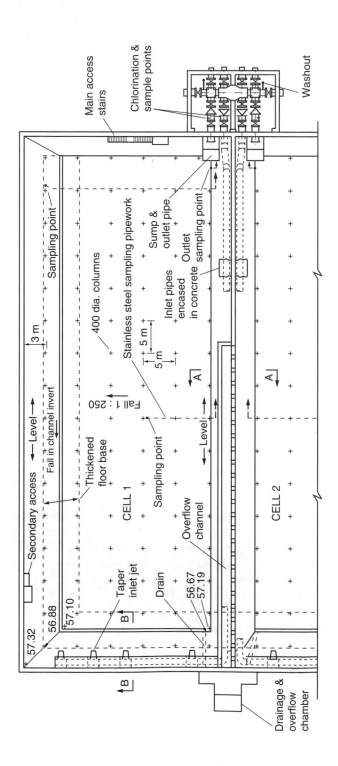

Fig 11.4   Monolithic concrete reservoir – plan.

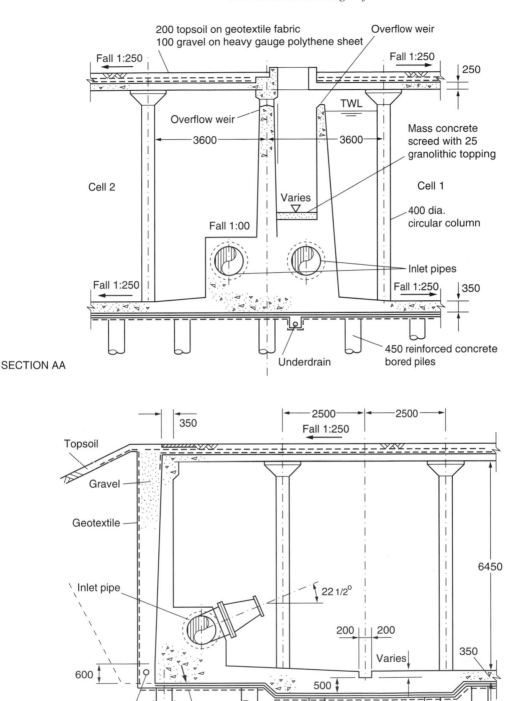

Fig. 11.5 Monolithic concrete reservoir – sections.

special circumstances apply. Some early reservoirs may still remain, built of brickwork with a rendered internal face and often had barrel vaulted roofs; but their construction was discontinued in the early 1900s. Steel tanks are occasionally used for small reservoirs if there is no amenity objection. The two main current alternatives to the traditional concrete reservoir are the following.

Pre-stressed concrete reservoirs offer the possibility of holding a greater depth of water than normally reinforced concrete walled reservoirs, but they need to be circular in plan. Hence they need a level (or levelled) site since they are not suitable for unbalanced earth surcharge backing. They are not usually earth embanked and often have an uncovered shell roof of concrete, so they are only possible where there is no amenity objection to their visual appearance.

Earth embanked reservoirs with columns and roof of reinforced concrete have been used for the construction of large reservoirs where the ground is suitable. The floor and inner face of the embankment have to be lined with asphaltic concrete or some other impervious membrane. The care necessary in the design and construction to prevent movement or settlement causing leaks at joints, and the different processes of construction involved (earth compaction, concrete construction, and extensive membrane water-proofing, etc.) means such reservoirs tend to be economic only in special circumstances.

Choice of type of reservoir

The practical choice of reservoir type will usually be between a jointed concrete reservoir and a wholly or largely monolithic reservoir. To enable this choice to be made, the differential movements likely to be experienced under all possible loading conditions must be assessed; in the case of the monolithic design, the need for piling or soil stabilisation may need to be considered. A jointed reservoir will be possible on any site where foundation movements will not cause movements in the structure which are too great to be accommodated by the joints. A monolithic reservoir requires any variations in the underlying ground to be within the limits that the monolith can withstand without cracking. However, bearing piles or ground stabilisation may sometimes be used to improve the foundation if the ground is weak or variable. The absence of movement joints eliminates a potential source of leakage or contamination by ground water.

11.7   Design procedures

Current British practice is to follow the procedures set out in BS 8007[2] for the design of liquid retaining structures by a method based on limit state philosophy in the same way as that adopted for the design of reinforced concrete structures to BS 8110.[3] The procedure involves the calculation of the maximum serviceability widths of cracks and the determination of reinforcement needed to restrict them to limiting values depending on the degree of exposure to which structural elements will be subjected. The Code requires checks to be made for safety or 'collapse' for other limit states, including the ultimate limit state, in order to ensure that the structure remains serviceable at working loads throughout its expected life. The design, detailing and workmanship of joints are discussed in some detail and the designer is left to choose, from a number of options, the best method of detailing joints to control shrinkage and initial thermal contraction.

The Code uses the term 'characteristic strength' for concrete and the same term is applied to the yield stress of the steel reinforcement. The characteristic strength is defined as the value of the concrete strength or steel yield stress below which not more that 5% of the test results may be expected to fall. A minimum cement content of 325 kg/m^3 is specified for structural concrete with a maximum water/cement ratio of 0.55 which should be reduced to 0.50 when pulverised fuel-ash cement replaces part or all of the ordinary Portland cement. The resulting concrete is classed as grade C35A to designate a requirement for a 28-day characteristic cube strength of 35 N/mm^2. The use of grade C20 concrete, containing a minimum of 220 kg/m^3 cement, is permitted for non-structural components of structures and C10 concrete, containing a minimum of 150 kg/m^3, can be used where there is no requirement other than to protect an exposed formation from the effects of weather or to fill an excavation with a stable material. The Code recommends that aggregates should normally be 20 mm maximum nominal size in reinforced members and the cover to all reinforcing steel should be not less than 40 mm. In practice aggregates should form a dense mix and inclusion of some 38 mm size stones can help to reduce shrinkage.

Since the difference in cost of high yield deformed bars over mild steel bars is small, and as economies result from the use of the former material, all reinforcement except that used for links and binders in beams and columns should be in high yield steel with a minimum characteristic yield strength of 460 N/mm^2. The Code sets out allowable steel tensile stresses for different design crack widths.

Where service reservoirs are to be built overseas, any relevant local codes of practice should be followed, and designs may have to take account of local constructional abilities and the quality of materials available. Often this involves using lower permissible design stresses for concrete whose quality control may not reach the standards possible in UK. Also design practices may need to be modified to take account of local prices prevailing so that an economic construction results.

11.8 Drainage and waterproofing requirements

Where the reservoir external wall is designed as a free-standing cantilever or is of mass concrete, the backfill against the wall should comprise a vertical drainage layer of gravel about 300 mm, extending the full height of the wall and continuing down over the wall heel to link with a drainage system. Where the wall is monolithic with the reservoir floor and roof, the granular wedge may – depending on the design assumptions – be omitted, but it will generally be advantageous to retain the vertical gravel layer to control the ground water level and also to transmit water draining from the roof. The wall and roof drainage systems must be kept separate from any underfloor drainage system, and it is an advantage to arrange the wall drainage pipework so that it can show from which wall (or pair of walls) there is a discharge into an observation chamber. This aids location of any wall leak that has developed.

The roof should prevent ingress of water and pollution by being covered with a waterproof membrane. For new reservoirs an adhesive membrane such as Bituthene DW is recommended, protected by heavy duty polyethylene sheet under the gravel drainage layer. The roof gradient should be no flatter than 1:250 for drainage, and the floor slope should be made parallel to it, so as to maintain constant wall and column heights. The simplest way to achieve this is to provide the slope in one direction only. Where a vertical drainage layer is provided behind the reservoir wall, the roof drainage can usually be allowed to flow over the roof edge into it. Where there is no vertical drainage layer, and stability of the embankment is essential for wall stability, a low peripheral kerb provided along the lowest edge of the roof can act as a collector for piping the water away. Inside the reservoir, the floor should have a shallow collecting channel leading to a drainage sump to aid cleaning of the floor of the reservoir.

The underfloor drainage system is usually laid to a herring bone or rectangular pattern, normally comprising porous pipes surrounded in gravel in a trench below the floor. The layout should make it possible to observe drainage or leakage flow from separate areas of the floor. The porous pipes from each area are continued in ductile iron piping laid in concrete below the wall and its embankment, to discharge individually to a collector manhole, from which there is a free outfall pipe to some lower point. This provision aids locating any floor leak. With jointed reservoir construction there must be no possibility of the outfall being submerged. If, on the other hand, conditions are such that uplift below the floor is unavoidable, then monolithic construction must be adopted to a special design, with the weight of the reservoir when empty being made greater than any possible uplift pressure.

11.9   Access provisions

Access to each reservoir compartment is needed for personnel, plant and materials. Personnel access is required comparatively frequently and openings for that purpose should be as small as practicable to minimise the weight of the cover. Access for plant and materials is less frequently required, and the opening has to be larger. Kerbing has to be provided around each opening to prevent surface water entering the reservoir. The covers to all access openings must be robust although they do not normally need to be designed to support heavy loadings. They must be secure to prevent unauthorised access, and must not allow rainwater to enter the reservoir. Hinged covers are usually preferable; lift-off covers introduce the risk that mud and debris might be accidentally carried into the reservoir after adhering to the underside of the cover. However, it is essential for the cover to have an effective system for holding it in the open position when the access is in use.

For personnel entry into the reservoir the preferred arrangement is an inclined ladder leading to a platform about 2.5 m below the roof and a stairway leading from the platform to the floor. Where a stairway exceeds 3 m, an intermediate landing is required. Reinforced concrete construction is recommended for platforms and stairways as this needs less long-term maintenance. The platforms can either be supported on columns or, in some cases, cantilevered from the walls. Alternatively the platforms and stairways can be fabricated in galvanised steel or anodised aluminium alloy. The same material should be used for the ladder. For reservoir compartments exceeding 2000 m^3 capacity, two separate human accesses should be provided into each compartment, near opposite corners to assist ventilation of the compartment when work is in progress and to provide an escape route for use in an emergency.

Access for plant and materials has to be unobstructed to allow items to be lowered vertically to the compartment floor. Provision for human access as well is impracticable without an unduly large roof opening. The clear opening needed for small plant and materials for normal maintenance should be not less than 1.5 × 1.0 m to allow a wheelbarrow to be lowered. Consideration should be given to the provision of removable handrailing around such openings, or of sockets into which it could be fitted. For reservoir compartments exceeding about 10 000 m^3, a second and larger access for plant and materials should be considered, if there is a possibility that larger mechanical equipment will be needed for cleaning or major repairs. If wheeled plant is to be used on a reservoir roof or within a reservoir, a suitable access ramp onto the roof will be desirable. However, it is important to ensure that appropriate arrangements are made to prevent unauthorised vehicles reaching the roof or plant being used outside any specially strengthened areas of the roof.

11.10 Pipework

Reservoir pipework will normally comprise, inlet main(s), outlet main(s), overflow pipe, drawdown/scour pipes, reservoir bypass and drainage pipes, and may include a suction main to a site pumping station. It is common practice to provide a small-diameter valved pipe through the division wall of a two-compartment reservoir, so that a copious supply of water is available for hosing down a compartment when taken out of service for cleaning. An 80 mm diameter pipe is usually sufficient for this purpose, and accords with standard fire hydrant sizes. Unless separate connecting pipes are used for flow in each direction, the control valve on the connection must be operable from outside the reservoir. Flexible joints should be incorporated between embedded or rigid pipes and external pipelines to accommodate differential settlement.

Inlet and outlet mains should bifurcate to serve each reservoir compartment equally. The inlet main can discharge at TWL or near BWL. One of the disadvantages of the latter is that, in the event of a burst on the incoming main, the reservoir contents will be lost unless a suitable non-return valve is provided. On the other hand, if the incoming supply is pumped, a high-level entry will forfeit the energy savings potentially available when the reservoir is operating below TWL.

Inlet pipework

The inlet piping arrangement needs to achieve complete mixing of the inflow with the stored water and ensure there will be a flow of water through all parts of the reservoir to avoid build-up of stagnant water areas. This involves suitable siting of the inlet and outlet pipes and, if necessary, the use of baffle walls. For the design of large reservoirs and where there are water quality problems, it is becoming more common to use 3-D models to optimise inlet and outlet arrangements and where baffle walls (if any) should be placed. Options for encouraging circulation comprise – placing the inlet and outlet at opposite ends of the compartment (see Fig. 11.4); distributing the incoming flow as evenly as possible along an end wall by the use of a long inlet weir; using a tapered diffuser pipe with several openings, or delivering to a semi-circular terminal box with slotted outlets.

Outlet pipework

The most common (and simplest) outlet system uses only one draw-off point per compartment, but this is likely to leave some potentially stagnant areas in one (or both) corners at the outlet end of the compartment. To avoid this the outlet may draw water from a number of points along an end wall if flow distribution is used as the sole means of avoiding stagnation.

If the inlet and outlet pipelines are to terminate at opposite sides of the reservoir compartment, it may be appropriate to have separate inlet and outlet valve houses.

However, for economy and ease of operation, a single valve house containing controls for both inlet and outlet is usually preferable. With this arrangement, one pipeline (usually the inlet) is normally laid within the reservoir compartment to feed water to the far end of it. This pipeline should be placed alongside the wall and encased in concrete to avoid 'dead' spaces and to inhibit external corrosion of the pipes. If the reservoir is of the jointed design, the internal pipework (and its surround) must have flexible joints corresponding with the joints in the structure.

The outlet main can be laid horizontally, either through the reservoir compartment wall, or under the floor with a 90° vertical bend. It is usual to provide an entry bellmouth, to reduce hydraulic losses. If the outlet main serves, or has a connection to, a pump suction, the outlet bellmouth must be sufficiently submerged at BWL to prevent the entrainment of air into the main. For a bellmouth in the horizontal plane (i.e. vertical exit), a safe rule for minimum submergence of the bellmouth lip is:

$$S/D = 1.0 + 2.3F$$

where S = submergence below BWL; D = bellmouth diameter at lip; and F is the Froude number $V_D/(gD)^{0.5}$, where V_D = the average flow velocity through the exit plane of the bellmouth. With a bellmouth in the vertical plane (horizontal exit) the same equation may be used, but S is measured from the bellmouth axis.

For a gravity supply outlet main, the submergence requirement can usually be relaxed depending on the acceptability of air entry into the pipeline, but should not be less than D. The required submergence may create an uneconomic depth of 'dead' water unless the reservoir outlet is lowered by means of a sump in the floor. The sump should be generously sized to avoid undesirable hydraulic turbulence. The bottom of the sump will collect floor deposits and should be not less than 300 mm below the bellmouth lip. Safety features for maintenance personnel will need to be considered in the detailed design of a sump. The outlet sump can also serve as the drain sump.

Where a service reservoir has a common inlet/outlet main, circulation inside the reservoir can be achieved by dividing the common main into inlet and outlet pipes before these pipes bifurcate to each compartment. If a low-level entry design is used, both inlet and outlet must be fitted with non-return valves. With the high-level type entry, a non-return valve is required on the outlet pipes only.

Overflow and draindown pipework

Adequate overflow arrangements must be made in case of an inflow malfunction, such as the failure of supply pumps to shut down when the reservoir water level reaches TWL. The overflow capacity from each compartment must not be less than the maximum uncontrolled inflow possible into that compartment. The usual provision of a vertical pipe with bellmouth attached as an overflow has the defect of limited capacity, and a horizontal weir is usually required. A convenient arrangement in a two-compartment reservoir is a weir box in the central division wall with weir entries from each compartment. The weir box often discharges to a pipe laid through the valve house, which can also receive the washout pipework connections.

The combined overflow/washout system should preferably discharge into an open watercourse, which must be of adequate capacity. It may be necessary to consult with the land drainage authority or sewerage agency concerned if drainage is to a sewer. A break pit must be provided before final discharge to provide a means of monitoring chlorine levels after disinfection and applying any necessary treatment. Drainage pipes should be connected to a drainage sump in each compartment and their size will be dependent on the compartment storage volume, the drawdown time from full that is acceptable, and any limitations applied to rate of discharge to a watercourse or sewer.

Drawdown times of 8–12 hours for reservoir compartments under 2000 m^3, and between 24 hours and 3 days for larger compartments, may be considered appropriate.

A reservoir bypass (between inlet and outlet pipes) is necessary in the case of a single-compartment reservoir, or where the whole reservoir may need to be taken out of service. It is normally possible to accommodate the valve on this within the valve house.

11.11 Use of baffle walls or curtains

Baffle walls or curtains aim to achieve plug flow through the reservoir and serve to direct the flow of water through the compartment from inlet to outlet by a circuitous route. The optimum arrangement of baffles will depend upon the shape of the compartment and the most convenient positions for the inlet and outlet. For construction convenience, baffles are normally installed between the columns supporting the reservoir roof.

Solid baffle walls are normally made of reinforced concrete, brickwork or blockwork. Lightweight, hollow blockwork should be avoided because the free chlorine normally present in a reservoir air-space has been known to cause deterioration of some blocks of this type. Plastic curtains are sometimes used. Where appropriate, openings must be provided along the bottom of the baffle walls to allow all areas of the reservoir floor to drain and to facilitate cleaning and maintenance activities. Openings should also be provided at the top of the walls to assist with ventilation. Joints must be provided in the walls at all points where they bridge movement joints in the floor or roof. Additional joints may be needed to allow for thermal movements or for the flexing of the reservoir walls as water levels rise and fall.

Baffle curtains may be less robust but easier to install. Care must be exercised in the selection of the material used, ensuring that it is suitable for long-term use inside a reservoir and is unlikely to be torn. The recommended method for supporting the top edge of the curtain is to build a dove-tail steel channel into the roof soffit for the subsequent attachment of hangers. A similar arrangement can be used for connecting to the vertical edges to walls and columns although, for columns, it may be easier to provide a system of lashing or strapping around the column. The bottom edge must also be firmly anchored and this is best done by fixing it to steel or GRP angles bolted to the floor. The use of pre-cast concrete blocks which are not fixed to the floor is not a recommended method of anchoring a curtain. The alignment of the curtains should be chosen so that they are not subject to high velocity or turbulent flow such as may occur near an inlet weir or inlet pipe discharge. A problem commonly arises near the inlet when a reservoir is filling – velocities of flow may initially be high until the depth of water has built up. In such circumstances, it is recommended that a dwarf concrete wall, bonded to the floor, is provided below the curtain. This wall should be about 500 mm high and the bottom of the curtain should be anchored to the top of the wall. Dwarf walls must contain drainage openings in the same way as full-height walls.

11.12 Valves

Stop valves must be provided on inlet mains, outlet mains, scour pipes and the reservoir bypass, but must not be provided on the overflow or any wall or underfloor drainage systems. Stop valves may be either gate (sluice) valves or butterfly valves. Gate valves become impracticable for normal reservoir use above about 600 mm diameter, when resilient-seated butterfly valves should be provided. The valve size can be less than that of the pipeline, though the saving in cost of the valve is at least partly offset by the need for tapers and the increased space occupied by the pipework in a valve house. If a smaller size of valve is selected, a check should be made that the maximum velocity through the valve does not exceed that recommended by the valve manufacturer.

Over-velocity valves, designed to close automatically when the water velocity in the pipeline exceeds a predetermined rate, have fallen out of general favour because of their high cost and infrequent use. They may still be appropriate in special circumstances, for example where a large reservoir provides the major supply to a distribution area, or where the loss of water from a failed outlet main would be severe because of high head. The possible need for such valves should therefore be reviewed in reservoir planning.

It is common practice to group the various valves in a valve chamber or valve house. However, in some circumstances, it may not be practicable to do this and alternative arrangements need to be made. In general, all butterfly valves and special control valves should be installed within chambers so they are accessible for maintenance. Important gate valves (such as the isolating valves on any pipes connecting into the reservoir) should also be placed within chambers but others can be buried in fill.

Special consideration needs to be given to the siting of isolating valves on pipes leading into or out of the reservoir. It is essential that such valves are installed so that they are connected directly to the reservoir structure without any possibility of differential movement occurring. For example, a valve separated from the reservoir by a length of pipework including flexible joints would not be acceptable because, if earth movements occurred, the pipe joints could fail and there would then be no way of preventing the entire contents of the reservoir being released. Thus, unless such valves are within a valve house forming an integral part of the reservoir structure, they should be placed outside the reservoir wall and its embankment but close to it, so that the connections to the pipe components built into the wall are flanged joints. The same principles apply to outlet or drain pipework built into the reservoir floor – all joints between the pipe end and the valve must be flanged and (except where exposed in chambers) surrounded in concrete.

Valve house requirements

It is often convenient for all reservoir control valves to be concentrated in one chamber or valve house. For security of supply, the valve house should be as close as possible to the reservoir and is usually part of the reservoir structure. Access to the valve house may be by top entry – through the roof – or side entry through a wall. Top entry may result in the whole of the interior of the house being classified as a 'confined space', with consequent safety constraints on entry. These could give rise to unacceptable delays in gaining access if, for example, it becomes necessary to isolate the reservoir in an emergency. Side-entry valve houses are to be preferred from an operational viewpoint although they may give rise to unacceptable visual impact in environmentally sensitive areas.

The pipework within the valve house must be arranged so that it is possible to install, maintain or remove any valve without great difficulty. If a valve is too heavy to be manhandled, it is important that there is clearance for a straight vertical lift out of the building (if top entry is provided) or to a position where it can be transferred to a trolley or road vehicle (if side entry is provided).

In addition to pipework, valve houses are often used to accommodate sampling pumps and pipework, level recording and indicating equipment, telemetry and site monitoring equipment for flowmeters on inlet and outlet pipework, ventilation and dewatering equipment. Provision must also be made for dealing with any water resulting from spillages during maintenance work or leakage from pipework components. As a minimum, this should comprise a sump into which the suction hose of a portable pump can be inserted.

11.13 Reservoir ventilation

Ventilation of reservoir compartments is needed to maintain a fresh supply of air above the water surface, for temperature control of that air and to admit or release air displaced by varying water levels in the compartment. The cross-sectional area of ventilation ducts or openings should be based on an air speed of 15 m/s. Allowance should be made for the reduction in effective air passage area caused by insect screens. In many cases, traditional mushroom-type roof ventilators have been found unsatisfactory in long-term service and to be a potential source of pollution. 'Vented' access covers have sometimes been used for small reservoirs but are also vulnerable to pollution. Alternatives include piped systems above the reservoir roof, leading to one or more ventilation chambers or ventilation ducts.

11.14 Sampling and instrumentation requirements

Monitoring of water quality in service reservoirs and water towers is required in UK under the Water Supply (Water Quality) Regulations 1989.[4] At sites with more than one service reservoir, each reservoir should be sampled separately unless they are inter-connected when the final outlet point should be sampled. Likewise where a reservoir is divided into compartments, each compartment should be sampled as a separate reservoir unless the compartments are inter-connected. The sampling arrangements at service reservoirs with combined inlet/outlet mains should always ensure that the water sampled is from the body of the reservoir or that leaving the reservoir. If there is no suitable location for a sample tap at the reservoir site, then a tap should be provided on the outlet main at the nearest possible point to the reservoir. Dip sampling is not recommended and should only be used as an emergency measure.

The following instruments and equipment are normally required.

(a) A stilling well, either within or external to the reservoir, in which the water level measuring equipment is installed. If the reservoir is divided into two compartments, an externally-mounted stilling well with valved connections to each compartment should be provided. The well and valving are normally sited in the valve house. In some cases it may be desirable to have separate stilling wells for each compartment.

(b) Equipment for displaying the level measurement in either digital or analogue form in the valve house or in a weatherproof and vandal-proof enclosure installed adjacent to the stilling well.

(c) Level electrodes or float switches at TWL (high) and BWL (low) for normal on/off pump control and alarm purposes. Additional switches at intermediate levels may be required for some pump control schemes.

(d) Level electrodes or float switches at OWL (extra high) and just below BWL (extra low) for alarm purposes and emergency control of pumps feeding the reservoir.

(e) A display panel (wall- or floor-mounted as appropriate) for level and alarm indications, power supplies and the like. Provision should be made for interfacing with a telemetry outstation which, if required, may be mounted inside the panel.

(f) A telemetry link to the pumping station or control centre to transmit water levels, flow rates through the inlet and outlet pipes, valve positions, remote control of valves, power supply failure, telemetry failure and, when provided, an intruder alarm.

Short- and long-term monitoring of movements of the reservoir structure can be accomplished by installing reference points at selected external locations on the structure, e.g. walls, roof and embankments, and by employing surveying techniques. Datum points will also usually need to be established well clear of the zone of influence of the reservoir. If required, strains within the reservoir may also be monitored remotely by a suitable instrumentation system.

11.15 Testing service reservoirs

Service reservoirs are usually required to be tested for watertightness before being put into service. The test should be carried out before placing any backfill or banks against the outside walls unless the wall design relies on the embankment to resist hydraulic forces. The roof should be in place. In the case of reservoirs of the monolithic design, the second-stage concrete at the roof/wall joint must also be completed. Each reservoir compartment should be tested separately, with the other compartment empty.

The compartment to be tested should be filled, to a test level about 75 mm below the overflow sill, with treated water at a uniform rate not exceeding 2 m vertical rise in water level per 24 hours. It should then be left to stand for at least 7 days to allow for absorption into the concrete. Longer periods (up to 21 days) may be required by some specifications. The water level should then be measured and recorded using a hook gauge with vernier control, or by other approved means of no less accuracy, and the water allowed to stand under test for 7 days. At least once each day during this period, the water level should be measured and recorded. During the 7-day test period, the effects of evaporation from the water surface can be reduced by closing all air vents and access openings (except for one vent left open for pressure balance).

Any flows in the underdrain and wall drain systems should be measured and recorded throughout the test, from a time at least 24 hours before beginning to fill, until 24 hours after emptying or on completion of a final water level measurement. Taking such measurements in chambers on the drain systems normally require safety precautions appropriate to confined spaces. The outfalls of all pipes connected to the reservoir should be inspected during the test to ensure that all isolating valves are shut tight. Any significant leakage through them should be measured. In some circumstances it may also be necessary to keep records of evaporation losses from the water surface.

The test may be deemed successful if the drop of water level over the 7-day test period does not exceed the lesser of $1/500 \times$ average water depth or 10 mm, after deducting any measured leakage through valves and making allowance for any evaporation or condensation.

If the test fails, any increase in underdrain or wall drain flow during the test period should be investigated to identify, if possible, the part of the reservoir that leaked. The test compartment should then be emptied and closely inspected for faults likely to cause the leakage.

Investigating reservoir leakage can be troublesome and time consuming. The interior of the reservoir – especially any joints – should be closely inspected before filling with water, and care is needed in setting up and using the flow measurement devices.

11.16 Searching for leaks

It is not easy to make a large reservoir fully watertight and the following notes as to what steps might be taken to track down the point of leakage may be found helpful.

(1) Flows from underdrains should be examined. These drains should preferably have been so designed that they can give some idea as to the location of the leakage; they should not all join to one common outlet point before being measurable.

(2) The inlet and outlet valves must be tested to ensure that they are not passing water into or out of the reservoir. The only secure way of knowing this is to have short removable sections of the mains on the outer side of the valves so that sections can be removed and the outflow, if any, through the valve can be observed and measured. If testing can be done before making all the pipe connections this will avoid the need for removable sections in them. Valves ought not to leak but many do, not necessarily through faulty design but most commonly because the gate is not shut properly because of dirt in the gate groove.

(3) The rate of leakage at full depth, half depth, and with about 0.6 m of water in the reservoir should be measured so that some idea is obtained as to the possible height of the leakage point within the reservoir. It is not likely that any revealing mathematical relationship between rate of leakage and depth of water in the reservoir will be established because any 'crack' through which water is leaking may be vertical or horizontal, long or short, and there may be several such cracks. It is usual to find that leakage is less when the depth of water is less, but occasionally one may find that there is no leakage at all below a certain level and this is a useful piece of knowledge.

(4) If the leakage is not traced by the above methods then the reservoir must be emptied and subjected to the most careful internal inspection. Emphasis must be placed on careful inspection. Good lights, adequate ladders, plenty of time, and a consistent pattern of examination should be adopted. It is extremely easy to miss a faint crack in the wall or floor. Walls (particularly the joints next to the corners) should receive special attention for it is here that there is most likelihood of movement having occurred. After emptying a reservoir the walls should be kept under observation when drying off because there is a certain stage of drying when leakage is evidenced by a dark patch on the surface of the concrete, even though this is the waterface. This patch will be short lived, but it may give a clue as to the whereabouts of a poor area of concrete through which leakage is taking place.

(5) The floor joints should be inspected. Jointing material should be examined to see if it has sunk, has holes in it, or has come away from or failed to bond to the concrete of

the sealing grooves. The majority of leakages arise from defects of this kind. Wall joints should similarly be examined.

(6) If failure still results, about 0.6 m of water should be put into the reservoir and be left to stand until the water is quite still. Then crystals of potassium permanganate may be dropped into the reservoir, widely spaced, and left for a considerable time. Then, descending into the reservoir with a good light and walking over a pre-arranged walkway so as not to disturb the water, streaks of colour may be noticed from the permanganate crystals showing some definite flow towards a point of leakage.

(7) As an alternative to method (6), about 150 mm of water can be put into the reservoir, a hole or several holes bored through the floor, and compressed air can be introduced under the floor. In certain conditions of floor foundation air bubbling upwards through the water may indicate where faulty floor joints occur.

(8) If, despite all these attempts, the cause of leakage is still unaccounted for then more drastic measures may have to be undertaken, such as digging pits in the bank to inspect the rear of the wall joints, placing further sealing strips (such as glass fibre embedded in bitumen) over joints, or even rendering wall face areas. Sealing of leakage through a reservoir floor has been achieved by gravity grouting. About 450 mm of thin grout mix is put into the reservoir and the cement is kept in suspension by continually sweeping the floor and disturbing the water with squeegees for two successive days. Thus grout passes into the unknown paths of leakage and the cement sets. It should not be necessary, however, to adopt these measures unless poor construction has taken place. Where ordinary care has been taken with construction, failure to find the cause of leakage should be taken as an indication that the interior inspection has not been carried out carefully enough and this should be repeated. In the long run most troublesome leaks are discovered to be in some rather obvious place which has not been thoroughly examined in a first or even a second examination.

11.17 Water towers

Water towers are necessary in areas of flat topography in order to provide sufficient pressure for delivery into the distribution system. A typical design for a reinforced concrete reservoir on a cylindrical shaft and a raft base is shown in Fig. 11.6. Many shapes and design features are possible; generally the designer will aim to produce a structure that is aesthetically acceptable to the water undertaker and the planning authorities, bearing in mind that it will become a landmark in the community which it serves. Ancillary equipment including pipework, ladders, instrumentation and booster pumps, if required, can all be hidden in the cylindrical shaft. Examples are illustrated in Plate 26.

Supplies to water towers are likely to involve pumping and energy consumption. The optimum depth/diameter ratios should be determined for each location having regard to any benefits available from use of power at off-peak charges and a need to avoid large pressure fluctuations in distribution that may be caused by drawdown or filling in excessively deep tanks. Typical dimensions adopted for the design shown in Fig. 11.6 are:

Size (m^3)	Depth of water (m)	Internal diameter (m)
1200	7.5	17.0
2000	9.1	19.4
3000	10.2	22.6

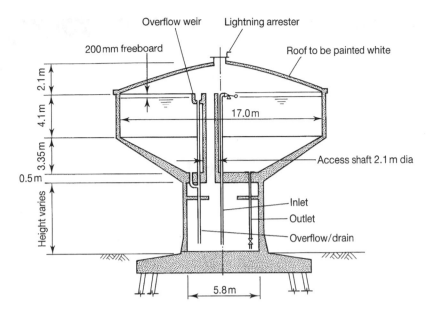

Fig. 11.6   Typical layout for RC water tower.

11.18 Steel and other reservoirs

Welded steel ground level or elevated tanks are often provided as storage reservoirs in capacities up to 9000 m³, particularly on pumped pressurised systems. Elevated storage tanks with conical or ellipsoidal bottoms can be manufactured in a variety of shapes and sizes for support on tubular columns or towers. Service reservoirs fabricated from modules of bolted pressed steel plates, e.g. Braithwaite tanks, are rarely constructed nowadays due to the high cost of protecting the metalwork against corrosion and their limited service lives. The corrosion problem can be overcome by spraying a special borosilicate glass onto a pre-coated steel plate which is then fired causing the glass slurry to fuse with the steel and produce a hard impervious finish. Service reservoirs are then formed by bolting together curved modular units to form cylindrical structures which can be mounted on steel or concrete floors at ground level or onto a concrete frame structure. Diameters of up to 17 m have been constructed without columns for supports of roof panels. Reservoirs have been constructed of rigid plastic panels bolted together but they have not proved satisfactory in hot climates because repeated flexing of plates has led to ultimate cracking of them near their fastenings.

References

1. The UK Reservoirs Act 1975 and associated Regulations dated 1985 and 1986.
2. BS 8007: Code of Practice for Design of Structures for Retaining Aqueous Liquids.
3. BS 8110: Structural Use of Concrete Part 1 – Code of Practice for Design and Construction
4. The UK Water Supply (Water Quality) Regulations 1989.

12

Pumping plant: electrical, control and instrumentation systems

12.1 Pumping plant

Using machines to lift water is a very ancient art, developed to satisfy the most basic of human needs, water for domestic use and water for irrigation to grow crops for food. Fortunately for the present-day pump engineer, designs of pumping machinery have progressed a long way from the hand, foot, or animal-driven shadufs and water-wheel pumps of ancient Egypt, India, and China. Most modern pump designs however still bear a strong resemblance to their predecessors of 100 years or so ago, although incorporating many improvements. Examples are better materials of construction, improved bearings, added protective coatings, better designed hydraulic passages and better methods of drive and control. All these have contributed to improvements in reliability and performance.

The laws of physics dictate the minimum power requirement for lifting a given mass of water through a given distance in a given time. The challenge of doing this seemingly-simple task reliably and efficiently, that is with the minimum consumption of energy, is one which has exercised the ingenuity of the pump engineer ever since the need for minimising power consumption to save cost was first appreciated. This need is almost always of great importance in water supply, since most water supply installations operate for long continuous periods and pumping costs generally dominate the running costs of water supply systems.

12.2 Centrifugal pumps

Pumps which operate by rotary action are called rotodynamic pumps and the centrifugal pump is the first type to be considered. Although other types of pump still have their uses, the centrifugal pump is the most commonly used because of the wide range of duties which it can cover and its comparatively high efficiency and low cost. Centrifugal pumps are available in a great variety of arrangements, as single or multistage units, and arranged vertically or horizontally to suit particular needs. The principal part of a centrifugal pump is the impeller, which is rotated at high speed. The impeller usually consists of two discs with a number of spiral blades between them. A pump with five impellers is shown in Plate 27. The impeller is manufactured as a single casting, and different materials are used in different applications. For use with fresh water, bronze is a satisfactory material for meeting the most important criteria. These include resistance to corrosion, abrasion, and cavitation damage, combined with ease of casting, good machining properties, and moderate cost. One of the discs is fixed to the shaft of the pump, and the other has a central hole in it making an annular space around the shaft – the 'eye' of the impeller. When the impeller rotates, water is drawn into it through the eye, passes through the impeller, and is flung radially off the tips of the vanes, which adds high kinetic energy. In the diffuser chamber around the impeller, part of the energy is converted to pressure energy, part to forward movement of the water through the connected system, and part is lost in turbulence and friction. The efficient conversion to useful energy – pressure rise and forward movement of the water – is accomplished by careful design of the impeller and diffuser chamber.

With good design the maximum efficiency of the pump can be around 80%, including all the energy lost in bearing friction as well as the hydraulic losses within the pump. Efficiency depends upon several factors including the size of the pump, with larger sizes generally more efficient than small, and the efficiency of a centrifugal pump is bound to vary at different flow rates. Maximum efficiencies of more than 90% are possible with special designs and large machines, but these may be obtained, amongst other things, by fine clearances between the moving and static parts of the pump so that efficiency may fall sharply with wear. In practice, all centrifugal pump efficiencies are likely to fall with time. The operating efficiency is a useful criterion in judging whether the time has come for a pump to be refurbished or replaced, since the efficiency inversely affects the running cost by influencing the power consumption for a given delivery.

12.3 Types of centrifugal pump

While all centrifugal pumps work on the principle set out above, their construction varies greatly according to the duty required from the pump. Multistage pumps consist of several impellers and diffuser chambers arranged in series, the impellers being fixed to one shaft. The water from the first diffuser chamber is led to the second impeller, from the second diffuser to the third impeller, and so on. The pressure developed by the pump increases stage by stage. For general waterworks duties the maximum pressure normally developed by one impeller may be between 80 and 100 m head. Higher heads can be produced by higher speeds of rotation and larger impellers, although these measures increase the cost.

A 'split casing' centrifugal pump is shown in Plate 27. The great advantage of this design is that the upper half of the casing is easy to remove, giving access to the impellers and diffuser chambers for inspection and any needed maintenance without the need to disconnect the pipework or the driver. An advantage of the multistage pump is that it is possible to have a 'dummy stage' for one or more of the stages. This is simply a diffuser chamber without an impeller, allowing for the addition of impellers if the pump is later required to develop more pressure. The efficiency of a pump is not much altered by the dummy stage, and its introduction at the time of installation when future increases in head are expected can prolong the useful life of a pump. The driving motor must of course have enough power to drive the pump when the impeller is added.

One of the problems in designing a centrifugal pump is accommodating the resulting end thrust. The pressure of the water at the inlet side of the impeller is low relative to that at the back, which is almost equal in surface area, so that axial thrust results. Several means are adopted for balancing this thrust, which would otherwise quickly cause wear on the pump and shorten its life. Small pumps can absorb the end thrust by the use of purpose-designed thrust bearings. For larger pumps a double-entry, back-to-back impeller design may be used, with the water entering the impeller from both sides. By this means the end thrusts can be effectively balanced. A multistage pump does not normally have double entry arranged for each impeller, although special designs have been successful for very high duties in which there is more than one double-entry impeller on the same shaft, with sometimes a balanced arrangement of single-entry impellers as well. Another common device for overcoming end thrust on smaller multistage centrifugal pumps is to incorporate a balancing disc on the shaft, high pressure water being led to one side of it so that most of the end thrust is taken by the disc.

The vertical spindle pump is frequently used for pumping water from a well, and for river intakes. The driving motor, Fig. 12.1(a), is at the surface, mounted above flooding level, but the pump, Fig. 12.1(b), is immersed in the water. The spindle rotates within a tube or sleeve, perhaps 75–125 mm diameter, and is held centrally in the rising main by 'spider' bearings. The pumped water is delivered to the surface via the annular space between the sleeving and the rising main. A typical arrangement would be a 250 mm diameter rising main in 3 m lengths bolted together with flanged joints, the sleeve tube being 100–125 mm diameter, with bearings for the spindle at every joint in the rising main. These bearings are nearly always water lubricated (oil or grease lubrication risks contamination of the pumped flow) the water being fed through the sleeving after being taken off the pumping main and filtered before being supplied to the bearings. The whole weight of the spindle and the pump impellers, and the hydraulic thrust generated, is taken by a Michell thrust bearing at the top of the shafting, just below the coupling to the motor. Vertical spindle machines of this type are very reliable, being robust and suitable for continuous heavy duty running. However they are expensive, and take time and skill to dismantle or erect when repairs are necessary. Their capital cost may be double that for a horizontal spindle pump, and they are now less common with the increased use of cheaper submersible pumps.

Submersible pumps

Submersible pumps should strictly be termed 'submersible-motor' pumps or 'submersible pumpsets'. The motor design (see Plate 28) is the main difference from more conventional designs. The pump, driven by a submersible motor, will be very similar to a pump driven by a vertical spindle 'dry' motor as described above, although some differences are given below. Submersible pumps gained in popularity because they usually result in a cheaper installation than one using dry motors. The disadvantages of having a submerged motor – out of the sight and hearing of any attendant, and certainly less reliable than a dry motor when the submersible machine was first introduced – have been largely overcome by improvements in the motor design, particularly in the insulation. Properly selected, modern submersible pumps have now been proved reliable in service over many years, and submersible designs are available from specialist manufacturers for a very wide range of duties.

Many submersible pumps in water supply are installed in boreholes. Since modern boreholes are invariably drilled, and no longer hand-dug wells as in the past, and since the high cost of drilling is affected by the borehole diameter, the diameter of the submersible pump has become of great importance. Designers have therefore concentrated on producing pumps and motors of small diameter, so that where high pumping heads are

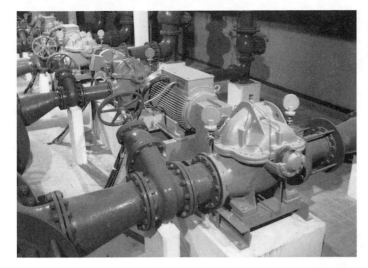

Single stage, split casing pumps. (Weir Pumps Ltd., Glasgow, Scotland)

Cut away diagram of a five-stage centrifugal pump. (Weir Pumps Ltd., Glasgow, Scotland)

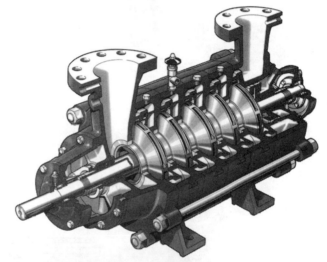

Two-stage split casing pumpsets in a water pumping station. (Weir pumps Ltd., Glasgow, Scotland)

Plate 27

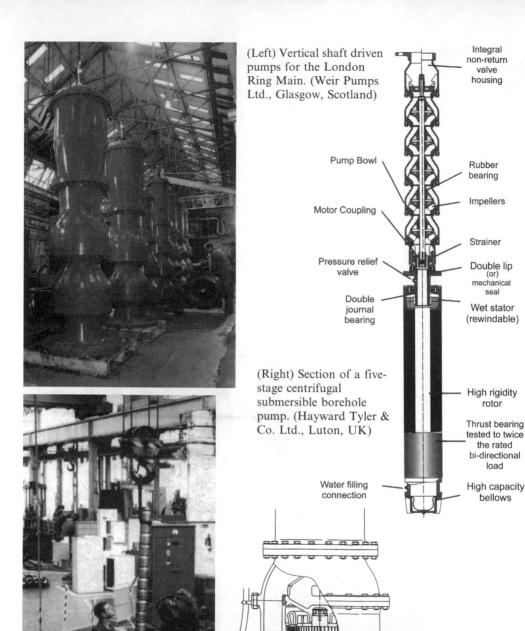

(Left) Vertical shaft driven pumps for the London Ring Main. (Weir Pumps Ltd., Glasgow, Scotland)

Integral non-return valve housing

Pump Bowl

Motor Coupling

Pressure relief valve

Double journal bearing

Rubber bearing

Impellers

Strainer

Double lip (or) mechanical seal

Wet stator (rewindable)

High rigidity rotor

Thrust bearing tested to twice the rated bi-directional load

Water filling connection

High capacity bellows

(Right) Section of a five-stage centrifugal submersible borehole pump. (Hayward Tyler & Co. Ltd., Luton, UK)

11-stage centrifugal submersible borehole pump. (Hayward Tyler & Co. Ltd., Luton, UK)

Submersible axial or mixed flow pump for pumping from a wet well. (Bedford Pump Co. Ltd., Kempston, UK)

Plate 28

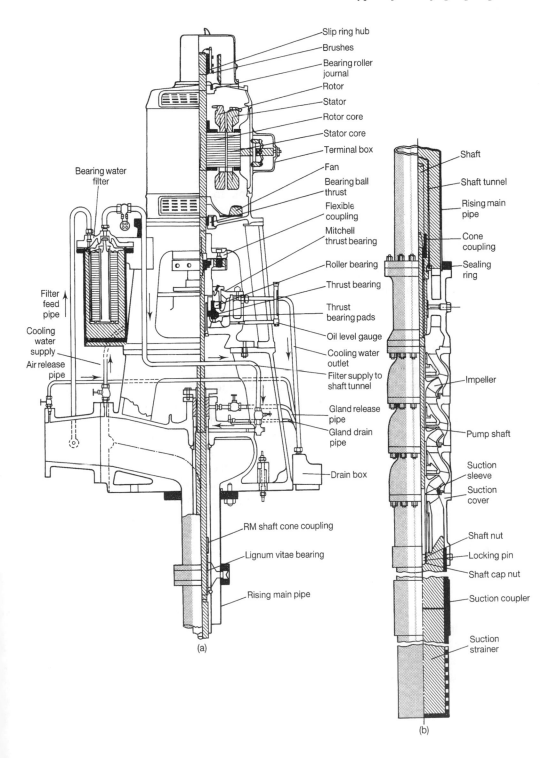

Fig. 12.1 Vertical spindle borehole pump.

needed many pump stages are required. This results in pumps which are longer and narrower than more conventional designs. For the same reasons, the submersible motors are longer than equivalent dry motors in order to develop the needed driving powers with a limited diameter available, and they are nearly always two-pole designs enabling them to run at the highest available speed. By this means, more power can be developed from a given size motor, which reduces the overall cost. Similarly the output of any given pump will be increased by running it at the two-pole speed in comparison with its performance at lower speeds. Naturally the mechanical design of the pump, especially its bearings, must be appropriate for the chosen speed. The disadvantage of having the higher speed pumpset may sometimes be important, particularly if there is any likelihood of abrasive particles suspended in the pumped water. Faster running pumps will also have reduced suction capability, so that deeper submergence may be required. Each installation needs careful consideration before the type and speed of the pumps is finally decided.

In a typical borehole installation, the pump is directly coupled to the submersible motor, which is underneath, and power is supplied to the motor through waterproof cables clipped to the outside of the rising main. The water inlet is in the centre between pump and motor with the outlet from the final pump stage leaving axially, vertically upwards. The motor is normally a fixed-speed caged induction motor, specially designed for underwater running. To ensure the motor is properly cooled by water passing over its surface, if there is any risk that inflow to the borehole could be predominantly from a higher level than the pump inlet, or if the pump is installed in a large body of water so that the pumped flow does not pass over the motor, a motor shroud should be used. This device is a simple canister, an annular shroud open at the bottom, which fits over the motor and is secured to the pump just above the inlet, thus ensuring that the pump inlet flow passes over the motor.

Submersible pumps are relatively quick and easy to install. The rising main is free of the spindle and sleeving needed with the vertical spindle pump, and a large thrust bearing to support the heavy rotating parts is not required. Submersible pumps need not be installed truly vertically, which may be a big advantage in very deep wells. They are sometimes used horizontally as booster pumps in distribution mains. Submersible-pump reliability in non-corrosive waters has been proven over the years, and even in corrosive waters when they may need attention they can be withdrawn and replaced more easily than pumps of the vertical spindle design. Some modern borehole installations are now designed without any surface housings, although provision still needs to be made for access for a mobile crane or sheerlegs for installation and withdrawal of the pumpset and its rising main. This simplification can make substantial cost savings.

Submersible pumpsets may be less efficient than the vertical spindle design, partly because of the special design of the motor but also because of the higher number of stages needed to achieve a given duty, as described above. This can be specially significant if the pumping duty is wrongly estimated, because of the pronounced peak in the efficiency curve with the multistage unit. However submersibles do gain by avoiding the transmission shaft losses of the vertical spindle design.

12.4   Characteristics of centrifugal pumps

The characteristics curves for a typical true radial flow centrifugal pump are shown in Fig. 12.2. The head–flow curve is relatively flat up to the design duty point and the power at zero flow is only about 40% of that required at the design duty. This illustrates two

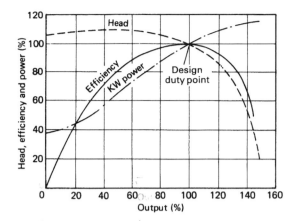

Fig. 12.2 Characteristic curves for a radial flow centrifugal pump at constant speed. Note that the head developed shows unstable characteristics up to about 90% output and a continuously falling head curve would be preferable and in some instances necessary (see main text).

useful characteristics of the centrifugal pump: it can be started against a closed valve, and the power required to start it is much less than the power required at the duty point. It is common to start a centrifugal pump against a closed delivery valve, and to close the valve before the pump is shut down. The pump is unaffected provided that the valve is opened (or the pump stopped when shutting down) before the pump becomes overheated. The reduced power required for start-up is also beneficial in reducing the starting current when an electric motor is used as the prime mover, and the pressure rise in the delivery system on starting the pump can be controlled by controlling the rate of opening of the valve.

The maximum head generated by the pump, for a given speed, is not greatly in excess of the design head. However in the particular instance shown, this head–output curve is unsatisfactory since for heads higher than the duty head there are two possible outputs. The pump is therefore unstable which could cause trouble if operated in parallel with another pump. For preference, the head–output curve should fall continuously with increasing output, and the pump is then stable. If the curve is steep, which is often specified for water supply pumps, the pump output will not vary much if the head alters somewhat during operation. The maximum head at zero flow – the 'closed valve' head – should not be excessive. The efficiency curve should be reasonably flat about the design duty point so that there is no great reduction in efficiency if the actual head is slightly different from the duty expected.

When the rotational speed (*N*) of a centrifugal pump is changed, there is little change in efficiency but the output (*Q*), head developed (*H*), and power required (*P*) are altered according to the following relationships, known as the 'affinity laws':

$$Q_1/Q_2 = N_1/N_2$$
$$H_1/H_2 = (N_1)^2/(N_2)^2$$
$$P_1/P_2 = (N_1)^3/(N_2)^3$$

The theoretical effect of changing the speed of the pump is illustrated in Fig. 12.3.

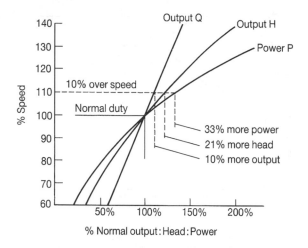

Fig. 12.3 Effect of changing the speed of a centrifugal pump.

Air in a pump can reduce its efficiency substantially and may also induce severe corrosion from cavitation (see Section 12.20). Air within a pump is evidenced by a hard crackling noise heard from the pump, almost as if the pump had some gravel inside it. If such a sound does not disappear shortly after starting, the cause should be investigated and eradicated.

Each pump has a minimum 'net positive suction head requirement', sometimes abbreviated to NPSHR, which is the head that causes water to flow into the eye of the impeller and is the minimum suction conditions required to prevent cavitation. Its value varies with the speed and capacity of the pump and will normally be given by the manufacturer, based on the results of tests. This must not be confused with NPSHA, i.e. the NPSH *available*, provided by the system in which the pump is installed, which is given by:

$$NPSHA = Z + \frac{(P_a + P_{vp})}{\gamma} - h_f - V_s^2/2g$$

where Z is the difference between the pump impeller eye level and the suction water levels, P_a is the absolute atmospheric pressure, P_{vp} is the absolute vapour pressure of the liquid at the pumping temperature, γ is the specific weight of liquid at pumping temperature, h_f is the head lost in friction in the suction pipework and $V_s^2/2g$ is the suction velocity head. (Although water supply pumps normally pump cold water at modest altitudes, terms are included in the above expression for NPSHA which take account of both atmospheric pressure and water temperature.) The value of NPSHA must always be greater than NPSHR, and a safety margin of perhaps 1 m is often specified. This allows for any minor differences between calculated and actual figures as well as changes with time.

Specific speed

To classify geometrically similar pumps, the numerical quantity 'specific speed' has been adopted. Specific speed is the speed required for delivery of unit flow against unit head; it will vary in accordance with the system of units used.

$$N_s = \frac{NQ^{1/2}}{H^{3/4}}$$

where N_s is specific speed; N is pump impeller speed in rpm; Q is output at maximum efficiency (m^3/s); H is delivery head at maximum efficiency (m). Specific speeds (in metric units) fall approximately into the following categories:

Type of pump	Specific speed
Radial flow	10–90
Mixed flow	40–160
Axial flow	150–420

12.5 Axial flow and mixed flow pumps

Axial flow pumps are of the propeller type, in which the rotation of the impeller forces the water forward axially, and therefore they strictly do not qualify as centrifugal pumps. Mixed flow pumps act partly by centrifugal action and partly by propeller action, the blades of the impeller being given some degree of 'twist'. However in practical terms there are no precise dividing lines between radial flow (centrifugal), mixed flow, and axial flow pumps. In general, axial and mixed flow pumps are primarily suited for pumping large quantities of water against low heads, whilst centrifugal pumps are best for pumping moderate outputs against high heads. Axial flow pumps have poor suction capability and must be submerged for starting. They are most often used for land drainage or irrigation, or for transferring large quantities of water from a river to some nearby ground-level storage. A mixed flow pump is shown in Plate 28.

Characteristic curves for typical mixed flow and axial flow pumps are given in Fig. 12.4. The starting power required by the mixed flow pump shown is about the same as the duty power, but for the axial flow pump the starting power is substantially in excess of the duty power. Axial flow pumps are therefore not started against a closed valve, which would overload a motor correctly sized for the expected duty. They are either started against an open valve to minimise the starting power and current required, or may be installed in systems specially designed to ensure no delivery valve is needed.

Even the smallest of pumps may be mixed flow rather than truly centrifugal. The borehole pump shown in Fig. 12.1 is really a mixed flow multistage pump. The designer has had to design a pump which is restricted in diameter so that the pump can be inserted in small boreholes. The quantity required has forced him to adopt a mixed flow impeller design, and because one impeller does not develop enough head, several similar impellers in series are needed.

12.6 Reciprocating pumps

Most pumps used for water supply nowadays use a rotating impeller, but the reciprocating pump still has its uses. The ram pump is the most common form of reciprocating pump, and it consists of a piston reciprocating within a cylinder provided with water inlet and outlet valves. Water is drawn in by one stroke of the pump and forced out by the next.

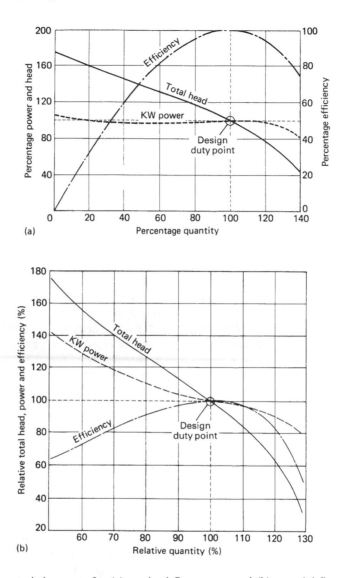

Fig. 12.4   Characteristic curves for (a) a mixed flow pump and (b) an axial flow pump.

With all ram pumps the output must fluctuate cyclically, but if three or more cylinders are used, a reasonably constant flow is maintained. The type of pump with three cylinders, is termed a 'triple throw ram pump', the pistons being connected to the operating crank 120° apart. Ram pumps were common for high-head duties in water supply before the widespread introduction of centrifugal pumps. Their slow speed suited the speed of the reciprocating steam engines which usually drove them. Driven nowadays by an electric motor, the triple-throw ram pump may sometimes be used where exceptionally high heads must be produced, since no special design is necessary other than making the working parts strong enough to sustain the head. Efficiencies of more than 90% can be obtained, falling off with piston and cylinder wear. However since very high head centrifugal pump designs are now available, the use of new ram pumps for high-head duties in water supply has almost completely ended.

Small ram pumps, usually driven by variable-speed electric motors, are however still in widespread use for the injection of the chemical solutions needed for dosing in water treatment. Their property of delivering a constant volume of liquid per stroke makes them well suited to the metering duty required. With appropriate design of the pump and its driving system, both the pump speed and its stroke are varied, so that any desired rate of chemical injection per unit volume of water to be treated can be maintained.

The bucket pump is another form of the ram pump, arranged to operate vertically for drawing water from a well or borehole. Instead of a piston a 'bucket' is given a vertical reciprocating motion in the rising main. The bucket incorporates a valve, which opens as the bucket descends below water level. When the bucket rises, the valve closes and water is lifted. An alternative arrangement makes the descent of the bucket force water up the rising main through a series of valves. This type of pump is known as a 'lift and force' pump. The village pump is a single-throw bucket pump operated manually, and it illustrates one of the chief advantages of the bucket pump apart from that of its low cost, that it is always ready for use and able to function even if erratically operated. Modern versions of these machines are still being installed for village water supplies in parts of the world where an electro-submersible borehole installation would be too costly. Small single-throw bucket pumps are also still used, for example to provide water supplies for remote farmland, sometimes with modern wind turbines driving them, again illustrating the reliability of these pumps under intermittent operating conditions.

The hydrostat or *hydraulic ram* pump is another example of reciprocating pump still in occasional use. A large volume of water flowing in one pipe or waterway is used to drive a ram which is connected to a smaller ram pump, which pumps part of the water to a higher elevation through a branch pipe. Of course it is not possible to raise the pressure of all of the water, and the flow which continues down the main pipeline suffers a loss of head. Hydraulic rams are not convenient in public water supply systems because any increase of demand on either the high or low pressure side will quickly invalidate the set-up. However they do find use in special situations, such as when a small volume of water is to be lifted in a remote location without an accessible power supply.

12.7 Choice of pumps for water supply

The horizontal centrifugal pump is suitable for nearly all waterworks duties, except those of handling very large volumes of water against low heads, and pumping from wells and boreholes. The main advantages of the horizontal pump are that it is relatively low in first cost, and it can readily be arranged to provide easy access for maintenance. A great variety of designs are available to meet a range of pumping conditions, but the most common arrangement is to use the horizontal, split-casing double-suction design, which has been developed over many years of water supply duties. For a single unit the output can range from 50 Ml/day by 60 m head to 10 Ml/day by 200 m head. The most common waterworks duty is from 10 to 25 Ml/day per unit by 30 to 120 m head, and in this range the horizontal pump is cheapest.

The pump should have stable characteristics and should be 'non-overloading', i.e. the power absorbed should not increase much if the delivery head drops. This is not always possible to arrange, and protection against overload must be provided for the motor. A low head could result if the delivery main from the pump were to burst near to the pumping station. The efficiency curve should also indicate no severe fall in efficiency for moderate variations of flow and head about the duty point. When specifying a pump, the manufacturer must be told of the complete range of duties the pump is intended to meet, including whether series or parallel operation is required (see Section 12.15). To specify the duty point only, i.e. the theoretical normal conditions under which the pump will operate, could lead to large efficiency loss if the actual running conditions vary from the theoretically calculated duty.

For wells and boreholes the choice lies between vertical spindle pumps and submersible pumps. The characteristics of these two types of pumps have already been given. The vertical spindle pump may be regarded as a 'heavy duty' pump. It may be driven by a synchronous electric motor which enables the power factor then to be brought to unity. The consequent cost of power can so be reduced, although the motor is more expensive initially and costs more to maintain. Large outputs are often handled by vertical spindle pumps driven by synchronous electric motors. Variable speed is another advantage that can be more easily obtained with the vertical spindle pump than with the submersible pump, which requires a variable frequency supply to run it at variable speed. (However recent developments with variable frequency drives are increasing their use.) Variation in speed is often necessary when pumping from a well or borehole because the output from the well may have to be kept in step with the demand, irrespective of seasonal fluctuation in water levels in the well. A thorough appraisal of all the possible operating conditions must be made before choosing the right pump for the duty.

For pumping large quantities of water against low heads, the vertical spindle motor driving a mixed-flow pump is also suitable, with the pump immersed in the water and the motor sited above the highest flooding level. A vertical spindle pump is not always needed for pumping water from a tank or main, because the cheaper horizontal centrifugal pump can be sited in a dry well, with its centreline below the bottom water level in the tank so that the pump is always primed. Centrifugal pumps should preferably not be sited so that they are subject to any suction lift, because their efficiency drops and cavitation may be experienced. But if essential and depending upon the design, up to 4 m lift can be managed, with 5.5 m the maximum possible.

For high lift pumping the most common choice is the fixed-speed, horizontal multistage centrifugal pump. Submersible pumps are also used for this duty, either for pumping direct from a well to a high-level tank, or for inserting as boosters in a pipeline in a pit below ground level. This is advantageous when it would otherwise be impossible or expensive to erect a building to house the normal arrangement of horizontal pump and motor.

In water supply pumping stations, a single pump is seldom relied upon for the full output. Adequate standby is essential to ensure continuity of supply. If the full duty can be handled by a single pump then a duplicate of equal capacity should be installed and this is a common arrangement when the unit sizes are small. However if pumps each sized for 50% total required output are chosen, the installation of three similar pumps ensures 100% output on the breakdown of any one machine, and 50% in the much more unlikely event of two pumps failing at the same time. The cost of the standby plant is then less than when two pumps of 100% capacity are installed. When pumps of different sizes are needed, for example when pumping into a system with limited storage or when large fluctuations in demand must be satisfied, the normal requirement is to ensure the availability on standby of at least one pump of each size.

12.8   Prime movers

Pumping machinery may be driven by almost any prime mover of suitable power and speed. Most water supply pumps are nowadays driven by electric motors, but some still exist that are directly driven by a diesel engine. The widespread availability of more reliable and economically priced electricity supplies, and the comparative cheapness, choice of design and sizing, ease of operation and control of electric motors make them clear favourites in most circumstances. Two disadvantages of motor drives nevertheless need mentioning. Firstly, the electricity supply must be secure to ensure the constant availability of pumping, which is nearly always important in water supply systems. Secondly, the cheaper motor designs are inflexible since they are normally able to run only at a fixed speed.

Security of electrical supplies. Although in many places the reliability of electrical power supplies has greatly improved, so that often no consideration need be given to outages of any long duration, there are still installations where power failure must be taken into account. A typical example is where severe weather can affect transmission lines to remote pumping stations. Some form of standby generating plant will then be needed. Except for very large unit sizes such as 5 mW and greater, for which gas-turbine generators are usually used, the choice nowadays is almost always to use diesel-driven generators. Where a single large water sourceworks provides a major part of the water supply, there may be no escaping the costly provision of a fixed, dedicated standby generating station, possibly serving high- and low-lift pumps and a water treatment works as well. Careful consideration then needs to be given to the generating capacity needed, which will depend on such fundamentals as the duration of the outage to be covered which affects the fuel storage required, how much of the plant must be kept operational, what size is the largest motor to be started, and so on. If providing a large generating plant is essential, substantial capital investment in plant will be needed, but it could stand idle for most of its life. The option should then be considered of operating the generating plant continuously to provide the base load required, since unused generating installations are prone to deterioration unless regularly operated, and the operative staff need to have experience of running the plant.

An alternative which may sometimes be possible when a number of sourceworks contribute to the total water supply, is to set up a pool of mobile generating sets, which can be transported to any site where a power outage has occurred, if the same outage is not expected to affect all works simultaneously. This reduces the amount of standby generating plant needed, and makes for more cost-effective use of the generators. But other difficulties can arise. Transporting large generators may be awkward and expensive, and may be impossible if, for example, snow-storm conditions causing the original electrical outage also make roads impassable. Simple and speedy arrangements for receiving and connecting the generators, and for fuelling them, must be made at each point of use, and more than one size of generator may be needed. Arrangements must also be made for safe storage and maintenance of the generators to ensure they are always ready for emergencies.

If pumping stations and treatment works have to be built where there is no public electricity supply available within an economic distance, power may have to provided by a diesel engine on site. If more than one pump has to be operated the diesel engine would drive a generator to make electrical power available to all pumps and also provide power for lighting and instrumentation, etc. In rare cases of small isolated pumping stations with only one pump, a small diesel engine can be direct coupled to a pump. Normally, however, the choice is between building a power station to supply the works, which will have to include some standby capacity, or negotiating with the electricity supply authority to make new supply arrangements. The economics of these situations can be complex, being dependent on the charge the electricity supplier makes for bringing a power line to site, the terms of the supply, and the need to take into account likely future developments in both power and water needs. The capital, operating and maintenance costs of diesel-generating power plant, and the need to provide adequate fuel storage, tend to be high, so that such a set-up is rarely economic in developed countries such as the UK where electrical grid supplies are widely available.

12.9   Electric motors for pump drives

The three-phase alternating current (AC) motor is the most common type, and can be classified as follows:

- AC induction motors of two types – (i) 'cage' or 'caged' (formerly known as 'squirrel cage' motors); and (ii) 'wound rotor' motors (sometimes called 'slip ring' motors) – see below;
- synchronous motors;
- commutator motors.

AC induction motors of the caged type are the most widely used because of their simplicity, robustness, reliability and low cost. They are inherently fixed-speed machines when connected to conventional fixed-frequency supplies. This is a handicap for centrifugal pumps, as the pumps themselves are capable of a wide range of duties without modification if the speed can be varied. Since many water supply systems require pumping installations of output varying at different times, fixed-speed pumps are often a disadvantage. Various methods of altering motor speed have been available for many years, but most of these have had limited application because of reduced overall efficiency, limited speed range, limited power capability, or high first cost. Fortunately the continuing development of various methods of variable frequency drive have overcome most of these disadvantages, although cost is still an important consideration. Motors designed for operating at two different fixed-speeds are also available and sometimes can provide an economical method of altering pump performance.

The *synchronous motor* is more expensive because of the need for more complex control equipment and a DC supply is required (supplied by an 'exciter' driven by the motor itself) for the rotor which results in the rotor turning at the same speed as the rotating magnetic field created by the stator. The synchronous motor (which has to be started up by short circuiting the rotor windings so it acts as an induction motor) is normally used for applications requiring constant speed operation under varying load conditions, or where its ability to apply a power factor correction or improve the voltage regulation of the supply system is justified on economic grounds. If adopted, it is usually only for pumping large steady outputs of water.

The *commutator motor*, which provides variable speed, is now rarely adopted because it has largely been replaced by the electronic variable frequency drive using the cage type motor, or the slip energy-recovery type using the wound rotor induction motor.

Because by far the most commonly used motors for driving pumps are the AC induction motor of the caged, or wound rotor type, only these are described in more detail below.

12.10 The AC induction motor

The induction motor comprises a stator which incorporates a distributed winding which is connected to the three-phase electrical supply, and a laminated steel rotor which in its simplest form has embedded large section bars which are short-circuited at each end. The motor thus consists of two electrically separate windings which are linked by a common magnetic field forming a transformer with an air gap magnetic circuit. The three-phase current in the stator winding produces a smoothly rotating magnetic field whose rotational speed is given by the equation:

$$\frac{\text{Rotation speed of magnetic field}}{\text{(synchronous speed) (revs per minute)}} = \frac{\text{supply frequency (Hertz)} \times 120}{\text{number of poles}}$$

Thus for a supply frequency of 50 Hz a two-pole motor will have a field speed of 3000 rpm, a four-pole motor 1500 rpm, and a six-pole motor 1000 rpm. The rotating magnetic field cuts the rotor bars and induces a current in them. The rotor current produces a magnetic field which interacts with the stator field producing an accelerating torque. The motor will run up to a speed at which the developed torque is equal to that of the driven load and that required to overcome friction and windage losses. In practice, the rotor cannot attain synchronous speed with the magnetic field because under such a condition no current would be induced in the rotor conductors and hence no magnetic flux and torque would be produced. The difference between the actual speed of rotation and synchronous speed with the magnetic field is termed the slip. The slip increases with the load and is normally expressed as a percentage of the synchronous speed. The slip at full load typically varies from about 6% for small motors to 2% for large motors. The starting torque and speed characteristics of the cage induction motor of basic standard design are shown in Fig. 12.5.

The *cage induction motor* has a rotor core which is made up of laminations and conductors of aluminium, copper or copper-alloy non-insulated bars in semi-enclosed slots, the bars being short-circuited at each end by rings. For the smaller motor ratings, the rotor bars and end rings are often cast. The advantages are simple construction, low cost and low maintenance. The limitations of the cage induction motor of basic standard design are low breakaway torque and high starting current, the former typically ranging from 0.5 to 2.0 times and the latter three to seven times rated values, depending on motor rating and number of poles. These limitations can be improved by the use of motors with multi-cage rotors. The multi-cage rotor in its simplest form, comprises a low reactance and a high resistance outer or starting cage which predominates during the starting period and results in increased torque and reduced current, and an inner cage exhibiting low resistance which is dominant under running conditions.

The *wound rotor induction motor* design incorporates a three-phase, star configuration rotor winding brought out to slip rings to which external resistance is connected for starting. This type of motor is used where high starting torque, reduced starting current and controlled acceleration characteristics are required. The magnitude of starting torque and current, and acceleration period are determined by the value and stages of external

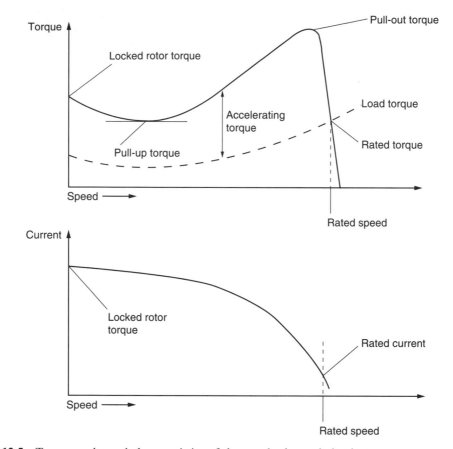

Fig. 12.5   Torque and speed characteristics of the standard cage induction motor.

resistance. Under running conditions the slip rings are shorted-out. Disadvantages of the wound rotor motor are higher cost and additional slip ring and brush gear maintenance.

Rated output, starting torque and start frequency

The *rated output* of an induction motor is for the designated duty when the ambient temperature of the coolant air does not exceed 40°C or, in the case of a water-cooled motor, the temperature of the water entering the heat exchanger, does not exceed 25°C – both at a height above sea level not exceeding 1000 m. The supply voltage may deviate between 95 and 105% of the rated voltage of the motor without affecting the rated output. When the motor is operated in conditions different from the reference values, the motor rated output will be affected, and these factors must be considered at the time of motor selection,[1] as must the requirement for the motor to operate continuously at voltage tolerances differing from the rated value by ±5%. For applications where the motor is required to operate under varying and cyclic load conditions, which may include periods of either no-load or standstill, reference should be made to the manufacturer giving details of the load inertia and the load–time duty sequence.

   Starting torque. The standard cage induction motor is designed to produce a starting torque between standstill and that at which pull-out (minimum) torque occurs, of not less than 1.3 times a torque characteristic varying as the product of the square of the per unit speed, and rated torque. The square law torque is representative of the run-up characteristic of the centrifugal pump. The factor of 1.3 is chosen to allow for a voltage of 90% rated value at the motor terminals during the starting period. The load and motor torques characteristic during the starting period should be considered at the time of motor selection. This is of particular importance where the motor is required to start with a voltage drop greater than 10%, typically for other than direct-on-line starting control.

Frequency of starting. The standard cage induction motor is designed to allow two starts in succession (running to rest between starts) from cold conditions, or one start from hot after running at rated conditions. Further starting is permissible only if the motor temperature does not exceed the steady temperature at rated load. Because the number of starts directly affects motor service life, they should be kept to a minimum. More onerous starting requirements must be considered at the time of motor selection. For the wound rotor induction motor the starting frequency is generally dictated by the short-time rating of the starting resistor.

12.11 AC induction motor starting methods

Factors influencing the starting method. The starting current drawn by an AC induction motor depends on its type, rating, voltage and starting method. When starting a motor the resulting current causes a voltage dip which has to be kept within defined limits stipulated by the electricity supply company to avoid annoyance to other consumers and the mal-operation of connected equipment. The limits will depend on the degree of voltage variation and frequency of occurrence. While a detailed electrical system analysis is required in special cases, experience has shown that it is possible to apply general guidelines based on permitted transient voltage dip at the point of common coupling, that is the busbar from which other consumers are supplied, and the frequency of motor starting, to assess whether annoyance is likely to be caused. Typical voltage dip limits are 3% for motors started infrequently, say at intervals longer than 2 hours, and 1% for motors started frequently. In special cases, for example motors which are either infrequently started or are located in remote areas, the electricity supply company may permit a transient voltage dip greater than 3%. Where large motors are to be installed it is important to establish at the design stage the transient voltage dip and starting frequency criteria to determine the method of starting. In addition to complying with voltage dip criteria at the point of common coupling, it is necessary to assess the ability of the motor to start and accelerate if the voltage drop at the motor terminals is greater than 10%, and the effect of motor starting on other parts of the consumer's installation. These aspects can be significant if the electricity supply is taken from a low fault level rural distribution system or small capacity transformers, and it is important to take account of these factors at the time of selecting the motor and its starting method.

The starting method is determined by motor starting torque and current requirements, and in some applications, the need to control acceleration. Starting methods for the cage induction motor can be classified as:

- full voltage or direct-on-line;
- reduced voltage (star-delta, auto-transformer and electronic soft start);
- rotor resistance starting (for wound rotor motors only).

Direct-on-line starting is the cheapest, simplest, and most reliable, and is therefore the most widely used. However, direct-on-line starting causes a high starting current, and may therefore not be suitable if the electricity supply company requires a reduced starting current (or limited voltage dip), or shock-free controlled starting is required. With direct-on-line starting, the applied voltage, starting torque and current are 100% rated values, and rapid run-up at maximum available torque is achieved. An advantage is that direct-on-line starting simplifies motor construction, insofar as only one end of each phase winding need be brought out to terminals. For low voltage motors, the size which can be started by the direct-on-line method is often arbitrarily limited by the designer to the smaller ratings, typically up to 15 kW. However, unless the maximum size of motor is stipulated by the electricity supply company, or is limited by supply transformer transient loading considerations, there is no restriction on the size of motor which can be direct-on-line started, provided transient voltage dip and the mechanical impact loading criteria are met. These design criteria also apply to high voltage motors.

Star-delta starting is the most usual method used for reduced voltage starting, and involves connecting the stator winding of the motor initially in star until an optimum speed is achieved, when it is switched to delta. In star connection the stator windings are connected phase to neutral; in delta connection the windings are connected phase to phase – for a 3-phase a.c. supply. When connected in star the voltage across each phase winding is 58% ($1/\sqrt{3}$%) of the supply voltage and the starting current and torque are reduced to 33.3% of the full voltage values. The starting method is relatively simple and inexpensive but its use is limited to low inertia drives because of the reduced starting torque. The disadvantage of the starter type is the possible high transient torque and current which could occur when switching from star to delta. The number of starts per hour is not normally restricted by the starter, although consideration has to be given to the characteristics of the main circuit, short circuit and thermal overload protection. To eliminate or reduce the high transient torque and current when switching from star to delta, the closed transient star-delta (Wauchope) type starter can be used, in which resistance is inserted when changing over from star to delta to provide a no-break transition. The closed transition type starter provides three steps of acceleration against the two steps provided by the standard star-delta starter. Because of the additional resistors and control equipment, the closed transition starter is more expensive than the standard starter. The number of starts per hour is limited by the short-time thermal rating of the transition resistors.

Auto-transformer starting provides more flexibility than the star-delta starting method because the applied voltage and hence the starting torque and current can be varied by changing the voltage tapping. The motor starting torque and current are a function of the square of the transformer tapping with respect to rated voltage. For example, at 50% voltage tapping, the torque and current are 25% of the full voltage values, and at 80% voltage tapping, the torque and current are 64% of the full voltage values. The ability to adjust the voltage of the auto-transformer provides a convenient means for closely matching the starting torque to the driven load, and reducing the starting current. For this reason, the auto-transformer method has the potential for starting larger motors than would be possible with the star-delta method. It also permits 'push button' starting so that pumps can be safely started by unskilled personnel. The disadvantages of the auto-transformer starter is cost, being considerably more expensive than the star-delta starter, and the possibility of high transient torque and current when switching from reduced to full supply voltage. To eliminate or reduce the high transient torque and current when switching from reduced to full voltage, a closed transition configuration whose operating principle is similar to that of the closed transition star-delta starter, can be used. This involves the use of a more costly transformer and additional control equipment. The number of starts per hour is limited by the short-time thermal rating of the auto-transformer.

Electronic soft starting. The disadvantages of the previous electro-mechanical starting methods can be mitigated in part by electronic soft starting. With this method, the motor supply voltage is gradually increased linearly up to rated value, providing smooth acceleration, controlled and reduced starting current typically between 200 and 300% of full load value, and controlled motor torque. At full speed, the electronic controls would normally be by-passed, and the motor connected directly to the mains supply. The electronic soft starter is more complex and expensive than the electro-mechanical direct-on-line and star-delta motor starting methods.

Rotor resistance starting of wound rotor induction motors is used where the applied load has a large moment of inertia requiring a high starting torque, or the starting current needs to be limited for supply system voltage drop considerations, or controlled acceleration is required. With the connection of external resistance into the rotor circuit via the slip rings, high starting torque and low starting current can be obtained. Typically, for full load torque at standstill, the starting current will be in the order of 125% full load value. By selection of resistance, starting torques of 200–250% can be attained with corresponding currents of 250–300%. The resistance is either the multi-stage metallic non-inductive grid, or the liquid type, the latter providing smooth control of acceleration. The number of starts per hour is limited by the short-time thermal rating of the rotor resistance.

12.12 AC induction motor protection

The function of the protection is to initiate disconnection of the motor from the supply to prevent or limit damage caused by overheating due to abnormal load or failure of the winding insulation. For low voltage motors, the protection provided needs to be determined by consideration of factors such as motor cost, and the characteristics and importance of the drive. The degree of protection provided can range from the thermal overload relay, to a motor protection relay providing high-set overcurrent, overload, earth fault, negative phase sequence (unbalance) and stall protection. The motor protection relay would not normally be used for motors rated below 50 kW. High voltage motors – irrespective of rating – are provided with high-set overcurrent, overload, earth fault, negative phase sequence (unbalance) and stall protection. For the larger motors, typically 1000 kW and above, high-speed differential protection is often provided to minimise damage to the stator core in the event of a stator winding fault. For motors where a reversal in the direction of rotation could cause damage to the driven load, for example to a screw pump, phase reversal protection should be provided.

Additional motor winding and bearing protection can be provided by thermistors, thermocouples and resistance elements. This method gives protection against faulty conditions which are not reflected in the line current of the motor. For applications requiring temperature indication, in addition to alarm and motor tripping initiation, the thermocouple or resistance element is used. Thermocouples can be located in stator slots, stator end windings, cooling air circuits and bearings. The location of resistance elements is the same as for the thermocouple with the exception of the stator end winding. The degree of protection provided by the three devices is good and response to temperature change is fast.

12.13 Speed control of AC induction motors

When speed control is needed for pump drives there is no simple solution which is of general application, and the capital cost and benefits provided must be assessed carefully in each case. The standard AC induction motor is a constant speed machine, the speed being determined by the number of poles and frequency of the supply on which they operate. The two most commonly used methods of speed control are based on changing these parameters, although it is to be noted that limited speed variation can be obtained by using the wound rotor type induction motor and varying the rotor resistance. The latter method of speed control is inefficient because of the power dissipated in the resistance and, for this reason, would not normally be considered, and has been replaced by slip-energy recovery systems whereby, instead of wasting power in the rotor resistance, it is returned to the supply thereby improving efficiency. The most commonly used methods of speed control are summarised below.

(a) *Pole changing motor.* The simplest form of pole changing motor has a single-tapped winding, also known as a Delanger winding, which provide two speeds in a ratio of 2:1, for example 3000:1500 rpm for a motor connected to a 50 Herz supply. A variant on the single-tapped winding is a motor with two separate windings, which can provide any combination of two speeds. By combining the tapped and two winding arrangements up to four speeds can be obtained, for example 3000, 1500, 1000, 500 rpm at a supply frequency of 50 Hz.

(b) *Pole amplitude modulated motors.* The pole amplitude modulated (PAM) motor is a development of the single-tapped 2:1 speed ratio winding, and speed change is achieved by reversing one-half of each phase which changes the magnetic flux distribution and produces a resultant field of different polarity. Various combinations of speed ratios can be obtained with this motor design. Other than speed ratio, the advantages of PAM motor design over the pole changing type are better utilisation of active materials resulting in smaller physical size for a specific speed ratio and rated power, and improved efficiency and power factor.

(c) *Variable frequency drives.* This form of drive uses a standard cage induction motor, and speed regulation is achieved by varying the supply frequency. The variable frequency drive controller is of the static design, the basic components of which are a rectifier connected to the AC supply, an inverter to provide a variable frequency supply to the motor, and a DC link between the rectifier and the inverter. The motor speed can be regulated typically over the range 10–200% motor rated speed. Harmonic currents are produced by the rectifier which reflect into the electricity supply system, and can cause interference with other consumers or be detrimental to connected plant such as capacitors, generators and motors. The electricity supply companies lay down guidelines for the permitted magnitude of individual harmonic currents and/or harmonic voltage distortion, and this aspect needs to be considered carefully at the design stage. Methods available for reducing harmonic currents are to increase the rectifier from 6 pulse to 12 or 24 pulse, depending on the power rating of the drive, install filters, or use phase shifting supply transformers. Increasing rectifier pulse number or installing filter will considerably increase the cost of the drive. When selecting an induction motor for use with a variable frequency drive controller, the following factors need to be considered:

- increased losses and hence heating in the magnetic circuit caused by the harmonics in the inverter output waveform;
- power and torque requirements throughout the speed operating range, and
- impaired cooling at low speed operation, an important consideration for constant torque loads.

(d) *Slip energy recovery*. The slip energy recovery variable speed drive, also known as a Kramer drive, uses a wound rotor type induction motor and operates on the principle of recovering rotor energy and feeding it back into the supply. The mechanism for doing this is to convert the slip ring frequency power to DC and, in the case of the static type drive, return the power to the mains supply via an inverter. An alternative to the inverter would be a DC motor driven asynchronous generator. A direct connection between the rotor and the mains is not possible because both the rotor voltage and frequency vary with motor speed. Harmonic currents are produced by the inverter, and the considerations summarised for the variable frequency drive apply.

12.14 Savings from use of automatic controls

The extra capital expenditure required to provide automatic controls can be quickly repaid by the resulting saving in labour costs. As a result, modern automatic control systems have been intensively developed in recent years and their reliability when properly designed and installed is now extremely high. Other cost benefits can follow, for example by optimising plant operating regimes. The application of automatic controls has been aided by the predominant use of electric motor drives, which are readily adapted to this means of control. Control may be completely automatic, i.e. both starting and stopping being controlled, or semi-automatic, when it is usual to arrange only for shutdown to be automatic.

The ever-increasing use of computers in control applications has also been beneficial, particularly in complex pumping systems. Regimes have been developed to regulate the operation of pumping stations in accordance with pre-planned programmes, for example, designed to minimise power costs. This can be done by ensuring that best use is made of electricity supply tariffs – concentrating pumping at times of low-cost electricity, reducing pumping when possible at times of higher cost, and avoiding incurring high maximum demand charges. The calculations involved in such optimisations may be complex, and may involve negotiation with the electricity supply authority to achieve the best results. Such computerised control systems can also be readily designed to take account of a wide variety of conditions. Some examples are: preventing the over-frequent starting of electric motors; sharing the hours run equally between a number of pumpsets; co-ordinating the operation of different pumping stations serving the same supply area; and regulating the operation of pumping stations arranged in series along a long transfer main.

12.15 Boosting

Boosting is a pumping arrangement which augments the pressure or quantity of water delivered through an existing system. (The term is sometimes wrongly used to mean simple pumping.) There are many possible arrangements, but the following are the most important:

(1) the addition of a fixed extra flow to an existing supply;
(2) the addition of a fixed extra pressure to an existing supply,
(3) the maintenance of a given pressure, irrespective of the flow.

Addition of fixed extra flow or pressure

To increase the flowrate, two similar pumps can be connected in parallel; to increase the pressure two similar pumps can be connected in series. For lesser increases the pump added in parallel or series will be of a smaller rating. In either example, the characteristics of the system into which the pumps are to deliver must also be considered. Figure 12.6 shows the characteristic flow–pressure curve for a pump A. For series working of two such pumps, the joint characteristic curve (A + A) is obtained. The characteristic curve S, which indicates the head–flow relationship for the system into which the pumps are delivering, must also be drawn, which shows how the friction losses increase with flow. The point of intersection of the curves (A + A) and S indicates the joint output of the two pumps A in series, which is clearly less than the sum of the individual outputs. If the pumps are connected in parallel, a pump characteristic curve (A ∥ A) can be drawn by adding the individual pump flows at the same head. Again, the point where this curve cuts curve S indicates the joint output of the pumps, this time working in parallel delivering into the same system.

The foregoing is not enough to confirm whether it is practicable to connect the two pumps in series or in parallel. A hydraulic gradient must now be drawn for the system in which the pumps are to work to confirm that the siting of the added pump is correct. Referring to Fig. 12.7, suppose pump A initially draws water from reservoir R_1 and pumps it to reservoir R_2. When pump A is delivering water the hydraulic gradient for the output

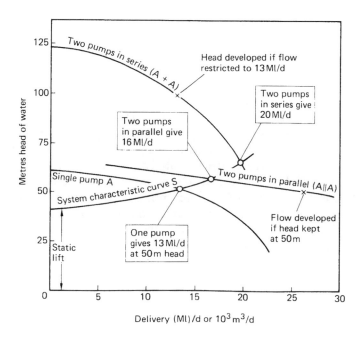

Fig. 12.6 Output of two pumps connected in series or in parallel.

of pump A will be the line *abcd*. If the flow is increased by an amount Q then the hydraulic gradient must change to some line $ab'c'd$ where b' is lower in elevation than b, and c' is higher in elevation than c. The difference cb was the pumping head which had to be developed by pump A alone. The difference $c'b'$ is the new pumping head which must be developed by pumps (A + A). The following must then be checked:

(1) The position of b' must not be so low in relation to the elevation of the pumps that suction troubles result. Ideally, there should be no negative pressure on the suction side of the pumps because this could cause cavitation and will reduce the pump efficiencies. (Negative pressures in water supply systems are also undesirable anyway, because of the risk of in-leakage of contaminated groundwater when buried mains are used.)

(2) The enhanced pressure represented by c' must not be beyond the safe rated working pressure on the body of pump A, or of the valves, fittings, and delivery main to reservoir R_2.

(3) If the pumps are to be operated in series the joint duty must not cause pump A to be overloaded (if this is so then the difficulty might be overcome by changing its impellers).

(4) Provided that pump A is not overloaded, its efficiency at its new duty point must be checked. The joint efficiency of the two pumps working together must also be studied; it may be concluded that it would be better to scrap pump A altogether and have an entirely new pump, capable of managing the whole of the enhanced duty.

(5) If the pumps are to be run in parallel they must have stable running characteristics.

If consideration (2) indicates that an extra pump cannot be sited alongside A, one solution might be to place pump A' somewhere along the line between the pumping station and R_2. Overpressurising the main would have to be overcome by replacing some or all of it. If suction conditions could cause trouble, then some or all of the delivery main from R_1 to the pumping station may have to be duplicated to reduce the friction losses at the augmented flow.

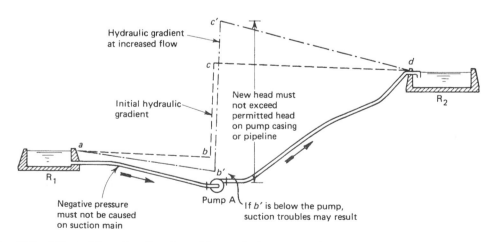

Fig. 12.7   Effect of boosting flow in a rising main to a reservoir.

Maintenance of a given pressure

One of the most frequent uses of a booster is to increase the pressure of water in a distribution system at times of high drawoff. Drawoff normally varies both with time of day and season. At low drawoff, the pressure may be adequate, but when the drawoff is high the terminal pressures may be too low. Instead of laying additional feeder mains into the distribution area, it may be more economical to boost the pressure at times of high flow. Figure 12.8 shows the hydraulic gradients that may apply before and after boosting.

An immediate consequence of pressure boosting will be that the flow increases. At the design stage, the amount of the increase can only be estimated, although modern network analysis techniques, aided by computer, have reduced the likely errors. Since the area is starved of water, the flow records will not indicate the true value of maximum demand when the pressure is raised. The design duty required from the booster pumps must therefore be judged by the engineer who will take into account that some margin will have to be kept in hand for any future demands on the system. The value of the pressure needed to give the flow required must also be investigated, and from these considerations the characteristics of the pumps required can be decided.

This kind of booster will probably be arranged to start automatically when the pressure downstream of the pump reaches a certain low value. Safeguards will have to be included in the pump controls to prevent hunting and ensure correct interpretation of events: such things as a burst main near the pumping station, which would increase flow and reduce pressure, must not be misinterpreted as a normal demand for more water. The duties required from a booster pump of this kind may be too wide for a fixed speed machine. For instance, if the demand only slightly exceeds the supply given by the unboosted system, then the pump is called upon to raise the head of the whole flow of water by only a few metres, i.e. flow is large against a small head. The flow may later increase substantially so that the duty required from the pump is a large flow against moderate to substantial head. If these conditions are more than a fixed speed pump can handle, then a variable speed drive may be needed. Alternatively a range of pumps could be provided so that these can be brought into use one by one. Pressure surges must be prevented by slow opening and closing of valves, or the slow starting of variable speed pumps, and the control system may become quite complex. A simpler solution to the problem is considered in Section 12.16.

All automatic control gear must be as robust and reliable as the equipment it controls, and must not be so complicated that time is wasted finding a fault that has caused a system

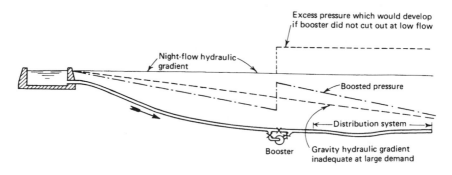

Fig. 12.8   Effect of boosting the pressure of water direct into a distribution system.

failure. Fortunately, developments with modern electronic control systems using solid state equipment and programmable logic controllers have greatly improved the reliability and life of such plants. Modern diagnostic systems and repair-by-replacement techniques have also greatly improved the speed and ease with which faults can be detected and cured. Careful design and selection of the right plant is still needed however to get the best from these innovations.

12.16 Use of balancing storage

Much of the complication of controlling a booster pumping station can be avoided if balancing storage, such as a high-level tank or water tower, can be built into the system at an appropriate point. This can add greatly to the safety and economy of the supply. Once such a tank is connected to the system the booster pumps may be operated by the simple use of level switches in the balancing tank. When peak drawoffs occur, the water level in the balancing tank falls and this may be arranged to start a pump in the booster station. With further lowering of the water level, further level switches may bring additional pumps into operation or increase the speed of the first pump. This system has many valuable characteristics:

(1) Sudden large increases of drawoff which last only for a short time may be handled by the balancing reservoir without causing the booster pumps to start. Such sudden increases can be caused by opening of washouts or by fire-fighting.
(2) The maximum capacity required from the booster pumps is reduced.
(3) The head range against which the pumps have to work is diminished.
(4) Pumping when started can continue until the balancing reservoir is refilled, and the load factor and efficiency of the pumps are thereby increased.
(5) Control of the pumps is simple and positive and the possibility of repeated stopping and starting of the pumps is almost entirely eliminated.
(6) Future extensions to the boosting system are easier to arrange because there are a number of possible ways of further development once there is balancing storage on the delivery line.

The size of balancing tank or water tower required is often small because the periods of peak drawoff on a water distribution system are quite short. If a 5 ml/day supply suffers from a peak drawoff of 50% above normal for a period of 3 hours, the theoretical size of tank required would be about 312 m^3, but a tank of one-half or even one-third of this size would greatly reduce the maximum duty required from the booster pumps.

12.17 Plant layouts

Pumping stations may be arranged in a great many different ways, depending on such things as land availability, the station duty, the required standards, and the preferences of the engineers responsible. Cost will always be a consideration, but compromising by saving initial cost at the expense of future benefits from reduced running costs or inaccessibility for maintenance should be avoided. Some principles to be considered at the design stage are as follows.

Accommodating the plant needed to meet the required duty comes first; this takes account of the site, the type and number of pumpsets, their drives, valves and pipework, their power supplies, methods of starting and control, access for maintenance, and any needed surge protection. Once the plant detailed requirements are decided, design of the pumping station buildings can proceed.

The main pump hall must leave adequate working space about the machinery. There should always be a clear space big enough for complete stripping down of one of the units, which can be the same space as a loading area big enough for access by a vehicle suitable for carrying the largest plant. A travelling crane is essential for all but the smallest pumping units, and power operation can usually be justified. The crane should be arranged to serve all the heavy plant and to cover the loading area. For borehole installations, a mobile crane may sometimes be judged adequate.

High-voltage (HV) switchgear and transformers are usually needed for all but the smallest pumping stations. HV switchgear should be sited in a locked room separate from the main rooms, with access restricted to authorised personnel. Transformers are usually mounted outside the building, although in cold climates use can be made of the heat they generate to provide some background heating to the station if they are inside. Main switchgear, both high and low voltage, and control panels, must have adequate space both in front and behind them. About 1 m should be regarded as the minimum at the back, and 2 m in the front. All switchgear should as far as possible be in line so that the main bus bars are kept short, and cable routes from switchgear to motors should also be kept as short as practicable. Instrumentation and control panelling is nowadays often combined with the low-voltage switchgear panels to achieve a neat installation, and this needs similar accessibility.

All cabling and pipework piping should be in conduits of ample size, and runs must be carefully laid out. When designing cable ducts, the minimum radius to which cables can be bent must be considered. When designing pipework, T-junctions or 90°-bends in the line of flow should be avoided. Radiused junctions and large radius bends should be used to reduce friction losses which add to station energy consumption. All moving machinery must be securely bolted to adequate foundation blocks. If pumps are sited over a well or tank they should preferably be fixed on steel joists, rather than bedded on concrete slabs.

Heaters are usually essential to prevent damp lowering the resistance of electrical insulation in switchgear and motors. Electric motors are frequently fitted with built-in heaters which are energised automatically when the motor is stopped. These are very effective and take little power – perhaps 0.25 kW for motors under 50 kW, and up to 1.5 kW for very large motors.

If air vessels for surge protection are needed, these can be placed outside the pumping station if the climate permits, or inside in an unheated part of the station. Outside vessels may affect the appearance of the station, and in severe climates will need frost protection. This can be done with insulation (which may only delay and not entirely prevent the onset of freezing), or by arranging a trickle flow of warmer water from the supply main to keep the temperature above freezing, or in extreme cases by incorporating electric heaters. If the vessels are inside, condensation may occur on their surfaces and cause staining of floors and rusting of the vessels.

Facilities for chlorination may be needed, but these must have no connection with the main parts of the building and should preferably be in a separate building with restricted access (see Section 9.12).

Some facilities must usually be provided for the station attendants, even if the site is not permanently manned; a small workshop may be justifiable, and a messroom containing a sink and means for cooking light meals. Sanitation of the highest standard is needed, together with hot water and washing facilities.

12.18 Transient pressures: water hammer and surge

If a valve in a pipeline is suddenly closed the water immediately upstream of the valve will be brought to an abrupt stop and will be compressed by the momentum of the upstream water column. This results in a sudden large increase in pressure known as water hammer. As the flow is progressively brought to a halt further and further upstream, so the increased pressure moves upstream as a positive pressure wave. This wave is transmitted up the pipeline at the speed of sound in water. Similarly, downstream of the valve or downstream of a pump which has been suddenly turned off, a rapid drop in pressure will occur and this will be transmitted down the line as a negative wave.

Secondary pressure waves will be generated as the initial wave passes any fitting or change to the pipeline, such as an enlargement or tee, but of greater importance is that when a pressure wave reaches the closed end of a pipeline (such as a shut control valve at entry into a reservoir) it will be reflected as a wave of the same type, i.e. a positive wave is reflected as a positive wave, a negative wave as a negative one. Conversely, a pressure wave reaching an open end to a pipeline discharging to a reservoir or tank will be reflected as a wave of the opposite type. Even in a relatively simple system, therefore, the pattern of waves can become complicated and although, in some instances it is the passage of the initial pressure wave which gives rise to the most critical pressures, that is not always the case. For instance, an initial positive wave may be reflected from a reservoir as a negative wave which may then cause negative (i.e. sub-atmospheric) pressures at high points in a pipeline and allow air valves to open, drawing in air. Later, as the pressure rises again, that air is expelled and the air valves may slam shut generating shock pressures, which may be much greater than in the initial event. Similarly, if the pressure falls low enough, vacuum cavities may form, which may also cause high shock pressures when they subsequently collapse on a rising pressure. The analysis of transient conditions can thus be very complicated and the results depend not only on the elements of the system, but also on the profile of the pipelines. Although methods have been developed in the past for hand or graphical computations, these days the use of computer programs for the analysis of the transient conditions is almost universal.

The main concern with transient pressures is that they should not be high enough to cause bursting of the pipes or fittings. To put the potential for damage into perspective, it is worth noting that the rise in pressure head, on a sudden change of velocity ΔV in a pipeline, is given by $a.\Delta V/g$, where a is the velocity of wave propagation and g is the acceleration due to gravity. In a ductile iron pipeline the wave speed may be as much as 1200 m/s although it is normally between about 900–600 m/s as a result of small quantities of air dissolved in the water. With this high value for the wave speed, the slamming of a valve in a pipeline operating at a velocity of 1.5 m/s could lead to a pressure rise of 180 m. Even with a wave speed of 600 m/s the pressure rise would be 90 m. Such surge pressures do occur, but are often not recorded to the full extent by the ordinary Bourdon pressure gauge which is too 'sluggish' in operation to record the peak of the transient pressure rise.

Codes of practice for most pipe materials allow some transient overstressing above the allowable operating pressure, which is defined as the internal pressure, exclusive of surge, that a component can safely withstand in permanent service. However, the designer must also consider all elements of the system including valves and jointing systems and the resistance of thrust blocks, particularly if an existing system is being uprated. There are, however, other considerations that may be more critical in a particular installation.

In potable-water systems it is normal practice to avoid any negative pressures and hence to eliminate any risk of contamination being drawn in through open-air valves (particularly if located in chambers below ground in areas of high water table) or through joints designed primarily to prevent leakage from a high internal pressure. With large diameter, thin-walled pipes there may a risk of collapse if negative pressures fall to near vacuum; and certain plastic materials, particularly uPVC, may suffer from fatigue failure if there are repeated excessive transient pressure fluctuations above a certain magnitude over the life of the system.

In the majority of systems the most likely causes of transient pressures are valve closure and pump stoppage, particularly the latter if a power failure causes simultaneous stoppage of all operating pumps. Valve closure can be controlled and it is always advisable to ensure that valves cannot be slammed shut and have a closure time long enough to limit the pressure rise to within an acceptable range. The minimum time of closure should, at least, be greater than the time it takes for the pressure wave to be reflected back to the valve from the far end of the line, $2L/a$, where L is the length of the pipeline and a is the wave speed as defined above. A longer closure time, or closure that allows the last 10–20% of the movement to be much slower than over the first 80%, may be required (see Fig. 13.5 in Chapter 13). However, the most critical case is usually a power failure causing simultaneous stopping of the pumps. Most modern low-inertia pumps stop producing forward flow of water in a few seconds when turned off.

Once the potential problems are identified, it is necessary to consider their alleviation. It is rarely economic to increase the strength of a pipeline solely to cope with surge pressures, and it is therefore necessary in most cases to provide some other form of protection for the system. There are a number protective measures that can be adopted as listed below.[2] An indication of suitable locations for installation of some of the devices mentioned is shown in Fig. 12.9.

(1) *Slower valve closure* as already mentioned, obtained by various mechanical means.
(2) *Increased pump inertia.* Fly-wheels fitted to the pumps reduce the rate of deceleration of the pump and the corresponding rate of change of flow.
(3) *Air vessels* (also referred to as 'surge vessels') which comprise pressure vessels connected directly to the pipeline, part of their volume being occupied by compressed air. They are commonly used to feed water into the pipeline when the pumps stop but they also provide a cushion to absorb high pressures on the returning wave and on pump start up. A drawback is that the air is gradually absorbed into the water and compressor facilities are required to provide occasional topping up of air in the vessel. For this reason they are generally installed only at pumping stations.
(4) *Accumulators* are identical in concept and operation to air vessels except that the air is separated from the water in the vessel by a flexible rubber membrane thus greatly reducing the loss of air by absorption. This eliminates the need for compressor facilities and allows the use of a gas such as nitrogen in place of air, topped up periodically from a portable cylinder.
(5) *Surge shafts* can be constructed, if the topography permits, but they must extend above the hydraulic grade line. Flow is diverted into the shaft or drawn from it, thus slowing down the rate of change of flow in the main pipeline.

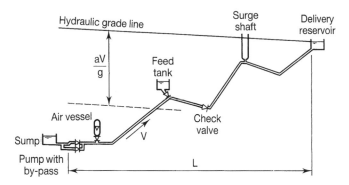

Fig. 12.9 Pipeline profile illustrating suitable locations for installation of surge protection devices (not all would usually be installed on one pipeline).

(6) *Feed tanks* operate by feeding water into the line to relieve low pressures. They can be located at high points below the hydraulic grade line as they are isolated by non-return valves which only allow flow into the pipeline.
(7) *Air valves* of the large orifice type are also used to prevent low pressures in the line by opening to admit air when the pressure falls below atmospheric. Their use for this purpose is not generally permitted on potable water schemes because of the risk of contamination as mentioned above, and there are potential drawbacks including slamming, unless special non-slam valves are used.
(8) *Pressure relief valves* can be set to open at a given pressure or operate in response to an initiating event thus limiting the maximum pressures in a system.
(9) *By-pass pipework* can be fitted around the pumps to allow water to be drawn from the sump provided the pressure on the delivery side falls low enough. However, this may be too late to prevent other low pressure problems occurring down the line on a simple system, but they may be effective at booster stations where the pressure on the suction side of the pumps rises significantly on pump stoppage.
(10) *Non-return check valves* can be used along a pipeline to reduce the effect of the returning positive pressure or water column, but they may give rise to adverse effects themselves so must be analysed carefully and used with care.

With the emphasis of pump manufacturers on producing lighter pumps, fly-wheels and added inertia are out of fashion but in the right circumstances are the most reliable and effective form of protection. More commonly on pumping systems a surge vessel or accumulator is used. Feed tanks may be necessary where the pipeline passes over a high point near the elevation of the discharge point, but must be designed to prevent the possibility of contamination entering the chamber and will need a small sweetening flow, (almost certainly having to be discharged to waste) to prevent stagnation of the water. Air valves, particularly of the non-slam type, can be effective; but careful consideration must be given to their use as the primary form of protection of a pipeline because they require regular maintenance and have often to be sited in remote locations.

Pump delivery non-return valves need particular attention to ensure they are suitable for the system and its transient response; especially if a surge vessel is also provided because the flow in the connecting pipe to the air vessel may reverse very quickly. Ideally, the non-return valve should shut at the moment of flow reversal but if it reacts more slowly the reversed flow may slam the valve shut with the generation of a high shock pressure. The dynamic response of the non-return valves should thus be matched to the transient characteristics of the pipeline system.

12.19 Pump suction design

Pumps should be sited so that negative pressure does not develop on the suction side. Negative pressure can cause a reduction in the performance of the pump, and may prevent the pump from being automatically primed. If negative pressure is really unavoidable a 'self-priming' pump must be specified, but this should be avoidable in many cases by a suitable design of suction. Figure 12.10 shows a bad design and a good design for a pump suction. The aim is to minimise pressure losses in the suction system and so prevent air being released from the flow of water. Another fault that can develop at the inlet to a suction pipe is vortexing, which can be reduced by fitting flow-straightening vanes at the inlet.

The diameter of pump suction pipes is usually larger than the delivery pipe diameter to reduce friction losses by reducing the flow velocity on the suction side. Many suction pipes for raw water need to be equipped with a strainer, and it is important to keep this in good condition. If strainers are used, they must be of ample size and should be specified as having a total area of opening at least double that of the suction pipe. Foot valves, i.e. non-return valves, fitted at the inlet to vertical suction pipes, are a potential source of

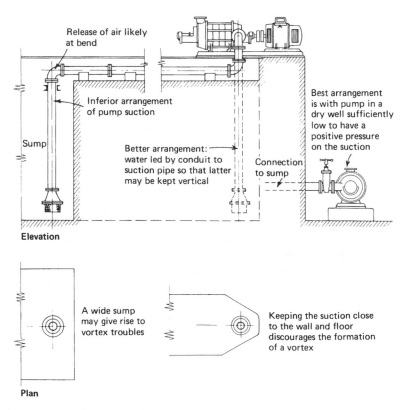

Fig. 12.10 Arrangement of pump stations.

trouble and should be avoided where possible. If installed they must be of high quality and be well maintained or they may tend to stick open. They are sometimes installed for keeping a pump primed when it is idle, but often they are not effective at this if the pump stands idle over a long period and other priming methods are better. Foot valves are also sometimes used to prevent reverse rotation of a vertical spindle pump which has a long rising main, when the pump is stopped. A high speed of reverse-rotation can be caused by the falling column of water, and to restart a pump when it is rotating in reverse may cause breakage of the shaft. Instead of a foot valve, a pump may be specified as suitable for being 'turbined' under reverse flow, and a time delay switch is then incorporated in the switchgear to prevent restarting before the reverse rotation has stopped.

12.20 Cavitation

Cavitation occurs when the absolute pressure in the pumped water falls to its vapour pressure, and this can happen if suction conditions are poor. Air or water vapour may then be released from the water and the entrained bubbles, which may be of minute size, will be carried into the pump, where entering a higher pressure zone they collapse. A stream of such bubbles continuously collapsing on the tip of an impeller blade quickly erodes the blade material. The blade tips become pitted and will be worn away in a short time. Cavitation can be prevented by good design of the pump, and by maintaining a positive pressure on the suction side by good design of the pipework and, wherever possible, siting the pump below or as near as possible to the suction-side water level when pumping is occurring.

12.21 Corrosion protection

Corrosion protection can be provided for most pumping plant by the correct choice of pump materials. A more recent development has been the application to pump internal parts of specially developed coatings. One commonly applied material is the glass-filled resin-based coating. These coatings have been successfully used for application to the internal surfaces of pipework, valves, pump bowls, and pump impellers to restore performance following corrosion or erosion attacks. They are also sometimes applied to new pumps to prevent damage and improve efficiency by presenting smoother surfaces to the flow. Metal surfaces require careful surface preparation to obtain the full benefits of the coating material, and since the coatings have significant thickness, care is needed to ensure that pump passages, particularly at the pump inlet, are not reduced in area so much that impaired performance results. Clearly this is more significant with the smaller sizes of pump, and for this reason combined with the cost, coatings cannot normally be justified on pump sizes of less than about 300 mm branch diameter.

12.22 Efficiencies and fuel consumption

Quoting efficiencies and fuel consumption are difficult because they vary so widely and every individual case has something special about it. The figures in Tables 12.1 and 12.2 are intended to act as guides to the efficiencies and fuel consumption normally to be expected. There are wide variations according to the power rating and type of pump and motor.

12.23 Effect of electricity tariffs

The charges made by the electricity supply companies are designed to favour the consumer who takes a steady supply of electricity at high power factor. The consumer who wants large power intermittently is penalised. Tariffs usually comprise two, three, or even four separate charges as listed below. If the supply is required for a new installation in a remote area, the supply company will usually also ask for a connection charge to recover the cost of providing the supply.

(1) A 'maximum demand charge' per kW or kVA of maximum demand in that month above a given figure (and sometimes, more onerously, per kW or kVA of maximum demand for the last 12 months) The charge may differentiate between summer and winter maximum demand, and sometimes may be varied month by month.
(2) A 'unit charge' which is a monthly charge for the number of units of electricity consumed, the charge per unit being substantially higher for daytime consumption. It is usually on a sliding scale allowing reductions when large numbers of units are consumed.
(3) A 'fuel clause' which increases the unit price of electricity according to the increase of basic fuel cost.

Table 12.1   Efficiencies

Pumps:	
Horizontal centrifugal	Medium size 80–82%, perhaps 85% large size. Even higher with special construction but at higher price
Vertical spindle shaft driven	Tending towards about 3% less than the horizontal centrifugal
Submersible	75–81% and can be lower, to about 70% for small sizes. Generally about 3% less again than the vertical spindle pump, the reason being that the pump is restricted in diameter
Electric motors:	
For horizontal pumps	93–95%. Fixed speed AC induction
For vertical pumps	90–94%. Fixed speed AC induction
For submersibles	85–89%. Less than the above because of the restrictions imposed on the design
Variable speed	About 3–5% less than with a caged AC motor

Table 12.2   Overall fuel consumption

Electrically driven pumps	About 1.0 kW for every 0.75 kW of water power output, this implies an overall efficiency of about 75% which would be usual. Up to 1.3 kW per 0.75 kW water power output or higher for small pumps or variable speed pumps
Diesel engines	0.21 kg of diesel fuel oil consumed per kWh of engine power exerted would be considered good, 0.28 kg per kWh being not unusually high. For lubricating oil add 5% to fuel oil cost

(4) A 'power factor clause' which increases the maximum demand charge if the power factor drops below a certain figure (usually 0.90).

The effect of the maximum demand charge (especially when based upon twelve months) can be onerous. The larger the gap between the maximum demand and the average demand, the more penal the maximum demand charge becomes and the more costly the charges per unit. Hence it is a matter of great importance for pumping stations to be run at a high 'load factor', i.e. the average power consumed should be as close as possible to the maximum power required at any time during a month (or year), and pumps should be run for long periods rather than short, so that the units consumed are as cheap as possible because their price reduces as more units are consumed per kW of maximum demand during the billing period. Pumping for short intermittent periods at high outputs is therefore an expensive way of operating. Similarly, the occasional running of pumps at full load when they normally run only at part load, or bringing in extra pumps for short periods can also be expensive. Even the testing of pumps when first installed – perhaps running several together on test – can bring a heavy extra cost for so brief a run as 30 minutes.

Choosing the right duty for the pumps, the right hours of working, and the right amount of standby to be provided for any pumping station are matters that must be thoroughly investigated. The benefits due to adequate storage in the supply system can be very significant. These include safeguarding continuity of the supply, levelling out pumping rates and so reducing friction losses and the maximum power demanded of the motors, and permitting long steady running at a high load factor to minimise electricity charges. With modern competitive electricity supply arrangements, the water supply company may be in a strong position to obtain specially favourable rates from the electricity supplier. Unlike many users, water companies are usually large electricity consumers with a steady load for most of the year.

12.24 Thermodynamic pump performance monitoring system

Pumps must be run at the highest possible efficiency to minimise operating costs. Most pump energy losses result in heating the pumped water by a small amount. Meters have been developed which rely on very accurate sensors to detect the resulting very small temperature rises across pumps, which are proportional to the wasted energy. An advantage of this method is that there is no need to measure the pumped flow directly to determine the pump efficiency. However these devices are not suitable for all installations, and they work best with high head machines because the resulting temperature rise is then higher for the same pump efficiency and power. They are not at present very effective when used with submersible borehole pumps. Development work is continuing however, and the performance of these devices continues to improve.

In a typical installation, temperature probes and pressure transducers are mounted on the pump pipework to detect the inlet and outlet conditions, as shown in Fig. 12.11.[3] A micro-processor, usually mounted in portable equipment, is used for analysis of the data and to allow display of efficiency, flow, head and power consumption. Thermodynamic pump testing is accepted in Pump Test Standard ISO 5198[4] as a precision-class test. This method may be most valuable if repeated regularly to monitor changes in pump performance with time.

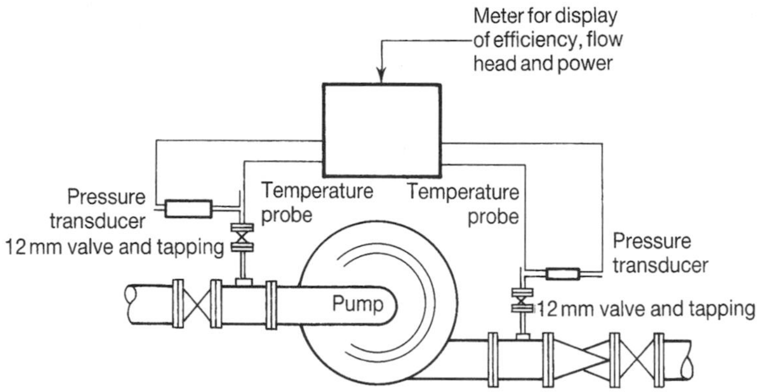

Fig. 12.11   Pump performance monitoring system.

12.25 Control and instrumentation (C and I)

The C and I of pumping plants, distribution network flows, and water treatment plant, has changed rapidly in the water industry with the introduction of stricter requirements for levels of service to consumers, maintenance of water quality through the distribution system, and the development of many specialised and advanced water treatment processes. The latter alone have necessitated the processing of thousands of signals and large amounts of data. A typical modern water treatment works control system will include 10 000–15 000 signals to be processed. The automation of such plants is becoming essential due to this complexity and the large amount of data to be handled. Also automatic control systems are seen as a way to reduce labour operating and supervisory costs, whilst at the same time maintaining a high quality product.

Current C and I development has evolved from changes in the electronics industry, which have reduced the capital cost of providing complex monitoring and control systems. Hence the industry has moved away from traditional 'hard wired' control systems to the use of more adaptable programmable electronic systems. Three levels of control are practised.

- Although manual control may be impracticable in the more complex cases, it must still be available as a 'fall-back' method for emergencies and when other systems are unsuitable or fail.
- Semi-automatic control aids the operator cope with treatment processes which are too complicated for the operator to control without assistance. Examples of such processes are ozone production, or the overall start-up of plant applying numerous chemicals. Some operations which were traditionally manual are now frequently specified as semi-automatic.
- Full automation is used to control processes with the minimum need for operator intervention. This type of control must ensure a high quality, safe product, and take corrective action when failure occurs of plant and equipment. Operator intervention is usually limited to entering key set point parameters related to output quality and quantity.

The degree of automation to be adopted is the most critical decision to be made. In the information technology industry the following terms are in common use.

Hardware means the device (sometimes programmable) which controls and monitors the operation of an item of plant.

Software is the programming logic or set of instructions which designates the task(s) to be carried out by an item of hardware or a group of hardware. These tasks can range from starting a pump, to calculating from monitored process variables the appropriate hardware action or combination of actions to be taken.

Firmware is a term used in the industry to designate a permanent form of software (embedded in hardware) which is normally not available for alteration by the operator, but may have features permitting selection from a standard menu suited to the application.

Software which is custom designed for a specific application is often referred to as 'application' or 'bespoke' software. Most control systems involve the use of some bespoke software, the precise specification of which is important. A 'functional design specification' (FDS) must be prepared in detail setting out all the function requirements for a piece or set of hardware, and the controls to be applied. Drawing up an FDS is a critical link in providing a satisfactory monitoring and control system, which must satisfy all parties involved that it meets their requirements.

Where two pieces of instrumentation equipment are required to operate as a system, the interface between them is very important to ensure all necessary data is communicated. After selecting the correct component for the application, the designer will be concerned primarily with component compatability – the ability for two or more items of equipment to operate together as a system. The system designer must also consider the operating life and maintenance associated with each piece of equipment selected to form a system, a task which becomes increasingly difficult as the size and complexity of monitoring and control systems increases.

Systems

A 'system' may involve only one measurement and a controller to maintain a set point of a single parameter; or it can handle data from many monitoring points and propagate many plant control commands. A 'distributed control system' (DCS) is now the type most commonly used (see Fig. 12.12). In it the control logic is distributed within outstations around the plant being controlled. Until relatively recently the computing power required to monitor and control plant was not available at outstations, so the entire computing power was held centrally, relying on a communication network to monitor and control an outstation and its associated plant. These centrally based systems are now, in almost every case, obsolete. An up-to-date DCS comprises a number of discrete process control units, i.e. outstations or remote terminal units (RTUs), linked by a 'state-of-the-art' communication network as Fig. 12.12. The automatic control logic for each process area is contained in the RTU and can therefore function independently of the overall monitoring and control system. A DCS links discrete process blocks to achieve:

- transfer of data round the system, including changes to control parameters at RTUs, such as adjustment of set points or duty plant selection;
- monitoring of individual RTUs;
- display and recording of data, events and alarms;
- safe reaction to changes or failure of individual RTUs which could affect product criteria, process performance, or plant safety.

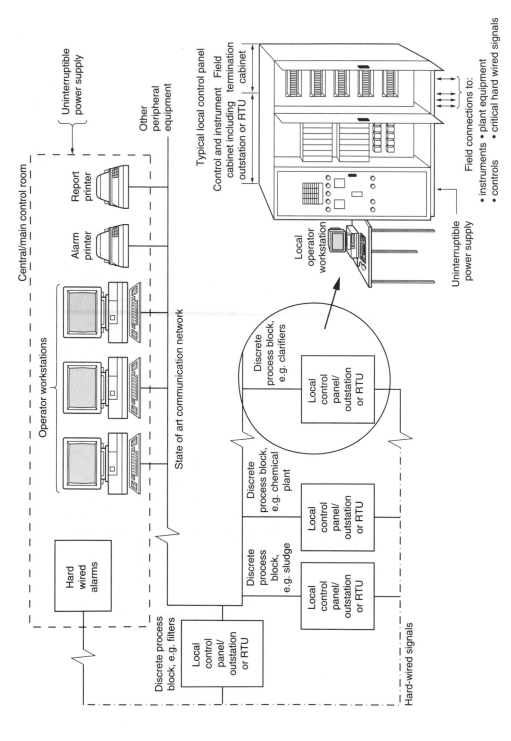

Fig. 12.12 Typical distributed control system linking a number of discrete process block control units.

Telemetry is the system for transmitting data from one location to another. The distance between the two locations can vary from a few metres to thousands of kilometres; virtually between any two locations with only the limitations of the transmission medium applying (e.g. hardwired data communication networks, telephone networks, microwave systems or radio systems, etc.). Telemetry systems used by most monitoring and control systems include DCS, and SCADA systems referred to below.

Data acquisition, control, and backup requirements

The term SCADA (supervisory control and data acquisition) refers virtually to any data acquisition system used, but usually one which exercises monitoring and supervisory control of a number of sites from a central control centre. Such systems are widely used in the water industry so that a 24-hour manned control centre can react to any problems arising at sources or throughout the distribution system.

Measurement of process parameters can be digital or analogue. Digital is used for reporting the status of a plant, e.g. whether running or stopped. Analogue measurement is used to represent variable values, such as chlorine residual or pump output. There are several standard analogue interface standards. The water industry uses the 4–20 milliamp loop as the *de facto* standard. The 4 milliamp value usually represents the lowest end of the value range and the 20 milliamp the highest. The 4 milliamp value is referred to as 'live zero', which is used to detect a problem, such as a zero milliamp value which could indicate a broken connection. The instrumentation industry is working towards standardising inter-instrument communication with the development of standard protocols (the communication terms of engagement between two components). Two such protocol standards are HART (highway addressable remote transducer) and 'Field Bus'. Also, intelligent instruments are being developed to operate on their own standard data communication network: 'Profibus' and 'ControlNet' are two such standards. Although instruments with these capabilities have been available for some time, the standards have yet to be used extensively in the water industry.

Uninterruptible power supplies (UPS) are now often used to support water industry monitoring and control systems, to maintain system operation for a predefined period during a power supply failure. A plant must always be reviewed for power failure; each piece of equipment must be controlled into a fail-safe condition when power supply fails. An ancillary plant may be necessary to fulfil these requirements. Also plant start-up following a power failure is another key consideration; and automatic start-up or manual start-up must be reviewed accordingly and specified clearly.

Modern control systems still frequently utilise hard-wired monitoring methods, primarily as a back up, fail safe system for a critical process or piece of plant. Shutdown systems have to be considered carefully; they have to prevent damage and danger to plant or personnel and ensure there is no failure to maintain output quality. *Alarms* notify the need for operator attention when plant, process, personnel, or output quality and quantity may be at risk. *Alerts* draw operator attention to any state that may develop into an alarm if left unattended to (e.g. tank water level high or process flow low, etc.). *Events* inform the operator of a 'normal' operating occurrence that may or may not require attention (e.g. a pumping sequence completed, etc.)

Operation and maintenance

Operators of automated treatment facilities need an understanding of electronic control and software programs, as well as a full knowledge of the treatment processes. Maintenance of automatic control systems is a specialist's field: this aspect of modern system installation is often overlooked. The system purchaser must fully appreciate the maintenance commitment that is involved. The shift from a traditional plant with a high level of operator input, to plant which has complex software and electronics control, has not significantly reduced overall operational and maintenance costs of water supply, because maintenance costs associated with C and I systems for a treatment works tend to offset the savings due to reduction in plant operational personnel needed. On the other hand, the wide variety of specialist processes now involved in water treatment require more complex systems of monitoring and control than would be possible manually.

References

1. BS 4999, General Requirements for Rotating Electrical Machines. British Standard Institution, 1987/1988.
2. Thornley A. R. D. and Enever K. J. *Control and Suppression of Pressure Surges in Pipelines and Tunnels.* CIRIA, 1984
3. Yates M. A. A Meter for Pump Efficiency Measurement. *World Pumps*, January 1989.
4. ISO 5198: *Specification for Acceptance Tests for Centrifugal, Axial and Mixed Flow Pumps: Precision Class Tests*, 1987 (BS 5316, Part 3: 1988).

13

Pipes, pipeline construction and valves

13.1 Introduction

Pipes, lining materials and joints must not cause a water quality hazard. From 1 July 1999, materials in the UK on customers' installations are covered by the Water Supply (Water Fittings) Regulations 1999 – and their equivalents in Scotland and Northern Ireland. Materials used in public supply systems are covered by Regulation 25 of the Water Supply (Water Quality) Regulations 1989 and amendments – and their equivalents in Scotland and Northern Ireland. Materials considered to comply on the water supply side are those (a) approved by the UK Government (DETR) Committee on Chemicals and Materials of Construction for use in Public Water Supply and Swimming Pools and (b) products verified and listed under the UK Water Regulations Advisory Scheme (formerly the Water Byelaws Scheme), administered by WRc plc, UK. From 1 April 2000 approval of materials which have been in use for some time without apparent problems to health, will no longer be permitted.

13.2 Types of pipes and organisations setting standards

Pipes found in waterworks systems are generally of the following materials:

(1) cast or 'grey' iron (now superseded by ductile iron);
(2) ductile iron (DI);
(3) steel;
(4) polyethylene (PE);
(5) PVC (polyvinyl chloride);
(6) GRP (glass reinforced plastic); and
(7) pre-stressed concrete, cylinder or non-cylinder (PSC);
(8) reinforced concrete cylinder (RC);
(9) asbestos cement (AC, no longer produced in the UK).

Other materials include galvanised iron, copper and lead which tend to be found in service pipes, plumbing, common connections and other small diameter mains. Lead was extensively used in the past for service pipes but is no longer installed because of the danger of plumbo-solvency (see Section 6.27); however many lead house connection pipes and plumbing still remain in service. Copper pipe can also give rise to plumbo-solvency problems from lead solder. In the UK these materials now tend to be superseded by polyethylene and, where lead in water is a problem, lead pipes are generally being replaced by polyethylene.

A number of organisations have set standards for pipes and their fittings, using various terms to define pipe characteristics or pressures applied to pipes or pipe systems. Table 13.1 lists the terms most widely used (as in this chapter) and references for their definitions. Other specific terms are explained in their context. A short bibliography at the end of this chapter gives the main range of standards applying.

Table 13.1 Abbreviations for pipe and pipe system standards

Standards issued by various organisations	
BS EN	denotes a British Standard according with a Comité Européen de Normalisation (CEN) standard
BS	denotes a British Standard issued before a CEN standard was issued and not necessarily complying with a subsequent CEN standard
prEN	CEN standard in preparation
CP	a British Standard Code of Practice, discontinued in mid 1970s (later Codes were given a BS or BS EN number)
ISO	a standard issued by the International Standards Organisation
Some abbreviations used in standards	
For pipes	
DN:	nominal diameter in millimetres, shown as DN followed by a number (ISO 6708)
PN:	nominal pressure in bar or metres of water (ISO 7268 and 7268/1). PN for flanges refers to the maximum allowable working pressure, excluding surge. 1 bar = 10.19 m head of water
For pipe pressures	
PFA:	allowable operating pressure, excluding surge (prEN 805)
PMA:	allowable maximum operating pressure, including surge (prEN 805)
PEA:	maximum allowable hydrostatic test pressure after installation (prEN 805)
For pipe system pressures*	
DP:	design pressure – maximum operating internal pressure of a system or zone fixed by the designer but excluding surge (BS EN 1295–1)
MDP:	as DP but including surge – designated MDPa when there is a fixed allowance for surge or MDPc where surge is calculated (BS EN 1295–1)
STP:	system test pressure as applied after installation (BS EN 1295–1)

*The pressure rating for a pipeline system may well be limited by fittings and particularly flange pressure ratings.

13.3 Structural design of pipes

Pipelines are required to withstand internal hydrostatic pressure, external pressure from soil, surcharge and traffic loads and to be safe against buckling.[1-3] Soil load increases with depth: surcharge and traffic loads reduce with depth. Main road traffic load depends on the wheel load and configuration considered but typically accounts for less than 10% of the total at 6 m depth and the minimum total load under main roads typically occurs at about 1.8 m depth. BS EN 545 gives traffic loads for ductile iron pipes and tabulates allowable heights of cover for different combinations of traffic load and backfill compaction.

Pipes can be grouped as rigid, flexible and semi-rigid. Rigid pipes deflect little and carry external loads in ring bending. Their load carrying capacity is derived from ring bending strength (as designed or measured by crushing tests) – which can be increased by bedding factors appropriate to various standardised bedding and surrounds. Flexible and semi-rigid pipes deflect under load in inverse relation to the stiffness of the pipe and overall soil modulus (comprising both embedment and native soil modulus). Flexible pipes derive their support primarily from passive soil pressure which develops as the pipe ovalises under vertical load and deflects horizontally into the trench backfill. The contribution of pipe stiffness is small. Ring bending stress is traditionally not taken into account for steel pipes, but is addressed as combined hoop and bending stress for thermoplastic (PE and PVC) pipes. Bending strain is calculated for GRP pipelines. Overall soil modulus is a function of the backfill type and compaction and depth. It also depends to some extent on the modulus of the native soil, which, if weak, may require the trench width to be increased or the backfill material and compaction to be improved or both. Safety against buckling of flexible pipes also depends on overall soil modulus. Flexible pipes require more care in selection and control of backfill than other pipe types but offer economical solutions and are therefore widely used. Other pipe types also require attention to backfill materials and workmanship. Semi-rigid pipes derive their support partly from the soil and partly from the pipe stiffness. Their design takes into account ring bending stress. Permissible deflection limits for ductile iron pipes are set to limit ring bending stress in the pipe wall and for joint and lining performance.

Rigid pipes include concrete and asbestos cement. Flexible pipes include thin walled steel (diameter-to-thickness ratio more than about 120) and plastics. Ductile iron pipes are classed as semi-rigid. Flexibility is determined by ring bending stiffness (specific stiffness), $S\,(\text{kN/m}^2) = EI/D^3 = 1000E\,(t/D)^3/12$, where E is the modulus of elasticity (N/mm^2), t is the mean pipe wall thickness and D is the mean diameter of the pipe (outside diameter less thickness). Ring bending stiffness may also be expressed in units of N/m^2. This diametrical value is distinct from and is one-eighth of the stiffness based on radius, used in Germany.

Typical values for semi-rigid and flexible pipes in the range DN 100–2000 are 1350–16 kN/m^2 for ductile iron; 580–4 kN/m^2 for steel; and, irrespective of diameter (except for small diameters) 80–16 kN/m^2 for PVC, 17–8 kN/m^2 for MOPVC (see Section 13.15), 10–5 kN/m^2 for GRP and 80–4 kN/m^2 for PE. In comparison, reinforced concrete pipe stiffness is typically about 500–600 kN/m^2 for DN 800 and above and stiffer at smaller diameters. Very flexible pipes (stiffness less than about 4 kN/m^2) are at risk of deflection to a 'squared' shape (as the pipe is deflected first inwards and upwards and then downward and outward as backfill is built up in layers): they are not recommended without great care in construction and specifically not for plastics pipes.[4] This value is also adopted under the European standards being developed for plastics pipes. A stiffness of 4 kN/m^2 (4000 N/m^2) is equivalent to a diameter/thickness (D/t) ratio of about 160 for steel pipes. This is similar to the D/t range of 160–180 quoted as the maximum suitable for centrifugal cement mortar or spun concrete lining. (Above this range, pipes need flexible lining or can be lined with cement mortar lining *in situ*.) Despite case histories of much greater values, D/t for steel pipes should in no case be more than 200 (stiffness 2 kN/m^2) to retain sensible control during installation.

Characteristics of plastics pipes change with time, in particular the modulus of elasticity, *E*, reduces very considerably with increasing time. 'Short term' is defined as the 1 hour modulus. Long-term values are obtained by extrapolation of tests over various periods to obtain 50-year values. Long-term values must be taken into account in design and for PVC and PE, respectively, may typically be 50 and 25% of the short-term value – but this depends on specifics of the actual material used. 'Ultra short-term' (10 second) values are significantly greater than the 1-hour values and are applied when calculating pressure surges.

Structural design in general is based on the pipe cross-section. Longitudinal bending arising from uneven trench support or differential movement tends only to be specially considered where pipes are designed to be supported at discrete intervals – as on piles or above ground. Similarly ring bending tends not to be included specifically in steel pipe design. These elements therefore are implied to be included for in the factors of safety – but where necessary can be separated and combined as equivalent stresses according to the appropriate design codes.

Work by the European standards organisation, CEN (Comité Européen de Normalisation), to develop a common calculation approach found that the several European methods (and American methods) were similar: principal differences were in coefficients and in values used for soil stiffness and assumptions for soil support. Differences may also arise from design strategy: adopting low factors of safety to upper and lower bound loads and strength, or higher factors of safety to mean loads and strength. BS EN 1295–1: 1998 summarises the various European design methods. National annex A of this standard sets out procedures established in the UK for the three pipe categories. BS EN 1295–2, due to be published at the end of 1999 will give details of the European methods. Work towards a common European design methodology is in hand but there is no date yet set for publication.

Cast iron pipes
13.4   Cast or 'grey' iron pipes

Iron pipes made towards the end of the nineteenth century were cast in vertical sand moulds. Many waterworks still have some of these pipes in use. Their walls were relatively thick and not always of uniform thickness because of occasional misplacement of the central core mould.

'Spun' grey iron pipes were first produced in the UK in the 1920s, the molten cast iron being poured into cylindrical moulds rotating at high speed so that the pipe walls were formed by centrifugal force. After a few moments of spinning the metal solidified and the pipe contracted slightly so that it could be drawn in a red hot state from the mould. It was then sent for reheating and slow cooling to reduce stresses arising from cooling. By this process a denser iron resulted and the pipes were of uniform wall thickness. These 'spun' iron pipes were in widespread use from the 1930s to the 1960s and now comprise a large proportion of the distribution mains of water undertakers in many countries. Three classes of such pipes were in common use: B, C, and D for working pressures of 60, 90, and 120 m respectively. Classes B and C were predominantly used, but because their outside diameters in the larger sizes were different, and also to reduce stocks and possible confusion in use, many water undertakers used only class C pipes rather than a mixture of classes B and C. Carbon is present in the iron matrix substantially as lamellar or flaky form: therefore the pipes are brittle and relatively weak in tension and liable to fracture. The manufacture of grey iron pipes has now been discontinued in most countries, except for the production of non-pressure pipes for rainwater and soil drainage.

13.5   Ductile iron pipes

Ductile iron pipes are normally cast by centrifugally spinning molten iron in high quality steel moulds; fittings are cast by pouring the metal into steel or sand moulds. Small quantities of magnesium are added to transform the lamellar form of carbon into a spheroidal form, thereby increasing the tensile strength and ductility of the iron. Pipes are heat treated (annealed) after casting. The resulting pipe is stronger and less liable to fracture than grey iron. Ultrasonic tests may be carried out at selected locations on the pipe body. Ductile iron pipes and fittings up to DN 2000 for potable water are covered by BS EN 545. This also covers design for external loads. BS 8010 Section 2.1 covers installation.

Ductile iron pipes are manufactured in the UK in sizes up to DN 1600 with socket and spigot ends suitable for forming push fit type joints, plain ends suitable for jointing with flexible couplings, or with flanged ends formed by welding on ductile iron flanges complying with BS EN 1092–2. Pipes larger than DN 2000 are manufactured elsewhere but sizes exceeding DN 2000 are jointed with other than the push fit type. Standard lengths vary depending on the diameter of the pipe, the type of ends required, and where the pipe is manufactured. In the UK standard lengths for socket and spigot ended pipes are 5.5 m up to DN 800 and 8 m for DN 900 and larger.

BS EN 545 requires wall thickness, e in mm for pipes and fittings to comply with the formula

$$e = K(0.5 + 0.001DN)$$

where DN is in mm; $K = 9$ for socket/spigot, plain ended and welded flange pipes; $K = 12$ minimum for fittings without branches, e.g. bends and tapers; $K = 14$ minimum for fittings with branches, e.g. tees; a negative tolerance is allowed not exceeding $(1.3 + 0.001 DN)$, subject to a minimum wall thickness of 6 mm for pipes and 7 mm for fittings.

The allowable operating pressure excluding surge (PFA), including surge (PMA), and the maximum allowable test pressure after installation (PEA) vary with the diameter of pipes as shown in the following table for selected sizes.

PFA, PMA and PEA values for K9 ductile iron pipes, based on BS EN 545

Nominal diameter DN (mm)	PFA value (bar)	PMA value (bar)	PEA value (bar)
<200	64	77	96
200	62	74	79
300	49	59	64
400	42	51	56
600	36	43	48
900	31	37	42
1200	28	34	39
1600	27	32	37

The PFA values allow for a safety factor of three on the ultimate tensile strength of ductile iron of 420 N/mm^2. PMA, the allowable maximum operating pressure including surge is approximately 20% more than the PFA. PEA, the maximum allowable hydrostatic test pressure after installation is, in general, PMA plus 5 bar; except for pipes smaller than DN 200 where PEA is 1.5 times PFA.

When ordering pipes and fittings, specification of the operating and test pressures is vital, so that the pipes and fittings, particularly tees, can be made to suit. For very high pressures (PFA typically 25 bar or more, but depending on the diameter and type of fitting) availability should be checked.

Joints are required to be watertight at PEA and to durably withstand without leakage the PMA pressure under service conditions, including angular, radial and axial movement.

Works test pressures on individual items are also specified in BS EN 545 and are generally similar to, or a little less than PFA values.

BS EN 545 sets out a design methodology and gives a table of allowable external loads for selected backfill conditions.

13.6   External coatings and internal linings

Current UK practice for external coating of ductile iron pipe comprises spray coating with zinc onto the exterior of the pipes, followed by a coating of bitumen paint. For aggressive soils polyethylene sleeving, either factory or site applied, is often adopted, plus imported backfill where appropriate. Minor damage and puncturing of the sleeving does not impair the efficiency of the protection, although any such tear should be patched with adhesive tape. The effectiveness of the wrapping, even though the sheeting is not watertight, is ascribed to insulation of the pipe from uneven soil contact (which can produce galvanic cells) where there is an uneven pipe bed and surround, particularly in the case of clay soils which are hard to compact near the bottom of the pipe; also the sleeving also holds a virtually static body of water against the pipe. For highly aggressive soils the pipes can be wrapped with a heavy duty PVC backed bitumen adhesive tape, overlapped 25 mm or 55%. Cathodic protection is not normally recommended but, if used, requires bonding across the joints to provide electrical continuity and tape wrap to reduce current and anode consumption. A points system, depending on soil resistivity, ground water level and other characteristics, can be used to judge the potential severity of aggression in a given location.

Earlier ductile iron pipes were coated with cold or hot applied bitumen or hot applied bitumen-based material sprayed or brushed onto the pipe metal. The former use of coal-tar products for coatings and linings was discontinued because coal-tar can give rise to PAHs in potable water (see Section 6.36).

Internal lining of ductile iron pipes now adopted in the UK comprises spun mortar lining, using sulphate resisting cement to BS 4027 plus, for pipes DN 800 and smaller, an epoxy seal coat on top of the cement mortar. Details of cement mortar lining are given in Section 13.10.

13.7   Joints for iron pipes

Types of joints in use are shown in Fig. 13.1.

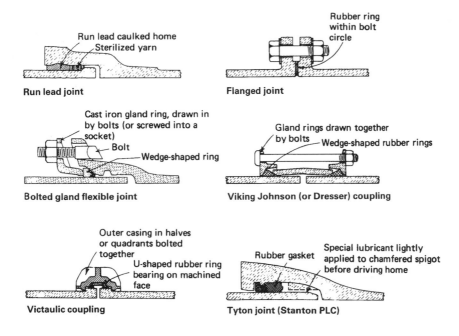

Fig. 13.1   Joints for iron pipes.

The run lead joint for use with spigot and socket pipes is now superseded by the various proprietary mechanical joints, but many mains still in use have been laid with this joint. Skilled workmanship is required to make the joint properly. The lead has to be heated to 400°C at which temperature the molten lead shows strong rainbow colours when the surface scum is drawn aside. A clip is placed around the pipe against the annular space of the socket and the lead must be poured in one continuous pour through an opening left at the top of the clip. The operation is dangerous if not undertaken with care and the socket space must be quite dry to avoid blowback of the lead by steam. The lead solidifies almost immediately after pouring and is then caulked up using a series of chisels. The joint is rigid and slight movement brings a tendency for such joints to weep; however, countless mains have been satisfactorily laid with this type of joint.

Flanged joints are covered by BS EN 1092–2. Flanges must be carefully aligned before the bolts are inserted and the flanges pulled together. The alignment must be almost as precise as that adopted for aligning motor couplings. To pull up misaligned flanges is likely to cause fracture of the pipe or flange. A rubber ring is inserted between the flanges, of such diameter that it lies inside the bolt circle but does not intrude into the pipe bore. The faces of the flanges and the rubber ring must be perfectly clean before assembly and the bolts must be tightened up in sequence little by little so that an even pressure is maintained all round. No grease, bitumen paint, oil, dirt, grit, or water should be permitted on the flange or rubber ring faces. The contact should be between clean dry metal and clean dry rubber. When making joint faces which are vertical, some difficulty may be experienced in keeping the rubber ring flat against the vertical flange face and, to counteract this, a little clear rubber solution may be used to tack the rubber ring on to the metal face. This is the only material whose use can be permitted in connection with rubber rings. If a greasy material such as a bitumen-based adhesive is used, tightening the flanges will cause the rubber ring to extrude into the pipeline and, however much further the flanges are squeezed together, an imperfect joint will result and be liable to leak. Rubber rings should be manufactured from high-grade, non-biodegradable natural or synthetic rubber to a thickness of 3.2–4.8 mm and must lie perfectly flat against the faces of the flanges without distortion.

Bolted or screwed gland joints in general are no longer used but worked on the principle of forcing a rubber ring into an annular space formed between a specially shaped socket and the plain spigot of the pipe. The rubber ring is forced into the annular space by a cast iron pressure gland which is drawn by bolts or screwed into the socket. The pipe barrel, the socket, and the rubber ring must all be scrupulously clean before erection and the pressure gland must be tightened up uniformly. Joints of this type were developed principally for gas mains.

Viking Johnson (or Dresser) couplings are another form of patent rubber ring joint used for connecting together lengths of pipe which are straight-barrelled, i.e. no socket is used and no spigot beads are necessary. This joint can be used with ductile iron pipes and is very widely used with steel pipes. A cross-section of such a joint is shown in Fig. 13.1 where it will be seen that it acts on the same principle as the patent ring type of joint. Slight angular deviation at joints is possible with these couplings; also, if the coupling is specified as without a central register it can be moved along the barrel of the pipe, thus permitting removal of a section of main.

The Victaulic coupling is used in conjunction with shouldered ends of pipes, thus holding them together longitudinally. The joint can be unmade and remade without difficulty. Hence this type of joint is most often used for temporary pipelines laid above ground. It is also sometimes used to connect meters in pipelines, thus facilitating their removal.

The push fit joint includes a specially shaped dual hardness synthetic rubber ring gasket fitted into the socket of a pipe before the spigot of the next pipe is pushed in. A little lubricant must be used on the inside of the gasket and on the outside of the pipe spigot before forcing it into the socket. Complete cleanliness of the socket, gasket, and spigot is essential. These joints are flexible: BS EN 545 requires a minimum possible angular deflection of 3.5° up to DN 300, 2.5° for DN 350–600 and 1.5° for larger sizes. Allowable deflections for pipes produced in the UK are 5° up to DN 300 and 4° for larger sizes. To allow for movement after laying, deflections at installation should be limited to 50% of the figures quoted. These joints are the norm for ductile iron pipes due to their simplicity and the ease and speed with which they may be made.

Steel pipes
13.8   Steel pipe manufacture and materials

BS 534 covers steel pipes, joints and specials (bends and other fittings) ranging from 60.3 to 2235 mm outside diameter. BS 534 also covers coating and lining. It applies to pipe to BS 3601 and dimensions to BS 3600 but pipe to API 5L and other standards can be used. BS 534 will be superseded by BS EN 10224 which was due to be published near the end of 1999: this standard will cover pipes only, separate standards will be issued for joints and for coatings and linings. The standards for pipe set out manufacturing tolerances, chemical composition and minimum values for yield stress and ultimate tensile strength.

BS 8010 Section 2.8 covers installation of steel pipes for oil, gas and associated products: Section 2.2, for water and associated products has not been published: the older CP 2010 Part 2 for steel pipes is not withdrawn.

BS 3601 covers four principal methods of manufacture:

- butt welded (B);
- electric resistance (including induction) welded (ERW);
- seamless (S);
- submerged arc welded (SAW).

Outside diameter ranges given in BS 3601 in mm are: (B) 8–50 (nominal); (ERW) 12.7–508; (S) 12.7–1016 and (SAW) 114.3–2220.

Steel pipes used for trunk mains are generally SAW. These pipes are fabricated from steel plate bent to a circular form, typically either (a) by rolling and butt welding the longitudinal seam; or (b) by 'U–O', first forming a U shape by hydraulic press, then curving over the straight parts and butt welding the longitudinal seam. Alternatively (c), they may be continuously produced from a coil of steel strip bent to a spiral and butt welded along the spiral seam. Joints between coil ends of spiral welded pipes are known as skelp end welds. Pipe may also be made by rolling and jointing plate into short 'cans' which are then joined by butt welding, end to end, to form the required lengths. Lengths are usually in the range of 9–12 m dependent on manufacture, transport and project requirements. Weld beads must be machined flush with the pipe surface at pipe ends to make them suitable for joint couplings. Spigot and socket ends, where shaped, are formed by die. Weld bead height needs to be limited for coating and lining.

Butt welded pipes are made from rolled strip with a longitudinal seam furnace butt welded by a continuous process. Electric resistance welded pipes are made up to about 610 mm outside diameter, also from rolled strip: the longitudinal weld is made by an electric current (by induction or direct contact) across the edges which are joined under pressure, without filler metal. Heat treatment at least of the weld zone is usual in sizes larger than DN 200. ERW pipes now tend to be known as HFI (high frequency induction) pipes. Seamless pipes are made, up to 508 mm outside diameter, from billet bars or ingots of hot steel which are pierced and rolled to the required dimensions.

Inspection typically includes chemical and mechanical material tests, ultrasonic inspection of plate and welds, radiography of welds and hydraulic pressure tests.

There are no standard classes for steel pipes: wall thickness above about DN 750 is designed for handling, for internal pressure, for buckling under external pressure and internal sub-atmospheric pressure and to limit deflection when buried. External load carrying capacity in trunk mains is mostly a function of the backfill and compaction design.

Steel grade in BS 3601 is designated by ultimate tensile strength in N/mm^2. Grade 430 is specified for SAW pipe: yield stress depends on thickness and is 275 N/mm^2 for thickness 16 mm and less, 265 N/mm^2 for thickness over 16 mm and up to an including 40 mm and 255 N/mm^2 over 40 mm.

pr EN 10224 grades steel by yield stress, options are 235, 275 and 355 N/mm^2.

BS 534 sets out nominal wall thicknesses considered to be the minimum for normal conditions. Selected data from BS 534 are given below: hydraulic test pressures are the factory test pressures according to BS 3601, using steel grade 430 and do not necessarily

Table 13.2   Steel pipes to BS 534: selected diameters

Outside diameter (mm)	Thickness (mm)	Test pressure (bar)	Diameter/thickness (D/t)	Ring bending stiffness, EI/D^3 (kN/m²)
168.3	3.6	70	47	183
323.9	4.0	54	81	34
610	6.3	45	97	20
914	7.1	34	129	8
1219	8.0	29	152	5
1626	10.0	27	163	4
2032	12.5	27	163	4
2235	14.2	28	157	5

relate to working pressures. Hydraulic test pressures quoted are based on hoop stress of 80% yield or a maximum of 70 bar.

Special pipes, e.g. bends, tees, etc., can be made to any dimensions required, bends being made by cutting and welding together mitred sections of pipe.

Steel grades as designated in the American Petroleum Institute standard API5L are designated by grade (A and B) and by yield stress in thousands of psi, as follows.

Table 13.3   Steel grades to API 5L

Grade		A	B	X42	X46	X52	X56	X60	X65	X70	X80
Yield	psi	25 000	30 000	42 000	46 000	52 000	56 000	60 000	65 000	70 000	80 000
stress	N/mm²	207	241	289	317	358	386	413	448	482	551

Grades less than grade B would not normally be used. Grades up to about X52 can normally be welded without special heat treatment. Their price is only marginally above that for grade B or BS 3601 grade 430 and provide good economy where high pressure or (typically for pipes above ground or installed underwater) significant longitudinal bending resistance is required.

The American Waterworks Association (AWWA) manual M11 gives a range of thicknesses and pressures and steels for diameters up to 4000 mm. Sizes in M11 are designated by outside diameter below 30 inches (762 mm), otherwise by inside diameter.

Pipe wall thickness, t (mm) for internal pressure is determined by hoop stress, as follows:

$$t = \frac{pd}{2a\sigma e}$$

where p is the internal pressure (N/mm²), d is the external diameter (mm), a is the design or safety factor, σ is the minimum yield stress (N/mm²), and e is the joint factor. The design factor, joint factor and definition of wall thickness depend on the design code.

Design factors for hoop stress typically range from 0.4 to 0.8; the joint factor is 1.0 for SAW pipes and certain codes require the negative tolerance to be deducted from wall thickness. BS 8010 Section 2.8 quotes a design factor of 0.72 and requires thickness tolerance ($\pm 7.5\%$ for BS 3601 SAW pipes) to be added to the design thickness. This can then be rounded up to the next standard nominal wall thickness. ASME codes B31.4 and B31.8 quote a basic design factor of 0.72 and state that this includes for thickness tolerance. For water supply under normal conditions, it is suggested here that the design factor is taken as 0.5 (as given in AWWA M11 and the WRc pipes selection manual) and that no deduction is made for the thickness tolerance.

Further consideration can be given where particular conditions warrant: for example the American Society of Mechanical Engineers (ASME) code B31.8 quotes design factors for a variety of laying conditions. Where necessary the analysis can be elaborated to include ring bending, longitudinal bending, longitudinal stress from temperature changes, Poisson's ratio effects on buried (and thus restrained) pipe under hoop tension, combined (equivalent) stresses – and where appropriate, for example for underwater pipes – can include strain based design.

13.9 External and internal protection of steel pipe

External and internal surfaces of pipes and specials are protected against corrosion by a coating and lining. It is not normally economic in design of water pipelines to provide a corrosion allowance (although, for comparison, corrosion allowance and corrosion inhibitors may be adopted in petrochemical practice). Corrosion protection may not always be carried out at the same place or by steel pipe supplier: pipe out of roundness and straightness can affect application of coatings and linings and must be specified to match the process of application of the corrosion protection.

Principal options for *external coating* of steel water pipes are:

● bitumen sheathing;
● fusion bonded epoxy (fbe);
● three-layer polyethylene (PE) and
● paints.

In all cases cleaning and preparation by grit blasting to BS 7079 second quality (or Swedish Standard Sa 2.5) is a prerequisite.

Bitumen sheathing is included in BS 534 and consists of a hot applied bitumen with an inert filler, reinforced if required with a woven glass cloth, to a thickness of 3 mm for small diameter pipes rising to 6 mm for diameters exceeding 350 mm. Bitumen enamel wrapping consists of a hot applied bitumen containing a mineral filler with an inner wrapping of glass tissue and an outer wrapping of bitumen impregnated reinforced glass tissue or composite glass fibre fabric to the same protection thickness as for bitumen sheathing. All reinforcement materials are spirally wound onto the pipe with adequate overlaps. The pipes are painted with emulsion to act as a reflective and non-stick surface: alternatively, sacrificial galvanised steel sheet may be used for both solar reflection and protection against damage during transit. Bitumen can biodegrade and as an alternative, coal tar has been used, particularly for underwater pipelines, but is losing favour in view of health and safety issues. Both bitumen and coal tar lose volatiles on exposure and crack. Bitumen still presents a viable option for buried pipelines but has tended to lose out in favour of fbe and PE coatings. Alternatives should be used for exposed pipelines.

Fusion bonded epoxy has been in use since the early 1970s. The pipe surface is prepared by grit blasting and by phosphate or chromate pre-treatment. The fbe coating is applied as a powder which is fused onto the pre-heated pipe surface and cured chemically to form a layer thickness typically about 450 ± 100 microns thick for pipelines onshore and 650 ± 100 microns for pipelines offshore. Fbe coatings generally have excellent resistance to cathodic disbonding when used on cathodically protected pipelines. Protection at joints is completed by heat shrink sleeves, tape wrap, by paint or by fusion bonded epoxy.

Three-layer polyethylene systems have been in use since the early 1980s, more in Europe than in the USA, and supersede the two-layer system. The three-layer system is more expensive than fbe but tends to be considered cost effective in view of potential damage during shipping over long distances. The system comprises the following typical thicknesses:

fusion bonded epoxy:	150–200 micron;
copolymer adhesive:	200–350 micron and finally;
polyethylene:	2.5 mm up to DN 750 and 3.0 mm for larger sizes.

Total thickness for the three-layer system is thus between about 3 mm up to DN 750 and 3.5 mm for larger sizes. After surface preparation by grit blasting and phosphate or chromate pre-treatment, the pipe is pre-heated and application of all three components takes place while the pipe is rotated about its axis and moved forward through a specially designed booth. Timing is vital and curing takes place before the pipe reaches supporting rollers. Completion at joints is by heat shrink sleeves or by tape wrap. Heat shrink sleeves tend to be favoured for the larger pipe diameters.

Paint options include epoxy and polyurethane and may receive a zinc rich base layer. Thickness typically is similar to that for fbe, with application by airless spray. Coal tar epoxy is not generally favoured. Care is required to ensure good quality preparation and protection at fittings, which tend to be anodic and more prone to corrosion due to the additional welding and working. Small diameter steel pipes are galvanised. Polypropylene is used for high temperature applications (not for water supply): other developments continue, including blends of fbe, adhesive and PE and some manufacturers may also provide a thin additional coating of concrete for mechanical protection. Cement mortar external coating is covered by American standards.

The continuity of the applied protection (other than cement-mortar) is checked with a 'holiday' detector. This comprises passing a scanning electrode in the form of a brush containing a high voltage electrical charge over the coating and lining; the voltage is set so as to produce a spark length of 10 mm or double the specified minimum thickness of the protective material whichever is the greater. Pinholes or breakages are disclosed by an electrical discharge to the steel of the pipe and this can be arranged to cause a buzzer to sound.

Internal surfaces of steel pipes can be protected with concrete or cement-mortar (described later in this chapter), or epoxy. Market forces and perceived quality issues in the UK currently favour epoxy. Epoxy coatings may be hot applied fbe or airless spray liquids and high solids systems. Thickness is typically about 0.5 mm and surface preparation is by grit blasting to a high standard. Bitumen was a popular and successful option but is no longer used in the UK. A hot applied bitumen with an inert filler was sprayed or brushed onto the pipe, after first priming it with or without a compatible priming coat; usual practice was to apply a minimum thickness of 1.5 mm for diameters up to 300 mm and 6 mm for diameter exceeding 1000 mm.

Laying 1.0 m diameter GRP pipeline, sand bedded and surrounded, within a 30 m easement. Part of the 350 km North-South Carrier pipeline for Botswana, year 2000. (Engineers: Binnie, Black & Veatch)

Cross-connection on a twin ductile iron 40 km pipeline in Middle East, 1989. At each cross-connection pipe diameters are reversed from 2300 mm to 2100 mm to permit nesting of pipes for delivery from Japan. (Engineers: Binnie & Partners. Contractors: Continental Construction, India)

Plate 29

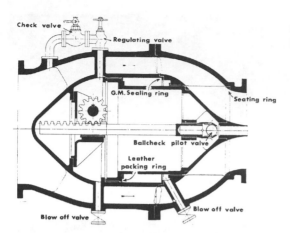

(Left) A 'needle' or 'streamline' cone valve, frequently used for flow control

A metal seated, hand operated 840 mm diameter butterfly valve. (Biwater Industries Ltd., Dorking, UK)

A pipe cutting machine. The driving shaft on the right can be rotated by a compressed air operated motor. (Pipeline Technology Ltd., Fareham, UK)

Making an under-pressure connection on an existing main. A valved split collar is clamped onto the main; the valve is opened for the drilling machine to trepan a hole in the side of the main; and the trepanned section is withdrawn and the valve closed. (Aqua-Gas (Valves & Fittings) Ltd., Northampton, UK)

Plate 30

13.10 Mortar and concrete linings

Steel and ductile iron pipes can be lined with cement-sand mortar or spun concrete and these form some of the best and most cost effective types of interior protection. The cement mortar or concrete lining generates a high pH environment which passivates and prevents corrosion. Cement mortar lining (cml) may be either factory lined by centrifugal spinning or may be lined *in situ* by spraying, (orange peel finish) followed, at large diameters and if required at small diameters, by trowelling to a smooth (washboard) finish. BS 534 covers both mortar and spun concrete and AWWA C205 and C602 cover shop applied and *in situ* mortar, respectively. Concrete lining has been chosen in favour of cml on some projects because it is thicker (typically 25 mm instead of 12 mm at DN 1400 and larger) and because the aggregate is larger, thus presenting a more robust and durable surface.

Difficulties in small diameter mains particularly are that aggressive waters (as indicated by the Langelier index, see Section 8.40) will dissolve the cement and increase pH, which may raise water quality problems. The cement lining may be sealed with an approved thin (150–250 micron) layer of epoxy paint. The cement mortar or concrete provides the corrosion protection layer by passivation: the seal coat provides a barrier to solution of the cement mortar by aggressive low pH, low calcium waters. It is not particularly robust. Where water cannot be dosed to make it less aggressive – for example with lime to raise pH – or where dosing cannot be guaranteed due to materials supply or operational uncertainties, epoxy or other lining should be used.

Cement mortar linings and concrete linings are thicker than epoxy. They therefore result in a slight reduction in pipe diameter (or require a slightly larger external diameter to obtain the same finished internal diameter). Although cement mortar or concrete linings may, after a while in service and with biofilm addition, attain a hydraulic roughness similar to epoxy, they will in general tend to have a greater initial hydraulic roughness than epoxy. These factors may lead to epoxy being a more cost-effective solution if energy costs are taken into account, dependent on the specific circumstances. The current trend away from concrete and mortar lining and towards epoxy lining for new steel pipe may be linked to market forces, and also to the need for material in contact with water used for the public supply (including raw water mains) to conform with the required approval process. The approval process is time-consuming and entails continued control of materials sources, including cement, as in concrete for water retaining structures and linings for ductile iron pipes. Epoxy lining on its own would be an expensive solution for ductile iron due to the surface roughness after grit blasting: the pipes are cement mortar lined at all sizes and in the UK have a seal coat on pipes only of DN 800 and smaller.

The bare metal pipe is spun at high speed on rollers and the mortar is poured as a slurry into the interior. The lining builds up by centrifugal force and when the required quantity has been poured the speed of rotation of the pipe is increased. This causes the mortar to compact further and surplus water, rising to the surface of the lining, then runs off as the pipe is tilted very slightly. The mortar mix comprises sand or fine broken rock aggregate mixed with Portland or sulphate-resisting cement in the ratio of 2.5:1 or 3:1 by weight. The lining is usually so well compacted after spinning that the pipe can be immediately taken off the spinning bed and be placed in a damp warm atmosphere for a curing period lasting 21 days. The stipulated lining thickness may vary and some specifications are given in Tables 13.4(A–C). The mortar lining thickness for steel is generally thicker than for ductile iron pipes of the same diameter because of the flexibility of steel pipes as compared with

Table 13.4A   Recommended thickness of cement-mortar lining (AWWA)

| Pipe diameter (mm) | AWWA C602: pipelines in place AWWA C205: shop applied – as for new steel | | | |
| | Type of pipe | | | |
	Old steel (mm)	Old and new ductile iron (mm)	New steel (mm)	Tolerance (mm)
100–250	8	4.8	6.3	−1.6, +3.2
280–580	10	6.4	8.0	−1.6, +3.2
600–900	11.1	8.0	10	−1.6, +3.2
>900	14.3	8.0	13	−1.6, +4.8

Note. For badly deteriorated pipe or abnormal use the purchaser should specify such greater thickness of lining as engineering judgement dictates.

Table 13.4B   Recommended thickness of cement-mortar and concrete lining – steel pipes (BS 534)

Pipe outside diameter (mm)	Minimum thickness of cement mortar lining (mm)	Tolerance (mm)	Minimum thickness of spun concrete lining (mm)	Tolerance (mm)
≤168.3	6	+2, −0	6	+3, −0
193.7–323.9	6	+2, −0	10	+3, −0
355.6–610	7	+2, −0	13	+3, −0
660–1219	9	+2, −0	19	+6, −0
1422–2235	12	+3, −0	25	+6, −0

Note. Thicker linings may be specified.

Table 13.4C   Recommended thickness of cement-mortar lining – ductile iron pipes (BS EN 545)

| Pipe nominal diameter (DN) | BS EN 545 | | BS EN 545 UK National annex | | |
	Nominal value (mm)	Tolerance (mm)	Nominal (mm)	Minimum arithmetical mean value (mm)	Individual minimum (mm)
80 (and DN 40 for BS EN 545)	3.5	−1.5	5	3.5	2.5
100–300	3.5	−1.5	5	4.5	3.5
350–600	5	−2	5	4.5	3.5
700–1200	6	−2.5	5	5.5	4.5
1400–2000	9	−3	5	8.0	7.0

To avoid excessive thicknesses, a minimum clear bore should also be specified.

ductile iron pipes. With regard to ductile iron pipes BS EN 545 national annex recommends that the thicknesses should be as given in the previous standard, BS 4772, for ductile iron pipes. All fittings associated with mortar lined pipes are also mortar lined, but this has to be trowelled on. The discontinuity of the lining at joints in steel pipes should also be filled if the pipe is of large enough diameter to give inside access. Steel pipes which are not large enough for this should either be jointed using flexible joints or be jointed externally using welded collar joints: the heat input is unlikely to damage the concrete lining: the lining will be brought flush with the pipe ends and the pipe ends will butt together. The collar will be protected internally at the gap and any water in the joint parts will be largely stagnant, without risk of causing corrosion.

13.11 Joints for steel pipes

The following types of joints are in common use:

(1) welded sleeve joints (see Figs 13.2a and b);
(2) butt welded joints;
(3) Viking Johnson type slip-on couplings (see Fig. 13.1);
(4) flanged joints (see Fig. 13.2c)
(5) push fit joints.

The BS 534 Type 1 taper sleeve joint (Fig. 13.2a) is common in the UK. It permits a small angular deviation (up to 1°) as long as the joint is detailed so that there is a penetration of at least four times the pipe wall thickness after deflection. The increased gap around the outside the pipe requires the weld to be buttered: filler rod can be used. The internal joint gap should be fairly constant as the mating surfaces are brought together. Pipes of DN 700 and larger should be welded internally and externally. The internal weld is the strength weld and the outer is a seal weld. The joint can then be tested by pressurising the annular space between the two welds (usually with nitrogen at a pressure of 200 kPa), which permits pipe joints to be tested before backfilling the trench. On pipes smaller than DN 700 the weld should be made on the outside only. Consideration needs to be given to access and the completion and inspection of the internal protection. On pipes DN 600 and smaller, the alternative of using mechanical or other joints can be considered.

BS Type 2 collar joints comprise a short sleeve usually in two parts which is slipped around the pipe end and joined by fillet welding. It is used for jointing small diameter pipes where internal access is not possible and for jointing closing lengths on larger diameters. Where access is possible welds should be made on the inside as well and each side should be air tested.

The BS 534 Type 3 spherical joint (Fig. 13.2b) was developed to permit deviation without unduly increasing the size of the external weld and has been found practicable and more satisfactory than the short sleeve joint because of the smaller welds required. It is common and successfully used on large diameter pipes overseas and allows larger deflections than the Type 1 joint, thus allowing gentle curves to be followed without use of gussetted (lobster back) bends. The design must provide an overlap of at least four times the pipe wall thickness.

For butt welded joints the ends of pipes are prepared by forming a 30° bevel on the full thickness of the pipe wall except for the inner 1.6 mm; the resulting V-groove between pipes is then filled with weld metal which is finished to stand proud of the external pipe

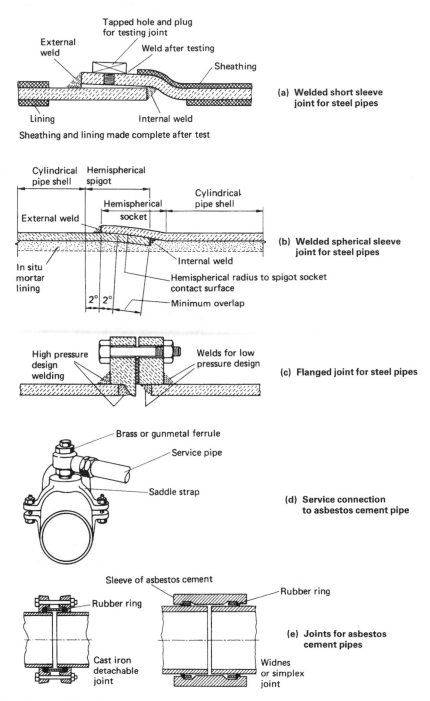

Fig. 13.2 (a)–(c) Joints for steel pipes. (d) Service pipe connection to an asbestos cement pipe. (e) Joint for an asbestos cement pipe.

surface. Butt welding is usually only adopted when the pipe wall thickness is substantial or when full longitudinal continuity of strength is required – such as in high pressure applications, or in underwater pipes, or if pipes are to be snaked into the trench.

Viking Johnson couplings permit slight angular deviation of pipes at joints. Longitudinal movement is also possible but is restricted to the amount allowed by shear movement within the rubber. Further movement will drag the rubber over the pipe surface eventually damaging the joint. Once filled and conveying water, temperature movements should be small even on an exposed pipeline in tropical climates, so it is possible to use these couplings, spaced at say two pipe joints (24 m centres) with an anchored joint between, to avoid the need for special contraction or expansion (bellows or slip) joints.

The continuity of the lining needs to be completed inside the pipe and this can be done by entry into the pipe when its diameter exceeds 600 mm, and from outside when its diameter is smaller. In the latter case a former is pushed into the pipe on a rod and expanded to form a temporary inside collar at the joint while protection material is injected through a hole in the coupling, thus filling the annular space between the pipes. Alternatively a lining train can be used to apply a measured quantity of paint by roller or spray. The outside coating is completed by heat shrink sleeves or by tape wrap with an overlap of at least 50%. Alternatively for bitumen sheathed pipes the external protection is completed by 'flood coating' – putting a mould over the coupling, filling it with bitumen and allowing it to cool before stripping off the mould.

Pipes for jointing with couplings must have true end diameters and if it is intended to cut pipes these must be ordered as 'true diameter throughout', otherwise they will only be truly circular at their ends. Some difficulty may therefore be experienced when inserting a branch by couplings into an existing main at a point where its diameter may not be true.

13.12 Cathodic protection

Cathodic protection provides an additional measure of security to cover breakdown and deficiencies in the external coating. Clays, sulphate-bearing soils, moorland acid waters, and saline ground-waters are the principal causes of aggressive ground conditions.

Cathodic protection uses the principal of electrolytic chemical action to protect pipes by providing the pipeline at intervals with sacrificial anodes or, alternatively, by impressing a (negative) voltage on the pipeline so that it becomes cathodic with respect to buried anodes in the medium of the wet soil which acts as a weak electrolyte. Corrosion takes place at the anode and not at the cathode. Sacrificial anodes are usually made of zinc, magnesium or aluminium. They are connected electrically to the pipe and cause a weak current to flow in the right direction since they are 'anodic' with respect to iron. They are buried a few metres away from the pipeline at 50–100 m intervals and are connected to it by insulated cables. If impressed current is used a DC negative voltage is applied to the pipeline and the anodes wired to it can be made of iron as the impressed voltage is sufficient to drive current in the right direction. Impressed current anodes are silicon iron, titanium and more currently, mixed metal oxide. Graphite is no longer used. This type of cathodic protection is used where the soil resistivity or where long stretches (typically more than 5 km) of pipe need protection since it can provide much higher potential driving forces than the sacrificial system. The lengths of pipe protected by a given anode must be electrically connected together. If the pipe joints use rubber rings then each pipe must be electrically connected across its joints by means of an insulated conductor welded onto two adjacent pipes.

Cathodic protection is not by itself complete protection of a pipeline against corrosion: it is an additional defence. Also it cannot be used without care where there are many other services, such as at works areas and urban areas, since it may reduce the effectiveness of the system by draining current to the other services. It may also increase corrosion of these services at points where the current picked up is returned to the soil.

Where a steel pipeline passes parallel to and up to about 50–200 m away from overhead high voltage lines, stray electric potential may be induced on the pipeline. This will require the pipeline to be earthed, using either zinc anodes or through diodes which limit the current flow to one direction and only above a predetermined limit. In addition, care must be taken to earth the pipeline during construction – not only to prevent current from welding damaging the pipe coating but also to earth induced currents. Length of pipe strung above ground should also be limited typically to about 400 m – and a strict safety regime implemented. Exact precautions need to determined according to circumstances.

13.13 Flexible pipe design

Flexible pipe design principles apply to steel and plastics and to some extent, particularly in the case of large diameters and large loads, to ductile iron also.

For steel pipes greater than about DN 750 (or D/t about 120–140, depending on conditions) the theoretical pipe thickness to meet usual internal pressures is frequently less than the thickness required to limit deflection under backfill load. To save adding extra steel thickness for stiffening the pipe against backfill load, the pipe may be laid under 'controlled backfill' conditions which, in practice, means bedding the pipe on a thin layer of uncompacted sand or onto a preformed circular invert (60° width) in sand or fine gravel and selecting and carefully compacting the backfill in shallow equal layers either side of the pipe to achieve a required soil stiffness.

At large diameters, temporary jacks may be inserted (in steel pipes) to maintain circularity or pre-deflect the pipe upwards: care needs to be taken to spread the jack load and not to damage any lining. As backfill is built up, lateral pressure increases deflection inward and upward; the jacks are therefore removed and as backfill progresses further the vertical load deflects the pipe downward and outward. Deflection is measured and if it exceeds the permissible value, the backfill should be removed and replaced to obtain the necessary circularity. Measurements should be continued periodically after installation and checked at critical sections before the pipe is filled for testing.

Deflection of flexible pipes is calculated using the Spangler formula or similar as follows:

$$\Delta/D = k(P_e D_l + P_s)/(8S + 0.061E')$$

where Δ = pipe deflection (assuming horizontal and vertical deflections are equal); D = mean diameter of the pipe; k = a constant, dependent on the angle, between contact points, over which the trench bed supports the pipe (typically 0.1 for 65°, 0.083 for 180°); P_e = soil load per unit area (kN/m^2); P_s = surcharge or traffic load per unit area, (kN/m^2); D_l = deflection lag factor, dependent on soil type and compaction; S = diametrical ring bending stiffness, = $1000E(t/D)^3/12$, (kN/m^2), where E = modulus of elasticity of the pipe wall; E' = soil stiffness (kN/m^2) and 0.061 is derived from assumed (parabolic) loading over a 100° lateral support angle.

For granular soils, D_l is unity. Long term, in clay, the value may be 3 or more. Where the cover is less than 2.5 m and the pipeline will be under sustained pressure within a year of installation, long-term deflection may be reduced by a re-rounding factor $D_R = 1 - (p_i/40)$, where p_i is the internal pressure in bar. Diurnal variations in pressure in distribution mains should be taken into account in deciding the re-rounding factor.

Tables in various publications and standards, including BS 1295–1 national annex A, link soil stiffness to soil type, depth of cover and compaction of the fill. Compaction is quoted either as per cent Proctor (modified Proctor density, M_p – which corresponds to the heavy compaction test to BS 1377) or, for granular materials, to relative density. Soil stiffness can also be derived from laboratory tests. The overall soil modulus is modified to take account of the native soil where the native soil modulus is less than 5 MN/m^2 and the trench width is less than 4.3 times the pipe diameter. AWWA M11 quotes accuracy of predicted deflections for different soil compaction: it does not cover native soil modulus.

Ring bending stress is traditionally not addressed in steel pipe design but is included for in PVC and PE (thermoplastics) pipe design and GRP (thermosetting) takes account of bending strain. Pipe thickness for plastic pipe is usually expressed as a minimum: thickness tolerance is not covered.

Bending stress is given by $\sigma_{bs} = E\,D_f\,(\Delta/D)\,(t/D)$, where E is the flexural modulus of elasticity of the pipe material and D_f is a strain factor, dependent on pipe and soil stiffness and is given in BS EN 1295–1 and other references. For PVC and PE pipe the sum of bending stress and hoop stress is required to be less than the design value. The approach is similar for GRP but using a criterion of strain. Designs for plastics pipes need to take into account both the initial short-term and the long-term characteristics.

Deflection limits vary according to pipe material. Deflection limits quoted in AWWA M11 for steel pipe are as follows:

Mortar lined and coated	2% of diameter
Mortar lined and flexible coated	3% of diameter
Flexible lined and coated	5% of diameter

The deflection limit of 2% frequently quoted can be exceeded where pipes are lined with cement mortar after installation or for flexible linings. However, if large deflections are to be permitted then for steel pipes it seems consistent that consideration should be given to a more detailed analysis, including ring bending stress, but adopting a design factor greater than the traditional value of 0.5 and taking into account the stress strain characteristics which differ from those applying to plastics.

Plastic pipes

13.14 Polyethylene pipes

Polyethylene is a thermoplastic and has been widely used for the production of pipes by an extrusion process for over 40 years. Pipe is produced in the UK up to DN 1000 and is available from Europe up to DN 1600. PE pipe is specified by nominal (minimum) outside diameter (OD), by standard dimensional ratio (SDR – the ratio of nominal outside diameter/minimum wall thickness, typically 11, 17.6 and 26) and by material. Material classifications in current use are PE 80 and PE 100, where the number refers to the minimum required strength (MRS) in bar (80 bar = 8 N/mm^2). The MRS is the pipe wall burst stress at 20°C at 50 years, extrapolated from shorter duration tests.

Materials terminology is confusing: current trend is to refer to grade (PE 80 or PE 100). PE 80 includes both medium density polyethylene (MDPE) and high density polyethylene (HDPE). MDPE was brought into service in the UK in the 1980s to replace low density polyethylene (LDPE) and high density polyethylene (HDPE). To take account of the risk of crack propagation (RCP) reported for gas pipes, PE 80 pipes DN 250 and larger were derated by about 20% and, above DN 315, were further derated for pipes containing more than 10% free air. This derating has been reviewed in the light of recent research: derating for fatigue is no longer considered necessary for modern PE 80 pipes which meet the stress crack requirements of WISs (Water Industry Standards) 4–32–03, –09 and –13 but is required for pipes which do not meet these high toughness requirements.

PE 100 has been produced since 1990 and is also known as high performance polyethylene (HPPE). Compared with PE 80 it has greater resistance to and is not derated for RCP. The material price is more than for PE 80 but its greater strength allows for higher pressures or thinner walls and hence provides economy under certain conditions.

Maximum operating pressures are controlled by the sum of hoop stress and ring bending stress which, as given in EN ISO 12162 and proposed in prEN 12201, is limited to 0.8 times MRS. Maximum operating pressure (PMA) includes surge. Recommendations were that the range of surge pressure should be limited to half the maximum operating pressure. However, because of the low wave speed in the PE pipe wall, surge pressures in PE pipes are typically about 25% of those in steel for the same event. Recent research shows that PE pipes will sustain surge pressures at least twice their static rating (the actual amount depending on the material and increasing with rate of strain). Attention must be given to derating for operating temperatures above 20°C.

Long-term deflection is allowed up to 6% of the diameter. However, this will be calculated taking into account creep in the soil and using long-term pipe stiffness – which may be one-sixth of the initial stiffness.

Advantages of PE pipes are that they are light and easy to handle, flexible but strong and resistant to cracking, do not corrode, are chemically resistant, have a low frictional resistance to water and can easily be cut to length. SDR 11 and 17.6 pipes can be bent safely to 25 times OD, increasing to 35 times in cold weather and reducing to lower values in warm weather. For SDR 26 and 33 pipes these radii need to be 50% greater. Small diameter pipe can be supplied in coils and straight lengths can be joined above ground and snaked into narrow non-man-entry trenches. Disadvantages are that the strength of pipes, defined as their ability to withstand hoop stress, decreases with time and reduces with increasing temperature. They are also liable to UV degradation if exposed overlong to sunlight. In the UK blue PE (for water supply) can be stored above ground for up to a year: for longer periods they should be covered. Under certain conditions they may be degraded or may be slightly permeable (and therefore give rise to taste problems) due to oils, organic solvents and by strong concentrations of halogens or acids. CP312 Part 1 gives a list of chemicals and suitability with plastics. Where soil is polluted by hydrocarbons, typically in former gas works sites, fuel stations, scrap yards and other industrial areas, it may be enough to use imported backfill and impermeable membranes: alternatively other measures or other pipe materials may be needed. The coefficient of linear expansion of PE 80 is about 1.5×10^{-4}, which is more than ten times greater than for steel and is equivalent to about 9 mm for 10°C in a 6-m pipe length. PE pipes are not traceable underground with metal detectors and do not transmit leak noises like metal pipes. BS 5336, BS 6437, BS 6572 and BS 6730 set out general requirements, respectively, for thermoplastic pipes, polyethylene pipes, and for blue (below ground) and black (above ground) polyethylene pipes up to 63 mm OD. Water Industry Standards WIS 4–32–03, –08, –09, –13, –14 and –15 specify requirements for polyethylene pipe and for joints and fittings from 90 to 1000 mm OD. DIN standards also cover polyethylene pipes and fittings and prEN 12201 is in preparation. For distribution mains PE pipes are supplied in 6 or 12 m straight lengths; their flexibility makes the use of small angle bends largely unnecessary.

PE pipes can be jointed mechanically or with push fit joints (with longitudinal strength) but are more usually jointed by butt fusion or electrofusion which form a continuous string and do not need thrust blocks. Butt fusion requires a purpose-designed powered machine in which the prepared and cleaned pipe ends are clamped and held under longitudinal pressure against a thermostatically controlled heated plate: after a given heating time the ends are pulled back, the plate is removed, and the two melted ends of the pipe are brought together under longitudinal pressure to fuse and cool. Growth of the bead of melted material is observed during heating: after jointing the external bead is removed, numbered with the joint and twisted at several positions. Any split in the bead requires the joint to be cut out and made again; and a further defect requires investigation and cleaning of the equipment and new trial joints before production continues again. Care is required to put in place good quality assurance and quality control procedures, with tracking to monitor performance of individual and gang workmanship. Experienced and disciplined inspection is essential. Internal weld beads can be removed if required. Electrofusion joints are used at fittings and comprise a spigot and socket with in-built electrical heating coils which are energised to locally melt and fuse the fitting and pipe together. Power supply can be by mains or generator. Equipment can be provided to read a bar code fixed to the fitting which contains information to control the weld. The equipment can then store weld data for downloading onto a computer database for traceability and quality assurance. Flow of melt material from indicator holes in the fitting gives visual evidence of weld completion.

Pressure testing must take into account the viscoelastic properties of the pipe material. Traditional pressure test techniques are not suitable due to creep and stress relaxation: pressure in a closed PE pipe will fall with time even when the pipe is leak free. The accepted method is to fill the pipe with water, raise the pressure, measure the pressure decay with time, derive decay parameters using a 'three-point analysis' (more points are strongly suggested) and compare with a range of standard values according to pipe restraint. The pipe must be sensibly free of air and not above 30°C. Pressure should be released slowly. If a retest is required, a rest time, typically of about 5 times the previous test duration, must be allowed beforehand for the pipeline to recover from the previous conditions.

Recommended test pressures are:

- for systems rated up to 10 bar, 1.5 times **rated pressure**;
- for 12 and 16 bar systems, 1.5 times **working pressure**.

The test pressure should not exceed 1.5 times the maximum rated pressure of the lowest rated component.

13.15 Polyvinyl chloride (PVC) pipes

Three types of PVC pipes available in the UK are:

- PVC-U: unplasticised PVC;
- MOPVC: molecular-orientated PVC; and
- PVC-A: PVC with added impact modifiers – also known as PVC-M (modified PVC).

MOPVC is first formed at about half the diameter and double the final thickness and is then heated and expanded to the final diameter. The result increases the strength particularly in the hoop direction and allows wall thicknesses slightly more than half of those for PVC-U. It also increases significantly the resistance to failure from cyclic loading. MOPVC pipes are about 30% of the stiffness of PVC-U pipes due to their reduced thickness but taking into account their greater modulus. PVC-A is an alloy of PVC-U, polyethylene (PE), and acrylics: wall thicknesses are about 80% of those for PVC-U.

A number of failures of early PVC-U pipes led to lack of confidence in their use. This is now understood to result from local bending stress from point loads, typically uneven bedding. Developments have now led to renewed confidence but

(a) pipe should be derated for temperatures above 20°C;
(b) due to notch sensitivity, pipe with significant internal or external scratches should be rejected;
(c) care is required with installation;
(d) care is required during tapping and not to overtighten saddles; and
(e) derating is required for repeated cyclic (fatigue) loads (distinct from diurnal variations), particularly for pumping mains.

PVC pipes are classified by nominal outside diameter. PVC-U pipes are available in the UK up to DN 630 and are supplied in 6 m lengths with spigot and socket rubber ring push fit joints. Solvent joints have not been used since the mid-1980s and are no longer permitted for underground use (primarily due to problems arising from movement during bedding and backfilling): they are permitted for above ground use. Classes C, D, and E (9, 12 and 15 bar operating pressures, respectively) are stipulated in BS 3505. Water Industry Standard WIS 4–31–06 covers PVC-U pipe for 8 bar and 12 bar operating pressures and has generally superseded BS 3505 for water supply. Pr EN 1452–2 will replace BS 3505 and WIS 4–31–06. WIS 4–31–08 covers MOPVC pipe from DN 110 to 400: sizes are only available from DN 110 to DN 200 at 12.5 bar rating and DN 250 and 315 at 16 bar rating. There is no national standard for PVC-M pipe: sizes are available from DN 110 to 630 at 10 and 16 bar ratings.

The pipes are suitable for use in hot climates but attention should be paid to derating for temperature above 20°C. Typical PVC-U pipe material has a relative density of 1.4 and a softening point of about 80°C. The coefficient of linear expansion is about 6×10^{-5} per °C which is more than five times the value for steel.

Wall thickness for PVC-U is based on an allowable design stress of 12.5 N/mm^2, which compares with a minimum required bursting stress of 26 N/mm^2 at 50 years, as extrapolated from shorter term tests. The design stresses for MOPVC and typically for PVC-M are 22.5 and 18 N/mm^2, respectively. Full structural design would not normally be needed for DN 315 and smaller.

Maximum operating pressures are controlled by the sum of hoop stress and ring bending stress which should not exceed the design stress. Maximum operating pressure includes surge. Recommendations were that the range of surge pressure should be limited to half the maximum operating pressure: however, because the wave speed in the pipe wall is lower, surge pressures in PVC pipes are less (typically 30% less) than those in steel for the same event. Recent research shows that PVC pipes will sustain surge pressures about twice their static rating (about 1.8 times for PVC-A and 2.2 times for MOPVC and PVC-U, the actual amount increasing with rate of strain). Care should be taken for pumping mains larger than DN315 and fracture toughness tests should be specified as set out in pr EN 1452.

The main advantage offered by PVC is its resistance to corrosion, hence its use for chemical transfer lines in water treatment works. It is not suitable for use in ground contaminated or likely to be contaminated by detergents or solvents or from oil storage areas. It is light in weight, flexible, and has easily made joints. Small size pipe can be joined above ground and snaked into narrow non-man-entry trenches. Care should be taken when handling pipe at temperatures below freezing due to reduced impact resistance.

Fittings can be made in PVC-U or metal, usually ductile iron. PVC-U pipes are degraded by ultraviolet light, the effect increasing with temperature so that the pipes must not be exposed to sunlight in hot climates for more than a day or two. In the UK, MOPVC, PVC-U and PVC-A can be stored above ground for up to a year: for longer periods they should be covered.

13.16 Glass reinforced plastics (GRP)

GRP pipes are lightweight and corrosion resistant and are made from a polyester resin, glass fibre reinforcement and usually a filler of silica sand. Pipes have been in use, primarily for trunk mains, since 1970. The sand increases the wall thickness and therefore the rigidity of pipes which can be an advantage for laying purposes. By adjusting the amount of glass fibre and the wall thickness, the pressure rating and stiffness can be varied independently, unlike other flexible pipes. GRP is thermosetting (sets irreversibly under heat) as opposed to thermoplastic (PE and PVC, which melt on heating).

Pipes may comprise layers of different resins and different quantities and qualities of glass through the pipe wall: typically using resin rich and corrosion resistant qualities for the surface layers. Where different resins are used, the inner surface layer comprises a more flexible and chemically-resistant resin, e.g. bisphenol, and the inner and outer layer is a cheaper, stronger resin, e.g. isophthalic or orthophthalic. Several formulations are used, different for each resin supplier and maybe peculiar to each pipe manufacturer. Care is required to specify and ensure that the resins are suitable for encasement in concrete, if required: alternatively fittings may be made of ductile iron or steel. Materials and the process must be suitable for potable water.

Pipes can be made by the Hobas centrifugal casting process or may be filament wound on a mandrel. Most pipes in the UK are made by the Hobas process in which metered weights of liquid resin plus filler and random orientated chopped glass strand are fed into a rotating mould. This produces a dense pipe wall with similar characteristics in the hoop and longitudinal direction – although fibre orientation can be adjusted if required to increase tensile capacity in a particular direction. The constant outside diameter allows pipes to be cut and joined at any position.

Filament wound pipes in the UK are made in discrete lengths on a rotating mandrel. For reasons of production economy, this arguably allows more choice in design of the pipe wall but may require more glass fibre to retain (fluid) material as the mandrel rotates. The Drostholm process comprises a cantilevered mandrel wrapped by a continuous steel strip which advances helically around and along the mandrel, providing a continuous rotating surface before returning up the inside of the mandrel and commencing the helix again. Pipe materials are built up on this surface, being added sequentially along the pipe as the helix advances. Material is heated so that the pipe cures on the mandrel and can be cut into discrete lengths shortly after it leaves the mandrel. The Drostholm process is not used in the UK but pipe made by this process has been used on some major projects.

There is less experience with long, large diameter high pressure trunk pipelines in GRP than in other materials. Manufacture is a complicated process which must come under skilled control and should preferably be automated. Potential problems can arise in the form of delamination and cracking due to imperfect manufacture and from strain corrosion if pipes are unduly stressed by deflection. Inspection using ultrasonic techniques is possible and should be adopted where there is concern. Pipes must be laid under strictly controlled backfill conditions to prevent unacceptable distortion of the pipe wall or bending – and must be designed for both for short-term and long-term conditions and typically using 50-year pipe characteristics, extrapolated from shorter duration tests.

In the UK pipes are available from DN 300 to 2500 for pressures up to 25 bar. Pressure rating and stiffness can be tailored independently to suit project specific requirements – which benefits economy. The pipes are flexible and typically are designed to BS 5480 and installed to BS 8010 Section 2.5. BS 5480 covers stiffness classes from 1250 to 20 000 N/m^2 and pipes in the range DN 100–4000. Stiffness should not be less than 5000 N/m^2 and typically is not more than 10 000 N/m^2. Lengths preferred in the BS are 3 and 6 m but lengths up to 18 m are produced.

Joints generally are the flexible push fit spigot and socket or collar type. Alternatives are resin adhesive or flanges: screw threads are possible. Deflection limits for push fit joints are at least 3° for DN 500 and less, 2° for DN 500–800, 1° for DN 900–1700 and 0.5° for DN 1800 and greater. Actual values at installation should not exceed half these values. Where required, rope can be threaded circumferentially into a pre-formed groove in push fit joints to lock and provide tensile strength across the joint, for anchorage and for installation under water.

Fittings can be made from GRP pipe which is cut and laminated with glass fibre across the joint both inside and outside. Any bend angle can be supplied as well as multiple fittings as a one piece unit – for example a duckfoot bend and integral bellmouth. Alternatively change pieces and metal fittings are used. Ductile iron fittings are compatible with centrifugally cast pipes up to and including DN 600.

The factory test pressure is 1.5 times their pressure rating.

GRP has a key advantage for carrying aggressive waters, or for laying in exceptionally aggressive ground conditions where both iron and concrete pipes would be severely attacked. Care is nevertheless required in selecting the resin and glass type to suit the fluid carried, ground conditions and any concrete surround.

A balance has to be drawn between resins with the best chemical resistance, such as vinyl ester, but which may be brittle and crack due to deflection in service – and those with less chemical resistance but better physical flexibility. Whilst it is possible to specify a 'recipe', a better option may be to specify performance and leave resin choice to manufacturer, who will have links to particular resin suppliers and will tailor resin according to pipe manufacturing process.

Concrete pipes

13.17 Pre-stressed concrete pressure pipes

Pre-stressed concrete pressure pipes are covered by BS EN 639 and BS EN 642: installation is covered BS 8010 Section 2.4. The nominal size range is from DN 200 to 4000 and is the design internal diameter. American Standards AWWA C301–92 and C304–92 also cover large diameters and recent designs for North Africa are typically DN 4000 up to 20 bar and DN 3600 up to 26 bar. Design limits are restricted in principle only by material properties and capability of the structural section. Pipes are made by tensioning high tensile wire wound spirally around a cylindrical core. This core may consist either of concrete which is pre-stressed longitudinally or of a thin steel cylinder which has a thick spun concrete lining to the interior as shown in Fig. 13.3. For large diameter pre-stressed concrete pipes the core is cast vertically around the inside and outside of the steel cylinder. Pipes with steel cylinders in the core are called 'pre-stressed concrete cylinder pipes', the others with longitudinal pre-stressing are termed 'non-cylinder'. When the wires have been wound on to the core, stressed, and anchored, a relatively thin but dense cement rich mortar coating (20 mm minimum) is applied externally to the pipe over the pre-stressing wires. This coating is applied pneumatically or by machine at high velocity in even layers as the pipe is rotated, resulting in a hard and very dense mortar coating. This mortar coating provides cover to the pre-stress wires ensuring a high performance mechanical and corrosion protection.

The chief advantage of pre-stressed concrete pressure pipes is that they can offer a cost advantage over other pipes, particularly for large diameter and higher pressures. Pre-stressed pipes can be made to withstand high pressures simply by increasing the number of

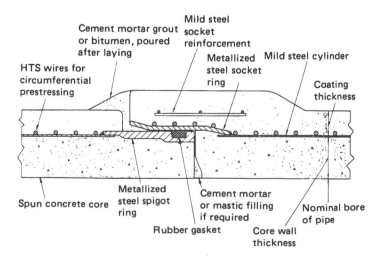

Fig. 13.3 Pre-stressed concrete pipe, with metal cylindrical core.

turns of pre-stressing wire per unit run of pipe or by doubling up the layer of pre-stressing wire. A second advantage of pre-stressed concrete pressure pipes is that they are proof against certain corrosive conditions that would attack iron and steel, although they may need special protection if the groundwater is high in sulphates and chlorides or otherwise aggressive to concrete. In very aggressive cases the pipes may need to be cathodically protected. This would involve the provision of electrical continuity to the pre-stressing wires, the steel cylinder and joint rings at each end of the pipe so that these can be bonded together electrically after the pipe is laid. Alternatively, or in addition, the pipes can be coated with urethane, coal tar enamel, or coal tar epoxy. The pipes can also be made using sulphate-resisting cement. The pipes are used in the marine environment where buried and permanently submerged. Attention is needed to the pre-stressing steel quality – and in the tidal and surf zone where the water level varies and there is a good oxygen supply, special attention is necessary and needs to be pursued according to detailed circumstances.

The pre-stressed pipe steel joint rings provide a high performance O-ring seal allowing joint rotations of up to 0.9° in any direction at each joint while maintaining water tightness. Deep joints are also available which can accommodate up to 1.4° of rotation at any joint. Full bevel and half bevel joints which allow several degrees of deflection can be built into standard sections of pipe, enabling pre-stressed concrete pipes to deal with the severe changes of grade and alignment (up to about 5°) before having to use pre-formed bends. Joints can be welded if required and will then carry longitudinal force.

The pipes are rigid thus providing a reasonable degree of resistance against rough handling and poor backfilling techniques.

There are some disadvantages to pre-stressed concrete pressure pipes. In the larger diameters the pipes are very heavy and difficulty will occur in aligning them in very soft ground unless they are placed on a prepared bed of granular material of adequate thickness to prevent uneven settlement. Connections can be made after the pipeline has been laid and while the pipe is in operation although to avoid shutdown, tees have to be incorporated in the line as it is laid.

Under normal operating conditions (up to the design operating pressures and external loads) cracking of the mortar coating will not occur. Cracking of the mortar coating may occur if the design operating conditions are accidentally exceeded through pipeline operational errors. However the elastic behaviour of prestressed concrete pipe is such that the cracks will close up after the overpressure or overloads have been corrected and the pipeline has returned to its normal operating condition. It should be noted that the mortar coating cannot be pre-stressed directly. However because the manufacturing process requires the coating to be applied immediately after the pre-stressing operation the mortar coating does receive some compressive strain as the concrete core continues to shrink after pre-stressing, taking the freshly applied mortar coating, which has been mechanically anchored by the pre-stressing wires, with it. This feature together with a feature known as tensile softening allows the mortar coating to take much more strain than the standard calculated plain strain before cracking. If pre-stressed concrete pipes are to be tested for long durations (say 24 hours), at a pressure greater than the design operating pressure, then the pipe must be designed accordingly: that is, the sustained test pressure should be considered as the design pressure. This would allow the pipe to be tested to the required pressures and for the specified duration without overstressing the pipe and unnecessarily subjecting the mortar coating to cracks.

The works hydrostatic proof pressure is the design working pressure plus surge (MDP) or 1.5 times the design working pressure (1.5 times PFA), whichever is the greater. Works hydrostatic proof pressure is required to be applied for at least one minute.

Joints for prestressed concrete pressure pipes are usually of the socket and spigot O-ring push fit type. The socket is normally mortared up afterwards and the joint then becomes rigid. Where ground conditions are known to cause differential settlement the joints may be filled with a bitumen mastic which provides the necessary protection while allowing joint rotation to occur.

13.18 Reinforced concrete cylinder pipes

Reinforced concrete cylinder pipes are similar to pre-stressed concrete pipes, except that there is no pre-stress and instead of using high tensile steel wire for circumferential reinforcement, mild steel rod is used. Size range is from 250 to 4000 mm internal diameter: corresponding wall thicknesses are 78 to 320 mm, giving external diameters from 406 to 4640 mm. Effective lengths in the UK are about 6 m up to DN 2100, reducing to 2 m at DN 4000 and are available up to 10 m from DN 400 to 1000. Up to DN 1000 in the UK the reinforcing rod is wound spirally under low tension on to a rotating metal cylinder, being welded to the cylinder at both ends. The reinforced cylinder is then covered internally and externally with concrete, the internal lining being spun on and the external coating being applied by impact and smoothed in layers. At larger sizes the pipe is cast vertically (wet mix) using a pre-formed spot welded reinforcement cage with either one or two layers of reinforcement according to design requirements. In the design of the pipe both the steel cylinder and the reinforcement are assumed to resist hoop tension stresses. The principal advantages of reinforced concrete cylinder pipes are as follows: (1) they can be designed for relatively high heads, (2) they require less sophisticated manufacturing techniques than do pre-stressed concrete pipes, (3) the concrete acts as a good protection to the steel and so permits some saving over the amount of steel that would be used in equivalent steel pipes, (4) the pipes are rigid thus providing a reasonable degree of resistance against rough handling and poor backfilling techniques. They can be used in the marine environment: care is needed in the tidal and surf zone. Disadvantages are as mentioned for pre-stressed concrete pipes. Some cracking is inherent as in any concrete structure. Cracks in reinforced concrete cylinder pipes caused by operational overload may not close up in the same way as for pre-stressed concrete pipes as the concrete does not have the same opportunity to recover after being subjected to large strains. Some degree of shrinkage cracking is normal after manufacture and can be accepted typically up to 1 mm in surfaces which will be continually wet after installation (internal surfaces and surfaces constantly below water level) as they will close after wetting and seal by the process of autogenous healing. Reinforcement size and spacing are designed to limit structural crack widths under load, typically to 0.3 mm, as in water retaining, maritime and other codes. Joints generally are welded which can carry longitudinal thrusts: the joints are rigid and the pipes require careful bedding to limit differential movement. Rubber ring push fit joints and other joints can be made if required. Fittings are coated internally and externally with reinforced concrete.

BS EN 639 covers reinforced concrete pipes and BS EN 641 covers reinforced concrete cylinder pipes. Reinforced concrete non-cylinder pipes are produced in the UK but are used for gravity, drainage, flow.

AWWA standard C300 addresses reinforced concrete cylinder pipes that are not pre-stressed or pre-tensioned. It covers pipe from DN 760 to DN 3600, typically with wall thicknesses between 89 and 305 mm (3.5–12 inches). This type of pipe has a relatively thick welded cylinder towards the inside and a layer of bar reinforcement towards the outside separated by concrete which renders the pipe rigid.

AWWA standard C303 covers reinforced concrete cylinder pre-tensioned pipes. The pipe is reinforced with a steel cylinder that is helically wrapped with mild steel bar reinforcement (using moderate tension in the bar), in sizes ranging from DN 250 to DN 1520 (10–60 inches) inclusive and for working pressures up to 27 bar. The lining thickness is 13 mm minimum for DN 250–410 and 19 mm minimum for DN 460–1520. The cement mortar coating thickness is 19 mm over the reinforcing bar or 25 mm over the cylinder, whichever results in the greater thickness of coating. Compared with reinforced concrete cylinder pipes, they have thicker steel shells and thinner concrete and are semi rigid. They are not used in the UK.

13.19 Asbestos cement pipes

Asbestos cement pipes are made of Portland cement and asbestos fibre mixed into a slurry and deposited layer upon layer onto a cylindrical mandrel. When the required thickness has been built up, pipes are steam or water cured, cleaned, the ends turned down to an accurate diameter for some 150 mm, and then usually dipped into cold bitumen. Although they have been in use for over 50 years and continue to be produced in some countries, the danger to health posed by the handling of asbestos in their manufacture and on site, particularly due to the risks of inhaling asbestos dust during cutting and handling pipes, has resulted in cessation of production in many countries. There is however no evidence that their use for conveyance of drinking water presents a danger to health.

UK standards warn against the hazards (asbestosis, lung cancer and mesothelioma) of breathing asbestos dust and refer to prohibitions on the supply of products containing amosite (brown asbestos) and crocidolite (blue asbestos). AC pipes made in the UK since 1982 are said to contain only chrysotile (white asbestos). Production of AC pipes in the UK ceased in 1986. BS EN 512 applies to two types: AT (Asbestos Technology), containing chrysotile asbestos and NT (Non-asbestos Technology) containing fibres other than asbestos.

Asbestos cement pipes have the advantages that they can often be produced from local resources more cheaply than other types of pipe, especially in countries which have to import steel or iron. They are also resistant to internal and external corrosion except in the case where sulphated soils occur which attack the cement of the pipe unless it is protected with bitumen. The principal disadvantage of asbestos cement pipes is that they are brittle; hence they need careful handling in transit, their bedding and backfill must comprise soft material containing no large stones, and they should preferably not be laid beneath roads subject to vibration from heavy traffic unless surrounded by concrete. Because a screw thread tapped into asbestos cement is fragile, a service pipe connection is made using a metal saddle clamped onto the pipe which is drilled and tapped for the insertion of a service pipe ferrule (see Fig. 13.2d). Ferrules tapped directly into a main without the use of a saddle, or the use of unsuitable saddles which corrode, are frequent causes of leakage.

BS EN 512 and BS 8010 Section 2.3 give standards for fibre cement pipes up to DN 2500. Pipes up to DN 1000 are classified according to nominal pressure up to 20 bar.

Preferred classes are 4, 6, 10 and 16 bar but pipe classes greater than 20 bar can be supplied if required. The required nominal pressure rating is decided in relation to hydraulic and operating conditions and external load. Pipes exceeding DN 1000 are designed to suit particular requirements of the pipeline. Tests include works pressure test, burst tests, crushing strength, beam bending strength (pipes DN 150 and smaller) and joint performance (including deflection and shear). The works test pressure (PT) is specified as double the nominal pressure up to DN 500 and 1.67 times the nominal pressure for larger sizes. Works test pressure duration is 30 seconds, or if the test pressure is increased by 10%, 10 seconds. BS 8010 Section 2.3 states that the design pressure (PFA) should be no more than half the works test pressure (PT) and that the maximum pressure with surge (PMA) should not exceed 60% of PT. Sizes for water supply are commonly DN 100–900 for working pressures of 75, 100 and 125 m. Nominal lengths are typically up to 5 m for DN 300 and smaller and up to 6 m for larger sizes. Although some fittings are asbestos cement, common practice is to use steel or standard ductile iron fittings with spigots to match the class of pipe used. Asbestos cement pipes and fittings are all plain ended and hence are jointed using collars of asbestos cement or by use of Viking Johnson or similar couplings (see Fig. 13.2e). Great care is needed to ensure that O-ring joints have no twist in them when placed, otherwise the joint will leak. A leak under pressure at a joint may in time cut right through the pipe wall. A gap must be left between the ends of pipes to allow for deflection and, where appropriate, thermal movement.

Pipeline construction
13.20 Choice of pipes

Principal factors affecting choice of material are technical considerations, price, local experience and skills, ground conditions, preference and standardisation. Only an indication can be given here of general patterns in the choice of pipes.

For new service pipes (DN 50 and less) in the UK, over 99% of material now used is MDPE. The remainder tend to be copper (for ground conditions).

For the smaller pipe sizes in water distribution systems (DN 51–300), in the UK over 80% of new pipes are plastic (about 10% PE 100, 50% PE 80 and over 20% PVC-U and MOPVC) and over 15% ductile iron. Plastic pipes offer cost advantages at small diameters and – as they can be joined above ground to form continuous lengths which can be snaked into narrow trenches – they also offer advantages of speed and consequently less social and environmental impact during construction. This advantage disappears with increasing diameter and pressure class.

For the middle range of pipe sizes used in water distribution systems (DN 350–800) ductile iron is most widely used because of stiffness, strength, toughness and durability in many kinds of ground. In addition a complete range of compatible, standard dimensioned ductile iron fittings and valves are available to make up a homogeneous pipeline, which simplifies both the design and construction processes. Service pipe connections can also be tapped directly on to the main, under pressure if necessary. Almost any other pipe type can be used as an alternative. Asbestos cement pipes in some countries form the main alternative to ductile iron pipes, their advantages being that they are usually cheaper (especially if ductile iron must be imported). However, asbestos cement pipes are more liable to breakage problems than are ductile iron pipes, both in transport and when laid, fittings must be mainly of ductile iron, and gunmetal or cast iron saddles have to be strapped to the pipes before service pipe connections can be made. Deterioration of these saddles can bring about severe leakage problems. Asbestos cement pipes have now been phased out of production in many countries.

For trunk mains and large diameter pipes no general rules of choice can be laid down since steel, ductile iron, pre-stressed concrete, concrete cylinder, and GRP pipes may each be used in any particular case according to circumstances applying. Steel is predominantly used for trunk mains or high pressure mains, the welded joints offering a distinct advantage in the latter case. Steel pipes are also particularly useful in congested and urban areas where welded joints provide longitudinal strength and avoid the need for large (or any) thrust blocks.[5] Steel pipes do not normally require rocker pipes at junctions with chambers and minor structures but may be required to accommodate major settlement where temporary excavation has to be backfilled at large structures. Anchorage points are nevertheless required at terminations and at connections to chambers where flexible couplings will be installed at valves. Alternatively, the pipeline may be tied across flexible joints to provide longitudinal continuity of strength: the pipeline can then in principle be allowed to move longitudinally within the chamber and need not be anchored into the chamber walls: leakage of groundwater into the chamber would have to be addressed but would not normally be an issue in dry conditions. To appreciate the positioning and design of joints and anchorages the pipeline can be considered as a potential mechanism limited only by the pipe structure, friction and soil stiffness. Ductile iron pipes in the largest diameters tend only to be used instead of steel if their price is competitive or if it is expected that there will be difficulty in getting the skilled welders for steel pipes. The alternatives of pre-stressed concrete, concrete cylinder, or GRP pipes tend to be used because of circumstances such as local preference and practice, price competitiveness, tied funding, in-country manufacture as opposed to importing, aggressive ground conditions or aggressive water to be conveyed, or (in the case of concrete pipes only) where a greater margin of safety is required against rough handling and backfilling. GRP pipes have the principal advantage that they are not attacked by ground conditions or by waters, such as desalinated water, which are severely aggressive to both iron and concrete.

13.21 Laying of pipes

Strict control needs to be exercised over the laying of pipes because of the high capital cost of pipelines and their swift deterioration if not properly laid. In this section the key factors to which attention should be paid to produce a well laid pipeline will be considered.

Typically a 'working width' for construction is defined in the contract documents, to allow space for access, pipe stringing and separate stockpiles for topsoil and excavated spoil. In addition space will be required for pipe storage and office and support facilities. The pipe 'easement' as finally granted relates to the pipe as laid and will set out rights for access and will limit subsequent construction and other works. A 'wayleave' is an agreement similar in nature but less specific than an easement, granted by the landowner, permitting the pipeline owner to carry out the works.

Care should be taken to ensure that pipe supports on vehicles transporting pipes are adequate to prevent damage to the external coating (especially in hot climates where the coating may soften) and that flexible pipes particularly those with concrete or cement mortar linings are supported to maintain circularity and avoid damage to the lining. The stock dumps for pipes should be properly planned and pipes should only be stacked one above the other if they are properly provided with timber supports and packers. Care should be taken not to damage pipe ends: bevel protectors should be fitted to pipes to be joined by butt welding. Pipes should be stacked in accordance with the manufacturers recommendations and should not be stacked in areas where long grass may grow: in a dry period this grass can catch fire, thus ruining the exterior protection of the pipes. All pipes should be handled by using purpose made lifting slings of a wide fabric material so that the external coating is not damaged. The practice of lifting pipes by means of chains or wire ropes 'packed off' the barrel of the pipe by pieces of wood should be forbidden. This not only damages the pipe, but can be a dangerous practice. When pipes are delivered, and again just before they are lowered into the trench, they should be inspected for flaws. The Holiday detector should be passed over steel pipes. Any coating or lining flaws detected should be made good. The interior of each pipe should be inspected as the pipe is lifted and any debris must be brushed out.

General practice is that pipes are buried for safety and security, for protection against weather, fire and other physical damage; so that they do not obstruct surface drainage or access and do not present an obvious visual environmental impact or target for vandalism or source for theft of water. In certain circumstances, for example in sustained lengths of rock or swampy ground, where excavation would be expensive and where the issues arising from the above can be accepted, pipes can be laid on supports above ground. Design of the supports and selection of joints must allow for longitudinal movement due to thermal effects and provide anchorage for thrusts due to internal pressure. For the above reasons pipes for water supply in most circumstances are buried.

At small diameters, flexible PE and PVC pipes can be joined on the surface and snaked into narrow non-man-entry trenches. Pipes are then usually installed on 100 mm granular bedding and backfilled to 100 mm above the pipe with free flowing granular material, above which the main backfill is placed. This allows trenches to be dug by chain excavator, rock-wheel or narrow bucket. If the pipe is laid directly on the soil then the trench bottom must give sensibly uniform support to the pipe: it must be stable, fine grained and free from flints and large stones or other material which may cause point loads. Additional excavation is required at pipe sockets. If the trench bottom is unsuitable a bedding should be used. A bedding must be provided for support in soft ground. A nominal minimum of 150 mm granular bedding must be provided in rock. The trench width for non-man-entry trenches should be at least 300 mm greater than the outside of the pipe. Trench fill must be brought up evenly either side of the pipe with soil selected and compacted to give the required support.

Steel, plastics and to a large extent ductile iron, are flexible conduits which when buried rely on the pipe/soil-structure interaction for their load carrying capacity. Deflection under vertical load is limited by support obtained from the trench sidefill, which in turn transfers load to the trench sides. It is therefore essential that the pipes are bedded evenly and are surrounded in material which is well compacted and can transmit the lateral thrusts from the pipe to the trench sides and that the undisturbed ground does not become overstressed. Detail design will determine whether the pipe can be laid on the trench bottom after trimming or whether a bedding must be used.

Successful implementation requires care in handling the pipe and in placing and compaction of backfill – which in turn requires competent labour and dedicated supervision and quality control. During backfill pipes must not be supported on timber or hard blocks. Pegs should not be used in the trench bottom for setting out bedding levels. The danger otherwise is that the pipe will, in time, settle down on hard points in the trench bottom and damage the coating and may overstress the pipe locally, both in ring and longitudinal bending. The pipe must be bedded evenly and all voids beneath the pipe must be filled with compacted fill. It is particularly important to achieve compaction in the 'upper bedding', in the narrow section of fill above the pipe invert: 'pogo stick' type rammers may be needed for this purpose. Compaction equipment must not damage or come into contact with the pipe or within typically 150 mm of the top of the pipe. Material selection and compaction should initially be specified as minimum requirements, with detail left to implementation. The bedding, pipe surround and initial backfill 300 mm above the pipe must be free from large stones (larger than say 15 mm) which may damage the pipe coating. Similarly, voids left by removal of trench supports must be filled – typically by withdrawing trench supports as fill is raised. Failure to observe the above may result in loss of the soil side support on which the pipe design relies.

It is vital to make an even bed for pipes, with joint holes previously excavated in the positions required. All large stones must be removed from the bed and no hard bands of rock should straddle the trench. The use of boning rods and sight rails for every pipe is essential and work should be stopped until these are provided. What constitutes a 'large stone' which should be removed in the base of the trench depends upon the pipe and the nature of the material forming the base of the trench. Generally any stone or stones which might give uneven bedding to the pipe invert or puncture its coating, over one-third of the width of the trench, i.e. along the central band of the trench where the pipe will bed, will need to be removed. In all cases where rock is encountered, and rock includes chalk with hard bands in it, the base of the trench must be filled with 150 mm of granular material. After trimming the trench bottom or compacting the bedding a depth of 50 mm of uncompacted soft sand may be left loose to form a uniform support for the bottom arc of the pipe. Alternatively the bedding must be shaped and compacted to match the curvature of the bottom of the pipe: the method adopted will depend on *in situ* soils and workmanship. Either way the work is required such that the specified deflections are not exceeded.

Where rock – or unduly soft soil – is expected there should be liberal provision for bedding in the bills of quantities. If this is not adequately provided for, the supervising engineer on site may be reluctant to order the proper bedding the site conditions demand because of the additional cost involved. He or she should not be put in such a position. The provision of bedding or fully surrounding pipes can solve many pipe-laying problems in a sound manner. The bedding not only gives full support to the pipe, protecting it from settlement on hard points and from excessive overburden pressure or traffic loading, but it can also give protection against corrosion in aggressive soils.

Lengths of pipeline should be laid to even grades: good practice is to limit grades to not less than 1:500. This gradient can be readily achieved and monitored during construction. Flatter gradients are possible but have greater risk of backfall (reverse gradient) either due to limitations of control during construction or due to movement during backfill. Backfalls will produce minor ponding – which may not be important – and air traps, which can affect pressure tests. For distribution pipes which have service connections a flat grade does not particularly matter as air will be drawn off via the service connections. On trunk pipelines, however, it is important to arrange even rises to air valves.

Care should be taken during pipe-laying to ensure that pipes cannot float in the event of the trench flooding. Trenches should be backfilled as soon as possible after laying and jointing: long lengths of uncovered pipe should not be permitted. This will also help limit thermal movement, which can be particularly important for plastics pipes.

Considerable trouble is experienced when laying pipelines through urban districts where many other services, e.g. gas, electricity, etc., will have to be negotiated. No amount of pre-planning work will ever reveal all the problems that will be encountered since records of these other services are seldom perfect or to the accuracy necessary. Close liaison with the utility operators is required; utilities location by CAT and Genny surveys are essential, followed where possible by trial pitting to find such services ahead of pipe-laying. The trench should preferably be excavated well ahead of pipe-laying, if the road authority will permit this so that the line and level can be adjusted in good time. Small angle deviations can be accommodated at joints and if need be by introducing shorter lengths. Stocks of pre-formed bends should also be held available for use when negotiating the pipeline around obstacles. Gussetted bends (otherwise termed mitred, or lobster back bends) in steel pipes can be fabricated from straight pipe to suit any combination of vertical and horizontal angles.

If two pipelines are to be accommodated in one trench, a minimum spacing of 300 mm should be kept between the two lines. If a parallel pipeline is to be constructed at a future time enough space must be allowed in the pipeline reserve to construct a second trench without disturbing the original works – and consideration must be given to unbalanced thrusts as it will be unlikely that the original pipeline can be shut down for an extended period.

Backfilling to pipes should be placed in even layers either side of the pipe up to soffit level and, in addition, for non-rigid pipes the backfill must be carefully compacted to keep the pipe in a true cylindrical form. The backfill material adjacent to the pipe and for 150 mm above its crown must be free of large sharp stones that could puncture the sheathing of the pipe. When the material from the trench is being excavated it should be inspected and instructions should be given for setting aside material that should not be used against the pipe. This material can be used in refilling the trench once the pipe has been properly covered with soft material.

A principal requirement for satisfactory pipe-laying is care in making the joints. Achieving cleanliness in a muddy trench is far from easy and the men should be provided with the facilities required, such as clean water and buckets, plenty of wiping rags, and enough room to work and time to make the joint properly. The reward for taking care with each joint is a pipeline which passes the test requirements at the first test: this can save weeks of extra work.

Cover to pipes should normally be not less than 1.0 m to provide some protection against physical (third party) damage; for frost protection and to limit the effect of seasonal ground movement. Depths of cover will need to be increased in climates with very low and prolonged low temperature. Cover can if necessary be reduced (for example to limit temporary environmental impact or to reduce excavation in rock) where flow in the pipe is continuous even in areas liable to frost – but requires calculation of heat transfer. Cover may be increased to reduce loading under heavily trafficked roads (minimum load is typically at about 1.8 m, depending on pipe and traffic load). Cover may also be increased for flexible pipes which rely on soils support, which may be removed by excavation for other services – alternatively an easement must be created (and enforced) to prevent encroachment of other excavations on the pipeline soil support (which typically requires no soil movement within two pipe diameters each side of the pipe, total five diameters width). Where shallow burial is necessary below roads, a reinforced concrete slab across and bearing on undisturbed ground each side of the trench can be used to transfer imposed load away from the pipe. Thrust blocks must be designed having regard to both the pipeline thrust developed under operating and surge pressure or pipeline testing conditions and to the soil resistance available. Thrust blocks do not resist pipeline thrust by themselves: they transfer the thrust to the ground and hence the soil resistance that can be mobilised to resist this must be carefully analysed. All such blocks must bear against undisturbed ground. Support blocks are required where pipes are laid at gradients of 1:6 or steeper and may be needed on slopes between 1:6 to 1:12, depending on ground conditions. Differential settlement at thrust and anchor blocks needs attention: rocker pipes may be required. Care is needed in backfill compaction and in selection of backfill material and grading, particularly on steep gradients, to ensure that the pipe trench does not become a drain, washing out trench fill and collecting at low points, softening the trench fill and weakening the combined pipe and soil structure. In weak soils particularly, backfill grading should be selected using normal filter rules: in certain circumstances it may be necessary to contain the pipe backfill within a membrane of filter fabric. Care should be taken during installation to ensure that there is no relative movement at joints: good bedding and care in backfill placing and compaction are important. Pipelines with flexible joints at close spacings provide inherently more capacity for natural minor movement after installation: flexible joints at wide spacings are fewer but can increase local movement and shear forces.

13.22 Testing of pipelines

The test usually consists first of filling the pipeline with water and allowing it to stand and stabilise under working pressure. The test pressure is then applied slowly by pumping, typically to about 50% in excess of the design or working pressure or of some fixed amount, such as 30 or 50 m water head, in excess of the maximum working pressure. The pump is then disconnected and the amount of fall of pressure in a given time (usually 1 hour) is measured. After this time has elapsed water is again pumped into the line under test, to bring the pressure back to its initial value, and the amount of water pumped in is measured. Water is then bled off to reduce the pressure to the value at the end of the standing period and this quantity is checked against the quantity pumped in and is recorded as the loss.

BS 8010 gives the standard for field testing of pipelines with non-porous linings as a loss not exceeding 2 litres of water per metre of nominal internal diameter of pipe per kilometre of pipeline per m head of applied water pressure per 24 hours. The standard is fairly rigorous. The allowable loss does not set out to permit leakage of any value but is intended to represent inaccuracies in measurement, temperature changes, small quantities of dissolved air and other elements. All exposed parts of the pipeline, including in chambers should be visually checked and all leakages rectified. A loss of 3.3 l/m diameter/m pressure/km length/24 h is permitted for new concrete pipes where the lining is absorbent and is likely to require uneconomic and probably unjustified time in reaching the recommended 2 litre figure. This higher value is permitted provided that a minimum of four and preferably at least six tests are made and a plot of leakage shows that the loss is reducing and that the 2 litre value would be likely to be reached. No such adjustment is mentioned but might be reasonable for steel pipes with concrete or cement mortar linings – and for asbestos cement pipes the 2 litre value is retained but the soak time prior to test is extended to at least 4 hours or until stabilisation. It would be reasonable to expect zero leakage on a buried welded steel pipeline tested over a 24-hour period and to 90% yield, provided correction is made for dissolved air and for temperature changes.

The test pressure recommended varies between standards as follows:

Steel:	pipe stress should not exceed 75% yield (AWWA M11);
Ductile iron:	allowable operating pressure (PFA) plus 5 bar or maximum operating pressure (including surge) if less (BS 8010, Section 2.1);
AC:	75% of the works hydraulic test pressure measured at the lowest point of each section: sections should be 500–1000 m long and selected so that the test pressure is not less than 60% of the works hydraulic test pressure or 1.5 times the pipeline design pressure if the above is not economic (BS 8010, Section 2.3);
PSC:	1.5 times working pressure except if the pipeline is strengthened for surge, in which case it should be working pressure plus surge (BS 8010, Part 2.4);
RC:	as for PSC (BS 8010, Section 2.7);
PVC:	1.5 times nominal working pressure of the lowest rated component or the rated pressure if greater (WRc manual);
GRP:	1.5 times working pressure subject to a minimum of working pressure plus 5 bar (BS 8010 Section 2.5);
PE:	1.5 times rated pressure for 6 and 10 bar systems and 1.5 times working pressure for 12 and 16 bar systems (WRc manual)

Where the alternative test pressure relates to surge, an additional margin of at least 20 m would be prudent. Pipe testing should proceed quickly after installation and results should be linked to payment for the pipeline. Initial tests should be made starting with short lengths for each mainlaying gang and pipe size and material. Test lengths can gradually be increased typically to 5 or 10 km or between section valves as a successful track record is developed. Test lengths will depend on topography and availability of water.

Fluctuating test pressure results are likely to be caused by air locks in the pipe. To avoid air locks there must be suitable air valves on the pipeline. Filling must proceed slowly, particularly on falling gradients, to vent air and avoid hydraulic jumps and entraining air. BS 8010, Section 2.3 quotes a filling rate equivalent to a water velocity about 0.05 m/s. This is very low. There is no hard and fast rule: time of the operation is also important and is linked to test lengths. An equivalent velocity between about 0.2–0.5 m/s is sensible but can be varied to suit circumstances. Source of water and pumping arrangements should be designed to allow any entrained air to escape and to prevent air entrainment into the pipe.

Air must not be used for testing water mains. The test must be hydrostatic and take place between blank flanges, bolted or welded to pipe ends, or caps may be used if fully supported by anchor blocks. Where pipes have flexible joints the end pipe must be fully anchored. Testing should not take place between closed valves because if the valve is already inserted in the line it will not be possible to detect any leakage past the valve, and if the valve is exposed at the end of a section of the line it will be in the 'open end condition' and will leak unless it is designed for the 'closed end condition' (see Section 13.26).

When a pipeline fails its test and it has been backfilled, searching for leaks can be troublesome. It is best to leave the pipeline under pressure for a day or two so that, possibly, a wet patch on the surface of the ground will indicate the whereabouts of failure. The pipe may also be sounded for leakage using normal waste detection methods (see Section 15.16) and it is possible to use some kind of tracer element in the water, although this is a skilled matter requiring a specialist. In practice, therefore, it is usual to dig down to the joints until the leakage point is found. Usually this is not as time-consuming as one might expect because wet ground is easily found if the pipeline is left under pressure.

13.23 Making connections

If a socket and spigot tee has to be inserted into an existing pipeline the length cut out of the latter must be slightly greater than the overall length of the tee. The socket of the tee is pushed up to fit one end of the cut pipe and the resulting gap between the two spigots at the other end is joined by using a collar. If a double-socketed tee is used this must be inserted using a plain piece of pipe on one side, again joined by a collar to the pipeline.

An alternative is to use an 'under-pressure connection', as shown in Plate 30. A split collar is clamped on to the main, the collar having a flanged branch on it to which is bolted a valve. A cutting machine is attached to the valve, the latter is then opened, and the cutter is moved forward through the valve and trepans a hole in the side of the pipe. The cutter is withdrawn with the trepanned piece of pipe wall and the valve is closed. The cutting machine is then removed and the branch connection can be made. Steel and iron pipes can be cut *in situ* using a rotating cutting tool which is clamped onto the main. A manually operated wheel cutter can be used on small diameter cast iron mains of 80 or 100 mm size. Oxyacetylene cutting of steel and iron pipes can be used, but the cut is ragged and difficult to make exactly at right-angles to the axis of the pipe. Cast and ductile iron pipes can be cut above ground using a hammer and chisel. The pipe is placed on a timber baulk below the line of cut and is rolled back and forth as the chiselling proceeds: first to 'mark' the cutting line and then to deepen the chiselled groove. At a certain stage the pipe will come apart at the chiselling line. When asbestos cement pipes are cut, or sawn, a water jet must be directed on to the cutting point to prevent asbestos fibre being raised in the air.

13.24 Underwater pipelines

Steel and PE are materials commonly used for pipelines laid below water for the crossings of rivers or estuaries. Both materials can be welded, thus providing longitudinal continuity of strength. Other materials can also be used, including concrete, GRP, PVC and ductile iron but require particular attention to tolerance on bed preparation, practicalities of level and position measurement underwater, plus the strength and deflection limits of flexible joints. Pipe-laying is normally carried out by one of the following methods:

(1) laybarge;
(2) reel barge;
(3) bottom pull;
(4) float and sink.

The laybarge is a 'factory' for progressively adding pipes to a string whilst winching the barge along the pipeline route so that the string hanging in a catenary from the back of the barge is gradually lowered onto the river or sea bed or into a pre-dredged trench. The reel barge method is similar but is used for unreeling lengths of plastic tubing as the barge moves along the pipeline route. In the bottom pull method lengths of pipe, prefabricated onshore, are joined to form a pipe string which is progressively pulled into the water by a winch, mounted on a pontoon or on the far shore, until the crossing is complete. In the float and sink method lengths of pipe are made up into strings at a remote fabrication yard, the string is towed at or below the water surface to the crossing location where it is aligned into position and sunk by removing supporting buoyancy tanks or filling with water.

Internal operating conditions are similar to those described for land pipelines but special attention needs to be given to external conditions as, once laid, access to a pipeline for remedial work is unlikely to be available. Attention has to be paid to stresses during laying, soil conditions, the effects of currents and waves (including soil liquefaction), sea bed topography and morphology, protection against corrosion and protection against damage by ships anchors, dropped objects, fisheries (trawl boards) and other activities. Physical protection in shallow water typically requires burial with some cover below the lowest likely level of bed movement and depth of penetration of any ship's anchor. In deep water where wave and current action is small, the pipeline may be laid without burial, subject to limitations of profile and hazards.

Because of their weight, pipelines are in general installed empty. Positive submerged weight ('negative buoyancy') is required to prevent the pipes from floating and for stability under the action of waves and currents. (Pipes installed by float and sink would be towed in strings and manoeuvred into position using temporary buoyancy.) Weight is added typically, for steel pipes, as a continuous concrete cladding and, for polyethylene pipes, concrete collars are used, spaced, shaped and sized to give the required underwater specific gravity and stability. Design considerations include buckling under external hydrostatic pressure, pull forces, longitudinal bending as the pipe is towed or conforms to the underwater profile – and ring bending under backfill loads. The cladding thickness is typically about 10% of the pipe diameter. It is sometimes considered simply as temporary weight coat but it does serve to protect the external corrosion coat and therefore can act as part of the permanent structure. The cladding can contribute significantly to bending stiffness particularly in ring bending: differences at joints can raise stresses locally. Above about DN 1500 the cladding is stiff and thick enough to be designed as reinforced concrete structural ring. This can be used to support and reduce the thickness required to withstand buckling of the steel shell – typically by 40% – and carry external backfill loads. The combined structure can also be designed to withstand ships anchors, thus saving considerably on the costs and environmental impacts of deep burial.[6] Backfill must be designed not to liquefy and to be stable under wave action. Rock armour can also be chosen and designed to save on burial costs. With care, considerable economies can therefore be achieved.

Design standards include BS 8010 Part 3; BS 6349; DNV rules; ASME B31.8, PIANC, US Shore Protection Manual and CIRIA/CUR recommendations for rock armour (see Bibliography).

Valves
13.25 Valve types

Valves used in waterworks may be listed as stop valves, non-return valves, flow control valves, pressure control valves, energy dissipaters, and air valves.

Stop valves are of several different varieties, namely:

(1) sluice valves which consist of a gate shut down into the pipeline;
(2) butterfly valves which consist of a pivoted disc in the pipeline, turned through 90° to block the flow of water;
(3) needle valves which consist of a cone, moved longitudinally or expanded in diameter so as to stop annular flow around the cone;
(4) screwdown plug valves which consist of a plug or diaphragm which is forced on to a circular seating through which the water is flowing (used only for small pipes or low heads).

Valves may be operated by hand or by actuator. Actuators generally are electric (but may be hydraulic or pneumatic) and may be operated locally or by remote control. Where there is no local power supply, portable power packs may be used. An inching facility for fine control and impulse facility for offseating valves which have not been regularly operated are useful.

13.26 Sluice valves

Sluice valves were invented over one hundred years ago. The metal seated type has not materially altered in design except that toroidal (O ring) seals have in general replaced gland packing in the stuffing box at the top of the valve. Figure 13.4 shows a section through a large valve. The gate is wedge shaped and is lowered into a groove cast in the

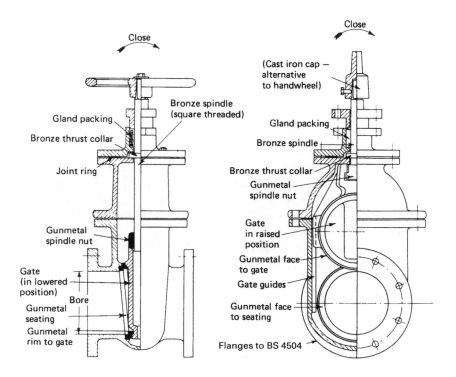

Fig. 13.4   Metal seal sluice valve for 300 mm diameter pipe.

body of the valve, the meeting faces being made of gunmetal. The design is simple and relies for watertightness upon the gunmetal faces being forced together to make a watertight joint. Resilient seat valves offer an alternative and are in general use up to DN 300, in which the gate is encapsulated with rubber and seals against a clear full bore typically without grooves. A further development is the boltless design for the valve body: this stems from the cost benefits of mass production and the hypothesis that it is cheaper to replace rather than refurbish valves up to and including DN 300. Sluice valves have a low head loss coefficient, hence their use is ideal for maximising pump efficiency on pumping installations. However they do have a number of troublesome characteristics.

(1) When closed, an unbalanced head of water will press the gate heavily against the seating; to re-open the valve against any large unbalanced head requires a great force. Above certain unbalanced heads for the larger sizes of valve it is impossible to open such valves manually, even if gearing is used. Where high unbalanced heads must be met, anti-friction devices, such as ball-bearing thrust collars and anti-friction rollers and external rising screws which can be greased, must be used – although these measures are not usually required. The most effective method and first step in meeting high unbalanced heads is to provide a bypass to the valve.

(2) The amount of work required for full opening or closing of a large valve is also great. The time taken by two men to open one of the larger valves of DN 700 or more may well be more than 1 hour. This is not a defect hydraulically since large flows of water should not be suddenly stopped or severe surge pressures will result, but it does mean that in waterworks shutdowns, as for instance for repairs of mains, considerable time must be allowed for operating valves.

(3) Sluice valves which are left shut for a long time tend to stick shut and require even greater force to get the gate off the seating. Similarly sluice valves which have been left open for a long time may not close properly because of collection of dirt in the gate groove which prevents proper closure of the gate.

(4) A sluice valve is only designed to close drop tight at a given unbalanced pressure; it may not close drop tight at greater or lesser unbalanced pressures.

(5) Sluice valves must be specified as designed for the 'open end test' or the 'closed end test'. A valve designed for drop tight closure under open end conditions may not close drop tight for closed end conditions. A closed end siting of a valve is when that valve is so fixed that it cannot expand when pressure is put on it, as in the case of a double-flanged valve connected into rigidly held pipework. An open end situated valve is one where the valve is free to expand, as for instance where it is held to the pipeline by one flange only or where the joint on the downstream side is a slip joint, such as a flange adapter coupling, which permits expansion of the valve body along the line of pipework.

(6) A sluice valve is not the proper device for controlling the rate of flow of water through a pipe. Figure 13.5 shows how the waterway on a sluice valve decreases with travel of the gate. Only the last 10% travel of the gate towards closure has any substantial effect on the flow rate, depending upon the pressure of the water in the pipeline, and this means that when controlling flow the water will be passing under the gate at high velocity. The gate may develop vibration as a consequence and cavitation resulting from the reduction in pressure of the high velocity water flowing under the gate may cause pitting and erosion.

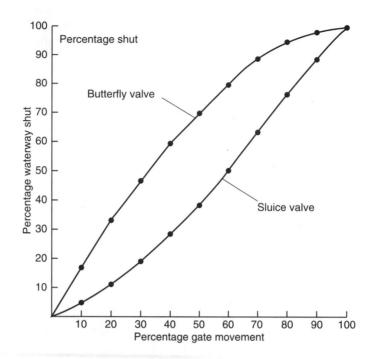

Fig. 13.5   Valve waterway area in relation to gate travel.

(7) Repairs to a sluice valve (other than repacking of glands) involve emptying the main in which the valve is situated.
(8) Sluice valves are expensive and, in the larger sizes, they are heavy single items of equipment to handle.

Some of these difficulties can be minimised by making the valve about five-eighths to three-quarters the size of the pipeline. If the valve is connected in line by properly designed tapers, particularly if a long taper is given on the downstream side, very little headloss will occur when the valve is fully open. The amount of head lost will be a few cm of water pressure only. The cost of the valve will be less (by as much as 50% in the larger sizes), its insertion into line will be made easier because it will have less weight, and the force required to open it will be substantially reduced. This practice should always be followed where possible. Difficulty with sticking valves and dirt on the gate groove can be greatly reduced by routinely operating valves. To leave valves unoperated for years on end is an unwise policy as they may not then close in some emergency. The problem of opening large valves against unbalanced heads is greatly eased by installing a bypass to the valve, on which bypass a second smaller valve is fixed as shown on Fig. 13.13. Bypasses need not be very large in diameter, a DN 100 bypass would normally be suitable for a DN 800 valve. The effect of the bypass depends upon the downstream conditions of the pipeline. In general, however, even if the pressure is not substantially raised downstream of the main valve when the bypass is opened, pressure is reduced on the upstream side of the main valve and, accordingly, the force required to unseat the gate of the main valve is reduced. A bypass is also useful for filling a main since this can be opened first and the flow into the empty downstream section more carefully controlled than if the main valve were opened.

A power driven actuator may be used to operate a valve; it may be directly coupled to the valve spindle or be mounted on a headstock and coupled to the valve through an extension spindle.

13.27 Butterfly valves

At and below about DN 300–400 sluice valves are cheaper than butterfly valves due to the quantity required and the benefits of mass production. Butterfly valves tend otherwise to be cheaper because they are smaller and require less material and less civil works. The use of the butterfly valve in water distribution systems has increased; it is easy to operate such a valve against unbalanced water pressures because of the partially balanced pressure against the disc. Butterfly valves can be metal seated or resilient seated: in the latter case the seat is usually made of natural or synthetic rubber and can be fixed to the disc or to the body of the valve.

Resilient seated valves can be specified as 'droptight', meaning virtually watertight when shut against unbalanced pressures up to the design head. Hence such valves are the type usually specified for isolating valves in distribution systems. They do, however, prevent the passage of foam swabs used for cleaning mains, but this does not usually pose a problem if the valves are spaced reasonably far apart, since the main can then be cleaned in sections. Short lengths of pipe of between 1.5 and 2.0 m on either side of the valve are made removable so that the cleaning apparatus can be inserted and removed.

Metal seated butterfly valves do not have tight shut-off characteristics and are mainly intended for flow control purposes where they need to be held in the partially open position. (Resilient seated valves are used for this purpose also.) Metal seated valves are not necessarily more durable than resilient seated valves since scouring of the seating can occur if silty water flows at high velocity past a disc which is held only slightly open. Solid rubber seatings are the type of resilient seating most usually adopted. Inflatable seals have been used on very large valves, but not always with success. One advantage of the butterfly valve is that closure of the valve in flowing water causes loose debris below the disc to be swept away. Disc position indicators are useful and strong disc stops integral with the body should be specified, so that the operator can feel with certainty when the disc is fully closed or fully open.

Butterfly valves have been made to very large diameters (10 m or more) operating under very high heads and at high water velocities (20 m/s or more) and have proved successful in use. However when a valve is to be used for flow control purposes the maximum velocity through the valve should be limited to 5 m/s. Permissible leakage under test should not exceed about 0.06 l/h per 100 mm of nominal diameter for resilient seated valves under pressures of up to 100 m. 'Wafer thin' butterfly valves fitted with simple lever operating systems are available for uses such as isolation of air valves to enable maintenance to be carried out *in situ* whilst the pipeline remains in service. Because they do not provide a clear waterway, butterfly valves do not allow pipeline pigging.

13.28 Screwdown valves

Screwdown plug valves are normally made only in the smaller sizes, of which the bibtap is a typical example. The body of the valve is so cast that the water must pass through an orifice which is normally arranged in the horizontal plane. A plug or diaphragm, or in the case of a bibtap or stopcock a 'jumper', can then be forced down on to this orifice by a screwed handle, shutting the water flow off as shown in Fig. 13.6. This principle is used in all sorts of valves for shutting off or controlling flow.

The same principle applies to ball valves, to pressure or flow control valves, to hydrant valves and to many other types of valves. Where the pressure is low the size of valve can be quite large, as for instance with simple plugs for stopping flow into a drain. When the size of pipe, and therefore of orifice, is small then high pressures can be controlled, as in the case of the ordinary domestic tap. The defects of these particular types of stop valves are that their seatings need renewal from time to time if they are in frequent use and that, even when wide open, they cause a considerable loss of pressure.

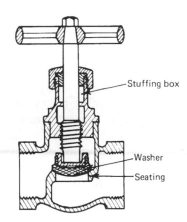

Fig. 13.6   Screwdown stopcock for service pipe.

13.29 Non-return valves

The usual type of non-return valve consists of a flat disc within the pipeline pivoted so that it is forced open when the flow of water is in one direction and forced shut against a seating when the flow tries to reverse (see Fig. 13.7) and is often termed a 'swing check valve'. The seating is arranged slightly out of perpendicular when the valve is to be inserted into a horizontal pipe so that the flap will close by gravity when there is no flow. The speed with which the flap acts is often important, especially on pump delivery lines where the valve is required to be fully shut the moment forward flow of the water ceases. The valve can be assisted to shut quickly at the appropriate time by bringing its pivoting spindle out through the body of the valve and putting a counterbalancing weight upon this. Conversely a damping device may be required to prevent slamming of the gate which might cause high surge pressures. The defect of non-return valves is that they cause a substantial headloss because the weight of the gate must be lifted and kept suspended by the pressure of the flowing water. They are normally designed for flow velocities of 3–5 m/s. If the gate is very large and heavy the headloss created can be large and this objection can sometimes be met by providing several smaller and lighter flaps in a single bulkhead.

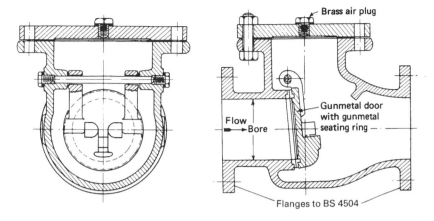

Fig. 13.7   Non-return valve.

13.30 Control valves

The operation of a typical control valve is best illustrated by taking as a first example the standard pressure reducing valve shown in Fig. 13.8. This maintains a pre-set pressure in the main downstream of the valve irrespective of the pressure upstream. Of course, if the pressure upstream is insufficient the valve simply opens wide and no more pressure is available downstream than the upstream pressure less the loss through the valve, which is about $5v^2/g$, where v is the mean velocity at the valve inlet (taking a typical 'k' value of 10 in $kv^2/2g$). Pressure reducing valves of this sort should be sized so that their full open capacity is more than adequate for the desired maximum flow.

A pressure sustaining valve is shown in Fig. 13.9 and operates to keep the pressure upstream of the valve to a given amount. This sort of valve is used, for instance, where a distribution area A feeds a second distribution area B. With a sustaining valve in the feedline, adequate pressure can be maintained in the distribution area A without excessive

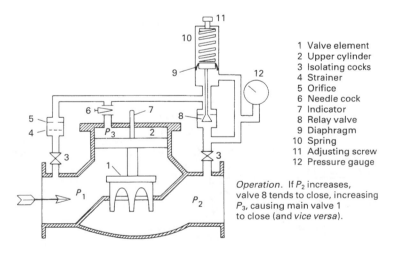

1 Valve element
2 Upper cylinder
3 Isolating cocks
4 Strainer
5 Orifice
6 Needle cock
7 Indicator
8 Relay valve
9 Diaphragm
10 Spring
11 Adjusting screw
12 Pressure gauge

Operation. If P_2 increases, valve 8 tends to close, increasing P_3, causing main valve 1 to close (and *vice versa*).

Fig. 13.8   Pressure reducing valve.

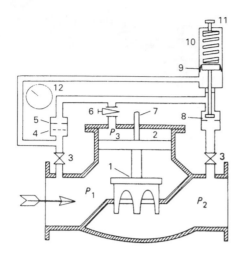

1 Valve element
2 Upper cylinder
3 Isolating cocks
4 Strainer
5 Orifice
6 Needle cock
7 Indicator
8 Relay valve
9 Diaphragm
10 Spring
11 Adjusting screw
12 Pressure gauge

Operation. If P_1 increases, valve 8 tends to open, reducing P_3, causing main valve 1 to open (and *vice versa*).

Fig. 13.9   Pressure sustaining valve.

demand in B pulling the pressure down in A. (In effect A's demand takes precedence over B's demand.) When the valve is full open, however, the pressure upstream passes through to the downstream side, less the loss of head in the valve ($5v^2/g$). This valve can also be used as a pressure relief valve if connected on a branch to a main where the branch, downstream of the valve, discharges to waste.

A constant flow valve is shown in Fig. 13.10. An orifice plate in the main upstream of the valve measures the flow and creates a differential pressure. The valve adjusts so as to maintain a constant differential pressure across the orifice, thus maintaining a constant flow in the pipe.

A variable flow control valve works on the same principle as the constant flow valve illustrated, except that the differential pressure set across the orifice is adjusted manually,

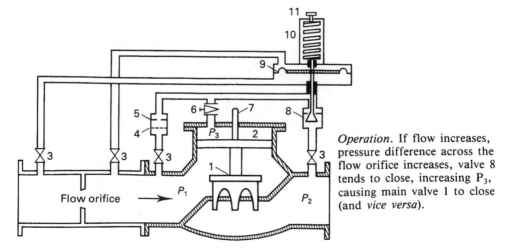

Operation. If flow increases, pressure difference across the flow orifice increases, valve 8 tends to close, increasing P_3, causing main valve 1 to close (and *vice versa*).

Fig. 13.10   Constant flow valve.

or altered to follow some signal received from a controller, so that the flow in consequence is altered. The valve can be fitted with a motor for control of the outlet pressure from a remote point via a telemetry system or for automatic closure in the event of the outlet pressure falling below a pre-set value, i.e. due to a pipe burst.

13.31 Needle (or hollow jet) valves

In a typical needle valve (Howell Bunger type) flow passes around the housing for a conical shaped piston located centrally within the valve body. The flow can be gradually reduced by advancing the position of the piston into a tapered nosing on the upstream side until complete closure is achieved. Plate 30 shows a cross-section through such a valve which is suitable for pressure regulation of flow in situations of high pressure and is free from cavitation and vibration. The valves are particularly suitable for 'free discharge' high pressure duties such as occur during draw-off at the base of a dam. A variant is the submerged discharge valve which is fitted to carry flow down to a ported and sleeved discharge at the base of a concrete pit: water then flows up the annular space between the valve body and the pit, before flowing to the outfall.

13.32 Jet dispersers

All control valves are energy destroying devices, but there is a particular kind of fitting whose sole aim is to destroy the energy of fast flowing water before it falls or is ejected into a river or a basin. These have no moving parts and are known as 'jet dispersers' and are used at the ends of pipelines under high head where water must be given a free discharge. If the full bore jet of water were to be left to fall upon the soft bed of a river then a large hole would be quickly scoured in the river bed. The aim of the jet disperser is therefore to break up the solid flow of the water and disperse it into numerous fine jets or droplets. This is assisted by angled fins in the jet disperser which give the water a 'twist' so that it sprays out over a wide area and the discharge cannot be sustained as a sheet, and breaks down into droplets whose fall velocity is reduced by friction through the air. Jet dispersers are usual on scour pipes, including washouts from mains, when the head on the pipe is likely to be high.

13.33 Air valves

Air trapped in a pipeline will reduce the area of flow, increasing the friction loss and hence reducing the flow capacity. Air trapped at a series of high points on a pipeline profile will have a cumulative effect and may seriously restrict flow. For air to be a problem under normal flow conditions there must be a source of air (for example at poorly designed intakes and pumps), the air must separate or come out of solution (typically as internal pressure reduces along a pumped system) and must accumulate and not be moved along the pipe (which is a function of pipe gradient, diameter and flow velocity). A well-designed system will therefore aim to prevent entry of air and will provide air release points (a) to allow filling and emptying of the main, (b) to avoid flow instability and surge complications due to air pockets, (c) to limit subatmospheric pressures and pipe buckling in the event of a burst, (d) to allow for development of flow up to the ultimate design flow and (e) as a safeguard so that if air does enter, it is not swept along the pipe.

Before a pipeline can be filled with water, means must be provided for releasing air from it. Once the pipe is full of water, however, any aperture for release of air must close so that no water is lost. An air release valve is designed to meet this condition. The orifice through which the air is to pass is circular and below it is a ball of special composition which will float in water (Fig. 13.11). So long as air exists below the orifice the ball will be floating on the water surface below, but as soon as water begins to be released the ball will rise and close the orifice. The design must, however, take several other factors into account, for example the ball must not be sucked on to the orifice by high velocity air and nor must the ball vibrate up and down. It is usual to provide any large pipeline with two sorts of air release valve, one being a 'large orifice' air valve designed to release large quantities of air when a pipeline is being filled with water and the other a 'small orifice' type which is designed for continuous operation, releasing small quantities of air as they collect in the main. Double orifice air valves comprise one of each type in the same unit. Their price is not materially different from a single air valve and after allowing for the cost of a tee, an isolating valve, installation and a chamber, it is sometimes common to install only 'double air valves' except at particular locations such as reservoirs and pump manifolds.

Apart from air which must be discharged to permit the filling of a pipeline, air can be introduced from pumps and from the release of air in solution in the water as its pressure reduces. The latter amount depends upon the pressure reduction and other factors, but water at standard conditions contains 2% dissolved air by volume. Whilst it is obvious that some air will collect at high points, there is also always a likelihood of air collecting at changes of gradient in the pipeline (see Fig. 13.12). The air is not necessarily shifted by the water flow, especially on a downhill gradient where air at the soffit of the pipe may not be able to travel backwards against the flow. Test data[7-12] show that air tends to move as

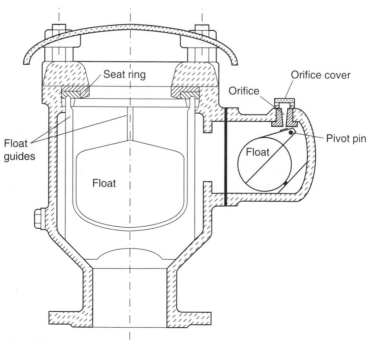

Fig. 13.11   Double orifice air valve.

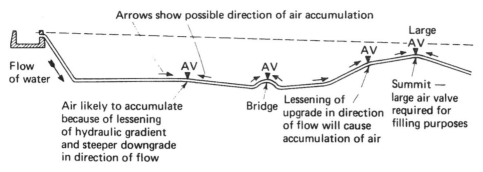

Fig. 13.12 Diagram showing desirable positions for air valves on a length of pipeline.

pockets, which can be carried forward on a downward gradient but will rise against the flow when the gradient is steeper than a critical value, which can be calculated for gradients less than about 1° as about 1 in 120 D/V^2, where D is the pipe diameter in metres and V is the water flow velocity in m/s. For example, an air pocket in a DN600 pipeline laid at 1 in 500 downhill would travel uphill against a flow of about 0.38 m/s. This implies that air can accumulate and therefore requires air release valves at quite shallow changes of gradient, particularly where flows may be less than design peak flows. Caution is required in assessing gradient, to allow for tolerances and pipe movement. Air bubbles will be carried forward at gradients typically about seven times steeper. Air pockets imply a reduction in pipe carrying capacity and potential for instability during any surge. Any movement of air along a pipeline is slow: when filling a 2 km length of main, for instance, although the flow from some downstream hydrant may soon be steady, it will be interrupted by occasional bursts of air for perhaps half an hour or more.

When filling a main it is common practice to open all fire hydrants along the length of line to be filled, shutting them only when they cease to discharge air. Whilst large orifice air valves could be used instead, the fire hydrants permit the discharge of dirty water to waste and also permit sampling of water quality. Therefore it is a debatable point as to whether hydrants or large orifice valves should be provided for filling a main.

It is seldom necessary to put small orifice air valves on distribution mains since air will be removed via the service pipe connections which should be soffit connected. It is only necessary to put small orifice valves on distribution mains where there is a sudden hump in the pipeline, such as when it is laid over a bridge. Where the location of an air valve is not possible in the road (as cover at high points is usually a minimum), consideration can be given to taking a tee off the high point to an air valve in the verge. For similar reasons of cover, air valve chambers may need to be raised above ground level to provide space.

Bearing the above comments in mind, positions for air valves might be summarised as follows.

(1) Double orifice air valves are required:
    (a) at high points where air must emerge to permit filling – or air enters to allow emptying – and at high points relative to the slope of the hydraulic gradient;
    (b) where a falling pipeline steepens its gradient;
    (c) where a rising pipeline flattens its gradient;

(d) at intervals on long lengths of pipeline (with no fire hydrants) where necessary to permit the release of air for filling the pipeline in convenient sections, generally at 2 km intervals or thereabouts;

(e) in each section where a long pipeline is divided into sections by intermediate line valves.

(2) Single orifice valves may be required additionally:

(f) on access manhole covers and other local 'humps' where air may collect;

(g) at some suitable point a short way downstream of any pump;

(h) after a pressure reducing valve if there is a substantial reduction of pressure;

(i) (possibly) at 0.5 km intervals on downward legs of pumping mains if the pump may have entrained air and on substantially horizontal mains at 1–2 km intervals, especially when the flow velocity tends to be low, account being taken of small orifice valves provided under (1) and the presence or absence of service pipes connected to the soffit.

Not so many small orifice valves are required towards the ends of a system as over the beginning lengths where the aim is to ensure early release of air. Large orifice valves are usually sized to limit sub-atmospheric pressures in the event of a burst – and according to the estimated rate of filling or emptying of a main. Fast filling is inadvisable. For a large main over 10 km long a filling time of 3 hours would be reasonable, with shorter lengths filled in not less than 1 hour: time also being linked to resources and supervision. The flow capacity of an air valve needs to be taken from manufacturers' catalogues, bearing in mind that the air pressure within the main will be low. Small orifice valves have an orifice size which is related to the operating pressure in the main: the higher the pressure the smaller must be the diameter. It would be usual to ensure that the total small orifice capacity provided on the system is sufficient to release air at 2% by volume of the water flow. Again capacities of valves should be taken from manufacturers' catalogues, but in this case the pressure taken is the working pressure in the main at the location where the air valve is fixed.

An isolating valve (sluice valve, butterfly valve or stopcock) should be sited below each air valve, thus making it possible to remove the air valve for repairs without shutting down the main. The restriction effect of valves and fittings between the main and the air valve must be taken into account in assessing air valve capacity as installed. Orifice sizes may be smaller than the connection size and this coupled with differences in valve design and test conditions or performance assumptions may give different values for valves of the same apparent size. It is therefore important to specify performance. On large pipes, air valves may be fitted to a blank flange on a DN 600 or larger tee to allow access to the pipeline. Care should be taken to site all air valves above the highest possible groundwater level that can occur in any pit; if this is not done polluted water may enter the main via the air valve when the main is emptied. On steel pipelines anti-vacuum valves may be essential to admit air to prevent the build-up of sub-atmospheric pressures which may cause the pipeline to collapse when water is drawn off by emptying the main from a lower point. It is important also to check the need for this type of protection in analyses for surges on large diameter steel pipes. An air valve of the large orifice kind can be used as an anti-vacuum valve.

Numbers and sizes of air valves require calculation: typical large air valve sizes for various pipe diameters are:

Pipe diameter (mm)	Large air valve branch diameter (mm)
DN ≤ 250	DN 50–65
DN 250–600	DN 80–100
DN 600–900	DN 150
DN 900–1200	DN 200
DN 1400–1800	2 number DN 200

13.34 Valve operating spindles

A valve must often be placed so deep that it cannot be reached with the normal length of tee key so that an extension spindle must be arranged for the valve, running in brackets rigidly attached to the chamber walls. These extension spindles may be obtained for the exact length required. It is worthwhile making spindles which are to be immersed in water, such as those for operating valves inside a reservoir, of manganese bronze so that they are proof against corrosion. There is, unfortunately, no standard direction of rotation for operating valves, but the most usual method is for clockwise rotation of the handwheel or tee key to cause the valve to close. Clear labelling of the operating direction is essential.

13.35 Washouts

Despite the name, a washout is seldom used for scouring or 'washing out' a main because its diameter is usually too small to create sufficient flow velocity in the main to wash out debris; its principal use is for emptying a main or for the removal of stagnant or dirty water.

Primary washouts may be installed to drain to a watercourse the majority of the length between section valves. Secondary washouts of smaller diameter can then be used to empty undrained subsections as required. Sizes, particularly primary washouts, need to be calculated according to the draindown time – which is decided by the number of washouts, limits on discharge to watercourses, valve spacing, access and resources, but typically to allow a section to be substantially drained within a working shift.

Typical minimum sizes adopted for secondary washouts are:

Main pipeline diameter	Washout branch diameter
Up to 300 mm	80 mm
400–600 mm	100 mm
700–1000 mm	150 mm
1100–1400 mm	200 mm
1600 mm and larger	250 mm

In open country it is usual to install washouts at every low point with additional washouts being provided in each section where a main is sub-divided into sections by stop valves. Each washout should discharge, wherever possible, by gravity to the nearest watercourse. The discharge should be to a concrete pit with overflow to the watercourse in order to prevent scour from the high velocity discharge. The washout branch on the main should be a 'level invert tee'. In flat country washouts should be spaced 2–5 km apart, depending on pipeline gradients and valving. Where it is not possible to get a free discharge to a watercourse the washout will have to discharge to a manhole from which the water can be pumped out to some other discharge point.

In distribution systems principal feeder mains are usually provided with washouts wherever this is convenient, regard being paid to the position of valves on the main and any branch connections and to the need to be able to empty any leg of the main in a reasonable time of one or two hours. On small mains, washouts are not normally provided because fire hydrants can be used for this purpose, but it would be usual to lay a specific washout to empty a part of the system where a convenient watercourse exists. Washouts are also placed at the end of every spur main; these usually comprise fire hydrants even though they may not be officially paid for by the fire authority and designated as such. They are operated regularly to sweeten the water at the end of the main.

Care is essential in the design of washouts since, under high heads, the velocity of discharge can be very high and the consequent jet discharge destructive and dangerous. Manholes receiving the jet should be of substantial construction and valves should be lockable and slow opening for high heads. Two valves, one guard and one operating may be installed where prolonged throttling at high head is required.

13.36 Valve chambers

A typical simple valve chamber is shown in Fig. 13.13. It should be noted that the valve is anchored on the upstream side having a Viking Johnson flange adapter on the downstream side which permits the valve to be removed. The valve body is therefore free to expand downstream when the valve is closed under pressure and it must therefore be specified for the open end conditions (see Section 13.26). The chamber cover can be made of pre-cast reinforced concrete beams or slab which can be removed for lifting out the valve. Removal of a valve should not be necessary for many years so it is frequent practice to pave or earth over the cover to the chamber, leaving only the access manhole and the spindle manhole exposed. Every effort should be made to avoid siting the chamber in a road; it should preferably be in the road verge, but even there the cover should be strong enough to take heavy vehicles which sometimes run off roads. The thrust or anchor block must be designed according to the ground conditions so that it is capable of taking the full thrust when the valve is closed. Two flexible joints, with an intervening 'rocker pipe' at least 1 m long, should be provided on either side of the chamber to avoid damaging pipework if any differential settlement of the pipeline, relative to the chambers, should occur.

Small valves, i.e. those DN 300 and smaller, are not usually installed in purpose built chambers. The valve is backfilled around the body to the spindle and then a pre-cast concrete or brick pit is set round the upper part, or a pipe (DN 80/100) is fitted as a sleeve for the spindle, and accessed through a surface box for tee key operation. Valves up to and including DN 600 may also be buried, depending on location and preference: the choice being on maintenance and the civil works required in accessing the valve.

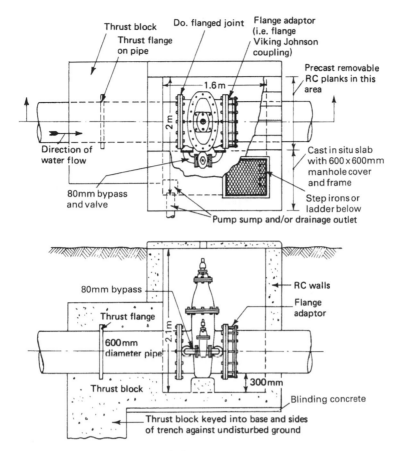

Fig. 13.13 Typical sluice valve chamber design.

Bibliography

WRc publications

Pipe materials selection manual – water supply, 2nd edition, June 1995
Manual for PE pipe systems, 2nd edition, July 1994
Manual for PVC pressure pipe systems, 2nd edition, July 1994
Service pipes manual, (Farrage, R.), WRc report FR 0355, WSA/FWR, 1993

Water Industry Standards (WIS)
IGN 4–08–01–1994: Bedding and sidefill materials
WIS 4–08–02–1994: Bedding and sidefill materials
IGN-4–31–01–1986: PVC-U pipes
WIS-4–31–06–1994: PVC-U pipe
WIS-4–31–07–1994: PVC-U fittings
WIS-4–31–08–1991: MOPVC pipe
WIS 4–32–03–1987: Specification for blue polyethylene (PE) pressure pipe for cold potable water (nominal sizes 90–1000 for underground or protected use)

610 Pipes, pipeline construction and valves

WIS 4–32–06–1989: Specification for polyethylene electrofusion couplers and fittings for cold potable water supply for nominal sizes up to and including 180
WIS 4–32–08–1989: Specification for site fusion jointing of PE80 and PE100 pipe and fittings
WIS 4–32–09–1991: Specification for black polyethylene pressure pipes for above ground or sewerage (nominal sizes 90–1000)
WIS 4–32–13–1993: Specification for blue higher performance polyethylene, HPPE/PE100, pressure pipes nominal sizes 90–1000 for underground or protected use for the conveyance of water intended for human consumption
WIS 4–32–14–1995: Specification for PE80 and PE100 electrofusion fittings for cold water supply for nominal sizes up to and including 630
WIS 4–32–15–1995: Specification for PE80 and PE100 spigot fittings and drawn bends for nominal sizes up to and including 1000
IGN 4–37–02: Design against surge and fatigue conditions for thermoplastic pipes

British Standards, London

BS EN 512: Asbestos cement pipes
BS EN 545: Ductile iron pipes, fittings, accessories and joints for water pipelines – Requirements and test methods
BS EN 593: 1998 Butterfly valves
BS EN 681–1: Elastomeric seals – material requirements for pipe joint seals
   Part 1 Vulcanised rubber, including amendment A1, 1998
BS EN 1092: Flanges
BS EN 1295–1: Structural design of buried pipelines under various conditions of loading, 1998
BS EN 60534–2–3: 1998, Industrial process control valves
BS 534: Steel pipes, joints and specials for water and sewage, 1990
BS 3505: Unplasticised PVC pressure pipes for cold potable water, 1986
BS 3600: Specification for dimensions and masses per unit length of welded and seamless steel pipes and tubes for pressure purposes
BS 3601: Steel Pipes and Tubes for Pressure Purposes, 1987
BS 4504: Section 3.1 Flanges and Bolting for Pipes,
BS 4625: Pre-stressed concrete pressure pipes, 1970
BS 5163: Sluice valves for waterworks purposes
BS 5480: Glass Reinforced Plastics (GRP) Pipes, Joints and Fittings, 1990
BS 5500: Specification for unfired fusion welded pressure vessels
BS 6349: Maritime structures
   Part 1: General criteria
   Part 7: Guide to the design and construction of breakwaters
BS 7079: Preparation of steel substrates before application of paints and related products
BS 7361: Cathodic protection
BS 8010: Code of Practice for Pipelines
   Part 1: Pipelines on land: general, 1989
   Part 2: Pipelines on land: design, construction and installation:
      Section 2.1 Ductile Iron, 1987
      Section 2.2 Steel (for water and associated products) (not published)
      Section 2.3 Asbestos Cement, 1988
      Section 2.4 Pre-stressed Concrete Pressure Pipelines, 1988
      Section 2.5 Glass Reinforced Thermosetting Plastics, 1989

Section 2.6 PVC *(in draft)*
Section 2.7 Pre-cast concrete, 1989
Section 2.8 Steel (for oil, gas and associated products), 1992
Part 3: Pipelines subsea: design construction and installation, 1993
CP 312: Code of Practice for Plastics Pipework
Part 1: General principles and choice of material
Part 2: Unplasticised PVC pipework
Part 3: Polyethylene pipes
CP 2010: Code of Practice for Pipelines
Part 2: Design and construction of steel pipelines on land, 1970

European standards

EN ISO 12162: 1995, Thermoplastics materials for pipes and fittings for pressure applications – Classification and designation – Overall service (design) coefficient

European standards in preparation

pr EN 805: Water supply – Requirements for systems and components outside buildings
pr EN 1452: Plastics piping systems for water supply – unplasticised poly vinyl chloride (7 parts)
pr EN 10224: Steel pipes, joints and fittings for the conveyance of aqueous liquids including potable water
prEN 12201: Plastics piping systems for water supply, polyethylene.

American Waterworks Association (AWWA)

AWWA M9: Concrete pressure pipe
AWWA M11 Manual: Steel Pipe – A Guide for Design and Installation
AWWA M45: Fiberglass pipe design
AWWA C205: Cement-mortar protective lining and coating for steel water pipe – 4 in (100 mm) and larger – shop applied
AWWA C602: Cement-mortar lining of water pipelines in place – 4 in (100 mm) and larger
AWWA C906–90: PE pressure pipe and fittings
AWWA C950: Fiberglass pressure pipe

American Society of Mechanical Engineers (ASME)

ASME B31.4: Liquid transportation systems for hydrocarbons (etc.), 1992
ASME B31.8: Gas transmission and distribution piping systems, 1992

American Petroleum Institute (API)

API 5L: Specification for line pipe

Construction Industry Research and Information Association (CIRIA), London

Thrust blocks for buried pressure pipelines, Report 128, 1994
Design and construction of buried thin walled pipes, Report 78, 1978

Manual on the use of rock for in coastal and shoreline engineering, CIRIA special publication 83/CUR report 154, 1991

US Army Corps of Engineers

Shore Protection manual, 1984, Vols I and II. Coastal Engineering Research Centre. US Government Printing Office, Washington, DC 20402

Permanent International Association of Navigation Congresses (PIANC)

Guidelines for the design of armoured slopes under open piled quay walls, Report of Working Group No. 22 of the Permanent Technical Committee II, Supplement to Bulletin No. 96, 1997 (October).

Det Norske Veritas (DNV), *Oslo*

Rules for submarine pipeline systems, 1996

References

1. Young O. C. and O'Reilly M. P. *Simplified Tables of External Loads on Buried Pipelines*. HMSO, London, 1986.
2. Young O.C. and O'Reilly M. P. *A Guide to Design Loadings for Buried Rigid Pipes*. HMSO, London, 1983.
3. Clarke N. W. B. *Pipelines – A Manual for Structural Design and Installation*, 1968.
4. Janson L.-E. *Plastics Pipes for Water Supply and Sewage Disposal*. Borealis, Stockholm, 1995.
5. Little M. J. New Pipelines on Land and Across Hong Kong Harbour. *JIWES*, June 1986, pp. 271–287.
6. Little M. J. and Duxbury J. A. Tolo channel submarine pipelines, Hong Kong. *Proc ICE*, Part 1, 1989, pp. 395–412.
7. Ervine, D. A. Air Entrainment in Hydraulic Structures: A Review, *Proc ICE, Water, Marit & Energy*, 1998, 130, Sept., pp. 142–153.
8. United States Department of the Interior. Engineering monograph No. 41, *Air-Water Flow in Hydraulic Systems*.
9. Lescovitch J. E. Locating and Sizing Air Release Valves. *JAWWA*, Water Technology/Distribution, July 1972.
10. Edmunds R. C. Air Binding in Pipes. *JAWWA*, Water Technology/Distribution, May 1979.
11. Kalinske A. A. and Robertson J. M. Closed Conduit Flow, *Transactions ASCE*, **108**, pp. 1435–1516.
12. Liseth P. Luft i utsslipsledninger (Air in outfalls), Prosjektkomiteén for rensing av avløpsvann, PRA report 8, NIVA (Norwegian Water Institute), Oslo, 1975 (in Norwegian).

14

Pipeline and distribution system design and analysis

14.1 Introduction

The pipelines of a distribution system can be divided into three functional categories:

(1) *trunk mains* which convey water in bulk from the source, usually to a service reservoir;
(2) *principal feeder mains*, or principal mains, which convey relatively large quantities of water from the service reservoir into demand areas;
(3) *distribution mains* which supply water to consumers' connections.

Trunk mains supply the daily quantity of water required and have relatively constant flow rates because the service reservoir storage is used to even-out the hourly changes of demand over 24 hours. However there will be day-to-day variations of demand and trunk pipelines must be designed to carry the maximum day demand because this often extends with little change over a week or more (see Figure 1.5). Such maximum demands are frequently in the range 120–140% of the annual average daily demand. Consideration may also have to be given to the possible use of a trunk main for re-routing supplies should some other trunk main suffer a burst, and for use of the main for transfer of water between zones of the distribution system to achieve a balanced storage in service reservoirs on a weekly cycle of demand.

Principal feeder mains are designed for peak hour demands but, because they normally convey water to large areas, the peak hourly factor applying to them is less than that applying to individual distribution mains because a 'diversity factor' applies, related to the size of the area served (see Section 1.24).

Distribution mains are designed to meet the hourly variation of consumers' demand, which, as Section 1.25 indicates can vary between 200 and 300% or more of the annual average annual daily demand. A typical hourly variation of demand over 24 hours is shown in Section 11.2. Urban, industrial and rural distribution systems exhibit different diurnal demand patterns which also vary from one year to the next. The system must have adequate pressures to meet the peak hour demand, but at night, during low demand periods, pressures need to be kept as low as practicable to minimise leakage. A good interconnection of distribution mains is desirable with dead ends eliminated wherever practical, in order to avoid long detention times for the water within the mains, which can result in water quality deterioration. Exceptionally high flow rates and reversals of flow in mains also need to be avoided where possible, because they can cause water quality deterioration through scour and suspension of interior water main deposits.

14.2   Asset management of pipeline

The pipelines of the distribution system normally comprise the largest capital asset an undertaking possesses. The cost of maintaining this asset in good condition has now become a major financial consideration for water undertakings for several reasons. First, the long life of pipelines, 60–100 years, has meant the problem has not been urgent in the past, but with many mains now approaching their serviceable life, the high cost of their rehabilitation or replacement is now seen as a large financial liability lying ahead. Second, the privatisation of undertakings in England and Wales has meant that, instead of relying on public funding as and when necessary to replace mains, the companies must fund asset renewals from their own resources. Third, stricter requirements concerning levels of service that must be provided to consumers imposed under privatisation (see Section 14.3), and stricter water quality requirements under EU and UK legislation, have meant extra measures have been necessary to ensure the condition of the distribution system does not cause the quality of the water to deteriorate before it reaches consumers. These requirements apply equally to the growing number of cases where private companies have been engaged to operate water undertakings under concessionary agreements (see Section 2.10).

It was a condition of the appointment of private water companies in England and Wales in 1989, when the water industry was privatised, that they should base their capital expenditure programme on an 'asset management plan'. Such plans were required to estimate the yearly expenditure necessary to maintain the underground assets in a condition sufficient for the company to meet its obligations for a period of not less than 15 years ahead. In practice a 20-year period divided in 5-year periods is used, the first 5-year plan being produced in detail. As a consequence the companies now produce asset condition assessments, which forecast the cost liability such assets pose as their performance or condition deteriorates, or their current hydraulic capacity is insufficient to meet foreseeable rises of demand. The assessment process takes into account the type, age, condition, leakage and burst frequency of mains; their flow, pressure and water quality performance, and hence the potential length of asset life remaining and when the need for rehabilitation or replacement is likely to occur. For the distribution system alone, the work of asset assessment represents a task of considerable magnitude because of the thousands of kilometres of mains involved and also the need to take account of the state of control valves, pumping and treatment equipment, etc..

To aid the work, comprehensive mains record drawings are increasingly held on electronic mapping systems. Pipework data and system performance information may also be held electronically in one or more databases. Data management and processing activities can be simplified and analysis conclusions more easily derived if the mapping and databases are integrated within a single geographic information system (GIS). Further comments on this use of computer software for analysing the behaviour of a distribution system are given in Sections 14.14 and 14.15 below.

Asset management planning of a distribution system necessitates the study of many options. Water mains can be refurbished in a number of ways (see Sections 15.8 and 15.9); but it may be more economical in the long term to replace an old main with a larger one which will meet forecasted rising demand and last for a longer period. Re-routing of pipeline flows may offer other alternatives. New mains can bring benefits in reducing leakage, improving pressures, and reducing bursts. Economic studies can also lead to other solutions, such as upgrading or automating pumping plant, or adopting some re-zoning of supplies. The latter can promote better pressure regimes, aid leak detection measures, and reduce problems of water quality deterioration through the distribution mains. All this work has to repeated at intervals to keep the asset renewal plans and costs up-to-date.

14.3 Service levels

Service levels are the standards of supply which a water undertaking affords its customers. The standards can be targets for achievement set by a water undertaking for itself, or set by some outside authority, such as the service targets set by the Director General of Water Services (OFWAT) for the privatised water companies in England and Wales. Levels of service are often set by, or agreed with, international funding agencies such as the World Bank, to define what some programme of rehabilitation or improvement of an undertaking should achieve. Records of how far such levels of service have been achieved in any year can act as performance indicators.

The three principal levels of service which relate to the performance of distribution systems are the following.

Hydraulic performance defines the minimum pressure and flow domestic consumers should experience. The reference level of service required by OFWAT when demand is not abnormal is a flow of 9 l/min at a pressure of 10 m head on the customer's side of the main stop tap (at the property boundary). In practice the pressure is difficult to measure at this point and therefore a 'surrogate' pressure of 15 m in the main supplying the property may be used. The use of lower surrogate pressures is acceptable only where justified with supporting evidence. Where two properties are supplied through a common service pipe the pressure reference level is the same for twice the flow. For common services feeding more than two properties, the increase in minimum mains pressure is linked to the size of the common service, number of properties supplied and the concept of 'loading units' defined in BS 6700.[1,2]

Continuity of supply is measured by the number, duration and circumstances relating to interruptions or deficiencies of supply. OFWAT uses a scoring system for assessing water companies' performance in England and Wales. The undertaking is required to report the number of properties affected by interruptions to supply of more than 6, 12 and 24 hours by the four categories of:

- unplanned interruptions due to bursts, etc.;
- planned and warned interruptions due to planned maintenance, new connections, etc.;
- unplanned interruptions caused by a third party, for example another utility damaging a water pipe while excavating for their own service; and
- unplanned interruption due to overrun of a planned and warned interruption.

Under certain circumstances customers affected are entitled to financial compensation, typically £10 for each event plus a further £10 for each 24-hour period the supply remains interrupted.

The Director General of OFWAT has also stated the maximum frequency he thinks is acceptable for restrictions of supply to consumers by water companies in England and Wales, e.g. hosepipe bans not more than once in 10 years, restriction on non-essential uses of water once in 20 years, and so on – details are given in Section 3.15. However these targets are not mandatory and during recent droughts hosepipe bans have been imposed by a number of the undertakings.

It should be appreciated that the requirements quoted above apply to privatised water companies in England and Wales. Similar targets can be appropriate where the operation and maintenance activities of a public water supply system have been contracted out to a company or private firm for a term of years. Where water supply undertakings are publicly owned (as in the majority of cases throughout the world), similar target service levels may be the aim, but their achievement is often not possible in many instances because of lack of resources. For the many water undertakings in countries which are not able to afford a 24-hour supply, a frequent aim is to achieve 4-hour supply morning and evening, with standpipes for low income householders being widely used.

Water quality standards. Standards for the quality of water supplied, together with sampling and reporting requirements are set by legislation in most of the developed countries (see Section 6.47). They universally apply to the water reaching consumers' taps. Hence though the quality of water leaving a treatment works may be acceptable, any decline of quality as it passes through the distribution system must be taken into account. All waters contain nutrients and long retention in mains can promote 'aftergrowth', i.e. increase of bacteria and small organisms in the mains. Taste, odours, and suspended matter in the water can occur. Thus, the design of the distribution system has to minimise these effects by ensuring good circulation of water, flow velocities kept within an acceptable range, short retention times, and cleanliness of the interior of mains. Sometimes it is found necessary to add a further disinfectant (usually chlorine) to the water at service reservoirs in order to inhibit aftergrowth in mains.

To assist in meeting the above requirements, particularly for the maintenance of water quality, water companies in UK are increasingly reorganising their distribution networks into separate 'stable' supply areas. Each area is preferably fed by only one source; but if two sources of different waters have to be used, the supplies need first to be blended in a predefined ratio to obtain a consistent water quality. Flows into an area and pressures within it need to be monitored and controlled to ensure adequate supplies to consumers and to aid leakage and waste detection measures. The operation of boundary valves connecting to adjacent areas and key valves may need to be put under control, in order to prevent unauthorised changes of valve setting affecting the quality of the supply.

14.4   Factors in pipeline design

It is often necessary to decide what size of main to lay to meet an estimated new or extra demand and its possible increase in the future. If the choice of diameter proves too small, a second main may have to be laid some years later; hence one needs to know for how many years ahead a new main should be laid.

For a given hydraulic gradient, the flow, Q, through a circular pipe is proportional to d^x, where d is the diameter and x has a value of 2.65 according to the Hazen–Williams formula (see Section 10.5). Hence increasing the diameter of a pipe by only 30% doubles its flow capacity. Consequently if it is thought the initial demand, Q, on a pipeline might double in n years' time, the question is whether to lay a pipeline for $2Q$ now, or lay one for Q now and duplicate it in n years' time. Since pipeline costs are roughly proportional to diameter up to about 600 mm diameter (provided wall thickness are also proportional), it is only worthwhile duplicating a main if the 'present value' (see Section 2.16) of duplication is less than the present value of the main for $2Q$ now. For this to be so

$$\text{(Present value) } £X + £X/(1+r)^n \leq 1.3£X$$

where £X is the cost of a main for Q; n is the number of years before the duplicate main is required; and r is the discount rate. This gives:

$$1/(1+r)^n \leq 0.3$$

which shows that, for a discount rate of 8% n must > 15 years; for 10% $n > 12$ years; and for 12% $n > 10$ years.

Such a calculation is only one factor to be taken into account when deciding what policy to follow. Many other considerations apply such as:

- the difficulty of forecasting future demand and its rate of rise;
- whether routes are possible for two pipelines, or the second might have to be longer;
- whether two pipelines would offer better security for maintenance of supplies;
- and various further factors discussed in Sections 14.5–14.7.

14.5 Pipeline planning

In pipeline planning to meet future demand increases it may be advantageous to lay a pipeline in two lengths of different diameter. For example in the case illustrated in Fig. 14.1, existing sources (1), (2) and (3) feed water into the principal network shown. A new source (4) is proposed to bring, say, another 80 ml/day into the urban area via pipeline AB. Possible areas of substantial growth of demand are labelled D1, D2, ..., D5 but, as often occurs in practice, it is uncertain to what extent each will develop and in what order.

If it is not known whether the most of the future extra demand will arise to the east or the west of the present system, it could be a false move to lay the pipeline for the full

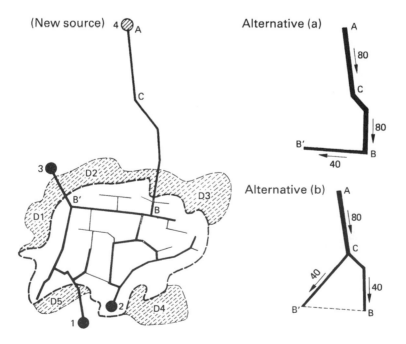

Fig. 14.1 Trunk mains alternatives.

80 Ml/day to point B on the east (alternative (a) in Fig. 14.1). In that case, extra demand on the west side (areas D1, D2, D5) would have to be met through existing main BB', increasing the loss of head to B' and perhaps making it necessary to lay an extra main along BB'. A better strategy is to bifurcate the main at C, so that additional supplies can be provided at both B and B', the existing pipeline BB' need then only carry the imbalance of demand on either leg. A ring main is formed which is always advantageous in ensuring at least some supply if there is burst on either part of the ring, and pressures are maintained better.

Pipeline planning comprises finding a strategy that best covers all the reasonably feasible demand options, maintaining flexibility in case development does not follow the current predictions. For this purpose ring main formations, and avoidance of any need to duplicate mains back to the source later, are useful precautions. The various alternatives have to be worked out, taking into account such matters as proposed development plans for new housing and industry, where uncontrolled developments might occur, the peak outputs of existing sources and differences in their production costs, possible locations for future sources, service reservoir capacities and fire risks. There is also the need to plan for separate or 'discrete' zones or district areas, fed only by one type of water so that water quality problems can be better controlled, and preferably supplied only by one or two mains which are metered so that distribution losses can be monitored. The exercise can be complex, involving a logical analysis of all the options to devise the best solution.

14.6 Design of a pumping main

A pumping main commonly delivers water from a pumping station to a service reservoir, and is therefore usually operated at a constant flow rate, the output being managed by varying the hours of pumping. The designer aims to find the diameter of main for which the total of capital expenditure on pumping plant and pipeline together with the cost of power consumption and other operating costs is a minimum. The following simplified example shows the type of approach used.

Example. A main is to convey 20 Ml/day (average) and 22.5 Ml/day (maximum) against a static lift of 55 m through 16 km of main (see Fig. 14.2). Costs are: pipeline £0.35 per mm diameter per m laid in rural conditions; pumps £3000 per kW installed with 50% standby required; power 4.5 pence/kWh. Capital repayment charges assumed 12%, and

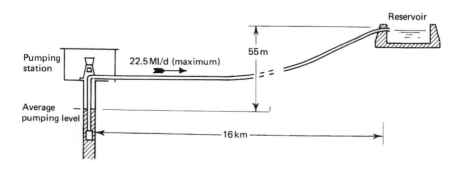

Fig. 14.2 Example for pumping main calculations.

friction coefficient of the pipeline k = 0.5 mm and water viscosity at 15°C in the Colebrook–White formula.

	Diameter of main (mm)			Notes
	500	600	700	
(a) Static lift (m)	55	55	55	
(b) Friction loss at 20 Ml/day (m)	46	18	8	
(c) Friction loss at 22.5 Ml/day (m)	59	22	10	
(d) Maximum water power (kW)	287	197	166	*22.5 Ml/day at head $(a + c)$
(e) Installed motor power (kW)	590	406	341	Allow for 73% efficiency of (d) + 50% standby
(f) Average power used (kWh)	313	226	196	*20 Ml/day at head $(a + b)$ at 73% efficiency
Capital costs: (£1000)				
– pipeline	2800	3360	3920	
– pumps	1771	1217	1024	
Total capital cost	4571	4577	4944	
Annual costs: (£1000 per annum)				
– 12% on capital	549	549	593	
– power at 4.5p/kWh	123	89	77	$(f) \times 24 \times 365$ at £0.045
Total annual costs	672	638	670	
Present value cost for 25 years at 12% discount: (£1000)				
– Initial capital	4574	4577	4944	
– Renewal of pumps at 20 years	184	126	106	Note (1)
– Power costs for 25 years	1016	735	636	Note (2)
Total present value	5774	5438	5686	

Notes.
*Power in kW = $0.1135Qh$ where Q is in Ml/day and h is in metres.
(1) PV = Cost $\times 1/(1 + r)^{20}$ where r = 0.12 (see Sections 2.14 and 2.16).
(2) PV = annual power cost $\times [(1 + r)^{25} - 1]/r(1 + r)^{25}]$. The limit is taken as 25 years because longer forecasts are of doubtful validity.

Both methods of comparison, annual cost and present value cost, show the 600 mm main is cheapest, although the difference is not large. In practice a more detailed analysis would be adopted to allow for:

● the growth of flow over years 1–25;
● the probable increase of pipeline friction with time;
● more accurate costing of pipelines, pump, valves, electrical and surge protection equipment, etc. and inclusion of their maintenance costs;
● the effect of any possible future increased price of power;
● actual expected terms for capital repayment.

Using a spreadsheet or a programmable calculator, the calculations can be extended to evaluate the optimum year for possible phasings and for a range of different parameter values.

There will be other strategic options to consider, such as alternative demand forecasts, possible phasing of capital expenditure, whether high efficiency pumps, or full automation would be worthwhile, and so on. The choice of the preferred solution may only emerge after consultation with the client and specialists to ensure that no hidden engineering or operational problems are overlooked, the route for the proposed pipeline should be walked to see that it is practicable, and some preliminary layout drawings may be necessary to ensure the costing is realistic.

14.7   Comments on the use of discounting

Sections 2.16 and 2.17 set out some of the problems of discounting. A high rate of discount devalues future expenditure more than a low rate of discount. Hence high discount rates tend to favour schemes that have low initial capital cost even though this may increase later running costs. This can be justified on the basis that high discount rates occur when capital is scarce; and when capital is scarce it is not unreasonable to require projects to be economical in their use of capital funds. However, discounting is becoming less frequently used in the UK because the method makes assumptions about future costs which are not borne out by events. For example the relationship between capital and operating costs may be different in the future from that pertaining at present. New manufacturing techniques can keep rises in plant and building costs down, but social changes can cause labour costs to rise proportionately more. Thus a project which is chosen because it is the 'most economical' in terms of its 'present value' at present-day costs, can prove less economical when relative prices change.

Of course present value calculations can be based on a range of discounts, and possible future changes in relative costs can also be tried out. Although this results in a less precise answer, it can reveal projects which are unlikely to be economic under any reasonably foreseeable circumstances and which can therefore be eliminated from further consideration. It can also show how some policy of rephasing or redesign can offer a financial advantage. Thus discounting is an aid to decision-making, not the basis of it. In practice a legion of other factors usually determine the choice of scheme, such as its environmental consequences, or in many cases 'how much can be afforded now?' (another form of 'present value' estimating). Discounting is useful, but not the arbiter of choice. Over-riding it are many technical matters that need to be taken into account.

14.8   Design of a gravity main

This section illustrates the design of a gravity trunk main connecting an impounding reservoir to a service reservoir.

Example. The proposed impounding reservoir 30 m deep will have a top water level of 410 m OD. From this reservoir 60 ml/day is to pass through rapid gravity filters located at the foot of the dam and then into a balancing reservoir beneath. From the balancing reservoir the water is to be conveyed 35 km to a service reservoir from which a distribution system of 4 km extent is fed. The ground elevations of the distribution system vary from 10 to 50 m OD. The distribution pipes are suitable for 100 m head working head. The proposed pipe profile and leading dimensions are given in Fig. 14.3.

The conditions at the headworks are therefore:

Minimum level in impounding reservoir		380 m OD
less		
– loss through filters	5 m	
– loss through balancing tank	4 m	
– allowance for other local losses	4 m	
Total losses	13 m	
Head at inlet of trunk main		367 m OD

At the downstream end of the gravity trunk main the water level in the service reservoir needs to be between the following limits

A	*Minimum water level*	
	Minimum service level pressure required	35 m
	Maximum ground level in supply area	50 m OD
	Assume distribution losses @ 1.5 m/1000 m for 4 km	6 m
	Hence service reservoir minimum bottom water level	91 m OD
B	*Maximum water level*	
	Maximum pipe working pressure	100 m
	Minimum ground level in supply area	10 m OD
	Hence service reservoir maximum top water level	110 m OD

A site for a service reservoir must therefore be found where the top water level can be not higher than 110 m OD and the bottom water level is not below 91 m OD. An elevation nearer the upper limit would be prudent in order to provide spare head on the distribution system to allow for unexpected growth in demand. Hence a service reservoir top water level of 110 m OD is taken.

The trunk main conditions are therefore:

Head at inlet to gravity main		367 m OD
Inlet level to service reservoir	110	
Allow for service reservoir inlet losses, say	6	116 m OD
Hence allowable loss in 35 km trunk main		251 m

This amount of head loss is very high, raising the following problems:

- the lower end of the pipeline would come under 251 m static head, necessitating expensive pipes;
- the control and stop valves at the reservoir inlet would be expensive because of the high pressure;
- a back-up inlet control valve would be necessary for safety;
- a break on the pipeline would cause a high velocity discharge posing a significant risk to public safety and property.

A break-pressure tank is therefore necessary. To choose a location for it the hydraulic gradient for maximum flow must be drawn on a profile of the main for an assumed diameter of pipe. For a flow of 60 Ml/day and an available head of 7.2 m/km (251 m/

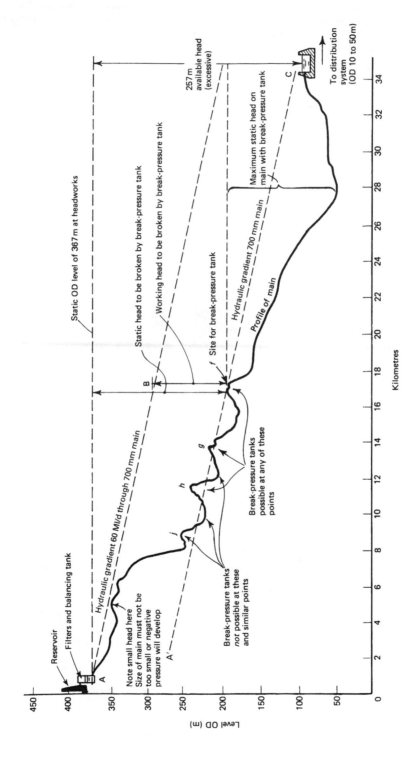

Fig. 14.3 Considerations in the design of a gravity main.

35 km) a pipeline of at least 700 mm diameter would be required.

$$\frac{\text{Loss of head through 700 mm pipeline at 60 Ml/day}}{\text{at } k_s = 0.75 \text{ mm (in Colebrook–White formula)}} = 5.0 \text{ m/km}$$

$$\frac{\text{Required head at service reservoir inlet 110 twl}}{+6 \text{ m loss through inlet valve}} = 116 \text{ m OD}$$

The hydraulic gradient at a slope of 5.0 m/km drawn upstream from 116 m OD at the service reservoir inlet shows that a break-pressure tank could be sited at *h*, *g*, or *f*, (but not at *j*). Which of the possible locations to choose depends on the profile of the main. In the case shown, *f* is chosen because it most nearly equalises the maximum static heads on the mains upstream and downstream of the tank. This means that pipes and valves of the same class strength can be used in both sections of main.

The investigation may need to be repeated for another size of main and to assess the costs of various alternatives. Cost will probably favour the smallest practicable diameter main, but the high velocity of flow in it means that it has practically no further carrying capacity so that a larger diameter main may be preferred. Other alternatives can be considered, such as relocating the treatment works closer to the service reservoir. However the main would then carry raw water between the impounding reservoir and the treatment works, and it might need to be slightly larger diameter or be concrete lined to allow for any reduction in carrying capacity due to carrying raw water. The raw water pipeline would also need to be sized for the additional usage at the treatment works for process water and cleaning.

An example of a break-pressure tank is shown in Fig. 14.4. One point that should be mentioned is that a break-pressure tank should not be provided with a by-pass, to make it impossible to bring too a high a pressure on the downstream length of pipeline which could have disastrous results. Instead the tank should have duplicate compartments to permit maintenance.

14.9   Source and pipeline layouts

The most economic layout for a supply system is to locate a service reservoir as near as technically feasible to the distribution area it serves, as shown in Fig. 14.5(a). Because the service reservoir evens out the peak demands for water, the further the service reservoir is from the distribution area, the longer must be the lengths of main designed for peak hourly rates of flow and, as Fig 14.5(b) shows, the more costly is the system. A service reservoir close to the distribution area provides advantages when maintaining supplies under emergency conditions and for fire-fighting. It also helps to reduce pressure fluctuations in the distribution system and aids economic development of the system. Nevertheless there are occasions where the configuration of the land makes it impossible to comply with this arrangement.

If the service reservoir cannot be sited close the distribution area, the most desirable layout in these circumstances is to have at least two major delivery mains from the reservoir, connected together at their extremities to form a ring main through the distribution area (Fig. 14.5c). By this means not only will the system be able to cope more effectively with peak rates of flow but, in addition, should a repair be necessary on one main, at least some flow can be maintained to the distribution system.

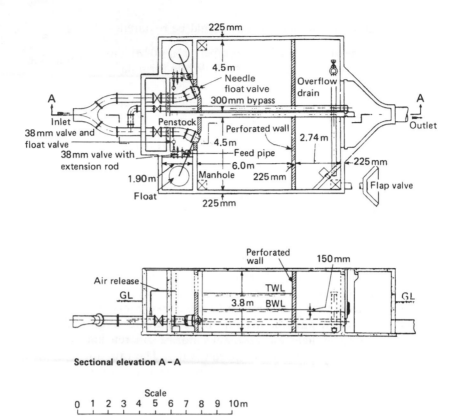

Fig. 14.4 A break-pressure tank.

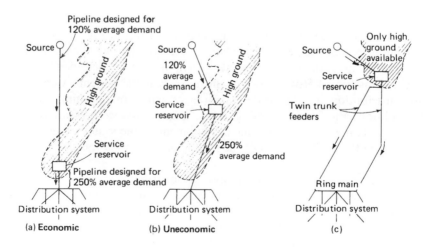

Fig. 14.5 Distribution system feeds: (a) single main feed and (b) use of ring main.

A *'rise-and-fall' main* can be adopted where the only possible siting of a service reservoir is further away from the distribution system than is the source of water. Figure 14.6(a) illustrates such a scheme. If the rate of output from the source is equal to the average daily supply, water flows out of the service reservoir whenever the demand rate exceeds the average and flows into the reservoir when demand is less than the average. Alternatively, if the source output is made sufficiently large to fill the reservoir during the day shifts, (typically from 06.00 to 22.00), when pumping ceases the distribution area is fed only by the service reservoir. This lowers distribution pressure during the hours 22.00–06.00, thus reducing night-time losses. However the design needs to ensure an adequate turnover of water in the reservoir to prevent water quality problems arising, particularly during periods of low demand. This type of system can be analysed using extended period (dynamic) hydraulic modelling to optimise the pumping and storage requirements, and ensure the age of the water, particularly at the reservoir, does not become excessive and hence require additional local disinfection facilities.

Elevated storage is often necessary on flat ground and an arrangement similar to that shown in Fig. 14.6(b) can then be adopted. As shown, the main to the elevated storage is a rise-and-fall main. If, however, the pump delivers water to the top of the tank the distribution system is at all times fed by gravity from the tank. In either case a float-operated valve can prevent the elevated tank from overflowing and water level sensors in it can control the number of pumps running, or their output, to keep the supply roughly in step with demand. The advantage of the rise-and-fall main is that, during pumping, a higher head can be maintained in the distribution system than is possible by gravity alone from the elevated tank which is costly if constructed more than about 30 m high.

A pressure sustaining valve could, in fact, be used to achieve the same effect without the use of an elevated tank, but even a small elevated tank is to be preferred because it permits water level control of the pumps which is simple and reliable in operation. Without storage, the pump output must be controlled by monitoring the flow or pressure in the mains feeding the distribution system. This is more difficult to set up and manage, and

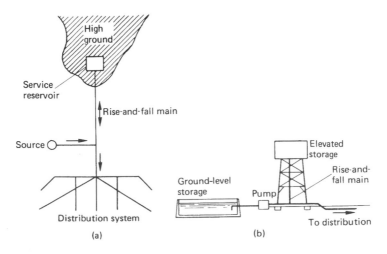

Fig. 14.6 (a) A rise-and-fall main supply, (b) use of elevated storage.

requires considerably more regular maintenance and understanding of demands and how they vary diurnally if the system is to be operated efficiently.

Elevated storage is expensive to construct, hence it is seldom possible to provide the same amount of this type of storage as is usually provided when ground-level storage can be constructed. As a consequence both ground-level and elevated storage may be provided at a source in flat country, with the ground-level storage providing the larger capacity. To ensure that the reserve supply in the ground-level tank can be used at all times the transfer pumps to the elevated storage usually have 50% standby capacity or more; an independent standby source of power, such as a diesel generator, may also be provided.

Many utilities, particularly overseas, prefer to pump directly into distribution. Although there are significant disadvantages as outlined above, the need to limit capital investment often outweighs the control advantages of elevated storage. It is true that systems for control of direct pumping into a distribution system are becoming increasingly sophisticated, but these require an appropriate level of technical maintenance which may not be available to water undertakings with limited resources or in remote places where simple fail-safe systems are preferable.

14.10 Distribution system characteristics

The principal performance requirements for a distribution system are the following.

(1) Minimum pressures at peak demand times on 24-hour supply systems need to be not less than 15–20 m in the main at the highest supply point a system can serve. Higher pressures may be necessary in some areas where there are a significant number of dwellings exceeding three-storey height; but high rise blocks will normally be required to have their own booster.

(2) Static pressures under low night-time demand should be as low as practicable to minimise leakage. For flat areas a maximum static pressure in the range 40–50 m is desirable. For undulating areas a higher static pressure may be unavoidable. If the distribution system contains older mains laid in Class B, C or D pipes, the maximum static pressure must not exceed 60, 90 and 120 m, respectively, which are the working pressures for such pipes.

(3) In the UK, the ideal fire demand requirements range from 8 l/s from a single hydrant in a one- or two-storey housing development to up to 75 l/s from pipework infrastructure to an industrial estate (see Section 15.2). Owners of property requiring supplies for sprinklers or hydrants on their premises may need to enter special arrangements with the water supplier. In the USA higher fire flows are required.

(4) Consistent water quality must be maintained through the system by establishing discrete hydraulic and source water quality areas. Wherever possible dead ends, long retention times, mixing of different waters within the distribution system, and diurnal reversals of flow in mains should be avoided.

(5) Spare flow capacity must exist in the system sufficient to meet foreseeable rises of demand over the next few years.

Occasionally static pressures have to be high because of widely fluctuating ground levels over short distances, making division of the distribution system into different pressure zones difficult. Pressure reducing valves can be adopted to reduce pressures in low lying areas, but they must be regularly maintained if they are to operate reliably. To guard against mishaps, the low level system must be strong enough to resist the maximum possible static pressure if the valve fails or needs to be by-passed for operational reasons.

Lower supply pressures of about 10 m head are desirable if only standpipes have to be supplied. This makes it possible to use low pressure, easily operable standpipe taps which give less maintenance problems and less wastage. However, standpipe-only systems are rare and it is not practicable to supply water over large areas at very low pressure.

Network mains usually have diameters in the range 100–250 mm. For 300 mm diameter pipes and above, the work of making service pipe connections to mains increases substantially, because it is necessary to excavate below the pipe to accommodate the chain which holds the tapping machine in place. Also a 300 mm main is of such large capacity that it will normally be used to convey water in bulk to an area for further distribution. Ideally service connections should be restricted to the smaller network mains. In areas where there are large industrial consumers, or where the per capita consumption for domestic purposes is exceptionally high, some network mains may be 400–450 mm diameter. Small mains of 80 mm and below should comprise only spur or rider mains feeding small groups of properties so they do not form part of the network. In practice, however, the older systems often have a significant proportion of 80 mm mains.

Principal feeder mains vary in diameter according to the size of area served: some may supply properties en route. For instance a principal main in a rural area may be as small as 100 or 150 mm diameter and would supply properties en route, but in most urban areas 200 mm would be regarded as the smallest feeder main, and many are larger than this. The proportions of mains laid in the different sizes in typical water undertakings are given in Table 14.1. The preponderance of mains of 80–150 mm size indicates that these comprise the majority of mains from which properties are served and thus form the network mains. The proportion of mains over 450 mm size usually reflects the distance between sources or reservoirs and the focal points of the area served.

Table 14.1 Percentage of mains laid in different sizes in five typical urban water undertakings

Diameter		UK (A)	UK (B)	UK (C)	Overseas (D)	Overseas (E)
mm	inches	%	%	%	%	%
80	3	29.9	39.3	10.5	18.0	6.0
100–150	4, 6	52.6	43.5	53.0	53.0	72.0
200–225	8, 9	9.2	4.7	9.8	12.0	3.0
250–350	10, 12, 14	5.1	4.9	8.0	⎫	
375–450	15, 18	1.7	3.8	6.0	⎬ 17.0	11.0
500–600	20, 21, 24	1.5	1.9	6.6	⎭	
Over 600	Over 24	0.2	1.9	6.1	Nil	8.0
		100.0	100.0	100.0	100.0	100.0

Note. The 225 mm (9 in) and 375 mm (15 in) diameter pipes are no longer in production, but many were laid and still exist. The 175 mm (7 in) mains are found in some older distribution systems.

14.11 Design of networks

There are few occasions when the design of a completely new distribution system is required, hence the majority of hydraulic studies and designs involve analysing the performance of an existing system and where necessary improving its performance to meet demands from new developments and increased consumption by existing consumers. Due to the complex nature of existing systems and the need to meet more exacting performance criteria for service levels and water quality, distribution network analysis has become increasingly complex requiring the use of computers and hydraulic network analysis techniques as discussed in Section 14.14 below.

However the design of new local distribution pipework is largely empirical. A main must be laid in every street along which there are properties requiring a supply. Mains most frequently used for local distribution are 100 or 150 mm diameters, as Table 14.1 shows. Smaller mains are still laid, but it is not the best practice to use pipes smaller than 100 mm diameter because they are unlikely to have adequate capacity to meet the minimum fire demand at hydrants.

The diameter of local distribution pipes depends primarily upon the population density of the area to be served and how the pipe is supplied. Although guidelines can be developed for sizing new pipes when extending an existing system, the designer needs an understanding of the site specific conditions and performance of the system. At any time, t, of the day, the demand from an area comprises:

$$\frac{\text{Average day}}{\text{domestic demand}} \times D_{t1} + \frac{\text{Average day}}{\text{non-domestic demands}} \times D_{t2} + \frac{\text{Allowance for leakage}}{\text{at time } t}.$$

where D_{t1} and D_{t2} are the diurnal factors applying at time t to the domestic and non-domestic demands, and the leakage allowance varies with the system pressure at time t as suggested by the 1994 WRc formula given in Section 1.16. The peak hour demand occurs at different times for different types of demand in different parts of the system. The designer therefore needs to develop the demand characteristics from field measurements of flow and pressure. If not a system demand profile must be built up from its constituent parts using available records of demand and patterns of usage, making assumptions only where better information is not available. A similar approach needs to be applied when deriving seasonal variations.

Pipe velocities at high demands should be limited to keep friction head losses within acceptable limits. Minimum velocities should also be achieved wherever possible to ensure the age of the water in the pipes does not become excessive and that loose deposits in the mains are not cyclically deposited and lifted into suspension, causing dirty water problems. The designer can draw up basic tables for sizing pipes manually, based on limiting velocities according to diameter of pipe, or to maximum headloss per km.

Where dwelling numbers and occupancy ratios cannot be readily assessed from utility records and census data, figures for the average number of people per hectare or per kilometre of mains may be used, but with caution. Table 14.2 gives typical average number of connections per km of main for large utilities. They include both domestic and trade connections. The number of people served by each street main will be influenced not only by the type and arrangement of the dwelling units and their occupancy ratio, but also by the kind of water supply given. In densely populated low income areas of overseas cities a large proportion of the population may be served by street standpipes. High-rise buildings result in high population densities, but each such building is generally fed by a single large connection. Table 14.3 gives some indication of the densities that can be met in small areas of cities.

Table 14.2 Population served per kilometre of mains laid

Large areas; UK only (50–2500 km^2)	Gross density (People/ha)	Average no. people per km of main	Average no. connections per km of main
Built-up urban area	40–60	370–450	150–170
Predominantly residential urban area	20–40	290–370	120–150
Well spread urban area with parks, etc.	7–20	200–290	85–120
Mixed urban and rural	4–7	160–200	70–85
Populated rural areas with villages	1.5–4	110–160	50–70
Predominantly rural areas	0.5–1.5	65–110	25–50

Note. The number of connections includes usual small trade supplies.

Table 14.3 Population densities in urban areas

Small areas (5–25 ha)	Net density (people/ha)	People per household	People per km of street
Overseas			
Densely populated low income areas	800–1200	Over 6	3000–4000
Planned high occupancy dwellings	1000	5–6	3000
High density – flats	750	3–5	
High density – condominiums	510	3–5	
Medium density housing	360	3–5	
Low density housing	60	3–5	
Dormitory accommodation	750	n/a	
UK			
Four-storey dwellings	450	2.5	2250
Residential areas with gardens	45	2.75–3	120–140

City centres (averages for 1000–1500 ha)	Gross density (people/ha)	People per household	People per km of street
Densest areas	100–125	2.5	700–800
Next densest areas	70–100	2.5	600–700

Notes.
Gross density – on total area inclusive of open spaces, etc.
Net density – on area reserved for housing, inclusive of service roads.
The population per km of street is given because the population per km of mains will depend upon the size of main and whether dual mains are laid.
The range of 1998 UK figures for household size is given in Section 1.8.

To meet fire demands the whole head at a hydrant can be utilised. A single fire hydrant flow of 2 m³/min can be relatively easily met by a 150 mm main fed from both ends; the flow rate in the main being 0.94 m/s and a rate of loss in the main of about 11.5 m/km. If only fed from one end, the velocity increases to 1.9 m/s and the head loss becomes 46 m/km which may be acceptable in urban areas if the lengths of pipe between cross connections are relatively short. Further away from the hydrant being used, the number of mains contributing to the flow will increase so that the headloss in them will be very much less. The same flow in a 100 mm pipe fed from both ends would produce both high velocity (2.12 m/s) and head loss (98 m/km). Hence although the full system head can be taken into account under fire demand conditions, sizing distribution mains should take account of the fire risk along the pipe alignment and hence the minimum fire demand associated with the risk. Fire flows are discussed further in Section 15.2.

The layout must form a sensible hierarchy, in which the larger mains feed the smaller ones, and also the designer may need to take into account a water undertaking's preferred range of pipe diameters and material, usually adopted to limit the numbers of spares an undertaking must hold. Care needs to be taken to ensure the chosen network of preferred pipe diameters does not restrict potential growth or create unacceptable velocities, high and low.

Where a network is open and highly interconnected, there will be opportunities to optimise the layout to maintain acceptable operating conditions under emergency flows while reducing retention times and minimising pressure variations. However, as discussed above, increasingly designs need to take into account the desirability of establishing stable supply areas which optimise operational efficiency, facilitate control of leakage, and aid maintenance of consistent water quality within the distribution system.

Computer software need not be used to analysis flows in small and straightforward street mains layouts. It is, however, an appropriate engineering design tool for more complex pipe arrangements, the analysis of existing networks, and where the design process is integrated with a base mapping and computer aided drafting package (CAD) or a GIS. It is also useful for testing the capacity of a network as a whole to meet a major fire demand.

14.12 Design of principal feeder mains

The matters to be considered when designing principal feed mains and trunk mains include:

- the optimum engineering or most physically practical route;
- proposed mode of operating the system; available resources versus seasonal and diurnal variations;
- location and capacity of storage;
- provision of system flexibility without creating diurnal reversal of flow in mains;
- capital versus operating cost;
- confidence in demand projections.

The design process must include decisions as to which source should supply each area, which route is practical for new mains and how peak day demands are to be met when some sources have spare capacity and others have not. Oversized mains will be advisable in some areas to cater for possible additional future demand, and ring mains will be needed to transfer water across the distribution system under different phases of development. The design should achieve the optimum combination of pipe sizes and alignments to satisfactorily meet the various demand conditions. The first step will be to identify route options. The process will involve desk studies of maps, records of physical obstructions including the locations of other utilities, and site visits to assess construction constraints and practicability of route.

Once realistic routes have been identified for the mains, pipe sizing can proceed with reference to a summation of the flows in mains working back from the extremities of the system towards the source(s). This process is similar to that used to assess demands in distribution pipework except that the demands on feeder mains are the summations of the demands from the pipes they supply.

Manual calculations using spreadsheets and programmable calculators are useful tools when analysing options which cover alternative routes of different lengths entailing a different pattern of interconnections. Quick approximations of the required diameters of new mains for a given flow, and the flow capacity of existing mains according to their diameter, can be made if a table of appropriate head losses and consequent flow capacity for each size of pipe is first drawn up. Flow velocities are so chosen that losses through a typical distribution system under peak hour demands will be about the maximum practicable. The table is drawn up to show mains' capacities to meet the maximum day demand because this is the output which must be matched by sources. The method permits the planning of routes for new mains with a rough knowledge of the size of pipe required. Later the pipe sizes chosen can be checked in more detail.

As in the case of designing networks, hydraulic modelling is used to analyse more complex networks and resolve operational management issues such as blending different sources in order to maintain water quality, or optimising use of storage capacity to maintain diurnal reliability.

14.13 Manual calculation of network flows

Manual calculations are cheap where they take only a day or two to complete, and give the engineer an invaluable understanding of the way a system must work. They can show which principal mains are overloaded or have spare capacity; why areas of low pressure occur; and what new mains must be laid to meet additional demands or improve pressures.

For manual analysis the network is reduced to a 'skeleton layout' of principal mains, and the total average daily demand is divided out over areas and allocated to nodes. There are various ways of analysing the system but the simplest way is to divide the network into ring mains and 'tree branches'. To analyse flow in ring mains, a trial division of flows at the input point can, after two or three attempts, be sufficiently adjusted to disclose the 'point of balance' in the ring main where friction losses in either leg are equal. If two ring mains are conjoined by a common leg (a 'figure of eight' formation), then a further process of trial and error can bring the flows and pressure losses of the two rings into balance. Flows taken by 'tree branch mains' can be calculated directly. The analysis can be repeated using a peaking factor applied to the average daily nodal demands.

Pressures in the distribution system are easily measured in the field for average and peak hourly conditions and, from them, pressure contour maps can be drawn for each case. These contours are reliable indicators of loss because pressure in principal mains cannot vary abruptly from one location to the next if valves are kept open. The pressures can be used to check, or adjust if necessary, the friction coefficients used in the analysis. These analyses will reveal where inadequate mains capacity exists, evidenced by high flow velocities and large friction losses. The effect of additional mains to improve pressures or cope with unsatisfied or future demands can easily be ascertained. A manual analysis will be a reliable basis for the planning of distribution improvements and extensions, but a detailed computer hydraulic analysis can provide extra information about parts of a system, the behaviour of reservoir rise and fill mains, and the effects of a burst main or major fire demand. It has to be remembered however that the accuracy of any computer model is not greater than the accuracy with which nodal demands can be estimated. Also every distribution system must have spare capacity in it to meet future increases of demand, and often the magnitude and location of such future demands is not known with any accuracy. Hence planning the development of a system is more often the choice of a strategy that best meets a range of future options than using precise calculation to meet some fixed situation ahead. The small mains of 200, 150 and 100 mm diameter are mostly ignored in a manual analysis, because they are normally designed according to local demand conditions.

14.14 Analysing existing systems by computer modelling

Network analysis by computer-based mathematical models can be used to analyse the hydraulic performance of existing systems and the design of new systems. Models representing both trunk mains and distribution pipework are used to assess system performance under a variety of supply and operating conditions.

With an increasing need to achieve operational efficiency, maintain service levels, and produce most economical rehabilitation and reinforcement proposals, models need to be constructed and validated to more exacting criteria. As hardware has developed, so it has been possible to build larger models in greater detail and also to use them to analyse water quality parameters. Both trunk main (strategic) and distribution models are built, but current practice is to construct 'all mains models' that comprises all the pipework between source and the service connection.

The physical characteristics of the system are represented in the model by nodes and pipes (or elements) in a set of linked data files. The nodes, joined together by pipes, represent pipe junctions and changes in pipe diameter, and include other system attributes such as flow or pressure control valves. Water demand on a main is allocated to the nodes either end, taking account of the distribution of the drawoffs along it. The node and pipe data files contain geographic co-ordinates, ground levels, basic demand information, internal diameter and friction coefficients, pump curves, service reservoir geometry, and valve performance characteristics.

As software has developed, so model construction has become more sophisticated. It is now common to abstract the data required for modelling from other databases used for company records. Recent and ongoing software developments can link the network analysis software directly to GIS, customer information systems (CIS), and telemetry data for the necessary input. The distribution of demands to nodes can be related to the area served from each and the make up of each demand. Where all consumers are metered, the demand at a node can be readily assigned as a measured consumption by category of user. Typically consumption is based on postal address or code and a geographic reference system, or use is made of data from the billing system and GIS. Where customers are unmetered, nodal demands have to be assessed from unit consumption estimates applied to property count or population figures. In practice demands at a node are made up of a number of categories of demand including, metered and unmetered domestic demand, and commercial, institutional and industrial demands. The different diurnal patterns of demand associated with each type of demand have to be taken into account. For example, industrial demands may be 24 hour or occurring during daytime working hours only. Different industries, such as food processing, dairies and laundries, will exhibit different diurnal demand patterns. The number of demand types taken will depend on the capacity of the software and the accuracy of the data used to derive the base demands. The modeller needs to keep in mind the accuracy of the output required from the model and the quality and availability of the data from which the model will be built.

The structures of most packages follow similar data formats for the physical system attributes and base demand information. Variations are usually associated with model size and how network apparatus (pumps, valves, reservoirs) are represented mathematically. Each package has its strengths and weaknesses and the choice of software will depend on the user's personal preference, the application for which it is required, the availability and quality of source data, the type of output required, and whether the software is one component of an integrated management information system. However a model constructed to the format of a particular package need not be a constraint because, in practice, provided access is possible to the data files, it is relatively quick and easy to convert a model from one format to another. If the hydraulic model construction procedures are carefully developed from the outset, future updating, particularly of the demand data, can become a relatively straightforward and mechanical process.

Although some software packages impose physical size restrictions, it is the ability of the engineer to assimilate and interpret complex analytical results which represents the overriding constraint on the size of model which can be built. Water quality analyses also impose analysis run-time constraints which need to be considered when deciding how large a model to construct

The actual mathematical analysis conducted by the computer comprises the iterative solution of a collection of algorithms in order to simulate the behaviour of the flow through the system. The computer solves the equations to specified tolerances by successive approximations subject to the following rules:

(1) the algebraic sum of the flows entering and leaving a node must be zero;
(2) in any closed loop in the system the algebraic sum of the pressure losses must be zero;
(3) the combined inputs to the system must equal the total of the nodal demands.

The operational control and dynamic (extended period) characteristics of the distribution system are contained in a time or control file which imposes particular supply/demand and operational conditions on the network. It may cover a single 'snapshot' or a series of prescribed time periods for each of which the changing activities of the system are listed. These include factors to represent the diurnal demand patterns, system losses and other demand related matters such as times at which pumps start and stop or when valves are opened and closed.

The model operates by applying operational conditions to the network data. Starting with initial pre-set reservoir levels, it establishes the status of pumps, pipes and valves and calculates demands, pressures, and flows in the network over the first time period. The net reservoir inflows/outflows and, from the reservoir geometry, the rise or fall in reservoir levels is also calculated at the end of the time step. These new reservoir levels are then used as the starting values for the second time period together with the diurnal demands and activities corresponding to the next analysis step. The analysis repeats this for each subsequent time step. At the end of each time period a 'snapshot' of flows, pressures, demands, pump performance and other relevant data is stored; thus by the end of the analysis the sequence of snapshot results can be displayed both graphically and in tabular form to illustrate the changing performance of the network and individual elements of the system over the period of the analysis. For hydraulic design, typical simulation times are over 24 hours, but for operational management and predictive planning, they may be extended over 7 days, or be further extended to analyse seasonal variations or long-term planning changes. Fixed and variable factors can be redefined for each run, to represent different operating conditions, and permitting an optimum sequence of results to be selected.

Complex design and operational efficiency simulations should only be undertaken if the model can first be validated against the results of field calibration tests in either steady state (snapshot) or, preferably dynamic (extended period) mode. This requires the collection and analysis of contemporary flow and pressure measurements from the distribution system over a period of time, typically 7 days, and comparing these measurements with the calculated flows and pressures obtained by the model which has been set up to represent the same supply and demand conditions. The quality of a calibrated model is dependent on the quality of the data derived from the field test. Field test activities include:

- flow and pressure measurements;
- establishing pump head/discharge and efficiency characteristics;
- valve status checking;
- reservoir drop tests;
- internal diameter and velocity profiles at insertion probe flow measurement points;
- estimating pipe friction characteristics by some form of internal visual inspection;
- interviewing metered consumers to establish their diurnal patterns of demand;
- resolving anomalies.

In order to limit the number of unknowns before the calibration field testing takes place, thereby avoiding an abortive test, extensive preliminary field investigations are required. Flows should be recorded at all sources into the system and internal flow meters may be used to determine area or sub-area demand, including inflows and outflows to storages and the usage by the larger metered consumers. Pressures are measured simultaneously at typically between 20 and 35% of the number of demand nodes. The quality of the calibration improves and the time required to complete the validation process reduces with the number of available reference points. Where the field recorded pressures and flows do not agree with those computed by the model within specified tolerances, pipe characteristics are adjusted using engineering judgement and the analysis is repeated until the calculated flows match those measured in the field. The model is then considered to be calibrated.

Successful construction of a validated model requires a methodical approach to both model construction and the fieldwork, and should include the development of procedures to simplify future model updating. The calibration process itself needs to be systematic and all adjustments to pipe characteristics must reflect known information of the condition and performance of a pipe. Where it is found necessary to adjust pipe roughnesses beyond reasonable values for the type, age and condition of the pipe, the anomaly should be recorded, and data searches and field investigations will be needed to justify the use of the anomalous value. Similarly conclusions that flows within a pipe or demands are incorrect, require investigation to substantiate the figures. Demand data should never be arbitrarily reallocated to 'fit the field measurements'. All adjustments must be justified or identified as an unresolved anomaly to be investigated later or taken into account when using the model for predictive work. Full documentation of the model, including the modelling techniques used, the assumptions made and anomalies remaining, is also required.

14.15 Water Quality Modelling

As water quality issues have become more significant in the UK, software developers have created water quality analysis modules, either as additional features to their hydraulics network analysis packages or as an integral part of a new suite of programs. These packages are now being used to model water quality parameters within a distribution system.

The majority of programs used in the water industry now include water quality modelling to a greater or lesser degree. As with hydraulics packages each has developed a slightly different approach, with different water quality features and results presented in different ways. Water quality models can be built to analyse at any point in the system:

- the age of water and hence the retention time within the network;
- the amount of water from a specified source, where more than one source is used;
- concentrations of substances which are independent of time but may be altered by mixing;
- decline or increase of substance concentrations which are time dependent.

The calculated ages and relative proportion of source waters at model nodes, are used to correlate with the results from water quality sampling programmes, or where consumer complaints have occurred. Source contribution analyses can be used to track and predict the effects and extent of pollution incidents. Chlorine residual models can be calibrated against recorded data and field measurements, using the time of travel and age of water analysis mode. Chlorine models used in conjunction with chemical and bacteriological sampling data and statistical predictive modelling techniques, can be used to improve the efficiency of dosing controls.

Experience to date shows that water quality modelling must be based on a validated hydraulic model to ensure the real time of travel of water is represented. Where, due to software constraints or client requirements the hydraulic model is a simplification of the network, modelled velocities usually represent an under-estimate of the time of travel or age of water at any point because the model tends to compute higher flow rates in the reduced number of pipes. It has therefore been found that water quality modelling generally requires more detailed (all mains) hydraulic models which replicate more closely the velocities in mains. With the current trend to build all mains models this problem is being overcome. Age of water models frequently form part of current model construction specifications.

The key to all mathematical modelling is verification. The equipment required to calibrate hydraulic models has developed over a number of years to meet the increasing accuracy and reliability requirements associated with the greater sophistication of hydraulic models. However, the field test equipment for continuous sampling and analysing of water quality needed to provide the data for calibrating water quality models is still being developed. At present the equipment is relatively expensive and not available in the numbers necessary to achieve the coverage necessary. As modelling techniques are devised and developed, so the monitoring equipment will evolve. There is no question that water quality modelling represents a potentially powerful tool for the future, but until the field sampling equipment is available, validation will be constrained by the quality of data obtainable using existing equipment and techniques.

14.16 Updating of network models

Since the network file is a static database, it should not be necessary to make any alterations to the data file after the model has been proved except when changes are made to the network, or base demands change. It is essential that all significant changes should be recorded in order to maintain the accuracy of the model. Changes are likely to be associated with:

● demand changes (new developments, changes in metered consumption rates);
● extensions of mains and rehabilitation, replacement and renewals;
● operational changes (alterations of zonal boundaries, opening/closing of valves or replacement of pumps).

Domestic demand changes can be assessed by reviewing seasonal variations, reductions in the level of leakage, increased per capita consumption, or changes in consumption volumes recorded in metering records. Trade, industrial and institutional consumers will generally be metered and any changes can be derived from billing records. Changes in mains layouts or mains condition, e.g. by scraping and lining to improve the internal roughness, will require amendments to the network file and schematic maps. Where permanent changes to zonal boundaries are made it will be necessary to revise the base demands in each of the distribution zones affected.

The increasing use of GIS, CIS, and associated system performance databases, provides opportunities for building and updating hydraulic models electronically, thereby eliminating the costly manual process which has inhibited model revision in the past. Further developments with genetic algorithms should bring additional model construction cost savings and improve optimisation facilities.

Hydraulic analysis software which can be integrated with the different business databases is currently being developed by a number of suppliers and associated organisations. Although some packages available are designed to deal with particular construction problems, none have yet resolved the problem of repeatability or only updating where there have been changes. The debate centres around whether to rebuild completely each time a new generation of model is required, or whether to update by exceptions; that is only revise sections of the model where there are known changes to the data. A second debate centres on the frequency of re-validating the model and whether to undertake a full or partial field test. Both these issues are influenced by the type of software adopted and the purpose for which the model is required.

Using information from mains record, customer billing, telemetry, levels of service, and water quality data accessed through an integrated management information system, combined with on line 'real-time' hydraulic analysis will provide future planners and operations engineers with a powerful suite of engineering analysis tools.

References

1. Office of Water Services. 1998 July Return Reporting Requirements and Definitions Manual, 31 January 1998.
2. BS 6700: 1987 *Design, Installation, Testing and Maintenance of Services for Supplying Water for Domestic Use*.

15

Distribution practice

15.1 Distribution organisation

Distribution organisational systems vary, reflecting division into operational functions or into separate regions of supply where the area served is large. Figure 15.1 shows the predominantly functional division adopted by a very large undertaking. Distribution activities normally controlled by head office departments comprise:

(1) management of meter reading, customer billing and income collection;
(2) negotiating agreements for extension of mains and new services;
(3) response to customer complaints;
(4) engineering planning and design;
(5) training;
(6) all legal, budgetary and financial control measures.

Activities delegated to regional or local divisions of the distribution system usually comprise:

(1) maintenance of supplies;
(2) customer relationship and investigation of customer difficulties;
(3) repair and maintenance of mains;
(4) monitoring of levels of service, flows and pressures;
(5) leak detection and repair, and reduction of waste;
(6) inspection of plumbing systems and enforcement of water byelaws.

Water undertakings in the more industrialised countries are increasingly making use of telemetry to set up central or regional Control Centres manned 24 hours a day. Data from key monitoring points throughout the distribution system, and the output of sources of supply, is monitored and processed so that deficiencies of supply, burst mains, fire demands, or consumer complaints, can be handled swiftly by calling out the necessary staff to deal with them. With an increasing trend towards the adoption of automated or semi-automated plant, the output of source works or pumping stations may be remotely controlled from the Centre.

For undertakings with no central telemetering control, it is usual for local distribution managers to keep in touch with the source output personnel, so that source outputs are altered as necessary to meet demand variations. Frequently regular times of reporting are arranged – at 09.00 h to consider what source output is required according to the previous day's consumption and the water levels in service reservoirs; and at 16.00 h to agree any further adjustment of source output in the light of the current day's demand. There will also be a need for communication when a fire demand arises or a burst main occurs.

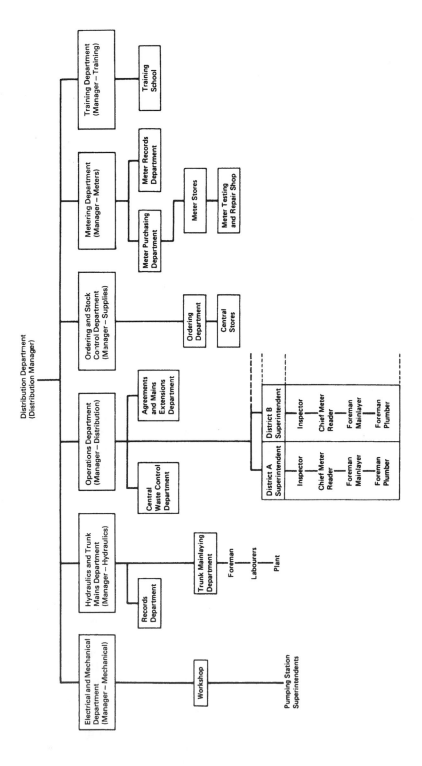

Fig. 15.1 The Distribution Department for a large waterworks undertaking.

A large distribution network is typically divided into separate areas termed 'zones', which are further subdivided into 'district meter areas' (districts or DMAs). Zones are usually based on the supply from one source, or a group of sources giving similar water, in order to provide a consistent quality of water. DMAs are used for monitoring consumption and evidence of leakage, and are chosen to achieve stable flows through their mains under normal conditions, to avoid flow reversals or excessive velocities in mains which could cause water quality deterioration (see Section 14.3). The measured flows into zones and districts permit monitoring of consumption in each part of the system, so that areas showing unusually high consumption can be scheduled for investigation. These flows, together with pressures taken at critical points in the network, can be transmitted by telemetry to the Central Control where they can be monitored and recorded to assess the operational performance of the system.

The key operative at local level for a distribution system is usually designated the 'District Inspector'; though sometimes the earlier term of 'Turnkey' is used. The inspector liaises with the public and investigates customer complaints, monitors levels of service, inspects properties for waste, and keeps the area under surveillance for visible signs of leakage. The inspector should also take part in waste detection activities. His or her intimate knowledge of the local mains system, its consumers and operational anomalies is invaluable in diagnosing the cause of troubles and directing where remedial measures are necessary. When burst mains occur or a large fire demand for water arises the inspector initiates the necessary valve operations. The inspector's principal task is to maintain supplies in his or her area with the minimum of waste.

Where a network is large and service laying and small mains repairs are undertaken by direct labour gangs, the distribution inspectors, waste and repair gangs may be backed by facilities provided in local depots and stores. Where, however, these activities are now largely contracted out to private contractors, some organisations have used the advantages of information technology to give their district inspectors a greater degree of autonomy. They may now operate out of fully equipped vans and, provided with ready access (via portable computers) to up-to-date mains plans and asset data, they can manage their area independently, and are able to respond swiftly to any operational requirements. This procedure has proved an efficient method of working and is a growing practice in UK.

Difference under intermittent supply conditions

In many countries, water undertakers are only able to provide intermittent supplies due to lack of source capacity and limited financial resources. Water rationing has then to be practised by operating distribution valves daily to apportion water in rotation to different areas, or the whole supply to an area has to be turned on and off, twice a day, to provide water to consumers morning and evening. Under these circumstances many difficulties arise, and maintaining water quality may be difficult. The operation of valves will need to minimise the risk of subjecting parts of the network to vacuum pressures which could cause groundwater infiltration and contamination of the supply. However, since valves have to be operated manually, it is usually only the branch mains off principal feed mains that are operated to shut-off or re-open supplies. Most consumers will use times of supply to fill storage receptacles for use during periods of non-supply and many may not turn off their taps so that such receptacles often overflow and waste water. If the quality of water is suspect, it is quite common for consumers in such intermittent supply areas to boil water for drinking – and they may even be advised to do so by the supply undertaker. Drinking water may also be purchased from vendors, though unless the water undertaker controls such vendors, the quality of water they supply can be very variable.

Distribution network extensions

The design and supervision of construction of network extensions and reinforcements, or the rehabilitation of mains (see Sections 15.8–15.10) is normally managed by the central engineering department of a water undertaking. In countries, however, where local water undertakings have limited technical resources, a separate regional organisation may be responsible for the design and construction of network extensions for the water undertakings in its region. Where water companies come under regulatory control, as in England and Wales, a company's in-house design team may be required to be competitive against outside designers, and its direct labour section competitive against construction contractors.

Some water undertakings still retain direct labour gangs to carry out routine mains repairs, distribution mains extensions, and the laying of service pipes to consumers. Alternatively 'term' contractors may be employed to do this type of work for a fixed period, typically for 1–3 years duration. As discussed in Section 2.10 there is increasing use being made of private contractors to provide management and maintenance functions of this type which, previously, were traditionally resourced in-house. The extent to which these 'facilities management contractors' will be employed depends on the efficiency and savings in costs such outside contractors are able to provide, as compared with those which can be achieved using in-house resources.

15.2   Fire-fighting requirements

Under current UK legislation[1] a water undertaker must install fire hydrants on mains where required by the fire authority, who must pay for the hydrant's installation and maintenance. The only exception is that the undertaker is not required to install a fire hydrant on a trunk main. There are no requirements on the water undertaker to provide any given pressure or flow at a hydrant, except that when the fire authority is dealing with a fire it can call upon the water undertaker to provide a greater supply or pressure by shutting off water from mains or pipes in any area. The water undertaker has, however, a duty to maintain a constant supply of water in its mains at a pressure sufficient to reach the topmost storey of every building in its area – but not to a greater height 'than that to which it will flow by gravitation through its water mains from the service reservoir or tank from which that supply is taken'.

Water UK and the Local Government Association have issued a 'National Guidance document on the provision of water for fire fighting'[2] which recommends the flow provisions set out in Table 15.1.

From Table 15.1 it will be seen that the minimum requirement is 0.5 m^3/min for two-storied housing, with up to 2.1 m^3/min for larger dwellings. This agrees with the specified minimum capacity of fire hydrants of 2 m^3/min at a pressure of 1.7 bar at the inlet as BS750.[3] When larger fires occur, suction hose lines will be put out to additional hydrants further afield, thus spreading the water demand over more mains. In the USA the National Board of Fire Underwriters stipulates minimum fire flow capacity of mains according to size of area and nature of property. These requirements often exceed the peak domestic rate of demand and therefore dominate the design of distribution mains. Some

Table 15.1 Guidelines on flow requirements for fire fighting issued by the Local Government Association and Water UK[a]

Category	Description	'Ideal' requirements
Housing	Detached or semi-detached houses of not more than two floors	8 l/s through a single hydrant
	Multi-occupied housing with units of more than two floors	20–35 l/s through a single hydrant
Transport	Lorry/coach parks – multi-storey car parks – service stations	System capable of delivering 25 l/s minimum from any single hydrant on site or within 90 m.
Industry	Recommended for industrial estates:	Site mains normally at least 150 mm diameter.
	– for up to 1 ha	Supply – 20 l/s
	– 1–2 ha	– 35 l/s
	– 2–3 ha	– 50 l/s
	– over 3 ha	– 75 l/s
Commercial	Shopping, offices, recreation, and tourism development	System capable of delivering 20–75 l/s
Institutional	Village halls	15 l/s through any single hydrant on site or within 100 m
	Primary schools and single storey health centres	20 l/s through any single hydrant on site or within 70 m
	Secondary schools, colleges, large health and community	35 l/s through any single hydrant on site or within 70 m

[a]Water UK is the trade association for water companies in England and Wales.

engineers question whether such high rates are necessary and, although not yet formalised, discussions on guidelines are taking place in the industry in the USA with the objective of limiting supply pressures; thereby also reducing unaccounted for water.

Fire appliances in common use have built-in pumps of capacities 2.3 and 4.5 m^3/min. Nozzle sizes are commonly 13 and 19 mm, discharging flows of 0.16 and 0.45 m^3/min, respectively. Larger nozzles of 25 mm diameter with capacities up to 1.1 m^3/min represent the largest practicable size for hand-held branches.

Table 15.2 summarises hydrant spacing and flow fire requirements within Europe and the USA.

15.3 Service pipes

Service pipe connections from a main to a property are usually laid as shown in Fig. 15.2. In the UK the length of service pipe from the main to the company stop tap (or consumer's meter if fitted) is termed 'the communication pipe'; the balance of service pipe to the consumer's internal stop tap being termed 'the supply pipe'. The communication pipe is maintained by the company and the supply pipe by the customer.

Table 15.2 Fire hydrant and fire flow requirements

	England and Wales	USA	Continent
Hydrant characteristics	2 m³/min at 1.7 bar	2 m³/min	1.0–1.5 m³/min minimum
Initial flow in close proximity to fire	1.4 m³/min upwards	Not stated	Generally 1.0 m³/min
Subsequent rates of flow (as required) for a single fire	9 m³/min or more, up to 25 m³/min for major fires	4 m³/min for 1000 population, rising to 45 m³/min for 200 000 population*	3.6–6.0 m³/min in high risk areas, or more
Minimum residual pressure in mains	Preferably 0.7 bar but not less than zero	Not stated	Not stated
Possible quantity of water used in a fire	Maximum flow times 'several hours' say 6 hours	Maximum flow for 4–10 hours	Maximum not stated; average incidents last less than 2 hours (France)
Spacing of hydrants	Generally 100–150 m but 30 m in high risk areas	100–150 m	80–100 m in urban areas; 120–140 m in residential arcas; widcr in rural areas

Notes.
Information for USA and continental requirements as reported by J Bernis in Water Requirements for Firefighting, Special Subject 7, Proc. IWSA Congress, 1976, and for UK requirements from BS 750.
*These flows are National Board of Fire Underwriters requirements; actual fire requirements according to records are 6.3–12.5 m³/min in built-up areas and up to 25 m³/min in high risk areas.

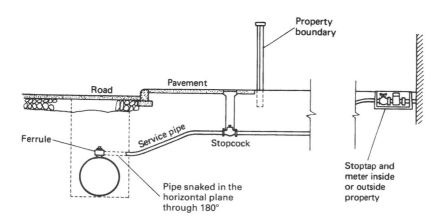

Fig. 15.2 Typical service pipe construction.

The ferrule is normally inserted into the main by means of an 'under-pressure' tapping machine as shown in Fig. 15.3. The machine bores a hole into the main and taps a screw-thread in the hole. By rotating the head of the machine, the ferrule is brought into position over the hole and screwed into place. The ferrule has a plug in it which can be screwed down, so cutting off the supply.

Service pipe connections for houses are usually to a standard size according to the practice of the water undertaker, 13 mm being the typical size for a one-family house or flat which is within 30 m of the distribution main. The pressure in the main would depend on the system supply characteristics. In the UK where utilities are required to maintain minimum levels of service, the lowest mains pressure equivalent to the minimum supply standard is 12–15 m. This 'surrogate' pressure is equivalent to maintaining 10 m available head at the consumer's stop tap when there is a flow of 9 l/min from the first tap within the consumer's property.

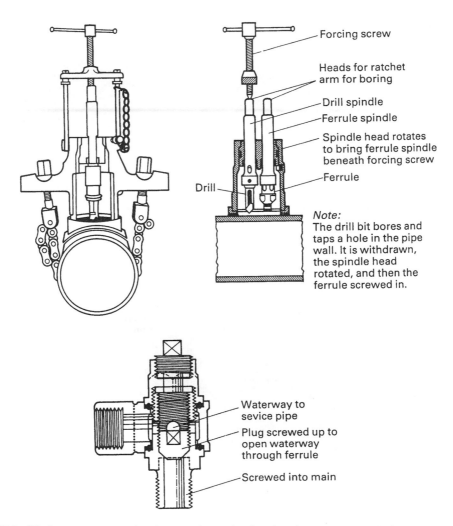

Fig. 15.3   Under-pressure tapping in a service main showing the type of ferrule inserted.

Higher operating pressures are maintained in the USA, where there are buildings greater than three storeys and where a water utility is contracted to provide higher mains pressures for contractual or fire-fighting reasons.

Service pipe materials

Materials used for service pipes comprise: (1) lead and lead alloys, (2) copper, (3) steel, (4) uPVC, (5) polyethylene.

(1) Lead pipes

Lead and lead alloy pipes are now no longer installed, but many older service pipes in the UK are of lead and still continue in service. As an EC Directive requires the maximum value for lead in drinking water to be reduced to 10 μg/l by year 2013 (see Section 6.27) UK water utilities have started to replace lead communication pipes whenever they have to undertake other rehabilitation and service improvement work. However they have recognised that where treatment solutions alone will not achieve the EC standard, they will need to implement significant lead service pipe replacement programmes over the next 10 years if they are to meet the new standard. If a consumer wishes to replace his lead supply pipe with another type of pipe, the water undertaker must likewise replace any lead piping used in the communication pipe.

In the USA the Environmental Protection Agency has set a level of 0.015 mg/l (15 μg/l) for lead. If this is exceeded the water has to be treated to reduce its plumbosolvency and, where this treatment is not effective, lead pipe replacement must be undertaken.

(2) Copper pipes

Copper pipes are widely used for plumbing and sometimes also for service pipes where ground conditions have been corrosive to iron. They are strong, durable, resistant to corrosion, easily jointed, and capable of withstanding high internal pressures, but are expensive. Pipes and fittings are jointed either by compression or solder/capillary joints. However as a consequence of the changes to the standards for acceptable concentrations of lead in drinking water in Europe and the USA, there is increasing discussion about the contribution of soldered joints to the lead concentrations in water, particularly in the first draw after water has been standing in the service pipe overnight.

(3) Steel pipes

Although no longer used for service connections in the UK, steel pipes have been widely used in the UK in the past and are still used in many countries because they are one of the cheapest forms of service pipe and can sustain high pressures. They may be supplied 'black' (i.e. untreated) or galvanised and with a range of internal and external protection systems. However, they are susceptible to structural failure under heavy traffic and, when laid unprotected in aggressive ground conditions, their life expectancy can be as little as 2–5 years. Poor installation workmanship, particularly associated with the rigid screw-thread joints, can result high leakage from an early stage in the life of the pipe. Many older properties have steel internal plumbing pipes. If steel pipes must be used for underground service connections, heavy grade steel tubes to BS 1387[4] should be used, galvanised and if necessary wrapped with protective adhesive tape if laid in corrosive ground. Permissible

Table 15.3   Steel pipes – heavy grade steel tube to BS 1387:1985 (Table 5)

Nominal diameter (mm)	Outside diameter (mm)	Min. wall thickness (mm)	Mean bore (mm)
15	21.1–21.7	3.2	14.9
20	26.6–27.2	3.2	20.4
25	33.4–34.2	4.0	25.7
32	42.1–42.9	4.0	24.4
40	48.0–48.8	4.0	40.3
50	59.8–60.8	4.5	51.3

working pressures are ample for the highest waterworks distribution pressures likely to be met in practice. Dimensions of steel service pipes are given in Table 15.3.

(4) uPVC pipes

uPVC or PVC-u, unplasticised polyvinyl chloride piping is described in Section 13.15. It is used for service piping in temperate climates and is increasingly being installed for cold water domestic plumbing pipework. It cannot be used for hot water plumbing and is not wholly suitable for hot climates because the maximum working pressure reduces by about 2% per 1°C above 20°C. However it does represent a low-cost pipe where locally made. The pipes should be stored under cover to protect them from the ultraviolet rays of sunlight. The pipes are corrosion-resistant, light to handle, and easy to joint using either solvent cement or compression joints. Solvent cement joints are not now favoured; if a solvent joint is broken it cannot be remade. uPVC pipes suffer from the same damage from traffic as steel service pipes but, being less strong than steel, are more susceptible to damage. Dimensions of uPVC service pipes to BS 3505[5] are given in Table 15.4.

Table 15.4   uPVC and polyethylene pipes for cold potable water up to 20°C

Nominal bore (mm)	uPVC pipes to BS 3505:1986 (Table 1)				Polyethylene pipes to BS 6572:1985 (Blue) and BS 6730:1986 (Black) (Tables 1)			
	Mean OD		Wall thickness		Mean OD		Wall thickness	
	Min. (mm)	Max. (mm)	Class D* (mm)	Class E* (mm)	Min. (mm)	Max. (mm)	Min. (mm)	Max. (mm)
15	21.2	21.5	–	2.1	20.0	20.3	2.3	2.6
20	26.6	26.9	–	2.5	25.0	25.3	2.3	2.6
25	33.4	33.7	–	2.7	32.0	32.3	3.0	3.4
32	42.1	42.4	2.7	3.2	–	–	–	–
40	48.1	48.4	3.0	3.7	50.0	50.4	4.6	5.2
50	60.2	60.5	3.7	4.5	63.0	63.4	5.8	6.5
75	88.7	89.1	5.3	6.5				

*For working pressures Class D 120 m, Class E 150 m.

(5) Polyethylene pipes

Medium density polyethylene piping (MDPE) (see Section 13.14) is now the principal material used in the UK for service pipes and is being installed increasingly internationally, where locally manufactured or where material costs are comparable with other service pipe materials. Piping can be supplied as straight pipes, sawn to length if necessary, or in coils up to 150 m long suitable for jointing electrothermally or with mechanical couplings. BS 6572[6] and BS 6730[7] specify requirements for the five most commonly used sizes for cold potable water services at pressures of up to 120 m and temperatures of up to 20°C – the former standard is for use of blue pigmented pipes in below ground systems or where the pipes are protected from sunlight by enclosure in ducts or buildings, and the latter standard is for black pigmented pipes laid above ground. Dimensions of polyethylene service pipes are given in Table 15.4.

15.4 Supply meters

Supply meters in common use on consumers' service pipes are of two types: semi-positive and inferential. These terms describe the manner in which they measure flow. The semi-positive meter, typically sized in the range 15–40 mm, is almost universally used in the UK for metering domestic and small trade supplies. Several different designs are available, the most usual type in UK being the rotary piston (or rotary cylinder). This has an eccentrically pivoted, light weight, freely moving cylinder inside the cylindrical body of the meter, which is pushed around by the water, opening and closing inlet and outlet ports as it turns (see Plate 31). This movement operates a counter mechanism which summates the total flow.

Another type of semi-positive meter uses a nutating disc which wobbles around in a circle as water is drawn through the meter and endeavours to pass above or below the disc. This meter will not measure low flows to the same degree of accuracy as the rotary piston meter, but is widely used in the USA where domestic flows are generally higher than in UK. All semi-positive meters should incorporate a strainer upstream as the meter is only suitable for measuring water free from grit or other suspended matter.

The inferential meter has a bladed turbine inside which is turned by the flow of water, i.e. it 'infers' the quantity of water passing by counting the revolutions of the turbine (see Plate 31). Any particular design must therefore be calibrated at the maker's works. It is primarily used on industrial supplies, being suitable for measuring large flows. It can summate the total flow either through counter-gearing, or each revolution of the propeller may initiate an electrical pulse so that an electrical logger can summate the total flow, and also record the flow rates over short time intervals. This data can be locally recorded or transmitted elsewhere by telemetering. For the measurement of widely fluctuating flows, beyond the range of any single meter, past practice has been to use two meters of different sizes connected in parallel, an automatic device ensuring the smaller flows pass through the smaller meter, and vice versa. However such combination meters were expensive to install, and have now been superseded by the electromagnetic meter (see Sections 10.17 and 15.15) which gives a good accuracy over a wide range of flows and is cheaper and easier to install.

The required accuracy of semi-positive and inferential meters is given in Table 1.8 of Section 1.13.

15.5 Domestic flow requirements and design of service pipes

Flow requirements

The maximum demand rate from a house will depend upon the amount of storage, if any, provided on the premises. In most modern UK plumbing systems the only storage provided is that on the hot water system and in WC flushing cisterns. In such cases all cold water supplies to taps, showers, WC ball valves, and the cold water feed tank to the hot water system will be fed directly from the mains. In other cases, generally older properties, only the cold water taps in the kitchen and to a bathroom washbasin will be direct off the mains; the WC cisterns and showers being fed via the cold water storage tank. The maximum rate of flow required to a property therefore depends upon the type of plumbing arrangement adopted.

BS 6700[8] recommends that systems are designed to provide a total demand not exceeding 0.3 l/s (18 l/min). Simultaneous discharges are likely to cause reduced flows and hence adverse or malfunctioning of appliances. If the pressure in the mains is high this effect may be less noticeable. However it is when pressures are below 30 m that an observable effect will be caused. A householder does not normally expect to get the maximum flow at all his water consuming facilities simultaneously: he knows that fully opening of the kitchen tap will tend to reduce the cold flow to washbasin and bath taps on the floor above. Conversely, the recent trend to install pump assisted (power) showers has significantly increased the potential peak demand rate from individual dwellings. However in many cases the capacity of the existing service pipework will restrict the maximum flow rate to the unit. Table 15.5 gives the British Standard recommended design and minimum flow rates from different appliances.

For 'high-rise' flats the cold water supply from the mains will be boosted by a small pump to a roof tank which then feeds all the supplies to the flats below. The peak demand from the group of flats will therefore depend upon the amount of storage provided and the pump characteristics and controls.

Service pipe design

For the design of a service pipe the peak demand rate must be estimated and the minimum mains pressure must be known. The flow and headloss through the piping can be calculated using one of the formulae described in Sections 10.4 and 10.5. Section 10.4 suggests some friction coefficients that can be used in the Colebrook–White formula for new small diameter service pipes. For service pipes that have been in use for some (or many) years, friction coefficients are too variable to be able to quote any value that does not run the risk of being misleading. One can only recommend that a 'realistic' friction value be taken, bearing in mind the k values in the Colebrook–White formula are meant to represent 'equivalent sand grain size' of the pipe's internal roughness, and that allowance must also be made for the relatively high number of joints and fittings occurring on service pipes and internal plumbing.

Connections to the distribution main are usually made under pressure, the size of the tapping being 6 mm less in diameter than the service pipe diameter, except in the case of a 13 mm service pipe diameter (the minimum size) which has a tapping of the same diameter as itself. There are usually two stopcocks on the service pipeline: one at the boundary to the consumer's property, which may be operated only by the water undertaker, and one just inside the consumer's property for his own operation.

Table 15.5 gives headlosses expected through typical fittings. Pressure losses in the tapping to the main, in stopcocks, and at ball valves and taps may represent the major pressure losses on the delivery line. Losses at fittings which have been in use for a number of years may be very high. It must be appreciated that no formula gives consistently identical results to those obtained even in laboratory tests and, in practical terms, discrepancies will be not less than ± 10%. Hence for design purposes actual flows should be taken as something less than the calculated value and headlosses as something more.

The headloss through supply meters of sizes 15–150 mm is required by BS 5728 not to exceed 2.5 m at the nominal flow rate of the meter. However it is best to consult manufacturers' literature for actual headlosses.

All supply meters need to be regularly tested for accuracy every few years. A typical small meter-testing bench is shown in Fig. 15.4. Large industrial supply meters may be tested *in situ* using a turbine or electromagnetic flow probe as described in Section 15.15.

Table 15.5 BS 6700 Design flow rates and headlosses through fittings

	Rate of flow (l/m)		Headloss (m) at flow rates		
	Design	Minimum	7.5 l/min	15 l/min	25 l/min
Kitchen tap 1.2 inch (13 mm)	12	6	0.7	1.7	4.9
Kitchen tap 3/4 inch (19 mm)	18	12	0.3	0.7	2.4
Washing machine	12	9			
Dish washing machine	9	6			
Handbasin (pillar taps)	9	6			
Handbasin (mixer/spray taps)	3	1.8			
Bath 3/4 inch (19 mm)	16	12			
Shower (typical)	12	6			
WC cistern to fill in 2 minutes	7.8	6			
Ferrule 13 mm			0.3	0.7	2.4
Ferrule 19 mm			0.15	0.4	1.1
Float valve to tank – 6 mm orifice			1.7	4.7	–
Float valve to tank – 9 mm orifice			0.7	1.8	4.0

Other data:

Service pipe fittings headloss	Values of K in $V^2/2g$
Joints in 13 mm pipes	1.2
Joints in 19 mm pipes	0.8
Elbows	1.0
Short radius bends	0.8
Equal tees – flow to branch	1.3
Screwdown stopcock – fully open	8.2
Gate type stopcock	0.2

BS flows in l/s have been converted to l/min. Loss for service pipe fittings are for average condition.

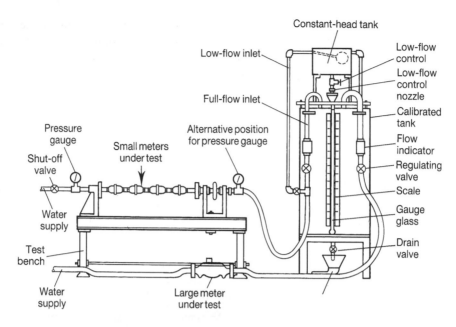

Fig. 15.4   Meter testing bench.

15.6   Waterworks byelaws

Nearly all water undertakers have regulations or byelaws setting out a variety of requirements with respect to consumers' use of water and the materials and design of plumbing systems. In England and Wales all the water companies' individual byelaws were repealed and replaced by the Water Supply (Water Fittings) Regulations 1999. These new Regulations, (issued by the Secretary of State for the Environment, Transport and the Regions), follow requirements laid down by the EC and, although not as prescriptive in detail as some of the previous UK byelaws, they maintain many of the previous byelaw provisions, such as the requirement that – 'No water fitting shall be installed, connected, arranged or used in such a manner which causes or is likely to cause waste, misuse, undue consumption or contamination of water supplied by a water undertaker'. Among the new provisions of the 1999 Regulations are the following.

Section 5 of the Regulations requires notification to the water undertaker of any proposal to install:

- a bath of capacity over 230 litres;
- a single shower unit;
- a pump or booster drawing more than 12 l/min;
- a reverse osmosis plant;
- a garden watering system 'unless designed to be operated by hand';
- 'construction of a pond or swimming pool of capacity greater than 10 000 litres ... designed to be replenished by automatic means and is to be filled with water supplied by the undertaker'.

Schedule 2 to the Regulations gives details of requirements for plumbing fittings for cold and hot water supply systems. Among the principal new provisions is that WC flushing cisterns installed after 1 January 2001 must not give a single flush exceeding 6 litres. Until then cisterns of 7.5 litres (the current requirement) can be installed; and there is no time limit to the replacement of cisterns installed before 1 July 1999 by ones of similar type and capacity. Both flushing cisterns and *pressure* flushing cisterns are permitted; but pressure flushing *valves* may only be installed in a building which is not a house. As siphonic flushing cisterns are no longer mandatory, flushing cisterns equipped with flap valves are permitted (see Section 1.6).

Another new provision is that domestic water-using machines are required to be economical in the use of water, and will be regarded as complying with this requirement if they do not use more than the following:

horizontal axis washing machines	27 litre per kg of washload;
washer-driers	48 litre per kg of washload; both for a standard 60°C cotton cycle; and
dishwashers	4.5 litres per place setting.

An overall requirement is that fittings and their installation must conform to – 'an appropriate British Standard or some other national specification of an EEA [European Economic Area] State which provides an equivalent level of protection and performance'.

Distribution system maintenance

15.7 Checking network performance

To assess the performance of the distribution system and plan repair and rehabilitation measures, water undertakers now maintain asset records in linked or integrated databases which comprise:

(1) mains record drawings, and files recording pipe material, age, internal and external condition and rehabilitation history;
(2) valve status records;
(3) telemetry data logging flows, pressures and water quality parameters at critical monitoring points in the distribution system;
(4) estimates of water losses derived from zonal or district meter recordings;
(5) burst and leakage recurrence rates, including repairs, ground conditions, and cause of failure;
(6) consumer complaints concerning poor pressure, reduction or loss of supply, and dirty water;
(7) results of water quality tests on random samples of water from the distribution system.

System performance analysis is necessary for long-term planning of the expenditure required to maintain the overall condition of assets. This is usually associated with and required by industry regulation of privatised water companies. Day-to-day system performance assessments usually arise as a consequence of consumer complaints, which are investigated in the first instance by a distribution inspector. Where a complaint warrants further action, pressure and flow data would be retrieved from the nearest critical monitoring point and district meter, and would also be measured at the consumer's tap from which a sample of water might be taken for testing. Further investigations may then include analysing the system's performance with a hydraulic model. The objective is to determine whether the problem is an operational one, for which immediate remedial measures can be initiated, or whether due to a gradual deterioration of the main, in which case longer term asset renewal may be needed. Dirty water complaints are often related to the removal of corrosion deposits from old cast iron mains caused by some change in flow conditions and are normally rectified by mains cleaning or, if persistent, by mains renewal or relining. Some complaints may concern the presence of animals or insects due to larvae having entered the system at source works.

15.8   Mains rehabilitation and cleaning

Rehabilitation techniques can be divided into short- and long-term measures. Immediate actions include, re-zoning supply areas to improve pressures and prevent interruption to supply, repairing bursts, and mains flushing to resolve dirty water problems. Valve maintenance should also be included, since for relatively small cost, air and gate valves can be repaired and leaking glands re-packed. Longer term techniques include mains rehabilitation or replacement. Rehabilitation techniques can comprise non-aggressive and aggressive cleaning methods or use of non-structural and structural linings; whilst pipe replacement[9,10] may utilise trenchless technology.[11]

Mains cleaning methods

Mains cleaning techniques include flushing, air scouring, swabbing, jetting and scraping, of which the first three are non-aggressive. Flushing, air scouring and jetting are only effective for pipes up to 300 mm diameters.

Flushing is a well-established technique for small diameter mains where there is adequate mains water pressure. The system hydraulics has to be managed to ensure the velocity at the pipe invert is sufficient to pick up material and hold it in suspension. Whether this can be achieved depends on the specific gravity of the material, water velocity and pipe diameter. In practice pipes up to about 100 mm diameter can be flushed through one or two hydrants supplied from both ends provided there is adequate pressure in the main. For larger diameter pipes, flushing generally needs to be through three or more adjacent hydrants, supplied from one direction only at a pressure of at least 4 bar. Achievement of this may be difficult due to the system hydraulic conditions. It is unlikely to be effective on mains over 300 mm diameter. When planning a flushing programme the following issues need to be taken into account:

● the real source of the problem;
● the characteristics of the material to be removed;
● the maximum achievable pipe velocity under realistic hydraulic conditions;
● the discharge capacity and locations of the hydrants or washouts in relation to the size of pipe to be flushed;
● acceptable methods of disposing of the flushing water.

Any flushing programme must be done systematically, taking into account likely secondary water quality problems created by isolating lengths of pipe. Also it must be expected that flushing is ineffective in a heavily turberculated main. Use of Flowjet equipment currently under development may be of assistance. The device relies on mains pressure and discharge from a hydrant to generate high velocities around the circumference of a telescopic device, so that these high velocities are directed at the pipe wall. The device is held against the water pressure to control movement towards the point of application required.

Air scouring can be used to generate higher velocities during the flushing process without using as much water. The injection of filtered compressed air will force slugs of water along the pipe which will cause more disturbance of loose deposits on the pipe walls as the slugs form and collapse. The procedure is suitable for pipes up to 200 mm diameter and for lengths of up to 1000 m; its effectiveness relies on the skill of the operator in forming suitable air/water mixtures.

Foam swabs can be used for removing soft or loose material, such as organic debris, iron and manganese deposits, sand, and stones. The process usually involves using a series of increasingly hard plastic swabs. The swab should have a diameter 25–75 mm larger than the bore of the pipe and a length of 1.5–2.0 times the bore. The softer swabs can pass through butterfly valves. They can be inserted into the main via a hydrant branch, but for harder or larger swabs the main must be opened. The optimum velocity of the swab is generally about 1 m/s. Its speed can be controlled by regulating the outflow at the discharge end, the speed of the swab being usually one-half to three-quarters of the water velocity because of flow past the swab. Swabbing can greatly improve the flow characteristics of a slimed main and is frequently used to clean out a newly laid main before being put into service.

Scraping and relining is applicable to old tuberculated cast iron mains which are structurally sound. Pipe scraping methods include drag scraping, power boring and pressure scraping. Pigs or aggressive swabs comprise plastic swabs incorporating grades of wire brush or studs. For harder encrusted materials, several passes may be necessary and it is essential to have a good flow of water past the device to prevent debris accumulating in front of it. An electrical transmitter incorporated in the swab or pig will assist in tracking its progress along the pipe. Scraping with pigs is effective in removing hard deposits, but can also damage and remove linings. The latter may increase PAH levels in the water by exposing old coal tar linings (see Section 6.36). Aggressive cleaning is therefore usually followed by applying a secondary lining to the pipe.

Long lengths of main can be aggressively cleaned in 'one go' if the run is fairly straight and no obstacles to the passage of the device exist. However, because of the past practice of using tapers either side of reduced diameter valves, the length is frequently limited to the distance between valve positions. Service connections off the main must be isolated and, after cleaning and relining, must be blown back with fresh water. The cleaning and relining processes is also likely to dislodge, damage or block ferrule connections, so excavation to these and their repair may also be required, increasing the cost of rehabilitating the main.

15.9   Pipe lining methods

Cement mortar and epoxy resin linings applied to the cleaned internal surfaces of cast iron mains can improve the hydraulic capacity of a pipe and reduce discolouration caused by corrosion. *In situ* linings can be applied to pipes from 75 mm diameter upwards. Cement mortar linings are typically 4–6 mm thick. They tend to increase the pH of the water. For pipes up to 150 mm diameter, the reduction in pipe bore and relative roughness of the mortar surface, may not provide adequate hydraulic capacity.

Epoxy resins are used for relining mains up to 450 mm in diameter, although it is technically feasible to line 600 mm diameter mains (see Plate 32). The finished lining is 1–2 mm thick, has a projected life of up to 75 years, and is very smooth providing a good hydraulic performance with no water quality problems. The lining can be applied to most pipe materials, not only iron and steel mains; but care has to be taken to ensure the epoxy lining makes proper adherence to the wall of the pipe.

Structural lining methods are adopted where the structural integrity of the main has deteriorated. These lining methods involve inserting a soft insertion lining into the main by methods denoted as – 'sliplining', 'rolldown', or 'swagdown'. Rolldown involves inserting a polyethylene pipe, which has previously been reduced in diameter by up to 10% by squeezing in a 'rolldown' machine. When in position the pipe's 'elastic memory' is activated by internal pressure and it reverts to its former larger diameter to make a close-fit with the existing pipe wall. The process is suitable for re-lining pipes of 75–600 mm diameter. All these techniques reduce the internal bore of the host pipe, but can significantly improve its hydraulic performance because of the smoothness of the lining. They are quicker to install and less disruptive than relaying the pipe in open cut, but require each ferrule connection to be redrilled through the lining.

15.10 Pipe replacement

The traditional replacement method is to lay a new main adjacent to the main to be replaced, later transferring the service connections to the new main as a separate operation to minimise interruption of supply to consumers. Apart from traditional methods of trenching, newer techniques include narrow bucket conventional excavation, rockwheels, chain trenching machines and mole ploughing. Narrow trenching for laying pipes up to 500 mm diameter avoids the need for an operative to enter the excavated trench and results in less handling of excavated material and reinstatement; but it can only be used for piping which is sufficiently flexible for pipe lengths to be jointed before lowering into the trench. It is quicker and cheaper than traditional open cut excavation. However it does not reduce the risk of disrupting other services and is therefore only suitable for uncongested locations, or to meet restricted site possession or when rapid laying and reinstatement is required. Mole ploughing is only used for pulling small diameter pipes through soft ground.

The high cost of installing, replacing or renovating small diameter underground pipes and services by traditional methods including surface reinstatement, has resulted in the increased use of trenchless construction methods. In addition to the relining methods discussed above, there is a range of 'no dig' or 'low dig' pipe replacement and tunnelling techniques, which include pipe bursting, guided boring, thrust boring, and auger boring (see Plate 32). These techniques are applicable in urban areas, where the presence of many other underground services or much surface traffic makes trenching difficult and expensive. In pipe bursting a tapered expandable impact mole is winched through an existing pipe to break it up and displace the fractured material into the surrounding fill material. A new PVC liner pipe, pulled in behind the mole provides external protection to a replacement MDPE pipe installed in a second stage operation. Pipes up to 300 mm diameter can be installed. Extensive prior site preparation may be necessary to excavate or remove steel repair collars, pipe bends and fittings, and concrete surrounds, and to disconnect existing mains and service connections to prevent their damage. Ground disturbance may also occur and be a problem.

Micro-tunnelling machines for 150–900 mm diameter tunnels have been developed in which articulated shields, adjustable by means of remotely controlled jacks in the launch chamber, allow the machines to be accurately steered on course. Muck removal is carried out by auger or by fluid transport using bentonite or water. The permanent pipe liner is jacked into position behind the shield. When complete the water pipe is installed and the annular space filled by grout. A similar technique for pipes over 900 mm diameter is pipe jacking where a tunnelling shield and string of tunnel lining is jacked into the ground from the drive shaft to a reception shaft. Lengths of several hundred metres can be achieved. Directional drilling can be used to install pipes under an obstruction, such as a road, railway, or river, etc. The technique involves drilling a bore between two shafts sunk either side of the obstruction, using a remotely controlled articulated drilling head. When complete the hole is reamed out to the required size and the pipe is pulled through. Depending on the capacity of the pulling equipment, length of pull and size of pipe, pulls of up to 1500 m and 800 mm diameter of continuously welded steel pipes have been achieved.

Control of leakage and losses
15.11 Reduction of non-revenue water and leakage

The chief components of non-revenue water can be divided into the following.

1. *Unmeasured legitimate use*, which includes:
    *   legal connections with no payment requirements (connection recorded), e.g. standpipe use and (overseas) watering of green areas;
    *   legal connections but consumption not billed (connection not recorded);
    *   demands for fire-fighting (usage could be measured);
    *   street cleaning and sewer flushing (usage could be measured);
    *   old or unmaintained meters (equipment error);
    *   incorrectly read meters (employee error);
    *   incorrect billings (employee error).
2. *Losses and non-legitimate uses*, which include:
    *   leakage from distribution network;
    *   leakage and overflows from service reservoirs or water towers;
    *   illegal connections, meter tampering or by-passing;
    *   unreported third party damage to pipework.

Reducing non-revenue water losses needs to be a continuous activity required to maintain losses within acceptable limits. If not so regarded, overall losses will increase and the system will deteriorate. To reduce losses and non-revenue water an undertaking must allocate the necessary resources in terms of expenditure and the use of an adequate number of properly trained staff. Programmes to educate the consumer in the need to avoid leakage and wastage will also be necessary.

However the data forming the basis for such action needs to be validated. Meter error resulting from the under-recording of consumers' supply meters can represent a significant percentage of non-revenue water. Low flow consumer demands, particularly at night, are frequently below a supply meter's stalling speed and hence go unrecorded. Over a period of time they can, in total, represent an appreciable demand on the system. The accuracy of consumers' supply meters and the magnitude of their likely under-recording has to be assessed by methods described in Section 1.3. Where unmetered supplies are given free to government or other public bodies, or for watering green areas – as occurs in a number of countries – checking their consumption is essential. All meters measuring the output of sources have also to be checked for accuracy: unless regularly serviced they will almost certainly be in error. Probable meter errors were given in Section 1.13; consumer wastage was referred to in Section 1.14; and figures for distribution losses were given in Section 1.16.

15.12 Leakage levels experienced

The scale of the problem of locating leaks on the hundreds of kilometres of pipework that make up a typical water distribution system should not be under-estimated. Modern leak detection equipment can be used to assist in locating leaks and flow monitoring can assist in identifying components of system losses, but they are only tools which aid leak detection. Reducing leakage is therefore inevitably a labour intensive process which has to be pursued continuously. Table 15.6 presents some typical 'background' leakage levels which are estimated to occur on UK water distribution networks; and Table 15.7 shows the likely incidence of defects based upon figures from the annual reports of UK undertakings.

The level of leakage experienced depends on a water undertaker's loss reduction policy, and on the age and condition of its system and the system operating pressure. Policies depend upon the financial and manpower resources an undertaking can allocate to leakage control, both for 'one-off' exercises to reduce current levels of leakage to a satisfactory level and for continued application of leakage control to maintain that level. Either in-house resources can be used, or the work may be put out to a contractor who is set a target level of leakage he is required to achieve.

Analysing the record of bursts on mains may reveal those which are particularly prone to failure. Causes of failure include fracturing due to severe cold weather periods, prolonged dry spells causing ground movement, mining settlement, poor original workmanship when laid, age of pipes or their faulty manufacture, inadequate cover for increased traffic loading, original class strength too low for current pressures, or because

Table 15.6   Typical UK distribution system background leakage levels at 50 m pressure[12]

Infrastructure element	Estimated leakage level		
	Low	Average	High
Distribution mains (l/km/h)	20.0	40.0	60.0
Average for all metered service pipes:			
– meter at property boundary (l/connection/h)	1.50	3.00	4.50
– meter in-house (l/connection/h)	1.75	3.50	5.25
In house plumbing losses			
– average over all houses (l/property/h)	0.25	0.50	0.75

Table 15.7   Incidence of defects on mains and service per annum

	Average number per annum	
Type of defect	UK[13–16]	Germany[12]
Trunk main fracture rate		
– mains in good condition	1.5–2 per 100 km	
– below average condition	6–7 per 100 km	
Distribution main fracture rate		
– lowest rate	10–15 per 100 km	14 per 100 km
– average rate	20–35 per 100 km	
– above average rate	40–55 per 100 km	
Valve, hydrant, and joint defects	1–1.5 times number of distribution mains fractures	
Leakages and other defects on service pipes and stoptaps	1–2 per 100 services	0.5 per 100 services
Defects on consumers' premises causing wastage of water	About 2 per 100 services	

supply operating conditions have caused cyclic stressing of pipes. Where evidence indicates that a main fails frequently during the last 5 years, typically more than two or three times per annum per kilometre, it will probably need to be scheduled for immediate or early replacement.

15.13 Need for active leakage control and setting of targets

Increasingly financial and environmental pressures have forced water undertakings to abandon 'passive leakage control' where only visible and reported leaks are repaired, and instead adopt 'active leakage control'. This is a policy of regularly testing all parts of the distribution system for signs of leakage, searching for and repairing leaks, and aiming to achieve an economic level of leakage. The latter is the level at which the cost of saving additional units of water increases until it equals the cost of producing those units of water.

Each separate part of a distribution system will have characteristics which determine its economic level of leakage. But the parameters determining this level will change as total percentage losses on the whole system reduce. Each area will have a *natural rate of recurrence* of leaks. Left unmaintained the leakage will progressively rise, but the recurrence rate will vary with supply conditions. For example, on systems where leakage is so high that pressures are low, reduction of leakage will result in improved pressures which will be reflected in increased demand until it is satisfied. Only when supply and demand are in balance will the pressure regime, and hence system leakage targets, stabilise.

Once leakage has been reduced to a given target level – termed the *exit level* – active leak detection and repair measures can be suspended. The area will then need to be monitored to ascertain when leakage levels have risen to a level at which it is necessary to re-enter the area and repeat leak reduction measures. This level – termed the *intervention level* – is determined by the combination of a number of parameters. These include the resources available within the organisation, both manpower and financial; the unit cost of water in an area compared with the unit cost of reducing non-revenue water in that area; the other savings that can be made by reducing losses; and whether other areas should take priority. The aim of setting targets is to maximise savings by targeting the area(s) which will deliver the higher value or immediate return. As the characteristics of each area are better understood, targets and thresholds will need to be reviewed and revised, taking into account the latest cost information on water production and manpower, and the resources required to achieve the targets. Where a target has been exceeded or significant rehabilitation has taken place, it may be appropriate to reduce both the target and the threshold to reflect the improved physical condition of the asset. Political and environmental requirements will also influence intervention levels; an undertaking may need to show it has reduced leakage down to some regulating authority's requirements or to justify a proposal to develop a new source. Targets for leak reduction therefore require re-assessment from time to time to ensure they relate to the appropriate source supply conditions and meet the overall policy requirements.

15.14 Active leakage control

The process of active leakage control comprises monitoring for evidence of leakage; detection of leaks; and leak repairs. The most effective method of monitoring is to divide the system into three levels of control, as mentioned in Section 15.1:

● zones of 10 000 to 25 000 connections or perhaps more;
● DMAs of typically 500–3000 connections;
● waste districts of a few hundred connections.

The areas chosen to form these three divisions follow existing system divisions as far as possible for economy. Flow meters are installed on the feed mains to zones and district meter areas. Preferably there should be only a single feed main for each zone or DMA, although this is not always possible. For the waste districts, a by-pass in a chamber is constructed around a valve on the main feed to the district, so that a waste meter can be temporarily inserted in the by-pass when waste metering by step-testing takes place (see Section 15.16). The boundary valves to zones and DMAs are normally kept closed unless, for operational reasons, some DMAs need to be combined. Pressure monitoring equipment is also installed at selected critical points in the system.

Using an integrated telemetering system, zonal and DMA flow and pressure measurements can be transferred to a control centre where they are monitored and recorded. Alternatively, on-site data loggers can be used which can be interrogated, and their data later transferred to the control centre for analysis. Where the integrated system is used, data assessments can be prepared over-night, the output being a summary report which identifies areas where flows or pressures are outside pre-defined parameters, indicating where investigation may be necessary. The daily demand and minimum nightflows can also be monitored and compared with historic data or calculated theoretical flows, to identify differences that may need investigation. If on-site data logging is used, the data is typically downloaded on a 28-day cycle. Hence there is inevitably a delay before such information is processed.

A full system as outlined above is not always adopted. Practices vary. Not all undertakings use waste meter districts, they may rely on DMA flow data to monitor

Semi-positive rotary piston meter, Kent KVM, Class C or D for 15 mm or 20 mm supply pipe. (Manufacturers: ABB Kent Meters, Luton, UK)

A

B

Sequence of operation of semi-positive rotary piston meter

C

D

Manufacturer's data		Class C		Class D	
	Size	15 mm	20 mm	15 mm	20 mm
Q_{max}	m³/h	3.0	5.0	2.0/3.0	5.0
Q_t ±2%	l/h	22.5	37.5	11.5/17.25	28.75
Q_{min} ±5%	l/h	15.0	25.0	7.5/11.25	18.75
Starting flow	l/h	5.7	9.5	3.75/5.63	9.4

E

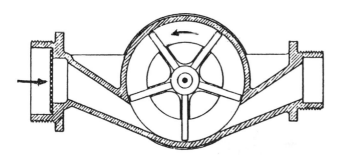

Single jet meter, 15 or 20 mm

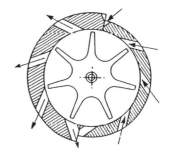

Multi jet meter, 15 to 50 mm

Inferential meters Class B or C

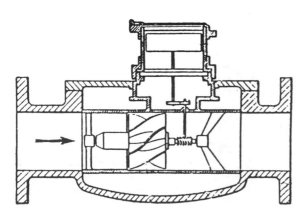

Inferential (Woltman type) vane meter used for measuring large flows. Class B or C, 50 to 300 m

Plate 31

(Left) Mercol pipe cleaning equipment and (right) epoxy resin lining machine. (Mercol Products Ltd., Bolsover, Chesterfield, UK)

Relining 150 mm (6 inch) diameter cast iron main with 160 mm MDPE coiled pipe using the pipe bursting method

Plate 32

One of two 660 mm casings thrust bored below river, to house 450 mm flanged ductile iron pipes for Yorkshire Water, UK. (Allen Watson Limited, Horsham, UK)

A 750 mm diameter auger bore under railway for a new 450 mm diameter pumping main for Sutton and E. Surrey Water plc. (Allen Watson Limited, Horsham, UK)

Plate 33

AquaMaster electromagnetic water meter, 15–600mm size. Accuracy ±0.25% at flows above 5% of nominal flow rating Q_n. (Manufacturers: ABB Instrumentation Ltd., Stonehouse, Glos., UK)

For district metering an *AquaMaster* can be buried directly in the ground with the electronics housed in a street bollard as illustrated above

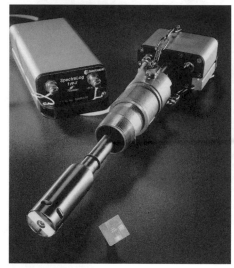

SpectraCense continuous on-line water quality monitoring apparatus for measuring, recording, and transmitting – pH, conductivity, dissolved oxygen, chlorine or free residual chlorine, turbidity and colour. (Manufacturers: Biwater SpectraScan, Waterlooville, Hants, UK)

Quadrina turbine insertion flow meter. (Manufacturers: Quadrina Limited, Letchworth, UK)

Aquaprobe electromagnetic probe flow meter. (Manufacturers: ABB Instrumentation Ltd., Stonehouse, Glos., UK)

Plate 34

consumption and levels of leakage because modern meters are able to measure small changes of flow more accurately than previously. Anglian Water was reported as having 2500 district meters each covering typically 1500 properties.[17] South West Water had 65 water supply zones and 471 district meter areas 'at each of which the diurnal flow is logged by a permanently installed data logger'.[18]

For many large water undertakings in under-developed countries, the setting up of many DMAs or waste districts where no provision for this has previously been made is very expensive, because of the need to purchase and install the meters and probably replace or repair many boundary valves that need to be in working order for flow monitoring to be valid. The approach has to be a long-term one, starting with the zonal metering of supplies, and progressively developing DMAs starting where zonal flows show losses are probably highest. For undertakings only able to provide intermittent supplies, quite different methods for checking levels of consumption and losses have to be adopted, as described in Section 15.17.

15.15 Measurement of pipeline flows

To monitor consumption and check leakage and losses it is desirable to install permanent meters on the flows to zones and district areas of supply. On large supplies to zones, the venturi meter or its shorter form the 'Dall tube' (see Section 10.16) has been used in the past, but these are no longer installed because of their high cost. Instead the electromagnetic meter (see below) would now be used. For the smaller diameter mains the inferential meter may continue to be used; equipped with electrical pulse recording equipment it can log either the total flow or rates of flow at stated intervals. But the tendency is to replace it with an electromagnetic meter when it needs renewal to save on long-term maintenance costs.

Temporary flow measurement devices

For the temporary measurement of flows when conducting consumption surveys or when flow data is required for distribution network analyses, 'probeflow insertion meters' are useful. These probeflow devices come in two types – the turbine type, and the electromagnetic type; the latter being of more recent development. Either is inserted through a tapping in the main at the location where measurement of flow is required. The turbine type (see Plate 34) has a small vane at its end the rotation of which records the velocity of flow past the vane. The electromagnetic probe has an electromagnet at its end which applies a magnetic field to the water, and electrodes either side of the probe at the same level pick up the induced EMF in the water which is proportional to the velocity of flow past the electrodes. The vane probeflow has the disadvantage that if its vane is damaged (or thought to be damaged) the instrument has to go back to the makers for recalibration and repair if necessary.

Either type of probe is used to record the centreline velocity of flow, but in order to derive from this the mean or total flow through the pipe, the velocity profile across the pipe diameter has to be determined. The probe has to be set at different points across the diameter to measure the velocities at these. The measurements taken are modified for each measurement point to take account of the disturbance caused by the instrument in the water (e.g. increased velocity), and then the velocity ratios to the centreline velocity are calculated to give the mean pipe velocity and the consequent total rate of flow according to the internal diameter of the pipe. This correction also takes some account of wall friction and any abnormal flow profile. For a 'perfect' profile the velocity ratio – meanflow to centreline flow – is 0.833, but in practice it can range from 0.7 to 1.0. The further the ratio is from 0.833 the more caution must be used in applying the result, e.g. the ratio may vary with mean velocity. The internal diameter of the pipe has to be measured accurately, and the number of flow profile readings required are five for pipes of 150 mm diameter; nine for 300 mm; and 13 for larger diameters. The measurements need to be repeated at least three times to ensure the ratio is consistent and repeatable, and the flow must be relatively constant during each profile run. In practice there may be difficulty in obtaining consistent results so that several attempts may be necessary, the poor field conditions often applying making it difficult to conduct precise measurements. Once the profile is satisfactorily established, the instrument is set at the centreline and a data logger is attached. The tapping for a turbine probeflow is 40 mm diameter and the minimum pipe diameter in which it can be installed is 200 mm. The tapping for an electromagnetic probe is 20 mm diameter, usually installed on pipes of not less than 150 mm diameter. At present electromagnetic probes are only made up to 1 m long, so they can only just be used in 900–1000 mm maximum diameter pipes where the flow rate does not exceed 1.75–2.0 m/s because of the probe's flexibility. These limitations, however, will rarely be significant in the usual distribution system.

Permanent flowmeter installations

The permanent type of electromagnetic flowmeter (see Plate 34) is now widely used for monitoring and recording flows from sources, pumping stations and reservoirs, and for the measurement of zonal and district metering areas. They can also be used for step-testing in waste metering (see Section 15.16). They are available in a full range of sizes from 15 mm to 2400 mm and can be mains or battery powered. They have the advantages of ease of installation, low maintenance, accuracy, and negligible head loss. They can be equipped with full telemetry interface capability to meter requirements for district metering, customer billing, leakage control and treatment works applications. They must however be carefully installed, requiring a minimum of five diameters of straight pipe upstream and two diameters downstream. Frequently they are sited too close to fittings, valves or tees, or are affected by protruding gaskets and other factors such as vibration, flooding, or large ambient temperature swings.

When purchasing a magnetic flowmeter, the manufacturer usually gives a written specification for the instrument and a flow calibration certificate from a recognised test bench. However, nearly all flowmeters that are tested in a flow laboratory are installed in pipelines where the conditions are different. This invariably leads to a shift in performance between laboratory and site, due to installation effects. Hence without *in situ* testing it is likely a meter will possess unknown metering errors. If it is then used to estimate leakage or water imbalances, the basic data coming from such meters may be open to question. Some types of meters and some specific designs do experience installation effects and the only means of establishing these is to perform a traceable calibration *in situ* against a known field standard. The field standard may have to comprise some master meter of proved accuracy, the use of volumetric measurement, or merely using other flowmeters temporarily installed as a cross check of the meter under test. Recently 'digital fingerprinting' has been added to these techniques. This tests the electrical characteristics of a magnetic meter which should be basically stable with time. Provided the electrical field remains constant and other key electrical parameters within the circuitry are also stable, it is possible to relate the electrical 'fingerprint' of a meter and transmitter back to a change in meter calibration. This method should also be traceable *in situ* and is not a calibration, but a check that the calibration has not changed within known limits. All such *in situ* or other testing should be conducted by an experienced engineer.

Despite the growing use of electromagnetic meters, many older meters of the venturi or Dall type remain in operation, most usually at source works. If properly maintained they can continue to give satisfactory readings. In particular, however, the throat of a venturi meter needs inspection from time to time to ensure that it is clean and free of slime or deposits that would otherwise affect its accuracy. For this purpose a hatch is usually provided over the throat of a venturi. The recording equipment, usually mechanical, also needs keeping in good order. To be certain that source outputs are known as accurately as possible, *in situ* volumetric testing of such meters is always advisable.

15.16 Leak location

The leak detection methods that are available include: (1) waste metering/step-testing; (2) 'sounding'; (3) leak noise correlation and other location equipment, and (4) visual observation.

(1) Waste metering/step-testing

Waste metering and step-testing comprise metering the night flow to a small part of the distribution system, and then shutting down the mains within the area, one by one, so the meter registers the drop in flow with each main shut down. A larger drop in flow than would be expected from the estimated night consumption by properties connected to the main may indicate the main has a leak. This 'step-testing' as it is called, is only practicable on 24-hour supply systems. For each waste area a valved by-pass on the feed main to the district is constructed in a chamber and, for step-testing, all flow to the area is passed through a 'waste meter' installed on this bypass with all other feeds shut off. The mains within the area are then shut down according to a pre-arranged timetable. Alternatively step-testing may be applied to a district metered area if it is of suitable size and the district meter (usually electromagnetic) is sufficiently sensitive to measure changes of flow as each main is shut down. The results of the test are interpreted and passed to the district inspector to arrange for further investigations to locate a possible leak. After all leaks suspected have been found and repaired the night test is repeated to confirm that there has been an improvement.

Technology developments have reduced the time required to analyse the flow data during step-testing. After shutting down a main, the waste inspector may use a mobile phone to interrogate the waste meter and find its drop in recording. Comparing this with data held in a laptop computer he or she can judge whether the drop in flow relates reasonably to the number of domestic connections on the main and any expected non-domestic night-time demand.

The cost of step-testing tends to be high. A chambered by-pass can be expensive to construct, all boundary valves have to tested to ensure they shut off properly, if not, they must be repaired; and the shutting and opening of mains isolating valves is time-consuming. Also a distribution manager may be reluctant to permit step-testing if there is any risk of dirty water contamination resulting from the inevitable re-routing of supplies or from high velocities caused in mains when reopening valves.

(2) Sounding

The sound made by water leaking from a pipe is the basis of the majority of leak location techniques and electronic equipment. Under traditional 'sounding' an inspector uses a 'listening stick', which may be a light bar of solid metal about 1.5 m long or the bar of a valve key. Purpose-made listening sticks are sometimes equipped with an earpiece. One end of the stick is placed on an exposed part of the main or service pipe, such as a valve or hydrant spindle or stopcock, and the other is placed against the ear. The sound emitted by a leak, if audible, is 'a low drumming noise' or 'a continuous buzzing sound' and tends to be continuous without any change of audibility or quality. It stops abruptly when, and if, the water can be turned off. An experienced waste inspector using a listening stick can detect even a small leak at a distance of 10–15 m, if it is making a sound. By listening at another point of contact on the main, the inspector can judge by the difference in sound volume the probable location of the leak.

Sounding is frequently done at night when background noise should be low. However an experienced operator can still be successful during the day away from main roads and where there are lulls in traffic noise. Daytime sounding may also be preferred because, at night, parked cars may prevent access to valves and stopcocks. Sounding all pipes in a system is not usually adopted because it is very labour intensive, and relatively ineffective on non-metallic pipes and not efficient at identifying new or increased leakage as it occurs. In the UK where a high percentage of service connections do not leak, an inspector can sound up to 200 connections per shift. Overseas where pressures are lower and background conditions are less favourable, an experienced inspector should be able to sound 80–120 connections per day, on average. Ground microphones which amplify the sound of a leak are effective for leaks from non-ferrous pipes. It must be borne in mind, however, that not all leaks emit a sound, and the volume of sound emitted is mostly not related to the size of the leak.

(3) Use of leak noise correlation and other equipment

Leak noise correlators are electronic devices for amplifying the sound of a leak and judging its location. The software analyses the sounds from two points of contact with a main, matching the sound traces by time-delaying one, and so judging the position of the leaks between the two points of contact by measuring the time-delay used. Recent developments in the electronics for the sound filters allow finer location determination, particularly for multiple leaks, and uses lower frequencies for testing for leaks on non-metallic pipes including plastic pipes.

Acoustic loggers are deployed in groups, typically three to five, in areas where there is a high continuous transient noise. They record the noise on the system over a period of time, typically a week. When retrieved, the data is analysed to filter out the background and 'normal' system noises and hence locate leaks. A group of acoustic loggers can successfully detect and locate a number of leaks at different locations within the triangulation of the deployed loggers. When used regularly within a system, system noise profiles can be developed and used to assess system changes and predict the need for intervention.

Pipeline integrity management systems use continuously monitored flow and pressure data linked to hydraulic modelling software to provide on-line active leak detection on a pipeline. The technology, developed from the oil and gas industry, is suitable for application to lengths of trunk main where pipe failure could be damaging and would need immediate, automatic shutdown of a section of the main. The technology uses flow and pressure measurements to continuously assess the performance of the main between the sensing points and compares the measured values against normal pipe performance characteristics. Where the software detects abnormal measurements, the software calculates the location of a leak and initiates the closure of appropriate valves in order to isolate the fractured length of pipe. The equipment is expensive to install and maintain but cost effective in terms of the consequential damage that might otherwise have occurred if a large pipeline were to develop a serious leak.

(4) Visual observation

Many leaks are reported first by the public. However more can be found by searching for surface signs of them. The district inspector should visit the whole of his or her supply area regularly and will know where leaks are possible. Damp patches, trickles of running water and extra vegetation growth close to pipe alignments, valves or fire hydrants, or above ferrules or stop taps, may be indicative of a leak. The inspector will also be able to detect signs of consumer wastage, such as overflows discharging outside properties. However, beneath metalled roads, or where mains are laid in freely draining ground, or adjacent to or below a water course, even large leaks may give no surface sign of existence. Trunk main routes through open country should be inspected from time to time for signs of leakage.

15.17 Repairing leaks

Confirmation of the existence of a leak is only obtained when a pipe is exposed and the leak located. However, the repair can be expensive, especially if the leak is under a heavily trafficked road or road junction, and special planning of the repair operation may be necessary to minimise traffic disruption. The relevant road authorities will have to be prior informed and all their requirements met. If the leak is known to be small and unlikely to increase, and its repair is expensive and would cause serious traffic delays resulting in strong adverse publicity, the water undertaking may be faced with the problem of how to repair the leak – if at all. Various 'no-dig' possibilities outlined in Sections 15.9 and 15.10 may assist, and repair from inside the pipe may offer a solution if internal access is possible.

Where the work of leak detection is let out to a contractor, the contract should require the contractor to confirm the existence of leaks detected by excavating down onto them. Where the contractor is also required to repair leaks, payment should be made against reductions in assessed leakage flows, not the number of excavations necessary to locate leaks, which would involve payment for abortive excavations where no leak is found. Also payment by reduction of leakage flows avoids the problem of deciding what size of leak justifies being repaired leaving this decision to the contractor.

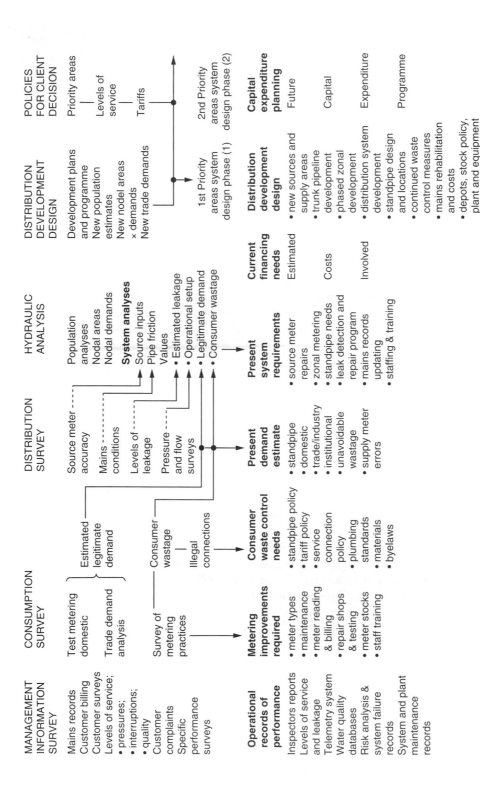

Fig. 15.5 Schematic for starting a distribution system rehabilitation and development project.

15.18 Rehabilitation, leak detection and development of distribution systems in disrepair

Many large public water supply undertakings in parts of the world have distribution systems in need of rehabilitation and repair. Total losses or unaccounted-for water of 50–55% of the supply can be reported, made up of unknown proportions of distribution leakage, consumer wastage, and failure to meter or bill all consumers taking a supply. It is not easy to rectify a large system of this sort in a short time; the approach has inevitably to be in progressive stages of improvement. An indication of the work involved in starting off a rehabilitation project is given in Fig. 15.5.

The difficulties of rehabilitation are increased in the many undertakings where supplies have to be intermittent because of the scarcity of supplies, often exacerbated by the large losses of water (and revenue) through leaks, wastage, and unpaid supplies. Traditional methods of leak detection suitable for 24-hour supplies, such as waste metering, step-testing and sounding, become impracticable because many consumers leave taps open, ready to discharge water into storage receptacles to cover periods of non-supply. Even the procedure of valving off lengths of main to put them under a pressure test for leakage after shutting down all service pipes, may be vitiated by the lack of stop-valves on service pipes, or leaking ferrule connections or tappings to the main, and many illegal or unknown connections. Often a first attempt at a pressure test results in a burst main because the system has long been on intermittent supply at low pressures, or it may be difficult to gain any pressure because of leaking boundary valves or the existence of some unknown connection which has not been marked on the mains record plans.

A more productive approach can be the direct one of exposing the soffit of lengths of main in selected areas to find the principal causes of loss. Often it will be found that high losses are due to one or more of the following – badly made service pipe tappings on the main; illegal and unknown connections; illegal by-passes to customer meters; leaking service pipes not fitted with a stopcock; service pipes continuously taking water because of waste on consumers' premises. Where the groundwater table level is high, or the ground is very free draining, or a pipe is laid close to or under a river, high rates of leakage from a pipe joint or a cracked pipe may be found which have given no surface indication of their existence.

Although the remedial measure of exposing the soffit of mains is time consuming and relatively expensive, it has the merit of directly reducing leakage, and can aid the finding of leaking joints in the main. It is not as expensive for many countries as relaying a main since it is primarily labour-intensive work involving the use of less expensive materials and plant. Attempts to find all leaks on a pipe network known to be in poor condition by using various methods of surface detection equipment and flow monitoring, can be very frustrating because of repeated failure to get an acceptable result when a system is re-tested. This non-success after so much work reduces motivation to pursue leak detection. The direct approach gives positive results and, if pursued, can bring increased efficiency in finding leaks through experience and increasing areas of the system put into a good condition.

References

1. Fire Services Act 1947, and the Water Industry Act 1991.

2. *National Guidance Document on the Provision of Water for Fire Fighting*, Water UK, 1999
3. BS750: 1984 Underground fire hydrants and surface box frames and covers.
4. BS 1387: 1985 Screwed and Socketed Steel Tubes and Tubulars and for Plain End Steel Tubes Suitable for Welding or for Screwing to BS21 Pipe Threads.
5. BS 3505: 1986 Unplasticised Polyvinyl Chloride (PVC-U) Pressure Pipes for Cold Potable Water.
6. BS 6572: 1985 Blue Polyethylene Pipes up to Nominal Size 63 for Below Ground Use for Potable Water Supply.
7. BS 6730: 1986 Black Polyethylene Pipes up to Nominal Size 63 for Above Ground Use for Potable Water Supply.
8. BS 6700: 1987 Design, Installation, Testing and Maintenance of Services for Supplying Water for Domestic Use.
9. *Planning the Rehabilitation of Water Distribution Systems*, WRc, 1989.
10. IWEM Monographs of Best Practice No. 1, *The Rehabilitation of Water Pipelines*, 1996
11. *Trenchless Technology Manual*, WRc, 1995.
12. Lambert, Myers & Trow, Managing Water Leakage, *Financial Times Energy*, 1998.
13. Reed E. C. Report on Water Losses. *Aqua, JIWSA*, 1980, p. 178.
14. Lackington D. W. Survey of Renovation of Water Mains, Paper No. 4, *Proc Symposium on the Deterioration of Underground Assets*, IWES, 1983.
15. Clark P. G. Factors Relating to the Deterioration of Water Mains, Paper No. 2, *Symposium* referred to in Reference 13.
16. Newport R. Factors Influencing the Occurrence of Bursts in Iron Water Mains. *Aqua, JIWSA*, 1980, p. 274.
17. *Water and Environmental Manager. JCIWEM*, July 1997, p. 20.
18. *Water and Environmental Manager. JCIWEM*, May 1996, p. 8.

Conversion factors

Metric to British units

British to metric units

Length:

1 m = 39.37 in	1 in = 25.40 mm
1 m = 3.2808 ft	1 ft = 304.80 mm
1 m = 1.0936 yd	1 yd = 0.91440 m
1 km = 0.6214 mile	1 mile = 1.60934 km

Note: In nominal sizing 300 mm is taken as equivalent to 1 ft and 25 mm as equivalent to 1 in.

Area:

1 m$^?$ = 1.196 yd^2 = 10.764 ft^2	1 ft^2 = 0.0929 m^2
1 ha = 2.471 acres	1 yd^2 = 0.8361 m^2
1 km^2 = 0.386 square mile	1 acre = 0.4047 ha
	1 square mile = 2.5900 km^2

Notes:

1 km^2 = 100 ha (hectares) and 1 ha = 10000 m^2.

1 square mile = 640 acres and 1 acre = 4840 yd^2.

Volume:

1 m^3 = 35.314 ft^3	1 ft^3 = 0.02832 m^3
1 m^3 = 1.3079 yd3	1 ft^3 = 28.32 litres
1 m^3 = 219.97 gallons (British)	1 yd^3 = 0.76456 m^3
1 litre = 0.21997 gallon	1 British gallon = 4.546 litres
1 Ml = 0.21997 Mgallon	1 US gallon = 3.785 litres

Notes:

1 m^3 = 1000 litres and 1 Ml = 1000 m^3.

1 US gallon = 0.83267 British gallon; also 1 ft^3 = 6.2288 British gallons.

Mass:

1 kg = 2.2046 lb	1 lb = 0.453 59 kg
50 kg = 0.9842 cwt	1 cwt = 50.802 kg
1 tonne = 19.684 cwt	1 ton = 1.01605 tonne
1 tonne = 0.9842 ton	

Pressure:

1 metre head of water   1 ft head of water
$= 1.422\ \text{lb/in}^2$ $= 0.03048\ \text{kgf/cm}^2$
$1\ \text{kgf/cm}^2 = 14.223\ \text{lb/in}^2$ $= 0.00299\ \text{N/mm}^2$
$1\ \text{N/mm}^2 = 145.038\ \text{lb/in}^2$ $1\ \text{lb/in}^2 = 0.0703\ \text{kgf/cm}^2$
$= 0.006895\ \text{N/mm}^2$

Notes:
$1\ \text{N/mm}^2 = 10.197\ \text{kgf/cm}^2$, $1\ \text{kgf/cm}^2 = 10$ metres head of water, and
1 bar $= 10.197$ metres head of water.
$1\ \text{lb/in}^2 = 2.3067$ ft head of water.
1 pascal $= 1\ \text{N/m}^2$.

Density:
$1\ \text{kg/m}^3 = 0.06243\ \text{lb/ft}^3$ $1\ \text{lb/ft}^3 = 16.018\ \text{kg/m}^3$

Flow rates:
$1\ \text{m}^3/\text{s} = 35.31\ \text{ft}^3/\text{s}$ $1\ \text{ft}^3/\text{s} = 0.0283\ \text{m}^3/\text{s}$
$1\ \text{m}^3/\text{s} = 19.00\ \text{mgd}$ $1\ \text{mgd} = 0.05262\ \text{m}^3/\text{s}$
$= 4.546\ \text{Ml/day}$

1 litre/s $= 13.20$ gpm $= 0.019$ mgd   1 gpm $= 0.0758$ litre/s

Notes:
mgd $=$ million gallons per day; gpm $=$ gallons per minute.
$1\ \text{m}^3/\text{s} = 86.4 \times 10^3 \text{m}^3/\text{day} = 86.4\ \text{Ml/day}$.
$1\ \text{ft}^3/\text{s} = 86400\ \text{ft}^3/\text{day} = 0.53817\ \text{mgd}$.
$1\ \text{Ml/day} = 1000\ \text{m}^3/\text{day} = 0.22\ \text{mgd}$.

Hydrological units:
1 litre/s per km^2 $1\ \text{ft}^3/\text{s}$ per 1000 acres
$= 0.09146\ \text{ft}^3/\text{s}$ per 1000 acres $= 6.997$ litres/s per km^2
1 mm rainfall per km^2 $1\ \text{ft}^3/\text{s}$ per square mile
$= 1000\ \text{m}^3$ $= 10.933$ litres/s per km^2
$= 0.220\ \text{Mg}$ 1 in rainfall per square mile
$= 65786\ \text{m}^3$
$= 14.471$ million gallons.

Filtration rate:
Note:
100 gallons per ft^2 per hour $= 117.44\ \text{m}^3$ per m^2 per day
$= 4.89\ \text{m/h}$

Power:
1 joule (J) $= 0.73756$ ft lb   1 horsepower (hp) $= 0.74570\ \text{kW}$
1 kW $= 1.3410$ hp

Notes:
1 J/s $= 1$ watt (W).
1 Ml/day of water raised through 8.81 m $= 1\ \text{kW}$ (at 100% efficiency).
1 hp $= 550$ ft lb/s.

Index